Mesozoic Birds

Mesozoic Birds

Above the Heads of Dinosaurs

EDITED BY

LUIS M. CHIAPPE

AND

LAWRENCE M. WITMER

University of California Press

Berkeley   Los Angeles   London

Title page illustration: Archaeopteryx lithographica by
Charles Knight (courtesy Los Angeles County Museum of
Natural History).

University of California Press
Berkeley and Los Angeles, California

University of California Press, Ltd.
London, England

© 2002 by the Regents of the University of California

Library of Congress Cataloging-in-Publication Data

Mesozoic birds : above the heads of dinosaurs / edited
    by Luis M. Chiappe and Lawrence M. Witmer.
        p.   cm.
    Includes bibliographical references and index.
    ISBN 0-520-20094-2 (alk. paper)
    1. Birds, Fossil.   2. Paleontology—Mesozoic.
3. Birds—Evolution.   I. Chiappe, Luis M.   II. Witmer,
Lawrence M.

QE871 .M47 2002
568—dc21   2001044600

Printed in China

10   09   08   07   06   05   04   03   02
10   9   8   7   6   5   4   3   2   1

The paper used in this publication meets the
minimum requirements of ANSI/NISO Z39.48-1992
(R 1997) (*Permanence of Paper*).♾

To Pat
and to Patty and Sam

Contents

Preface

Less than two years after the 1859 publication of Charles Darwin's *Origin of Species,* the first skeletal specimen of *Archaeopteryx* was discovered—a nearly perfect evolutionary "missing link"—and since that time, the origin and early history of birds have been prime topics of debate among natural historians and evolutionary biologists (De Beer, 1954; Desmond, 1982; Elzanowski, Chapter 6 in this volume). Not much later, the paleontological exploration of the American West resulted in the discovery of multiple specimens of *Hesperornis, Ichthyornis,* and their kin (Marsh, 1880), toothed birds that, although much less primitive than *Archaeopteryx,* strongly influenced discussions about early avian evolution. For many decades, this handful of Mesozoic taxa was all we had. It was the only fossil evidence available for understanding this fascinating chapter of vertebrate history. In fact, it was not until the early 1980s that significant new data started to fill the extensive evolutionary gap separating the basal birds discovered in the previous century. The first evidence of a previously unknown radiation of Mesozoic birds, Enantiornithes, was announced by Cyril Walker in 1981 (Chiappe and Walker, Chapter 11 in this volume), and soon other taxa (e.g., the Mongolian *Gobipteryx*) were recognized as part of this radiation (Martin, 1983). In addition, the discovery in the 1980s of some fossil-rich Early Cretaceous lake beds in Spain (Sanz et al., Chapter 9 in this volume) and intensified collecting in previously underexplored sites in Spain (Chiappe and Lacasa-Ruiz, Chapter 10 in this volume) and China (Zhou and Hou, Chapter 7 in this volume; Sereno, Rao, and Li, Chapter 8 in this volume) led to many other discoveries not much younger than *Archaeopteryx.* The two-dimensional preservation of this Early Cretaceous avian diversity limited anatomical studies, but three-dimensional bird skeletons began to be found in other Mesozoic fossil sites (e.g., Molnar, 1986; Chiappe, 1991). At the same time, other Cretaceous discoveries (e.g., Kurochkin, 1985; Olson and Parris, 1987; Hope, Chapter 15 in this volume) afforded in-

formation about the closest relatives of the living lineages of birds (Neornithes). If the abundant discoveries of the 1980s made a significant contribution to our understanding of the early evolution of birds, the 1990s brought an embarrassment of riches. In just that decade alone, the number of species of Mesozoic birds probably tripled those discovered in all previous years (Chiappe, 1997). These fossils were unearthed all around the world (Chiappe, 1995; Feduccia, 1996; Chatterjee, 1997; Padian and Chiappe, 1998), although primarily in China, Mongolia, Argentina, Spain, Madagascar, and the United States.

A volume entitled *Mesozoic Birds* faces the sometimes sticky problem of just what constitutes a "bird." In our narrow time plane of the present day, birds are so markedly distinct from other vertebrates that our perception of what is or is not a bird is patently obvious. But the Mesozoic era witnessed the dawning of birds, or, in scientific parlance, the evolutionary transition to birds. As a result, the line between bird and not-bird is often a fuzzy one, and there are many taxa whose avian status is highly controversial. This volume deals with several of these controversial taxa: for example, *Protoavis* (Chapter 1), *Caudipteryx* (Chapters 1, 2, and 7), *Avimimus* (Chapter 3), *Mononykus* and its alvarezsaurid kin (Chapters 4 and 5). Decisions about where to draw the line could be made on the basis of particular features, such as feathers or a furcula (wishbone), but, as it turns out, virtually all the attributes that characterize modern birds were acquired sequentially in the Mesozoic.

Thus, the line between bird and not-bird is essentially arbitrary and must be defined. In this modern phylogenetic era, taxonomic definitions are based on relationships, and we have adopted for this volume the convention of regarding Aves (and the colloquial terms "birds" and "avian") as pertaining to the group comprising the most recent common ancestor of *Archaeopteryx* and neornithine ("modern") birds and all its descendants (Witmer, Chapter 1 in this volume; Clark, Norell, and Makovicky, Chapter 2 in this volume; Sereno, Rao, and Li, Chapter 8 in this volume; see also Padian and Chiappe, 1998; Sereno, 1998, 1999). This convention, while not entirely satisfactory, conforms to traditional usage. Chapter 2, by Clark, Norell, and Makovicky, however, breaks with this convention and speaks cogently for another convention; as editors, we respect their position and have not forced them to conform.

Just as phylogenetic systematics (i.e., cladistics) has transformed taxonomy, it has also changed the way we approach just about every question in comparative biology. Fundamentally, cladistics is a means of discovering the genealogical relationships of organisms, and in their chapter Clark, Norell, and Makovicky provide a very useful guide to the objectives and methods of cladistics. A few of the chapters in *Mesozoic Birds* indeed have detailed phylogenetic analyses with character-taxon matrices (e.g., Chapter 2, by Clark, Norell, and Makovicky; Chapter 11, by Chiappe and Walker; Chapter 20, by Chiappe), but "phylogenetic thinking" pervades almost the whole volume. Most noteworthy perhaps is Gatesy's innovative phylogenetic treatment of the evolution of the avian locomotor system in Chapter 19. Likewise, Witmer in Chapter 1 discusses the role of cladistics in the debate on avian origins and the relationship of phylogenetics to theories on the origin of flight. At the same time, some chapters have relatively little overt phylogenetic focus, either because the authors were dealing with a diverse regional avifauna (e.g., Chapter 7, by Zhou and Hou) or because the authors are not cladists (e.g., Chapter 14, by Galton and Martin). Nevertheless, phylogenetics is the underpinning of the volume, and a simplified cladogram of Mesozoic birds is presented in Figure P.1.

The study of Mesozoic birds is diverse, and this volume has sought to characterize this diversity. *Mesozoic Birds* presents a collection of essays covering a wide range of topics bearing on the origin of birds, their Cretaceous morphological and osteohistological diversity, their genealogical history, and their functional transformations during 85 million years of Mesozoic avian evolution. We have divided the volume into four parts. Part I deals with larger and more conceptual issues, such as those surrounding avian origins (Chapter 1) and the broader phylogenetic relationships of birds (Chapter 2). Part II provides a treatment of some of the controversial taxa mentioned previously. We have separated these more contentious taxa out because, although some analyses have placed them within Aves, others have suggested placement outside birds. Part III presents the undisputed members of the Mesozoic aviary and is devoted to chapters dealing with the anatomy, systematics, and paleobiology of the various groups of Mesozoic birds. As workers struggle to keep pace with the seemingly endless new discoveries, a great deal of alpha-level description and systematic work is required, and this part of the volume presents these findings.

Some of the chapters deal with a single clade or even a single species (e.g., Chapter 6, by Elzanowski, on *Archaeopteryx* and its kin; Chapter 8, by Sereno, Rao, and Li, on *Sinornis;* Chapter 10, by Chiappe and Lacasa-Ruiz, on *Noguerornis;* Chapter 12, by Forster et al., on *Vorona;* Chapter 13, by Chiappe, on *Patagopteryx;* and Chapter 14, by Galton and Martin, on *Enaliornis* and other Hesperornithiformes). However, other chapters have a more geographic flavor, dealing with the fossil birds from a particular region (e.g., Chapter 7, by Zhou and Hou, on the Chinese birds; Chapter 9, by Sanz et al., on the Spanish birds).

In all cases, our intent was to have the primary experts who had the actual fossils in their laboratories write the chapters, ensuring that the most up-to-date and authoritative treatments would be available. This principle of seeking primary workers also resulted in members of one taxon, Enantiornithes, being covered in five chapters (Chapter 7, by Zhou and Hou, on the Chinese birds; Chapter 8, by Sereno, Rao, and Li, on *Sinornis;* Chapter 9, by Sanz et al., on the Spanish birds; Chapter 10, by Chiappe and Lacasa-Ruis on *Noguerornis;* and a summary, phylogenetic chapter [Chapter 11] by Chiappe and Walker). We worked with these authors to minimize the amount of overlap, and, although some minimal redundancy was perhaps introduced, Enantiornithes is such a new and important clade—recognized only since 1981 yet having greater known species diversity than any other Mesozoic avian clade—that we felt that extensive coverage was merited.

Although virtually all the taxa discussed in *Mesozoic Birds* were previously described elsewhere, the original descriptions are often very brief parts of short papers in journals such as *Science* and *Nature.* We are very pleased to be able to present in this volume new, definitive descriptions and illustrations of a number of taxa, including the enigmatic *Avimimus* (Vickers-Rich, Chiappe, and Kurzanov), the alvarezsaurid *Shuvuuia* (Chiappe, Norell, and Clark), the enantiornithines *Sinornis* (Sereno, Rao, and Li) and *Eoalulavis* (Sanz et al.), the basal ornithuromorphs *Vorona* (Forster et al.) and *Patagopteryx* (Chiappe), and the hesperornithiform *Enaliornis* (Galton and Martin). Thus, although our intent was to assemble and synthesize available data, *Mesozoic Birds* exceeds this goal by offering descriptions, figures, and other data that cannot be found elsewhere.

Lamentably, our rule of using only primary experts resulted in the almost total absence from the volume of one major clade of Mesozoic birds (Ichthyornithiformes) and only partial coverage of another major clade (Hesperornithiformes); both are marine clades known best from the Late Cretaceous Niobrara Chalk of Kansas. Although we had an agreement from the main researcher on these two clades, it was not fulfilled. Rather than enlist a secondary author, we choose to refer the reader to Marsh's 1880 monograph, *Odontornithes: A Monograph on the Extinct Toothed*

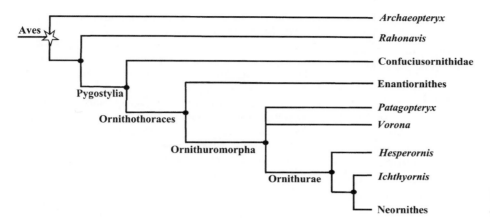

Figure P.1. Simplified cladogram of Mesozoic birds.

Birds of North America, which remains a useful source. Nevertheless, *Ichthyornis* and its kin are given passing treatment in Hope's chapter on Neornithes, and Chiappe (Chapter 20) includes *Ichthyornis* as a terminal taxon in his phylogenetic analysis. Chapter 14, by Galton and Martin, is largely devoted to the basal hesperornithiform *Enaliornis,* but they also discuss the anatomy and biogeography of other hesperornithiforms and provide a diagnosis of Hesperornithiformes; again, Chiappe includes this clade within his phylogenetic analysis presented in Chapter 20. For a detailed treatment of *Ichthyornis,* refer to Clarke (2002).

While we are on the subject of what might be perceived to be "missing" chapters, *Mesozoic Birds* lacks significant treatment of *Rahonavis,* the recently discovered Late Cretaceous bird from Madagascar (Forster et al., 1998). This discovery was so recent that a chapter for this volume was not feasible, and we regard the 1998 paper as sufficient for the time being (once again, Chapter 20, by Chiappe, includes *Rahonavis* in the phylogenetic analysis). But, truth be told, there are many such newly discovered Mesozoic birds that are absent from the pages of this volume and that, at this writing, remain unpublished. Discoveries are coming faster than scientists can write them up. With several new Mesozoic birds being discovered every year, the task of assembling exhaustive coverage of their early diversity became an impossible one. Although as editors we would prefer that the volume were "complete" (whatever that means), as avian paleontologists we are thrilled at the prospects that these new discoveries hold. We always knew that this volume would represent but a snapshot of this fast-moving field.

Part III concludes with two chapters on nonskeletal remains of Mesozoic birds, in particular an account of avian feathers by Kellner (Chapter 16) and a discussion of the footprint record by Lockley and Rainforth (Chapter 17). Part IV deals with issues surrounding the functional morphology, physiology, and evolution of birds. Chinsamy (Chapter 18) reviews the often contentious bone histology data for Mesozoic birds and its physiological implications. Gatesy (Chapter 19) provides an exciting new way of looking at locomotor evolution in birds that presents a true departure from previous studies. Finally, Chiappe (Chapter 20) presents a comprehensive phylogenetic analysis of basal birds and discusses its implications for avian evolution. Part IV would have been an appropriate place for a chapter on the origin of avian flight, and its absence may constitute, for some people, another "missing" chapter. In fact, Witmer's and Gatesy's chapters (Chapters 1 and 19, respectively) provide some discussion of the origin of flight, but, in general, our position is that so much has been written on the subject, with so little positive outcome, that another dedicated review was not warranted (see Chapter 1).

Mesozoic Birds has been a long and difficult project, but we believe the resulting volume is very satisfying. For the first time, most of the major primary workers on Mesozoic birds have been assembled to provide authoritative treatment of the subject from a variety of angles. The authors hail from ten different countries and all six inhabited continents—a truly international effort. The volume is more than a reference for the anatomy and systematics of Mesozoic birds—although we think that it performs that important task quite well. It also presents new ideas, new approaches, and new perspectives. As a result—although we always knew it would be valuable to those interested in birds and dinosaurs—we are confident that all vertebrate paleontologists and indeed most evolutionary biologists will find much of interest in its pages.

This volume would have not been possible without the help of a number of individuals and institutions. Special acknowledgment is given to Stacie Orell, a former volunteer at the American Museum of Natural History, who devoted many days to editing many of the manuscripts and bringing them to the required style and format. We are also grateful to many colleagues for acting as chapter reviewers and to two anonymous referees who painstakingly reviewed the whole volume and made numerous valuable suggestions.

The Frick and Chapman Funds of the American Museum of Natural History, the J. S. Guggenheim Foundation, the Dinosaur Society, and the National Science Foundation (DEB-9407999; DEB-9873705) were the major sources of funding supporting Luis Chiappe's contributions to this volume. Lawrence Witmer was supported by the National Science Foundation (BSR-9112070; IBN-9601174), the Dinosaur Society, and grants from the Ohio University College of Osteopathic Medicine.

Finally, we are grateful to the staff of the University of California Press, especially Elizabeth Knoll and Doris Kretschmer, who over the years provided continuous support and assistance in seeing this project through to publication, and to George Olshevsky for his efforts in performing the daunting task of indexing the book.

Literature Cited

Chatterjee, S. 1997. The Rise of Birds. Johns Hopkins University Press, Baltimore, 312 pp.

Chiappe, L. M. 1991. Cretaceous avian remains from Patagonia shed new light on the early radiation of birds. Alcheringa 15(3–4):333–338.

———. 1995. The first 85 million years of avian evolution. Nature 378:349–355.

———. 1997. Aves; pp. 32–38 in P. Currie and K. Padian (eds.), The Encyclopedia of Dinosaurs. Academic Press, New York.

Clarke, J. A. 2002. The morphology and systematic position of Ichthyornis Marsh and the phylogenetic relationships of basal Ornithurae (dissertation). Yale University, New Haven, 532 pp.

De Beer, G. 1954. Archaeopteryx lithographica: A Study Based upon the British Museum Specimen. British Museum (Natural History), London, 68 pp.

Desmond, A. 1982. Archetypes and Ancestors. University of Chicago Press, Chicago, 287 pp.

Feduccia, A. 1996. The Origin and Evolution of Birds. Yale University Press, New Haven, 420 pp.

Forster, C. A., S. D. Sampson, L. M. Chiappe, and D. W. Krause. 1998. The theropod ancestry of birds: new evidence from the Late Cretaceous of Madagascar. Science 279:1915–1919.

Kurochkin, E. 1985. A true carinate bird from the Lower Cretaceous deposits in Mongolia and other evidence of early Cretaceous birds in Asia. Cretaceous Research 6:271–278.

Marsh, O. C. 1880. Odontornithes: a monograph on the extinct toothed birds of North America. Memoirs of the Peabody Museum of Natural History 1:1–201.

Martin, L. D. 1983. The origin and early radiation of birds; pp. 291–338 in A. H. Bush and G. A. Clark Jr. (eds.), Perspectives in Ornithology. Cambridge University Press, New York.

Molnar, R. 1986. An enantiornithine bird from the Lower Cretaceous of Queensland, Australia. Nature 322:736–738.

Olson, S. L., and D. C. Parris. 1987. The Cretaceous birds of New Jersey. Smithsonian Contributions to Science 63:1–22.

Padian, K., and L. M. Chiappe. 1998. The origin and early evolution of birds. Biological Reviews 73:1–42.

Sereno, P. C. 1998. A rationale for phylogenetic definitions, with application to the higher-level taxonomy of Dinosauria. Neues Jahrbuch für Geologie und Paläontologie Abhandlungen 210:41–83.

———. 1999. Definitions in phylogenetic taxonomy: critique and rationale. Systematic Biology 48:329–351.

Walker, C. A. 1981. New subclass of birds from the Cretaceous of South America. Nature 292:51–53.

Part I

The Archosaurian Heritage of Birds

1

The Debate on Avian Ancestry
Phylogeny, Function, and Fossils

LAWRENCE M. WITMER

 Ever since the time of Huxley (1868), the debate on the place of birds within vertebrate phylogeny has been one of the highest-profile and hotly contested of all evolutionary debates. What makes the origin of birds so contentious? Why can we not agree and then move on to other issues? Will we ever regard the problem as solved? It would seem that establishing the ancestry of birds would be a relatively simple, empirical task, but, despite abundant data, the debate rages like never before. I previously published a fairly exhaustive review of the history of the controversy (Witmer, 1991), which has been updated and expanded by Feduccia (1996, 1999b) and Padian and Chiappe (1998b). Rather than repeat the historical chronicle, this chapter seeks to explore the impact of more recent developments on the debate and also to touch on the allied debate on the origin of avian flight. At the time of completing my previous review, it is fair to say that the theropod hypothesis was the most widely held notion, and other ideas (e.g., basal archosaurs, crocodylomorphs) were on the decline. There simply was little substantial opposition to birds being dinosaurs.

In the intervening years, however, a great deal has happened. *Protoavis texensis* moved from the popular arena to the scientific arena with the publication of Sankar Chatterjee's 1991 monograph (see also Chatterjee, 1995, 1997a, 1998b, 1999; Witmer, 2001b). Alick Walker (1990) published a large monograph on *Sphenosuchus acutus* in which he renewed his support for the relationships of birds to crocodylomorphs. Relationships of the basal archosauriform *Euparkeria capensis,* previously little more than a historical footnote, resurfaced in a paper by Johann Welman (1995; but see Gower and Weber, 1998). Various conferences provided a forum for discussion of new hypotheses and perspectives, such as the 1994 meeting of the International Ornithological Congress in Vienna (Bock and Bühler, 1995) and the 1992, 1996, and 2000 meetings of the Society of Avian Paleontology and Evolution in Frankfurt (Peters, 1995), Washington (Olson, 1999), and Beijing (Shi and Zhang, 2000), respectively. The late 1990s saw the publication of two important and high-profile reviews of early avian evolution in the science weeklies (Chiappe, 1995; Feduccia, 1995), a seminal and wide-ranging review (Padian and Chiappe, 1998b), a major symposium with subsequent volume (Gauthier and Gall, 2001), and a host of books directed toward a broader audience (Feduccia, 1996, 1999b; Chatterjee, 1997a; Dingus and Rowe, 1997; Shipman, 1997a; Paul, 2002).

Most important, however, are the many newly discovered testaments of the avian transition coming from the fossil record. For example, as documented in this volume, many new fossil birds relevant to avian origins have recently been described, primarily from Spain, China, Argentina, Madagascar, and Mongolia, and, from Germany, even new specimens of *Archaeopteryx*. Likewise, a number of new birdlike theropods have come to light, such as *Sinornithoides youngi* (Russell and Dong, 1993), *Unenlagia comahuensis* (Novas and Puerta, 1997), *Sinornithosaurus millenii* (Xu et al., 1999b), *Caudipteryx zoui* (Ji et al., 1998), *Bambiraptor feinbergi* (Burnham et al., 2000), *Nomingia gobiensis* (Barsbold et al., 2000a,b), *Sinovenator changii* (Xu et al., 2002), and *Microraptor zhaoianus* (Xu et al., 2000), as well as important new specimens of previously known taxa such as *Troodon formosus* (Currie and Zhao, 1993), *Deinonychus antirrhopus* (Witmer and Maxwell, 1996; Brinkman et al., 1998), *Velociraptor mongoliensis* (Norell and Makovicky, 1997, 1999; Norell et al., 1997), oviraptorids (Clark et al., 1999), and ornithomimosaurs (Pérez-Moreno et al., 1994; Makovicky and Norell, 1998; Norell et al., 2001), among others.

The debate heated up considerably with the publication in 1996 of Alan Feduccia's *The Origin and Evolution of Birds.* In this book, Feduccia launched a vehement attack on the prevailing consensus that birds are nested within Theropoda, opening the door to other studies purporting to cast doubt on the theropod relationships of birds (e.g., Burke and Feduccia, 1997; Ruben et al., 1997; Feduccia and Martin, 1998; Martin, 1998; see also Thomas and Garner, 1998). Some of the scientific points discussed by Feduccia and his colleagues will be taken up in this chapter, but it is worthwhile to examine briefly the tone the debate has taken and direction to which it has turned because they cannot help affecting the science. The initial response to the volume (e.g., Norell and Chiappe, 1996; Padian, 1997) was highly critical of Feduccia's apparent disregard for the recent cladistic analyses that argue for the theropod relationships of birds, although several later reviews were either mixed (Sereno, 1997a; Witmer, 1997b; Steadman, 1998) or highly favorable (Bock, 1997; Mayr, 1997; Ruben, 1997). The popular press was quick to provide an opportunity for the players to voice strong opinions. Among the many barbs slung back and forth to reporters were "paleobabble," "total garbage," "pure Fantasyland," "the greatest embarrassment of palaeontology of the 20th century," "absurdity," "poisoning his own discipline," "as impervious to evidence as the fundamentalists," "beyond ridiculous," "just hot air," "bombastic rhetoric and armwaving," "like taking candy from a baby," and "that's always the problem, these paleontologists just don't know birds." In order not to further the ad hominem tone of the discourse, these remarks will remain unattributed here, but the reader may consult McDonald (1996), Zalewski (1996), DiSilvestro (1997), Morell (1997), and Shipman (1997b). Clearly, the vituperative and combative turn that the debate has taken is likely to do little to advance the science—bombast is a poor substitute for evidence.

Perhaps a more important question is, Why is the origin of birds so important that professional scientists would make such strong public statements? In other words, what is the larger meaning of the debate? It is impossible to escape the fact that, as with any form of human endeavor, personalities may collide (and the media will be there with a microphone). But the *scientific* stakes really are quite high, and for at least two reasons: crownward inferences and stemward inferences. First, the origin of birds is critical if we are to gain a deeper understanding of birds themselves. Establishing the phylogenetic relationships of Aves is the logical first step in a wide variety of inferences such as the evolution of flight, feathers, metabolism, and various ecological and physiological parameters. All these attributes have a phylogenetic history, and thus we must know what came before birds in order to truly comprehend birds. If we are interested in, say, tracing the evolution of avian craniofacial kinesis, it makes a great difference whether birds are viewed as being derived from theropod dinosaurs (Chatterjee, 1991) or from crocodylomorphs (Walker, 1972).

Second, there is a sense that if we can sort out the origin of birds, we will *automatically* know a great deal about the biology of their antecedents. For example, if birds are descended from theropod dinosaurs, then we might feel justified in reconstructing nonavian theropods with a whole suite of avian attributes, from feathers and endothermy (Paul, 1991) to reproductive biology (Varricchio et al., 1997; Clark et al., 1999) and locomotor attributes (Gatesy, 1990; Gatesy and Dial, 1996a). Indeed, there may be compelling reasons to do so, and, according to the inferential hierarchy of Witmer (1995a), these might be reasonable level II inferences. If the prevailing orthodoxy is correct, then all dinosaurs are *not* extinct, and we thus have the potential to say a lot about even the extinct clades of dinosaurs. If, however, birds have no close relationship with dinosaurs, then all of a sudden dinosaurs seem more remote, less familiar, and perhaps even less interesting. Feduccia correctly noted that the theropod hypothesis provides "a mechanism by which you can vicariously study dinosaurs by stepping into your backyard. There's a real emotional investment here" (quoted in Zalewski, 1996:24). In fact, for studies that employ the extant phylogenetic bracket approach to make inferences about extinct archosaurs (e.g., Witmer, 1995a,b, 1997a, 2001a; Rowe, 2000; Hutchinson, 2001a,b), it is particularly fortunate and useful for birds to be dinosaurs because then extant birds and crocodilians together bracket a huge diversity of archosaurs. If birds are instead more closely related to, say, crocodylomorphs, then our inferential base with regard to other archosaurs is weakened significantly. Thus, resolution of the problem is critical for studies not just of birds and theropods but really of all archosaurs.

This chapter examines a variety of issues surrounding the debate. In all cases, the goal is to discover how that particular issue relates and contributes to the *resolution* of the question of avian origins. The next section takes up the central role that *Archaeopteryx* has played in the debate, and the following section briefly examines the controversial taxon *Protoavis*. The discovery of an incredible fossil deposit in China holds great importance for understanding avian origins and is discussed here. The relationship of the origin of flight to the origin of birds, which has reemerged as a critical topic with the publication of Feduccia's 1996 book, will then be taken up. The question of alternatives to theropod dinosaurs will then be explored, followed by an assessment of the status of the theropod hypothesis itself.

The Centrality of *Archaeopteryx* in the Debate

The first sentence of the abstract of John H. Ostrom's 1976 landmark paper states: "The question of the origin of birds can be equated with the origin of *Archaeopteryx,* the oldest known bird" (Ostrom, 1976:91). This statement reflects a very common sentiment, with, for example, Martin (1991: 485) regarding *Archaeopteryx* as occupying "the center stage" and Feduccia (1996:29) calling it an "avian Rosetta

Stone." Certainly, virtually all the ancillary issues surrounding avian origins—from the origin of flight (Padian, 1985; Rayner, 1991; Feduccia, 1993, 1996; Herzog, 1993; Gatesy and Dial, 1996b) to feathers (Parkes, 1966; Dyck, 1985; Griffiths, 1996), endothermy (Ruben, 1995, 1996), and others—either take *Archaeopteryx* as their starting point or use it as the ruler against which particular scenarios are measured. A high-profile international meeting was devoted entirely to *Archaeopteryx*, and, as the editors of the subsequent volume noted (Hecht et al., 1985:7), it probably was "the first time that a scientific conference was devoted to a single fossil species." Martin (1995:33) is no doubt correct in noting that "there are probably more individuals who have worked on *Archaeopteryx* than all other palaeoornithologists put together."

But *Archaeopteryx* is more than a series of scientific specimens (see Elzanowski, Chapter 6 in this volume). *Archaeopteryx* has reached iconic status, partly because of the beauty of the specimens and partly because they have been important and prominent documents in establishing the fact of organic evolution; certainly creationists have regarded *Archaeopteryx* as a serious challenge (e.g., Cousins, 1973). *Archaeopteryx* is a celebrity. A respected scientific periodical bears its name. It is the logo for museums and graces the covers of numerous books and magazines. As a fossil celebrity, *Archaeopteryx* is more similar to *Australopithecus* in inspiring a sense of respectful awe and reverence than to "pop favorites" like *Tyrannosaurus*. It is indisputable that *Archaeopteryx* has historically been the central focus of virtually all studies on the origin and early evolution of birds, and this is likely to continue for the foreseeable future.

The obvious next question is: Is this reliance—perhaps even overreliance—on *Archaeopteryx* justified and wise? The avian status of *Archaeopteryx* is often an unquestioned assumption of many analyses, and in some phylogenetic studies (e.g., Padian, 1982; Holtz, 1994; Weishampel and Jianu, 1996), *Archaeopteryx* alone stands as a proxy, in a sense, for all birds. The concern obviously is that if we have hung all our conclusions about avian evolution on a fossil that is peripheral to avian origins, then much of the research for the past century may have been misguided (Witmer, 1999). The stakes could not be higher. Looking carefully back over the history of this debate reveals a small but continuous thread of dissent (see Witmer, 1991). P. R. Lowe (1935, 1944), for example, considered *Archaeopteryx* to be too specialized to be ancestral to later birds. Moreover, he went so far as to claim that *Archaeopteryx* "represented the culminating attempt of the reptiles toward flight, that is to say, it was a flying dinosaur. This, of course, implies a belief in the diphyletic origin of feathers—a zoological transgression for which I expect no mercy" (Lowe, 1935: 408–409). Lowe (1935:409) denied any features that were "definitely avian as opposed to dinosaurian," and he spent much of the 1944 paper attempting to refute the most bird-

like characters (e.g., the boomerang-shaped element could not be a furcula because *Hesperornis, Palaelodus,* and embryonic birds have unfused clavicles). Lowe's views on avian origins were complex and somewhat idiosyncratic (Witmer, 1991) and were refuted to most people's satisfaction by Simpson (1946) and de Beer (1954, 1956).

Although it is tempting to marginalize Lowe's views because they were expressed so long ago, a number of modern workers have questioned the avian status of *Archaeopteryx*. For example, in a series of papers from 1975 to 1984, R. A. Thulborn moved progressively closer to the conclusion that *Archaeopteryx* was a theropod dinosaur of no particularly close relationship to birds (Thulborn, 1975, 1984; Thulborn and Hamley, 1982); his 1984 paper presented a cladogram in which tyrannosaurids, an ornithomimosaur-troodontid clade, and *Avimimus* were closer to birds than was *Archaeopteryx*. He denied virtually all characters that unite *Archaeopteryx* with true birds to the exclusion of other theropod groups, noting that furculae were present in a variety of theropods and that the presence of feathers in *Archaeopteryx* "probably signifies nothing more than a rare circumstance of preservation" (Thulborn, 1984:145). Thus, for Thulborn (1984:151), "*Archaeopteryx* is not a convincing 'intermediate' between reptiles and birds, nor is it an ancestral bird" because there are other theropods even more birdlike than *Archaeopteryx*. Likewise, Kurzanov (1985, 1987) was so struck by the birdlike features of the Cretaceous theropod *Avimimus portentosus* (see also Vickers-Rich, Chiappe, and Kurzanov, Chapter 3 in this volume)—not least of which was the inference of feathered forelimbs—that he regarded the supposed avian features of *Archaeopteryx* as insufficient evidence to establish it as a true bird.

A similar theme was elaborated somewhat earlier by the Mongolian paleontologist Barsbold Rinschen, who articulated a notion that has major implications for the interpretation of the avian attributes of not just *Archaeopteryx* but really all theropods. Barsbold's (1983) concept of "ornithization" in theropod evolution suggests that various lineages of theropods independently evolved birdlike attributes but with no clade possessing the entire suite of avian apomorphies. Thus, *Archaeopteryx* is seen by Barsbold (1983) as potentially just one of several parallel, "ornithized" lineages, "aberrant" in and of itself yet otherwise showing the underlying affinity of birds and theropods. In fact, it was conceivable to Barsbold (1983) that perhaps more than one group of "ornithized" theropods crossed the line into "birds" and that neornithine birds may actually be diphyletic (reminiscent of Lowe) or even polyphyletic. It is not clear what kind of evolutionary process Barsbold envisioned that would produce the recurrent evolution of avian features, but the fact that avian features have arisen repeatedly and independently in theropod evolution now seems to be an inescapable conclusion.

The preceding discussion, of course, begs the question of just what is a bird: How do we recognize birds? How do we define them in both a scientific and a colloquial sense? The foregoing has presumed a more colloquial sense of birds, that is, a sense of "birds" being feathered vertebrates that fly or had flight in their ancestry. This idea generally works in the modern time plane but breaks down when evaluated over the fossil record of theropods because the attributes of extant birds were acquired sequentially. Thus, there is no sharp line demarcating bird and nonbird—the distinction has become entirely arbitrary. Defining taxa has emerged as a major focus for phylogenetic taxonomists, and *Archaeopteryx* in relation to the definition of "Aves" and "birds" has likewise become a central test case (see Gauthier, 1986; de Queiroz and Gauthier, 1990, 1992; Padian and Chiappe 1998b; Sereno, 1998, 1999b; Padian et al., 1999; see also Clark, Norell, and Makovicky, Chapter 2 in this volume). The issues surrounding this matter are lengthy and complex, and they have been well discussed in the references just cited.

Traditionally, *Archaeopteryx* has been regarded as both a bird and a member of Aves. For example, Richard Owen, in his 1863 description of the London specimen, "declare[d] it unequivocally to be a Bird, with rare peculiarities indicative of a distinct order in that class" (Owen, 1863:46). Likewise, Haeckel (1866, 1876), Huxley (1867), and other early taxonomists referred *Archaeopteryx* and other Mesozoic birds to Aves. All modern ornithology texts regard *Archaeopteryx* as within their purview. This traditional sense can be captured by modern phylogenetic taxonomy by defining the name "Aves" using a node-based definition: *Archaeopteryx,* Neornithes ("modern birds"), and all descendants of their most recent common ancestor (see Chiappe, 1992, 1997; Padian and Chiappe, 1998b; Sereno, 1998, 1999b; Padian et al., 1999). The colloquial term "birds" is usually applied to this same group. These definitions of Aves and birds are the ones adopted generally for this volume. However, an alternative nomenclature pioneered by Gauthier (1986) advocates a crown-group definition for Aves that encompasses just the clades of living birds: ratites, tinamous, neognaths, and all descendants of their common ancestor. Gauthier (1986) defined the term "Avialae" as more or less equivalent to the traditionally conceived Aves, yet he applied the colloquial term "birds" to Avialae. The Avialae convention has seen a fair amount of use, and with good reason (see Clark, Norell, and Makovicky, Chapter 2 in this volume). Nevertheless, at this writing, preferences seem to be tending toward the Aves convention adopted here (see Padian and Chiappe, 1998b; Sereno, 1998; and Padian et al., 1999, for justification). Interestingly, Padian et al. (1999) retained the term "Avialae" but redefined it as a stem-based taxon comprising Neornithes and all taxa closer to them than to the dromaeosaurid *Deinonychus.* Thus, if we regard the terms "Aves" and "birds" as being more or less synonymous, then *Archaeopteryx* is a bird, because we have defined it as such, as a member of the clade Aves.

Nevertheless, nomenclature aside, given the number of birdlike theropods and theropodlike birds, what is it about *Archaeopteryx* that leads us to conclude that it is truly a bird (i.e., by definition part of Aves) and thus worthy of all the attention? Long before the recent flurry of discoveries, workers recognized that *Archaeopteryx* possessed few uniquely avian characters that, in modern parlance, would contribute to a diagnosis of Aves (see Owen, 1863; Heilmann, 1926). For example, de Beer (1954:44) listed only four "features which differ completely from the condition in reptiles and agree with that of modern birds": (1) a retroverted pubis, (2) a furcula, (3) an opposable hallux, and (4) feathers. Of these, a pubis with at least some measure of apomorphic retroversion is now well documented for dromaeosaurids, basal troodontids, and therizinosauroids (Barsbold, 1979; Barsbold and Perle, 1979, 1980; Norell and Makovicky, 1997, 1999; Rasskin-Gutman, 1997; Xu et al., 2002), and furculae are turning out to be very widely distributed indeed among theropods (Barsbold, 1983; Bryant and Russell, 1993; Chure and Madsen, 1996; Norell et al., 1997; Dal Sasso and Signore, 1998; Ji et al., 1998; Makovicky and Currie, 1998; Norell and Makovicky, 1999; Xu et al., 1999a). Although the opposable hallux may still stand as an avian apomorphy (Gauthier, 1986; Chiappe, 1995; Feduccia, 1996; Sereno, 1997b; Forster et al., 1998), it is, of course, the presence of feathers that, since its discovery in 1861, has garnered for *Archaeopteryx* a seemingly unshakable position within Aves. However, even the uniqueness of feathers to birds was cast into doubt in 1996 with reports of nonavian theropod dinosaurs with feathers or featherlike filamentous structures from Liaoning, China (Ji and Ji, 1996, 1997 [see also Gibbons, 1996, 1997a, reporting on the work of Ji and Ji]; Currie, 1997; Chen et al., 1998; Ji et al., 1998; Xu et al., 1999a,b). Although the true identity of the filamentous structures has called into question (Brush et al., 1997; Geist et al., 1997; Gibbons, 1997b) and has been highly controversial, the reports fueled the persistent speculation, if not expectation, that feathers might have been present in taxa outside birds (Gauthier, 1986; Paul, 1991). The startling announcement in 1998 (Ackerman, 1998; Currie, 1998; Ji et al., 1998) of nonavian dinosaurs with indisputable feathers truly has shaken the foundations of just what it takes to recognize something as a bird (Padian, 1998). The importance of feathered dinosaurs is taken up in a later section.

Without the uniqueness of opisthopuby, furcula, and even feathers, is there any basis for uniting *Archaeopteryx* with birds to the exclusion of other archosaurian taxa? Definitions can be constructed to suit personal tastes, and, regardless of whether one defines Aves to exclude (Gauthier, 1986) or include (Chiappe, 1992) *Archaeopteryx,* it is relevant to ask just what features suggest that *Archaeopteryx* is the outgroup to less controversial birds. As it turns out, there are many such features. In addition to the reflexed hallux, other oft cited synapomorphies (see also Elzanowski, Chapter 6 in this volume) include (1) an elongate prenarial portion of the premaxilla (Gauthier, 1986; Sereno, 1997b); (2) the breaking down of the postorbital bar (Chatterjee, 1991, 1997a; Sanz et al., 1997; Sereno, 1997b); (3) the absence of dental serrations and the presence in the teeth of a characteristic constriction (Martin et al., 1980; Gauthier, 1986; Chiappe, 1995; Chiappe et al., 1996); (4) the enlargement of the cranial cavity (Gauthier, 1986; Chatterjee, 1997a); (5) the caudal tympanic recess opening within the columellar recess (Witmer and Weishampel, 1993; Chiappe et al., 1996); (6) the presence of a caudal maxillary sinus (Witmer, 1990; Chiappe et al., 1996; Chatterjee, 1997a); (7) fewer tail vertebrae, with the prezygapophyses reduced distally (Gauthier, 1986; Chiappe, 1995; Chiappe et al., 1996; Chatterjee, 1997a; Sereno, 1997b; Forster et al., 1998); and (8) various modifications of the shoulder girdle (Gauthier, 1986; Feduccia, 1996; Chatterjee, 1997a), although some of the shoulder characters may have a broader distribution (Novas and Puerta, 1997; Forster et al., 1998; Norell and Makovicky, 1999; Xu et al., 1999b, 2002).

Some of the foregoing characters may seem a bit subtle, even trifling, but such is the nature of any phylogenetic transition as it becomes better and better known. In fact, we should *predict* that the number of characters per node should decrease (and that the characters themselves may well seem more trivial) as sampling of the fossil record improves. What is remarkable is that so many characters do affirm the avian status of *Archaeopteryx*, thus providing ample justification for the sharp focus placed on *Archaeopteryx*. Although many of the characters listed previously are far from "clean," with homoplasy and missing data complicating the picture such that Chiappe (Chapter 20 in this volume) found only three unambiguous synapomorphies for the node Aves, *Archaeopteryx* is indeed avian. Moreover, it has very few autapomorphies (Gauthier, 1986; Chatterjee, 1991; Sereno, 1997b) and thus, in a strictly phylogenetic sense, may legitimately serve as a model for an avian ancestor.

However, there is a danger here. Even if *Archaeopteryx* is the best available *model* for an ancestral bird, the worry is that we might come to regard it as *truly* the first bird. Even the German common name for *Archaeopteryx*—Urvogel—carries this sense of being the very first bird. But *Archaeopteryx* had an evolutionary history, and, as with any organism, its phylogenetic heritage has an impact on not only its form and function but also our *interpretation* of its form and function. Thus, although *Archaeopteryx* may be our best and oldest evidence for, say, feathers, there is no guarantee that *Archaeopteryx* gives us any *direct* glimpse into the origin of feathers and flight because both pre-

sumably predate *Archaeopteryx* by perhaps millions of years. For example, whether or not there were trees in the Solnhofen landscape suitable for *Archaeopteryx* to perch upon (Peters and Görgner, 1992; Feduccia, 1993, 1996; Padian and Chiappe, 1998a,b) bears little relevance for the arboreal versus cursorial origin of flight. On the one hand, the presence of Solnhofen trees would not automatically mean that *Archaeopteryx* either used them or evolved from an arboreal ancestor. On the other hand, the absence of Solnhofen trees would not dictate that flight arose in a terrestrial context, because *Archaeopteryx* could well have evolved from fully arboreal ancestors and apomorphically became terrestrial or just visited Solnhofen seasonally. We can never know these things. The point here is that we have tended to run all hypotheses through the filter of *Archaeopteryx*, almost as if we believed that it was truly the *first* bird, not just the *oldest known* or *most basal* bird.

The historical centrality of *Archaeopteryx* in the debate is quite understandable given that, until very recently, it was all we had—that is, it provided the only solid evidential basis for hypothesis testing. Although most analyses continue to support the basalmost avian position for *Archaeopteryx*, recent discoveries of Cretaceous birds and bird-like theropods have considerably lightened the load that *Archaeopteryx* must bear in teasing apart the details of early avian evolution (see discussion in Witmer, 1999). From *Unenlagia*, *Rahonavis*, and *Microraptor* to *Confuciusornis*, *Sinornis*, and *Concornis*, new finds are documenting the details of both the phylogenetic and functional transition to birds. *Archaeopteryx* will always merit a special place in the minds (and hearts) of scientists and the public in general. Only recently, however, could *Archaeopteryx* assume its proper role in the drama of the avian transition as one of a number of important players in an ensemble cast.

The Significance of *Protoavis*

A particularly difficult question is whether this ensemble cast should rightly include the Texas fossils known as *P. texensis*. These fossils, whose discovery was announced only in 1986, have had a troubled and controversial history. In a long series of published works, Sankar Chatterjee (1987, 1988, 1991, 1995, 1997a, 1998b, 1999) argued that the *Protoavis* fossils are not only those of a bird but from a bird that lived 75 million years before *Archaeopteryx*! The fossils are generally attributed to two individuals excavated from the Early Norian Cooper Member of the Dockum Group of western Texas (Chatterjee, 1991), although other material from a different formation and county also was later referred to *P. texensis* (Chatterjee, 1995, 1997a, 1999; justification for this referral has not been presented, and I will not consider that material here). Although the report of any new Mesozoic bird is greeted with great interest, the reception of *Protoavis* was unique, largely because of the implications that a Triassic bird holds. Perhaps surprisingly, despite the great age of the fossils, Chatterjee has never argued for any major changes in the general notion of avian ancestry from dromaeosaurlike coelurosaurian dinosaurs; in other words, *Archaeopteryx* remains the basal bird, and the Ostrom/Gauthier hypothesis of theropod relationships is not challenged. The irony that emerges is that *Protoavis* perhaps should have relatively little relevance for the origin of birds in that, according to Chatterjee's cladograms, *Protoavis* is nested well within Aves.

Skepticism about the avian status of *Protoavis* was immediate and did not necessarily follow along lines of allegiance to any particular theory of avian origins. For example, Feduccia (1996) and Martin (1998), on the one hand, and Ostrom (1987, 1991, 1996), Wellnhofer (1992, 1994), Chiappe (1995, 1998), and Sereno (1997b, 1999a), on the other hand, have all expressed doubt that the fossils of *Protoavis* are adequate to substantiate the claim of a Triassic bird. At the same time, Chatterjee has had some supporters, including Peters (1994), Kurochkin (1995), and Bock (1997, 1999). Why the controversy? A fuller critical appraisal of the status of *Protoavis* is presented elsewhere (Witmer, 2001), but a brief analysis is presented here.

Detractors of *Protoavis* have raised a variety of complaints, the most important of which relate to the taphonomy of the specimens and their preservation and preparation. With regard to the taphonomy, there is a widespread concern that *P. texensis* is a chimera, that is, a mixture of more than one species. Indeed, the quarry from which the specimens derive is a multispecific bonebed that has recorded many taxa (Chatterjee, 1985), and thus mixing is a possibility, as has already been suggested for other taxa from the same quarry (e.g., *Postosuchus kirkpatricki;* Long and Murry, 1995). Chatterjee (1991, 1998b) has steadfastly maintained the association of the holotype and paratype skeletons of *Protoavis,* and Kurochkin (1995) offered his support based on his study of the original material. Nevertheless, the specimens were collected inadvertently while removing overburden with a jackhammer, and hence we can never be completely sure of the taphonomic setting. The possibility that *Protoavis* is a composite of several species is commonly voiced but, even if true, does not rule out the chance that some of the included bones are avian (Witmer, 1997c). However, as Chiappe (1998) correctly pointed out, the chimera problem presents itself most insidiously during phylogenetic analysis, the ultimate arbiter of avian origins, in that the mixture of taxa means a mixture of characters, all of which leads to phylogenetic nonsense. Thus, the taphonomic question—Is *Protoavis* a species or a fauna?—is a critical one, and one that is not likely to go away until new material is discovered.

And, according to many of those who have studied the specimens, new material is needed desperately. In other words, a major concern has been that the *Protoavis* specimens are simply too poorly preserved, too scrappy, to be diagnostic. Unlike *Archaeopteryx* and the Cretaceous birds from Spain and China, *Protoavis* is not a "slab animal"; that is, it is not preserved in situ, but rather all the bones have been prepared free of the matrix. Thus, without the aid of the positional information that slab animals preserve, the identification of isolated elements is difficult and can lead to widely different interpretations. The other side of this coin is that all sides of the elements are available for study rather than being half entombed in stone, as is the case for slab birds such as *Archaeopteryx.* Nevertheless, it has been difficult to confirm not only many of the structures but even some of the bone identifications made by Chatterjee (Witmer, 2001b). Perhaps Padian and Chiappe (1998b:13) best characterized the situation by noting that the "material has become a paleontological Rorschach test of one's training, theoretical bias, and predisposition." Coupled with this is the problem that the specimens are extensively reconstructed with plaster and epoxy, and it often seems that the published descriptions are of these reconstructed composites rather than of the fossils themselves.

But ultimately—even given the gravity of these and other concerns—it comes down to the fossils and their structures. Are there clearly interpretable anatomical clues revealing the phylogenetic relationships of the beast? Again, a more comprehensive skeletal analysis is presented elsewhere (Witmer, 2001b), but it is worthwhile to examine here a few of the more important anatomical systems. For Chatterjee (1991, 1995, 1997a, 1998b), the skull, particularly the temporal region, is the most critical, because he regarded *Protoavis* as possessing the ornithurine condition: that is, loss of the postorbital bone, leading to confluence of the orbit, dorsotemporal fenestra, and laterotemporal fenestra. Given that *Archaeopteryx, Confuciusornis sanctus* (Peters and Ji, 1998; Chiappe et al., 1999; Hou et al., 1999), and at least some enantiornithines (e.g., the Catalan hatchling, Sanz et al., 1997; *Protopteryx fengningensis,* Zhang and Zhou, 2000) retain both the postorbital bone and its contact with the squamosal, presence of the advanced ornithurine condition in *Protoavis* indeed would be highly significant and would, in fact, argue for a higher phylogenetic position within birds than that advocated by Chatterjee (1998b). Unfortunately, the temporal region of the holotype skull has been assembled from disarticulated pieces, and thus the identity and positions of elements are not certain. The most telling clues are found in the squamosal and quadrate; the absence of the postorbital is negative evidence and hence hard to evaluate. The squamosal identification is key, because the element would lack an articular surface for the postorbital, which implies absence of the postorbital bone itself. Similarly, the putative quadrates are important, because they would be drastically modified along the lines of birds, presumably, according to Chatterjee, in association with avian craniofacial kinesis. The squamosal identification is defensible in that the element in question has a cotyla that could receive a quadrate, but other interpretations are possible. The quadrate identifications are less certain (see Witmer, 2001b). My general impression of the reconstructed temporal region is that Chatterjee's view is understandable and justifiable but is not sufficiently clear to merit drawing firm phylogenetic conclusions.

Without question, the braincase is the most easily interpreted part of the skull, at least with respect to bone identifications. It is doubtful that there are any indisputably avian apomorphies in the braincase. However, it is indeed the braincase of a coelurosaur in that it possesses cranial pneumatic recesses (including the caudal tympanic recess, which thus far is not known outside Coelurosauria), a large cerebellar auricular fossa, a metotic strut, and a vagal canal opening onto the occiput (Chatterjee, 1991, 1998b; Witmer, 1997d, 2001b), and thus the braincase may pertain to the oldest known coelurosaur.

Postcranially, little is unambiguously avian. Exceptions are the cervical vertebrae, which are truly heterocoelous, if only incipiently so; have prominent ventral processes (hypapophyses); and have large vertebral foramina. Of course, heterocoely has a fairly homoplastic distribution within birds (Martin, 1983; Chiappe, 1996), and *Protoavis* is not as heterocoelous as *Hesperornis* or most neornithines, but the vertebral structure nevertheless represents one of the few bona fide avian suites of *Protoavis*. There are many problems in the thoracic appendage (Witmer, 2001b), not least of which is the coracoid, which, although having a generally advanced avian shape, seems positively minuscule in comparison with the rest of the skeleton. I cannot confirm the remigial papillae on the ulna or manus. In fact, the identification of the four-digit manus itself has been called into question, with Sereno (1997b) regarding it as the foot of an archosaur. The pelvic appendage likewise is not particularly birdlike. Perhaps the most avian feature is a medial fossa within the os coxae regarded by Chatterjee (1995, 1998b) as a renal fossa; however, no other Mesozoic bird has a renal fossa, and the structure in *Protoavis* differs somewhat from that in neornithines (Witmer, 2001b).

It is probably fair to state that the case for the avian status of *P. texensis* is not as clear as generally portrayed by Chatterjee. The temporal configuration and vertebral morphology might argue for a position near Ornithurae, yet the long tail, the archaic ankle, the four-digit manus, and other features would argue for a basal position, probably well outside Aves. The braincase is more or less coelurosaurian. The taphonomic problems and the possibility that this is a chimera may make this an intractable problem. An option is simply to take Chatterjee's analyses at face value and proceed (as done by Dyke and Thorley, 1998), but this seems

naive at best. Critical study of the original material is absolutely necessary, but even here, in the absence of newly collected material of known association, firm conclusions will likely be elusive.

It would thus seem that *Protoavis* bears little significance for solving the riddle of the origin of birds. The specimens themselves are problematic, and so we rightly should be skeptical. But even assuming that Chatterjee has interpreted them 100% correctly, *Protoavis* would have little impact on the phylogenetic pattern of avian origins (Witmer, 1997c, 2001b; Dyke and Thorley, 1998). Then why the often vitriolic controversy? The reasons are complex, but probably a major component is that Chatterjee's acceptance of the Ostrom/Gauthier orthodoxy would require that virtually all theropod cladogenesis had taken place well back in the Triassic, at least in the Norian, if not earlier—that is, right at the very dawning of the dinosaurs. I have elsewhere (Witmer, 2001b) referred to this (somewhat whimsically) as the "Norian Explosion." The problem is that we have no real evidence of such an explosion: no Norian tyrannosaurids, no Norian oviraptorosaurs, etc. Thus, many people simply have not accepted this proposition. This incongruity has not gone unnoticed and has been exploited by opponents of the idea of theropod relationships. For example, Martin (1988), Tarsitano (1991), and Bock (1997) were receptive to the avian status of *Protoavis* and pointed out that a Triassic bird would essentially disprove the prevailing notion of theropod relationships.

But a long view is appropriate. The origins of many theropod groups are constantly being pushed back further in time. Therizinosauroids have been reported from the Early Jurassic of China (Zhao and Xu, 1998; Xu et al., 2001a), and Chatterjee (1993) reported an ornithomimosaur from the Late Triassic of Texas, although such claims generally are controversial (Rauhut, 1997). As mentioned, *Protoavis* itself represents a temporal range extension for Coelurosauria. Whether the idea of a Triassic bird will ever be more palatable is hard to predict, but stranger things have happened in the history of science, and *Protoavis* may yet prove to be a key player. For the present, however, it is probably both prudent and justifiable to minimize the role that *Protoavis* plays in any discussions of avian ancestry.

The Significance of the Feathered Chinese Dinosaurs

The 1990s will go down in history as a time when one of the most significant fossil deposits ever discovered was brought to light. The Lower Yixian Formation (Chaomidianzi Formation of some) and allied rock units in western Liaoning Province, People's Republic of China, have yielded a wealth of fossil vertebrates, preserving—in often astounding abundance—an entire fauna in all its diversity (Luo, 1999;

Swisher et al., 1999). The basal birds from these deposits are discussed in this volume by Zhou and Hou (Chapter 7). With regard to the origin of birds, several additional taxa are relevant, particularly because of the preservation of integumentary remains interpreted to be feathers or featherlike filaments. At this writing, six theropod taxa (other than the indisputable birds) have been reported to have "feathery" skin (with, no doubt, more taxa on the way): *Sinosauropteryx prima* (Ji and Ji, 1996; Chen et al., 1998), *Protarchaeopteryx robusta* (Ji and Ji, 1997; Ji et al., 1998), *Caudipteryx* spp. (Ji et al., 1998; Zhou and Wang, 2000; Zhou et al., 2000), *Beipiaosaurus inexpectus* (Xu et al., 1999a), *Sinornithosaurus millenii* (Xu et al., 1999b; Ji et al., 2001, if the juvenile dromaeosaurid pertains to this species), and *Microraptor zhaoianus* (Xu et al., 2000). The obvious significance for the debate on avian origins is that if feathers are truly present in nonavian theropod dinosaurs, then this should effectively close the door to any opposition to the theropod hypothesis. The debate, for all intents and purposes, will be over.

Thus, it is necessary to assess these claims carefully. See also the chapters in this volume by Clark, Norell, and Makovicky (Chapter 2) and Zhou and Hou (Chapter 7) for their assessments. The controversy began when Ji and Ji (1996: translation courtesy of Chen P.-J. and P. J. Currie) identified feathers in *Sinosauropteryx* and argued that (1) they were similar to modern down in lacking rachis and barbs, and (2) they were restricted to a median frill running from the head to the tip of the tail dorsally and onto the ventromedian surface of the tail. For Ji and Ji (1996), the presence of feathers required the referral of *Sinosauropteryx* to Aves. In the subsequent furor, there was a retreat from their interpretation as true feathers, being instead "protofeathers" (e.g., Brush et al., 1997). Moreover, the status of *Sinosauropteryx* as a bird was questioned, as it clearly had the skeletal anatomy of a small theropod dinosaur. Indeed, Chen et al. (1998), based on additional specimens, formally referred *Sinosauropteryx* to Compsognathidae, a clade of relatively basal coelurosaurs. These authors also presented the first in-depth morphological analysis of the integumentary structures, describing them as coarse, probably hollow, filaments up to 40 mm in length; a chemical or elemental analysis has not been published. The ultimate question, of course, is, What makes these feathers or even "protofeathers" (Unwin, 1998)?

Indeed, Geist et al. (1997) and Feduccia (1999b) suggested that the structures in *Sinosauropteryx* were in fact *not* feathers at all and, moreover, that they were not external, epidermal appendages of any kind. Rather, they argued that, based on comparative anatomy, the fossil structures more closely resembled collagenous fibers supporting a midsagittal dermal frill, that is, *internal* structures that became frayed in the process of decomposition. Although the apparent midline distribution of the filaments is indeed fully consistent with dermal frills, which are widely present in modern squamates and even well known in some dinosaur groups (e.g., sauropods: Czerkas, 1994; hadrosaurids: Lull and Wright, 1942), the filaments are not actually in the median plane in all regions but rather are in some places offset, such as the head region, which is not preserved in a straight lateral view (Padian et al., 2001). In fact, a routine finding with the Liaoning birds and dinosaurs is that the feathers or filaments are preserved as a halo around the skeletal remains. Thus, the "midline frill" is perhaps more safely interpreted as an artifact resulting from the animals being preserved lying more or less on their sides, such that the halo would roughly correspond to the median plane. Geist et al. (1997) suggested that another problem with the feather interpretation in *Sinosauropteryx* is that although some specimens may show a ruffle of fibers extending along the tail, another specimen shows a smooth outline along the tail.

It is valid to question whether these shortcomings falsify the feather hypothesis or may simply be ascribed to vagaries of preservation. Nevertheless, the inference of feathers in *Sinosauropteryx* has such profound implications— not only for the origin of birds but also for the origin of feathers and endothermy—that we should be compelled by the weight of evidence before accepting such momentous claims. In the acknowledged absence of calamus, rachis, and barbs, the identification of these structures as "true" feathers in *Sinosauropteryx* is clearly unjustified. Also problematic is the inference of "protofeathers." Although true feathers certainly had epidermal precursors that lacked such definitive attributes as rachis and barbs, how would we recognize them? Chemical analysis showing unique feather proteins might provide valid evidence, but again such studies have not been performed. Significantly, according to Prum's (1999, 2000) developmental model of feather evolution, the filaments of *Sinosauropteryx* are entirely consistent with an early stage of feather evolution. Moreover, Padian et al. (2001) argued that these filaments have enough morphological attributes in common with feathers that it is fair to accept that the filaments pass the similarity test of homology with avian feathers. Finally, the notion of feather precursors in *Sinosauropteryx* is significantly enhanced by the feathered theropods from the Yixian discussed later, leading Padian (1998:729) to state that "doubts [raised by Geist et al. (1997)] can now be put to rest." Thus, in effect, the filaments of *Sinosauropteryx* might be regarded as passing the congruence test of homology, as well (Padian et al., 2001). Nevertheless, the evidence in *Sinosauropteryx* obviously should be judged on its own merits, and stemward inferences based on crownward observations require considerable justification (i.e., level II inference; Witmer 1995a).

Shortly after the announcement of *Sinosauropteryx*, Ji and Ji (1997) announced the discovery of another feathered creature, *Protarchaeopteryx robusta*. They regarded it as the sister group of *Archaeopteryx*, even placing it within Archaeopterygidae, but Ji et al. (1998) removed it from a position within Aves. Unlike the case of *Sinosauropteryx*, the feathers attributed to *Protarchaeopteryx* are absolutely indisputable, with clear rachis and barbs. Thus, if a phylogenetic placement outside birds is justified, then a feathered nonavian theropod would be at hand. Unfortunately, the unique specimen of *Protarchaeopteryx* (NGMC 2125) is quite poorly preserved, and many attributes either are open to interpretation or beyond reliable observation (e.g., about 50% missing data according to Ji et al., 1998). I was unable to confirm the two plesiomorphies identified by Ji et al. (1998) that would deny *Protarchaeopteryx* a higher position—a short frontal process of the premaxilla and serrated teeth. The premaxilla is badly damaged, and the teeth seemed to lack clear serrations; Ji et al. (1998) regarded the serrations as so small (7–10/mm) that they were not visible even with my hand lens, but one then wonders if something so small can be regarded as truly a "serration." It may lack a reversed hallux, which would be an important plesiomorphy, but neither foot of the holotype is well preserved. Given the current state of our knowledge of *Protarchaeopteryx*, it is difficult to predict whether better specimens will show it to be outside or within Aves. It is even conceivable that a sister group relationship with *Archaeopteryx*, as originally suggested by Ji and Ji (1997), will be borne out (see Elzanowski, Chapter 6 in this volume). Certainly, *Protarchaeopteryx* is very close to the transition to birds, which makes its state of preservation all the more frustrating.

Much better preserved, however, is the material of *Caudipteryx* (Ji et al., 1998; Zhou and Wang, 2000; Zhou et al., 2000). *Caudipteryx* in many ways seems to be the perfect "feathered dinosaur." It possesses clearly "avian" feathers (i.e., with calamus, rachis, and barbs), yet, unlike those of *Archaeopteryx*, these feathers are not part of a flight apparatus, and hence *Caudipteryx* obviously did not fly. Moreover, *Caudipteryx* lacks many of the derived bony features unique to "proper" birds and hence has justifiably been hailed as the first animal to be discovered that is both indisputably feathered and indisputably *not* a bird. Indeed, *Caudipteryx* truly begs the question of just what may be called a "bird" in the colloquial sense of the word.

As mentioned, feathers are known for the two widely studied specimens described by Ji et al. (1998); the several new specimens reported by Zhou and Wang (2000) and Zhou et al. (2000) confirm a consistent pattern. In their preserved state, well-developed feathers are largely restricted to the manus and distal portion of the tail. As far as can be discerned, the inner and outer vanes are symmetrical about the rachis. Ji et al. (1998) and Zhou and Wang (2000) reported preservation of filamentous structures in the body regions, but the real question is whether the distribution of true feathers was more extensive in life or actually restricted to the tips of the hands and tail.

The association of true feathers with the skeletons of *Caudipteryx* is beyond any doubt, which is important because the skeleton is decidedly nonavian—that is, this is no chimeric association. The following discussion is not intended to be a description of the bony anatomy of *Caudipteryx* but rather a tabulation of its *primitive, nonavian* attributes (see also Zhou and Wang, 2000). Although it is more customary in this cladistic age to enumerate *derived* characters, documentation of the *primitive* characters of this feathered creature is necessary to counter claims that *Caudipteryx* is in fact "a secondarily flightless bird, a Mesozoic kiwi" (Feduccia, 1999a:4742; 1999b; see also Jones et al., 2000b). In addition to feathers, another significant avian apomorphy would be the shortened tail. In *Caudipteryx* there are only 22 caudal vertebrae, the same number as in *Archaeopteryx* and fewer than in any other known nonavian theropod (Ji et al., 1998). Moreover, the distal portion is clearly very stiff, although, as correctly noted by Ji et al. (1998; see also Zhou et al., 2000), definitely not fused into a pygostyle (or a "protopygostyle," as Feduccia [1999a] called it). Other than its short length and distal stiffening, nothing about the tail is particularly birdlike. Its distal caudal vertebrae have very short centra (Ji et al., 1998), rather than the elongate distal centra observed in *Archaeopteryx* and *Rahonavis* (Forster et al., 1998). Moreover, the proximal caudal haemal arches (chevrons) are very long and spatulate, again unlike those of basal birds and unlike those of even most derived nonavian coelurosaurs. The closest match to the tail of *Caudipteryx* may well be among oviraptorosaurs. As particularly well demonstrated by *Nomingia* (Barsbold et al., 2000a,b), oviraptorosaurs display the following derived characters: a reduced number of caudal vertebrae (24 in *Nomingia*—only 2 more than in *Caudipteryx*), rigid distal tail with short centra, relatively long transverse processes on the proximal caudals (Sereno, 1999a), and elongate and spatulate haemal arches. Although Barsbold et al. (2000a,b) regarded the tail of *Nomingia* as bearing a "pygostyle," it is certainly not homologous (or even that similar) to the avian structure, and I would tend to reserve that name for pygostylian birds (see Chiappe, Chapter 20 in this volume). In any event, the shortened tail of *Caudipteryx* is not particularly birdlike and is basically matched by the tails of oviraptorosaurs.

Zhou et al. (2000) advanced a few additional birdlike characters that, although they still regarded *Caudipteryx* as a nonavian dinosaur, "indicate that its phylogenetic position remains a debatable issue." Not having examined their new specimens firsthand, I cannot comment in detail on the birdlike attributes, but a few points are pertinent. Of the birdlike characters that they advance, some are clearly homoplasies (e.g., manual phalangeal formula of 2-3-2), some are more widely distributed in maniraptorans (e.g., tooth form, uncinate processes), and some are open to interpretation (e.g., the "partially reversed" hallux). They also pointed to the remarkably short trunk (only nine thoracic vertebrae) and elongate hindlimbs, birdlike features that had earlier attracted the attention of Jones et al. (2000b).

Jones et al. (2000b) argued that *Caudipteryx* possessed a strikingly birdlike attribute relating to the location of the center of mass and the proportions of the trunk and hindlimb; these parameters were entirely unlike those of any known nonavian theropods but indistinguishable from those of cursorial birds. They provided three alternatives to explain these data. First, perhaps simply *Caudipteryx* apomorphically and convergently developed a locomotor style similar to that of cursorial birds. Second, perhaps *Caudipteryx* was a nonavian theropod that had flight in its ancestry. And third, perhaps *Caudipteryx* was in fact "a secondarily, flightless, post-*Archaeopteryx*, cursorial bird" (Jones et al., 2000b). The authors clearly favor this third hypothesis. Testing all three hypotheses is firmly within the realm of phylogenetic analysis, and the paper of Jones et al. (2000b) was a functional analysis, not a comprehensive phylogenetic study.

Nevertheless, despite the presence of true feathers, birdlike hindlimb proportions, and perhaps other, less certain features, *Caudipteryx* displays a variety of plesiomorphic characters throughout the skeleton that, when taken together, clearly place it outside Aves. Ji et al. (1998) listed three such characters. Their first two characters are very similar and relate to the quadratojugal and its contact with the quadrate and squamosal. I concur that the quadratojugal of NGMC 97-9-A bears the primitive character of a relatively long dorsal (squamosal) process that probably is sutured to the quadrate, and the new specimens reported by Zhou et al. (2000) confirm this arrangement. *Caudipteryx* clearly lacks the small quadratojugal of birds, including such basal birds as *Archaeopteryx, Confuciusornis,* and enantiornithines.

The other plesiomorphic trait cited by Ji et al. (1998) involves the retention of a prominent, triangular obturator process of the ischium. Again, I fully agree, and I regard the shape of the ischium as one of the clearest manifestations of the position of *Caudipteryx* outside Aves. Basal birds have complex ischia (Forster et al., 1998) that generally are characterized by a small (or even absent) obturator process and instead a large, tablike (i.e., rectangular) proximodorsal process extending up toward the ilium. This is the condition in, for example, *Archaeopteryx, Confuciusornis,* and enantiornithines. The very birdlike theropod *Unenlagia comahuensis* (Novas and Puerta, 1997) presents the intermediate condition of possessing both a large obturator process and a proximodorsal process. The shape and orientation of the ischium

of *Caudipteryx* are clearly visible in the new material described by Zhou and Wang (2000; see also Zhou et al., 2000). Both ischia are well preserved and show only a single process that is large and triangular. This shape is exactly like that of the obturator process of, say, dromaeosaurids and oviraptorosaurs. Another primitive character thus would be the absence of the proximodorsal process.

A number of other primitive characters of *Caudipteryx* can be added to those discussed by Ji et al. (1998; see also Zhou and Wang, 2000, and Zhou et al., 2000). For example, the jugal is a typically nonavian theropodan jugal with a very large postorbital process. Birds, on the other hand, have lost the postorbital process of the jugal or, at most, have reduced it to a small process. Even taxa that retain a postorbital bone and a dorsotemporal arch (e.g., *Archaeopteryx,* the Catalan enantiornithine nestling, alvarezsaurids) lack a large postorbital process of the jugal and basically have a jugal bar. The postorbital bone of the enantiornithine *Protopteryx* has a long jugal process that might reach the jugal, but such a contact is not clear on the specimens (Zhang and Zhou, 2000). The only certain exception is *Confuciusornis,* which curiously possesses a complete postorbital bar formed by contact of the postorbital and jugal bones. But even in *Confuciusornis* most specimens have a relatively small postorbital process of the jugal (in some cases, little more than a bump), and the postorbital bone makes up almost all of the bar (Martin et al., 1998; Peters and Ji, 1998; Chiappe et al., 1999; Hou et al., 1999; Zhou and Hou, Chapter 7 in this volume). It may be noted here that the Eichstätt specimen of *Archaeopteryx* displays a somewhat bifid caudal extremity to the jugal. The dorsal prong of this bone could be interpreted as a postorbital process (e.g., Paul, 1988), but it seems to be situated too far caudally to reach the ventral ramus of the postorbital as preserved in the Berlin specimen (see also Chiappe et al., 1999); hence, I tend to agree more with the restoration of *Archaeopteryx* produced by Chatterjee (1991).

Another primitive character of *Caudipteryx* is the relatively very deep mandibular fenestra, as evidenced by the deep caudal embayment of the dentary of the paratype skull. Absence of a mandibular fenestra had been thought to characterize Aves because such an opening is absent in *Archaeopteryx,* hesperornithids, *Ichthyornis,* and neornithines, but the discovery of mandibular fenestrae in *Confuciusornis* (Martin et al., 1998; Chiappe et al., 1999; Zhou and Hou, Chapter 7 in this volume) makes this assessment a bit problematic. Nevertheless, the fenestra in *Confuciusornis* is not nearly as deep as in *Caudipteryx* and has an unusual form and thus may well be a reversal. The mandibular fenestra of *Caudipteryx,* on the other hand, is very comparable to that of dromaeosaurids, oviraptorosaurs, and other nonavian coelurosaurs and thus represents the primitive condition.

The thoracic girdle of *Caudipteryx* is also quite primitive and has none of the avian apomorphies seen in members of Aves. For example, the scapula has a relatively broad blade with a pronounced distal expansion, indicating the retention of a broad suprascapular cartilage. The blade clearly is not the slender and elongate structure seen in all basal birds. The shape of the coracoid is more or less that of a conventional nonavian coelurosaur coracoid, with a quadrilateral shape, proximal supracoracoidal nerve foramen, and moderate biceps tubercle. The coracoid certainly is not the elongate "straplike" bone seen in ornithothoracine birds. Another primitive trait here relates to the orientation of the girdle in that it is located on the lateral aspect of the thorax with the scapula at an angle to the axial column rather than on the dorsal aspect of the thorax with the scapula parallel to the column. The former condition is the primitive condition, whereas the latter condition is observed in all birds (Jenkins, 1993), including *Archaeopteryx, Confuciusornis,* and other basal birds. One hesitates to make too much of the orientation of elements in two-dimensional specimens, but, taken at face value (and all specimens agree on this point), *Caudipteryx* again displays the primitive condition.

The pelvic girdle of *Caudipteryx* presents primitive, nonavian characters beyond the ischiadic shape noted earlier. For example, the ilium is relatively very tall directly above the acetabulum, and its preacetabular portion is not expanded cranially; this is the typical condition for most nonavian coelurosaurs. In birds, on the other hand, the ilium is relatively low, with a greatly elongate preacetabular portion (see Elzanowski, Chapter 6 in this volume; Zhou and Hou, Chapter 7 in this volume). The pubic apron is extensive in *Caudipteryx,* measuring about 56% of total pubic length in the holotype. This is considerably more than the 45% measured in the London *Archaeopteryx,* the bird with the longest known pubic apron, and may even exceed that of some dromaeosaurids (Norell and Makovicky, 1997, 1999). Finally, Zhou and Wang (2000) and Zhou et al. (2000) noted that the pubis is not retroverted (as argued by Feduccia, 1999b) but rather is directed cranially, as in most nonavian theropods.

The picture that emerges from this brief survey of *Caudipteryx* is of a feathered theropod dinosaur that is probably well outside the avian lineage. I have not performed a more extensive formal analysis, but it seems readily apparent that it would be much less parsimonious to include *Caudipteryx* within Aves. And this point leads to the question of the phylogenetic position of *Caudipteryx.* The analysis of Ji et al. (1998) was not very inclusive, using only *Velociraptor* as an outgroup to the Chinese taxa and birds. My initial study of the specimens suggested a number of derived features pointing to oviraptorosaur relationships for *Caudipteryx,* including the following. The jaws are almost completely edentulous in *Caudipteryx,* which is indeed a re-

semblance to oviraptorosaurs, but even more striking is the conformation of the jaws. In both groups, the dentary is very deep between the mandibular fenestra and the symphyseal portion, which is deflected ventrally; moreover, the symphyseal portion is medially inflected, and the caudal processes of the dentary diverge widely around the mandibular fenestra, both of which are attributes of oviraptorosaurs (Makovicky and Sues, 1998). The premaxilla has an extensive prenarial portion (which is also an avian apomorphy), and the naris itself is retracted (extensively in oviraptorosaurs). Finally, the maxilla is very characteristic, being a relatively small, rostrally displaced triangular element. Many of the postcranial elements compare well with oviraptorosaurs but also with other clades of coelurosaurs. However, the tail of *Caudipteryx,* as detailed previously, is quite similar to that of oviraptorosaurs in that both are short and proximally very thick. Given that my observations were not part of a comprehensive phylogenetic analysis, it was gratifying to see these impressions of an oviraptorosaurian *Caudipteryx* borne out by numerous cladistic analyses presented at the Ostrom Symposium at Yale University in 1999 (e.g., by P. C. Sereno, T. R. Holtz, M. A. Norell, and P. J. Currie), suggesting broad independent discovery of these relationships (and a heartening affirmation of phylogenetic systematics). Barsbold et al. (2000a,b) also regarded *Caudipteryx* as a basal oviraptorosaur. More significant, *Caudipteryx* was included in the very extensive phylogenetic analysis of Sereno (1999a). Sereno scored *Caudipteryx* for 204 characters (only 16% missing data) and found not only that *Caudipteryx* is well outside Aves but also that it is indeed a basal oviraptorosaur, sharing a dozen characters with oviraptoroids.

Thus, feathers of essentially modern structure do indeed predate the group conventionally known as "birds." This finding may seem shocking, but it is to be expected. This surprise again may relate to the pervasive sense of *Archaeopteryx* as truly the Urvogel, or "first bird." Common sense, of course, dictates that the elaborate feathers of *Archaeopteryx,* arranged as they are in their "modern" array of primaries and secondaries, must have had predecessors. However, *Caudipteryx* will likely remain difficult for some to accept, perhaps because it is such a dramatic repudiation of opposition to the theropod origin of birds.

In fact, more "feathered dinosaurs" are likely to come to light. For example, Xu et al. (1999a) described *Beipiaosaurus,* a new therizinosauroid theropod from the Lower Yixian Formation that bears filamentous dermal structures that are perhaps similar to those of *Sinosauropteryx.* These structures lack the unambiguous feather structure seen in *Caudipteryx* (i.e., they lack calamus, rachis, and barbs), but they are clearly present on areas of the body that cannot be explained away as remnants of a median frill. In *Beipiaosaurus,* filamentous structures are associated with elements of both fore- and hindlimbs. The best-preserved filaments are attached to the ulna, where some approach 70 mm in length. Xu et al. (1999a) describe some filaments as distally branched and with hollow cores. These filaments have the same "protofeather" problems as did those of *Sinosauropteryx* (i.e., Are filaments truly the evolutionary precursors of feathers?), but their association with the limbs and their considerable length clearly indicate that they are some kind of epidermal appendage rather than an artifact of desiccating dermal collagen (see also Prum, 1999).

Another nonavian theropod with preserved integumentary filaments is the Yixian dromaeosaurid *Sinornithosaurus millenii* (Xu et al., 1999b). Unfortunately, the filaments are not in their natural positions, and thus, for example, the cluster of filaments adjacent to the skull cannot be reliably attributed to the head region. A very significant finding of *Sinornithosaurus* is a negative one, and that is the absence of hand and tail feathers. No true feathers (i.e., with rachis and barbs) of the sort seen in birds and *Caudipteryx* have been recovered with *Sinornithosaurus*. This is a bit troubling because the phylogenetic hypothesis of Sereno (1999a) predicts that, minimally, hand and tail feathers should be found in dromaeosaurids. However, given that the integumentary structures are not in life position and that the *Sinornithosaurus* specimen is generally jumbled somewhat on the slab, it is probably best not to make too much of this absence and assume that it is preservational. Close examination reveals some details suggesting that the filaments of *Sinornithosaurus* are more structured than those of *Sinosauropteryx* and *Beipiaosaurus* and hence more similar to avian feathers. Xu et al. (2001b) documented branching of some filaments, the compound construction of filamentous bundles, and even the occurrence of basal tufts of filaments, all features indicative of a more structurally complex integumentary covering. This report was followed shortly by the announcement by Ji et al. (2001; see also Norell, 2001) of a new specimen of a juvenile dromaeosaurid that is very similar to *Sinornithosaurus* and may even be the same species. The specimen preserves the integument in place and affirms the complex nature of the integument in dromaeosaurids. Not only does the juvenile specimen show branching and tufted filaments, but it also shows fibers branching off of a central axial filament—that is, it shows structure that could be interpreted as being the rachis and barbs of a "true" feather. Moreover, Ji et al. (2001) argued that the structures were so well ordered that birdlike barbules almost certainly had to have been present. As in *Caudipteryx,* the tail and forelimbs have the best-organized integumentary structures.

Microraptor, the tiny dromaeosaurid reported by Xu et al. (2000), lacks hand and tail feathers but has the now typical filamentous coat. As in the juvenile dromaeosaurid, some integumentary impressions bear a rachislike structure, suggesting that true feathers might have been present

in this animal, although this finding awaits confirmation with better-preserved material.

It is also relevant at this point to mention the findings of Schweitzer et al. (1999) on the biochemistry and morphology of fibrous integumentary structures recovered from the head region of the alvarezsaurid *Shuvuuia* from Mongolia (see also Chiappe, Norell, and Clark, Chapter 4 in this volume). Schweitzer et al. (1999) reported two important observations about these structures. First, biochemical studies are consistent with their being composed of beta keratin, a protein found in the feathers and scales of sauropsids. Second, the structures were apparently hollow. At present, the only structures known to be both hollow and composed of beta keratin are avian feathers. These findings are more provocative than conclusive, and, given the controversial phylogenetic position of alvarezsaurids (see Novas and Pol, Chapter 5 in this volume), one should be hesitant to make too much of these findings. Nevertheless, they may be legitimate evidence for feather or featherlike structures outside Aves.

In sum, the significance of these Chinese (and Mongolian) fossils for the debate on avian origins, in one sense, should be minimal. That is, we should not be surprised at the identification of feathers in a group of animals that a broad consensus had always thought was close to avian ancestry. The discovery of feathers in, say, *Caudipteryx* simply adds one more apomorphy to the long list of derived characters linking birds with theropod dinosaurs. The disproof of feathers in any of these Chinese forms would simply remove one character; all the others would remain. Likewise, forcing *Caudipteryx* to be within Aves because of its possession of true feathers (Cai and Zhao, 1999) would not automatically strip it of its clear theropod heritage. Thus, in this context, feathered dinosaurs are not that important. But, of course, in this high-profile, high-energy debate, rhetoric, regrettably, sometimes seems paramount to evidence. Feathers—that quintessentially avian trait—have always been the great definer of birds. The presence of unambiguous feathers in an unambiguously nonavian theropod has the rhetorical impact of an atomic bomb, rendering any doubt about the theropod relationships of birds ludicrous.

The Relationship of the Origin of Flight to the Origin Of Birds

The origin of birds is, at its core, a matter of genealogy. That is, regardless of your systematic philosophy—whether it be cladistic, phenetic, or eclectic—avian ancestry is a question of phylogeny, or, more precisely, phylogenetic reconstruction. For most biologists, phylogenetic reconstruction has become more or less synonymous with phylogenetic systematics or cladistics, whereby the distribution of attributes among taxa forms the primary raw data used to develop hypotheses of relationship. The debate on the origin of birds, however, has been unusual in that a different approach has been applied by a minority of workers for many years (Witmer, 1991, 1997b, 1999). This approach is (1) to create the most likely scenario for the origin of avian flight, (2) to deduce from this scenario the morphological features likely to be present in the hypothetical "proavis," and then (3), as I have said before (Witmer, 1997b:1209), "to search the animal kingdom for a match." Thus, *functional* hypotheses on the origin of *flight* are being used to test *phylogenetic* hypotheses on the origin of *birds*. This distinction between the functional and phylogenetic approaches to avian origins has not been widely appreciated.

For decades, of course, discussions on the origin of flight in birds have been dominated and dichotomized by the arboreal hypothesis (a.k.a. the "trees down" theory) and the cursorial hypothesis (a.k.a. the "ground up" theory). The cursorial hypothesis has been closely associated with the notion of relationships to theropod dinosaurs, whereas the arboreal hypothesis has been tied to the "alternative ancestry" hypothesis (that is, the origin of birds from a usually poorly defined group other than theropods, most often basal archosaurs). It is my intention in this section neither to evaluate these ideas nor to provide a historical account; these are beyond the scope of this chapter (see Hecht et al., 1985; Feduccia, 1996; Shipman, 1997a). Rather, my goal is to examine how these two functional hypotheses relate to the phylogenetic question of avian origins. Moreover, it is worthwhile to question the strict coupling of the cursorial hypothesis with theropods, on the one hand, and the arboreal hypothesis with alternative ancestors, on the other hand (Witmer, 1999). For example, is the arboreal hypothesis truly inconsistent with theropod relationships?

Opponents of theropod relationships have argued strongly for a tight linkage between these functional and phylogenetic issues. For example, Feduccia (1996:viii) stated that "a dinosaurian origin of birds is inextricably linked with the cursorial, or ground-up origin, of avian flight, which is a biophysical impossibility." Martin (1998:40) characterized the debate on avian origins exclusively in functional terms, claiming that "in the great bird-dinosaur debate, the participants huddle in two camps, which paleontologists have nicknamed 'ground up' and 'trees down.'" Bock so intertwined the functional and phylogenetic questions that he entitled a paper "The Arboreal Theory for the Origin of Birds" and used "origin of birds" and "origin of flight" almost interchangeably (Bock, 1985). Tarsitano (1985, 1991) also strongly advocated this approach. The basic premise here is that flight began in animals that lived in high places (trees, in most formulations) and made use of gravity and expanded body surface area to slow descent during falls and leaps; hence, these animals should have been small, quadrupedal, and with arboreal adaptations—

attributes not typical of theropod dinosaurs (Martin, 1983, 1991, 1998; Tarsitano, 1985, 1991; Feduccia and Wild, 1993; Feduccia, 1996, 1999a).

Again, my intent is not to explore this model or its theoretical premises but rather to evaluate the validity of the approach. We are faced with two interesting and obviously related issues: the genealogical ancestry of birds, and the evolution of flight in birds. The question then becomes, Is resolution of one issue logically prior to resolution of the other? The answer is yes, and most theorists would argue that workers such as Martin, Feduccia, and Tarsitano have the logical order reversed. In other words, the phylogenetic question of avian ancestry must precede the functional question of how flight arose. There is a fairly extensive literature on the relationship of functional inference to phylogenetic inference, most prominently discussed by Lauder (1981, 1990, 1995; Lauder and Liem, 1989), although others have commented on the issue (e.g., Padian, 1982, 1985, 1995; Liem, 1989; Bryant and Russell, 1992; Weishampel, 1995; Witmer, 1995a). These authors all agree that functional hypotheses and scenarios are best tested within the context of a strict hypothesis of phylogenetic relationships, primarily for the simple reason that evolutionary history constrains functional systems and their evolution. The evolutionary "starting point" for any functional transition is absolutely critical. The evolutionary trajectory from a *Megalancosaurus*-like form to a flying bird will be much different from the trajectory from a *Deinonychus*-like form to a flying bird, regardless of whether these "starting points" are arboreal, terrestrial, aquatic, or whatever. Determining this evolutionary "starting point" is a matter of genealogy, that is, of phylogenetic inference, not functional inference. Thus, the details of any functional transition, such as the origin and refinement of flight, can be best dissected with the tool of a well-resolved phylogenetic hypothesis (see Cracraft, 1990; Chiappe, 1995; Sereno, 1997b, 1999a).

Bock (1965, 1985, 1986), on the other hand, has argued vigorously and persuasively for the validity of what amounts to a "function-first," scenario-based kind of approach, couched in the philosophical terms of historical-narrative explanations. Nevertheless, most current opinion has found such scenario building in the absence of a strict phylogenetic hypothesis to fall short on the grounds of testability. More to the point, it seems unjustified to believe that such scenarios—no matter how intuitively appealing, such as is the case with the arboreal theory—can overturn as well substantiated a phylogenetic hypothesis as is the theropod hypothesis. The hypothesized steps in the functional transition from an arboreal proavis to a flying bird are generally tested by only plausibility or modeling rather than hard data. On the other hand, phylogenetic hypotheses are much better grounded in tangible evidence—in this case, actual objects (bones) that can be observed, measured, and compared—and hence cladograms are subject to more rigorous tests.

But, in many respects, it is the role that cladistic analysis has played in the theropod hypothesis that has elicited the opposition. Opponents of theropod relationships have simultaneously waged a war against phylogenetic systematics, because, for authors like Feduccia (1996), Martin (1998), and Bock (1999), "cladistic analysis . . . lies at the core of the debate concerning bird origins" (Feduccia, 1996:59). Such statements are a little difficult to reconcile with the fact that, say, John Ostrom (1973, 1976), who is not a cladist, formulated the theropod hypothesis using precisely the systematic methodology these authors advocate. Nevertheless, more recent authors have indeed employed phylogenetic systematics, and some reviewers of Feduccia's 1996 book agreed that differing systematic philosophies are part of the source of the conflict (Norell and Chiappe, 1996; Sereno, 1997a; Witmer, 1997b; see, in particular, Padian, 1997, for an analysis of this issue). What is pertinent here for the "function versus phylogeny" debate is that, given the role of cladistic analysis in (1) modern functional inference in general and (2) the theropod hypothesis for avian ancestry in particular, it seems unlikely that those in the Feduccia/Martin school will adopt the "phylogeny-first, function-second" approach to understanding the origin of flight advocated here.

Perhaps the greatest irony for this whole issue is that there probably is no adequate justification for tightly coupling the cursorial theory with theropod relationships and the arboreal theory with alternative ancestry. It is conceivable that the Feduccia/Martin school is correct that the arboreal model for the origin of flight is the superior model—but the evolutionary starting point may in fact be a small theropod dinosaur. Why must these functional and phylogenetic models be coupled (Witmer, 1999)? The coupling of these models has more to do with the tactics of the debate than the debate itself. Advocates of the alternative-ancestry/arboreal pairing have pointed to the large size of such obviously terrestrial theropods as *Tyrannosaurus* or even *Deinonychus* in the hope of illustrating how ludicrous the notion of an arboreal/theropod origin of birds is (e.g., Tarsitano, 1985; Martin, 1991, 1997; Feduccia, 1996, 1999b). Advocates of the theropod/cursorial pair called attention to the same terrestrial attributes of theropods in arguing for their position (e.g., Ostrom, 1986). In general, each camp has chosen a single point on which to be immovable and hence forces the functional or phylogenetic issue to fall in line with that point. For the Feduccia/Martin school, the arboreal theory is unshakable, and hence all phylogenetic possibilities must be concordant—and theropods, they argue, are the height of discord. On the other hand, the theropod school has remained intransigent on the phylogenetic issue, and hence the functional transition to flight has been constrained.

As outlined previously, the theropod school is on much firmer theoretical ground in placing phylogeny logically prior to function. In fact, it was this kind of reasoning that helped bolster the cursorial origin of avian flight. That is, since the theropod outgroups of *Archaeopteryx* and other birds were more or less large animals, the origin of flight clearly is best understood in this terrestrial or cursorial context (Padian, 1982; Gauthier and Padian, 1985; Padian and Chiappe, 1998a,b). However, as noted by Sereno and Rao (1992), arboreality was apparently an early adaptation for birds. Hence, the debates about the arboreality versus terrestriality of *Archaeopteryx* have always been seen as critical. Arguments on both sides have been presented for decades (see Hecht et al., 1985; Paul, 1988; Feduccia, 1996; Padian and Chiappe, 1998a,b), and, once partisanship is eliminated, no clear consensus emerges. Interestingly, independent studies on pedal proportions in *Archaeopteryx* and other taxa (Hopson and Chiappe, 1998; Zhou, 1998) have agreed in showing that the feet of *Archaeopteryx* are basically intermediate between those of a terrestrial cursor and those of an arboreal bird. Thus, *Archaeopteryx* itself is inconclusive in establishing the phylogenetic level at which arboreality occurred (assuming for the sake of argument that it occurred only once).

The question ultimately comes down to the actual theropod ancestor of birds: what it looked like and how it lived its life. Although virtually all recent analyses put Dromaeosauridae or Troodontidae (or the two together as Deinonychosauria) as the sister group of Aves, neither is truly the ancestor, and hence known forms like *Deinonychus* or *Troodon* can only go so far as models for the true avian ancestor (see Gatesy, Chapter 19 in this volume, for an insightful discussion). Virtually all early birds are small animals, so, at some point in the transition to birds, miniaturization took place. Small size has many virtues. That is, in the absence of the constraints imposed by large mass, small animals can exploit a broad behavioral repertoire without necessarily having to develop novel morphological adaptations. This line of reasoning is obviously leading toward the possibility that the miniaturization took place within a lineage that we would probably recognize as "nonavian," that is, a lineage of little dinosaurs. If such a tiny theropod habitually used trees or other high places (and one can easily envision many sound reasons for doing so), then perhaps the arboreal model propounded by opponents of theropod relationships would apply equally well to theropods. This notion is not new, and a number of workers have argued for an arboreal origin of avian flight from tiny, dromaeosaur-like ancestors (Abel, 1911; Paul, 1988, 1996, 2002; Witmer, 1995c; Chatterjee, 1997a,b; Xu et al., 2000; Zhou and Wang, 2000). In particular, Chatterjee and Paul have developed fairly elaborate models and have identified a number of features of dromaeosaurlike theropods that may indicate arboreal capabilities. Significantly, a variety of tiny theropods, such as *Bambiraptor* (Burnham et al., 2000) and *Microraptor* (Xu et al., 2000), have begun to turn up in the fossil record.

It is not my aim here to evaluate models for an arboreal origin of avian flight from theropod dinosaurs. Although the idea has a lot of merit, virtually all models on the origin of avian flight are so speculative and so data-poor that any satisfactory resolution is unlikely any time soon. In fact, there are serious testability problems for all these models. For example, mathematical models for the origin of avian flight abound (e.g., Caple et al., 1983; Balda et al., 1985; Norberg, 1985; Rayner, 1985; Pennycuick, 1986; Herzog, 1993; Ebel, 1996; Burgers and Chiappe, 1999), but they all suffer to varying extents from testability problems—and this problem pertains to all models, regardless of the phylogenetic starting point. The fact is that we simply have paltry data on the functional capabilities of any of the principal taxa (e.g., dromaeosaurids, troodontids, Triassic archosauromorphs like *Megalancosaurus* or *Longisquama*). Despite numerous studies, even the basic lifestyle of *Archaeopteryx* is disputed. It is conceivable that the origin of flight—as a matter of scientific discourse—is out of reach. We may simply never have the appropriate data to adequately test any models. In fact, this is probably the reason that the debate on the origin of flight has raged uncontrolled for a century with no sign of resolution in sight. All ideas remain active because almost none can be falsified.

It is fair to regard the foregoing as overly pessimistic, but one thing that must be true is that modeling the origin of avian flight is a very poor research strategy for discovering the origin of birds. The origin of flight is logically and methodologically secondary to the phylogenetic origin of birds. There is currently no good reason to rigidly couple models of the origin of flight with particular phylogenetic clades. And, perhaps most troubling, the details (or even the broader pattern) of the functional transition to powered flight may be lost in time and virtually unrecoverable in any rigorous scientific sense.

The Status of Alternatives to the Theropod Hypothesis

There is no question that the theropod origin of birds is by far the most popular hypothesis on avian ancestry. The question then arises, Are there credible alternatives? In most previous reviews (e.g., Ostrom, 1976; Gauthier, 1986; Witmer, 1991; Feduccia, 1996; Padian and Chiappe, 1998b), the debate on avian origins was divided into three competing hypotheses: (1) the theropod hypothesis, (2) the crocodylomorph hypothesis, and (3) the basal archosauriform or "thecodont" hypothesis. However, in recent years it has become apparent that there really are just two major hypotheses: (1) the theropod hypothesis and (2) the "not-theropod" or, as I have termed it previously, "the alternative ancestry hypothesis." Relationship to crocodylomorphs, originally proposed by Walker (1972), seems to have simply faded away in that earlier advocates, such as Martin (1983, 1991), Walker (1990), and Tarsitano (1991), have not renewed their support. The basal archosauriform hypothesis received a significant boost from Welman (1995), who suggested that *Euparkeria* shares with *Archaeopteryx* to the exclusion of theropods and crocodylomorphs a large suite of derived characters in the cranial base. This new "thecodont" hypothesis has received to date no additional adherents and was severely challenged by the detailed analysis of Gower and Weber (1998). Thus, for the present, the crocodylomorph and basal archosauriform hypotheses no longer appear to merit serious consideration.

Indeed, opposition to the theropod origin of birds has become almost exclusively just that, an argument of opposition rather than an argument of advocacy. Criticism is a necessary and appropriate part of the scientific process, and opponents have published a number of papers taking issue with certain of the characters (Martin et al., 1980; Martin, 1983, 1997; Tarsitano, 1991; Feduccia, 1996). It is not my goal here to analyze these criticisms or to provide responses, although a few will be touched on in the next section. My main point here is that opponents have sought to destroy but not build in that they have lost sight of the goal of phylogenetically linking birds to actual taxa. The cladistic approach is more constructive in that a particular phylogenetic hypothesis is refuted not simply by criticizing the characters but rather by offering an alternative that better accounts for the available data, that is, an hypothesis that is more parsimonious, a shorter tree. In 1991, I stated: "At present, supporters of relationships of birds with crocodylomorphs, 'thecodonts,' or mammals have failed to produce a competing cladogram, and in this respect the coelurosaurian hypothesis is uncontested" (Witmer, 1991:457).

That statement still stands today, largely because there are no serious alternative phylogenetic hypotheses. For some time, opponents have offered a variety of small, generally poorly preserved Triassic forms as being relevant to the debate (Martin, 1983, 1991, 1997, 1998; Tarsitano, 1985, 1991; Feduccia and Wild, 1993; Feduccia, 1996). These Triassic taxa include *Megalancosaurus*, *Cosesaurus*, *Scleromochlus*, and *Longisquama*. These forms do not constitute a clade but are a hodgepodge of basal archosaurs or basal archosauromorphs. *Megalancosaurus* and *Cosesaurus* both pertain to the archosauromorph clade Prolacertiformes (Sanz and Lopez-Martinez, 1984; Renesto, 1994). *Scleromochlus* has been thought to be related to a variety of taxa, most commonly pterosaurs and dinosaurs (Padian, 1984; Gauthier, 1986; Benton, 1999). *Longisquama* has never been subjected to adequate phylogenetic scrutiny; Sharov (1970) placed it in "Pseudosuchia," and Haubold and Buffetaut (1987) agreed, although Charig (1976:9) argued that "the justification for this assignation is obscure." Tarsitano (1991:549) and Feduccia (1996:86) referred to these taxa as "avimorph thecodonts" in that they regarded them as basically birdlike. The resemblances, however, have never been particularly strong or numerous. Authors such as Feduccia, Martin, and Tarsitano generally have not considered these taxa to be truly ancestral to birds but rather as merely representative of what their hypothesized arboreal proavis was like. In other words, these taxa show that there were small, arboreal, quadrupedal animals running around before *Archaeopteryx* (although it should be pointed out that the preferred habitat and mode of life of these animals are perhaps not as obvious as commonly portrayed).

But even given that such animals existed and are consistent with the arboreal theory for the origin of avian flight, this does not constitute actual evidence relevant to the ancestry of birds. For example, unless *Megalancosaurus* is being considered as close to the ancestry of birds, whether or not it has a "straplike scapula" or a "birdlike orbit" (Feduccia and Wild, 1993) is of questionable significance. It is conceivable that viable candidates for avian ancestry could emerge from such a nexus of "avimorph" forms, but such hypotheses will continue to be relegated to the fringe unless they are framed in explicit phylogenetic terms and take head-on the theropod hypothesis on its own terms.

Nevertheless, one of these "avimorph" forms captured broad attention when Jones et al. (2000a:2205) pointed to a number of features of the integumentary appendages of *Longisquama* that led them to conclude that these structures represent "nonavian feathers, probably homologous to those in birds." The most compelling resemblances center on the presence of a calamuslike base wrapped in a presumably epidermal sheath, indicating that the appendages probably developed in a follicle, as is characteristic of feathers and unlike scales. However, their interpretation of the structures coming off the central axis as separate "barbs" seems overly generous at best. These structures unite distally, forming a continuous ribbon around the periphery of the appendage. This distal union is completely unlike the situation in avian feathers, and even if a few tolerably similar examples—all of which are specialized feathers—can be found among birds, it is clearly not the primitive avian condition (Kellner, Chapter 16 in this volume). It seems more likely that the "barbs" identified by Jones et al. (2000a) are in fact plications or corrugations in a continuous structure, which would be more consistent with a modified scale than a feather. Reisz and Sues (2000) also were critical of the hypothesis of Jones et al. (2000a), advancing many of the same arguments just articulated.

Still, Jones et al. (2000a) regarded the structures as feathers probably homologous to those of birds. The implications of such a hypothesis were not explored in the paper. For example, if feathers are a very basal innovation among archosaurs, then this would constitute strong support for the interpretation of the filaments of, say, *Sinosauropteryx* as feathers (which would be ironic given that Jones and colleagues were such vocal opponents of feathered dinosaurs). But if feathers are a basal character evolving in the Triassic, then where are all the Triassic and Jurassic fossil feathers? There are abundant unequivocal fossil feathers in the Cretaceous, but none prior to those of *Archaeopteryx* in the Late Jurassic (see Kellner, Chapter 16 in this volume).

The Jones et al. (2000a) paper carefully avoided any statement on the origin of birds, but the authors were very vocal in the associated media furor (e.g., see Stokstad, 2000), arguing that the finding of feathers in *Longisquama* refuted the theropod hypothesis and that *Longisquama* itself is "an ideal bird ancestor" (J. A. Ruben quoted in Stokstad, 2000:2124). This example of disparity between scientific and public statements is just the latest in the long history of the debate on avian origins, and it is best to focus on the scientific evidence. In this case, the paper of Jones et al. (2000a) offered no scientific statement on the ancestry of birds. In fact, their claims of homology of the integumentary appendages of *Longisquama* with avian feathers was an incidental point of the paper, based basically on their opinions and not on a careful phylogenetic treatment, which is the ultimate arbiter of homology. It is fair to say that Jones et al. (2000a) demonstrated that the integumentary appendages of *Longisquama* are more interesting and unusual than previously thought. Beyond that—and in the absence of a phylogenetic analysis—*Longisquama* and its appendages are as irrelevant to the debate on the origin of birds as are the other "avimorph" forms.

In sum, at present there remains no credible alternative to maniraptoran theropod dinosaurs for the origin of birds. Previous tangible alternatives (crocodylomorphs, basal archosauriforms such as *Euparkeria*) have been refuted or summarily dropped because of lack of interest. What has replaced these are intangible "models" that conform to preconceived notions on how bird flight evolved, that is, taxa that, although not truly related to birds, are "much like what we would expect" the true ancestors to be. In some ways, it seems as if the search for real avian relatives has been supplanted by the mission to discredit both the theropod hypothesis and the cladistic methodology that continues to corroborate the hypothesis. Fossils such as *Longisquama* may someday emerge as more relevant players in the debate, but if the media hype surrounding the Jones et al. (2000a) paper was any indication, even *Longisquama* will be just another attempt to develop a rhetorical weapon to attack the theropod hypothesis and cladistics.

The Status of the Theropod Hypothesis

The only explicit hypothesis for the phylogenetic relationships of birds states that avian ancestry is fully embedded somewhere within the nexus of maniraptoran theropod dinosaurs, probably nearest to Dromaeosauridae and/or Troodontidae among known groups. As such, it is "the only game in town." As mentioned previously, opponents of theropod relationships have regarded the hypothesis as basically an unfortunate outcome of sloppy application of phylogenetic systematics. However, the idea was formulated by Ostrom (1973) and initially supported (e.g., Bakker and Galton, 1974; Thulborn, 1975; Thulborn and Hamley, 1982) without cladistics. In large measure, the many cladistic studies that have followed have served mostly to support, clarify, and update Ostrom's original work. More important, they have repeatedly tested the hypothesis (although, to be fair, it must be pointed out that they rarely include nondinosaurian taxa in the analysis). Among the more important cladistic studies are those of Padian (1982), Thulborn (1984), Gauthier (1986), Holtz (1994), Novas (1996), Forster et al. (1998), Sereno (1999a), and Clark, Norell, and Makovicky (Chapter 2 in this volume).

Until the reemergence of Alan Feduccia in the debate in the mid-1990s, opposition to the theropod hypothesis had become basically mute, and, in my opinion, theropod advocates had become complacent (Witmer, 1997b). But Feduccia reenergized the opposition, enlisting new recruits (e.g., J. A. Ruben) and strengthening former alliances (e.g., with L. D. Martin). The warfare metaphor is intended to be lighthearted, but there is clearly a sense that this group feels as if it is fighting a holy war against a great oppressor; Feduccia (quoted in Shipman, 1997b:29) went so far as to regard Ruben as a "comrade in the war against hot-blooded dinos." This group has offered criticisms on a number of fronts. Much of the criticism has taken the form of comments to the media, book reviews (Martin, 1988, 1998), popular articles or books (Feduccia, 1994, 1996, 1999b), and other outlets outside normal peer review (e.g., Martin 1997). Chief among these criticisms is that relating to the functional problems of evolving flight in an arboreal context when theropods seem to have been such obligate terrestrial animals; having discussed the multiple fallacies in this general approach earlier, I will turn to other issues. More specific challenges can be grouped into three categories: (1) the time problem, (2) problems of morphological interpretation or homology, and (3) single problems of such significance that they alone would falsify the theropod hypothesis. Martin (1997:337) regarded these criticisms as causing a "collapse of various anatomical arguments for a bird-dinosaur connection followed by determined efforts [by advocates of theropod relationships] to bolster failing characters." However, I do not regard such efforts at "damage control" (as he later put it [Martin, 1998:42]) as inappropriate but rather as a normal part of the scientific process in which new data or claims are evaluated. In this light, let us briefly examine these three categories.

The "time problem"—or "temporal paradox," as it is often known—relates to the fact that the closest nonavian theropod sister groups of birds are all Cretaceous in age and hence younger than *Archaeopteryx*. The issue has been raised many times over the years, but Feduccia has wielded it as a bludgeon. For example, he stated that "to such workers [paleontologists] it is inconsequential that birdlike dinosaurs occur some 75 million or more years after the origin of birds" (Feduccia, 1996:vii). Elsewhere, Feduccia (1994:32, italics in original) painted an even worse picture, claiming that "most of the supposed similarities between the *urvögel* [meaning specifically *Archaeopteryx*] and dinosaurs are seen in birdlike dinosaurs that lived 80 to 100 million years later." The latter date in the second quote would actually put these "birdlike dinosaurs" in the Eocene (!) —perhaps a simple mistake on Feduccia's part, but it reflects a consistent hyperbolic exaggeration of the time discordance (Witmer, 1997b).

It is true that, say, *Velociraptor* is 70 My younger than *Archaeopteryx*, but other dromaeosaurids and troodontids are much closer in age to *Archaeopteryx*: *Deinonychus* is 35 My younger, *Utahraptor* is only 25 My younger, *Sinornithosaurus* is only about 20 My younger, and *Sinovenator* is less than 17 My younger (Swisher et al., 1999; Xu et al., 2002). There are much greater time discordances in the dinosaur fossil record (Sereno, 1997b, 1999a) than this one. But, moreover, there are a variety of fragmentary specimens (mostly teeth) of animals that closely resemble those of dromaeosaurids and troodontids recovered from Middle Jurassic deposits that *predate Archaeopteryx* by 20 My (Evans and Milner, 1994; Metcalf and Walker, 1994). Similarly, Zinke (1998) reported on an extensive collection of theropod teeth from deposits perhaps just slightly older than *Archaeopteryx*; Zinke made firm assignments of these teeth to Dromaeosauridae (29 teeth), Troodontidae (14 teeth), and Tyrannosauridae (3 teeth). Finally, Jensen and Padian (1989) described fragmentary but provocative skeletal material of maniraptoran theropods from the Late Jurassic Morrison Formation. Even if some of these precise taxonomic assignments do not stand scrutiny, they clearly indicate that there were nonavian maniraptorans that existed prior to *Archaeopteryx*. Moreover, Brochu and Norell (2000) pursued the temporal paradox issue by comparing stratigraphic consistency indices and other phylogenetic metrics among various hypotheses for avian origins, and they found that the theropod hypothesis actually compares *favorably* to the alternatives when considered globally across the cladogram. Thus, not only is the time problem not particularly severe, it does not even exist, and its perpetuation in the face of such data is untenable.

Opponents of theropod relationships have questioned either the interpretation or the homology of a number of characters (see Martin, 1983, 1991, 1997, 1998; Tarsitano, 1991; Feduccia, 1996, 1999a). Only one of the higher-profile characters will be discussed here as an example, and that is the semilunate carpal, one of the classic Ostrom characters. Ostrom (1973) originally—and erroneously—regarded the element in *Deinonychus* as a radiale (i.e., a proximal carpal element), even though it clearly was much more tightly articulated to the metacarpus than to the antebrachium. Gauthier (1986), without fanfare, corrected this error and regarded it as a distal carpal, which is the identity of the semilunate element in birds. Nevertheless, the homology has been vigorously questioned (see Martin, 1991, 1997; Feduccia, 1996, and references therein). Until relatively recently, the complete carpal structure was understood for relatively few theropods. However, the wrist is now known in many theropods, and the presence of a semilunate carpal characterizes a broad taxon (Neotetanurae), where it can be seen to be a distal carpal element (Sereno, 1999a). For example, within coelurosaurs there are specimens (e.g., *Scipionyx samniticus* [Dal Sasso and Signore, 1998] and *Sinornithoides youngi* [Russell and Dong, 1993; personal observation of IVPP V9612]) that clearly show the semilunate carpal to be *distal* to another carpal element (the radiale), clinching its identity as a distal carpal. Neither Feduccia (1996) nor Martin (1997) cited the new evidence from *Sinornithoides,* despite the fact that Russell and Dong (1993:2169) clearly identified both a radiale and a semilunate carpal. There seems to be little reason to doubt the homology of the carpal elements of nonavian maniraptorans, *Archaeopteryx,* and other birds. Some of the challenges to other characters have been addressed by other workers, for example, the furcula and sternum (Norell and Makovicky, 1997; Norell et al., 1997; Makovicky and Currie, 1998; Clark et al., 1999; Sereno, 1999a) and the pelvis (Norell and Makovicky, 1997, 1999).

Two issues have been proposed as being so important that they alone would have the power to overthrow the entire theropod hypothesis. These issues are the homology of the manual digits and the evolution of the lung ventilatory mechanism. The question of digital homologies has a fairly long and extensive history (see Hinchliffe and Hecht, 1984) but basically involves a conflict between paleontology and embryology. There is almost unanimous agreement that the pattern of digit reduction observed throughout theropod phylogeny indicates that the digits of maniraptorans are I-II-III (Ostrom, 1976; Gauthier, 1986; Tarsitano, 1991; Feduccia, 1996; Sereno, 1997b; Chatterjee, 1998a; Padian and Chiappe, 1998b). Thus, those who regard birds as dinosaurs accept that birds also retain digits I–III. The con-

flict arises because embryologists have repeatedly come up with the result that the avian hand skeleton has digits II–IV (Hinchliffe and Hecht, 1984; Hinchliffe, 1985; Shubin and Alberch, 1986). Opponents of theropod relationships (Tarsitano and Hecht, 1980; Martin, 1991; Tarsitano, 1991; Feduccia, 1996) seized on this as a potentially fatal flaw: if birds truly evolved from dinosaurs, then birds would have had to have lost one finger (I) and gained another (IV)—an unlikely proposition.

For more than a decade this debate was basically a stalemate, until a paper published by Burke and Feduccia (1997) reopened the issue. The paper offered few new data or insights; instead, it largely restated the conflict, updated the embryological component by integrating the primary-axis paradigm of Shubin and Alberch (1986), and asserted that these data refute the theropod relationships of birds. There are valid complaints that can be leveled at the Burke and Feduccia study (e.g., see Chatterjee, 1998a; Garner and Thomas, 1998; Padian and Chiappe, 1998a,b; Zweers and Vanden Berge, 1998), and I could add some additional ones. But rather than expand this discussion with elaborate counterpoints, a broader issue needs to be raised, and that is the relationship between different kinds of data, in this case paleontological and embryological. There is a sense among some that the embryological signal must be correct because it involves seemingly high-tech bench science and makes reference to hox genes, as opposed to the dust and dirt of paleontology. The fact is that both disciplines require a lot of interpretation of the data. The observed embryonic condensations and their pattern of connectivity do not come with unequivocal labels but rather require much interpretation before elements can be identified. For instance, the structure that Burke and Feduccia (1997) identified as a transient metacarpal V in chicken hand development could indeed be just that, but it is never part of the metacarpal arcade and could just as easily be regarded as a bifurcation of the adjacent carpal condensation. Given that these embryological data are no "cleaner" than paleontological data, I am reluctant to accept their implications, particularly when studies by Hinchliffe (e.g., 1985) clearly document that avian hand development is a complicated and unusual system (with the ulnare progressively disappearing and replaced by a mysterious "X element" of uncertain origin). Thus, I do not find compelling the paper of Wagner and Gauthier (1999) that argued that, in effect, both embryologists and paleontologists are correct. They suggested that there has been a "frame shift" in "developmental identities" over the course of theropod phylogeny such that embryonic condensations II–IV differentiate into definitive digits I–III somewhere near the origin of Tetanurae. It is a tidy and intriguing hypothesis, but it seems largely untestable (since the embryology of fossil taxa is unknowable) and probably is circular (they must "postulate a frame shift" and hence cannot then use these data to *deduce* a frame shift). In sum, unless some new line of evidence arises, I continue to find the I-II-III assessment to be the most conservative and the best supported.

Whereas the discussion on digital homologies often seems to be a tired topic, a fresh new challenge came from a paper by J. A. Ruben and colleagues (1997). In this paper, they argued that theropod dinosaurs lacked an avian-style flow-through lung (i.e., with abdominal air sacs, etc.) but rather had a crocodilianlike "hepatic piston" whereby the lungs were ventilated with the assistance of a hepatic-diaphragmatic complex that was retracted (like a piston) by diaphragmatic muscles attaching to the pubis. Despite obvious significance for dinosaur physiology, what has engendered more controversy is a statement in the Ruben et al. (1997) paper suggesting that the presence of such a lung ventilatory system in theropods would effectively deny them the possibility of being the progenitors of birds. Their point is that the hepatic piston is an evolutionarily canalized system that could never evolve into an avian system. Thus, despite all the evidence from cladistics, birds could not have had their origins among theropod dinosaurs.

Evaluating this proposition involves two separate issues. First, did theropods actually have a crocodilianlike hepatic piston? And second, even if they did, how do we know that such a system could not evolve into the avian system? The first issue is really beyond the scope of this chapter. Nevertheless, Ruben et al. (1997:1270) stated that "the hepatic-piston diaphragm systems in crocodilians and theropods are convergently derived." Thus, the theropod system would constitute, according to the inferential hierarchy of Witmer (1995a), a level III inference, that is, a relatively weak soft-tissue inference requiring exceptionally compelling morphological evidence. The second issue is more pertinent here in that it speaks directly to the origin of birds. The problem with the claim of canalization by Ruben et al. (1997) is that it is based not so much on evidence as on authority. That is, canalization is asserted rather than demonstrated. These authors have argued that, basically, "you can't get there from here." How could we test this hypothesis that the avian system could not evolve from the presumed hepatic piston of theropod? An obvious test is a phylogenetic one: integrate it with other data and let it play out on the cladogram. Given the considerable evidence that birds are embedded within Theropoda, it would seem that indeed "you *can* get there from here," even if the physiological or anatomical mechanism is at present obscure.

In sum, regardless of whether the inference of the hepatic-pump ventilatory system in theropods is sufficiently robust to be sustained, there seems to be little in the Ruben et al. (1997) paper that requires an overhaul of our views on avian origins. Significantly perhaps, in a more recent paper that sought to bolster the theropod hepatic-pump hypothesis, Ruben et al. (1999) made no mention of any relevance of this research to the origin of birds, although Feduccia (1999a) continued to tout such evidence as damning to the theropod hypothesis.

Despite such external challenges to the theropod origin of birds, this hypothesis has survived and continually gains news adherents. This final section will examine briefly the theropod hypothesis from within, particularly with regard to the diversity of opinion. In my 1991 review, I was able to report a fair amount of diversity within the general notion that birds were somehow closely related to theropods. At that time, there were active hypotheses that suggested that birds were closest to coelophysoids, troodontids, oviraptorosaurs, dromaeosaurids, and *Avimimus* (see Witmer, 1991, and references therein). At the dawn of the twenty-first century, there is remarkably little diversity. Instead, the phylogenetic analyses from all the sources cited earlier seem to be converging on close relationships to Dromaeosauridae, Troodontidae, or a Deinonychosauria clade (Dromaeosauridae + Troodontidae).

Still there is diversity. A recent development that has not received wide attention derives from the collaboration of G. A. Zweers and J. C. Vanden Berge (Zweers et al., 1997; Zweers and Vanden Berge, 1998). These authors have devised an elaborate and intriguing scheme for the evolution of the feeding or trophic apparatus in birds. Despite Feduccia's consistent portrayal (e.g., Feduccia, 1994, 1996) of the origin debate as basically a dichotomy between ornithologists and paleontologists, Zweers and Vanden Berge, two of the leading anatomical ornithologists in Europe and North America, respectively, firmly embedded birds within Theropoda. In fact, they not only "embedded" birds within Theropoda but actually "scattered" avian clades throughout Theropoda in that they, somewhat reminiscent of Lowe (1935, 1944), argued for the polyphyly of birds (Zweers and Vanden Berge, 1998). In their scheme, *Archaeopteryx*, Alvarezsauridae, and Enantiornithes form a clade with dromaeosaurids as the basal taxon; hoatzins, cranes, and palaeognaths form a clade with *Hesperornis*, *Ichthyornis*, and ornithomimids; and all other birds and *Confuciusornis* form a clade with troodontids as its basal taxon. It is impossible to do justice here to the complexity of the functional arguments presented in these papers, although they are very enlightening and engaging whether or not one accepts the authors' phylogenetic scheme. In fact, throughout both papers, particularly Zweers et al. (1997), it is not that clear whether the scheme is intended to reflect a hypothetical functional framework or a true depiction of phylogeny. But in Zweers and Vanden Berge (1998:183) it eventually becomes clear that they indeed regard birds as polyphyletic, describing "successive waves of avian radiation." Nevertheless, they recognized that "at several points this scenario does not coincide with the most recent avian phylogeny, which remains to be explained" (Zweers and Vanden Berge, 1998:183).

Indeed, it is likely that their scheme would be found to be less parsimonious than a cladogram produced by, say, Chiappe or Sereno. But their notion of "successive waves of avian radiation" is not entirely new and represents just the most recent version of the idea that birds and "conventional" theropods are more intertwined than commonly thought. For example, G. S. Paul (1984, 1988, 2002) suggested that perhaps the traditional ancestor-descendant relationships have been interpreted backward: perhaps some "conventional" theropods (such as dromaeosaurids, troodontids, oviraptorosaurs) are in fact secondarily flightless descendants of a persistent lineage of "protobirds." This protobird lineage would have had its origins in the Jurassic period with *Archaeopteryx*, becoming progressively more birdlike in the Cretaceous. Paul (2002) cited a variety of lines of evidence by which "neoflightless" taxa could be identified, pointing in particular to the shoulder girdle and thoracic appendage. In some ways, this hypothesis arises from the realization that, unlike in other groups of flying vertebrates (i.e., bats and pterosaurs), flightlessness has been a recurrent evolutionary theme of birds. Thus, what would a flightless form look like at about the *Archaeopteryx* stage? It might look very much like a small dromaeosaur. Paul (2002) acknowledged that the cladistic representation and discovery of such a pattern are problematic. Paul is not alone in deriving some "nonavian" theropods from birds. For example, Elzanowski (1995, 1999) and Lü (2000) regarded oviraptorosaurs as being not just very closely related to birds but potentially a clade of early flightless birds. If this is proven true, Feduccia (1999a), perhaps ironically, would then be correct that the basal oviraptorosaur *Caudipteryx* is just a Mesozoic kiwi after all! Finally, Olshevsky (1994) proposed the "Birds Came First" (BCF) theory, which maintains that the evolution of archosaurs is characterized by an arboreal "central line" of "dino-birds" that sprouted terrestrial branches, giving rise to the various clades of archosaurs. Throughout the Mesozoic, this central line would have gotten more and more birdlike, and thus their terrestrial offshoots also became progressively birdlike. As in Paul's hypothesis, the Cretaceous coelurosaurs would be secondarily flightless.

All these latter ideas are truly out of the mainstream of current thought and present some problems for testing by phylogenetic analysis. In a sense, they are similar to the scenario-based, "function-first" methodology criticized in an earlier section. Nevertheless, they merit the scrutiny that they have never adequately received. As a class, they are all very similar in that they propose an iterative process to the evolution of birds and theropods. For what it is worth, these proposals have the distinct advantage that all the supposed time discordances basically disappear yet all the anatomical

similarities remain homologous. The "successive waves of avian radiation" scheme described by Zweers and Vanden Berge is not far removed from the successive waves of theropod descent from an avian stem envisioned by Paul and Olshevsky. In all these formulations, the evolution of birds and theropods is hopelessly intertwined. The current orthodoxy (Ostrom, 1976; Gauthier, 1986; Sereno, 1999a) has produced a fairly tidy phylogenetic pattern. These nonstandard views are decidedly untidy, yet they still should receive serious consideration, and this will happen only when they are framed in explicit, phylogenetic terms.

Conclusions

The issues surrounding the origin of birds are wide ranging, and this chapter has attempted to capture this diversity. As a result, it is a bit of a hodgepodge and has sought to touch on those issues that, in particular, have controlled the debate. Throughout the chapter, I have focused on the debate itself, because *how* things are discussed affects *what* is discussed. That is, rhetoric and science are, lamentably, inextricably linked. As I am neither a sociologist nor a psychologist, I have tried not to delve into matters of motivation, politics, and ego. Nevertheless, in this modern media age, it would be naive to think that these factors have not helped shape the debate, and the sociology of the debate would be a very interesting study indeed. Science and the scientific method, however, will ultimately prevail and produce a broad consensus on at least the major issues.

For paleontologists faced with controversy, the traditional appeal is for more and better fossils. In this case, however, methodology—not fossils—will have to be the key to achieving agreement and converting dissenters. We already have abundant well-preserved fossils documenting the transition to birds among theropod dinosaurs. Yet those opposed to theropod relationships remain unwilling to accept and apply the cladistic methodology that has elucidated this transition. If the theropod ancestry of birds is the nonsense that some would have us believe, why is it that so many highly trained specialists seem to keep confusing birds for dinosaurs or vice versa? The list of taxa that have bounced back and forth between birds and theropods is quite long: Alvarezsauridae, *Archaeopteryx* (Eichstätt specimen), *Archaeornithoides, Avimimus, Avisaurus, Bradycneme, Caenagnathus, Caudipteryx, Limnornis,* Oviraptoridae, *Palaeocursornis, Protarchaeopteryx, Protoavis, Wyleyia.* It would seem to be simple common sense to think that birds and dinosaurs must have *some* close relationship if we have such trouble telling them apart. Of course, evolutionary convergence is the usual explanation invoked by opponents of theropod relationships to explain the resemblances. But this, too, seems to fly in the face of logic: on the one hand, we are told that the similarities have arisen because of convergence—the independent acquisition of similar attributes due to similar function and mode of life—but then, on the other hand, we are told that theropods and the ancestors of birds had totally different body plans, body sizes, preferred habitats, and modes of life (i.e., large, bipedal, terrestrial theropods versus small, quadrupedal, and arboreal alternative ancestors). How can they have it both ways? Cladistic methodology argues that convergence and homology are not asserted or assumed but rather are hypothesized and subsequently tested within a comprehensive phylogenetic analysis. I fully expect that, if we had the *true* and complete phylogenetic tree, we would indeed find that some of the characters shared by birds and maniraptoran theropods were convergently acquired—but certainly not *all* of them, considering that there is hardly a bone of the body that fails to show synapomorphies at some level.

The debate on avian ancestry has had a long and contentious history, and this is likely to continue for some time. It all started with *Archaeopteryx*, and *Archaeopteryx* remains the central figure, although, if the situation is as complicated as Paul (2002) or Zweers and Vanden Berge (1998) suggest, this may prove to have been folly. A remarkable development witnessed in recent years is that, as support for the theropod origin of birds mounts and becomes more refined, the opposition has taken on a shrill tone and has seemingly abandoned the search for cogent and articulate alternative hypotheses, instead choosing to attack the dinosaur hypothesis, its methodology, and even its purveyors. Ironically, virtually all recent thought on the relationship of functional to phylogenetic inference has challenged the "function-first" approach advocated by those arguing for an arboreal origin of avian flight from nontheropodan ancestors. But ultimately the debate will be resolved by both fossils and philosophy. For example, the recent discovery of nonavian theropods with true feathers, such as *Caudipteryx*, has already satisfied many people that the debate is over. History, however, would suggest that such a view is unwarranted. The irony of paleontological discovery has always been that the picture seems simpler and more understandable with fewer fossil taxa; as new taxa are discovered, the water often becomes muddier. Hypotheses fall, to be replaced by new ones. The theropod ancestry of birds has weathered many challenges, and I predict that it will continue to do so. But if new fossils are discovered (as we know they will) and if minds remain open (as we hope they will), the true nature of this relationship may be much more complicated than we can even envision today.

Acknowledgments

For comments on this chapter, I thank L. M. Chiappe, K. Padian, and J. C. Sedlmayr. For sharing unpublished information, I thank S. Chatterjee, L. M. Chiappe, P. J. Currie, T. D. Jones, G. S. Paul, J. A. Ruben, Xu X., and Zhou Z. Funding was provided by the National Science Foundation (IBN-9601174) and the Ohio University College of Osteopathic Medicine.

Literature Cited

Abel, O. 1911. Die Vorfahren der Vögel und ihre Lebensweise. Verhandlungen der Zoologischen-botanischen Gesellschaft, Vienna 61:144–191.

Ackerman, J. 1998. Dinosaurs take wing. National Geographic 194(1):74–99.

Bakker, R. T., and P. M. Galton. 1974. Dinosaur monophyly and a new class of vertebrates. Nature 248:168–172.

Balda, R. P., G. Caple, and W. R. Willis. 1985. Comparison of the gliding to flapping sequence with the flapping to gliding sequence; pp. 267–277 in M. K. Hecht, J. H. Ostrom, G. Viohl, and P. Wellnhofer (eds.), The Beginnings of Birds. Freunde des Jura-Museums, Eichstätt.

Barsbold R. 1979. Opisthopubic pelvis in the carnivorous dinosaurs. Nature 279:792–793.

———. 1983. Carnivorous dinosaurs from the Cretaceous of Mongolia. Transactions of the Joint Soviet-Mongolian Paleontological Expedition 19:5–119. [Russian, English summary]

Barsbold R. and Perle A. 1979. Modification of saurischian pelvis and parallel evolution of carnivorous dinosaurs. Transactions of the Joint Soviet-Mongolian Paleontological Expedition 8:39–44. [Russian, English summary]

———. 1980. Segnosauria, a new infraorder of carnivorous dinosaurs. Acta Palaeontologica Polonica 25:187–195.

Barsbold R., P. J. Currie, N. P. Myhrvold, H. Osmólska, Tsogtbaatar K., and M. Watabe. 2000a. A pygostyle from a nonavian theropod. Nature 403:155–156.

Barsbold R., H. Osmólska, M. Watabe, P. J. Currie, and Tsogtbaatar K. 2000b. A new oviraptorosaur (Dinosauria, Theropoda) from Mongolia: the first dinosaur with a pygostyle. Acta Palaeontologica Polonica 45:97–106.

Benton, M. J. 1999. *Scleromochlus taylori* and the origin of dinosaurs and pterosaurs. Philosophical Transactions of the Royal Society of London B 354:1423–1446.

Bock, W. J. 1965. The role of adaptive mechanisms in the origin of higher levels of organization. Systematic Zoology 14:272–287.

———. 1985. The arboreal theory for the origin of birds; pp. 199–207 in M. K. Hecht, J. H. Ostrom, G. Viohl, and P. Wellnhofer (eds.), The Beginnings of Birds. Freunde des Jura-Museums, Eichstätt.

———. 1986. The arboreal origin of avian flight; pp. 57–72 in K. Padian (ed.), The Origin of Birds and the Evolution of Flight. California Academy of Sciences, San Francisco.

———. 1997. Review of: The Origin and Evolution of Birds, by A. Feduccia. Auk 114:531–534.

———. 1999. Review of: The Mistaken Extinction: Dinosaur Evolution and the Origin of Birds. Auk 16:566–568.

Bock, W. J., and P. Bühler. 1995. Introduction to the symposium. Archaeopteryx 13:3–4.

Brinkman, D. L., R. L. Cifelli, and N. J. Czaplewski. 1998. First occurrence of *Deinonychus antirrhopus* (Dinosauria: Theropoda) from the Antlers Formation (Lower Cretaceous:

Aptian-Albian) of Oklahoma. Oklahoma Geological Survey Bulletin 146:1–27.

Brochu, C. A., and M. A. Norell. 2000. Temporal congruence and the origin of birds. Journal of Vertebrate Paleontology 20:197–200.

Brush, A. H., L. D. Martin, J. H. Ostrom, and P. Wellnhofer. 1997. Bird or dinosaur?—statement of a team of specialists. Episodes 20(1):47.

Bryant, H. N., and A. P. Russell. 1992. The role of phylogenetic analysis in the inference of unpreserved attributes of extinct taxa. Philosophical Transactions of the Royal Society of London B 337:405–418.

———. 1993. The occurrence of clavicles within Dinosauria: implications for the homology of the avian furcula and the utility of negative evidence. Journal of Vertebrate Paleontology 13(2):171–184.

Burgers, P., and L. M. Chiappe. 1999. The wing of *Archaeopteryx* as a primary thrust generator. Nature 399:60–62.

Burke, A. C., and A. Feduccia. 1997. Developmental patterns and the identification of homologies in the avian hand. Science 278:666–668.

Burnham, D. A., K. L. Derstler, P. J. Currie, R. T. Bakker, Zhou Z., and J. H. Ostrom. 2000. Remarkable new birdlike dinosaur (Theropoda: Maniraptora) from the Upper Cretaceous of Montana. University of Kansas Paleontological Contributions, New Series 13:1–14.

Cai Z. and Zhao L. 1999. A long tailed bird from the Late Cretaceous of Zhejiang. Science in China (Series D) 42(4):434–441.

Caple, G., R. P. Balda, and W. R. Willis. 1983. The physics of leaping animals and the evolution of preflight. American Naturalist 121:455–467.

Charig, A. J. 1976. Order Thecodontia Owen 1859; pp. 7–10 *in* O. Kuhn (ed.), Handbuch der Paläoherpetologie, Teil 13. Gustav Fischer Verlag, New York.

Chatterjee, S. 1985. *Postosuchus,* a new thecodontian reptile from the Triassic of Texas and the origin of tyrannosaurs. Philosophical Transactions of the Royal Society of London B 309:395–460.

———. 1987. Skull of *Protoavis* and early evolution of birds. Journal of Vertebrate Paleontology 7(Supplement to 3):14A.

———. 1988. Functional significance of the semilunate carpal in archosaurs and birds. Journal of Vertebrate Paleontology 8(Supplement to 3):11A.

———. 1991. Cranial anatomy and relationships of a new Triassic bird from Texas. Philosophical Transactions of the Royal Society of London B 332:277–342.

———. 1993. *Shuvosaurus,* a new theropod. National Geographic Research and Exploration 9:274–285.

———. 1995. The Triassic bird *Protoavis*. Archaeopteryx 13:15–31.

———. 1997a. The Rise of Birds: 225 Million Years of Evolution. Johns Hopkins University Press, Baltimore, 312 pp.

———. 1997b. The beginnings of avian flight; pp. 311–335 *in* D. L. Wolberg, E. Stump, and G. Rosenberg (eds.), Dinofest International: Proceedings of a Symposium Held at Arizona State University. Academy of Natural Sciences, Philadelphia.

———. 1998a. Counting the fingers of birds and dinosaurs. Science 280:355a.

———. 1998b. The avian status of *Protoavis*. Archaeopteryx 16:99–122.

———. 1999. *Protoavis* and the early evolution of birds. Palaeontographica, Abteilung A 254:1–100.

Chen P.-J., Dong Z.-M., and Zhen S.-N. 1998. An exceptionally well-preserved theropod dinosaur from the Yixian Formation of China. Nature 391:147–152.

Chiappe, L. M. 1992. Enantiornithine (Aves) tarsometatarsi and the avian affinities of the Late Cretaceous Avisauridae. Journal of Vertebrate Paleontology 12:344–350.

———. 1995. The first 85 million years of avian evolution. Nature 378:349–355.

———. 1996. Late Cretaceous birds of southern South America: anatomy and systematics of Enantiornithes and *Patagopteryx deferrariisi;* pp. 203–244 *in* G. Arratia (ed.), Contributions of Southern South America to Vertebrate Paleontology. Münchner Geowissenschaftliche Abhandlungen, Reihe A, Geologie und Paläontologie 30. Verlag Dr. Friedrich Pfeil, Munich.

———. 1997. Aves; pp. 32–38 *in* P. J. Currie and K. Padian (eds.), The Encyclopedia of Dinosaurs. Academic Press, New York.

———. 1998. Review of: The Rise of Birds: 225 Million Years of Evolution, by S. Chatterjee. American Zoologist 38(4):797–798.

Chiappe, L. M., M. A. Norell, and J. M. Clark. 1996. Phylogenetic position of *Mononykus* (Aves: Alvarezsauridae) from the Late Cretaceous of the Gobi Desert. Memoirs of the Queensland Museum 39(3):557–582.

Chiappe, L. M., Ji S., Ji Q., and M. A. Norell. 1999. Anatomy and systematics of the Confuciusornithidae (Theropoda: Aves) from the Late Mesozoic of northeastern China. Bulletin of the American Museum of Natural History 242:1–89.

Chure, D. J., and J. H. Madsen Jr. 1996. On the presence of furculae in some non-maniraptoran theropods. Journal of Vertebrate Paleontology 16(3):573–577.

Clark, J. M., M. A. Norell, and L. M. Chiappe. 1999. An oviraptorid skeleton from the Late Cretaceous of Ukhaa Tolgod, Mongolia, preserved in an avianlike brooding position over an oviraptorid nest. American Museum Novitates 3265:1–36.

Cousins, F. W. 1973. The alleged evolution of birds (*Archaeopteryx*); pp. 89–99 *in* D. W. Patten (ed.), Symposium in Creation III. Baker Book House, Grand Rapids, Mich.

Cracraft, J. 1990. The origin of evolutionary novelties: pattern and process at different hierarchical levels; pp. 21–44 *in* M. H. Nitecki (ed.), Evolutionary Innovations. University of Chicago Press, Chicago.

Currie, P. J. 1997. "Feathered" dinosaurs; p. 241 *in* P. J. Currie and K. Padian (eds.), Encyclopedia of Dinosaurs. Academic Press, New York.

———. 1998. *Caudipteryx* revealed. National Geographic 194(1):86–89.

Currie, P. J., and Zhao X. J. 1993 A new troodontid (Dinosauria, Theropoda) braincase from the Dinosaur Park Formation (Campanian) of Alberta. Canadian Journal of Earth Sciences 30:2231–2247.

Czerkas, S. 1994. The history and interpretation of sauropod skin impressions. Gaia 10:173–182.

Dal Sasso, C., and M. Signore. 1998. Exceptional soft-tissue preservation in a theropod dinosaur from Italy. Nature 392:383–387.

de Beer, G. R. 1954. *Archaeopteryx lithographica:* A Study Based upon the British Museum Specimen. Natural History Museum, London, 68 pp.

————. 1956. The evolution of ratites. Bulletin of the British Museum (Natural History), Zoology 4(2):59–70.

de Queiroz, K., and J. A. Gauthier. 1990. Phylogeny as a central principle in taxonomy: phylogenetic definition of taxon names. Systematic Zoology 39:307–322.

————. 1992. Phylogenetic taxonomy. Annual Review of Ecology and Systematics 23:449–480.

Dingus, L., and T. Rowe. 1997. The Mistaken Extinction: Dinosaur Evolution and the Origin of Birds. W. H. Freeman and Company, New York, 332 pp.

DiSilvestro, R. L. 1997. In quest of the origin of birds. BioScience 47(8): 481–485.

Dyck, J. 1985. The evolution of feathers. Zoologica Scripta 14(2):137–154.

Dyke, G. J., and J. Thorley. 1998. Reduced cladistic consensus methods and the inter-relationships of Protoavis, Avimimus and Mesozoic birds. Archaeopteryx 16:123–129.

Ebel, K. 1996. On the origin of flight in Archaeopteryx and in pterosaurs. Neues Jahrbuch für Geologie und Paläontologie 202:269–285.

Elzanowski, A. 1995. Cranial evidence for the avian relationships of Oviraptorosauria. Journal of Vertebrate Paleontology 15(Supplement to 3):27A.

————. 1999. A comparison of the jaw skeleton in theropods and birds, with a description of the palate in Oviraptoridae; pp. 311–323 in S. L. Olson (ed.), Avian Paleontology at the Close of the 20th Century: Proceedings of the 4th International Meeting of the Society of Avian Paleontology and Evolution, Washington, D.C., 4–7 June 1996. Smithsonian Contributions to Paleobiology 89, Washington.

Evans, S. E., and A. R. Milner. 1994. Middle Jurassic microvertebrate assemblages from the British Isles; pp. 303–321 in N. C. Fraser and H.-D. Sues (eds.), In the Shadow of the Dinosaurs: Early Mesozoic Tetrapods. Cambridge University Press, New York.

Feduccia, A. 1993. Evidence from claw geometry indicating arboreal habits of Archaeopteryx. Science 259:790–793.

————. 1994. The great dinosaur debate. Living Bird 13:29–33.

————. 1995. Explosive evolution in Tertiary birds and mammals. Science 267:637–638.

————. 1996. The Origin and Evolution of Birds. Yale University Press, New Haven, 420 pp.

————. 1999a. 1, 2, 3, = 2, 3, 4: accommodating the cladogram. Proceedings of the National Academy of Sciences 96:4740–4742.

————. 1999b. The Origin and Evolution of Birds, 2nd edition. Yale University Press, New Haven, 466 pp.

Feduccia, A., and L. D. Martin. 1998. Theropod-bird link reconsidered. Nature 391:754.

Feduccia, A., and R. Wild. 1993. Birdlike characters in the Triassic archosaur Megalancosaurus. Naturwissenschaften 80:564–566.

Forster, C. A., S. D. Sampson, L. M. Chiappe, and D. W. Krause. 1998. The theropod ancestry of birds: new evidence from the Late Cretaceous of Madagascar. Science 279:1915–1919.

Garner, J. P., and A. L. R. Thomas. 1998. Counting the fingers of birds and dinosaurs. Science 280:355a.

Gatesy, S. M. 1990. Caudofemoral musculature and the evolution of theropod locomotion. Paleobiology 16(2):170–186.

Gatesy, S. M., and K. P. Dial. 1996a. Locomotor modules and the evolution of avian flight. Evolution 50(1):331–340.

————. 1996b. From frond to fan: Archaeopteryx and the evolution of short-tailed birds. Evolution 50(5):2037–2048.

Gauthier, J. 1986. Saurischian monophyly and the origin of birds; pp. 1–55 in K. Padian (ed.), The Origin of Birds and the Evolution of Flight. California Academy of Sciences, San Francisco.

Gauthier, J., and L. F. Gall (eds.). 2001. New Perspectives on the Origin and Early Evolution of Birds: Proceedings of an International Symposium in Honor of John H. Ostrom. Yale University Press, New Haven.

————. 2001. The role of Protoavis in the debate on avian origins; pp. 537–548 in J. Gauthier and L. F. Gall (eds.), New Persepctives on the Origin and Early Evolution of Birds: Proceedings of an International Symposium in Honor of John H. Ostrom. Yale University Press, New Haven.

Gauthier, J., and K. Padian. 1985. Phylogenetic, functional, and aerodynamic analyses of the origin of birds and their flight; pp. 185–197 in M. K. Hecht, J. H. Ostrom, G. Viohl, and P. Wellnhofer (eds.), The Beginnings of Birds. Freunde des Jura-Museums, Eichstätt.

Geist, N. R., T. D. Jones, and J. A. Ruben. 1997. Implications of soft tissue preservation in the compsognathid dinosaur, Sinosauropteryx. Journal of Vertebrate Paleontology 17(Supplement to 3):48A.

Gibbons, A. 1996. New feathered fossil brings dinosaurs and birds closer. Science 274:720–721.

————. 1997a. Feathered dino wins a few friends. Science 275: 1731.

————. 1997b. Plucking the feathered dinosaur. Science 278: 1229.

Gower, D. J., and E. Weber. 1998. The braincase of Euparkeria, and the evolutionary relationships of birds and crocodilians. Biological Reviews 73:367–411.

Griffiths, P. J. 1996. The isolated Archaeopteryx feather. Archaeopteryx 14:1–26.

Haeckel, E. 1866. Generelle Morphologie der Organismen. Zweiter Band: Allgemeine Entwicklungsgeschichte der Organismen. G. Reimer, Berlin.

————. 1876. The History of Creation, Volume II (Lankester Translation). Appleton, New York, 544 pp.

Haubold, H., and E. Buffetaut. 1987. A new interpretation of Longisquama insignis, an enigmatic reptile from the Upper Triassic of Central Asia. Comptes Rendus de l'Académie des Sciences du Paris 305:65–70.

Hecht, M. K., J. H. Ostrom, G. Viohl, and P. Wellnhofer. 1985. Preface; p. 7 in M. K. Hecht, J. H. Ostrom, G. Viohl, and P. Wellnhofer (eds.), The Beginnings of Birds. Freunde des Jura-Museums, Eichstätt.

Heilmann, G. 1926. The Origin of Birds. Witherby, London, 208 pp.

Herzog, K. 1993. Stammesgeschichtliche Entwicklung des Flugvermögens der Vögel. Archaeopteryx 11:49–61.

Hinchliffe, J. R. 1985. "One, two, three" or "two, three, four": an embryologist's view of the homologies of the digits and carpus of modern birds; pp. 141–147 in M. K. Hecht, J. H. Ostrom, G. Viohl, and P. Wellnhofer (eds.), The Beginnings of Birds. Freunde des Jura-Museums, Eichstätt.

Hinchliffe, J. R., and M. K. Hecht. 1984. Homology of the bird wing skeleton: embryological versus paleontological evidence. Evolutionary Biology 18:21–39.

Holtz, T. R., Jr. 1994. The phylogenetic position of the Tyrannosauridae: implications for theropod systematics. Journal of Paleontology 68:1100–1117.

Hopson, J. A., and L. M. Chiappe. 1998. Pedal proportions of living and fossil birds indicate arboreal or terrestrial specialization. Journal of Vertebrate Paleontology 18(Supplement to 3):52A.

Hou L.-H., L. D. Martin, Zhou Z.-H., A. Feduccia, and Zhang F. 1999. A diapsid skull in a new species of the primitive bird *Confuciusornis*. Nature 399:679–682.

Hutchinson, J. R. 2001a. The evolution of pelvic osteology and soft tissues on the line to extant birds (Neornithes). Zoological Journal of the Linnean Society 131:123–168.

———. 2001b. The evolution of femoral osteology and soft tissues on the line to extant birds (Neornithes). Zoological Journal of the Linnean Society 131:169–197.

Huxley, T. H. 1867. On the classification of birds; and on the taxonomic value of the modifications of certain of the cranial bones observable in that class. Proceedings of the Zoological Society of London 1867:415–472.

———. 1868. On the animals which are most nearly intermediate between birds and reptiles. Annals and Magazine of Natural History 4(2):66–75.

Jenkins, F. A., Jr. 1993. The evolution of the avian shoulder joint. American Journal of Science 293-A:253–267.

Jensen, J. A., and K. Padian. 1989. Small pterosaurs and dinosaurs from the Uncompahgre fauna (Brushy Basin Member, Morrison Formation: ?Tithonian), Late Jurassic, western Colorado). Journal of Paleontology 63:364–373.

Ji Q. and Ji S.-A. 1996. On discovery of the earliest bird fossil in China and the origin of birds. Chinese Geology 233:30–33. [Chinese]

———. 1997. *Protoarchaeopteryx*, a new genus of Archaeopterygidae in China. Chinese Geology 238:38–41. [Chinese]

Ji Q., P. J. Currie, M. A. Norell, and Ji S.-A. 1998. Two feathered dinosaurs from northeastern China. Nature 393:753–761.

Ji Q., M. A. Norell, Gao K.-Q., Ji S.-A., and Ren D. 2001. The distribution of integumentary structures in a feathered dinosaur. Nature 410:1084–1088.

Jones, T. D., J. A. Ruben, L. D. Martin, E. N. Kurochkin, A. Feduccia, P. F. A. Maderson, W. J. Hillenius, N. R. Geist, and V. Alifanov. 2000a. Nonavian feathers in a Late Triassic archosaur. Science 288:2202–2205.

Jones, T. D., J. O. Farlow, J. A. Ruben, D. M. Henderson, and W. J. Hillenius. 2000b. Cursoriality in bipedal archosaurs. Nature 406:716–718.

Kurochkin, E. N. 1995. Synopsis of Mesozoic birds and early evolution of Class Aves. Archaeopteryx 13:47–66.

Kurzanov, S. M. 1985. The skull structure of the dinosaur *Avimimus*. Paleontological Journal 1985(4):92–99.

———. 1987. Avimimidae and the problem of the origin of birds. Transactions of the Joint Soviet-Mongolian Paleontological Expedition 31:31–94. [Russian, English summary]

Lauder, G. V. 1981. Form and function: structural analysis in evolutionary morphology. Paleobiology 7:430–442.

———. 1990. Functional morphology and systematics: studying functional patterns in an historical context. Annual Review of Ecology and Systematics 21:317–340.

———. 1995. On the inference of function from structure; pp. 1–18 in J. J. Thomason (ed.), Functional Morphology in Vertebrate Paleontology. Cambridge University Press, New York.

Lauder, G. V., and K. F. Liem. 1989. The role of historical factors in the evolution of complex organismal functions; pp. 63–78 in D. B. Wake and G. Roth (eds.), Complex Organismal Functions: Integration and Evolution in Vertebrates. John Wiley and Sons, New York.

Liem, K. F. 1989. Functional morphology and phylogenetic testing within the framework of symecomorphosis. Acta Morphologica Neerlandoscandinavica 27:119–131.

Long, R. A., and P. A. Murry. 1995. Late Triassic (Carnian and Norian) tetrapods from the southwestern United States. New Mexico Museum of Natural History and Science, Bulletin 4:1–254.

Lowe, P. R. 1935. On the relationship of Struthiones to the dinosaurs and to the rest of the avian class, with special reference to the position of *Archaeopteryx*. Ibis 5(2):398–432.

———. 1944. An analysis of the characters of *Archaeopteryx* and *Archaeornis*. Were they birds or reptiles? Ibis 86:517–543.

Lü J. 2000. Oviraptorosaurs compared to birds. Vertebrata PalAsiatica 38(Supplement):18.

Lull, R. S., and N. E. Wright. 1942. Hadrosaurian dinosaurs of North America. Geological Society of America, Special Papers 40:1–242.

Luo Z. 1999. A refugium for relicts. Nature 400:23–25.

Makovicky, P. J., and P. J. Currie. 1998. The presence of a furcula in tyrannosaurid theropods, and its phylogenetic and functional implications. Journal of Vertebrate Paleontology 18(1):143–149.

Makovicky, P. J., and M. A. Norell. 1998. A partial ornithomimid braincase from Ukhaa Tolgod (Upper Cretaceous, Mongolia). American Museum Novitates 3247:1–16.

Makovicky, P. J., and H.-D. Sues. 1998. Anatomy and phylogenetic relationships of the theropod dinosaur *Microvenator celer* from the Lower Cretaceous of Montana. American Museum Novitates 3240:1–27.

Martin, L. D. 1983. The origin and early radiation of birds; pp. 291–338 in A. H. Brush and G. A. Clark (eds.), Perspectives in Ornithology. Cambridge University Press, Cambridge.

———. 1988. Review of: The Origin of Birds and the Evolution of Flight, edited by K. Padian. Auk 105(3):596–597.

———. 1991. Mesozoic birds and the origin of birds; pp. 485–540 in H.-P. Schultze and L. Trueb (eds.), Origins of the Higher Groups of Tetrapods: Controversy and Consensus. Cornell University Press, Ithaca, N.Y.

———. 1995. A new skeletal model of *Archaeopteryx*. Archaeopteryx 13:33–40.

———. 1997. The difference between dinosaurs and birds as applied to *Mononykus*; pp. 337–343 in D. L. Wolberg, E. Stump, and G. Rosenberg (eds.), Dinofest International: Proceedings of a Symposium Held at Arizona State University. Academy of Natural Sciences, Philadelphia.

———. 1998. The big flap. Sciences 38(2):39–44.

Martin, L. D., J. D. Stewart, and K. N. Whetstone. 1980. The origin of birds: structure of the tarsus and teeth. Auk 97:86–93.

Martin, L. D., Zhou Z., Hou L., and A. Feduccia. 1998. *Confuciusornis sanctus* compared to *Archaeopteryx lithographica*. Naturwissenschaften 85:286–289.

Mayr, E. 1997. Review of: The Origin and Evolution of Birds, by A. Feduccia. American Zoologist 37(2):210–211.

McDonald, K. 1996. A dispute over the evolution of birds. Chronicle of Higher Education 25 Oct. 1996:A14–A15.

Metcalf, S. J., and R. J. Walker. 1994. A new Bathonian microvertebrate locality in the English Midlands; pp. 322–331 *in* N. C. Fraser and H.-D. Sues (eds.), In the Shadow of the Dinosaurs: Early Mesozoic Tetrapods. Cambridge University Press, New York.

Morell, V. 1997. The origin of birds: the dinosaur debate. Audubon 99(2):36–45.

Norberg, U. M. 1985. Evolution of flight in birds: aerodynamic, mechanical, and ecological aspects; pp. 293–302 *in* M. K. Hecht, J. H. Ostrom, G. Viohl, and P. Wellnhofer (eds.), The Beginnings of Birds. Freunde des Jura-Museums, Eichstätt.

Norell, M. A. 2001. The proof is in the plumage. Natural History 110(6):58–63.

Norell, M. A., and L. M. Chiappe. 1996. Flight from reason [Review of: The Origin and Evolution of Birds, by A. Feduccia]. Nature 384:230.

Norell, M. A., and P. J. Makovicky. 1997. Important features of the dromaeosaur skeleton: information from a new specimen. American Museum Novitates 3215:1–28.

———. 1999. Important features of the dromaeosaurid skeleton II: information from newly collected specimens of *Velociraptor mongoliensis*. American Museum Novitates 3282:1–45.

Norell, M. A., P. J. Makovicky, and J. M. Clark. 1997. A *Velociraptor* wishbone. Nature 389:447.

Norell, M. A., P. J. Makovicky, and P. J. Currie. 2001. The beaks of ostrich dinosaurs. Nature 412:873–874.

Novas, F. E. 1996. Alvarezsauridae, Cretaceous basal birds from Patagonia and Mongolia. Memoirs of the Queensland Museum 39:675–702.

Novas, F. E., and P. F. Puerta. 1997. New evidence concerning avian origins from the late Cretaceous of Patagonia. Nature 376:390–392.

Olshevsky, G. 1994. The birds first? A theory to fit the facts. Omni 16:34–38, 40–43, 80–84.

Olson, S. L. 1999. Avian Paleontology at the Close of the 20th Century: Proceedings of the 4th International Meeting of the Society of Avian Paleontology and Evolution, Washington, D.C., 4–7 June 1996. Smithsonian Contributions to Paleobiology 89, Washington.

Ostrom, J. H. 1973. The ancestry of birds. Nature 242:136.

———. 1976. *Archaeopteryx* and the origin of birds. Biological Journal of the Linnean Society 8:91–182.

———. 1986. The cursorial origin of avian flight; pp. 73–81 *in* K. Padian (ed.), The Origin of Birds and the Evolution of Flight. California Academy of Sciences, San Francisco.

———. 1987. *Protoavis*, a Triassic bird? Archaeopteryx 5:113–114.

———. 1991. The bird in the bush. Nature 353:212.

———. 1996. The questionable validity of *Protoavis*. Archaeopteryx 14:39–42.

Owen, R. 1863. On the *Archeopteryx* of von Meyer, with a description of the fossil remains of a long-tailed species, from the lithographic stone of Solenhofen. Philosophical Transactions of the Royal Society of London 153:33–47.

Padian, K. 1982. Macroevolution and the origin of major adaptations: vertebrate flight as a paradigm for the analysis of patterns. Proceedings of the Third North American Paleontological Convention 2:387–392.

———. 1984. The origin of pterosaurs; pp. 163–168 *in* W.-E. Reif and F. Westphal (eds.), Third Symposium on Mesozoic Terrestrial Ecosystems, Short Papers. Attempto Verlag, Tübingen.

———. 1985. The origins and aerodynamics of flight in extinct vertebrates. Palaeontology 28:413–433.

———. 1995. Form versus function: the evolution of a dialectic; pp. 264–277 *in* J. J. Thomason (ed.), Functional Morphology in Vertebrate Paleontology. Cambridge University Press, New York.

———. 1997. The continuing debate over avian origins [Review of: The Origin and Evolution of Birds, by A. Feduccia]. American Scientist 85:178–180.

———. 1998. When is a bird not a bird? Nature 393:729–730.

Padian, K., and L. M. Chiappe. 1998a. The origin of birds and their flight. Scientific American 278(2):28–37.

———. 1998b. The origin and early evolution of birds. Biological Reviews 73:1–42.

Padian, K., J. R. Hutchinson, and T. R. Holtz Jr. 1999. Phylogenetic definitions and nomenclature of the major taxonomic categories of the carnivorous Dinosauria (Theropoda). Journal of Vertebrate Paleontology 19:69–80.

Padian, K., Ji Q., and Ji. S. 2001. Feathered dinosaurs and the origin of avian flight; pp. 117–135 *in* D. H. Tanke and K. Carpenter (eds.), Mesozoic Vertebrate Life. Indiana University Press, Indianapolis.

Parkes, K. C. 1966. Speculations on the origin of feathers. Living Bird 5:77–86.

Paul, G. S. 1984. The archosaurs: a phylogenetic study; pp. 175–180 *in* W.-E. Reif and F. Westphal (eds.), Third Symposium on Mesozoic Terrestrial Ecosystems, Short Papers. Attempto Verlag, Tübingen.

———. 1988. Predatory Dinosaurs of the World. Simon and Schuster, New York, 464 pp.

———. 1991. The many myths, some old, some new, of dinosaurology. Modern Geology 16:69–99.

———. 1996. Complexities in the evolution of birds from predatory dinosaurs: *Archaeopteryx* was a flying dromaeosaur, and some Cretaceous dinosaurs may have been secondarily flightless; p. 15 *in* H. F. James and S. L. Olson (eds.), Program and Abstracts of the Fourth International Meeting of the Society of Avian Paleontology and Evolution. Smithsonian Institution, Washington.

———. 2002. Dinosaurs of the Air: The Evolution and Loss of Flight in Dinosaurs and Birds. Johns Hopkins University Press, Baltimore.

Pennycuick, C. J. 1986. Mechanical constraints on the evolution of flight; pp. 83–98 *in* K. Padian (ed.), The Origin of Birds and the Evolution of Flight. California Academy of Sciences, San Francisco.

Pérez-Moreno, B. P., J. L. Sanz, A. D. Buscalioni, J. L. Moratella, F. Ortega, and D. Rasskin-Gutman. 1994. A unique multitoothed ornithomimosaur dinosaur from the Lower Cretaceous of Spain. Nature 370:363–367.

Peters, D. S. 1994. Die Entstehung der Vögel. Verändern die jüngsten Fossilfunde das Modell? Senckenberg-Buch 70:403–424.

———. 1995. Acta Palaeornithologica. 3. Symposium SAPE. Courier Forschungsinstitut Senckenberg 18, Frankfurt, 361 pp.

Peters, D. S., and E. Görgner. 1992. A comparative study on the claws of *Archaeopteryx*; pp. 29–37 *in* K. E. Campbell Jr. (ed.),

Papers in Avian Paleontology Honoring Pierce Brodkorb. Natural History Museum of Los Angeles County, Los Angeles.

Peters, D. S., and Ji Q. 1998. The diapsid temporal construction of the Chinese fossil bird *Confuciusornis*. Senckenbergiana lethaea 78:153–155.

Prum, R. O. 1999. Development and evolutionary origin of feathers. Journal of Experimental Zoology 285:291–306.

———. 2000. Feather development and the homology of avian feathers and dinosaur integumental filaments. Vertebrata PalAsiatica 38(Supplement):25–26.

Rasskin-Gutman, D. 1997. Pelvis, comparative anatomy; pp. 536–540 *in* P. J. Currie and K. Padian (eds.), Encyclopedia of Dinosaurs. Academic Press, New York.

Rauhut, O. W. M. 1997. Zur Schädelanatomie von *Shuvosaurus inexpectatus* (Dinosauria; Theropoda); pp. 17–21 *in* S. Sachs, O. W. M. Rauhut, and A. Weigert (eds.), 1. Treffen der deutschsprachigen Palaeoherpetologen, Düsseldorf, Extended Abstracts—Terra Nostra 7/97.

Rayner, J. M. V. 1985. Cursorial gliding in proto-birds: an expanded version of a discussion contribution; pp. 289–292 *in* M. K. Hecht, J. H. Ostrom, G. Viohl, and P. Wellnhofer (eds.), The Beginnings of Birds. Freunde des Jura-Museums, Eichstätt.

———. 1991. Avian flight evolution and the problem of *Archaeopteryx*; pp. 183–212 *in* J. M. V. Rayner and R. J. Wootton (eds.), Biomechanics in Evolution. Cambridge University Press, New York.

Reisz, R. R., and H.-D. Sues. 2000. The "feathers" of *Longisquama*. Nature 408:428.

Renesto, S. 1994. *Megalancosaurus,* a possibly arboreal archosauromorph (Reptilia) from the Upper Triassic of northern Italy. Journal of Vertebrate Paleontology 14:38–52.

Rowe, M. P. 2000. Inferring the retinal anatomy and visual capacities of extinct vertebrates. Palaeontologia Electronica 3:1–43.

Ruben, J. A. 1995. The evolution of endothermy in mammals and birds: from physiology to fossils. Annual Review of Physiology 57:69–95.

———. 1996. Evolution of endothermy in mammals, birds and their ancestors; pp. 347–376 *in* I. A. Johnston and A. F. Bennett (eds.), Animals and Temperature: Phenotypic and Evolutionary Adaptation. Cambridge University Press, New York.

———. 1997. Review of: The Origin and Evolution of Birds, by A. Feduccia. BioScience 47:392–394.

Ruben, J. A., T. D. Jones, N. R. Geist, and W. J. Hillenius. 1997. Lung structure and ventilation in theropod dinosaurs and early birds. Science 278:1267–1270.

Ruben, J. A., C. Dal Sasso, N. R. Geist, W. J. Hillenius, T. D. Jones, and M. Signore. 1999. Pulmonary function and metabolic physiology of theropod dinosaurs. Science 283:514–516.

Russell, D. A., and Dong Z.-M. 1993. The affinities of a new theropod from the Alxa Desert, Inner Mongolia, People's Republic of China. Canadian Journal of Earth Science 30:2107-2127.

Sanz, J. L., and N. Lopez-Martinez. 1984. The prolacertid lepidosaurian *Cosesaurus aviceps* Ellenberger & Villalta, a claimed "protoavian" from the middle Triassic of Spain. Geobios 17:741–753.

Sanz, J. L., L. M. Chiappe, B. P. Pérez-Moreno, J. J. Moratalla, F. Hernández-Carrasquilla A. D. Buscalioni, F. Ortega, F. J. Poyato-Ariza, D. Rasskin-Gutman, and X. Martínez-Delclòs. 1997. A nestling bird from the Lower Cretaceous of Spain: implications for avian skull and neck evolution. Science 276:1543–1546.

Sanz, J. L., B. P. Pérez-Moreno, and F. J. Poyato-Ariza. 1998. Living with dinosaurs. Nature 393:32–33.

Schweitzer, M. H., J. A. Watt, R. Avci, L. Knapp, L. M. Chiappe, M. A. Norell, and M. Marshall. 1999. Beta-keratin specific immunological reactivity in feather-like structures of the Cretaceous alvarezsaurid, *Shuvuuia deserti*. Journal of Experimental Zoology 285:146–157.

Sereno, P. C. 1997a. Ancient aviary, featherweight phylogeny [Review of: The Origin and Evolution of Birds, by A. Feduccia]. Evolution 51(5):1689–1690.

———. 1997b. The origin and evolution of dinosaurs. Annual Review of Earth and Planetary Sciences 25:435–489.

———. 1998. A rationale for phylogenetic definitions, with application to the higher-level taxonomy of Dinosauria. Neues Jahrbuch für Geologie und Paläontologie 210:41–83.

———. 1999a. The evolution of dinosaurs. Science 284:2137–2147.

———. 1999b. Definitions in phylogenetic taxonomy: critique and rationale. Systematic Biology 48:329–351.

Sereno, P. C., and Rao C. 1992. Early evolution of avian flight and perching: new evidence from the Lower Cretaceous of China. Science 255:845–848.

Sharov, A. G. 1970. An unusual reptile from the Lower Triassic of Fergana. Paleontological Journal 1970:112–116.

Shi L. and Zhang F. 2000. 5th International Meeting of the Society of Avian Paleontology and Evolution and the Symposium of Jehol Biota (Abstracts). Vertebrata PalAsiatica 38(Supplement):1–63.

Shipman, P. 1997a. Taking Wing: *Archaeopteryx* and the Evolution of Bird Flight. Simon and Schuster, New York, 336 pp.

———. 1997b. Birds do it . . . did dinosaurs? New Scientist 153(2067):26–31.

Shubin, N. H., and P. Alberch. 1986. A morphogenetic approach to the origin and basic organization of the tetrapod limb. Evolutionary Biology 20:319–387.

Simpson, G. G. 1946. Fossil penguins. Bulletin of the American Museum of Natural History 87:1–95.

Steadman, D. W. 1998. Review of: The Origin and Evolution of Birds, by A. Feduccia. Wilson Bulletin 110(1):140–141.

Stokstad, E. 2000. Feathers, or flight of fancy. Science 288:2124–2125.

Swisher, C. C., III, Wang Y. Q., Wang Z.-L., Xu X., and Wang Y. 1999. Cretaceous age for the feathered dinosaurs of Liaoning, China. Nature 400:58–61.

Tarsitano, S. 1985. The morphological and aerodynamic constraints on the origin of avian flight; pp. 319–332 *in* M. K. Hecht, J. H. Ostrom, G. Viohl, and P. Wellnhofer (eds.), The Beginnings of Birds. Freunde des Jura-Museums, Eichstätt.

———. 1991. *Archaeopteryx:* Quo vadis?; pp. 541–576 *in* H.-P. Schultze and L. Trueb (eds.), Origins of the Higher Groups of Tetrapods: Controversy and Consensus. Cornell University Press, Ithaca, N.Y.

Tarsitano, S., and M. K. Hecht. 1980. A reconsideration of the reptilian relationships of *Archaeopteryx*. Zoological Journal of the Linnean Society 69:149–182.

Thomas, A. L. R., and J. P. Garner. 1998. Are birds dinosaurs? Trends in Ecology and Evolution 13(4):129–130.

Thulborn, R. A. 1975. Dinosaur polyphyly and the classification of archosaurs and birds. Australian Journal of Zoology 23:249–270.

———. 1984. The avian relationships of *Archaeopteryx,* and the origin of birds. Zoological Journal of the Linnean Society 82:119–158.

Thulborn, R. A., and T. L. Hamley. 1982. The reptilian relationships of *Archaeopteryx.* Australian Journal of Zoology 30:611–634.

Unwin, D. M. 1998. Feathers, filaments and theropod dinosaurs. Nature 392:119–120.

Varricchio, D. J., F. Jackson, J. J. Borkowski, and J. R. Horner. 1997. Nest and egg clutches of the dinosaur *Troodon formosus* and the evolution of avian reproductive traits. Nature 385: 247–250.

Wagner, G. P., and J. A. Gauthier. 1999. 1, 2, 3 = 2, 3, 4: a solution to the problem of the homology of the digits of the avian hand. Proceedings of the National Academy of Sciences 96:5111–5116.

Walker, A. D. 1972. New light on the origin of birds and crocodiles. Nature 237:257–263.

———. 1990. A revision of *Sphenosuchus acutus* Haughton, a crocodylomorph reptile from the Elliot Formation (late Triassic or early Jurassic) of South Africa. Philosophical Transactions of the Royal Society of London B 330:1–120.

Weishampel, D. B. 1995. Fossils, function, and phylogeny; pp. 34–54 *in* J. J. Thomason (ed.), Functional Morphology in Vertebrate Paleontology. Cambridge University Press, New York.

Weishampel, D. B., and C.-M. Jianu. 1996. New theropod dinosaur material from the Hateg Basin (Late Cretaceous, western Romania). Neues Jahrbuch für Geologie und Paläontologie 200(3):387–404.

Wellnhofer, P. 1992. *Protoavis:* der älteste Vogel? Naturwissenschaftliche Rundschau 45(3):107–108.

———. 1994. New data on the origin and early evolution of birds. Comptes Rendus de l'Académie des Sciences, Paris, Series II 319:299–308.

Welman, J. 1995. *Euparkeria* and the origin of birds. South African Journal of Science 91:533–537.

Witmer, L. M. 1990. The craniofacial air sac system of Mesozoic birds (Aves). Zoological Journal of the Linnean Society 100:327–378.

———. 1991. Perspectives on avian origins; pp. 427–466 *in* H.-P. Schultze and L. Trueb (eds.), Origins of the Higher Groups of Tetrapods: Controversy and Consensus. Cornell University Press, Ithaca, N.Y.

———. 1995a. The Extant Phylogenetic Bracket and the importance of reconstructing soft tissues in fossils; pp. 19–33 *in* J. J. Thomason (ed.), Functional Morphology in Vertebrate Paleontology. Cambridge University Press, New York.

———. 1995b. Homology of facial structures in extant archosaurs (birds and crocodilians), with special reference to paranasal pneumaticity and nasal conchae. Journal of Morphology 225:269–327.

———. 1995c. The Search for the Origin of Birds. Franklin Watts, New York, 64 pp.

———. 1997a. The evolution of the antorbital cavity of archosaurs: a study in soft-tissue reconstruction in the fossil record with an analysis of the function of pneumaticity. Memoirs of the Society of Vertebrate Paleontology, Journal of Vertebrate Paleontology 17(Supplement to 1):1–73.

———. 1997b. Flying feathers [Review of: The Origin and Evolution of Birds, by A. Feduccia]. Science 276:1209–1210.

———. 1997c. Foreword; pp. vii–xii *in* S. Chatterjee, The Rise of Birds: 225 Million Years of Evolution. Johns Hopkins University Press, Baltimore.

———. 1997d. Craniofacial air sinus systems; pp. 151–159 *in* P. J. Currie and K. Padian (eds.), The Encyclopedia of Dinosaurs. Academic Press, New York.

———. 1999. New aspects of avian origins: roundtable report; pp. 327–334 *in* S. L. Olson (ed.), Avian Paleontology at the Close of the 20th Century: Proceedings of the 4th International Meeting of the Society of Avian Paleontology and Evolution, Washington, D.C., 4–7 June 1996. Smithsonian Contributions to Paleobiology 89, Washington.

———. 2001a. Nostril position in dinosaurs and other vertebrates and its significance for nasal function. Science 293: 850–853.

———. 2001b. The role of *Protoavis* in the debate on avian origins; pp. 537–548 *in* J. A. Gauthier and L. F. Gall (eds.), New Perspectives on the Origin and Early Evolution of Birds: Proceedings of an International Symposium in Honor of John H. Ostrom. Yale University Press, New Haven.

Witmer, L. M., and W. D. Maxwell. 1996. The skull of *Deinonychus* (Dinosauria: Theropoda): new insights and implications. Journal of Vertebrate Paleontology 16(Supplement to 3):73A.

Witmer, L. M., and D. B. Weishampel. 1993. Remains of theropod dinosaurs from the Upper Cretaceous St. Mary River Formation of northwestern Montana, with special reference to a new maniraptoran braincase. Journal of Vertebrate Paleontology 13(Supplement to 3):63A.

Xu X., Tang Z.-L., and Wang X.-L. 1999a. A therizinosauroid dinosaur with integumentary structures from China. Nature 399:350–354.

Xu X., Wang X.-L., and Wu X.-C. 1999b. A dromaeosaurid dinosaur with a filamentous integument from the Yixian Formation of China. Nature 401:262–266.

Xu X., Zhou Z., and Wang X.-L. 2000. The smallest known nonavian theropod dinosaur. Nature 408:705–708.

Xu X., Zhao X., and J. M. Clark. 2001a. A new therizinosaur from the Lower Jurassic Lower Lufeng Formation of Yunnan, China. Journal of Vertebrate Paleontology 21:477–483.

Xu X., Zhou Z., and R. O. Prum. 2001b. Branched integumental structures in *Sinornithosaurus* and the origin of feathers. Nature 410:200–204.

Xu X., M. A. Norell, Wang X.-L., P. J. Makovicky, and Wu X.-C. 2002. A basal troodontid from the Early Cretaceous of China. Nature 415:780–784.

Zalewski, D. 1996. Bones of contention. Lingua Franca 6(6): 22–24.

Zhang F. and Zhou Z. 2000. A primitive enantiornithine bird and the origin of feathers. Science 290:1955–1959.

Zhao X. and Xu X. 1998. The oldest coelurosaurian. Nature 394:234–235.

Zhou Z. 1998. Origins of avian flight: evidence from fossil and modern birds. Journal of Vertebrate Paleontology 18(Supplement to 3):88A.

Zhou Z., and Wang X.-L. 2000. A new species of *Caudipteryx* from the Yixian Formation of Liaoning, northeast China. Vertebrata PalAsiatica 38:113–130.

Zhou Z., Wang X.-L., Zhang F., and Xu X. 2000. Important features of *Caudipteryx*—evidence from two nearly complete specimens. Vertebrata PalAsiatica 38:241–254.

Zinke, J. 1998. Small theropod teeth from the Upper Jurassic coal mine of Guimarota (Portugal). Paläontologische Zeitschrift 72:179–189.

Zweers, G. A., and J. C. Vanden Berge. 1998. Birds at geological boundaries. Zoology: Analysis of Complex Systems (ZACS) 100:183–202.

Zweers, G. A., J. C. Vanden Berge, and H. Berkhoudt. 1997. Evolutionary patterns of avian trophic diversification. Zoology: Analysis of Complex Systems (ZACS) 100:25–57.

2

Cladistic Approaches to the Relationships of Birds to Other Theropod Dinosaurs

JAMES M. CLARK, MARK A. NORELL, AND PETER J. MAKOVICKY

 For many years the study of bird origins focused on *Archaeopteryx*, with its feathers taken as conclusive proof of its avian status, its long, bony tail and toothed jaws proclaiming its intermediate position between birds and their reptilian ancestors, and its geological antiquity sufficient to qualify for ancestral status. Its evolutionary origin thereby became equated with the origin of all birds. This simplistic view was heuristic in the context of the "search for ancestors" prevalent during most of the past century, but it fails when a more critical view is taken of what constitutes the evidence for evolutionary relationships. As unique a phenomenon as birds may be, their study nevertheless must proceed using the same methods and assumptions required for understanding the evolution of any group of organisms.

This chapter reviews the rationale for using the system of comparative biology known as cladistics to address the question of bird origins, evaluates the results of cladistic analyses of extinct theropod dinosaurs and their implications for the origin of birds, examines the completeness of the fossil record in light of these hypotheses, and comments briefly on the implications these analyses have for the evolution of avian flight.

The Logic of Cladistics

Science endeavors to explain natural phenomena with the fewest untestable notions, and this methodological truism underlies the widespread acceptance of cladistic methods. The main premise of cladistic analysis, indeed any method of phylogenetic inference, is that characteristics shared by organisms provide the strongest evidence of their evolutionary relationships (Farris, 1983; Rieppel, 1988; Kluge, 1997). Many of the criticisms leveled at cladistic analysis either are true of any method of inferring evolutionary relationships or else assume the generality of particular patterns or processes of evolution. Such generalities require an understanding of evolutionary relationships before they can be deduced, and if based on evidence from other groups, such generalities are unwarranted extrapolations. However much we may believe a particular evolutionary process to occur among organisms, the evidence provided by the characters shared among organisms must be used to test such generalities (Brady, 1985).

Contextual evidence—where an organism is found or its geological age—does not provide as strong a test of the relationships of an organism as do the characteristics of the organism itself. If fossils were all identical, there would be no pattern in nature in need of an explanation; their differences and similarities provide the pattern. Stratigraphic information (i.e., the relative ages of fossils) is irrelevant without reference to the characteristics of the organisms—fossils do not have a temporal record beyond individual occurrences until they are grouped together, and shared characters are the only means of doing so.

A critical distinction, brought to prominent attention by Willi Hennig (1966), is that it is only the shared presence of particular characteristics, or characters, and not their shared absence, that provides the evidence of evolutionary groups. For example, having feathers is a feature of birds, but the absence of feathers in other organisms does not imply that, for example, jellyfish and mammals are more closely related to each other than either is to birds. In addition to the problematic dependence on absence of evidence to test a hypothesis (Kluge, 1997), methods that consider both the presence and absence of characters as evidence of relationships (i.e., phenetic methods) are viable only when the rate of evolution has been uniform (Farris, 1981). Because this regularity in rates of evolution is not a necessary assumption of cladistic methods, these methods are preferable.

Shared characters are most simply interpreted within a hierarchical framework. Characters that are stable among groups of organisms nest them into groups of greater and greater generality, so that a given character has a particular level of generality (i.e., breadth of taxonomic inclusion) at which it provides evidence for a close evolutionary relationship. For a selected group of taxa, called an ingroup, some similarities have a more general distribution than the group under study, and such similarities do not help in resolving relationships among the taxa within the group (and, indeed, will be misleading if misinterpreted). Furthermore, for every character reflecting evolutionary relationships that is shared among taxa in a selected group, there is a corresponding, more general character in unrelated forms that represents the ancestral condition. The correspondence between these characters is itself a hypothesis (Patterson, 1982; de Pinna, 1991; Hawkins et al., 1997), based on similarity in topology (i.e., position, such as relative positions of anatomical features or the position of nucleotides within a gene) or other qualitative characteristics (e.g., chemical composition). A character that is distributed beyond the group being studied is termed plesiomorphic, and the alternative condition, which provides evidence for the relationships of taxa, is termed apomorphic. A character that is apomorphic for a particular group is said to be synapomorphic among the taxa sharing it, and it will be plesiomorphic for groups within this group. Synapomorphy has been equated with the classical concept of biological homology (Patterson, 1982; Rieppel, 1994), although Kluge (1997) makes the distinction that synapomorphy is an approximation of homology, in the same way that the standard deviation of a sample approximates the variance of a population.

The simplest method for determining which characters in the ingroup have a more general distribution is by reference to a taxon outside the ingroup (Naef, 1931; Farris, 1970; Nixon and Carpenter, 1993). The presence of a particular condition in the outgroup—a straight humerus, for example—implies that alternative conditions, such as a curved humerus, provide evidence for groups of related taxa. Although the selection of one or more outgroups requires the acceptance of a particular evolutionary hypothesis before the analysis (i.e., that the ingroup taxa are more closely related to one another than any is to an outgroup taxon), this is open to test by examining further taxa.

Cladistic analysis uses the distribution of all relevant characters among taxa to test hypotheses of relationships among taxa. In so doing, the principle of parsimony is imposed to prefer the hypothesis that minimizes the number of groups to which a given apomorphy applies, or, in evolutionary terms, the number of times it evolved. The rationale for this derives directly from accepting a character as evidence that those organisms possessing it are related to one another and that ad hoc explanations of independent acquisition should be minimized. If a particular character—for example, feathers—is accepted as evidence that the taxa possessing them are related, then this group is supported, and feathers are considered to have evolved only once, unless there is stronger evidence indicating another set of relationships. This stronger evidence comes only from other apomorphic characters, so that if a larger number of characters were shared by penguins and turtles than by penguins and other birds, then a relationship with turtles would be supported, and we would infer that feathers evolved twice.

The relationships indicated by character evidence are most commonly displayed with a cladogram, although they could equally well be displayed by other diagrams indicating nested sets of taxa. This branching diagram indicates only the pattern of relationships inferred from the characters (the groups of related taxa) and does not include ancestors or attempt to incorporate time. A diagram indicating that some observed or inferred taxa are ancestors of others is called a tree.

Although patterns of direct ancestry can be traced among individuals, the concept of ancestry is problematic when applied to groups (Platnick, 1977; Gaffney, 1979) and is certainly illogical for taxa higher than the species level. The main problem is that characters alone do not identify single taxa as ancestors because it is the shared presence of characters in two or more taxa that provides the evidence of relationships. The shared possession of apomorphic characters identifies *inclusive* groups of individuals and taxa that are more closely related to one another than to some other group (Rieppel, 1988), whereas the evidential basis for identifying a group comprising an ancestor but *excluding* its descendants cannot be based solely on the shared presence of characters. Despite this incompatibility, some paleontologists argue that ancestors can be identified using the fossil record to test whether a species with no apomorphic characters (in comparison with their closest relatives) is from a time period allowing it to be the ancestor of later taxa (e.g., Smith, 1994), even though the absence of evidence (apomorphies) cannot be considered to test any hypothesis (Platnick, 1977). Even if ancestry is considered testable, however, it is simply a refinement of a cladogram and does not contradict the relationships indicated by shared characters alone.

Other paleontologists argue that the fossil record of taxa can proscribe certain phylogenies, because of the size or number of the gaps in the stratigraphic record they imply (e.g., Wagner, 1995). However, this depends on the fossil record providing stronger evidence (a stronger test of relationships) than character data, which is clearly not the case (Schoch, 1986; Norell, 1996). For example, unless character evidence is utilized, there is no basis for choosing a group on which to focus attention; all published studies attempting to incorporate stratigraphic data begin with a group chosen on the basis of morphological characters, because stratigraphic information provides no information about whether to group, say, a Miocene whale with a Miocene coral. The necessity of using the data of characters to "get you in the right ballpark" highlights the weakness of the evidence from the age or geographic distribution of a fossil and its inability to "move you to a different ballpark," one contradicted by character evidence. The fossil record is unquestionably incomplete, and unless the circumstantial evidence it provides can be shown to preclude the possibility of relationship between particular taxa (i.e., the lack of physical contact between populations), then gaps in it implied by particular phylogenies are better explained by lack of fossilization than by ad hoc hypotheses of convergent evolution.

The search for ancestors that preoccupied paleontologists for most of the twentieth century, epitomized by the overemphasis on fossils such as *Archaeopteryx,* failed because it discounted with no justification much of the evidence provided by characters. Instead of determining the pattern of relationships derived from the simplest explanation for the characters shared among different groups of organisms, certain taxa (e.g., *Archaeopteryx*) were exalted as ancestors, while other taxa, such as those with specializations supposedly precluding ancestral status, were explicitly disregarded. Phylogenetic hypotheses live or die on the character support mustered for them, and a given hypothesis supported by evidence can be refuted only by demonstrating that further character evidence changes the balance of support toward an alternative hypothesis, either by adding new characters, by extending the distribution of characters among taxa to demonstrate they are apomorphic at a different level, or by showing that a character was misinterpreted on the specimens. Scenarios involving selection pressures supposedly operating in the past, the adaptive utility (or lack thereof) of particular features, or the supposed effects of environmental changes on taxa are irrelevant to phylogenetic inference.

The importance of phylogenetic hypotheses to the study of evolution cannot be overstated. As the simplest explanation of the features shared among groups of organisms, a phylogenetic hypothesis must be the basis for any consideration of the evolution of these organisms and of their constituent features. Thus, with regard to the topic of this chapter, phylogenetic hypotheses must be the basis for answering such questions as: What are birds? What are the characteristics that identify them as a group? In what pattern did these characteristics evolve in the groups most closely related to birds? For example, questions about the origin of flight depend on a phylogenetic hypothesis and the identity of those characteristics related to flight. These characteristics must be studied not only in *Archaeopteryx*

but among all taxa relevant to the relationships of birds within Theropoda.

Reptiles, Birds, and the Meaning of Group Names

The development of cladistic methods in the 1960s and 1970s was fueled by the testability of phylogenetic hypotheses, and the untestable aspects of taxonomy as practiced at that time soon also became apparent. By using cladistic methodology, it is possible to test whether a group such as Aves is monophyletic, comprising organisms that are more closely related to one another than any one of them is to another group of organisms. (The term "monophyletic"— "one class"—has a long history of use in taxonomy, and as with the term "homology," its current, cladistic usage as defined here is a more explicit refinement of its traditional usage.) No comparable test exists for membership in traditional groups, which explains the many and creative definitions for "monophyly" that predate cladistics.

A group such as Reptilia is not monophyletic if it does not include birds, because the simplest explanation for the morphological and molecular evidence is that birds are more closely related to crocodilians than they are to any other living organism (Eernisse and Kluge, 1993; but see Hedges and Poling, 1999, for an alternative hypothesis). Advocates of traditional taxonomic methods argued that Reptilia without birds constitutes a natural group because it is adaptively unified (e.g., Mayr and Ashlock, 1991), but the only means ever outlined for assessing which groups are adaptively unified and which groups are not are either vaguely defined or arbitrary. Although it is possible to point to particular features present in birds that are absent in their closest living relatives, such as an endothermic metabolism or feathers, it is not possible to argue that these particular features should be more important in formulating a taxonomy—on what objective basis can we deduce the relative importance of such features? More to the point, if these features are used to change the hierarchical level at which a group is placed in the taxonomy, thereby contradicting the groupings indicated by the cladistic analysis, then this new hypothesis should have greater explanatory power than the phylogenetic hypothesis it replaces. However, the cladogram inferred from character data has the greatest explanatory power of any hypothesis of group membership because it requires the fewest ad hoc hypotheses (Farris, 1983), and any classification contradicting it will necessarily require more ad hoc hypotheses and therefore have less explanatory power.

The supposed adaptive unity of groups of organisms does not lend itself well to the hierarchical organization demanded by taxonomy, wherein every group is included within other groups. There is no necessary hierarchy of levels of "adaptive unity" in nature, and if a taxon such as Aves is removed from another taxon (i.e., Reptilia) to emphasize its adaptive unity, then it leaves behind a group whose adaptive unity is highly suspect under any definition. If there is some sense in which the group comprising turtles, lizards, and crocodilians is equivalent to the group comprising birds, it has yet to be explicated. Thus, in a traditional classification, the reasons that birds are a class within the subphylum Vertebrata, based on how distinct they are from other vertebrates, are not the same reasons given for why Reptilia is a class within the same group. On the other hand, the relationships among organisms inferred by cladistic analysis precisely mirror the hierarchy of groups within groups formalized in taxonomy, and there seems to be no scientific rationale for perturbing this natural hierarchy on the basis of subjective decisions about which groups appear to be more or less adaptively unified or distinct.

There is little question that extant birds form a monophyletic group, but the rich diversity of fossil archosaurs blurs the distinction between birds and their closest relatives. Because of the central role accorded *Archaeopteryx* historically in studies of the origin of birds (see Witmer, Chapter 1 in this volume), traditionally it has been included as the most primitive member of the formal taxonomic group that corresponds to birds—the class Aves. However, as the differences between birds and nonavian dinosaurs have dwindled, this emphasis on one species in defining birds has been called into question.

The definition of the class Aves is in one sense trivial, in that it is only a name arrived at by common consent among taxonomists to facilitate communication, but it is important to the extent that it influences research into topics such as bird origins. Several definitions of Aves have been suggested in recent years. Gauthier (1986) restricted the name Aves to the group comprising all living birds and only those extinct taxa belonging to a subgroup within Aves, a concept termed a "crown group." Thus, fossil woodpeckers and ostriches belong to this definition of Aves, but *Archaeopteryx* does not. Although somewhat arbitrary, crown groups are useful because they identify the taxonomic boundaries beyond which nonskeletal features apomorphic for living birds cannot be inferred to be present in fossil taxa. For example, an endothermic metabolism can be inferred to have been the ancestral condition for extinct taxa within the crown group, but where this feature evolved on the lineage leading to birds but outside the crown group is unclear. For the group comprising *Archaeopteryx,* Aves, and other extinct taxa within this monophyletic group, Gauthier (1986) erected the taxon Avialae.

An alternative definition of Aves has been proposed by Patterson (1993) in which Aves includes all taxa more closely related to birds than to their closest living relatives, the crocodilians. This definition is an extension of a system applied to fossil fish, in which it is not the crown group that receives the name of the living taxa but the entire group more closely related to one living group than to its closest living relative. This system has the advantage of reducing the number of names of groups, but it is unclear whether minimizing the number of groups to which names are attached is a desirable goal (as in this group, since there is no question that the crown group merits a name; Patterson et al., 1993). In any case, this disagreement highlights the need to define precisely which meaning is intended when taxonomic names are discussed.

Traditionally, the group Aves included *Archaeopteryx* and excluded those archosaurs more distantly related to living birds. More recently Aves has been defined explicitly to include "the common ancestor of *Archaeopteryx* and modern birds plus all its descendants" (Chiappe, 1992:348; see also Padian and Chiappe, 1998). Mandating that *Archaeopteryx* is included in Aves and that more distantly related taxa are excluded has the virtue of stability of usage, in that its content is similar to that used in the past.

We prefer to define Aves as the crown group rather than the traditional group (i.e., Avialae). Although crown groups are arbitrary, defining Aves to include *Archaeopteryx* and exclude other archosaurs invites attempts to recognize the "essence" of being a bird. The distinction between *Archaeopteryx* and other theropods has oversimplistically reduced the question of bird origins to the origin of one species, and drawing a class-level taxonomic boundary between them, as is often done, will only continue this unfortunate tradition.

[*Editors' note:* The authors of this chapter have chosen to use a different phylogenetic definition of Aves than that used elsewhere in this volume. The editors respect their choice. See the preface and the chapters in this volume by Witmer (Chapter 1) and Sereno, Rao, and Li (Chapter 8) for a justification for the taxonomic definitions used otherwise in this volume.]

Cladistic Analyses of Theropoda and the "Origin" of Birds

Character evidence is the basis for inferring phylogenetic relationships, so the question of the evolutionary origin of a particular taxon is really a question of what are its closest relatives as determined by the character evidence. Indeed, the word "origin" has the unfortunate connotation of a particular taxon being ancestral to another, whereas it is problematic whether character evidence allows for the identification of a direct ancestor. Furthermore, during most of the twentieth century, studies of the origin of birds, in the poorly defined essentialistic sense that was the hallmark of traditional classifications, often equated their evolutionary origin with the evolutionary appearance of feathered flight in this group (e.g., Heilmann, 1926). These are best considered two separate questions, one concerning the evolutionary relationships of birds as determined by character evidence and the other the pattern of appearance of the characters related to flight as inferred from these evolutionary relationships.

The evolutionary relationships of birds with theropod dinosaurs has received a great deal of attention in recent years. The evidence is overwhelming that birds are more closely related to some theropod dinosaurs than to others, but their precise relationships remain elusive. Gauthier (1986) applied the existing name Coelurosauria to a group of theropods, most of which are distinguished by their elongate forelimbs, and the evidence strongly supports the inclusion of Avialae within this group. In the following section we discuss recent work on coelurosaurians other than Avialae. For each group we list the valid taxa we consider to belong to the group and discuss the evidence for its monophyly, relationships among its members, its geographic distribution, similarities shared with other taxa of Coelurosauria, and recent work on the group, emphasizing data and interpretations published after the study of Gauthier (1986) and the reviews published in *The Dinosauria* (Weishampel et al., 1990). The oldest occurrence of each group is given in Appendix 2.1 and illustrated in Figure 2.2. In light of the differences between recent cladistic analyses (Fig. 2.1), it is not possible to identify all the characters diagnosing these taxa because those shared by more than one basic taxon may be either synapomorphies of more inclusive groups, convergently evolved autapomorphies of separate groups, or plesiomorphies that are uninformative in diagnosing monophyletic groups.

Dromaeosauridae

Dromaeosaurids are small to medium-sized bipedal carnivores, including *Deinonychus, Velociraptor, Dromaeosaurus, Adasaurus,* and *Utahraptor.* The group was most recently diagnosed by Currie (1995), who offered 17 characters in support of their monophyly. Although a few of these characters have distributions outside the group, this clade can be diagnosed on the basis of several characters, including a second pedal digit modified into a raptorial sickle claw (Colbert and Russell, 1969; Ostrom, 1969), a tail with extremely elongate prezygapophyses on the neural arch, and proximal processes of the haemal arches (Ostrom, 1969). The recent report of a basal avialan, *Rahonavis,* with a retractable, raptorial second pedal digit (Forster et al., 1998a,b) indicates that this feature is distributed outside Dromaeosauridae and may be diagnostic of a group including basal Avialae.

Dromaeosaurids have received much attention in both the scientific and popular arenas, owing to their putative close association with the ancestry of Avialae (Gauthier, 1986) as well as their role as movie villains. Important new specimens (especially of *Velociraptor mongoliensis* [see Norell and Makovicky, 1997; Norell et al., 1997; Barsbold and

Osmólska, 1999]) have pointed to a number of interesting features of the dromaeosaurid skeleton that are present or further modified in birds. These features include the presence of a furcula, the presence of paired sternal plates with attachment points for three pairs of ventral ribs (Norell and Makovicky, 1997, 1999), a dorsal tympanic recess (Norell et al., 1992), and a secondary palate formed by maxillary shelves and vomer. As in birds, the pubis is reversed (Barsbold, 1979; Norell and Makovicky, 1997) and the pelvis lacks a broad shelf for the origin of the M. cuppedicus, the antiliac hook is reduced, and the ischia are distally unfused (Norell and Makovicky, 1997). Contrary to many previous reconstructions, the hallux was not reversed but rather was in the typical nonavian theropod position on the side of the foot (Norell and Makovicky, 1997).

In his discussion of dromaeosaurids, Currie (1995) divided the Dromaeosauridae into two subfamilies, the Dromaeosaurinae (including only *Dromaeosaurus*) and the Velociraptorinae (for the remaining taxa, with some qualification). Interestingly, based on the cranial evidence, members of the Velociraptorinae share possible derived similarities with birds that are lacking in *Dromaeosaurus,* thus contradicting the monophyly of Dromaeosauridae. Unfortunately, nearly all the postcranial skeleton of *Dromaeosaurus* is unknown, which is where some of the most distinctive derived characters of dromaeosaurs are found (the retractile second pedal digit, specializations of the opisthopubic pelvis, and the superelongate vertebral articulations in the tail).

Most recently, the Dromaeosauridae has been greatly expanded by the addition of two taxa from the feathered dinosaur beds of China, *Sinornithosaurus millenii* (Xu et al., 1999b) and *Microraptor zhaoianus* (Xu et al., 2000). In addition to documenting the presence of simple featherlike structures in this family (in *Sinornithosaurus* and another dromaeosaurid of uncertain affinities; Ji et al., 2001), these specimens broaden the osteological diversity of the family. The small size, and possible arboreality, of *Microraptor* may be significant in the debate over the habits of the origin of bird ancestors, although the testability of these hypotheses is open to question (see later).

Troodontidae

The Troodontidae are especially difficult to diagnose because most of the taxa are known from very incomplete material, with the exception of *Sinornithoides youngi* (Russell and Dong, 1994b). The monophyly of Troodontidae is supported by a few unambiguous characters. Currie (1987) and Osmólska and Barsbold (1990) diagnosed the group, but many of the characters listed are plesiomorphies, and others are shared with other Coelurosauria, especially Ornithomimosauria. Russell and Dong (1994b) offered a list of characters differentiating troodontids from dromaeosaurids, and included among these are additional potential synapomorphies of Troodontidae. From these sources and our observations (Norell et al., 2001) the following characters appear valid: lacrimal with elongate cranial process dorsal to antorbital cavity, accessory antorbital fenestra large (more than half the length of the antorbital] fenestra), mental foramina in longitudinal groove on lateral surface of dentary, clavicles unfused (in *Sinornithoides*), and second metatarsal reduced in length and breadth.

A recently described troodontid from below the "feathered dinosaur" level of the Yixian Formation in Liaoning is significant in lacking features previously thought to ally troodontids with other theropod groups. *Sinovenator changii* (Xu et al., 2002) has teeth with small serrations, unlike the large serrations of troodontids and therizinosauroids, and these teeth are not reduced in size and numerous, as in other troodontids and some ornithomimosaurs, therizinosauroids, and alvarezsaurids. Furthermore, it lacks a bulla on the parasphenoid, as in some other troodontids and ornithomimosaurs.

Two instances of eggs associated with troodontid remains are known, although in only one is an embryo preserved within an egg, and that specimen is so incomplete that its identification is not definitive. The embryonic remains, from the Two Medicine Formation of Montana, previously were identified as those of the ornithopod *Orodromeus* but recently were recognized as troodontid (Horner and Weishampel, 1996). A second occurrence of troodontid remains with eggs, from Ukhaa Tolgod (?Djadokhta Formation) of Mongolia, involves small posthatchling individuals associated with a nest that also preserves an isolated tooth of an adult troodontid (Norell et al., in prep.).

Oviraptorosauria

Until recently Oviraptorosauria was among the most poorly known groups of theropods, but several spectacular discoveries in recent years now place it among the best known. In particular, discoveries of embryos and of adults overlying nests at Ukhaa Tolgod, Mongolia (Norell et al., 1994, 1995; Clark et al., 1999) and Bayan Mandahu, China (Dong and Currie, 1996), along with several dozen new skeletons of at least two taxa from Ukhaa Tolgod, provide a firm basis for interpreting the skeletal anatomy of the members of this group and several aspects of their development and behavior. More fragmentary but less specialized material from North America of the primitive taxa *Microvenator* and *Chirostenotes* has contributed significantly to understanding the relationships of several enigmatic taxa and the primitive conditions for this group (Sues, 1997; Makovicky and Sues, 1998). The recently described "feathered dinosaur" *Caudipteryx* also may be allied with this group (see later).

The most specialized members of the group, the Oviraptoridae (*Oviraptor, Conchoraptor,* and *Ingenia*), are easily diagnosed by their unusual cranial anatomy, including a brief, tall rostrum with large pneumatic pockets, a robust palate, and an edentulous beak, and their equally unusual mandible that is also edentulous, highly arched, and with a rostrocaudally movable articulation with the skull. Several avian features are now known to be present in oviraptorids, including a single ossified ventral rib segment, more than two ribs articulating with the sternum, ossified uncinate processes (Clark et al., 1999), and a quadrate articulating with both the squamosal and the braincase (Maryanska and Osmólska, 1997). However, the distribution of these features among other theropods is poorly determined.

Therizinosauroidea

The relationships of the Segnosauria, known exclusively from the Late Cretaceous of Asia, were enigmatic until recently, but the discovery of *Alxasaurus elesitaiensis* in Early Cretaceous deposits of China now clarifies their relationships and taxonomy (Russell and Dong, 1993a). It is now evident that segnosaurs are related to another poorly known taxon, *Therizinosaurus* Maleev, 1954, and that the valid name for the group is therefore Therizinosauroidea. These awkward-looking animals had unusually long arms with long, straight claws on their hands, and their skulls lacked teeth rostrally and apparently bore a beak. In addition to *Alxasaurus* and *Therizinosaurus,* four genera of Segnosauridae have been described (Barsbold and Maryanska, 1990), and recently a new monotypic genus was described (Xu et al., 1999a). This new species, *Beipiaosaurus inexpectus* from the Early Cretaceous of Liaoning, China, is notable in possessing simple, unbarbed featherlike structures and a foot in which the first metatarsal has a compressed proximal end, as in other theropods but unlike derived therizinosaurs and primitive dinosaurs. Most recently, new therizinosaurs have been described from North America (Kirkland and Wolfe, 2001) and Inner Mongolia (Zhang et al., 2001), and a jaw from the Early Jurassic of China has been tentatively assigned to this group (Xu et al., 2001a).

Although the cladistic analysis of therizinosauroid relationships presented by Russell and Dong (1994a) is flawed, they provided ample evidence for placing Therizinosauroidea in the Coelurosauria. Further support for the placement of Therizinosauroidea in Coelurosauria was provided by Clark et al. (1994b) based on the exceptionally well preserved holotype skull of *Erlikosaurus andrewsi* and by Xu et al. (1999a). Any future study of relationships among basal Coelurosauria must consider this group.

Ornithomimosauria

Ornithomimosauria includes the Ornithomimidae and the monotypic Garudimimidae and Harpymimidae (Barsbold and Osmólska, 1990), although other authors consider the family Ornithomimidae to be equivalent to the Ornithomimosauria (e.g., Smith and Galton, 1990; Sereno, 1999). *Elaphrosaurus* was considered an ornithomimosaurian by some authors, but Holtz (1994) provided evidence that it is not a member of this group. Seven valid genera were recognized by Barsbold and Osmólska (1990), and a new, primitive member of the group, *Pelecanimimus polyodon* Pérez-Moreno et al., 1994, was described from the Early Cretaceous of Spain. A detailed description of this form will provide important evidence for relationships of this group to other Coelurosauria.

Ornithomimosauria except *Harpymimus* and *Pelecanimimus* are toothless, and all but *Harpymimus* have a specialized hand in which the three metacarpals are equal in length (unlike other Coelurosauria). Several features are shared with Troodontidae, such as a hollow inflation of the parasphenoid (but see earlier), prompting Holtz (1994) to erect the taxon Bullatosauria to include both taxa, but troodontids also share features with birds and dromaeosaurids that are not present in ornithomimosaurians, such as the retractable second digit of the foot.

Compsognathus

This primitive coelurosaurian is represented by two specimens, one from the same formation in southern Germany as *Archaeopteryx* (Ostrom, 1978) and the other from slightly younger beds in France (Bidar et al., 1972). Although both are nearly complete, the specimens are crushed and not well preserved, and the forelimbs of the German specimen are disarticulated. Thus, many important anatomical details are obscured, such as the number of digits on the hand.

Coelurus and Ornitholestes

Two taxa from the Late Jurassic Morrison Formation of Wyoming are important in demonstrating that the diversification of Coelurosauria preceded the appearance of *Archaeopteryx* (see Appendix 2.1 for the ages of these taxa). *Coelurus fragilis* is a small carnivore known from a single, fragmentary specimen (Marsh, 1879). It is assignable to the Coelurosauria on the basis of the presence of a semilunate carpal (Ostrom, 1976; Norman, 1990). However, the cervical vertebrae lack hypapophyses, indicating that if *Coelurus* is a coelurosaur it may be basal. *Ornitholestes hermanni* Osborn, 1903, is a small theropod known from a single specimen (although a second manus has been referred to this taxon). Gauthier (1986) placed *Ornitholestes* within Coelurosauria on the basis of a curved ilium and proportions of the forelimb. However, because the specimen has not been adequately studied and is fragmentary, accurate forelimb proportions cannot be determined.

Tyrannosauridae

Tyrannosaurids are among the best-known group of theropods, with several taxa (e.g., *Tarbosaurus baatar*) known from nearly complete specimens. Monophyly of the family is supported by features such as the large foramen in the surangular and the diminutive forelimbs. The higher taxon Tyrannosauroidea was recognized by Sereno (1998) to distinguish a group including all tyrannosaurs except *Nanotyrannus*, but he has not yet described the character data in support of this hypothesis. The relationships of *Nanotyrannus* have been controversial, with some considering the relatively small skull of the single specimen to be that of a juvenile animal (Carr, 1999) and others (e.g., Bakker et al., 1988) considering it to represent a "pygmy" species. Recently, a new taxon was reported from the Early Cretaceous of Thailand, *Siamotyrannus* (Buffetaut et al., 1996). Unfortunately, the incomplete specimen provides only a few characters suggestive of tyrannosaurid affinities, and none that are unique.

Gauthier (1986) did not include Tyrannosauridae within the Coelurosauria, but several later analyses provided evidence that it was a member of this group, as reviewed subsequently.

Unenlagia

One of the more newsworthy discoveries of 1997 was the announcement by Novas and Puerta of a theropod from Patagonia, *Unenlagia comahuensis*, that exhibits several features of the shoulder and pelvis that are otherwise known only in Avialae among theropods. Unfortunately, *Unenlagia* is known only from a single partial skeleton. These features include the presence of a hypopubic cup on the caudal surface of the distal end of the pubis, the presence of a pubic apron, and a laterally oriented glenoid fossa on the scapulocoracoid. The reorientation of the glenoid from the primitive position on the caudal surface of the scapula, allowing *Unenlagia* to position its arm in an avian manner, is the most notable feature. Alternatively, however, a number of *Unenlagia* features thought at the time to be unique to it and Avialae are present in *V. mongoliensis* (Norell and Makovicky, 1999), including the reorientation of the glenoid fossa. Furthermore, two aspects of the vertebrae—stalked parapophyses on the dorsal vertebrae and the mediolateral expansion of the tip of the neural spine in the caudal dorsal vertebrae—suggest a possible relationship with dromaeosaurids (Norell and Makovicky, 1999).

Scipionyx

An unusually well preserved juvenile theropod, *Scipionyx samniticus* Dal Sasso and Signore, 1998, was reported recently from the Lower Cretaceous Pietraroia Plattenkalk of southern Italy. Remarkable is the preservation of soft tissues, especially the intestinal tract, preserving details of muscle structure. Skeletal features indicate that it is coelurosaurian and shares similarities with both dromaeosaurids and troodontids. Ruben et al. (1999) inferred that the soft tissues preserved on this specimen indicate that its lungs were ventilated by the M. diaphragmaticus, as in crocodilians, but the incomplete preservation of this structure and its position near the body wall do not contradict an alternative identification as hypaxial musculature. In any case, the contention of Ruben et al. (1999) that the possession of a crocodilian-type ventilatory system precludes a close relationship between theropods and birds is invalid for the reasons outlined at the beginning of this chapter—the hypothesis of homology between the lung structure of this fossil and that of either crocodilians or birds is tested by its congruence with other characters, not by a priori judgments about how evolution can and cannot occur.

Avimimus

An enigmatic form from the Late Cretaceous of Mongolia, *Avimimus portentosus* (Kurzanov, 1981), combines some unusually birdlike features with those of more primitive coelurosaurians, along with several specializations (Kurzanov, 1987). For example, the skull is similar to that of oviraptorosaurs in having an edentulous beak and a short rostrum, the fused pelvis includes a medially inclined ilium similar to that of birds, and the poorly known forelimb elements include metacarpals that appear to be fused and an ulna that may have evidence of feather attachments (see Vickers-Rich, Chiappe, and Kurzanov, Chapter 3 in this volume).

Feathered Dinosaurs of Liaoning

Recently, new and interesting theropod specimens are coming out of Liaoning, China, at a rapid pace, including primitive avialans (Sereno and Rao, 1992; Chiappe et al., 1999) and, rarely, nonavialans (e.g., Ji et al., 1998; Xu et al., 1999a,b; Ji et al., 2001). These specimens have fundamentally changed the way that we look at dinosaurs (including birds). Although the precise age of these beds within the Late Jurassic to Early Cretaceous time span is debated (Zhou and Hou, Chapter 7 in this volume; see Swisher et al., 1999), their importance is clear, and even if the later age is correct, they have important consequences for maniraptoran phylogeny (Sereno 1999; Norell et al., 2001).

Seven nonavialan theropods are particularly noteworthy—*Protarchaeopteryx robusta* (Ji and Ji, 1997), *Caudipteryx zoui* (Ji et al., 1998), *Sinosauropteryx prima* (Chen et al., 1998), the therizinosauroid *B. inexpectus* (Xu et al., 1999a), the dromaeosaurids *S. millenii* (Xu et al., 1999b) and *M. zhaoianus* (Xu et al., 2000), and the troodontid *S. changii* (Xu et al., 2002). The first of these animals to be discovered was *Sinosauropteryx*, a small, long-tailed animal of supposed compsognathid affinities (Currie et al., 1998). In addition to being a new taxon of a rare group, the specimens of *Sinosauropteryx* show an integumentary covering of small fibers that completely cover the body of the animal.

Both *Caudipteryx* and *Protarchaeopteryx* are small theropods with relatively long front limbs. *Protarchaeopteryx* is known from two specimens (Ji and Ji, 1997; Ji et al., 1998), while *Caudipteryx* is known from several individuals (Ji et al., 1998). The exciting thing about these specimens is that they possess true feathers, with remiges and rectrices, along both their forearms and tails. Both of these taxa lie outside Avialae, as indicated by, for example, the presence of a contact between the quadratojugal and squamosal in *Caudipteryx*. Subsequently discovered specimens of *Caudipteryx* have revealed other features, such as the unreversed first digit of the pes (pers. obs.). Although *Caudipteryx* was originally interpreted as a close relative of Avialae (Ji et al., 1998), more recent phylogenetic studies support a *Caudipteryx*-oviraptorosaur clade (Sereno, 1999; Holtz, 2001; see later). The position of *Protarchaeopteryx* is more ambiguous, owing to the incompleteness of the material, but it is certainly outside Avialae, as indicated by, for example, its serrated teeth.

The featherlike structures and feathers found in the nonavialan theropods of Liaoning occur in four types (Xu et al., 2001b): (1) simple filamentous structures (e.g., *Sinosauropteryx, Beipiaosaurus*), (2) simple filaments joined in a basal tuft (e.g., *Sinornithosaurus*), (3) filaments joined at their bases in series along a central filament (e.g., *Sinornithosaurus*), and (4) true feathers with a central rachis and barbs (e.g., *Caudipteryx, Protarchaeopteryx*). The distributions of these types among and within taxa are poorly known at present (but see Ji et al., 2001) but will undoubtedly become better known as the specimens from Liaoning continue to appear.

Study of these interesting animals is just beginning, and surely they and other specimens will become the focus for intense study for their bearing on the origin of birds and the evolutionary appearance of avian features (see also Witmer, Chapter 1 in this volume; Zhou and Hou, Chapter 7 in this volume). For example, the phylogenetic position of *Caudipteryx* as an oviraptorosaur relative has profound implications. The presence of feathers in this taxon indicates that feathers were present in other nonavialan dinosaur groups that are descended from the same common ancestor as oviraptorosaurs and modern birds. This group includes dromaeosaurids, troodontids, and therizinosauroids—all of which must now be interpreted in lieu of other evidence as being cloaked with a feathery body covering.

Protoavis

The Late Triassic species *Protoavis texensis* was described by Chatterjee (1991; see also Chatterjee, 1997), who considered it to be a bird. As described, it has an unusual combination of features; for example, the ulna supposedly has raised areas for the insertion of feathers, but the hand has four fingers rather than the three typical of Coelurosauria. However, questions have been raised about the accuracy of the published description and whether the material is from a single taxon (e.g., Chiappe, 1998; Witmer, 2001, Chapter 1 this volume).

"Archaeornithoides"

In 1992, Elzanowski and Wellnhofer described *Archaeornithoides deinosauriscus* on the basis of the incomplete rostrum and mandibular symphysis of a juvenile theropod collected by the 1965 Polish Mongolian expeditions from Bayn Dzak (the Flaming Cliffs), Mongolia (see also Elzanowski and Wellnhofer, 1993). They implicated *Archaeornithoides* as the sister group to Avialae. The type specimen is extremely fragmentary and possibly passed through the digestive tract of an organism before fossilization. Few details can be determined from this specimen, but it does seem to possess broad palatal shelves, lack interdental plates in the area preserved, and lack denticles on its teeth. It is mainly on the basis of the palatal shelves that Elzanowski and Wellnhofer defended its phylogenetic position as near the base of Avialae. However, broad palatal shelves (in some cases forming complete secondary palates) are now known from a number of theropods, including adult and juvenile troodontids (Norell et al., 2000) and dromaeosaurids (Norell and Makovicky, 1999). Furthermore, teeth without denticles and dentaries lacking interdental plates can also be observed in the juvenile troodontids from Ukhaa Tolgod. Considering these observations, it is likely that *A. deinosauriscus* represents a poorly preserved specimen of a juvenile of a nonavialan coelurosaurian taxon.

Cladistic Analyses of Coelurosaurian Theropods

Cladistics was developed as a formal analysis only within the past three decades, so the relationships of extant birds to fossil archosaurs were not subjected to cladistic analysis until the mid-1980s (e.g., Thulborn, 1984; Gauthier, 1986). But the groundwork of comparisons was laid by Ostrom (1976), who cataloged the many similarities of *Archaeopteryx* to theropod dinosaurs, dromaeosaurids in particular, and explicitly interpreted them in terms of synapomorphies (shared derived characters). The character evidence amassed by Ostrom was persuasive when he presented it anecdotally and became overwhelming when incorporated into cladistic analyses of birds with fossil archosaurs.

The study of Gauthier (1986) is a convenient starting point in discussing the relationships of birds to other dinosaurs because of the detailed discussion of the evidence presented therein. (The analysis by Thulborn [1984] reached similar results but found several theropods to be more closely related to extant birds than was *Archaeopteryx*.) Gauthier's study surveyed 84 characters in 18 taxa, with the results shown in Figure 2.1A. Noteworthy was the finding that several theropods with unusually long arms are the closest relatives of birds, together constituting Coelurosauria. One of these, Ornithomimidae (corresponding generally to what is now named Ornithomimosauria), is more distantly related to birds than the others, and Gauthier erected the group Maniraptora for the latter. Beyond this there was little resolution indicating which groups were most closely related to Avialae, as is evident from the lack of resolution in the consensus cladogram (which shows only those groups supported in all the equally most parsimonious cladograms). In the text, Gauthier hypothesized that Dromaeosauridae and Troodontidae formed a group that is the closest relative of Avialae, but in an addendum he pointed out evidence contradicting the placement of the two families together, thus leaving the relationships of nonavialan Maniraptora entirely unresolved.

Several recent cladistic analyses addressed the relationships of birds among theropods, and another analyzed relationships among several nonavialan theropod taxa that are closely related to Avialae. We do not include analyses without published data matrices in this summary. In a study of a new relative of "segnosaurs" from China, Russell and Dong (1994a) presented an analysis of theropods with several novel results (Fig. 2.1B). Segnosaurs are a highly specialized group of dinosaurs known with confidence only from the Late Cretaceous of Asia, and hypotheses of their affinities have ranged throughout the Dinosauria with little consensus (Barsbold and Maryanska, 1990). The study presented by Russell and Dong summarized important evidence for the relationships of segnosaurs and the new taxon, the Early Cretaceous *A. elesitaiensis*, but unfortunately they apparently restricted the data set, derived mainly from Gauthier (1986), to only those characters that could be determined in the new taxon (59 characters in 11 taxa). Thus, characters shared by, for example, dromaeosaurids and birds are not included unless the anatomical region in which they are expressed is preserved in the *Alxasaurus* material. The exclusion of this evidence renders the results inconclusive.

Holtz (1994) presented an analysis revising the characters summarized by Gauthier and adding numerous others (a total of 126 characters in 19 taxa), especially many cranial characters. An important aspect of this analysis is that characters considered to be correlated with large size were explicitly excluded; in other words, several characters present only in large taxa were considered to represent a single character—large size—rather than different characters. The results (Fig. 2.1C) differ from those of Gauthier (1986) in at least three significant respects: (1) there is much greater resolution; (2) Ornithomimosauria (including Ornithomimidae) is not as distantly related to birds as Gauthier hypothesized, so that the groups Coelurosauria and Maniraptora are redefined (see Holtz, 1996); and (3) Tyrannosauridae is

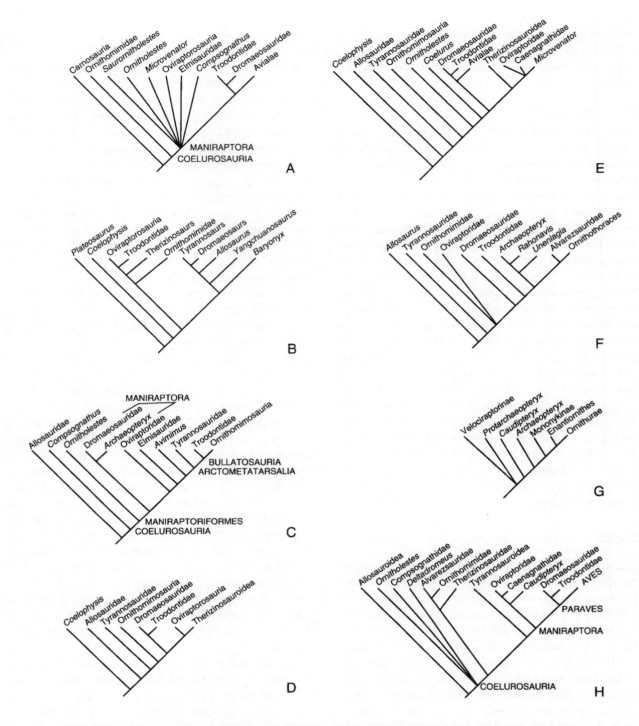

Figure 2.1. Cladograms of relationships among coelurosaurian theropods based on published data matrices. A, strict consensus cladogram of "numerous" equally parsimonious cladograms (84 characters for 18 taxa, CI = .85) presented by Gauthier (1986), who in an addendum to the paper questioned the close relationship of Troodontidae and Dromaeosauridae (treated as a single taxon, Deinonychosauria, in the analysis). Carnosauria includes Tyrannosauridae and Allosauridae. The original data matrix includes *Hulsanpes,* a taxon known only from the metatarsals, but Wilkinson (1995) reported that if this taxon is included then relationships of Coelurosauria are unresolved in the strict consensus. B, single most parsimonious cladogram (59 characters for 11 taxa, CI = .551) of Russell and Dong (1993a), who included *Microvenator* within Oviraptorosauria. C, single most parsimonious cladogram (126 characters for 19 taxa, CI = .470, RI = .684) of Holtz (1994) excluding noncoelurosaurians, with group names (but not relationships) modified following Holtz (1996). D, single most parsimonious cladogram (47 characters for 8 taxa, CI = .597, RI = .587) of Sues (1997), who included *Chirostenotes* within Oviraptorosauria and synonymized Elmisauridae with Caenagnathidae within this group. E, strict consensus of three equally parsimonious cladograms (95 characters for 14 taxa, CI = .562, RI = .635) of Makovicky and Sues (1998); F, single most parsimonious

hypothesized to be more closely related to birds than is Allosauridae (proposed also in a cladistic analysis by Novas [1992] but without a data matrix), whereas Gauthier's analysis placed the two together, as the Carnosauria, outside Coelurosauria.

However, several problems in the data matrix detract from Holtz's analysis (Clark et al., 1994b; Sereno, 1997), and the abbreviated descriptions of characters (in which only the derived state is described) often obscure the precise feature being specified in each case. Furthermore, it is not clear that removing characters that are correlated is justified, as correlation among characters is expected for those sharing a common evolutionary origin. There may be grounds for dismissing a few of the characters Gauthier cited in support of Carnosauria as redundant in some obvious way, but because ignoring evidence through ad hoc hypotheses is anathema in science, this should not be taken lightly. This study is therefore best considered a "work in progress" that is being refined (e.g., Holtz, 1996, and Padian et al., 1999), although the only published data matrix is that of Holtz (1994).

An important study of an enigmatic theropod dinosaur from the Late Cretaceous of Canada included an analysis of relationships among Coelurosauria, although it did not address the relationships of birds (Fig. 2.1D). *Chirostenotes pergracilis* had been poorly known until a partial skeleton was discovered in museum collections. Sues (1997) describes this new material and recognizes its affinities with taxa forming Caenagnathidae (indeed, Sues synonymizes *Caenagnathus* with *Chirostenotes,* as suggested tentatively by Currie and Russell [1988]). Caenagnathidae forms a group (Oviraptorosauria) with the highly specialized Oviraptoridae. A cladistic analysis of 8 taxa and 47 characters finds tyrannosaurs to be outside Coelurosauria, contra Holtz (1994), and provides evidence that Oviraptorosauria are most closely related to Therizinosauroidea. Dromaeosauridae and Troodontidae are sister taxa to each other, and together to the Avialae.

A related study by Makovicky and Sues (1998), of an equally enigmatic theropod dinosaur from the Early Cretaceous of Montana, expanded the analysis of Sues (1997) with the addition of vertebral characters surveyed by Makovicky (1995). In their redescription of *Microvenator celer* Ostrom, 1970, Makovicky and Sues pointed out several features indicative of oviraptorosaurian affinities, including a bone tentatively identified as an edentulous dentary. Their

analysis (Fig. 2.1E) found results similar to those of Sues (1997) and with *Microvenator* within Oviraptorosauria and Therizinosauroidea as the sister group to this group.

In the description of a new species of primitive avialan from the Late Cretaceous of Madagascar, *Rahonavis ostromi* (Forster et al., 1998a,b), an analysis of selected coelurosaurians was presented to explore its relationships (Fig. 2.1F). As in the analysis of Sues, Tyrannosauridae is farther removed from Avialae than indicated by the analysis of Holtz (1994). Noteworthy also is that Troodontidae, rather than Dromaeosauridae, were found to be the closest relatives of Avialae. *Rahonavis* and *Unenlagia* were found to be most closely related to *Archaeopteryx,* although an alternative placement of *Rahonavis* within Avialae and *Unenlagia* outside it was only a single step longer.

One of the most important discoveries in recent years has been new species of nonavialan coelurosaurs with feathers. In their description of *Protarchaeopteryx* and *Caudipteryx,* Ji et al. (1998) presented an analysis supporting the placement of both outside Avialae (Fig. 2.1F). The selection of nonavialan coelurosaurian taxa was limited, however.

Finally, an analysis of relationships among all dinosaurs was published by Sereno (1999). Notable elements of this phylogeny (Figure 2.1G) are the placement of Alvarezsauridae with Ornithomimidae (i.e., Ornithomimosauria of some other authors) and Therizinosauridae (contra Sereno, 1997), a clade comprising *Caudipteryx* and the Oviraptorosauria, and a monophyletic Deinonychosauria (Dromaeosauridae and Troodontidae).

In summary, the analyses of Coelurosauria published since the work of Gauthier (1986) have each resulted in more highly resolved relationships among the taxa, but with conflicting results. The problematic analyses of Russell and Dong (1993a) and Holtz (1994) aside, major conflicts are found in (1) the position of tyrannosaurs, either outside Coelurosauria or closer to birds than are ornithomimosaurs; (2) the relationships among Troodontidae and Dromaeosauridae relative to Avialae; (3) the relationships of therizinosauroids, either with oviraptorosaurs or ornithomimosaurs; and (4) the relationships of Alvarezsauridae, either with (or within) Avialae or with Ornithomimosauria. Given these conflicts, attempts to arrive at a stable nomenclature for coelurosaurians by defining groups impervious to changes in composition (Sereno, 1998; Padian et al., 1999) are more confusing than helpful. A notable shortcoming of all these analyses is that they considered

---

cladogram (113 characters for 14 taxa, CI = .579, RI = .712) of Forster et al. (1998a); four taxa of Ornithothoraces included separately in original data matrix. G, single most parsimonious cladogram (90 characters for 8 taxa, CI = .855, RI = .849) of Ji et al. (1998); two genera of Mononykinae, *Shuvuuia* and *Mononykus,* included separately in original data matrix. H, strict consensus of six equally most parsimonious cladograms (204 characters for 17 taxa, CI = .67, RI = .81) of Sereno (1999); Allosauroidea includes Allosauridae and Sinraptoridae, and four outgroup taxa are not illustrated.

either all higher taxa or a mixture of higher taxa and species, so that the character scores represent inferences of the primitive condition rather than observations.

A New Analysis of Coelurosaurian Relationships

The distribution of 197 characters among 37 species of coelurosaurian theropods and two outgroup taxa was analyzed by Norell et al. (2001), based on detailed study of specimens of nearly all taxa. The analysis presented here (see Hwang et al., in press) expands the data set to 208 characters and 46 theropod taxa (see Appendix 2.2 for the character list and data matrix). Recently discovered specimens of Troodontidae, Dromaeosauridae, Ornithomimidae, and Oviraptoridae from the Early Cretaceous of China and the Late Cretaceous of Mongolia provided important new data on character distributions.

Unlike previous studies, only species-level taxa were included. This is clearly preferable to the use of higher taxa, which require assumptions of monophyly. Composite taxa are often preferred because they have proportionately more data scored in a data matrix, because, for example, when one species is known only from a skull and another from a hindlimb, they can be combined into a single taxon scored for both. However, this assumes that the hindlimb and skull are uniform in both species and is a less accurate reflection of what is actually known of these taxa. Although missing data in species-level taxa may lead to an unresolved strict consensus cladogram (i.e., a cladogram that includes only those groups found in all equally most parsimonious cladograms will include few groups), this is no reason to prefer composite taxa a priori. We therefore hope that future studies that include composite taxa will present the species-level distribution of characters. We did not include the recently described *Scipionyx* and *Sinosauropteryx*, which we have not examined, nor did we include *Compsognathus* and *Coelurus*, for which we were able to determine few characters helpful in determining relationships.

All characters included in previous studies were considered. For several reasons many were excluded, however. In particular, many characters were found to vary continuously among taxa, making them problematic for phylogenetic analysis. None of the species considered here are known from sufficient specimens to assess sampling error (e.g., Farris, 1990), and the division of continuous variation into discrete character states in poorly sampled taxa can only be done arbitrarily. Many of these characters were expressed as ratios, with arbitrary limits to character state delimitation. For example, the shape of the antorbital fossa ("longer than tall" versus "taller than wide") varies considerably among archosaurs, and the down-turned dentary of segnosaurs and ornithomimosaurs is found to varying degrees in specimens of other taxa.

Several characters require comment. The distal carpals of coelurosaurs are notable for the large "semilunate" element in several taxa, but in other taxa the homologue of this element is unclear a priori. Thus, in the therizinosauroid *Alxasaurus* (and in *Beipiaosaurus)* there are two large distal elements, and it is unclear if both together or one alone is homologous with the single element of others. We therefore left this issue unresolved in the coding of carpal characters. Twelve of the 40 multistate characters (i.e., those with more than two states) were ordered (e.g., pleurocoels absent from sacral vertebrae, present on cranial sacrals, present on all sacrals), and some characters (e.g., glenoid fossa oriented caudally, caudolaterally, or laterally) were considered ordered in some runs and unordered in others.

The analyses were conducted in the computer program NONA (Goloboff, 1999) using heuristic search methods. One thousand replicates of the tree bisection and regraphing (TBR) algorithm were implemented retaining the 10 shortest cladograms for each replicate. The retained cladograms were then subjected to exhaustive branch swapping (MAX*). Branch swapping was extended to suboptimal cladograms up to 10% longer (JUMP 1).

Many (240) equally parsimonious cladograms resulted from the analysis, but a strict consensus of the equally parsimonious cladograms resulting from analysis is nonetheless highly resolved (Fig. 2.2). The results are generally concordant with those of Gauthier (1986) and most later studies except for the problematic studies of Russell and Dong (1994a) and Holtz (1994). Important aspects of the results include the following: (1) Tyrannosauridae is the sister group of other Coelurosauria; (2) Ornithomimosauria is at the base of Maniraptoriformes, and *Ornitholestes* at the base of the Maniraptora; (3) Therizinosauroidea and Oviraptorosauria are sister taxa; (4) *Caudipteryx* and *Avimimus* are members of Oviraptorosauria; (5) *Unenlagia* is a member of a largely unresolved Dromaeosauridae; and (6) Troodontidae is the sister group to Dromaeosauridae, and these two groups (Deinonychosauria) are the sister taxon of Avialae.

It is hoped that continuing phylogenetic studies of theropod dinosaurs will resolve several differences between these results and those of others. In particular, the results presented by Sereno (1999) differ mainly in the placement of Tyrannosauridae closer to birds than are ornithomimosaurs, the placement of alvarezsaurids as sister taxa to Ornithomimidae, and the grouping of Therizinosauroidea with the alvarezsaur-ornithomimid group. Among these differences the placement of alvarezsaurids requires the highest number of character changes. Although some of the character evidence provided by Sereno is convincing, we question several features proposed as homologies between the highly specialized skeleton of alvarezsaurids and ornithomimosaurs, especially those related to the fused, greatly abbreviated carpometacarpus. Furthermore, be-

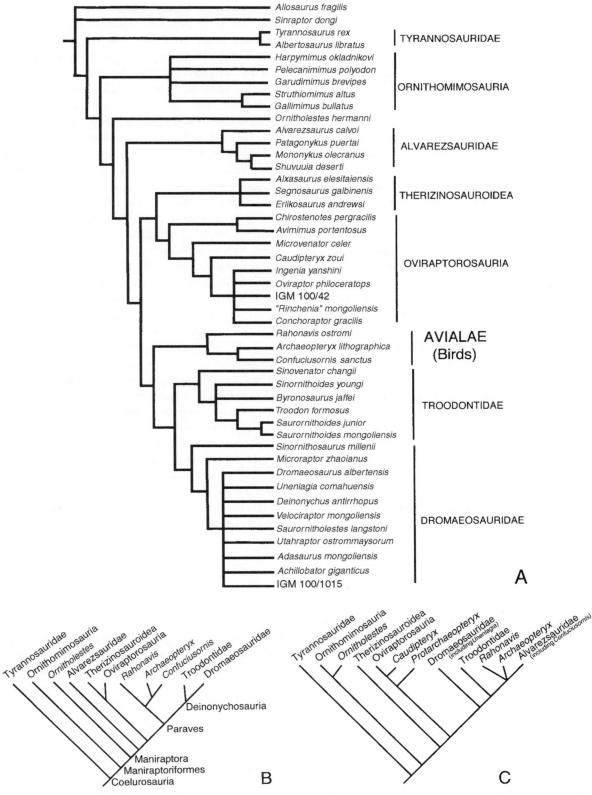

Figure 2.2. Results of an analysis of 208 characters among 44 taxa of coelurosaurian theropod dinosaurs and two outgroups (Hwang et al., in press). A, strict consensus of results from an analysis of all taxa resulting in 240 equally most parsimonious cladograms (length 599, CI = .41, RI = .69). B, cladogram derived from strict consensus cladogram showing relationships among major groups.

cause only the higher taxon Ornithomimidae was included, it is unclear to what extent the conditions of the basal ornithomimosaur *Harpymimus* were considered (e.g., its first metacarpal is shorter than II and III).

Cladistics and the Fossil Record of Theropods

Critics of the hypothesis that birds are theropod dinosaurs have claimed that the stratigraphic record supports their view, by indicating that the relevant theropod taxa occur too late in time. Feduccia (1996:90), for example, states, "Birds are supposed to be derived from the coelurosaurian dinosaurs, and the form initially used to relate dinosaurs to birds, *Deinonychus,* is from the early Cretaceous, some 40 million years after *Archaeopteryx,* and the most birdlike dinosaurs are from the late Cretaceous, some 75 or more million years after the appearance of the first bird. In fact, one could interpret the temporal evidence as indicating that birds and dinosaurs are indeed examples of convergent evolution." However, the only means of recognizing convergence —similarity that is not due to a common evolutionary origin—is through cladistic analysis of the characters shared by taxa, and in cladistic analysis the age of a taxon is irrelevant to determining its relationships. The lack of a substantive fossil record of chimpanzees and gorillas has not prevented anthropologists from recognizing the many similarities shared with humans as evidence of a close relationship, and such situations simply reflect the incompleteness of the fossil record. Feduccia's comments are therefore a criticism or misunderstanding of cladistic analysis rather than of this particular hypothesis. In any case, it is instructive to examine the stratigraphic distribution of these taxa to elucidate just how significant the gaps required by this hypothesis are.

The interpretation of stratigraphic occurrences of taxa in light of their phylogenetic relationships is controversial (see Smith, 1994), but several basic premises are common to all approaches. (1) The relevant age of a taxon is its first appearance in the stratigraphic record, the oldest occurrence of a specimen with the apomorphies of that taxon; (2) the branching order indicated by a phylogenetic hypothesis should reflect the order of appearance of these taxa in the fossil record; and (3) sister taxa (two taxa that are each other's closest relative) should appear at the same time in the fossil record unless one is potentially ancestral to the other (i.e., lacks autapomorphies). Discrepancies between phylogenetic hypotheses of taxa and their fossil record imply gaps in the fossil record (Norell, 1993), intervals during which fossils representing a clade should be present but are not. Note that predictions of precisely how long particular morphological changes took in the course of evolution depend on a "morphological clock" for which there is no evidence.

Figure 2.3 illustrates the first occurrence of each of the major groups of nonavialan coelurosaurs, as well as two other taxa suggested as bird relatives. All but one of the coelurosaurs first appear prior to the Late Cretaceous, a pattern that has been established by several recent discoveries in Early Cretaceous deposits. Second, several taxa predate the first occurrence of Avialae (i.e., *Archaeopteryx*), including *Ornitholestes* and *Coelurus,* and *Compsognathus* is the same age. Because the precise relationships of these taxa are ambiguous, especially *Ornitholestes* and *Coelurus,* it is unclear whether current hypotheses of relationships involving theropods and Avialae require any significant gaps in the fossil record. For example, if *Ornitholestes* is the closest relative of Dromaeosauridae and together they are the closest relative of Avialae (*Unenlagia* notwithstanding, for the moment), then with regard to the "origin" of Avialae, the only gap in the fossil record implied by this hypothesis is that between the Kimmeridgian (the occurrence of *Ornitholestes*) and Tithonian (the occurrence of *Archaeopteryx*) stages of the Late Jurassic, comprising only a few million years.

The first definitive occurrences of most of the major clades of Coelurosauria in the Early Cretaceous and Late Jurassic focus attention on the nonmarine fossil record of the Middle and Late Jurassic. A well-known bias of the geological record (Blatt and Jones, 1975; Blatt et al., 1991) is that older rocks are progressively less common—for example, in the United States there are 422,600 square miles of Cretaceous rocks exposed, but only 99,080 square miles of Jurassic rocks (Gilluly, 1949); calibrated for the somewhat shorter duration of the Jurassic, this is 5,336 square miles per million years for the Cretaceous versus 1,611 for the Jurassic. Middle Jurassic sedimentary rocks from nonmarine depositional environments are even rarer than those of the Early Jurassic and Late Triassic, and the fossil record of terrestrial vertebrates from this period is therefore among the worst in the fossil record (see, e.g., Benton, 1994). An attempt to illustrate this sampling bias is shown in the top of Figure 2.3, based on a compendium of fossil vertebrate localities with dinosaurs (Weishampel, 1990). Because dinosaurs occur throughout the world in nearly every nonmarine sedimentary formation with vertebrate fossils during this time span, this is offered as an admittedly flawed but nevertheless instructive approximation for how well sampled the terrestrial fossil record of these time periods is.

Late Jurassic terrestrial vertebrates are known primarily from a few localities with abundant fossils, especially from the Morrison Formation of western North America. The predominantly fluvial sediments of the Morrison Formation preserve few specimens of small size, but among them are several coelurosaurians. The best known of these are *Ornitholestes* and *Coelurus,* which deserve further study, but at least one other coelurosaur is represented by two isolated

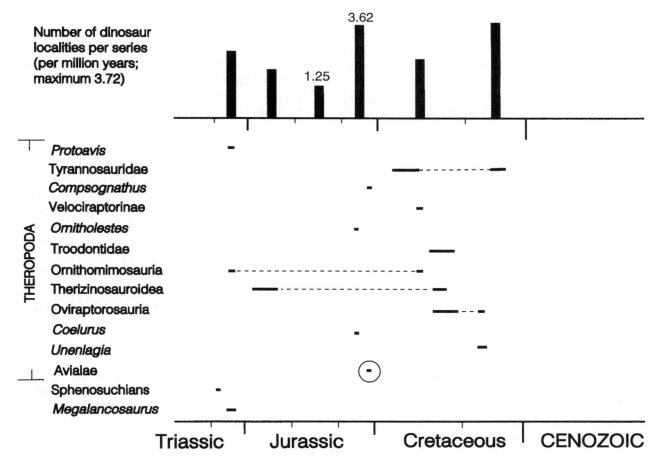

Figure 2.3. First occurrences in the fossil record of coelurosaurian theropods and two alternative bird relatives (bottom) and relative density of dinosaur localities (top). Dinosaur localities are plotted in number per million years per series (filled). Time scale is from Gradstein et al. (1995), dinosaur localities are from Weishampel (1990). Length of bars in the lower figure reflects uncertainties in the dating of single occurrences rather than durations. See Appendix 2.1 for references of first occurrences.

vertebrae (Makovicky, 1998), and other fragmentary material is known (Jensen and Padian, 1989). An isolated troodontidlike tooth also indicates the presence of coelurosaurs (Chure, 1994), but the affinities of this specimen must remain tentative until further material is discovered. In any case, these specimens indicate unequivocally that coelurosaurians had diversified prior to the Morrison, which is correlated with the Kimmeridgian marine stage, and therefore these specimens are older than the specimens of *Archaeopteryx*.

The low number of localities in the Middle Jurassic, scaled with time, reflects the relative paucity of the record from this time period. Despite this poor record, several fossils suggest that coelurosaurs occur in the Middle Jurassic. At two Middle Jurassic localities in Great Britain, isolated teeth similar to those of dromaeosaurids and troodontids have been reported (Evans and Milner, 1994; Metcalf and Walker, 1994). An incomplete theropod braincase from the

late Early Jurassic La Boca Formation of Mexico (Clark et al., 1994a) possesses a caudal tympanic recess, indicating affinities with coelurosaurians (Munter, 1999). Finally, a dentary from the Early Jurassic Lower Lufeng Formation of China shares derived features with therizinosauroids (Xu et al., 2001a), a group that is now considered to be coelurosaurian (Russell and Dong, 1993a; Clark et al., 1994b; Xu et al., 1999a).

Two occurrences from the Late Triassic may be extremely important for the timing of the diversification of Theropoda. *P. texensis,* from the Dockum Formation of Texas, was described as a bird (Chatterjee, 1991, 1997), but questions concerning the reconstruction of elements, the integrity of the specimen, and the interpretation of the anatomy have raised doubts about its importance. Nevertheless, several features of the specimens referred to this species indicate the presence of a coelurosaurian theropod, perhaps related to troodontids (Currie and Zhao, 1993), which moves the first occurrence of this group back significantly (see also Witmer, 2001, Chapter 1 in this volume). A second taxon from these same deposits, *Shuvosaurus inexpectatus,* shares interesting similarities with Ornithomimidae that may extend this particular coelurosaurian clade back in time.

The preponderance of specimens from the Late Cretaceous on which much of our knowledge of nonavialan coelurosaurian theropods is based deserves comment. Although the first occurrences of these groups in the stratigraphic record provide their minimum age, the most complete material is often from later deposits, perhaps because of the preponderance of sediments of later ages in the fossil record (Fig. 2.3). For example, many of the most complete articulated skeletons are from the Late Cretaceous Djadokhta Formation of Mongolia and China, a situation for which the unusual environment in which these sediments were deposited—sand dunes interspersed with streams and ponds—is presumably responsible (Eberth, 1994; Fastovsky et al., 1997; Loope et al., 1998). The later age of these specimens is of no consequence to their utility in understanding the relationships of these groups, just as the large amount of evidence provided by living organisms cannot be ignored when inferring relationships among taxa to which they belong (Rosen et al., 1981). Again, fossils do not have a stratigraphic record beyond individual occurrences until they are grouped together, and the only means of doing this is with shared characters. Ignoring particular fossils because of their age, as when they are considered a priori to be convergent with birds, is therefore an arbitrary dismissal of evidence, for which there is no justification.

The two most vocal critics of a close relationship between birds and theropod dinosaurs have proposed alternative hypotheses, although without providing sufficient character evidence to make compelling arguments. Martin (1983) hypothesized that birds and crocodilians share a closer relationship than do birds and dinosaurs and that the extinct group Sphenosuchia, usually allied with crocodilians, is in turn the closest relative of this group. Feduccia (Feduccia and Wild, 1993; Feduccia, 1996) has proposed instead that the poorly known *Megalancosaurus* is a closer relative of birds than are dinosaurs, arguing that it shares so few characters with birds because it is at the very base of a long evolutionary branch as yet unrepresented in the fossil record. Without addressing the deficiencies in the evidential support for both these hypotheses (see Currie, 1987; Sereno, 1991; Witmer, 1991; Gauthier, 1994; Renesto, 1994; Elzanowski and Wellnhofer, 1996), we point out that each requires either a much larger gap in the fossil record than does the fossil record of theropods or, if the Late Triassic and Early Jurassic coelurosaur records are valid, then a nearly identical gap (this has been quantified by Brochu and Norell, 2000, 2001).

The Evolutionary Origin of Avialan Features

Birds share many apomorphies that are lacking in their closest living relatives, as even the most cursory examination of a bird and a crocodilian will attest. Familiar features are feathers, an endothermic metabolism, a complex respiratory system in which pulmonary air flows in a single direction, a four-chambered heart completely separating arterial and venous blood, an enlarged forebrain, a syrinx, many features of the bony skeleton, and a variety of complex behaviors. The simplest explanation is that these features appeared after the divergence of birds from their common ancestor with crocodilians, and for the osteological characters the evolutionary appearance can be resolved more precisely among the taxa known only from the fossil record.

By documenting the distribution of osteological characters among theropod taxa, the fossil record provides information critical to determining which of these arose prior to the evolutionary appearance of flight. There is considerable disagreement about the precise capabilities for powered flight among basal fossil avialans, but the relatively large body size and weak forearms of dromaeosaurids, oviraptorosaurians, troodontids, and their closest relatives strongly suggest that none possessed these capabilities. Thus, the presence of a furcula, ossified ventral rib segments, and an ossified sternum in dromaeosaurids (Norell and Makovicky, 1999) and oviraptorids (Barsbold et al., 1990; Clark et al., 1999) indicates that these structures evolved before powered flight.

For the "soft" anatomical, physiological, and behavioral features, however, there is often only very weak evidence, so those rare occurrences hinting at their presence in fossils are the focus of a great deal of attention. The most important recent discoveries are the theropods from the Early Cretaceous of Liaoning, which preserve soft tissues including feathers and featherlike structures. The vaned feathers in *Protarchaeopteryx* and *Caudipteryx* (Ji et al., 1998) demonstrate that these structures appeared earlier in the evolution of birds than had previously been thought. Furthermore, the relationships of *Caudipteryx* with Oviraptorosauria imply that vaned feathers were broadly distributed among coelurosaurians (i.e., in troodontids, oviraptorosaurs, and dromaeosaurids). Because the forelimb anatomy of these two genera indicate that they were not capable of powered flight, the simplest interpretation of current evidence is that feathers evolved before powered flight.

Another important recent discovery is a skeleton of an oviraptorid overlying a nest in a posture almost identical to that taken by living birds (Norell et al., 1995; Clark et al., 1999). The eggs in the nest are identical in texture, size, and shape to another in which an oviraptorid embryo is preserved (Norell et al., 1994), and the adult skeleton directly contacts some of the eggs in the nest. Furthermore, among the first 17 oviraptorid skeletons collected from the Djadokhta Formation, 4 are preserved over nests of this kind (Osborn, 1924; Dong and Currie, 1996; Webster, 1996:80). These specimens strongly suggest that the adult was a parent of the nest, and they provide evidence that the behavior of sitting on a nest in this posture evolved prior to, or perhaps with, the common ancestor of oviraptorids and birds.

The dinosaurs most closely related to birds (or to Avialae) should offer the strongest evidence for the conditions ancestral to them, but our understanding of their behavior, especially the uses to which they put their forelimbs, is severely limited. For example, it is often contended that the closest relatives of Avialae (e.g., dromaeosaurids) were predominately cursorial, and therefore flight in birds evolved from cursorial animals (e.g., Ostrom, 1976). However, it is unclear whether we can be certain these taxa never wandered to the top of a cliff or clambered up trees with low

limbs during any stage of their ontogeny. Indeed, the strongly curved claws of the hand of dromaeosaurids and oviraptorids (Chiappe, 1997) could have been used in ways similar to those of living birds with strongly curved claws that climb trees or cliffs. Furthermore, although it is clear that birds are descended at some level from animals that were ground living (as there is no evidence that they evolved from "fish" independent of other tetrapods), it is unclear that our sample of nonavialan theropods is sufficient to address questions about the conditions *immediately* preceding the origin of flight in birds. The Mesozoic fossil record of theropod dinosaurs is clearly a poor sample of the true diversity of this group, and it is likely biased against taxa of smaller size. In light of this uncertainty, we feel that it has yet to be demonstrated that the fossil record is capable of testing critically either the "arboreal" or "cursorial" hypotheses of the origin of flight (see also Witmer, Chapter 1 in this volume).

APPENDIX 2.1
First Occurrences of Taxa in Figure 2.3

Theropods

Troodontidae—The oldest definitive troodontid specimen is *S. youngi* Russell and Dong, 1994, Ejinhoroqi Formation near Huamuxiao, Inner Mongolia, China (Russell and Dong, 1994b). Probably Barremian to Aptian (Eberth et al., 1994; Russell and Zhao, 1996). Isolated teeth similar to those of troodontids from Bathonian (Middle Jurassic) deposits in England (Evans and Milner, 1994) and a troodontidlike tooth from the Late Jurassic Morrison Formation of Utah (Chure, 1994) are not included.

Ornithomimosauria—*P. polyodon* Pérez-Moreno et al., 1994, is from the Upper Hauterivian–Lower Barremian Calizas de La Huérguina Formation, Las Hoyas, Spain. *S. inexpectatus* Chatterjee, 1993, from the early Norian Dockum Formation, exhibits several features of derived ornithomimosaurs but is poorly preserved.

Oviraptorosauria—*M. celer* Ostrom, 1970, recently has been identified as a primitive oviraptorosaur (Makovicky and Sues, 1998), as suggested earlier by Currie and Russell (1988). It is from the Aptian-Albian Cloverly Formation of Montana (Ostrom, 1970). Otherwise, the earliest member is *Caenagnathasia* from the upper Turonian Bissekty Formation of Uzbekistan (Currie et al., 1994).

Dromaeosauridae—*Utahraptor ostrommaysorum* Kirkland et al., 1993, Cedar Mountain Formation, Utah, cited as Bar-

remian in age. Possible dromaeosaurid or troodontid teeth from several Bathonian sites in Great Britain (Evans and Milner, 1994; Metcalf and Walker, 1994).

Therizinosauroidea—*A. elesitaiensis* (Russell and Dong, 1994a), Bayin Obo Formation, Alxa Desert, Inner Mongolia, China. Dated as Albian on the basis of pollen and tetrapods (Jerzykiewicz and Russell, 1991). A possible therizinosauroid has been reported from the Early Jurassic Lower Lufeng Formation of China (Xu et al., 2001a).

Coelurus—The only known specimens of *C. fragilis* Marsh, 1879, and *C. agilis* Marsh, 1884, are from the Morrison Formation of Como Bluff, Wyoming, dated by pollen as correlative with the Kimmeridgian marine stage (Kowallis et al., 1998). Ostrom (1980) suggested that they all are from a single specimen and species, *C. fragilis*.

Ornitholestes—The single specimen of *O. hermanni* Osborn, 1903, is from the Morrison Formation at Bone Cabin Quarry, Wyoming. The Morrison Formation is dated by pollen as correlative with the Kimmeridgian marine stage (Kowallis et al., 1998).

Tyrannosauridae—The oldest taxon is *Siamotyrannus isanensis* Buffetaut et al., 1996, from the Sao Khua Formation, Phu Wiang, Thailand, dated as pre-Albian Early Cretaceous. However, more complete, definitive tyrannosaurid remains do not appear in well-dated rocks until the Campanian (Molnar et al., 1990).

Compsognathus—The two specimens of *Compsognathus* are from Solnhofen (the holotype of *C. longipes* Wagner, 1861) and an unnamed unit at Canjuer, France (the holotype of *C. corallestris* Bidar et al., 1972, considered to represent *C. longipes* by Ostrom [1978]). The Upper Solnhofen Plattenkalk is dated by marine invertebrates as early Tithonian (Barthel et al., 1990); the French unit is dated by marine invertebrates as slightly later in the Tithonian (Bidar et al., 1972).

Protoavis—*P. texensis* Chatterjee, 1991, from the early Norian Dockum Formation of Texas, exhibits several coelurosaurian features, but the reconstructions of the specimen and the allocation of all the material identified with this taxon have been questioned.

Unenlagia—*U. comahuensis* Novas and Puerta, 1997, is from the Rio Neuquén Formation, Argentina, dated at Turonian-Coniacian (Cruz et al., 1989).

Nontheropods

Megalancosaurus—The five specimens of *Megalancosaurus preonensis* Calzavara et al., 1980, are from the "Dolomia di Forni" (Renesto, 1994), which is considered to be Norian by Tintori et al. (1985) on the basis of marine invertebrates.

Sphenosuchians—The first occurrence of sphenosuchians in the fossil record is *Trialestes romeri* (Reig, 1963), from the Ischigualasto Formation of Argentina (although the monophyly of this group has been questioned by Clark, in Benton and Clark, 1988). The Ischigualasto Formation is dated radiometrically as correlative with the mid-Carnian marine stage by Rogers et al. (1993).

APPENDIX 2.2
Character List and Data Matrix

Key: 0 = plesiomorphic character state; 1, 2, 3, 4 = apomorphic character states; ? = missing data; – = character not applicable (e.g., dental characters in *Oviraptor philoceratops*)

1. Vaned feathers on forelimb: symmetric (0); asymmetric (1).

Skull

2. Orbit round in lateral or dorsolateral view (0) or dorso-ventrally elongate (1). It is unclear that the eye occupied the entire orbit of those taxa in which it is keyhole-shaped.
3. Anterior process of postorbital projects into orbit (0) or does not project into orbit (1).
4. Postorbital in lateral view with straight anterior (frontal) process (0) or frontal process curves anterodorsally, and dorsal border of temporal bar is dorsally concave (1).
5. Postorbital bar parallels quadrate, lower temporal fenestra rectangular in shape (0) or jugal and postorbital approach or contact quadratojugal to constrict lower temporal fenestra (1).
6. Otosphenoidal crest vertical on basisphenoid and prootic and does not border an enlarged pneumatic recess (0) or well-developed, crescent-shaped, thin crest forms anterior edge of enlarged pneumatic recess (1). This structure forms the anterior, and most distinct, border of the lateral depressio of the middle ear region (see Currie and Zhao, 1994) of troodontids and some extant avians.
7. Crista interfenestralis confluent with lateral surface of prootic and opisthotic (0) or distinctly depressed within middle ear opening (1).
8. Subotic recess (pneumatic fossa ventral to fenestra ovalis) present (0) or absent (1).
9. Basisphenoid recess present between basisphenoid and basioccipital (0) or entirely within basisphenoid (1) or absent (2).
10. Posterior opening of basisphenoid recess single (0) or divided into two small, circular foramina by a thin bar of bone (1).
11. Base of cultriform process not highly pneumatized (0) or base of cultriform process (parasphenoid rostrum) expanded and pneumatic (parasphenoid bulla) (1).

12. Basipterygoid processes ventral or anteroventrally projecting (0) or lateroventrally projecting (1).
13. Basipterygoid processes well developed, extending as a distinct process from the base of the basisphenoid (0) or processes abbreviated or absent (1).
14. Basipterygoid processes solid (0) or processes hollow (1).
15. Basipterygoid recesses on dorsolateral surfaces of basipterygoid processes absent (0) or present (1).
16. Depression for pneumatic recess on prootic absent (0) or present as dorsally open fossa on prootic/opisthotic (1) or present as deep, posterolaterally directed concavity (2). The dorsal tympanic recess referred to here is the depression anterodorsal to the middle ear on the opisthotic, not the recess dorsal to the crista interfenestralis within the middle ear as seen in *Archaeopteryx lithographica, Shuvuuiau deserti* and Aves.
17. Accessory tympanic recess dorsal to crista interfenestralis absent (0) small pocket present (1) or extensive with indirect pneumatization (2). According to Witmer (1990), this structure may be an extension from the caudal tympanic recess, although it has been interpreted as the main part of the caudal tympanic recess by some authors (e.g., Walker, 1985).
18. Caudal (posterior) tympanic recess absent (0) present as opening on anterior surface of paroccipital process (1) or extends into opisthotic posterodorsal to fenestra ovalis, confluent with this fenestra (2).
19. Exits of CN X–XII flush with surface of exoccipital (0) or cranial nerve exits located together in a bowl-like basisphenoid depression (1).
20. Maxillary process of premaxilla contacts nasal to form posterior border of nares (0) or maxillary process reduced so that maxilla participates broadly in external naris (1) or

maxillary process of premaxilla extends posteriorly to separate maxilla from nasal posterior to nares (2).

21. Posterior process of premaxillary short and blunt (0) or elongate and extend along nasal-maxillary suture posterior to nares (1).
22. Internarial bar rounded (0) or flat (1).
23. Crenulate margin on buccal edge of premaxilla absent (0) or present (1).
24. Caudal margin of naris farther rostral than (0), or nearly reaching or overlapping (1), the rostral border of the antorbital fossa.
25. Premaxillary symphysis acute, V-shaped (0) or rounded, U-shaped (1).
26. Secondary palate formed by premaxilla only (0) or by premaxilla, maxilla, and vomer (1).
27. Palatal shelf of maxilla flat (0) or with midline ventral toothlike projection (1).
28. Pronounced, round accessory antorbital fenestra absent (0) or present (1). A small fenestra, variously termed the accessory antorbital fenestra or maxillary fenestra, penetrates the medial wall of the antorbital fossa anterior to the antorbital fenestra in a variety of coelurosaurs and other theropods.
29. Accessory antorbital fossa situated at rostral border of antorbital fossa (0) or situated posterior to rostral border of fossa (1).
30. Tertiary antorbital fenestra (fenestra promaxillaris) absent (0) or present (1).
31. Antorbital fossa without distinct rim ventrally and anteriorly (0) or with distinct rim composed of a thin wall of bone (1). A rim is most strongly developed in the therizinosauroid *E. andrewsi* (Clark et al., 1994b) but is nearly absent in ornithomimosaurs.
32. Narial region apneumatic or poorly pneumatized (0) or with extensive pneumatic fossae, especially along posterodorsal rim of fossa (1).
33. Jugal and postorbital contribute equally to postorbital bar (0) or ascending process of jugal reduced and descending process of postorbital ventrally elongate (1).
34. Jugal tall beneath lower temporal fenestra, twice or more as tall dorsoventrally as it is wide transversely (0) or rodlike (1).
35. Jugal pneumatic recess in posteroventral corner of antorbital fossa present (0) or absent (1).
36. Medial jugal foramen present on medial surface ventral to postorbital bar (0) or absent (1).
37. Quadratojugal without horizontal process posterior to ascending process (reversed L shape) (0) or with process (i.e., inverted T or Y shape) (1).
38. Jugal and quadratojugal separate (0) or quadratojugal and jugal fused and not distinguishable from one another (1).
39. Supraorbital crests on lacrimal in adult individuals absent (0) or dorsal crest above orbit (1) or lateral expansion anterior and dorsal to orbit (2).
40. Enlarged foramen or foramina opening laterally at the angle of the lacrimal, absent (0) or present (1).
41. Lacrimal anterodorsal process absent (inverted L-shaped) (0) or lacrimal T-shaped in lateral view (1) or anterodorsal process much longer than posterior process (2). Ordered.
42. Prefrontal large, dorsal exposure similar to that of lacrimal (0) or greatly reduced in size (1) or absent (2). Ordered.
43. Frontals narrow anteriorly as a wedge between nasals (0) or end abruptly anteriorly, suture with nasal transversely oriented (1).
44. Anterior emargination of supratemporal fossa on frontal straight or slightly curved (0) or strongly sinusoidal and reaching onto postorbital process (1) (Currie, 1995).

45. Frontal postorbital process (dorsal view): smooth transition from orbital margin (0) or sharply demarcated from orbital margin (1) (Currie, 1995).
46. Frontal edge smooth in region of lacrimal suture (0) or edge notched (1) (Currie, 1995).
47. Dorsal surface of parietals flat, lateral ridge borders supratemporal fenestra (0) or parietals dorsally convex with very low sagittal crest along midline (1) or dorsally convex with well-developed sagittal crest (2).
48. Parietals separate (0) or fused (1).
49. Descending process of squamosal parallels quadrate shaft (0) or nearly perpendicular to quadrate shaft (1).
50. Descending process of squamosal contacts quadratojugal (0) or does not contact quadratojugal (1).
51. Posterolateral shelf on squamosal overhanging quadrate head absent (0) or present (1).
52. Dorsal process of quadrate single headed (0) or with two distinct heads, a lateral one contacting the squamosal and a medial head contacting the braincase (1).
53. Quadrate vertical (0) or strongly inclined anteroventrally so that distal end lies far forward of proximal end (1).
54. Quadrate solid (0) or hollow, with depression on posterior surface (1).
55. Lateral border of quadrate shaft straight (0) or with lateral tab that touches squamosal and quadratojugal above an enlarged quadrate foramen (1).
56. Foramen magnum subcircular, slightly wider than tall (0) or oval, taller than wide (1). See Makovicky and Sues (1998).
57. Occipital condyle without constricted neck (0) or subspherical with constricted neck (1).
58. Paroccipital process elongate and slender, with dorsal and ventral edges nearly parallel (0) or process short, deep with convex distal end (1).
59. Paroccipital process straight, projects laterally or posterolaterally (0) or distal end curves ventrally, pendant (1).
60. Paroccipital process with straight dorsal edge (0) or with dorsal edge twisted rostrolaterally at distal end (1) (Currie, 1995).
61. Ectopterygoid with constricted opening into fossa (0) or with open ventral fossa in the main body of the element (1).
62. Dorsal recess on ectopterygoid absent (0) or present (1).
63. Flange of pterygoid well developed (0) or reduced in size or absent (1).
64. Palatine and ectopterygoid separated by pterygoid (0) or contact (1) (Currie, 1995).
65. Palatine tetraradiate, with jugal process (0) or palatine triradiate, jugal process absent (1).
66. Suborbital fenestra similar in length to orbit (0) or reduced in size (less than one quarter orbital length) or absent (1).

Mandible

67. Symphyseal region of dentary broad and straight, paralleling lateral margin (0) or medially recurved slightly (1) or strongly recurved medially (2).
68. Dentary symphyseal region in line with main part of buccal edge (0) or symphyseal end downturned (1).
69. Mandible without coronoid prominence (0) or with coronoid prominence (1).
70. Posterior end of dentary without posterodorsal process dorsal to mandibular fenestra (0) or with dorsal process above anterior end of mandibular fenestra (1) or with elongate dorsal process extending over most of fenestra (2).
71. Labial face of dentary flat (0) or with lateral ridge and inset tooth row (1).
72. Dentary subtriangular in lateral view (0) or with subparallel dorsal and ventral edges (1) (Currie, 1995).

73. Nutrient foramina on external surface of dentary superficial (0) or lie within deep groove (1).

74. External mandibular fenestra oval (0) or subdivided by a spinous rostral process of the surangular (1).

75. Internal mandibular fenestra small and slitlike (0) or large and rounded (1) (Currie, 1995).

76. Foramen in lateral surface of surangular rostral to mandibular articulation, absent (0) or present (1).

77. Splenial not widely exposed on lateral surface of mandible (0) or exposed as a broad triangle between dentary and angular on lateral surface of mandible (1).

78. Coronoid ossification large (0) or only a thin splint (1) or absent (2). Ordered.

79. Articular without elongate, slender medial, posteromedial, or mediodorsal process from retroarticular process (0) or with process (1).

80. Retroarticular process short, stout (0) or elongate and slender (1).

81. Mandibular articulation surface as long as distal end of quadrate (0) or twice or more as long as quadrate surface, allowing anteroposterior movement of mandible (1).

Dentition

82. Premaxilla toothed (0) or edentulous (1).

83. Second premaxillary tooth approximately equivalent in size to other premaxillary teeth (0) or second tooth markedly larger than third and fourth premaxillary teeth (1) (Currie, 1995).

84. Maxilla toothed (0) or edentulous (1).

85. All maxillary teeth serrated (0) or some without serrations anteriorly (except at base in *Saurornithoides mongoliensis*) (1) or all without serrations (2). Ordered.

86. All dentary teeth serrated (0) or some without serrations anteriorly (except at base in *S. mongoliensis*) (1) or all without serrations (2). Ordered.

87. Dentary and maxillary teeth large, less than 25 in dentary (0) or moderate number of small teeth (25–30 in dentary) (1) or teeth relatively small, and numerous (more than 30 in dentary) (2).

88. Serration denticles large (0) or small (1). Farlow et al. (1991) quantify this difference.

89. Serrations simple, denticles convex (0) or distal and often mesial edges of teeth with large, hooked denticles that point toward the tip of the crown (1).

90. Maxillary teeth constricted between root and crown (0) or root and crown confluent (1).

91. Dentary teeth constricted between root and crown (0) or root and crown confluent (1).

92. Dentary teeth evenly spaced (0) or anterior dentary teeth smaller, more numerous, and more closely appressed than those in middle of tooth row (1).

93. Dentaries lack distinct interdental plates (0) or with interdental plates medially between teeth (1). Currie (1995) suggests the interdental plates of dromaeosaurids are present but fused to the medial surface of the dentary, but in the absence of convincing evidence for this fusion we did not recognize this distinction.

94. In cross section, premaxillary tooth crowns suboval to subcircular (0) or asymmetrical (D-shaped in cross section) with flat lingual surface (1).

Axial Skeleton

95. Number of cervical vertebrae: 10 (0) or 12 or more (1).

96. Axial epipophyses absent or poorly developed, not extending past posterior rim of postzygapophyses (0) or large and posteriorly directed, extend beyond postzygapophyses (1).

97. Axial neural spine flared transversely (0) or compressed mediolaterally (1).

98. Epipophyses of cervical vertebrae placed distally on postzygapophyses, above postzygapophyseal facets (0) or placed proximally, proximal to postzygapophyseal facets (1).

99. Anterior cervical centra level with or shorter than posterior extent of neural arch (0) or centra extending beyond posterior limit of neural arch (1).

100. Carotid process on posterior cervical vertebrae absent (0) or present (1).

101. Anterior cervical centra subcircular or square in anterior view (0) or distinctly wider than high, kidney-shaped (1).

102. Cervical neural spines anteroposteriorly long (0) or short and centered on neural arch, giving arch an X shape in dorsal view (1).

103. Cervical centra with one pair of pneumatic openings (0) or with two pairs of pneumatic openings (1).

104. Cervical and anterior trunk vertebrae amphiplatyan (0) or opisthocoelous (1).

105. Anterior trunk vertebrae without prominent hypapophyses (0) or with large hypapophyses (1).

106. Parapophyses of posterior trunk vertebrae flush with neural arch (0) or distinctly projected on pedicels (1).

107. Hyposphene-hypantrum articulations in trunk vertebrae absent (0) or present (1).

108. Zygapophyses of trunk vertebrae abutting one another above neural canal, opposite hyposphenes meet to form lamina (0), or zygapophyses placed lateral to neural canal and separated by groove for interspinous ligaments, hyposphenes separated (1).

109. Cervical vertebrae but not dorsal vertebrae pneumatic (0) or all presacral vertebrae pneumatic (1).

110. Transverse processes of anterior dorsal vertebrae long and thin (0) or short, wide, and only slightly inclined (1).

111. Neural spines of dorsal vertebrae not expanded distally (0) or expanded to form spine table (1).

112. Scars for interspinous ligaments terminate at apex of neural spine in dorsal vertebrae (0) or terminate below apex of neural spine (1).

113. Number of sacral vertebrae: 5 (0) or 6 (1) or 8 or more (2). Ordered.

114. Sacral vertebrae wit unfused zygapophyses (0) or with fused zygapophyses forming a sinuous ridge in dorsal view (1).

115. Ventral surface of posterior sacral centra gently rounded, convex (0) or ventrally flattened, sometimes with shallow sulcus (1) or centrum strongly constricted transversely, ventral surface keeled (2). Note that in *Alvarezsaurus calvoi* it is only the fifth sacral that is keeled, unlike other alvarezsaurids (Novas, 1996).

116. Pleurocoels absent on sacral vertebrae (0) or present on anterior sacrals only (1) or present on all sacrals (2). A pleurocoel may be present on the first sacral in *A. elesitaiensis,* although this area is badly crushed (Russell and Dong, 1994a). Ordered.

117. Last sacral centrum with flat posterior articulation surface (0) or convex articulation surface (1).

118. Caudal vertebrae with distinct transition point, from shorter centra with long transverse processes proximally to longer centra with small or no transverse processes distally (0) or vertebrae homogeneous in shape, without transition point (1).

119. Transition point in caudal series begins distal to the 10th caudal (0) or at or proximal to the 10th caudal vertebra (1).

120. Anterior caudal centra tall, oval in cross section (0) or with boxlike centra in caudals I–V (1) or anterior caudal centra laterally compressed with ventral keel (2).

121. Neural spines of caudal vertebrae simple, undivided (0) or separated into anterior and posterior alae throughout much of caudal sequence (1).

122. Neural spines on distal caudals form a low ridge (0) or spine absent (1) or midline sulcus in center of neural arch (2).

123. Prezygapophyses of distal caudal vertebrae between one-third and whole centrum length (0) or with extremely long extensions of the prezygapophyses (up to 10 vertebral segments long in some taxa) (1) or strongly reduced as in *A. lithographica* (2).

124. More than 40 caudal vertebrae (0) or 25–40 caudal vertebrae (1) or no more than 25 caudal vertebrae (2). Ordered.

125. Proximal end of chevrons of proximal caudals short anteroposteriorly, shaft cylindrical (0) or proximal end elongate anteroposteriorly, flattened and platelike (1).

126. Distal caudal chevrons are simple (0) or anteriorly bifurcate (1).

127. Shaft of cervical ribs slender and longer than vertebra to which they articulate (0) or broad and shorter than vertebra (1).

128. Ossified uncinate processes absent (0) or present (1).

129. Ossified ventral rib segments absent (0) or present (1).

130. Lateral gastral segment shorter than medial one in each arch (0) or distal segment longer than proximal segment (1).

131. Ossified sternal plates separate in adults (0) or fused (1).

132. Sternum without distinct lateral xiphoid process posterior to costal margin (0) or with lateral xiphoid process (1).

133. Anterior edge of sternum grooved for reception of coracoids (0) or sternum without grooves (1).

134. Articular facet of coracoid on sternum (conditions may be determined by the articular facet on coracoid in taxa without ossified sternum): anterolateral or more lateral than anterior (0); almost anterior (1) (Xu et al., 1999b).

Pectoral Girdle and Forelimb

135. Hypocleidium on furcula absent (0) or present (1). The hypocleidium is a process extending from the ventral midline of the furcula and is attached to the sternum by a ligament in extant birds.

136. Acromion margin of scapula continuous with blade (0) or anterior edge laterally everted (1).

137. Anterior surface of coracoid ventral to glenoid fossa unexpanded (0) or anterior edge of coracoid expanded, forms triangular subglenoid fossa bounded laterally by coracoid tuber (1).

138. Scapula and coracoid separate (0) or fused into scapulacoracoid (1).

139. Coracoid in lateral view subcircular, with shallow ventral blade (0) or subquadrangular with extensive ventral blade (1) or shallow ventral blade with elongate posteroventral process (2).

140. Scapula and coracoid form a continuous arc in posterior and anterior views (0) or coracoid inflected medially, scapulocoracoid L-shaped in lateral view (1).

141. Glenoid fossa faces posteriorly or posterolaterally (0) or laterally (1).

142. Scapula longer than humerus (0) or humerus longer than scapula (1).

143. Deltopectoral crest large and distinct, proximal end of humerus quadrangular in anterior view (0) or deltopectoral crest less pronounced, forming an arc rather than being quadrangular (1) or deltopectoral crest very weakly developed, proximal end of humerus with rounded edges (2) or deltopectoral crest extremely long (3) or proximal end of humerus extremely broad, triangular in anterior view (4).

144. Anterior surface of deltopectoral crest smooth (0) or with distinct groove or ridge near lateral edge along distal end of crest (1).

145. Olecranon process weakly developed (0) or distinct and large (1).

146. Distal articular surface of ulna flat (0) or convex, semilunate surface (1).

147. Proximal surface of ulna a single continuous articular facet (0) or divided into two distinct fossae separated by a median ridge (1).

148. Lateral proximal carpal (ulnare?) quadrangular (0) or triangular in proximal view (1). The homology of the carpal elements of coelurosaurs is unclear (see, e.g., Padian and Chiappe, 1998), but the large, triangular lateral element of some taxa most likely corresponds to the lateral proximal carpal of basal tetanurans.

149. Two distal carpals in contact with metacarpals, one covering the base of metacarpal I (and perhaps contacting metacarpal II) and the other covering the base of metacarpal II (0) or a single distal carpal capping metacarpals I and II (1). In the absence of ontogenetic data, it is not possible to determine whether the single large semilunate carpal of birds and many other coelurosaurs is formed by fusion of the two distal carpals or is, instead, an enlarged distal carpal 1 or 2.

150. Distal carpals not fused to metacarpals (0) or fused to metacarpals, forming carpometacarpus (1).

151. Semilunate distal carpal well developed, covering all of proximal ends of metacarpals I and II (0) or small, covers about half of base of metacarpals I and II (1) or covers bases of all metacarpals (2).

152. Metacarpal I half the length of metacarpal II (0) or less than half the length of metacarpal II (1) or subequal in length to metacarpal II (2).

153. Third manual digit present, phalanges present (0) or reduced to no more than metacarpal splint (1).

154. Manual unguals strongly curved, with large flexor tubercles (0) or weakly curved or straight with weak flexor tubercles displaced distally from articular end (1).

155. Unguals on all digits generally similar in size (0) or digit I bearing large ungual and unguals of other digits distinctly smaller (1).

156. Proximodorsal "lip" on some manual unguals—a transverse ridge immediately dorsal to the articulating surface—absent (0) or present (1). In *V. mongoliensis* and *Deinonychus antirrhopus* a lip is present, contrary to previous contentions.

Pelvic Girdle and Hindlimb

157. Ventral edge of anterior ala of ilium straight or gently curved (0) or ventral edge hooked anteriorly (1) or very strongly hooked (2). Ordered.

158. Preacetabular part of ilium roughly as long as postacetabular part of ilium (0) or preacetabular portion of ilium markedly longer (more than two-thirds of total ilium length) than postacetabular part (1).

159. Anterior end of ilium gently rounded or straight (0) or anterior end strongly curved (1) or pointed at anterodorsal corner (2). Ordered.

160. Supraacetabular crest on ilium as a separate process from antitrochanter, forms "hood" over femoral head present (0) reduced, not forming hood (1) or absent (2).

161. Postacetabular ala of ilium in lateral view squared (0) or acuminate (1).

162. Postacetabular blades of ilia in dorsal view parallel (0) or diverge posteriorly (1).

163. Tuber along dorsal edge of ilium, dorsal or slightly posterior to acetabulum absent (0) or present (1).

164. Brevis fossa shelflike (0) or deeply concave with lateral overhang (1).

165. Antitrochanter posterior to acetabulum absent or poorly developed (0) or prominent (1).

166. Cuppedicus fossa formed as antiliac shelf anterior to acetabulum, extends posteriorly to above anterior end of acetabulum (0) or posterior end of fossa on anterior end of pubic peduncle, anterior to acetabulum (1).

167. Cuppedicus fossa deep, ventrally concave (0) or fossa shallow or flat, with no lateral overhang (1) or absent (2).

168. Posterior edge of ischium straight (0) or with median posterior process (1).

169. Ischium straight (0) or ventrodistally curved anteriorly (1) or twisted at midshaft and with flexure of obturator process toward midline so that distal end is horizontal (2) or with laterally concave curvature in anterior view (3).

170. Obturator process of ischium absent (0) or proximal in position (1) or located near middle of ischiadic shaft (2) or located at distal end of ischium (3).

171. Obturator process does not contact pubis (0) or contacts pubis (1).

172. Obturator notch present (0) or notch or foramen absent (1).

173. Semicircular scar on posterior part of the proximal end of the ischium, absent (0) or present (1).

174. Ischium more than two-thirds (0) or two-thirds or less of pubis length (1).

175. Distal ends of ischia form symphysis (0) or approach one another but do not form symphysis (1) or widely separated (2). Ordered.

176. Ischial boot (expanded distal end) present (0) or absent (1).

177. Tubercle on anterior edge of ischium absent (0) or present (1).

178. Pubis propubic (0) or pubis vertical (1) or pubis moderately posteriorly oriented (2) or pubis fully posteriorly oriented (opisthopubic) (3). The oviraptorid condition, in which the proximal end of the pubis is vertical and the distal end curves anteriorly, is considered to be state 1. Ordered.

179. Pubic boot projects anteriorly and posteriorly (0) or with little or no anterior process (1) or no anteroposterior projections (2).

180. Shelf on pubic shaft proximal to symphysis ("pubic apron") extends medially from middle of cylindrical pubic shaft (0) or shelf extends medially from anterior edge of anteroposteriorly flattened shaft (1).

181. Pubic shaft straight (0) or distal end curves anteriorly, anterior surface of shaft concave (1).

182. Pubic apron about half of pubic shaft length (0) or less than one-third of shaft length (1).

183. Femoral head without fovea capitalis (for attachment of capital ligament) (0) or circular fovea present in center of medial surface of head (1).

184. Lesser trochanter separated from greater trochanter by deep cleft (0) or trochanters separated by small groove (1) or completely fused (or absent) to form crista trochanteris (2).

185. Lesser trochanter of femur alariform (0) or cylindrical in cross section (1).

186. Posterior trochanter absent or represented only by rugose area (0) or posterior trochanter distinctly raised from shaft, moundlike (1). Cited by Gauthier (1986) as synapomorphy of Coelurosauria (his character 64), but he termed it the greater trochanter, which he equated with the posterior trochanter. Ostrom (1969) identifies the posterior and greater trochanter as separate structures, and we follow his terminology.

187. Fourth trochanter on femur present (0) or absent (1).

188. Accessory trochanteric crest distal to lesser trochanter absent (0) or present (1). This character was identified as an autapomorphy of *M. celer* (Makovicky and Sues, 1998), but it is more widespread.

189. Anterior surface of femur proximal to medial distal condyle without longitudinal crest (0) or crest present extending proximally from medial condyle on anterior surface of shaft (1).

190. Popliteal fossa on distal end of femur open distally (0) or closed off distally by contact between distal condyles (1).

191. Fibula reaches proximal tarsals (0) or short, tapering distally, and not in contact with proximal tarsals (1).

192. Medial surface of proximal end of fibula concave along long axis (0) or flat (1).

193. Deep oval fossa on medial surface of fibula near proximal end absent (0) or present (1).

194. Distal end of tibia and astragalus without distinct condyles (0) or with distinct condyles separated by prominent tendinal groove on anterior surface (1).

195. Medial cnemial crest absent (0) or present on proximal end of tibia (1).

196. Ascending process of the astragalus tall and broad, covering most of anterior surface of distal end of tibia (0) or process short and slender, covering only lateral half of anterior surface of tibia (1) or ascending process tall with medial notch that restricts it to lateral side of anterior face of distal tibia (2).

197. Ascending process of astragalus confluent with condylar portion (0) or separated by transverse groove or fossa across base (1).

198. Astragalus and calcaneum separate from tibia (0) or fused to each other and to the tibia in late ontogeny (1).

199. Distal tarsals separate, not fused to metatarsals (0) or form metatarsal cap with intercondylar prominence that fuses to metatarsal early in postnatal ontogeny (1).

200. Metatarsals not co-ossified (0) or co-ossification of metatarsals begins proximally (1) or distally (2).

201. Distal end of metatarsal II smooth, not ginglymoid (0) or with developed ginglymus (1).

202. Distal end of metatarsal III smooth, not ginglymoid (0) or with developed ginglymus (1).

203. Proximal surface of metatarsal IV subequal to II in size, proximal end of metatarsal III visible between metatarsals II and IV in anterior view (0) or metatarsal III pinched between metatarsals II and IV, the latter two contacting one another proximally in front of III (1) or metatarsal III does not reach proximal end of metatarsus (2). Ordered.

204. Ungual and penultimate phalanx of pedal digit II similar to those of III (0) or penultimate phalanx highly modified for extreme hyper-extension, ungual more strongly curved and about 50% larger than that of III (1).

205. Metatarsal I articulates in the middle of the medial surface of metatarsal II (0) or metatarsal I attaches to posterior surface of distal quarter of metatarsal II (1) or metatarsal I articulates to medial surface of metatarsal II near its proximal end (2) or metatarsal I absent (3).

206. Metatarsal I attenuates proximally, without proximal articulating surface (0) or proximal end of metatarsal I similar to that of metatarsals II–IV (1).

207. Shaft of metatarsal IV round or thicker dorsoventrally than wide in cross section (0) or shaft of metatarsal IV mediolaterally widened and flat in cross section (1).

208. Foot symmetrical (0) or asymmetrical with slender metatarsal II and very robust metatarsal IV (1).

Taxon	1–5	6–10	11–15	16–20	21–25	26–30	31–35	36–40	41–45	46–50
Allosaurus fragilis	?1100	0?100	00001	?0011	00010	00100	00001	11011	0010-	?0000
Sinraptor dongi	?1100	0??00	?0001	00010	00000	0010?	00000	01011	0010-	00?00
Ingenia yanshini	?01?0	?????	?????	?????	??1?1	?????	?1???	?????	?????	?????
Oviraptor mongoliensis	?01?0	????0	?????	?????	10111	?1???	?1?11	?0001	0??00	???00
O. philoceratops	?01?0	?????	?01?1	????0	1??1?	?111?	01011	???0?	??1?0	?11??
Conchoraptor gracilis	?0110	?????	?????	????0	10111	?1???	?1?11	?000?	?2100	0110?
Oviraptorid IGM 100/42	?0110	01001	001-?	22100	10101	111?0	01011	?0001	02100	01100
C. pergracilis	?????	1??01	?0110	???0?	???1?	110-0	1????	?????	?????	?????
Dromaeosaurus albertensis	?0??0	01100	00000	0010?	??0??	?0?01	???01	110??	??111	1????
D. antirrhopus	?0110	????1	?????	??1??	10000	?0111	00001	11000	11???	???0?
V. mongoliensis	?0110	01101	00001	20112	10000	10111	00001	11000	12111	?10?0
Mononykus olecranus	?????	?00??	?????	112??	?????	?????	?????	?????	?????	?????
S. deserti	?0110	10000	0000?	11201	01000	?0?0?	00-11	1-110	02100	00101
Patagonykus puertai	?????	?????	?????	?????	?????	?????	?????	?????	?????	?????
A. calvoi	?????	?????	?????	?????	?????	?????	?????	?????	?????	?????
O. hermanni	?0100	???0?	0?00?	1??0	1?010	?0111	00001	?1000	01?0?	?1000
M. celer	?????	?????	?????	?????	?????	?????	?????	?????	?????	?????
A. lithographica	101?0	100??	000??	11201	10010	??111	00?11	1?000	12100	?10?0
A. portentosus	?01?0	???10	011?0	0??0?	??1?1	?????	???1?	-?1??	???00	?11-?
C. zoui	00110	?????	?????	????0	0?111	??10?	00001	?10??	?2100	0???0
U. comahuensis	?????	?????	?????	?????	?????	?????	?????	?????	?????	?????
Confuciusornis sanctus	10110	?????	?????	????1	0-000	?00??	00001	???0?	?2??0	??1??
R. ostromi	?????	?????	?????	?????	?????	?????	?????	?????	?????	?????
Struthiomimus altus	?0101	0?010	??0??	10102	11000	10111	001-1	10000	10000	00000
Gallimimus bullatus	?0101	0?010	11010	10102	11001	10010	001-1	11020	10000	00000
Garudimimus brevipes	?0100	????0	1101?	????2	1?001	10101	1000?	?000?	10?00	?0?00
P. polyodon	?01??	?????	1????	????2	1?00?	???10	?0???	?????	?0??0	?????
Harpymimus okladnikovi	?0???	?????	?????	????2	??0??	?????	?????	?????	?????	?????
Troodon formosus	???1?	1102-	11010	00001	????0	?011?	10???	???20	22000	0210?
S. mongoliensis	?01??	1?0??	1101?	??0?1	01000	1?100	10???	?????	2????	?????
Byronosaurus jaffei	?????	100??	1101?	11001	01010	1011?	?0???	???20	220??	?????
Saurornithoides junior	?0110	1?02-	11010	0?001	0?000	??100	0000?	???20	22000	?21??
S. youngi	?0??0	1????	?????	????1	??000	??1??	?0???	?00??	?2???	????1
Segnosaurus galbinensis	?????	?????	?????	?????	?????	?????	?????	?????	?????	?????
E. andrewsi	?0110	??02-	0-1-0	??101	00011	100-?	10001	?1000	00100	00100
A. elesitaiensis	?????	?????	?????	?????	?????	?????	?????	?????	?????	?????
Tyrannosaurus rex	?1000	0?111	00001	00210	00000	10101	00000	00011	00001	02110
Albertosaurus libratus	?1000	0??0?	0????	???10	00000	101?1	00000	00011	0000-	?2110
Adasaurus mongoliensis	?????	?????	?????	?????	?????	?????	?????	?????	?????	?????
Utahraptor ostrommaysorum	?????	?????	?????	?????	1?0??	?????	?????	???00	1????	?????
Saurornitholestes langstoni	?????	?????	?????	?????	?????	?????	?????	?????	???11	1????
Achillobator giganticus	?????	?????	?????	?????	?????	?01?1	0????	?????	?????	?????
Dromaeosaurid IGM100/1015	?0110	01101	00001	20112	10000	10101	00001	?1000	1211?	11000
S. milleni	?011?	?????	0????	?????	??00?	??111	00???	?1000	11110	0?1??
S. changii	?0???	0012-	00001	11100	01010	1?111	10?01	1?0??	?2??0	011??
M. zhaoianus	0????	?????	?????	????1	0?0??	?????	?????	?????	?????	?????

Taxon	51–55	56–60	61–65	66–70	71–75	76–80	81–85	86–90	91–95	96–100
Allosaurus fragilis	00000	10000	10000	00000	01000	10000	00000	00101	?0100	10000
Sinraptor dongi	00000	00000	10000	?0000	0100?	10?00	00000	00101	10100	10000
Ingenia yanshini	?????	?????	?????	?2112	0-010	00?01	11?1-	---	?---?	?????
Oviraptor mongoliensis	?00??	????0	0??1?	12112	?–010	00??1	11-1-	---	---?	?????
O. philoceratops	??0?1	0????	0?11?	12112	0-01?	00?01	?1-1-	---	---?	?????
Conchoraptor gracilis	000??	???1?	0????	12112	0-010	?0?01	11-1-	---	---?	?????
Oviraptorid IGM 100/42	00011	00010	00110	12112	0-010	00101	11-1-	---	---1	01110
C. pergracilis	?????	10010	?????	?2112	0-000	00?01	???1-	---	---?	?????
Dromaeosaurus albertensis	01001	1?001	100??	00000	01001	11110	00?00	00101	1001?	?????
D. antirrhopus	?1???	?000?	11010	0000?	01001	11?10	00?01	10101	1000?	11000
V. mongoliensis	01001	00001	11010	00000	01001	111?0	00101	10101	10000	11?00
Mononykus olecranus	?????	??100	?????	?????	?????	?????	????2	2?--0	0????	????1
S. deserti	10001	0010?	-?1?1	00101	01000	00210	0?102	22--0	000??	01111
Patagonykus puertai	?????	?????	?????	?????	?????	?????	?????	?????	?1???	?????
A. calvoi	?????	?????	?????	?????	?????	?????	?????	?????	?????	?????
O. hermanni	00001	0?101	1????	00000	01000	?0?00	0010?	?0101	1001?	????0
M. celer	?????	?????	?????	?21?2	0?0??	?????	?????	?????	?????	011?0
A. lithographica	000?0	??100	111?1	?0000	01000	00??0	00002	20--0	00100	?1?1?
A. portentosus	?0001	00110	?????	?2?1?	??00?	0??01	11-?-	---	---?	01101
C. zoui	0?0??	?????	?????	?2112	0-0??	?????	?0?1-	---	---00	????0
U. comahuensis	?????	?????	?????	?????	?????	?????	?????	?????	?????	?????
Confuciusornis sanctus	100??	0?1??	?????	01001	0000?	10?00	01-1?	?-???	????0	?????
R. ostromi	?????	?????	?????	?????	?????	?????	?????	?????	?????	?????
Struthiomimus altus	001?0	??000	1?01?	?0001	00000	10200	01-1-	---	---0	01?10
Gallimimus bullatus	00100	00000	1?010	00000	00000	00200	01-1-	---	---0	01110
Garudimimus brevipes	?0???	??000	1??1?	0??0?	?0??0	?02?0	?1-1-	---	---?	?????
P. polyodon	?????	?????	?????	??00?	0000?	?????	?0002	22--0	0001?	?????
Harpymimus okladnikovi	?????	?????	?????	??1??	00?00	?????	????2	20--1	1-???	?????
Troodon formosus	00?1?	11100	????0	?10??	001??	?????	???01	11010	0100?	??111
S. mongoliensis	?????	?????	1?010	?100?	0010?	?1???	?0001	11010	01???	?????
Byronosaurus jaffei	?????	??100	?????	?0000	001??	11???	?0002	21--0	01?0?	0????
Saurornithoides junior	?0???	11100	?????	?100?	001??	?1???	?0001	11010	0100?	?????
S. youngi	0????	?????	?????	?00??	0010?	?????	?0001	110?0	01???	???11
Segnosaurus galbinensis	?????	?????	?????	??1??	10?00	00?00	0???0	0???0	01000	001??
E. andrewsi	?00??	00000	??111	12100	10000	00200	01-00	01000	001-?	?????
A. elesitaiensis	?????	?????	?????	?210?	100??	?????	?????	?10?0	001??	?????
Tyrannosaurus rex	00010	10000	1?000	00000	01001	10010	00000	00101	10110	10000
Albertosaurus libratus	?0010	10000	1????	00000	0100?	10010	00000	00101	10110	10000
Adasaurus mongoliensis	?????	?????	?????	?????	?????	?????	?0?0?	?????	?????	?????
Utahraptor ostrommaysorum	?????	?????	?????	?????	?????	?????	?0???	??101	1??1?	?????
Saurornitholestes langstoni	?????	?????	11???	?0???	?????	?????	????1	10101	1?00?	11000
Achillobator giganticus	?????	?????	?????	?????	?????	?????	???00	00101	1????	??0?0
Dromaeosaurid IGM100/1015	0100?	0?010	1?0??	00000	0100?	11110	00101	10101	1000?	???0?
S. milleni	0??01	?????	?????	?00?0	0100?	?1???	?0101	10??1	100??	?????
S. changii	0?11?	01110	?????	?000?	001??	????0	00001	1110?	?1???	??11?
M. zhaoianus	?????	?????	?????	??0??	010?0	?1???	?0?01	10000	00???	??001

Taxon	101–105	106–110	111–115	116–120	121–125	126–130	131–135	136–140	141–145	146–150
Allosaurus fragilis	00100	01000	00000	00000	10000	00?01	????0	00000	00000	00000
Sinraptor dongi	00010	01000	00000	00000	?????	?????	2?00?	0?0??	00??0	?????
Ingenia yanshini	?????	?????	??1??	?01-?	??2?0	0???0	01111	0011?	?0000	0??10
Oviraptor mongoliensis	?????	?????	?????	?0???	??210	?????	????1	??1??	?00?0	???10
O. philoceratops	?????	?????	?????	?0???	?????	?????	???11	??1??	???00	???1?
Conchoraptor gracilis	???01	0??1?	013?1	10??1	?????	0??1?	??1?1	00110	00010	?????
Oviraptorid IGM 100/42	11001	01111	001??	201-?	00210	0111?	01111	00110	10010	0?110
C. pergracilis	?1101	?11??	??1?1	2???0	?????	?????	?????	??01?	1????	?????
Dromaeosaurus albertensis	?????	?????	?????	?????	?????	?????	?????	?????	?????	?????
D. antirrhopus	1100?	11110	11???	?0011	011?1	11???	???1?	11011	10010	01110
V. mongoliensis	11001	11110	11111	00011	01111	11110	01110	11111	10010	01110
Mononykus olecranus	?1?11	10100	1????	?10?2	?????	?1???	2000?	00020	00301	10-11
S. deserti	11011	101?0	??1?2	01012	01221	0100?	2000?	00020	00301	1?-11
Patagonykus puertai	?????	111??	?????	?10?2	?????	?????	?????	?0020	0?301	10?11
A. calvoi	?100?	?????	????2	?10?2	?12??	?????	?????	0?0?0	0????	?????
O. hermanni	11?01	01100	00???	000?0	010??	1????	?????	?????	?01?0	0????
M. celer	?1100	01111	00???	0?1-1	002??	?????	?????	00?00	0?110	00???
A. lithographica	??00?	0??0?	0?0??	?0011	01221	1000?	1??10	11111	11100	0?110
A. portentosus	01101	01?0?	00??1	00???	?????	?????	?????	?????	?0100	?????
C. zoui	??00?	?????	??0??	?01-?	???20	?????	0???0	??0??	?0?0	???10
U. comahuensis	?????	1111?	11???	?0???	?????	?????	?????	1?0??	1001?	?????
Confuciusornis sanctus	?????	101??	0?2??	?0???	????2	-111?	21010	?-1-1	11400	?1111
R. ostromi	???01	?111?	??1?0	1?011	11221	1????	?????	?????	11??0	11???
Struthiomimus altus	11000	01000	00101	00000	00011	01001	?????	01120	00200	00000
Gallimimus bullatus	11000	01000	001?1	00000	00011	0100?	?????	01020	00200	000?0
Garudimimus brevipes	?????	??0?	??1??	0????	?????	?????	?????	?????	?????	?????
P. polyodon	?????	?????	?????	?????	?????	?????	0?00?	???2?	????0	0??00
Harpymimus okladnikovi	?????	?????	?????	?????	?????	?????	?????	?????	??2?0	0?000
Troodon formosus	11001	01101	111?1	000?1	020??	11???	?????	?????	1?010	????0
S. mongoliensis	?1???	0??0?	????1	00??1	?????	?????	?????	?????	?????	?????
Byronosaurus jaffei	???01	0110?	?????	??0??	02???	?????	?????	?????	?????	?????
Saurornithoides junior	?????	?????	??1?1	000?1	020?1	?????	?????	?????	?????	?????
S. youngi	??001	?????	?????	?0011	?1011	1101?	???1-	00?1?	101?0	???10
Segnosaurus galbinensis	?????	?????	?????	?0???	???????	?????	?????	??11?	?200?	?????
E. andrewsi	?????	?????	?????	?????	?????	?????	?????	?????	?????	??0??
A. elesitaiensis	???0?	01000	000?1	?001?	?02?0	?1???	?????	?????	?0000	0?100
Tyrannosaurus rex	0-000	01010	00000	10000	?0010	00??1	?????	00000	00100	00??0
Albertosaurus libratus	0-?00	01010	00000	10000	00010	000?1	00110	00000	00100	00000
Adasaurus mongoliensis	?????	?????	?????	?????	?????	?????	?????	?????	?????	?????
Utahraptor ostrommaysorum	11110	1????	?????	??0??	011??	?????	?????	?????	?????	?????
Saurornitholestes langstoni	11001	11110	11011	100?1	011?1	11???	1????	?????	?101?	?????
Achillobator giganticus	1100?	11110	?????	??0??	011?1	1????	?????	101??	?????	?????
Dromaeosaurid IGM100/1015	?1???	?????	?????	?????	?????	?????	?????	?????	?????	?????
S. milleni	?1??1	?????	??0??	?00?1	??1??	?????	0111?	11011	100?0	????0
S. changi	10100	01101	0?001	00011	0221?	1????	?????	11011	11???	???10
M. zhaoianus	?1??(01)	1?100	0?01?	??011	0?11?	1111?	01?10	1?1?1	11010	??100

Taxon	151–155	156–160	161–165	166–170	171–175	176–180	181–185	186–190	191–195	196–200	201–205	206–208
Allosaurus fragilis	?0000	01000	00010	00001	00000	00000	0000?	00110	00000	10000	00000	000
Sinraptor dongi	??000	01000	0?010	??001	00000	0000??00	00000	000?0	00000	10000	00000	000
Ingenia yanshini	00000	10002	1????	??032	01001	1010?	11011	01?00	00?00	010?0	00000	000
Oviraptor mongoliensis	??000	1100?	1????	11???	?????	?????	1????	????0	????0	??0?0	?????	??0
O. philoceratops	0??00	?00??	1??0?	?????	?????	?????	?????	????0	????0	?????	?????	??-
Conchoraptor gracilis	???00	11002	10001	-10??	?10??	?0100	1??11	010?0	?????	??0?0	000??	0?0
Oviraptorid IGM 100/42	01000	10002	11000	??032	010??	10101	11?21	01000	00000	0?000	00000	000
C. pergracilis	??00?	1?002	??00?	1?032	01011	10?01	?????	?????	???00	?100?	00100	000
Dromaeosaurus albertensis	?????	?????	?????	?????	?????	?????	?????	?????	?????	?????	???1?	??-
D. antirrhopus	01000	10022	11101	11022	0?011	11?01	00?11	11000	00000	01000	01000	010
V. mongoliensis	01000	10022	11101	11022	01011	11211	00111	1?000	00000	01010	11010	010
Mononykus olecranus	22011	0???1	????1	??000	-1???	??2??	0-?21	01001	11011	21100	00200	000
S. deserti	22011	000?1	−021	−2000	−1002	1022-	0-?21	01001	11011	21100	00200	000
Patagonykus puertai	0??11	0????	????1	?????	?1???	??10?	00?11	?1?00	???00	011?0	??0??	??-
A. calvoi	0????	??001	1-?0?	?2???	?????	?????	???1?	00???	0??0?	?1000	0000?	??0
O. hermanni	?????	?1000	1??1?	1?011	00000	101??	0????	????0	?????	???00	?00??	??0
M. celer	???00	10002	???00	10???	?????	???01	10?01	01100	01100	010??	?????	0?-
A. lithographica	01000	00112	11100	1?103	01012	-021?	01?11	11000	00?00	0?011	00001	0?0
A. portentosus	?????	?00?2	11?01	??032	01010	?010?	0?000	100?0	???00	01111	0010?	??0
C. zoui	01000	?100?	1????	??0?2	01???	1????	10?11	?????	0????	01000	??0?0	0?0
U. comahuensis	?????	?0122	1?10?	111?2	0101?	??211	00?1?	01??0	?10??	11??1	01001	??-
Confuciusornis sanctus	11000	00112	1?1??	-?1?0	?1?11	1032?	0112?	?10??	11??1	20111	0?100	0?0
R. ostromi	????0	?0112	1?10?	-?103	?1012	?02??	01?21	??000	0??0?	?1000	1101?	010
Struthiomimus altus	−2010	01000	00011	00011	00100	00000	00000	00110	00100	01000	00103	-00
Gallimimus bullatus	−2010	01000	00011	00011	00100	00001	00000	00110	00100	01000	00103	-00
Garudimimus brevipes	?????	???0?	0???0	?????	?????	?????	?????	????0	?????	???00	0000?	0?0
P. polyodon	−2010	?????	?????	?????	?????	?????	?????	?????	?????	?????	?????	??-
Harpymimus okladnikovi	00010	?10??	0????	?????	?????	?????	?????	?????	?????	???00	0000?	??0
Troodon formosus	?0000	10???	?????	??032	?10?1	1000?	0?011	11000	???00	01000	0011?	?01
S. mongoliensis	?????	?????	?????	??032	010?0	10??1	0??11	110??	?????	????0	??110	?01
Byronosaurus jaffei	?????	?????	?????	?????	?????	?????	????1	????0	?0??0	?????	???1?	???
Saurornithoides junior	?????	?????	?????	?????	?????	?????	?????	?????	?????	01???	?????	???
S. youngi	01000	????2	1??01	??0?3	0??1?	1?0??	00??1	110??	???0?	????0	00?10	001
Segnosaurus galbinensis	??010	?2002	1?000	0?102	11?0?	1020?	0????	????0	0??0?	00??0	00??2	1?0
E. andrewsi	?????	?????	?????	?????	?????	?????	?????	?????	?????	???0?	00?0?	??-
A. elesitaiensis	−0000	12002	1?00?	11002	11?02	1?2??	???01	????0	0????	????0	00002	100
Tyrannosaurus rex	-?100	01001	00010	00001	01101	10000	00?01	00100	00?00	01000	00100	000
Albertosaurus libratus	−0100	01001	00010	00001	01100	10000	00001	00000	00100	01000	00100	000
Adasaurus mongoliensis	?????	?1?22	111?1	1?0?2	01?1?	1?21?	0????	?????	????0	0?0??	???1?	??0
Utahraptor ostrmmaysorum	???0?	0????	?????	?????	?????	??10?	???0?	?????	??00?	0????	?????	??-
Saurornitholestes langstoni	??000	1????	?????	?????	?????	?????	?????	?????	?????	?????	?????	??0
Achillobator giganticus	??00?	?1022	1??11	11001	01?01	1010?	00?21	110?0	????0	0?000	?101?	??-
Dromaeosaurid IGM100/1015	?????	?????	?????	?????	?????	?????	?????	?????	?????	?????	?????	??-
S. milleni	?1000	?00?2	11?01	?11?3	?1?11	1021?	0????	1????	?????	11000	1?110	0?0
S. changi	01?0?	????2	11?01	111?3	01?11	10211	00?11	11000	0??11	01000	0101?	?11
M. zhaoianus	11001	11012	111?0	11103	?1?11	10(12)11	00?11	111??	00?00	01110	11111	0?1

Acknowledgments

We thank L. M. Chiappe and L. M. Witmer for inviting us to write this chapter and for help along the way. H.-D. Sues, Xu Xing, T. R. Holtz Jr., C. A. Forster, and M. Kearney provided useful comments and access to unpublished manuscripts and specimens. This research was supported by the National Science Foundation (DEB 9407999) and the George Washington University Facilitating Fund.

Literature Cited

Bakker, R. T., M. Williams, and P. Currie. 1988. *Nanotyrannus,* a new genus of pygmy tyrannosaur, from the latest Cretaceous of Montana. Hunteria 1:1–30.

Barsbold R. 1979. Opisthopubic pelvis in the carnivorous dinosaurs. Nature 279:792–793.

Barsbold R. and T. Maryanska. 1990. Segnosauria; pp. 408–415 *in* D. B. Weishampel, P. Dodson, and H. Osmólska (eds.), The Dinosauria. University of California Press, Berkeley.

Barsbold R. and H. Osmólska. 1999. The skull of *Velociraptor* (Theropoda) from the Late Cretaceous of Mongolia. Acta Palaeontologica Polonica 44(2):189–219.

Barsbold R., T. Maryanska, and H. Osmólska. 1990. Oviraptorosauria; pp. 249–258 *in* D. B. Weishampel, P. Dodson, and H. Osmólska (eds.), The Dinosauria. University of California Press, Berkeley.

Barthel, K. W., N. H. M. Swinburne, and S. Conway Morris. 1990. Solnhofen, a Study in Mesozoic Palaeontology. Cambridge University Press, New York. 236 pp.

Benton, M. J. 1994. Late Triassic to Middle Jurassic extinctions among continental tetrapods: testing the pattern; pp. 366–397 *in* N. Fraser and H.-D. Sues (eds.), In the Shadow of the Dinosaurs: Early Mesozoic Tetrapods. Cambridge University Press, New York.

Benton, M. J., and J. M. Clark. 1988. Archosaur phylogeny and the relationships of the Crocodylia; pp. 295–338 *in* M. J. Benton (ed.), The Phylogeny and Classification of the Tetrapods, Volume 1: Amphibians, Reptiles, and Birds. Systematics Association Special Vol. 35A. Clarendon Press, Oxford.

Bidar, A., L. Demay, and G. Thomel. 1972. *Compsognathus corallestris,* nouvelle espèce de dinosaurien théropode du Portlandian de Canjuers (Sud-Est de la France). Musee d'Histoire Naturelle Nice, Annales 1:1–34.

Blatt, H., and R. L. Jones. 1975. Proportions of exposed igneous, metamorphic and sedimentary rocks. Geological Society of America Bulletin 86:1085–1088.

Blatt, H., W. B. N. Berry, and S. Brande. 1991. Principles of Stratigraphic Analysis. Blackwell, Boston, 512 pp.

Brady, R. H. 1985. On the independence of systematics. Cladistics 1:113–126.

Brochu, C. A., and M. A. Norell. 2000. Temporal congruence and the origin of birds. Journal of Vertebrate Paleontology 20:197–200.

Brochu, C., and M. A. Norell. 2001. Time and trees: a quantitative assessment of temporal congruence in the bird origins debate; pp. 512–535 *in* J. Gauthier (ed.), New Perspectives on the Origin and Early Evolution of Birds. Peabody Museum of Natural History, Yale University, New Haven.

Buffetaut, E., V. Suteethorn, and H. Tong. 1996. The earliest known tyrannosaur from the Lower Cretaceous of Thailand. Nature 381:689–691.

Calzavara, M., G. Muscio, and R. Wild. 1980. *Megalancosaurus preonensis* n.g., n. sp., a new reptile from the Norian of Friuli. Gortania, Atti del Museo Friulano di Storia Naturale 2: 49–54.

Carr, T. D. 1999. Craniofacial ontogeny in Tyrannosauridae (Dinosauria, Coelurosauria). Journal of Vertebrate Paleontology 19:497–520.

Chatterjee, S. 1991. Cranial anatomy and relationships of a new Triassic bird from Texas. Philosophical Transactions of the Royal Society, London (B) 309:395–460.

———. 1993. *Shuvosaurus,* a new theropod. National Geographic Research and Exploration 9:274–285.

———. 1997. The Rise of Birds. Johns Hopkins University Press, Baltimore, 312 pp.

Chen P. J., Dong Z.-M., and Zhen S.-N. 1998. An exceptionally well-preserved theropod dinosaur from the Yixian Formation of China. Nature 391:147–152.

Chiappe, L. 1992. Enantiornithine (Aves) tarsometatarsi and the avian affinities of the Late Cretaceous Avisauridae. Journal of Vertebrate Paleontology 12:344–350.

———. 1995. The first 85 million years of avian evolution. Nature 378:349–355.

———. 1997. Climbing *Archaeopteryx?* A response to Yalden. *Archaeopteryx* 15:109–112.

———. 1998. Review of: The Rise of Birds, by S. Chatterjee. American Zoologist 38(4):797–798.

Chiappe, L., Ji S.-A., Ji Q., and M. Norell. 1999. Anatomy and systematics of the Confuciusornithidae (Aves) from the Mesozoic of northeastern China. Bulletin of the American Museum of Natural History 242, 89 pp.

Chure, D. J. 1994. *Koparion douglassi,* a new dinosaur from the Morrison Formation (Upper Jurassic) of Dinosaur National Monument; the oldest troodontid (Theropoda: Maniraptora). Brigham Young University Studies in Geology 40: 11–15.

Clark, J. M., M. Montellano, J. Hopson, R. Hernández, and D. Fastovsky. 1994a. An Early or Middle Jurassic tetrapod assemblage from the La Boca Formation, northeastern Mexico; pp. 295–302 *in* N. Fraser and H.-D. Sues (eds.), In the Shadow of the Dinosaurs: Early Mesozoic Tetrapods. Cambridge University Press, New York.

Clark, J. M., Perle A., and M. Norell. 1994b. The skull of *Erlicosaurus* [*sic*] andrewsi, a Late Cretaceous "segnosaur" (Theropoda: Therizinosauridae) from Mongolia. American Museum Novitates 3113:1–39.

Clark, J. M., M. Norell, and L. Chiappe. 1999. An oviraptorid skeleton from the Late Cretaceous of Ukhaa Tolgod, Mongolia, preserved in an avian-like brooding position over an oviraptorid nest. American Museum Novitates 3265:1–36.

Colbert, E. H., and D. A. Russell. 1969. The small Cretaceous dinosaur *Dromaeosaurus.* American Museum Novitates 1900: 1–20.

Cruz, C. E., P. Condat, E. Kozlowski, and R. Manceda. 1989. Análisis estratigráfico sequencial del Grupo Neuquén (Cretácico superior) en el valle del Rio Grande, Provincia de Mendoza. I Congreso Argentina Hidrocarbon 2:689–714.

Currie, P. J. 1987. Bird-like characteristics of the jaws and teeth of troodontid theropods (Dinosauria, Saurischia). Journal of Vertebrate Paleontology 7:72–81.

———. 1995. New information on the anatomy and relationships of *Dromaeosaurus albertensis* (Dinosauria: Theropoda). Journal of Vertebrate Paleontology 15:576–591.

Currie, P. J., and D. Russell. 1988. Osteology and relationships of *Chirostenotes pergracilis* (Saurischia, Theropoda) from the Judith River Formation of Alberta, Canada. Canadian Journal of Earth Sciences 25:972–986.

Currie, P. J., and Zhao X.-J. 1994. A new troodontid (Dinosauria, Theropoda) braincase from the Dinosaur Park Formation (Campania) of Alberta. Canadian Journal of Earth Sciences 30:2231–2247.

Currie, P. J., S. J. Godfrey, and L. Nessov. 1994. New caenagnathid (Dinosauria: Theropoda) specimens from the Upper Cretaceous of North America and Asia. Canadian Journal of Earth Sciences 30:2255–2272.

Currie, P. J., M. Norell, Ji Q., and Ji S.-A. 1998. The anatomy of two feathered theropods from Liaoning, China. Journal of Vertebrate Paleontology 18(Supplement to 3):36A.

Dal Sasso, C., and M. Signore. 1998. Exceptional soft-tissue preservation in a theropod dinosaur from Italy. Nature 392:383–387.

de Pinna, M. C. C. 1991. Concepts and tests of homology in the cladistic paradigm. Cladistics 7:367–394.

De Queiroz, K., and J. A. Gauthier. 1994. Toward a phylogenetic system of biological nomenclature. Trends in Ecology and Evolution 9(1):27–31.

Dong Z. M. and P. J. Currie. 1996. On the discovery of an oviraptorid skeleton on a nest of eggs at Bayan Mandahu, Inner Mongolia, People's Republic of China. Canadian Journal of Earth Sciences 33(4):631–636.

Eberth, D. A. 1994. Depositional environments and facies transitions of dinosaur-bearing Upper Cretaceous redbeds at Bayan Mandahu (Inner Mongolia, People's Republic of China). Canadian Journal of Earth Sciences 30(10–11):2196–2213.

Eberth, D. A., D. A. Russell, D. R. Braman, and A. L. Deino. 1994. The age of the dinosaur-bearing sediments at Tebch, Inner Mongolia, People's Republic of China. Canadian Journal of Earth Sciences 30(10–11):2101–2106.

Eernisse, D. J., and A. G. Kluge. 1993. Taxonomic congruence versus total evidence, and amniote phylogeny inferred from fossils, molecules, and morphology. Molecular Biology and Evolution 10:1170–1195.

Elzanowski, A., and P. Wellnhofer. 1992. A new link between theropods and birds from the Cretaceous of Mongolia. Nature 359:821–823.

———. 1993. Skull of *Archaeornithoides* from the Upper Cretaceous of Mongolia. American Journal of Science 293:A235–252.

———. 1996. Cranial morphology of *Archaeopteryx*: evidence from the seventh skeleton. Journal of Vertebrate Paleontology 16:81–94.

Evans, S. E., and A. R. Milner. 1994. Middle Jurassic microvertebrate assemblages from the British Isles; pp. 303–321 *in* N. C. Fraser and H.-D. Sues (eds.), In the Shadow of the Dinosaurs: Early Mesozoic Tetrapods. Cambridge University Press, New York.

Farlow, J. O., D. L. Brinkman, W. L. Abler, and P. J. Currie. 1991. Size, shape, and serration density of theropod dinosaur lateral teeth. Modern Geology 16:161–198.

Farris, J. S. 1970. Methods for computing Wagner trees. Systematic Zoology 19:83–92.

———. 1981. Distance data in phylogenetic analysis; pp. 3–23 *in* V. A. Funk and D. R. Brooks (eds.), Advances in Cladistics: Proceedings of the First Meeting of the Willi Hennig Society. New York Botanical Garden, New York.

———. 1983. The logical basis of phylogenetic analysis; pp. 7–36 *in* N. I. Platnick and V. A. Funk (eds.), Advances in Cladistics, Volume 2. Columbia University Press, New York.

———. 1990. Phenetics in camouflage. Cladistics 6:91–100.

Fastovsky, D. E., Badamgarav D., H. Ishimoto, M. Watabe, and D. Weishampel. 1997. The paleoenvironments of Tugrikin-Shireh (Gobi Desert, Mongolia) and aspects of the taphonomy and paleoecology of *Protoceratops* (Dinosauria; Ornithischia). Palaios 12:59–70.

Feduccia, A. 1996. The Origin and Evolution of Birds. Yale University Press, New Haven, 420 pp.

Feduccia, A., and R. Wild. 1993. Birdlike characters in the Triassic archosaur *Megalancosaurus*. Naturwissenschaften 80:564–566.

Forster, C. A., S. D. Sampson, L. M. Chiappe, and D. Krause. 1998a. The theropod ancestry of birds: New evidence from the Late Cretaceous of Madagascar. Science 279:1915–1919.

———. 1998b. Genus correction. Science 280:185.

Fraser, N., and H.-D. Sues (eds.). 1994. In the Shadow of the Dinosaurs: Early Mesozoic Tetrapods. Cambridge University Press, New York, 435 pp.

Gaffney, E. S. 1979. An introduction to the logic of phylogeny reconstruction; pp. 79–112 *in* J. Cracraft and N. Eldredge (eds.), Phylogenetic Analysis and Paleontology. Columbia University Press, New York.

Gauthier, J. A. 1986. Saurischian monophyly and the origin of birds; pp. 1–56 *in* K. Padian (ed.), The Origin of Birds and the Evolution of Flight. California Academy of Sciences, San Francisco.

———. 1994. The diversification of the amniotes; pp. 129–159 *in* Major Features of Vertebrate Evolution. Short Courses in Paleontology 7, Paleontological Society, Knoxville, Tenn.

Gilluly, J. 1949. Distribution of mountain building in geologic time. Geological Society of America Bulletin 60:561–590.

Goloboff, P. 1999. NONA (version 1.9). Published by the author, Fundación e Instituto Miguel Lillo, Tucuman, Argentina.

Gradstein, F. M., F. Agterberg, J. Ogg, J. Hardenbol, P. van Veen, J. Thierry, and Z. Huang. 1995. A Triassic, Jurassic and Cretaceous time scale; pp. 95–128 *in* W. A. Berggren, D. Kent, M.-P. Aubry, and J. Hardenbol (eds.), Geochronology, Time Scales and Global Stratigraphic Correlation, SEPM (Society for Sedimentary Geology), Tulsa, Okla.

Hawkins, J. A., C. E. Hughes, and R. W. Scotland. 1997. Primary homology assessment, characters and character states. Cladistics 13:275–283.

Hedges, S. B., and L. L. Poling. 1999 A molecular phylogeny of reptiles. Science 283:998–1001.

Heilmann, G. 1926. The Origin of Birds. Witherby, London, 208 pp.

Hennig, W. 1966. Phylogenetic Systematics. University of Illinois Press, Urbana, 263 pp.

Holtz, T. R., Jr. 1994. The phylogenetic position of the Tyrannosauridae: implications for theropod systematics. Journal of Paleontology 68:1100–1117.

———. 1996. Phylogenetic taxonomy of the Coelurosauria (Dinosauria: Theropoda). Journal of Paleontology 70:536–538.

Horner, J. R., and D. B. Weishampel. 1996. Correction: a comparative embryological study of two ornithischian dinosaurs. Nature 383:103.

Hwang, S., M. Norell, Ji Q., and K. Gao. In press. New specimens of *Microraptor zhaoianus*. American Museum Novitates.

Jensen, J., and K. Padian. 1989. Small pterosaurs and dinosaurs from the Uncompahgre fauna (Brushy Basin Member, Morrison Formation: ?Tithonian), Late Jurassic, western Colorado. Journal of Paleontology 63:364–373.

Jerzykiewicz, T., and D. Russell. 1991. Late Mesozoic stratigraphy and vertebrates of the Gobi Basin. Cretaceous Research 12:345–377.

Ji Q. and Ji S.-A. 1997. Protarchaeopterygid bird (*Protarchaeopteryx* gen. nov.)—fossil remains of archaeopterygids from China. Chinese Geology 238:38–41.

Ji Q., P. J. Currie, M. Norell, and Ji S.-A. 1998. Two feathered theropods from the Upper Jurassic/Lower Cretaceous strata of northeastern China. Nature 393:753–761.

Ji Q., M. A. Norell, Gao K., Ji S., and Ren D. 2001. The distribution of integumentary structures in a feathered dinosaur. Nature 410:1084–1088.

Kirkland, J., R. Gaston, and D. Burge. 1993. A large dromaeosaur (Theropoda) from the Lower Cretaceous of eastern Utah. Hunteria 2(10):1–16.

Kirkland, J. I., and D. G. Wolfe. 2001. First definitive therizinosaurid (Dinosauria: Theropoda) from North America. Journal of Vertebrate Paleontology 21:410–414.

Kluge, A. G. 1997. Testability and the refutation and corroboration of cladistic hypotheses. Cladistics 13:81–96.

Kowallis, B. K., E. H. Christiansen, A. L. Deino, F. Peterson, C. E. Turner, M. J. Kunk, and J. D. Obradovich. 1998. The age of the Morrison Formation; in K. Carpenter, D. Chure, and J. Kirkland (eds.), The Morrison Formation: An Interdisciplinary Study. Modern Geology 23:235–260.

Kurzanov, S. M. 1981. On unusual theropods from Upper Cretaceous of Mongolia. Transactions of the Soviet-Mongolian Paleontological Expeditions 24:39–50. [Russian]

———. 1987. Avimimids and the problem of the origin of birds. Transactions of the Soviet-Mongolian Paleontological Expeditions 31:5–95. [Russian]

Loope, D. B., L. Dingus, C. C. Swisher III, and Minjin C. 1998. Life and death in a Late Cretaceous dunefield, Nemegt Basin, Mongolia. Geology 26:27–30.

Makovicky, P. 1995. Phylogenetic aspects of the vertebral morphology of Coelurosauria (Dinosauria: Theropoda). Unpublished master's thesis, Copenhagen University. 230 pp.

———. 1998. A new small theropod from the Morrison Formation of Como Bluff, Wyoming. Journal of Vertebrate Paleontology 17:755–757.

Makovicky, P., and H.-D. Sues. 1998. Anatomy and phylogenetic relationships of the theropod dinosaur *Microvenator celer* from the Lower Cretaceous of Montana. American Museum Novitates 3240:1–27.

Maleev, E. A. 1954. New turtle-like reptile in Mongolia. Nature (Moscow) 1954:106–108.

Marsh, O. C. 1879. Notice of new Jurassic reptiles. American Journal of Science (Series 3) 18:501–505.

———. 1884. Principle characters of American Jurassic dinosaurs. Part VIII. The Order Theropoda. American Journal of Science (Series 3) 27:329–340.

Martin, L. 1983. The origin of birds and of avian flight. Current Ornithology 1:105–129.

Maryanska, T., and H. Osmólska. 1997. The quadrate of oviraptorid dinosaurs. Acta Palaeontologica Polonica 42:377–387.

Mayr, E., and P. D. Ashlock. 1991. Principles of Systematic Zoology, 2nd edition. McGraw-Hill, New York, 475 pp.

Metcalf, S. J., and R. J. Walker. 1994. A new Bathonian microvertebrate locality in the English Midlands; pp. 322–332 *in* N. C. Fraser and H.-D. Sues (eds.), In the Shadow of the Dinosaurs: Early Mesozoic Tetrapods. Cambridge University Press, New York.

Molnar, R., S. Kurzanov, and Dong Z.-M. 1990. Carnosauria; pp. 169–209 *in* D. B. Weishampel, P. Dodson, and H. Osmólska (eds.), The Dinosauria. University of California Press, Berkeley.

Munter, R. 1999. Two new theropod dinosaurs from Huizachal Canyon, Mexico. Unpublished master's thesis, George Washington University, Washington, D.C.

Naef, A. 1931. Phylogenie der Tiere. Handbuch der Vererbungswissenschaften III, 1 (Liefg. 13). Borntraeger, Berlin, 200 pp.

Nixon, K., and J. Carpenter. 1993. On outgroups. Cladistics 9:413–426.

Norell, M. A. 1993. Tree-based approaches to understanding history: comments on ranks, rules and the quality of the fossil record. American Journal of Science 293-A:407–417.

———. 1996. Ghost taxa, ancestors, and assumptions: a comment on Wagner. Paleobiology 22:453–455.

Norell, M. A., and P. Makovicky. 1997. Important features of the dromaeosaur skeleton: information from a new specimen. American Museum Novitates 3215:1–28.

———. 1999. Important features of the dromaeosaurid skeleton II: information from newly collected specimens of *Velociraptor mongoliensis*. American Museum Novitates 3282:1–45.

Norell, M. A., J. M. Clark, and Perle A. 1992. New dromaeosaur material from the Late Cretaceous of Mongolia. Journal of Vertebrate Paleontology 12(3):44A.

Norell, M. A., J. M. Clark, Dashzeveg D., Barsbold R., L. M. Chiappe, A. R. Davidson, M. C. McKenna, and M. J. Novacek. 1994. A theropod dinosaur embryo, and the affinities of the Flaming Cliffs dinosaur eggs. Science 266:779–782.

Norell, M. A., J. M. Clark, L. M. Chiappe, and Dashzeveg D. 1995. A nesting dinosaur. Nature 378:774–776.

Norell, M. A., P. Makovicky, and J. M. Clark. 1997. A *Velociraptor* wishbone. Nature 389:447.

———. 2000. A new troodontid theropod from Ukhaa Tolgod, Mongolia. Journal of Vertebrate Paleontology 20:7–11.

Norell, M. A., J. M. Clark, and P. Makovicky. 2001. Phylogenetic relationships among coelurosaurian theropods; pp. 49–67 *in* J. Gauthier (ed.), New Perspectives on the Origin and Early Evolution of Birds. Peabody Museum of Natural History, Yale University, New Haven.

Norman, D. B. 1990. Problematic Theropoda: "Coelurosaurs"; pp. 280–305 *in* D. B. Weishampel, P. Dodson, and H. Osmólska (eds.), The Dinosauria. University of California Press, Berkeley.

Novas, F. 1992. La evolución de los dinosaurios carnivoros; pp. 125–163 *in* J. L. Sanz and A. D. Buscalioni (eds.), Los Dinosaurios y su Entorno Biotico. Actas del Segundo Curso de Paleontologia en Cuenca. Unidad de Paleontologia, Universidad Autonoma de Madrid, Spain.

Novas, F. 1996. Alvarezsauridae, Cretaceous basal birds from Patagonia and Mongolia. Memoirs of the Queensland Museum 39:675–702.

Novas, F., and P. F. Puerta. 1997. New evidence concerning avian origins from the Late Cretaceous of Patagonia. Nature 387:390–392.

Osborn, H. F. 1903. *Ornitholestes hermanni,* a new compsognathoid dinosaur from the Jurassic of Wyoming. American Museum of Natural History Bulletin 12:161–172.

———. 1924. Three new Theropoda, *Protoceratops* zone, central Mongolia. American Museum Novitates 144:1–12.

Osmólska, H., and Barsbold R. 1990. Troodontidae; pp. 259–268 *in* D. B. Weishampel, P. Dodson, and H. Osmólska (eds.), The Dinosauria. University of California Press, Berkeley.

Ostrom, J. H. 1969. Osteology of *Deinonychus antirrhopus,* an unusual theropod from the Lower Cretaceous of Montana. Peabody Museum of Natural History Bulletin 30:1–165.

———. 1970. Stratigraphy and paleontology of the Cloverly Formation (Lower Cretaceous) of the Bighorn Basin area, Wyoming and Montana. Peabody Museum of Natural History Bulletin 35:1–234.

———. 1976. *Archaeopteryx* and the origin of birds. Biological Journal of the Linnean Society 8:91–182.

———. 1978. The osteology of *Compsognathus longipes.* Zitteliana 4:73–118.

———. 1980. *Coelurus* and *Ornitholestes:* are they the same?; pp. 245–256 *in* L. Jacobs (ed.), Aspects of Vertebrate Evolution. Museum of Northern Arizona, Flagstaff.

Padian, K., and L. Chiappe. 1998. The origin of birds and their flight. Scientific American 278:38–47.

Padian, K., J. R. Hutchinson, and T. R. Holtz Jr. 1999. Phylogenetic definitions and nomenclature of the major taxonomic categories of the carnivorous Dinosauria (Theropoda). Journal of Vertebrate Paleontology 19:69–80.

Patterson, C. 1982. Morphological characters and homology; pp. 21–74 *in* K. A. Joysey and A. E. Friday (eds.), Problems of Phylogenetic Reconstruction. Academic Press, New York.

———. 1993. Bird or dinosaur? Nature 365:21–22.

Patterson, C., M. Norell, J. Clark, and L. Chiappe. 1993. Scientific correspondence: naming names. Nature 366:518.

Pérez-Moreno, B. P., J. L. Sanz, A. D. Buscalioni, J. J. Moratalla, F. Ortega, and D. Rasskin-Gutman. 1994. A unique multi-toothed ornithomimosaur dinosaur from the Lower Cretaceous of Spain. Nature 370:363–367.

Platnick, N. I. 1977. Cladograms, phylogenetic trees, and hypothesis testing. Systematic Zoology 26:438–442.

Reig, O. 1963. La presencia de dinosaurios saurisquios en los "Estratos de Ischigualasto" (Mesotriasico superior) de las provincias de San Juan y La Rioja (Republica Argentina). Ameghiniana 3:3–20.

Renesto, S. 1994. *Megalancosaurus,* a possibly arboreal archosauromorph (Reptilia) from the Upper Triassic of northern Italy. Journal of Vertebrate Paleontology 14:38–52.

Rieppel, O. 1988. Fundamentals of Comparative Biology. Birkhauser, Boston, 250 pp.

———. 1994. Homology, topology, and typology: the history of modern debates; pp. 63–100 *in* B. K. Hall (ed.), Homology, the Hierarchical Basis of Comparative Biology. Academic Press, New York.

Rogers, R. R., C. C. Swisher, and P. C. Sereno. 1993. The Ischigualasto tetrapod assemblage (Late Triassic, Argentina) and 40Ar/39Ar dating of dinosaur origins. Science 260:794–797.

Rosen, D. E., P. L. Forey, B. G. Gardiner, and C. Patterson. 1981. Lungfishes, tetrapods, paleontology, and plesiomorphy. American Museum of Natural History Bulletin 167(4):163–275.

Ruben, J. A., C. Dal Sasso, N. R. Geist, W. J. Hillenius, T. D. Jones, and M. Signore. 1999. Pulmonary function and metabolic physiology of theropod dinosaurs. Science 283:514–516.

Russell, D. A., and Dong Z.-M. 1994a. The affinities of a new theropod from the Alxa Desert, Inner Mongolia, People's Republic of China. Canadian Journal of Earth Sciences 30(10–11):2107–2127.

———. 1994b. A nearly complete skeleton of a new troodontid dinosaur from the People's Republic of China. Canadian Journal of Earth Sciences 30(10–11):2163–2173.

Russell, D. A., and Zhao X.-J. 1996. New psittacosaur occurrences in Inner Mongolia. Canadian Journal of Earth Sciences 33(4):637–648.

Schoch, R. M. 1986. Phylogeny reconstruction in paleontology. Van Nostrand Reinhold, New York, 351 pp.

Sereno, P. C. 1991. Basal archosaurs: phylogenetic relationships and functional implications. Society of Vertebrate Paleontology Memoir 2, 53 pp.

———. 1997. The origin and evolution of dinosaurs. Annual Review of Earth and Planetary Sciences 25:435–489.

———. 1998. A rationale for phylogenetic definitions, with application to the higher-level taxonomy of Dinosauria. Neues Jahrbuch für Geologie und Paläontologie 210(1):41–83.

———. 1999. The evolution of dinosaurs. Science 284:2137–2147.

Sereno, P. C., and Rao C. 1992. Early evolution of avian flight and perching: new evidence from the Lower Cretaceous of China. Science 255:845–848.

Smith, A. B. 1994. Systematics and the Fossil Record. Blackwell Scientific Publications, Oxford. 223 pp.

Smith, D., and P. Galton. 1990. Osteology of *Archaeornithomimus asiaticus* (Upper Cretaceous, Iren Dabasu Formation, People's Republic of China). Journal of Vertebrate Paleontology 10(2):255–265.

Sues, H. D. 1997. On *Chirostenotes,* a Late Cretaceous oviraptorosaur (Dinosauria: Theropoda) from western North America. Journal of Vertebrate Paleontology 17(4):698–716.

Swisher, C. C., III, Wang Y., Wang X., Xu X., and Wang Y. 1999. Cretaceous age for the feathered dinosaurs of Liaoning, China. Nature 400:58–61.

Thulborn, R. A. 1984. The avian relationships of *Archaeopteryx,* and the origin of birds. Zoological Journal of the Linnean Society 82:119–158.

Tintori, A., G. Muscio, and S. Nardon. 1985. The Triassic fossil fishes localities in Italy. Rivista Italiana di Paleontologia e Stratigrafia 91:197–210.

Wagner, A. 1861. Neue Beiträge zur Kenntnis der urweltlichen Fauna des lithographischen Schiefers. V. *Compsognathus longipes* Wagn. Abhandlungen Bayerische Akademie der Wissenschaften 9:30–38.

Wagner, P. J. 1995. Stratigraphic tests of cladistic hypotheses. Paleobiology 21:153–178.

Walker, A. D. 1985. The braincase of *Archaeopteryx;* pp. 123–134 *in* M. K. Hecht, J. H. Ostrom, G. Viohl, and P. Wellnhofer (eds.), The Beginnings of Birds. Freunde des Jura-Museums Eichstätt, Willibaldsburg, Germany.

Webster, D. 1996. Dinosaurs of the Gobi. National Geographic 190(1):70–89.

Weishampel, D. B. 1990. Dinosaurian distribution; pp. 63–139 *in* D. B. Weishampel, P. Dodson, and H. Osmólska (eds.), The Dinosauria. University of California, Berkeley.

Weishampel, D. B., P. Dodson, and H. Osmólska (eds.). 1990. The Dinosauria. University of California, Berkeley, 733 pp.

Wilkinson, M. 1994. Common cladistic information and its consensus representation: reduced Adams and reduced cladistic consensus trees and profiles. Systematic Biology 43:343–368.

———. 1995. Coping with abundant missing entries in phylogenetic inference using parsimony. Systematic Biology 44: 501–514.

Witmer, L. M. 1990. The craniofacial air sac system of Mesozoic birds (Aves). Zoological Journal of the Linnean Society 100: 327–378.

———. 1991. Perspectives on avian origins; pp. 427–466 *in* H. P. Schultze and L. Trueb (eds.), Origins of the Higher Groups of Tetrapods. Cornell University Press, Ithaca, N.Y.

———. 2001. The role of *Protoavis* in the debate on avian origins; pp. 537–548 *in* J. A. Gauthier (ed.), New Perspectives on the Origin and Early Evolution of Birds; Peabody Museum of Natural History, Yale University, New Haven.

Xu X., Tang Z., and Wang X. 1999a. A therizinosauroid dinosaur with integumentary structures from China. Nature 399:350–354.

Xu X., Wang X. and Wu X. 1999b. A dromaeosaurid dinosaur with a filamentous integument from the Yixian Formation of China. Nature 401:262–266.

Xu X., Zhou Z., and Wang X. 2000. The smallest known nonavian theropod dinosaur. Nature 408:705–708.

Xu X., Zhao X., and J. Clark. 2001a. A new therizinosaur from the Lower Jurassic lower Lufeng Formation of Yunnan, China. Journal of Vertebrate Paleontology 21:477–483.

Xu X., Zhou Z., and R. O. Prum. 2001b. Branched integumental structures in *Sinornithosaurus* and the origin of feathers. Nature 410:200–204.

Xu X., M. Norell, Wang X.-L., P. J. Makovicky, and Wu X.-C. 2002. A basal troodontid from the Early Cretaceous of China. Nature 415:780–784.

Zhang X. H., Xu X., P. Sereno, Kuang X.-W., and Tan L. 2001. A long-necked therizinosauroid dinosaur from the Upper Cretaceous Iren Dabasu Formation of Nei Mongol, People's Republic of China. Vertebrata Palasiatica 39:282–290.

Part II

Taxa of Controversial Status

3

The Enigmatic Birdlike Dinosaur *Avimimus portentosus*
Comments and a Pictorial Atlas

PATRICIA VICKERS-RICH, LUIS M. CHIAPPE, AND SERGEI KURZANOV

The phylogenetic position of *Avimimus* is a puzzle that is yet to be solved. Since its announcement by Kurzanov in 1981, *Avimimus* has been alternatively regarded as a nonavian theropod close to the ancestry of birds (e.g., Thulborn, 1984; Norman, 1990; Holtz, 1994; Currie, 2000) or as a flightless, basal avian (e.g., Chatterjee, 1991, 1995, 1999). It is hoped that the comments and the high-quality stereophotographs provided in this short chapter (see Figs. 3.1–3.19) will yield some clues that will eventually help the relationships of *Avimimus* to be decoded.

Avimimus portentosus was first described by Kurzanov (1981) on the basis of three partial specimens from the Upper Cretaceous (Campanian; see Jerzykiewicz and Russell, 1991) Barun Goyot ("Burungoyotskaya" Svita) and Djadohkta ("Djadochtinskaya" Svita) Formations of Mongolia Omnogovi and Ovorkhangai Aimaks, respectively) (Weishampel, 1990). Kurzanov (1982, 1983, 1985, 1987) published a number of subsequent studies on *Avimimus*, several of them translated into English (Kurzanov, 1982, 1983, 1985). The specimen illustrated in this chapter (PIN 3907/1, Paleontological Institute, Russian Academy of Sciences) is from Udan Sayr in the eastern part of the Mongolian Gobi Desert and was made available for our study as a result of the organization of The Great Russian Dinosaurs Exhibition (1993–1997), a joint venture between the Paleontological Institute (Moscow), the Monash Science Centre, and the Queen Victoria Museum and Art Gallery (the last two institutions in Australia) (Vickers-Rich and Rich, 1993).

Kurzanov's opinion about *Avimimus* in 1993 was that it was "a [nonavian] dinosaur that had features in parallel with birds, perhaps even feathers" (Kurzanov, 1993). Norman (1990:286) concluded that *Avimimus* was an "unusual [nonavian] theropod from the evidence of the published description possessing, as it does, a very distinctive mixture of features, some of which are seen either in other groups of [nonavian] dinosaurs or in birds." Norman (1990) listed a few salient points: (1) The premaxilla (Fig. 3.10b) is reminiscent of lambeosaurine hadrosaurids in that it has a crenulate margin. (2) The braincase (Fig. 3.2) resembles that of birds and sauropods. (3) The vertebral column is typical for small, nonavian theropods, with the exception of several cervicals in the holotype specimen being radically different (notably, the lack of pneumatic foramina on the centra and the presence of large ventral processes—which are more commonly seen in birds more advanced than *Archaeopteryx*; see Chiappe et al., 1996). (4) The pelvis (Figs. 3.11, 3.12) is typical for nonavian theropods except for the unusual form of the iliac blades. (5) The hindlimb resembles that of nonavian theropods; the fusion of the distal end of the fibula with the tibia and proximal tarsals (Figs. 3.15, 3.16) occurs commonly in birds but occurs in some nonavian theropods as well. (6) The retention of the vestige of metatarsal V (Fig. 3.17) is unusual in nonavian theropods from the Late Cretaceous (although present in *Vorona* and *Rahonavis*, both birds from the Late Cretaceous of Madagascar [Forster et al., 1996, 1998, Chapter 12 in this volume], as well as the Late Cretaceous alvarezsaurids [Chiappe, Norell, and Clark, Chapter 4 in this volume]). (7) The forelimb is most puzzling; the humerus (Fig. 3.9) is typical for a bipedal, nonavian dinosaur; the metacarpal fragment (Fig. 3.10d–f) is similar to that of a bird.

Norman (1990:287) further suggested that "evidence can be brought forward to suggest [nonavian] theropod, sauropod, ornithopod, and avian affinities with some of the individual remains in this apparently associated material." Norman (1990:287) also pointed out that "using Gauthier (1986) as a guide it is clear that *Avimimus* displays some basal theropodan characters: spur-like metatarsal V, enlarged preacetabular portion of the ilium and large brevis fossa, and bowed femoral shaft and a fibular attachment crest on the lateral surface of the tibia. It also exhibits a small number of tetanuran and ornithomimid characters and one maniraptoran feature (development of hypapophyses on cervical vertebrae). However, it seems clear that our knowledge of this form is insufficient to establish precise relations with currently recognized higher taxa." *Avimimus* was also considered as a nonavian theropod by several other authors. Thulborn (1984) placed it as the sister group of his "Aves," which he used to name a group formed by Enantiornithes and other more advanced birds (a clade equivalent to Chiappe's Ornithothoraces; see Chiappe, Chapter 20 in this volume). Thulborn (1984) cited the presence of an intercotylar prominence and a straplike coracoid as the evidence supporting his proposed sister-group relationship. This support, however, is clearly weak at best, since *Avimimus* lacks an intercotylar prominence and its straplike coracoid appears to be of another taxon, something even Thulborn (1984) himself admitted.

In a discussion about the origin of birds, Molnar (1985) implicitly regarded *Avimimus* as a nonavian theropod, although he did not discuss its precise placement within the phylogeny of theropods. Paul (1988) followed Thulborn (1984) in placing *Avimimus* as the sister group of a clade equivalent to Chiappe's Ornithothoraces (see Chiappe, Chapter 20 in this volume). Paul also followed Thulborn (1984) in placing *Archaeopteryx* in a more basal position than dromaeosaurids, troodontids, and other nonavian theropod taxa. In Paul's (1988) opinion, *Avimimus* was secondarily flightless, a form derived from volant ancestors. Paul (1988) regarded *Avimimus* as the most birdlike of all nonavian theropods and allocated it within his "proto-birds," as the sister group of his "birds." The intimate relation between *Avimimus* and the origin of birds proposed by Thulborn (1984) and Paul (1988) was dismissed by Holtz (1994, 1996), who placed it within his Arctometatarsalia, not directly related to the origin of birds (see also Padian, 1997).

Even though, from a strict phylogenetic approach, Thulborn (1984) and Paul (1988) regarded *Avimimus* as a member of Aves (the common ancestor of *Archaeopteryx* and neornithine birds plus all its descendants), it is clear that this was not their intention. Chatterjee (1991, 1995), however, considered it to be a true, but flightless, bird. In 1991, Chatterjee placed it as the sister taxon of a group composed of *Protoavis* and Ornithurae (hesperornithiforms, ichthyornithiforms, and neornithine birds). Among the apomorphies cited by Chatterjee (1991) for *Avimimus* is the presence of teeth in the premaxilla and their absence in the maxilla. Yet, as pointed out by Kurzanov (1987) and Norman (1990) and confirmed by observations of one of us (Chiappe), the premaxilla of *Avimimus* lacks teeth, and the maxilla is not known for any specimen. (This fact appears to have been noticed by Chatterjee [1997], who mentioned the presence of "toothlike denticles" in the premaxilla of *Avimimus*.) Subsequently, Chatterjee (1995) supported the sister-group relationship of *Avimimus* and *Mononykus*, allocating them within Aves as the sister group of Ornithothoraces (see Chiappe, 1996, for the phylogenetic definition of this clade). Chatterjee (1995) cited the presence of a free orbital process of the quadrate, its ventral condylar articulation with the pterygoid, a lateral cotyla for the quadratojugal, and the presence of three condyles in the distal end of the quadrate as synapomorphies uniting *Avimimus* and *Mononykus*. Yet Norman (1990) remarked on the fusion of the suspensorium, and Kurzanov's (1987) illustrations show neither the free orbital process nor the pterygoid condyle of the quadrate (Figs. 3.2, 3.3a). Furthermore, the quadrate does not have a lateral cotyla or three condyles on its mandibular articulation. In his 1997 book, Chatterjee was more cautious about the avian identity of *Avimimus* and discussed it within a section entitled "Dubious Flightless Birds," although he concluded that it "may represent a flightless bird" (1997:121). This conclusion was more strongly reaffirmed in his recent monograph on *Protoavis* (Chatterjee, 1999). In this paper, Chatterjee (1999) regarded *Avimimus*, and its alleged sister-taxon *Mononykus*, as the sister-group of Ornithothoraces, although he did not include confuciusornithids (see Chiappe et al. [1999] and Chiappe [Chapter 20 in this volume] for the phylogenetic position of these birds) in the cladistic analysis.

Chiappe made the following observations, complementary to Norman's and Kurzanov's descriptions, when viewing the specimen at the New Jersey State Museum (Trenton) when The Great Russian Dinosaur Exhibition (Vickers-Rich and Rich, 1993) was in residence there during 1996 and 1997.

An axial centrum is preserved. It is very small compared with the remaining vertebral elements. This condition resembles that of alvarezsaurids (see Chiappe, Norell, and Clark, Chapter 4 in this volume). The axis has a small pneumatic foramen on the centrum and a short odontoid process.

Several thoracic vertebrae have open neurocentral sutures, which suggests that this specimen was not fully grown (see Sereno and Novas, 1993; Brochu, 1996). Some, but not all, have one small pneumatic foramen (sometimes two together) on the lateral side of the centrum. This is more evident in the caudal thoracic centra. The cranial thoracic vertebrae have prominent ventral processes (Fig. 3.4)—much larger than those in *Deinonychus* (Ostrom, 1969). The cranial and midthoracic vertebrae have a compressed and ventrally keeled centra. The caudal thoracic centra are more rounded. All the articular facets are platycoelous. The ratio between vertebral canal and centrum (in cranial view) is roughly 0.5, a condition typical for birds among theropods (see Chiappe, 1996). Hyposphene-hypantrum accessory articulations are well developed in the thoracic vertebrae.

The humerus (Fig. 3.9) appears to have a single condyle (as in Alvarezsauridae; see Chiappe, Norell, and Clark, Chapter 4 in this volume). There is neither a pneumatic fossa nor foramen. The head is flat when viewed proximally, and there is a very weak deltopectoral crest with a knoblike prominence toward the midshaft. The main planes of the proximal and distal ends are displaced by a fairly large angle.

The ulna is quite compressed (Fig. 3.10). The ulna was suggested to have quill knobs by Kurzanov (1987). Chiappe confirms that there are bumps on the caudal margin, but their function remains unclear. The proximal end is not very well preserved, but it does not seem to have two cotylae.

The carpometacarpus (as interpreted by Kurzanov [1987]) is a bone representing some sort of fusion of elements (Fig. 3.10). The semilunate carpal appears to be fused to metacarpals II and III in a cast displayed at the exhibition, but this is not so clear in the actual specimen.

The pubis has a large pubic foot (Fig. 3.12), primitively designed and not reduced cranially. It has a long pubic apron, compressed craniocaudally. The ventral surface of the pubic foot is flat with a central depression.

On the proximal end of the femur (Figs. 3.13, 3.14), the lesser trochanter has a globular-condylar appearance. The head does not have a fossa for the capital ligament. On the shaft, caudally and cranially, there is a prominent intermuscular line. The distal end, cranially, has no trace of a patellar groove. Interestingly, the caudal end has condyles that are connected below the popliteal fossa by a transverse ridge, a condition otherwise known only for birds among dinosaurs (see Chiappe, 1996). The lateral condyle is not distally projected, unlike that of alvarezsaurids (see Chiappe, Norell, and Clark, Chapter 4 in this volume). The fibular condyle is rather large.

The tibiotarsus is a true tibiotarsus, including astragalus, calcaneum, and tibia (Figs. 3.15, 3.16). The fibular crest is quite short. The cnemial crest projects only slightly cranially, and in medial view it has a crescentic shape. Distally, the calcaneum did not have a fossa for the articulation of the fibula, and it is uncertain whether the fibula reached the tarsals. Even though only a faint ridge and suture imply some portions of the ascending process, it is clear that it extended for roughly one-quarter of the length of the tibiotarsus.

The metatarsus (Figs. 3.17, 3.18) is fused proximally and includes the distal tarsals, as well. It has a metatarsal V, which is slender and small and runs along the caudal surface of metatarsal IV. There is no hypotarsus. Metatarsal IV is somewhat longer than metatarsal II. Metatarsal III is arctometatarsalian, yet it does not have the proximal expansion seen in tyrannosaurids or ornithomimids.

The pedal digits (Fig. 3.19) are short when compared with the metatarsus. The phalanges of digits IV and II are unique in that they are not flat ventrally but have a sharp ridge. Consequently, the cross section is oval, instead of triangular.

Final Comment

With these observations and the high-quality photographs made by Steve Morton (Monash University), it is hoped that this chapter will stimulate further research and discussion on this intriguing form. It should be kept in mind for further work on PIN 3907/1 that the association of all elements is not absolute, as at least three other taxa were found in the same concentration that yielded *Avimimus* (E. Kurochkin, pers. comm.). Yet a new articulated specimen from Udan Sayr, recently collected by the Mongolian-Japanese expeditions (M. P. Watabe, pers. comm.), supports the association of the holotypic hindlimb elements with the holotypic skull.

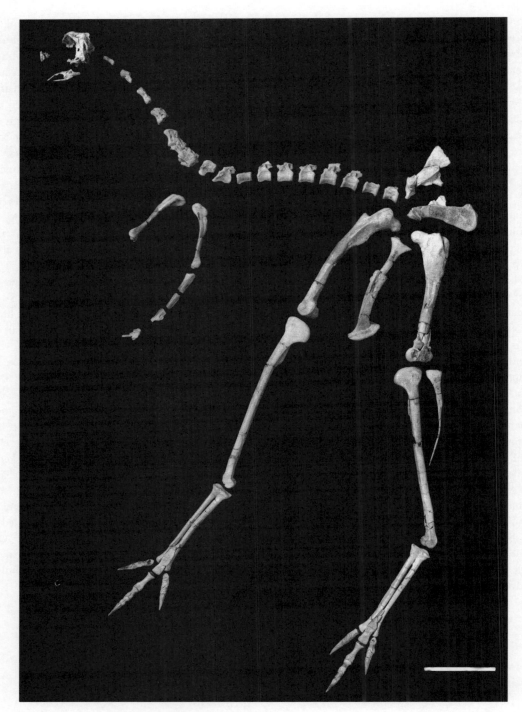

Figure 3.1. *A. portentosus* (PIN 3907/1). Scale bar = 10 cm.

PATRICIA VICKERS-RICH ET AL.

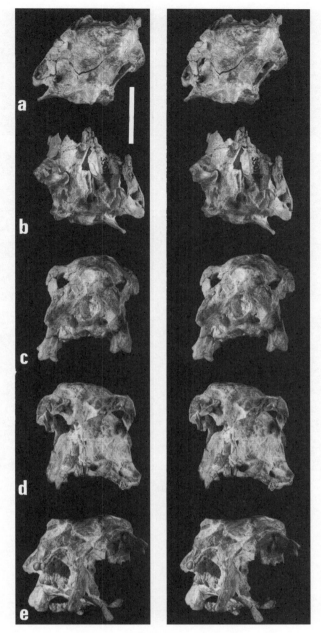

Figure 3.2. *A. portentosus* (PIN 3907/1). Stereo pairs of the skull in dorsal (a), ventral (b), occipital (c), rostral (d), and right lateral (e) view. Scale bar = 2 cm.

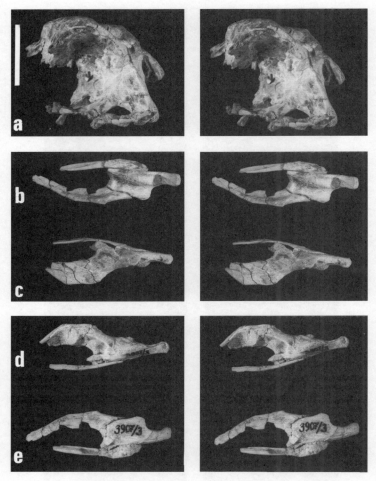

Figure 3.3. *A. portentosus* (PIN 3907/1). Stereo pairs of the skull in left lateral view (a). Stereo pairs of the caudal portion of the right mandible in dorsal (b), lateral (c), medial (d), and ventral (e) view. Scale bar = 2 cm.

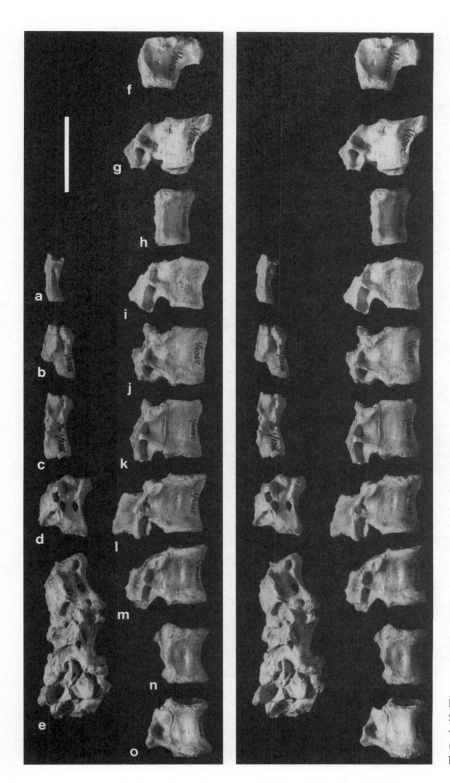

Figure 3.4. *A. portentosus* (PIN 3907/1). Stereo pairs of cervical (a–e) and thoracic vertebrae (f–o) in right lateral view (sequence follows Kurzanov [1982]). Scale bar = 3 cm.

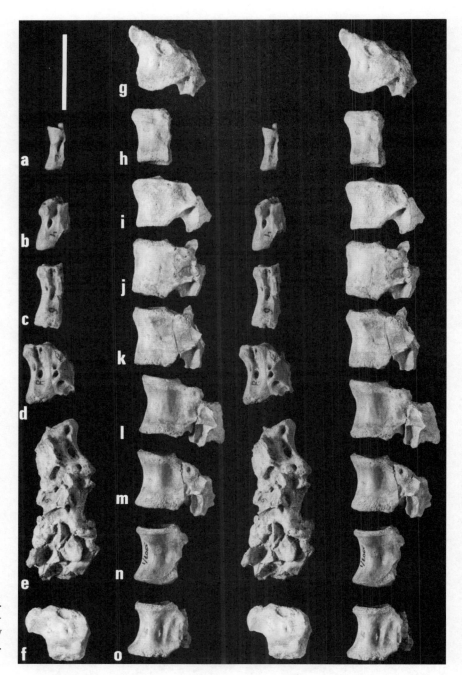

Figure 3.5. *A. portentosus* (PIN 3907/1). Stereo pairs of cervical (a–e) and thoracic vertebrae (f–o) in left lateral view (sequence follows Kurzanov [1982]). Scale bar = 3 cm.

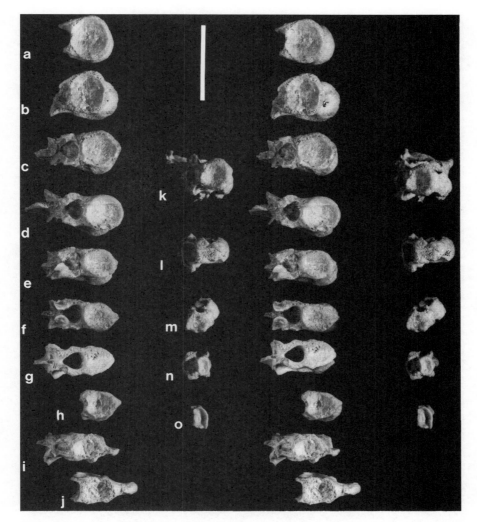

Figure 3.6. *A. portentosus* (PIN 3907/1). Stereo pairs of thoracic (a–j) and cervical vertebrae (k–o) in cranial view. Scale bar = 3 cm.

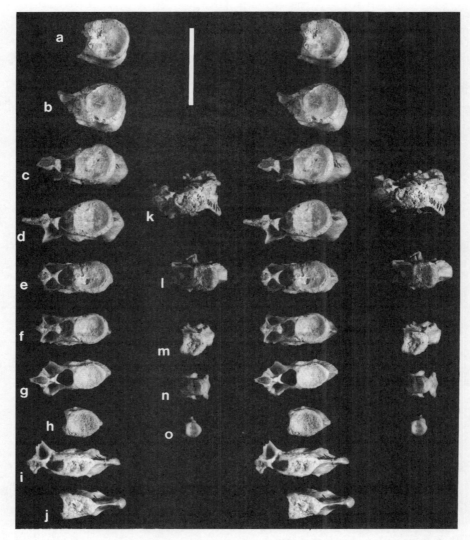

Figure 3.7. *A. portentosus* (PIN 3907/1). Stereo pairs of thoracic (a–j) and cervical vertebrae (k–o) in caudal view. Scale bar = 3 cm.

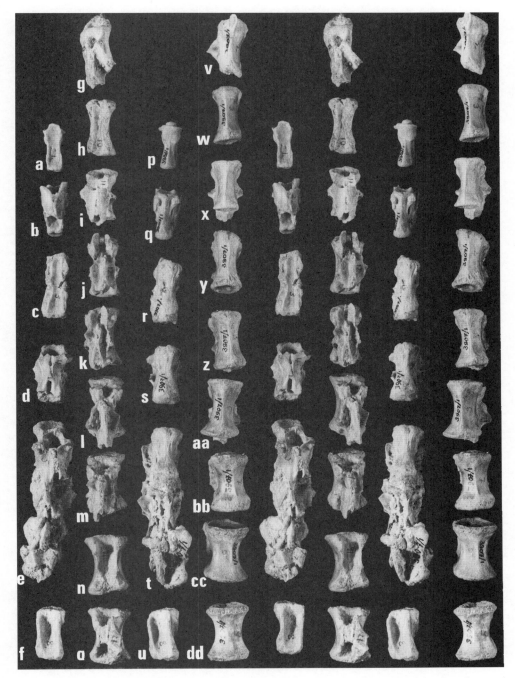

Figure 3.8. *A. portentosus* (PIN 3907/1). Stereo pairs of vertebral column in dorsal (a–o) and ventral (p–dd) view (sequence follows Kurzanov [1982]).

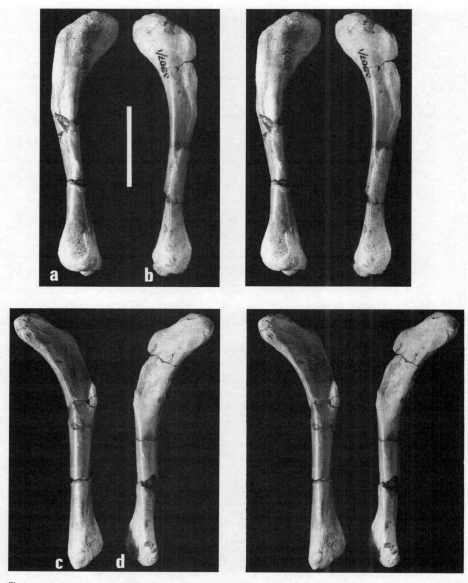

Figure 3.9. *A. portentosus* (PIN 3907/1). Stereo pairs of left (a, c) and right (b, d) humeri in dorsal (a, b) and ventral (c, d) view. Scale bar = 3 cm.

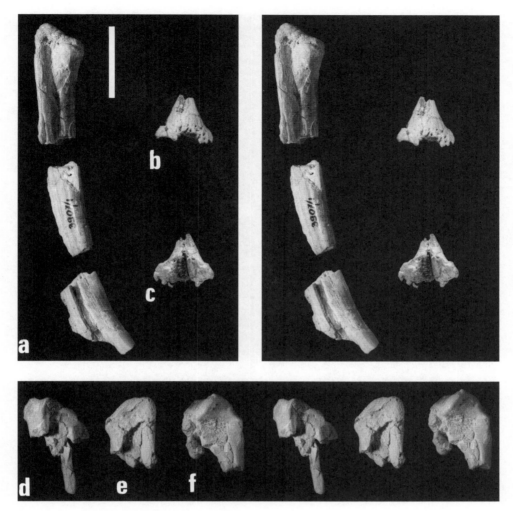

Figure 3.10. *A. portentosus* (PIN 3907/1). Stereo pairs of left ulna in dorsal view (a); stereo pairs of premaxilla in rostral (b) and caudal (c) view; stereo pairs of "carpometacarpus" (?) in several views (d–f). Scale bar = 2 cm.

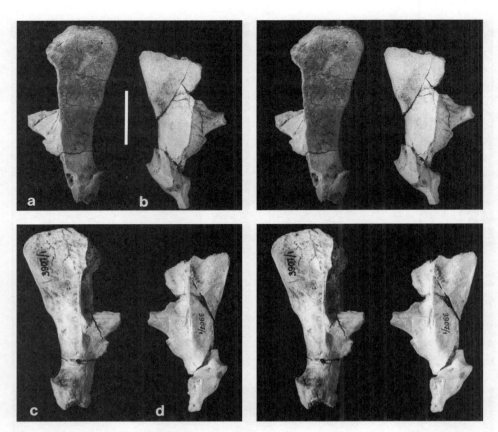

Figure 3.11. *A. portentosus* (PIN 3907/1). Stereo pairs of right (a, c) and left (b, d) ilia in ventral (a, b) and dorsal (c, d) view. Scale bar = 3 cm.

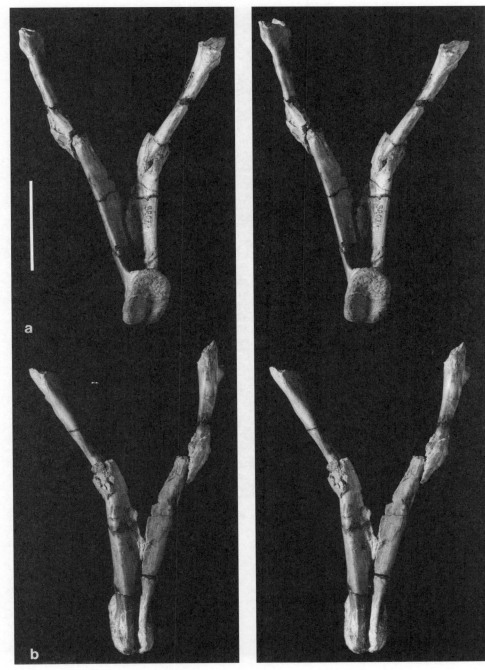

Figure 3.12. *A. portentosus* (PIN 3907/1). Stereo pairs of pubes in caudal (a) and cranial (b) view. Scale bar = 5 cm.

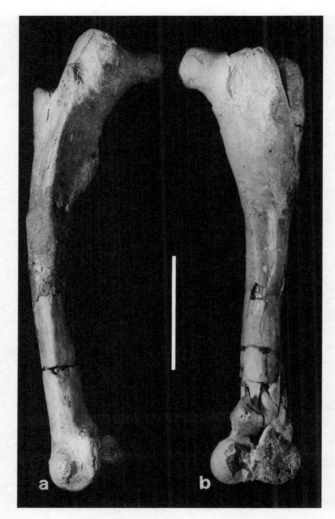

Figure 3.13. *A. portentosus* (PIN 3907/1). Left femur in caudo-lateral (a) view and right femur in caudal view (b). Scale bar = 5 cm.

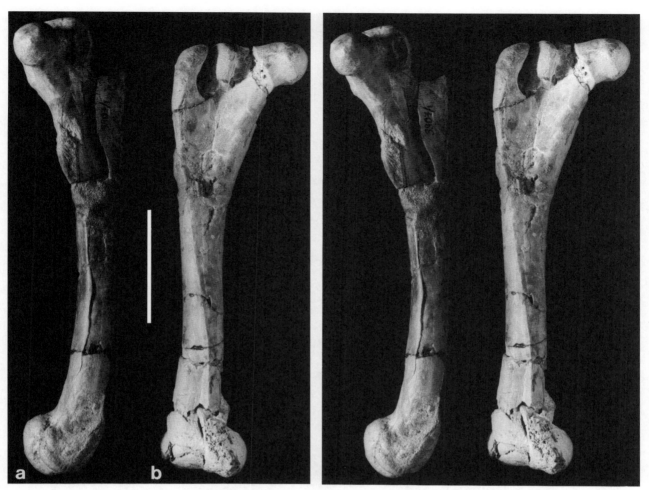

Figure 3.14. *A. portentosus* (PIN 3907/1). Stereo pairs of left femur in medial view (a) and the right femur in cranial view (b). Scale bar = 5 cm.

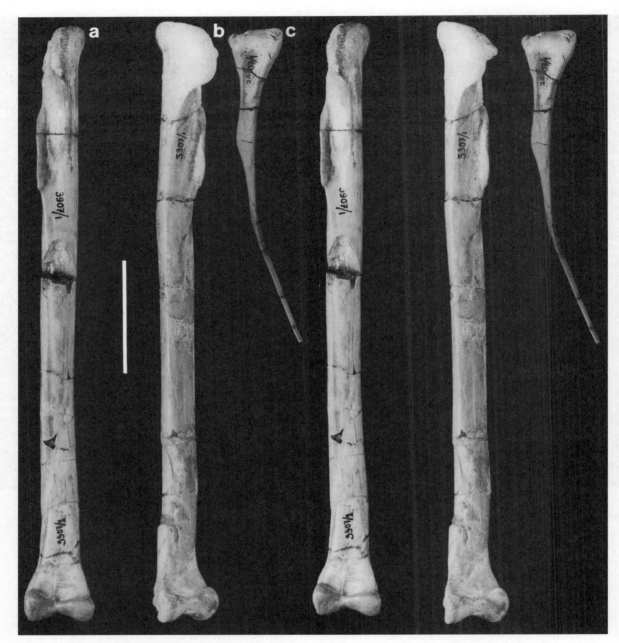

Figure 3.15. *A. portentosus* (PIN 3907/1). Stereo pairs of right (a) and left (b) tibiotarsi in cranial view and of right fibula in medial view (c). Scale bar = 5 cm.

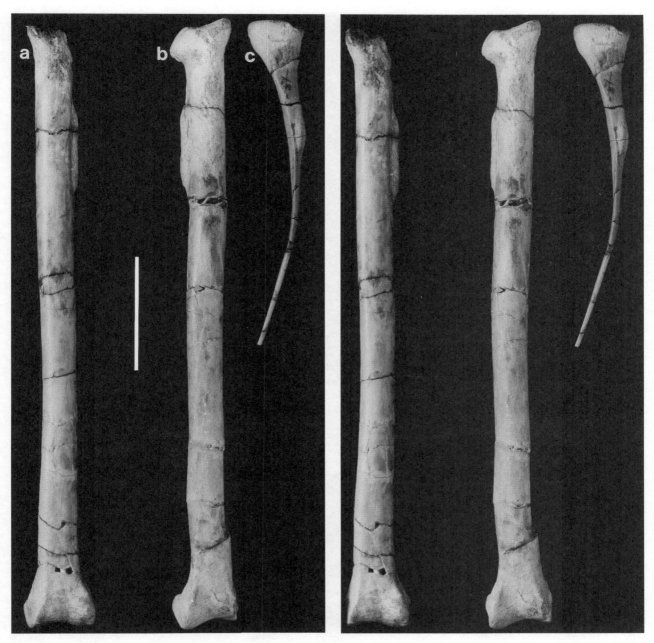

Figure 3.16. *A. portentosus* (PIN 3907/1). Stereo pairs of right (a) and left (b) tibiotarsi in caudal view and of right fibula in lateral view (c). Scale bar = 5 cm.

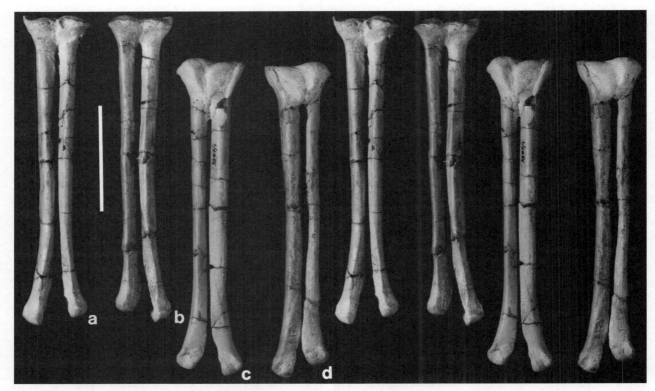

Figure 3.17. *A. portentosus* (PIN 3907/1). Stereo pairs of right (a, c) and left (b, d) metatarsal II and IV in dorsal (cranial) (a, b) and plantar (caudal) (c, d) view. Scale bar = 5 cm.

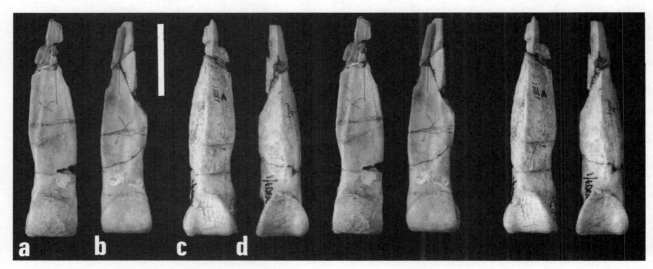

Figure 3.18. *A. portentosus* (PIN 3907/1). Stereo pairs of right (a, c) and left (b, d) metatarsal III in dorsal (cranial) (a, b) and plantar (caudal) (c, d) view. Scale bar = 2 cm.

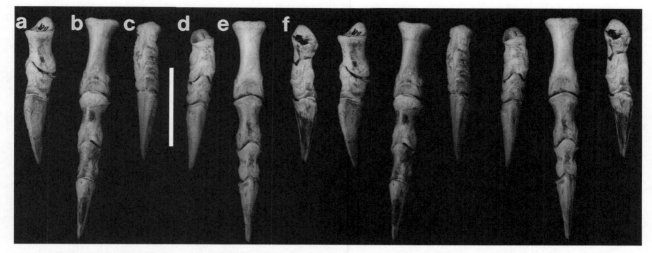

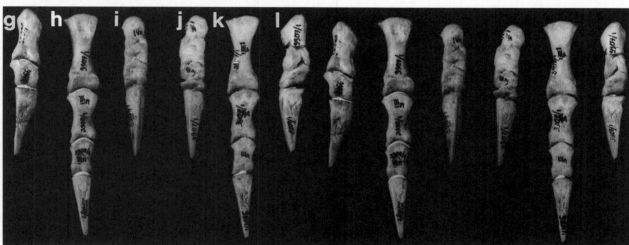

Figure 3.19. *A. portentosus* (PIN 3907/1). Stereo pairs of pedal digits (II–IV) in dorsal (a–f) and ventral (g–l) view; digit II (a, f, g, l), digit III (b, e, h, k), digit IV (c, d, i, j). Scale bar = 3 cm.

Acknowledgments

We are grateful to S. Morton (Monash University) for photographing PIN 3907/1 and to L. Meeker (American Museum of Natural History) and L. Rhoads (Natural History Museum of Los Angeles County) for mounting and labeling the photographs.

Literature Cited

Brochu, C. A. 1996. Closure of neurocentral sutures during crocodilian ontogeny: implications for maturity assessment in fossil archosaurs. Journal of Vertebrate Paleontology 16(1): 49–62.

Chatterjee, S. 1991. Cranial anatomy and relationships of a new Triassic bird from Texas. Philosophical Transactions of the Royal Society of London, Series B, 332(1265):277–346.

———. 1995. The Triassic bird *Protoavis*. Archaeopteryx 13:15–31.

———. 1997. The Rise of Birds. Johns Hopkins University Press, Baltimore, 312 pp.

———. 1999. *Protoavis* and the early evolution of birds. Palaeontographica 254(1–3):1–100.

Chiappe, L. M. 1996. Late Cretaceous birds of southern South America: anatomy and systematics of Enantiornithes and *Patagopteryx deferrariisi*; pp. 203–244 *in* G. Arratia (ed.), Contributions of Southern South America to Vertebrate Paleontology, Münchner Geowissenschaftliche Abhandlungen, Reihe A, Geologie und Paläontologie 30. Verlag Dr. Friedrich Pfeil, Munich.

Chiappe, L. M., M. A. Norell, and J. M. Clark. 1996. Phylogenetic position of *Mononykus* from the Upper Cretaceous of the Gobi Desert. Memoirs of the Queensland Museum 39:557–582.

Chiappe, L. M., Ji S.-A., Ji Q., and M. A. Norell. 1999. Anatomy and systematics of the Confuciusornithidae (Aves) from the late Mesozoic of northeastern China. Bulletin of the American Museum Novitates 242:1–89.

Currie, P. J. 2000. Theropods from the Cretaceous of Mongolia; pp. 434–455 *in* M. J. Benton, M. A. Shishkin, D. M. Unwin, and E. N. Kurochkin (eds.), The Age of Dinosaurs in Russia and Mongolia. Cambridge University Press, Cambridge.

Forster, C. A., L. M. Chiappe, D. W. Krause, and S. D. Sampson. 1996. The first Cretaceous bird from Madagascar. Nature 382:532–534.

Forster, C. A., S. D. Sampson, L. M. Chiappe, and D. W. Krause. 1998. The theropodan ancestry of birds: new evidence from the Late Cretaceous of Madagascar. Science 279: 1915–1919.

Gauthier, J. 1986. Saurischian monophyly and the origin of birds; pp. 1–55 *in* K. Padian (ed.), The Origin of Birds and the Evolution of Flight. California Academy of Sciences, San Francisco.

Holtz, T. R., Jr. 1994. Phylogenetic position of the Tyrannosauridae: implications for theropod systematics. Journal of Paleontology 68(5):1100–1117.

———. 1996. Phylogenetic taxonomy of the Coelurosauria. Journal of Paleontology 70:536–538.

Jerzykiewicz, T., and D. A. Russell. 1991. Late Mesozoic stratigraphy and vertebrates from the Gobi Basin. Cretaceous Research 12:345–377.

Kurzanov, S. M. 1981. On the unusual theropods from the Upper Cretaceous of Mongolia. Transactions of the Joint Soviet-Mongolian Paleontological Expeditions 15:39–50. [Russian]

———. 1982. Structural characteristics of the fore limbs of *Avimimus*. Paleontological Journal 1982(3):108–112.

———. 1983. New data on the pelvic structure of *Avimimus*. Paleontological Journal 1983(4):110–111.

———. 1985. The skull structure of the dinosaur *Avimimus*. Paleontological Journal 1985(4):92–99.

———. 1987. Avimimids and the problem of the origin of birds. Transactions of the Joint Soviet-Mongolian Paleontological Expeditions 31:5–95. [Russian]

Molnar, R. E. 1985. Alternatives to *Archaeopteryx*: a survey of proposed early or ancestral birds; pp. 209–217 *in* M. K. Hecht, J. H. Ostrom, G. Viohl, and P. Wellnhofer (eds.), The Beginnings of Birds. Freunde des Jura-Museums, Eichstätt.

Norman, D. B. 1990. Problematic Theropoda: "Coelurosaurs"; pp. 280–305 *in* D. B. Weishampel, P. Dodson, and H. Osmólska (eds.), The Dinosauria. University of California Press, Berkeley.

Ostrom, J. H. 1969. Osteology of *Deinonychus antirrhopus*, an unusual theropod dinosaur from the Lower Cretaceous of Montana. Bulletin of the Peabody Museum of Natural History 30:1–165.

Padian, K. 1997. Maniraptora; pp. 411–414 *in* P. J. Currie and K. Padian (eds.), The Encyclopedia of Dinosaurs. Academic Press, San Diego.

Paul, G. S. 1988. Predatory Dinosaurs of the World. Simon and Schuster, New York. 464 pp.

Sereno, P. C., and F. E. Novas. 1993. The skull and neck of the basal theropod *Herrerasaurus ischigualastensis*. Journal of Vertebrate Paleontology 13(4):451–471.

Thulborn, R. A. 1984. The avian relationships of *Archaeopteryx*, and the origin of birds. Zoological Journal of the Linnean Society 82:119–158.

Vickers-Rich, P., and T. H. Rich. 1993. The ICI Australia Catalogue of The Great Russian Dinosaurs Exhibition 1993–1995, Monash Science Centre, Melbourne.

Weishampel, D. B. 1990. Dinosaurian distribution; pp. 63–139 *in* D. B. Weishampel, P. Dodson, and H. Osmólska (eds.), The Dinosauria. University of California Press, Berkeley.

4

The Cretaceous, Short-Armed Alvarezsauridae
Mononykus and Its Kin

LUIS M. CHIAPPE, MARK A. NORELL, AND JAMES M. CLARK

 The discovery of *"Mononychus" olecranus* (Perle et al., 1993a; later emended to *Mononykus olecranus,* see Perle et al., 1993b) from the Late Cretaceous of the Gobi Desert stimulated a fruitful new discussion about the origin and early diversification of birds and the evolution of their flight. *Mononykus* not only led to the recognition of a previously unknown clade of theropod dinosaurs but also suggested, through its phylogenetic placement as the sister taxon of all birds except *Archaeopteryx,* the possibility (heretical to some) that flight may have evolved twice within birds (Perle et al., 1993a).

The study of *Mononykus* also helped to fine-tune the phylogenetic placement of certain enigmatic Cretaceous taxa. Novas (1996, 1997) convincingly argued for a common relationship between *Mononykus* and the Patagonian Late Cretaceous *Patagonykus puertai* and *Alvarezsaurus calvoi,* placing all of them within Alvarezsauridae (Bonaparte, 1991), a formerly monospecific taxon regarded by Bonaparte (1991) as a group of nonavian theropods. More recently, Karkhu and Rautian (1996) described *Parvicursor remotus* from the Late Cretaceous of the Gobi Desert. Although they placed it in a unique family, they recognized its strong similarity to *Mononykus.* We believe that this similarity is due to common descent and that Parvicursoridae is a junior synonym of Alvarezsauridae.

In addition to the taxa mentioned previously, several specimens (e.g., MGI N 100/99, MGI 100/975, MGI 100/977, MGI 100/1001; listed as *Mononykus* in Perle et al., 1993a, and Chiappe et al., 1996) that were collected between 1992 and 1995 by the American Museum of Natural History–Mongolian Academy of Sciences Paleontological Expeditions to the Upper Cretaceous red beds of southern Mongolia (Djadokhta and Barun Goyot Formations and their equivalents) display anatomical differences with the holotype of *M. olecranus.* These specimens can be diagnosed as a new taxon, *Shuvuuia deserti* (Chiappe et al., 1998a; Suzuki et al., 2002).

Recent findings and examination of existing collections have hinted at the presence of Alvarezsauridae in the Late Cretaceous of North America. Holtz (1994) suggested that the "cotype" specimen of *Ornithomimus minutus* (YPM-2042, originally 1049; Marsh, 1892), from the Lance Formation (Late Maastrichtian) of Wyoming, may be a member of the *Mononykus* lineage. In addition, and far more conclusive, is the occurrence of a pubis and ischium of nearly identical morphology to that of *Mononykus* and *Shuvuuia* in the Late Maastrichtian Hell Creek Formation of Montana (Hutchinson and Chiappe, 1998). Hutchinson and Chiappe (1998) placed this fragmentary specimen within the Mononykinae, a supraspecific taxon erected by Chiappe et al. (1998a) to include *Mononykus, Shuvuuia, Parvicursor,* and all the descendants from their most recent common ancestor, thus providing the first reliable record of Alvarezsauridae for North America.

Here we summarize the morphology of Alvarezsauridae, emphasizing the morphological differences between the included taxa. The present description is based primarily on *Mononykus* and *Shuvuuia,* which represent the best-known Alvarezsauridae. We then discuss differences between these two taxa as well as differences between them and the remaining Mongolian, North American, and Argentine forms. Specific diagnoses follow the guidelines of the Phylogenetic Species Concept (see Eldredge and Cracraft, 1980; Nixon and Wheeler, 1990, 1992), in which diagnostic characters do not necessarily represent autapomorphies but rather a unique character combination that differentiates specific taxa. This concept is used as an operational concept as suggested by Frost and Kluge (1994).

The following institutional abbreviations are used in this chapter: MGI, Mongolian Geological Institute, Ulaanbataar, Mongolia; MPD, Mongolian Paleontological Center, Ulaanbataar, Mongolia; MUCPv, Museo de Ciencias Naturales, Universidad Nacional del Comahue, Colección Paleontología Vertebrados, Neuquén, Argentina; PIN, Paleontological Institute, Moscow, Russia; PVPH, Museo Municipal "Carmen Funes," Plaza Huincul, Argentina; YPM, Yale Peabody Museum, New Haven, Connecticut, United States.

Geological Setting

Alvarezsaurid remains have been found in a variety of sedimentary rocks and paleoenvironments from fossil sites in the United States, China, Argentina, and Mongolia, although they are best known from the Gobi Desert (Mongolia) and Patagonia (Argentina) (Fig. 4.1).

Alvarezsaurid specimens from the Gobi Desert of Mongolia occur in two distinct environments. The holotype of *M. olecranus* is from Bugin Tsav, from a rich but poorly studied stratigraphic unit that lies in a small transverse basin north of the Altan Uul (Gradzinski et al., 1977). Rocks at Bugin Tsav have been described as belonging to the Nemegt Formation. However, major differences exist between these beds and more typical Nemegt rocks found in the Nemegt Basin. Characteristically, the Nemegt Formation contains fluvial rocks, predominantly channel fills with occasional overbank and lacustrine deposits (Gradzinski, 1970; Jerzykiewicz and Russell, 1991; Jerzykiewicz et al., 1993). Besides a typical "Nemegt" dinosaur fauna of *Tarbosaurus, Saurolophus,* ornithomimids, and occasional sauropods, fossils of aquatic animals (e.g., turtles, crocodyliforms, mollusks, and fishes) are common at Bugin Tsav. In some places, copious amounts of petrified wood are preserved. The sedimentation of the Nemegt Formation was considered by Jerzykiewicz and Russell (1991) to have occurred during the Maastrichtian, most likely in the middle Maastrichtian (~ 71 Ma).

Most of the Mongolian alvarezsaurid specimens, including those of *S. deserti* and *P. remotus,* have been collected from the Upper Cretaceous beds of southern Mongolia—the Djadokhta and Barun Goyot Formations and their equivalents (Dashzeveg et al., 1995). These rocks are generally considered slightly older than the Nemegt rocks, which superpositionally overlie Djadokhta-type rocks at some localities (e.g., Nemegt; Gradzinski et al., 1977). Jerzykiewicz and Russell (1991) considered the initial phases of sedimentation of the Djadokhta Formation to be of middle Campanian age. According to these authors, this sedimentation would have been centered around 78 Ma. The extraordinarily well preserved fauna of the Djadokhta and Barun Goyot Formations is remarkably diverse, including many species of mammals, lizards, turtles, nonavian dinosaurs, and birds (see Kielan-Jaworowska, 1969; Lavas, 1993; Dashzeveg et al., 1995; Novacek, 1996; Gao and Norell, 2000).

Various and disparate interpretations have been proposed concerning the environment during the time of deposition of the Djadokhta and Barun Goyot sediments. These range from interpretations as fluvial systems (Tverdokhlebov and Tsybin, 1974; Tumanova, 1987) to ideas that the massive sandstones are predominantly aeolian (Berkey and Morris, 1927; Eberth, 1993; Jerzykiewicz et al., 1993). Eberth (1993), whose perspective is based on work at Bayan Mandahu in Inner Mongolia (China), posits that some of these beds were part of a vast system of ergs—a point of view shared by Fastovsky et al. (1997) from their work at Tugrugeen Shireh (southern Mongolia).

However, several authors have noted that, while predominantly aeolian, these deposits contain intermittent lacustrine and fluvial episodes as suggested by occasional channel deposits, intermingled clays, and localized heavy bioturbation. Particularly enlightening has been the work of Loope et al. (1998) at Ukhaa Tolgod (southern Mongolia). While these authors found evidence of violent sandstorms in the cross-bedded deposits, their detailed study showed that the extraordinary fauna of Ukhaa Tolgod (Dashzeveg et al., 1995; Novacek, 1996) was entombed by massive sandstone of nonaeolian origin. Tracks found in the cross-bedded sandstones documented that dinosaurs and presumably other creatures frequented the dunes, though they were not buried in them. Loope et al. (1998) interpreted the rich fossil beds of Ukhaa Tolgod as sandy alluvial fans built at the margins of stabilized dunes during episodes of heavy rain.

Thus, the combined work of international expeditions in the Gobi Desert suggests that during the time *Shuvuuia* and *Parvicursor* lived, the region was covered by large dune fields with, at best, sparse vegetation, scattered with intermittent streams and small ponds and lakes of little depth (Novacek, 1996). Violent sandstorms may have occurred, but these cannot account for the exceptional preservation of the Mongolian Upper Cretaceous red beds. Most likely, occasional, heavy rains led to the development of the sandy alluvial fans that rapidly buried the organisms living at the edge of the dune slopes (Loope et al., 1998; Dingus and Loope, 2000). As stated by Loope et al. (1998:30): "The high, sparsely vegetated landforms were a sediment reservoir available for episodic gravity flows, capable of quickly burying intact, unscavenged skeletons, and, possibly, live animals."

According to Jerzykiewicz and Russell (1991:356), the "local onset of a more humid regime and the development of integrated drainage systems" resulted in the deposition of the Nemegt Formation. However, active fieldwork in the Gobi Desert has begun to show that the Nemegt and Djadokhta dinosaur faunas comingled with many forms common to both rock units and that faunas differ mainly in relative abundance rather than in composition. This seems to suggest that the Nemegt and Djadokhta faunas may be more environmentally than temporally controlled. Never-

Figure 4.1. Map indicating the central Asian, Patagonian, and North American localities that have furnished alvarezsaurid material.

1- Boca del Sapo, Neuquén
2- Sierra del Portezuelo, Neuquén
3- Tugrugeen Shireh, South Gobi
4- Bayn Dzak, South Gobi
5- Bugin Tsav, South Gobi
6- Khermeen Tsav, South Gobi
7- Khulsan, South Gobi
8- Ukhaa Tolgod, South Gobi
9- Iren Dabasu, Inner Mongolia
10- Sandstone Basin, Eastern Montana

theless, the common occurrence of extremely similar alvarezsaurids (i.e., *Mononykus* and *Shuvuuia*) in these deposits suggests that these animals were capable of existing in a variety of habitats.

Patagonian alvarezsaurids come from the reddish sandstones of the Neuquén Group, in the northwestern corner of Argentina's Patagonia. *A. calvoi* is from the Bajo de la Carpa Member (Río Colorado Formation), while *P. puertai* occurs in the Portezuelo Member of the underlying Río Neuquén Formation. Traditionally, these rocks have been considered as fluvial deposits, each formation representing successive sedimentary cycles (Cazau and Uliana, 1973; Ramos, 1981). Recent work at Boca del Sapo (vicinity of Neuquén city), the type locality of *Alvarezsaurus,* has suggested that these sandstones may be of aeolian origin, representing a small field of dunes (Heredia and Calvo, 1997). Nevertheless, only very localized rocks preserve evidence in favor of an aeolian type of deposition for the Bajo de la Carpa Member at this site (Clarke et al., 1998); these rocks appear to be predominantly fluvial (Clarke et al., 1998).

The fauna associated with *Alvarezsaurus* includes a variety of small vertebrates such as snakes, crocodyliforms, and other birds (Bonaparte, 1991; Chiappe, Chapter 13 in this volume), even though abundant sauropod remains have been found in other localities of the Río Colorado Formation (Powell, 1986). The Río Neuquén Formation has so far provided limited fossil material. The locality of *Patagonykus* has also yielded the remains of various nonavian theropods, sauropods, and crocodyliforms (Novas and Puerta, 1997; Novas, 1998; Novas, pers. comm.).

As for the Asian alvarezsaurids, no absolute dates are available for the Patagonian alvarezsaurid-bearing rocks. Recent stratigraphic and paleontological data have converged in placing the age of both the Portezuelo (Río Neuquén Formation) and Bajo de la Carpa (Río Colorado Formation) members within the Coniacian-Campanian interval (Cruz et al., 1989; Legarreta and Gulisano, 1989; Bonaparte, 1991; Calvo, 1991; Chiappe and Calvo, 1994). Recent magnetostratigraphic analyses at the sauropod nesting site of Auca Mahuevo (Chiappe et al., 1998b), within the Río Colorado Formation, placed this unit within the Campanian (Dingus et al., 2000).

Systematic Paleontology

Taxonomic Hierarchy

Theropoda[1] Marsh, 1881
Metornithes Perle et al., 1993
Alvarezsauridae Bonaparte, 1991

Diagnosis—Laterally compressed synsacral vertebrae; concave and convex cranial and caudal articular surfaces of synsacrum, respectively; short coracoid, wider than it is tall, lacking bicipital tubercle; sternum stout and subtriangular in section; humerus with prominent, proximally projected ventral tubercle and a single condyle; ulna with a hypertrophied olecranon process and a single cotyla; large and proximally expanded radiocarpal facet of radius; hypertrophied and strongly depressed metacarpal I; robust digit I with claw bearing two proximoventral foramina; pubic pedicel cranioventrally oriented; absence of both ischiadic and pubic distal fusion; laterally compressed and kidney-shaped proximal end of pubis; cone-shaped lateral condyle of femur.

Alvarezsaurus Bonaparte, 1991
A. calvoi Bonaparte, 1991

Holotype—MUCPv 54 (Bonaparte, 1991), including portions of several cervical and thoracic vertebrae, incomplete synsacrum, 13 caudal elements, right ilium and a fragment of the left ilium, portions of scapula and coracoid, segments of the hindlimb elements (including pedal phalanges), and a phalanx interpreted as a manual ungual (Novas, 1997).

Locality and horizon—Boca del Sapo, Neuquén city, Neuquén Province, Argentina. Bajo de la Carpa Member, Río Colorado Formation, Late Cretaceous (Campanian; Dingus et al., 2000).

Diagnosis—Amphyplatian cervicals with flat, paddle-shaped postzygapophyses; abrupt transition between cervicals and thoracic vertebrae; synsacrum less compressed than in *Patagonykus, Mononykus,* and *Shuvuuia;* spinous processes of proximal caudals absent or weakly developed; distal caudal vertebrae twice as long as proximal ones; scapula significantly smaller than in *Mononykus* and *Shuvuuia;* scapular blade curved in dorsal view; subequal pre- and postacetabular wings of ilium; postacetabular wing not depressed as in *Shuvuuia* and *Parvicursor;* metatarsal IV longer than metatarsal II; ungual phalanx of digit I with a short, ventral keel.

[1] The phylogenetic relationships of Alvarezsauridae within Theropoda are still a matter of debate. While several cladistic analyses have supported its inclusion within Aves, others have not. Their phylogenetic relationships are discussed in the section Phylogenetic Relationships.

Patagonykus Novas, 1997
P. puertai Novas, 1997

Holotype—PVPH 37 (Novas, 1997), assemblage of primarily disarticulated bones (presumably of a single individual) including a few thoracic and caudal vertebrae, a portion of the synsacrum, incomplete coracoids, segments of humeri and other elements of the forelimb, incomplete pelvic bones, and several portions of the hindlimb elements.

Locality and horizon—Sierra del Portezuelo, 12 km west of Plaza Huincul, Neuquén Province, Argentina. Portezuelo Member, Río Neuquén Formation, Late Cretaceous (Coniacian-Santonian; Legarreta and Gulisano, 1989).

Diagnosis—Weak hyposphene-hypantrum; thoracic postzygapophysis ventrally curved and with a tongue-shaped lateral border; tubercle on the caudal base of the neural arch of thoracic, synsacral, and caudal vertebrae; coracoidal shaft strongly convex ventrally and concave dorsally, with distinct ridge on ventral surface; transversely narrow humeral articular facet of coracoid; ventral tubercle of humerus proximally truncated; subcircular humeral condyle; humerus with sharp, ventral epicondyle, ventrally projected; olecranon not as large as in *Mononykus;* first phalanx of digit I with proximal articulation narrower than in *Mononykus* and *Shuvuuia,* and with prominent ventral ridge on its lateral margin; pubic foot; short, lateral projection between cnemial crest and lateral articular area of tibiotarsus; ascending process not indented.

Mononykinae Chiappe et al., 1998

Diagnosis—Alvarezsaurids with supracetabular crest on ilium extending only over the rostral half of the acetabulum; laterally compressed and kidney-shaped proximal end of pubis; absence of femoral posterior trochanter; medial border of tibiotarsus strongly projected proximally; fibula greatly reduced distally, without reaching the proximal tarsals; proximal end of metatarsal III reduced, not reaching the tarsals (arctometatarsalian condition).

Mononykus Perle et al., 1993
M. olecranus Perle et al., 1993

Holotype—MGI 107/6 (Perle et al., 1993a), including a fragmentary skull, an isolated tooth, several cervical and thoracic vertebrae, fragments of the synsacrum and one proximal caudal vertebra, forelimbs, scapulae and an incomplete coracoid, sternum, portions of ribs, incomplete pelvic elements, femora, tibiotarsi, portions of the metatarsals and pedal phalanges.

Locality and horizon—Bugin Tsav, South Gobi Aimak, Mongolia. Nemegt Formation, Late Cretaceous (mid-Maastrichtian; Jerzykiewicz and Russell, 1991).

Diagnosis—Cervical centra strongly compressed laterally, lacking pneumatic foramina; cranialmost thoracic vertebrae strongly compressed; fused ilium and ischium; pillarlike deltopectoral crest of humerus; supracetabular crest developed only in the cranial portion of acetabulum; subtriangular cross section of pubis; two cnemial crests in tibiotarsus; medial indentation of ascending process with deeply excavated base; ascending process arises from medial margin of astragalar condyle instead of from lateral margin.

Parvicursor Karkhu and Rautian, 1996
P. remotus Karkhu and Rautian, 1996

Holotype—PIN 4487/25 (Karkhu and Rautian, 1996), including three thoracic, three synsacral, and seven caudal vertebrae, portions of the pelvis, a nearly complete right hindlimb and several elements of the left hindlimb.

Locality and horizon—Khulsan, South Gobi Aimak, Mongolia. Barun Goyot Formation, Late Cretaceous (Campanian; Jerzykiewicz and Russell, 1991).

Diagnosis—Very similar to *Mononykus* but differs in having opisthocoelous caudal thoracic vertebrae; no biconvex thoracic vertebra; convex cranial margin of synsacrum; and a smaller adult size.

Shuvuuia Chiappe et al., 1998
S. deserti Chiappe et al., 1998

Holotype—MGI 100/975 (Chiappe et al., 1998a), including 9 cervical vertebrae, several thoracic vertebrae, synsacrum, and 19 caudals; portions of the forelimb and thoracic girdle; ilium and proximal portions of the ischium and pubis; portions of femur and tibiotarsus, complete metatarsals, and several pedal phalanges.

Referred specimens—MGI N 100/99 (juvenile), MGI 100/977, MGI 100/1001, including various postcranial elements and two complete skulls. MPD 100/120, a partial skeleton preserving a portion of the skull (Suzuki et al., 2002).

Locality and horizon—Ukhaa Tolgod and Tugrugeen Shireh, South Gobi Aimak, Mongolia. Djadokhta-like beds, Late Cretaceous (Campanian; Jerzykiewicz and Russell, 1991).

Diagnosis—Very similar to *M. olecranus* but differs in having cervical centra that are less compressed laterally and bearing large pneumatic foramina; deltopectoral crest of humerus continuous with humeral head; a pubis of subcircular section; femoral and tibiotarsal shafts bowed mediolaterally; medial border of distal end of tibiotarsus with a sharp ridge; medial margin of the ascending process of the astragalus less notched; much longer proximal phalanx of the hallux; an intermediate phalanx of digit II similar in length to its ungual (instead of much shorter as in *M. olecranus*), and longer and more slender intermediate phalanges of digit IV and lesser degree of co-ossification of

proximal tarsals and metacarpals, suggesting a size larger than *M. olecranus*. *S. deserti* differs from *P. remotus* because pedal digit IV is longer than half the length of metatarsal IV (in *P. remotus*, digit IV is less than half of metatarsal IV) and this digit is shorter than digit II (without counting ungual phalanges).

Anatomy

Skull and Mandible

Cranial remains are known only for the Asian alvarezsaurids. The holotype of *M. olecranus* preserves a braincase and a fragment of the maxilla, along with an isolated tooth found within the braincase (Perle et al., 1993a; Chiappe et al., 1996). Another fragmentary braincase and a partial skull are preserved in two juvenile specimens of *S. deserti* from Tugrugeen Shireh (MGI N 100/99 and MPD 100/120, respectively). By far, the best cranial remains are two nearly complete skulls of *Shuvuuia* (MGI 100/977 and MGI 100/1001) from the red beds of Ukhaa Tolgod (Figs. 4.2–4.6, 4.8; Chiappe et al., 1998a). The following description is mostly based primarily on these specimens.

The skull of *Shuvuuia* is lightly built with large orbits. The nares, surrounded by the premaxilla, nasal, and maxilla, are elongated, teardrop-shaped, and located at the tip of the rostrum (Figs. 4.2–4.4). The unfused premaxillae are very thin and are wedged in between the nasals caudodorsally. The dentigerous margin of the premaxilla is not preserved in any of the available skulls; two large neurovascular foramina are visible on its preserved rostroventral portion in MGI 100/977. The caudoventral end of the premaxilla forms a short, laterally concave process that abuts against the dorsomedial surface of a short, rostral process of the maxilla (Figs. 4.3A, 4.4A).

As in nonavian theropods and basal birds (Chiappe, 1996), the maxilla is the largest bone on the lateral surface of the snout. Its rostral half is subtriangular and vertically expanded, whereas the caudal half forms a slender bar, delimiting the ventral margin of the antorbital fossa (Figs. 4.2–4.4). The maxilla bears numerous tiny teeth; 14 are preserved in the right maxilla of MGI 100/977, but spaces between these teeth suggest that it probably had 25 or more. The maxillary teeth, however, are restricted to the rostral portion of this bone, being absent in its last two-thirds. The maxilla bears only a few external neurovascular foramina, which are parallel to the dentigerous margin. Ventromedially, the two maxillae abut, forming an extensive secondary palate that extends at least to the rostral level of the antorbital fossa.

The antorbital fossa is elongate and subtriangular (Figs. 4.5A, 4.8). Its is bounded by the maxilla rostrally and ventrally and by the lacrimal dorsally and caudally; the nasal is excluded from the antorbital fossa, but the rostral portion

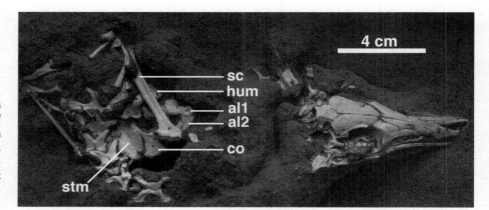

Figure 4.2. Articulated specimen with skull of *S. deserti* (MGI 100/977) from the red beds of Ukhaa Tolgod (southern Mongolia). Abbreviations: al1, proximal phalanx of digit I; al2, ungual phalanx of digit I; co, coracoid; hum, humerus; sc, scapula; stm, sternum.

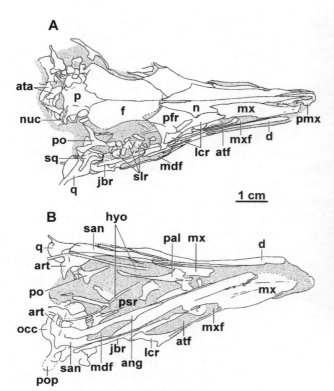

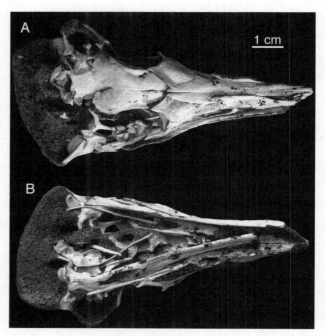

Figure 4.4. Skull of *S. deserti* (MGI 100/977) in dorsal (A) and ventral (B) views.

Figure 4.3. Camera lucida drawing of the skull of *S. deserti* (MGI 100/977) in dorsal (A) and ventral (B) views. Abbreviations: ang, angular; art, articular; ata, atlantal arches; atf, antorbital fenestra; d, dentary; f, frontal; hyo, hyoid; jbr, jugal bar; lcr, lacrimal; mdf, mandibular fenestra; mx, maxilla; mxf, maxillary fenestra; n, nasal; nuc, nuchal crest; occ, occipital condyle; p, parietal; pal, palatine; pfr, prefrontal (or ectethmoid; see text for discussion); pmx, premaxilla; po, postorbital; pop, paroccipital process; psr, parasphenoidal rostrum; q, quadrate; san, surangular; slr, sclerotic ring; sq, squamosal.

of the jugal contributes to the caudoventral corner of the fossa. A dorsal process of the maxilla separates a rostrally located fenestra (roughly one-third the length of the entire antorbital fossa) from the larger, antorbital fenestra (Figs. 4.3, 4.4). Given the preservation of this region in the two available skulls of *Shuvuuia*, it is not possible to discern whether this rostral fenestra is either the maxillary or the promaxillary fenestra of other maniraptoriform dinosaurs (Witmer, 1997). However, the presence of two fenestrae (i.e., the maxillary and promaxillary fenestra) rostral to the antorbital fossa cannot be discarded. Also unfortunate is the fact that none of the available specimens is conclusive on whether there is a dorsal maxillary wall lining medially the rostral portion of the antorbital fossa.

The nasals are long, rhomboid in shape, and slightly convex dorsally. United by a long, straight suture, these bones run the length of the snout (Figs. 4.2–4.4). They taper both rostrally and caudally. Rostrally, each ends in a very thin, sharp process that borders most of the dorsal margin of the naris.

In *Shuvuuia,* the rostral wall of the orbit is formed by a large ossification, which is broadly exposed on the dorsal surface of the snout (Figs. 4.3, 4.4, 4.6). In dorsal view, this bone has a rhomboidal shape and a sharp caudal apex. The combined caudal borders of the dorsal surfaces of the left and right elements of this ossification form a crescentic margin that overlaps the rostral end of the frontals. The orbital vertical wall of this ossification narrows ventromedially, contacting its counterpart in the midline. A wide space is left between the medial projections of these elements and the roof of the frontals; this space is comparable to the orbitonasal foramen of neornithine birds, which conducts both n. olfactorius and n. ophthalmicus (Baumel and Witmer, 1993). The homology of this ossification is unclear. On the one hand, it could represent a prefrontal. This bone, however, is small or reduced in maniraptoriform dinosaurs (Gauthier, 1986; Clark et al., 1994; Witmer, 1995). The prefrontal in ornithomimids is somewhat larger than in maniraptoriforms (e.g., Osmólska et al., 1972) but also exhibits

fundamental differences from the preorbital ossification of *Shuvuuia.* The prefrontal of ornithomimids, like the prefrontals of other nonavian theropods (e.g., *Erlikosaurus, Sinraptor, Allosaurus*), is essentially caudal to the ventral bar of the lacrimal; in *Shuvuuia,* the bone in question is mostly expanded rostral to this bar (Fig. 4.3, 4.4). Furthermore, a ventromedial contact between prefrontals, underneath the frontals, has not been reported for any theropod—the medial surface of this bone is typically sutured to the lateral margin of the frontal (see Madsen, 1976; Currie and Zhao, 1993a). Thus, if this element is truly a prefrontal, the condition in *Shuvuuia* is highly modified and autapomorphic. Alternatively, this ossification may represent an avian ectethmoid. In neornithine birds, this ossification forms a platelike surface that projects with different degrees from the interorbital septum and separates the orbit from the nasal cavity (Cracraft, 1968). Although the ectethmoid is present in most neornithine birds and in several instances it joins (or fuses) with the frontal dorsally (e.g., Podicipediformes, Falconiformes, Gruiformes, Piciformes), this ossification lacks a dorsal exposure on the cranial/snout roof (Cracraft, 1968). Thus, if this preorbital ossification is indeed an ectethmoid, it also represents a highly modified and autapomorphic condition (Suzuki et al., 2002).

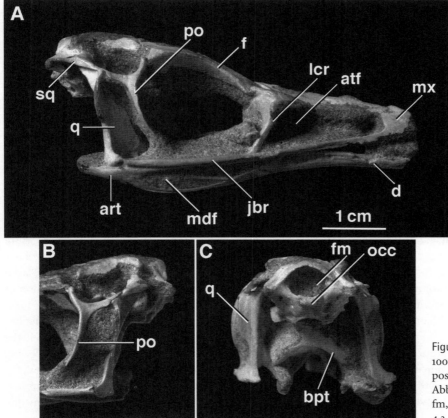

Figure 4.5. Skull of *S. deserti* (MGI 100/1001) in right lateral (A), left lateral (B; only postorbital region), and occipital (C) views. Abbreviations: bpt, basipterygoid process; fm, foramen magnum; others as in Figure 4.3.

Lateral to the prefrontal/ectethmoid ossification is the slender vertical bar of the lacrimal. This bone has an extremely long rostral portion, slightly recessed relative to its vertical bar, and a very short caudal process (Fig. 4.3B, 4.4B). Its inverted, L-shaped morphology resembles that of *Archaeopteryx* and certain nonavian theropods (Currie, 1995). The vertical bar of the lacrimal is subtriangular in cross section, and it slopes rostrodorsally. As in nonavian theropods and basal birds (including ratites) the nasolacrimal duct pierces the body of the lacrimal (Witmer, 1995, 1997). The frontals are large and long, forming virtually the entire roof of the orbit (Figs. 4.2–4.4, 4.6). These slightly vaulted bones are separated by a straight suture and a notch on their caudal margin.

The caudal margin of the orbit is formed by the postorbital (Fig. 4.5). This triradiate bone bears processes for the articulation to the frontal rostrodorsally and the squamosal caudally and a ventral, splintlike process that fails to reach the jugal. This gap between the postorbital and the jugal shows that in *Shuvuuia* the orbit was confluent with the laterotemporal fenestra, a uniquely avian condition among archosaurs (Zusi, 1993). Interestingly, the tip of the caudal process of the postorbital cooperates with the squamosal in the formation of a ventral facet for the articulation of the lateral head of the quadrate, a condition unknown in any other dinosaur (including birds) (Weishampel et al., 1990).

The ventral margin of the orbit is formed by a rodlike jugal (Figs. 4.3–4.6, 4.8), which is strongly depressed rostrally. This bone contacts the medial margin of the maxilla and lateral margin of the lacrimal rostrally, and caudally it is fused to the quadratojugal, leaving no sign of suture. While the contact with the maxilla is by means of a suture, that with the jugal appears to be by a ligamentous connection. This latter interpretation agrees with the displacement seen in some of the lacrimals of the two skulls of *Shuvuuia*. The jugal bar lacks a postorbital process for the articulation to the postorbital bone (Fig. 4.5A, B), again, a feature exclusive to birds among archosaurs. A distinct tubercle for the attachment of the postorbital ligament is also missing. The caudal tip of the quadratojugal (or the jugal bar) forms a tiny fork that must have been connected by ligaments to the ventrolateral corner of the quadrate (Fig. 4.5B). This condition, which agrees with that of most birds (with perhaps an exception in *Archaeopteryx*), contrasts with that of nonavian archosaurs, in which the quadratojugal has a distinct squamosal process forming a broad and firm articulation with the quadrate (Romer, 1956; Weishampel et al., 1990; Zusi, 1993).

The squamosal is triradiate and depressed (Figs. 4.3–4.6). Unlike neornithine birds but like nonavian theropods, *Archaeopteryx* (Elzanowski, Chapter 6 in this volume), *Confuciusornis* (Chiappe et al., 1999; Zhou and Hou, Chapter 7 in this volume), and certain enantiornithines (Sanz et al., 1997; Chiappe and Walker, Chapter 11 in this volume), alvarezsaurids exhibit the ancestral diapsid condition (Romer, 1956) in which the squamosal is not incorporated into the braincase but forms the external margin of a complete, subcircular dorsotemporal fenestra (Figs. 4.5A, B, 4.6A). This bone bears a short and robust medial process articulating with the parietal at the level of the nuchal crest. It has a rostrolateral process for the articulation to the postorbital; the tip of this process contributes to the ventral articular facet for the lateral head of the quadrate (Fig. 4.5A, B). The third process is directed caudolaterally, articulating with the paroccipital process of the braincase.

The quadrate is tall, slender, and rostrocaudally compressed (Fig. 4.5). As in *Archaeopteryx* and the enantiornithine *Gobipteryx* (Elzanowski and Wellnhofer, 1996), the length of the quadrate is nearly one-quarter the length of the skull. The proximal end of the quadrate of *Shuvuuia* exhibits two distinct articular heads (Fig. 4.5C). This suggests that our early interpretation of the presence of a single-headed quadrate in *Mononykus*, as derived from the study of the fragmentary braincase of the holotype specimen (Perle et al., 1994), is most likely incorrect. In the skull of *Shuvuuia*, the lateral quadrate head articulates at the joint between the squamosal and postorbital, while the medial head is caudodorsally directed toward the braincase, toward the region on the prootic on which the quadrate fragment articulates in the *Mononykus* holotype (Perle et al., 1994; Fig. 4.7). The orbital process (pterygoid flange of nonavian theropods) is broad and rostromedially oriented (Fig. 4.5A), a primitive condition retained in basal birds such as *Archaeopteryx, Confuciusornis,* and enantiornithines (Chiappe, 1996; Sanz et al., 1997; Chiappe et al., 1999). On its ventromedial corner, this process firmly abuts the caudal end of the pterygoid. Caudally, the shaft of the quadrate is excavated on its proximal half by a shallow, broad sulcus, while its distal half is flat (Fig. 4.5C). The medial margin of the caudal surface forms a ridge that becomes stronger proximally as it converges with the medial otic head. The lateral margin of the quadrate is very peculiar. In caudal view, this margin forms a convex, laterally projected ridge, while its distal half develops a distinct, crescentic notch.

The morphology of the jugal bar and the suspensorium region of the skull of *Shuvuuia* is very similar to that of neornithine birds. This suggests capabilities for craniofacial kinesis (Fig. 4.8), namely, the elevation and depression of the rostrum relative to the braincase (Simonetta, 1960; Bock, 1964; Zusi, 1985, 1993). The absence of a contact between the squamosal and the quadratojugal allowed the streptostylic quadrate to swing rostrocaudally. Likewise, without a connection between the jugal and postorbital, the jugal bar was able to act as a strut between the quadrate and the rostrum. Forces directed longitudinally from the quadrate would elevate or depress the rostrum around a

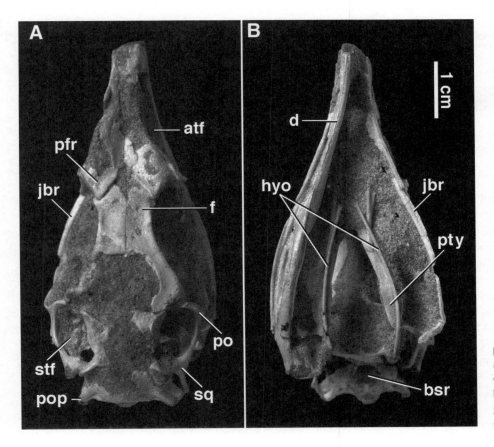

Figure 4.6. Skull of *S. deserti* (MGI 100/1001) in dorsal (A) and ventral (B). Abbreviations: bsr, basisphenoidal recess; pty, pterygoid; stf, supratemporal fossa; others as in Figure 4.3.

transverse axis at a bending zone immediately rostral to the orbit (Fig. 4.8). The existence of a thinning of the jugal just caudal to its lacrimal contact would have formed a ventral bending zone. This and the loose connection between the frontals and the preorbital ossifications (nasals and pre-frontals/ectethmoids) suggest that the snout may have moved as a unit, much in the manner of prokinetic neornithine birds (Simonetta, 1960; Bock, 1964; Zusi, 1985, 1993). The absence of a continuous, ossified septum, as indicated by the lack of a ridge, or even a trace of it, on the ventral surface of the frontals and nasals, agrees with this interpretation. The general design of the skull of *Shuvuuia* suggests that some motion, most likely of a prokinetic type, was possible (Chiappe et al., 1998a). Nevertheless, the prokinetic condition of *Shuvuuia* differs from that of prokinetic neornithines in that the dorsal bending zone is between the frontals and the nasals-prefrontals/ectethmoids and not between the frontals and the nasals-premaxillae. Furthermore, the dorsal and ventral bending zones of the skull of *Shuvuuia* are located in the rostral portion of the orbit, and not rostral to the orbit as in prokinetic neornithines. Although Zweers and Vanden Berge (1997:187) considered the skull of *Shuvuuia* as "almost avian-like prokinetic," the position of its bending zones led them to call this type of kinesis *mesokinesis*—a term functionally defined by the exis-

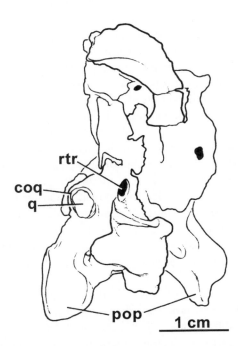

Figure 4.7. Skull of *M. olecranus* (MGI 107/6) (after Perle et al., 1994) in ventrolateral view. Abbreviations: coq, cotyla for the articulation of the lateral otic head of the quadrate; pop, paroccipital process; q, quadrate; rtr, rostral tympanic recess.

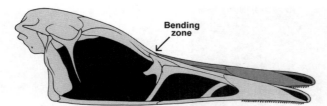

Figure 4.8. Reconstruction of the skull of *S. deserti*. Note the prokinetic movement of the rostrum.

tence of a flexible area in the roof of the orbit caudal to the lacrimal, specifically the frontoparietal area.

The braincase of alvarezsaurids is proportionally very small. Its roof is formed by the parietals, which are dorsally flat and completely fused to each other (Figs. 4.2, 4.3A, 4.4A). Caudally, the central portion of the parietals passes smoothly into the occipital table, while laterally, each parietal is separated from the occiput by a thin, but prominent, transverse nuchal crest (Fig. 4.6A). The bones surrounding the occiput, including the parietal, are co-ossified. Dorsal to the foramen magnum, the occiput is transversely vaulted. The angle between the central portion of the parietals and the dorsal surface of the occiput is very shallow; these two areas are almost coplanar.

The occipital condyle is significantly smaller than the foramen magnum (Fig. 4.5C). In this respect, the alvarezsaurid skull also resembles the condition in birds, differing from nonavian theropods, where the size of the condyle is comparable to that of the foramen magnum (Osmólska et al., 1972; Madsen, 1976; Weishampel et al., 1990; Currie and Zhao, 1993b; Clark et al., 1994; Currie, 1995). The large size of the foramen magnum is interesting considering the small size of the braincase. This proportion is not in agreement with the results of Mlikovsky (1990), who concluded that, in birds, the area of the foramen magnum was closely correlated to that of brain size. Two vertically aligned hypoglossal foramina (CN XII) lie on each side of the occipital condyle. Lateral to the dorsal hypoglossal foramen lies a larger vagus foramen (CN X); this foramen lies at the same level as the paroccipital process, just caudal to its base. Ventral and lateral to the hypoglossal foramina and the occipital condyle—lateral to the caudal portion of the basisphenoidal recess (see later)—is a large, ventrolaterally facing foramen interpreted as the entrance of the cerebral carotid artery.

The paroccipital process is short and slightly expanded distally (Fig. 4.7). Its dorsal surface faces up and slightly caudal, a condition comparable to that in *Archaeopteryx* and nonavian theropods such as *Erlikosaurus* (Clark et al., 1994). A weak, longitudinal ridge runs over its dorsal surface, and a small foramen—presumably for the entrance of the occipital artery (Chatterjee, 1991)—is present lateral to the dis-

tal end of this ridge. In ventral view, the distal half of the paroccipital process forms a subrhomboidal, slightly concave platform. The proximomedial border of this platform essentially limits the lateral margin of the columellar recess caudoventrally. The caudodorsal margin of this recess is formed by a strong ridge that projects from the proximal half of the paroccipital process. Dorsally, the columellar recess is bordered by the prootic, while a strong projected flange of the metotic strut forms its ventral floor. Inside the columellar recess there is a large, subtriangular vestibular fenestra separated from the cochlear fenestra by the crista interfenestralis. Caudodorsal to the cochlear fenestra is a large, cup-shaped cavity that bears the entrance of the caudal tympanic recess caudally. Thus, as in birds (Chiappe et al., 1996), the exit of the caudal tympanic recess opens inside the columellar recess instead of perforating the paroccipital process, as in several nonavian theropods.

The braincase wall of the holotype of *M. olecranus* (MGI 107/6) shows that the prootic articulation for the quadrate is located dorsal and rostral to the columellar recess (Perle et al., 1994). This facet is ventrolaterally oriented, with a slight caudal orientation. Just caudal to this facet is a small foramen leading to an air space inside the prootic, presumably the dorsal tympanic recess. Laterally, the prootic is perforated by a single facial foramen (CN VII) that is at the level of the vestibular fenestra. Ventral to the facial foramen is the entrance to the rostral tympanic recess (Fig. 4.7).

A portion of the medial braincase wall is also preserved in the braincase of the holotype of *M. olecranus*. The arcuate eminence projects medially from the braincase wall. The subarcuate fossa (= floccular recess) is large and deep. The vestibular eminence is swollen. The acoustic recess is shallow. Rostrally, a single facial foramen presumably transmitted the facial nerve (CN VII). Two small, rostral auditory foramina lie just dorsal to the facial foramen, between it and the inflated area for the vertical vestibular canal. These transmitted the rostral auditory ramus of the acoustic nerve (CN VIII). Caudal and slightly ventral to the acoustic foramen lies the fovea ganglii vagoglossopharyngealis (= metotic foramen). This foramen is a deep, nearly vertical slit that transmitted the glossopharyngeal (CN IX), vagus (CN X), and spinal accessory (CN XI) nerves through the wall of the braincase.

The basisphenoid is excavated by a subcircular recess, a condition known for several nonavian theropods (e.g., Currie and Zhao, 1993a; Currie, 1995). In neornithine birds, the basisphenoid is sheathed by the parasphenoid (i.e., parasphenoidal lamina), which in some instances it is distinctly recessed. Unfortunately, it is not clear whether the condition in *Shuvuuia* constitutes the primitive one of nonavian theropods (basisphenoid excavated by a recess) or the derived condition of neornithine birds (parasphenoidal lamina recessed by a fossa). Rostral to the basisphenoidal recess

(Figs. 4.5C, 4.6B), long basipterygoid processes project ventrolaterally. Their nearly vertical orientation in MGI 100/1001 (Figs. 4.5C, 4.6B) is most likely a preservational artifact, as suggested by (1) a crack in their base, (2) their caudal displacement with respect to the caudal end of the pterygoids, and (3) the subvertical orientation of the basipterygoid processes of MGI 100/977. In any event, the proportional length of the basipterygoid processes in *Shuvuuia* is unparalleled by any other bird or nonavian theropod, resembling more the development of diplodocid and dicraeosaurid sauropod dinosaurs (Romer, 1956). Rostral to the basipterygoid processes is a subtriangular, parasphenoidal rostrum (Figs. 4.3B, 4.4B). This is narrow and bears a strong ventral ridge, a condition contrasting with the swollen parasphenoidal rostrum of troodontids (Currie and Zhao, 1993b), *Erlikosaurus* (Clark et al., 1994), and ornithomimids (Osmólska et al., 1972).

The pterygoids are long and straplike, reaching near the rostral margin of the orbits (Fig. 4.6B). Caudally, a short, vertical quadrate ramus abuts firmly against the ventromedial corner of the orbital process of the quadrate. No ectopterygoid is preserved in the available two skulls of *Shuvuuia*. This does not mean, however, that these bones were absent. The palatine is long and slender (Figs. 4.3B, 4.4B). Sereno (2001) misidentified a portion of the maxillary secondary palate, visible through the antorbital fenestra of MGI 100/977, as an expanded rostral portion of the palatine. Examination of this specimen and of MPD 100/120 shows that Sereno's (2001) interpretation is erroneous. The caudal half of the palatine bears a prominent ridge that divides a lateral and somewhat depressed area from a smaller medial portion. Its caudal half forms a thin, straight bar that caudally widens into a transverse "foot" for its articulation with the pterygoid.

The lower jaw of alvarezsaurids, known only for *Shuvuuia*, is low and compressed (Figs. 4.3–4.6). In dorsal view, it has an elongate sigmoid shape. Laterally, the cranial half has subparallel margins, while the wider caudal half is convex both dorsally and ventrally. New information derived from MPD 100/120 shows that the two mandibular rami contact each other along a straight suture, forming a short symphysis (Suzuki et al., 2002). The dentary and splenial meet each other by means of straight sutures running both dorsally and ventrally throughout their length. The cranial half of the dentary bears a dorsal groove for implantation of the teeth—individual alveoli are clearly absent in the two known jaws of *Shuvuuia* (GMI 100/977, GMI 100/1001). This type of implantation has been reported for ornithomimids (Sereno, 2001), troodontids (Currie, 1987), and *Hesperornis* (Marsh, 1880). Dentary and splenial form the cranial margin of a large mandibular fenestra laterally and medially, respectively. This fenestra is laterally elongate and teardrop-shaped (Figs. 4.3B, 4.4B, 4.5A). It is bordered

dorsally by the surangular and ventrally by the angular. The minimal participation of dentary in the ventral margin of the mandibular fenestra is comparable to the condition in nonavian theropods (Ostrom, 1969; Osmólska et al., 1972; Currie, 1995) and certain basal birds (Sanz et al., 1997). Dorsally, the caudal two-thirds of the surangular are broad and strongly depressed. This bone also forms the caudal margin of the mandibular fenestra. The mandibular fenestra is lined medially by a wedged projection of the splenial and cranially by a thin prearticular. This latter bone is concave dorsally and convex ventrally. As in *Archaeopteryx* (Elzanowski and Wellnhofer, 1996), the rostral end of the prearticular abuts against the splenial without leaving any space for the coronoid. This suggest that *Shuvuuia*—like apparently all birds and some nonavian theropods—did not have a coronoid bone. The articular area is broken or covered by other bones in the available two specimens. The mandible bears a wide and depressed retroarticular process, which receives a lateral, thin contribution from the surangular. The caudomedial corner of the retroarticular process projects medially into a prominent, tapered process (Figs. 4.3B, 4.4B). Dorsally, the retroarticular process forms a subcircular basin. Given the available material, it is not possible to discern whether there is a pneumatic foramen perforating the articular.

Axial Skeleton

Differentiation between cervical and thoracic vertebrae in fossil vertebrates always involves a certain dose of arbitrariness. Typically, those vertebrae in which the ribs articulate with the sternum are considered thoracic vertebrae, but in most fossil vertebrates, the articulation between ribs and sternum is either poorly preserved or missing altogether. This is the case for *Mononykus* and *Shuvuuia*, in which the sternum lacks facets for the articulation of ribs, suggesting that they may have articulated to cartilaginous sternal facets. The distinction between the cervical and thoracic vertebrae of *Mononykus* and *Shuvuuia* is further complicated by the fact that, in contrast to the condition typical of the adults of ornithothoracine birds and certain nonavian theropods, their cervical ribs are not fused to the vertebral centra. Thus, our classification of these vertebrae is based exclusively on their overall morphology.

Cervical Vertebrae. The cervical series is well known for *Mononykus* and *Shuvuuia* (MGI N 100/99, MGI 100/975, MGI 100/977). In *Alvarezsaurus*, only the vertebral arch of a midcervical and portions of the caudal section of the neck are preserved. Novas (1997) included within the material of *Patagonykus* an isolated cervical along with an incomplete, isolated postzygapophysis. This short vertebra is quite different from the long and depressed cervicals of other alvarezsaurids. Because these remains do not show diagnos-

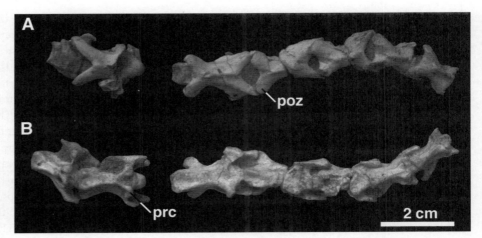

Figure 4.9. Cervical vertebrae of *S. deserti* (MGI 100/975) in dorsal (A) and ventral (B) views. Abbreviations: poz, postzygapophysis; prc, carotid process.

tic features of other alvarezsaurids, we prefer to leave them out of the present description.

The number of cervicals is not known for either *Mononykus* or *Shuvuuia*, but specimen MGI 100/975 of *Shuvuuia* preserves at least nine postaxial vertebrae of cervical morphology. The cervical vertebrae of alvarezsaurids are elongated and low. The longest elements in the series are the middle ones. The size reduction of the cranial cervicals, with respect to the middle ones, is quite remarkable. For example, in specimen MGI 100/975, the cranialmost preserved element is nearly half the length of a midcervical that is only four elements behind it (Fig. 4.9).

The vertebral arches are broad and depressed, with low spinous processes (Figs. 4.9, 4.10). The height of the vertebral arches increases in the caudal section of the series. The postzygapophyses bear small epipophyses. In *Alvarezsaurus* (Fig. 4.10A), the postzygapophyses of the preserved cervicals are nearly flat and paddle-shaped in dorsal view (Bonaparte, 1991). This apomorphic condition contrasts markedly with the rectangular morphology of the postzygapophyses of other alvarezsaurids (Novas, 1996).

Like the vertebral arches, the cervical centra of *Mononykus* and *Shuvuuia* are elongated and low. The strong, opisthocoelous condition present in these vertebrae (Fig. 4.9) seems to be absent in *Alvarezsaurus*. In an isolated centrum regarded by Bonaparte (1991) as a caudal cervical, both articular surfaces are concave (Figs. 4.10B, C). In *Mononykus*, the centra are strongly compressed laterally (Perle et al., 1994), differing in this regard from both *Shuvuuia* and *Alvarezsaurus* (based on the single isolated centrum). The centra of the cranial and middle cervicals of the Asian alvarezsaurids are cranioventrally excavated by a furrow bordered by prominences (Fig. 4.9B) comparable to the carotid processes of neornithine birds (Perle et al., 1994). In *Mononykus*, the preserved cervicals do not exhibit pneumatic foramina (i.e., pleurocoels). These, however, are present in *Shuvuuia* and apparently in *Alvarezsaurus*. A small,

oval pneumatic foramen is present behind the parapophysis (costal fovea) in the isolated cervical centrum of *Alvarezsaurus* (Fig. 4.10A; Bonaparte, 1991; Novas, 1996). In *Shuvuuia*, the pneumatic foramina of the caudal cervicals are very large. The cervical ribs of alvarezsaurids are not fused to the vertebrae.

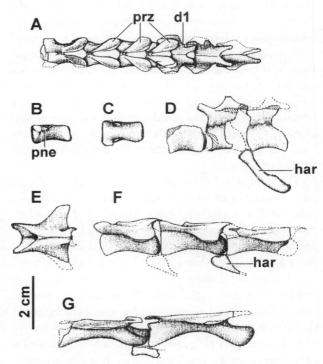

Figure 4.10. Vertebral elements of *A. calvoi* (MUCPv 54) (after Bonaparte, 1991). A, cervicothoracic neural arches in dorsal view. B, C, isolated cervical centrum in left lateral (B) and ventral (C) views. D, proximal caudals in left lateral view. E, dorsal view of the first vertebra illustrated in F. F, proximal middle caudals in left lateral view. G, distal middle caudals in left lateral view. Abbreviations: d1, first dorsal; har, haemal arch; pne, pneumatic foramen; prz, prezygapophysis.

Thoracic Vertebrae. The thoracic series is primarily known from the Asian alvarezsaurids. Only a few cranial vertebral arches, and a midthoracic vertebra and some portions of caudal thoracic vertebrae, are preserved in *Alvarezsaurus* and *Patagonykus*, respectively.

In *Mononykus* and *Shuvuuia*, the transition from typical cervicals to typical thoracic vertebrae is gradual (Perle et al., 1994). This is not the case in *Alvarezsaurus*, in which only an intermediate vertebra separates the caudal cervicals with paddled postzygapophyses from thoracic vertebrae with much smaller postzygapophyses (Fig. 4.10A) (Bonaparte, 1991).

In the transition from cervicals to thoracic vertebrae, the parapophysis (costal fovea) migrates from its cranioventral position in the centrum to a position that is lateral to, and at the same level as, the prezygapophysis (Fig. 4.11). This apomorphic condition is clearly seen in the cranial thoracic vertebrae of the Asian alvarezsaurids (Perle et al., 1994), although it seems to be absent in *Alvarezsaurus* itself (Bonaparte, 1991; Novas, 1996).

As in the cervical vertebrae, *Mononykus* differs from *Shuvuuia* in that its transitional and thoracic centra (particularly the cranial and middle ones) are strongly compressed laterally (Fig. 4.11). The lateral compression of the thoracic vertebrae of *Mononykus* is also greater than that of *Patagonykus* (Novas, 1996, 1997). The sharp ventral margins of the transitional centra of *Mononykus* bear small ventral processes; these processes are larger in the subsequent, cranial thoracic vertebrae (Fig. 4.11A; Perle et al., 1994). Prominent ventral processes are also present in the cranial thoracic vertebrae of *Shuvuuia*. Pneumatic foramina are absent in the thoracic vertebrae of all alvarezsaurids.

In *Mononykus* and *Shuvuuia*, the spinous processes of the cranial thoracic vertebrae are moderately low and caudodorsally oriented (Fig. 4.11). Toward the back (Fig.

4.12), these increase in height and become vertically directed (Perle et al., 1994). The spinous processes of the cranial thoracic vertebrae of *Alvarezsaurus,* the only known thoracic vertebrae, are also low (Bonaparte, 1991). The vertebral foramen of all alvarezsaurids is very large with respect to the centrum (Fig. 4.11C; Chiappe et al., 1996). Hyposphene-hypantrum articulations are absent in the Asian alvarezsaurids (Perle et al., 1994) but weakly developed in *Patagonykus* (Novas, 1996, 1997). These additional vertebral articulations are widespread among nonavian saurischian dinosaurs (Gauthier, 1986), and they have also been found in the basal bird *Rahonavis* from the Late Cretaceous of Madagascar (Forster et al., 1998a,b).

In *Mononykus* and *Shuvuuia*, the transitional elements as well as the cranial and middle thoracic vertebrae are opisthocoelous (Fig. 4.11). In the caudal section of the thoracic series, a biconvex element bridges the opisthocoelous morphology of its preceding vertebrae and the procoelous condition of the synsacrum (Perle et al., 1994). In this respect, *Mononykus* and *Shuvuuia* strikingly resemble *Patagopteryx*, a flightless bird from the Late Cretaceous of Patagonia that is more closely related to extant birds, although in this bird the biconvex element is the fifth thoracic vertebra (Chiappe, 1996, Chapter 13 in this volume). The biconvex vertebra of *Mononykus* and *Shuvuuia* may represent the last thoracic element, although it is unclear whether there were any (one or maybe two) procoelous thoracic vertebra caudal to this biconvex element. A biconvex thoracic vertebra is probably absent in *Parvicursor* (Karkhu and Rautian, 1996), in which the last element of the thoracic series is still opisthocoelous. The known thoracic centra of *Patagonykus* are not opisthocoelous; the preserved midthoracic centrum is amphicoelous, while the known portion of a thoracic centrum, of caudal position, possesses a slightly convex caudal surface (Novas, 1997).

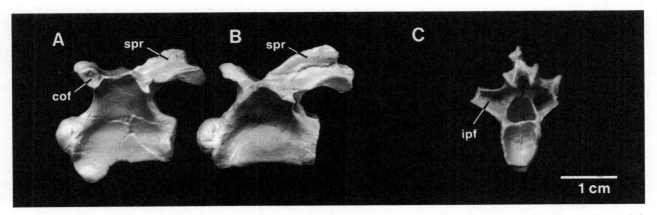

Figure 4.11. Thoracic vertebrae of *M. olecranus* (MGI 107/6) (after Perle et al., 1994). A, B, first two preserved thoracic vertebrae in left lateral view. C, caudal view of the second one. Abbreviations: cof, costal fovea; ipf, infrapostzygapophysial fossa; spr, spinous process of vertebral arch.

Synsacrum. The synsacrum is nearly complete in *Shuvuuia* (MGI N 100/99, MGI 100/975). Portions of it are known for all other alvarezsaurids.

The synsacrum of *Shuvuuia* is formed by seven vertebrae. In this taxon as well as in *Mononykus,* the cranial and caudal articular surfaces are concave and convex, respectively (Fig. 4.12). Four synsacral vertebrae are preserved in both *Patagonykus* and *Alvarezsaurus.* Novas (1997) considered that *Patagonykus* had at least five synsacral vertebrae, although we believe it may have had several more. In *Alvarezsaurus,* taking into consideration the length of the ilium, seven synsacral vertebrae seems to be a reasonable estimate. The fact that both *Alvarezsaurus* and *Patagonykus* are missing the first synsacral vertebra prevents determination of whether the cranial surface of the synsacrum was concave, as in *Mononykus* and *Shuvuuia.* In *Parvicursor,* however, the cranial surface of the synsacrum is convex (Karkhu and Rautian, 1996).

The synsacral vertebrae of alvarezsaurids are laterally compressed. This compression is greatest in the laminar, caudal portion of the synsacrum, which forms a prominent, ventrally projected keel (Fig. 4.12). In *Alvarezsaurus,* the synsacral centra are less compressed than those of *Patagonykus, Mononykus,* and *Shuvuuia.* The last synsacral vertebrae bear a pronounced ventral keel, although this condition is unknown for *Alvarezsaurus.*

The spinous processes of the cranial and central portion of the synsacrum of *Shuvuuia* form a continuous dorsal crest. It is not clear, however, whether this crest runs through the entire synsacrum. In *Patagonykus,* the spinous processes of the last two preserved synsacral vertebrae are clearly individualized (Novas, 1997).

Caudal Vertebrae. None of the alvarezsaurid specimens preserves a complete caudal series. Nevertheless, large portions of the tail are known for both *Alvarezsaurus* (Fig. 4.10) and *Shuvuuia* (Fig. 4.13). In specimens MGI N100/99 and MGI 100/975 of *Shuvuuia,* 19 and 20 caudal elements are preserved, respectively, starting with very proximal elements (unfortunately none of these caudal series were articulated to the synsacrum). A recently described specimen (MPD 100/120) of this taxon (Suzuki et al., 2002) provides the best estimate for the number of caudal vertebrae in alvarezsaurids. Some 35 vertebrae fit the preserved segment of the tail of MPD 100/120, and additional remnants distal to this series suggest a slightly higher number.

The caudals are laterally compressed and procoelous. In the two most proximal elements preserved in *Shuvuuia* (perhaps the first two caudals), the centra are ventrally sharp, a condition shared by the first preserved caudal vertebra (a proximal element) of *Alvarezsaurus* (Bonaparte, 1991) and apparently *Parvicursor* as well (Karkhu and Rautian, 1996). In the subsequent vertebrae of *Shuvuuia,* the ventral border of the centra bears a strong, longitudinal furrow that runs throughout its entire length. In both *Shuvuuia* and *Alvarezsaurus,* the transverse processes are subtriangular and laterodistally directed. This appears not to be the case in *Parvicursor,* however, in which the transverse process is slightly lateroproximally oriented. Transverse processes are present in the first 11 preserved caudals of *Shuvuuia.* The caudal spinous processes of this taxon are proximodistally narrow and located immediately above the short postzygapophyses (Fig. 4.13A, B)—a condition shared by *Parvicursor* (Karhku and Rautian, 1996). Yet *Parvicusor* differs from *Shuvuuia* in that the postzygapophyses of the proximal caudals are far removed proximally from the distal margin of the centrum (Karkhu and Rautian, 1996). By the ninth caudal vertebra of *Shuvuuia,* the spinous process is virtually absent. The spinous processes appear to be absent or weakly developed (in the proximal section) in *Alvarezsaurus* (Bonaparte, 1991).

In the proximal section of *Shuvuuia,* the prezygapophyses are dorsoproximally oriented, projecting beyond the proximal margin of the centrum. For example, in the first seven preserved vertebrae of specimen MGI 100/975, their proximal extension is roughly one-third the length of the centrum (Fig. 4.13A). In the distal section of the tail, the prezygapophyses are short and do not extend to the preceding vertebra (Fig. 4.13C). The prezygapophyses are short in all preserved vertebrae of *Alvarezsaurus* (Bonaparte, 1991).

The caudal series of *Alvarezsaurus* differs drastically from that of *Shuvuuia* in that its distal elements are much

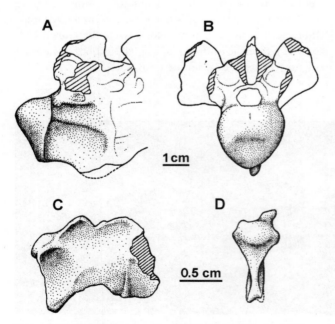

Figure 4.12. Caudal synsacral end of alvarezsaurids in right lateral (A, C) and caudal (B, D) views (after Novas, 1996). A, B, *P. puertai* (PVPH 37). C, D, *M. olecranus* (MGI 107/6).

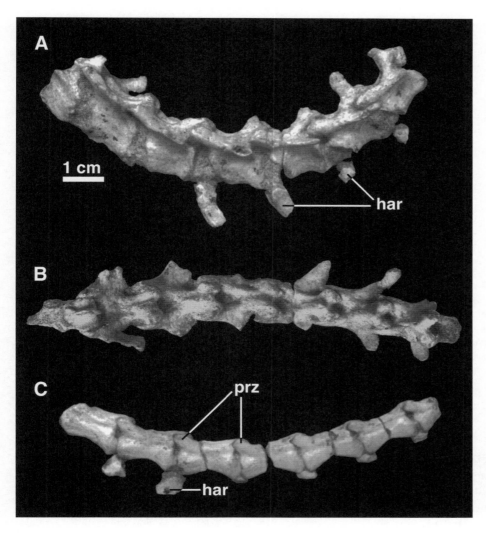

Figure 4.13. Caudal vertebrae of *S. deserti* (MGI 100/975). A, B, left lateral and dorsal views of the second? to eighth? vertebrae. C, left lateral view of 14th? to 20th? vertebrae. Abbreviations as in Figure 4.10.

longer than the proximal ones. For example, in the Patagonian form, the centrum of the most proximal vertebra (presumably the first caudal) is less than 60% of the length of the last preserved element (probably a midcaudal) (Bonaparte, 1991). This condition compares with that of *Archaeopteryx,* in which the midcaudal centra typically have more than twice the length of the centra of the proximal caudals (Wellnhofer, 1974). Interestingly, this condition is different from *Rahonavis* (Forster et al., 1998a,b), in which the length of the midcaudals is subequal to that of the most proximal vertebrae of the tail, even though the centra of the midcaudals are much more elongate than their most immediate preceding centra.

In *Shuvuuia*, the haemal arches of the 10 proximal vertebrae are higher than their proximodistal length (Fig. 4.13A). Similar haemal arches are present in the proximal portion of the tail of *Alvarezsaurus* (Fig. 4.10D). The haemal arch between the 10th and 11th caudals of *Shuvuuia* has an inverted T shape and is proximodistally longer than it is high. Subsequent haemal arches also have an inverted T shape, and

they gradually diminish in size (Fig. 4.13C). Inverted, T-shaped haemal arches are also present in the distal portion of the tail of *Alvarezsaurus* (Fig. 4.10F, G).

Thoracic Girdle and Sternum

The scapula, coracoid, and sternum are well known in both *Mononykus* and *Shuvuuia*. Portions of the scapulocoracoid and coracoid are preserved for *Alvarezsaurus* and *Patagonykus,* respectively. None of the available specimens of these taxa preserves a furcula. The fact that some of these (e.g., MGI 100/977) are exquisitely preserved and articulated suggests that this element was probably absent in Alvarezsauridae, but the recent discoveries of furculae in several nonavian theropods thought to lack them (Chure and Madsen, 1996; Norell et al., 1997; Makovicky and Currie, 1998) underscore the lack of certainty.

Scapula. The scapula is firmly attached to the coracoid, although in none of the known specimens are these bones fused to each other (Fig. 4.14). The scapular blade is slender

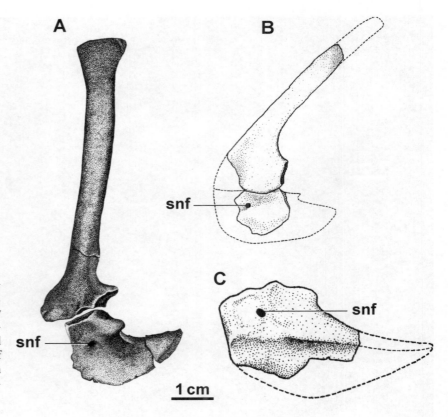

Figure 4.14. Thoracic girdle of alvarez-saurids. A, left scapulocoracoid of *M. olecranus* (MGI 107/6) in lateral view (from Perle et al., 1994). B, left scapulocoracoid of *A. calvoi* (MUCPv 54) in lateral view (after Novas, 1996). C, left coracoid of *P. puertai* (PVPH 37) in lateral view (from Novas, 1996). Abbreviation: snf, supracoracoid nerve foramen.

and compressed. This transverse compression is stronger in its distal third. In *Mononykus* and *Shuvuuia*, the blade is straight in both lateral and dorsal view, differing from that of *Alvarezsaurus*, which is distinctly curved in dorsal view (Bonaparte, 1991). Dorsal and ventral margins are sub-parallel for most of the blade. The distal portion expands dorsoventrally, and it ends abruptly.

The shoulder end is dorsoventrally expanded and slightly twisted medially. The acromion does not project be-yond the articular facet for the coracoid. The latter facet is subtriangular and much larger than the semioval and slightly concave humeral articular facet. The orientation of the scapular blade suggests that, in life, the humeral articu-lar facet of the scapulocoracoid faced caudolaterally. *Al-varezsaurus* also differs from *Mononykus* and *Shuvuuia* in that the scapula is proportionally smaller. The scapula of the Patagonian taxon is roughly half the length of the ilium (Bonaparte, 1991; Novas, 1996), a proportion much lower than in the Asian alvarezsaurids.

Coracoid. The coracoid is short, wider than it is tall, and has a generally crescentic shape (Fig. 4.14). In *Mononykus* and *Shuvuuia*, the shaft is strongly compressed and virtually flat. In contrast, in *Patagonykus*, the coracoid shaft is strongly convex ventrally (rostrally) and concave dorsally (caudally), and its medial portion is very thick (Novas, 1997). In the Patagonian taxon, a distinct, oblique ridge runs transversely on the ventral (rostral) surface of the shaft (Fig. 4.14C).

The bicipital tubercle is absent in all alvarezsaurid cora-coids. A round supracoracoid nerve foramen perforates the center of the shaft, somewhat below the scapular articula-tion (Fig. 4.14). The medial margin of the shaft is straight to slightly convex. Laterally, the shaft projects into a long, tri-angular, tapered flange. The sternal border of the coracoid is convex, providing an ample area of articulation for the sternum.

The humeral articular facet is suboval, slightly convex, and smaller than the articular surface for the scapula. As in *Archaeopteryx* and nonavian theropods (Chiappe, 1996), this facet is on the same plane of the humeral articular facet of the scapula. The general shape and the orientation of the scapulocoracoid, preserved in articulation with the sternum in specimen MGI 100/977 of *Shuvuuia*, suggest that the humeral articular facet probably faced caudolaterally.

Sternum. The sternum is a stout and caudally tapered bone, longer than it is wide, with a subtriangular cross sec-tion (Fig. 4.15). It is simple and small and without ossified processes or facet for the articulation of ribs. Ventrally, it bears a strong carina that splits rostrally in two lateral crests enclosing a median groove (Perle et al., 1994). Specimen MGI 100/977 of *Shuvuuia* (Fig. 4.2) shows that the long ster-

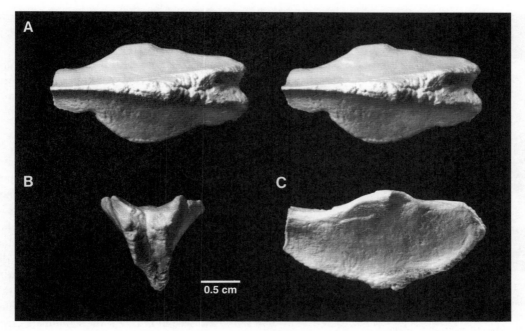

Figure 4.15. Sternum of *M. olecranus* (MGI 107/6) in ventral (A, stereo pairs), cranial (B), and right lateral (C) views (after Perle et al., 1994).

nal margin of the coracoid articulated with the craniolateral portion of the sternum.

Thoracic Limb

The bizarre structure of the thoracic limb of alvarezsaurids is their most characteristic skeletal feature (Fig. 4.16). Nearly complete thoracic limbs are known for both *Mononykus* and *Shuvuuia*, and portions of it are preserved in *Patagonykus*. The thoracic limb is, however, unknown in *Alvarezsaurus* and *Parvicursor*.

Humerus. The humerus is short and robust, with both extremities expanded in the same plane (Perle et al., 1994). In *Mononykus* and *Shuvuuia*, its proximal end bears a prominent, proximally projected ventral tubercle that tapers toward its tip (Fig. 4.17). A similarly projected ventral tubercle is present in *Patagonykus* (Novas, 1997), although it is shorter than in the Asian alvarezsaurids, and it has a truncated end (Fig. 4.16A). The humeral head is large. In *Mononykus* and *Shuvuuia*, it is cranially concave and caudally convex (Fig. 4.17B); in *Patagonykus*, it is cranially flat. The caudal surface of the proximal end lacks both a pneumatic fossa and a foramen. In *Mononykus*, the deltopectoral crest is a prominent, pillarlike, craniodorsally directed structure (Figs. 4.17, 4.18; Perle et al., 1994). In this respect, *Mononykus* differs from *Shuvuuia*, in which the deltopectoral crest forms a robust craniodorsal expansion continuous with the humeral head. The deltopectoral crest is not preserved in *Patagonykus* (Novas, 1997).

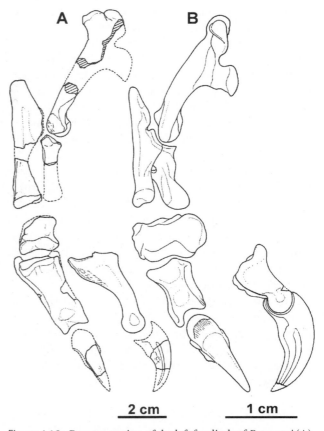

Figure 4.16. Reconstruction of the left forelimb of *P. puertai* (A) and *M. olecranus* (B) in dorsal view, with digit I shown in lateral view (after Novas, 1996).

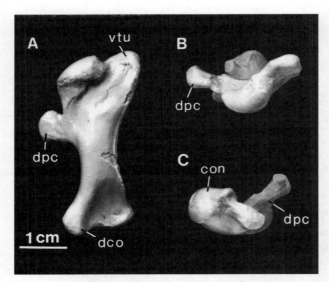

Figure 4.17. Left humerus of *M. olecranus* (MGI 107/6) in caudal (A), proximal (B), and distal (C) views (after Perle et al., 1994). Abbreviations: con, condyle; dco, dorsal epicondyle; dpc, deltopectoral crest; vtu, ventral tubercle.

Distally, the alvarezsaurid humerus bears a single condyle that is mainly developed on the cranial surface (Figs. 4.17C, 4.18A). In this view, the condyle is subtriangular in *Mononykus* (Perle et al., 1994) and in *Shuvuuia* but subcircular in *Patagonykus* (Novas, 1997). The dorsal epicondyle is prominently developed, projecting dorsodistally. In *Patagonykus*, there is also a sharp ventral epicondyle, projecting ventrally (Novas, 1997). The caudal surface of the distal end of the humerus is concave in the Asian taxa and nearly flat in *Patagonykus*; a distinct olecranon fossa is absent (Fig. 4.17A).

Ulna. The ulna of alvarezsaurids bears a characteristic hypertrophied olecranon process (Figs. 4.16, 4.18C), with a transversely expanded end. In *Mononykus*, the olecranon process forms nearly half the length of the ulna (Perle et al., 1994). In *Patagonykus*, it is much shorter (Fig. 4.16), although it may have formed one-third of the ulnar length, but a precise estimate of its olecranon length is uncertain, given the available material (Novas, 1997).

The ulna has a single, subtriangular cotyla. The ulnar shaft is short. Its caudal margin consists of a robust ridge, lacking papillae for feather attachment. Distally, the ulna has a distinct carpal trochlea. This is craniocaudally expanded and subtriangular from the distal view (Perle et al., 1994; Novas, 1996, 1997).

Radius. The radius is a short and robust bone (Fig. 4.18C). In *Mononykus*, it is roughly half the size of the ulna (Perle et al., 1994). Its proximal extremity is strongly co-ossified to the ulna. Its subtriangular cotyla is connected to

that of the ulna, forming an ample articular surface for the humeral condyle. The distal end bears a large, proximally expanded radiocarpal facet (Fig. 4.18C). This facet is craniodorsally oriented. Its ventral surface bears a distinct ligamental depression that separates the dorsal portion of the radiocarpal facet from a proximal extension of its ventral margin (Perle et al., 1994).

Carpometacarpus. The carpometacarpus of alvarezsaurids is formed by three metacarpals (alular, major, and minor, or I, II, III) and at least one carpal, similar to the semilunate carpal of *Archaeopteryx* and nonavian theropods. This latter element is fused to the proximal end of the metacarpal I. As discussed elsewhere (Perle et al., 1994; Chiappe et al., 1996), the logic behind the interpretation of the proximal part being a distal carpal lies first in the fact that the convex, pulleylike morphology of the proximal end of the carpometacarpus is remarkably similar to the condition in *Archaeopteryx* and nonavian maniraptoriform dinosaurs, and second in that if the semilunate carpal and the metacarpal I were not fused to each other, the proximal end of the latter would be flat or slightly concave, as in nonavian theropods.

The carpometacarpus is massive and quadrangular in shape (Fig. 4.19; Perle et al., 1994). The metacarpals are co-ossified to varying degrees. In *Mononykus* (holotype, MGI 107/6), they are firmly co-ossified, although lines of contact are still distinct. In *Shuvuuia*, in specimens (e.g., MGI 100/975) of comparable or larger size than the holotype of *Mononykus*, metacarpals II and III are not preserved—but the articular facet for metacarpal II is present on metacarpal I. This condition suggests that in these specimens metacarpals II and III were not co-ossified to metacarpal I (and its fused semilunate), suggesting the bones co-ossified at a larger size in *Shuvuuia*.

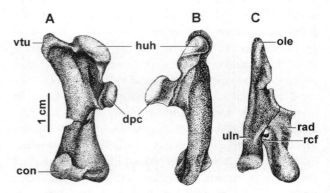

Figure 4.18. Left humerus (A, B) and right radius-ulna (C) of *M. olecranus* (MGI 107/6) in cranial (A), dorsal (B, C) views (after Perle et al., 1994). Abbreviations: huh, humeral head; ole, olecranon; rad, radius; rcf, radiocarpal facet; uln, ulna; other abbreviations as in Figure 4.17.

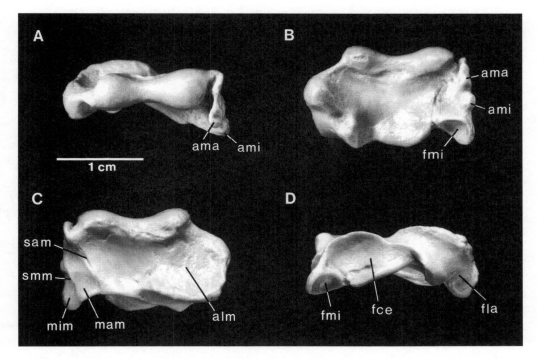

Figure 4.19. Left carpometacarpus of *M. olecranus* (MGI 107/6) in distal (A), ventral (B), dorsal (C), and proximal (D) views (after Perle et al., 1994). Abbreviations: alm, metacarpal I; ama, articular facet of metacarpal II; ami, articular facet metacarpal III; fce, central proximal articular facet; fla, lateral proximal articular facet; fmi, proximal articular facet of metacarpal III; mam, metacarpal II; mim, metacarpal II; sam, suture between metacarpals I and II; smm, suture between metacarpals II and III.

The alvarezsaurid metacarpal I is hypertrophied. In *Mononykus*, and presumably in all alvarezsaurids, metacarpals II and III are tiny and cling to the lateral margin of the latter, leaving no intermetacarpal space between them (Fig. 4.19).

Metacarpal I is dorsoventrally compressed (Fig. 4.19A), transversely broad, and subpentagonally shaped (Fig. 4.19C). Its distal articular facet forms a broad, well-developed trochlea. The distal ends of metacarpals II and III bear small, ball-shaped articular facets (Fig. 4.19A). The proximal articular surface of the carpometacarpus exhibits three distinct articular surfaces (Fig. 4.19D; Perle et al., 1994). The medial facet, formed by the semilunate carpal, is a pulley-shaped articulation. The central facet, lateral to the semilunate portion, is subtriangular and is mainly formed by metacarpal I, although it also receives a contribution from metacarpal II. The lateral articular facet of the carpometacarpus is small and subcircular and is formed exclusively by metacarpal III. The relationships of these facets with both the free carpal elements and the radius and ulna are not clear. Free carpals are not known, but the morphology of proximal facets of the carpometacarpus indicates that they were certainly present.

Manual Digits. Despite the robustness and hypertrophy of digit I, the hand of alvarezsaurids consists of three digits (Perle et al., 1994; Suzuki et al., 2002). Evidence in support of this initially came from the articular facets at the distal end of metacarpals II and III of *Mononykus* and the presence of a tiny phalanx, found in the small concretion containing the carpometacarpus of specimen MGI 100/975 of *Shuvuuia*, that could not represent anything other than an intermediate phalanx of either the major or minor digit. Since then, confirmation of the presence of a three-fingered hand in alvarezsaurids was provided by Suzuki et al. (2002), who described a specimen of *Shuvuuia* (MPD 100/120) with a completely articulated hand.

Digit I is composed of two phalanges (Figs. 4.16, 4.20). The proximal phalanx is dorsoventrally compressed, with a broad, proximal articular surface. In the Asian alvarezsaurids, this phalanx is strongly asymmetrical, with its dorsolateral corner projecting dorsally (Fig. 4.20G). This condition contrasts with its more symmetrical morphology in *Patagonykus* (Fig. 4.20I; Novas, 1997). *Patagonykus* also differs from its Asian relatives in that the proximal articular facet of this phalanx is proportionally narrower. In both Asian and Patagonian alvarezsaurids, the ventral surface of

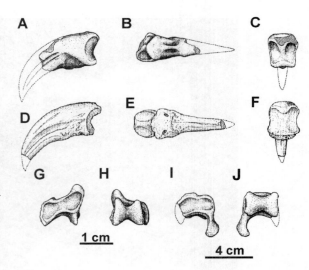

Figure 4.20. Phalanges of digit I of alvarezsaurids (after Novas, 1996). A, B, C, ungual phalanx of *A. calvoi* (MUCPv 54) in lateral (A), ventral (B), and proximal (C) view. D, E, F, ungual phalanx of *M. olecranus* (MGI 107/6) in lateral (D), ventral (E), and proximal (F) view. G, H, proximal phalanx of *M. olecranus* (MGI 107/6) in proximal (G) and distal (H) view. I, J, proximal phalanx of *P. puertai* (PVPH 37) in proximal (I) and distal (J) view. Four-centimeter scale applicable for I and J only.

the proximal phalanx of digit I bears a broad, axial furrow outlined by distinct, ventrally projected ridges (the medial ridge is not preserved in *Patagonykus*). The lateral ridge is much more prominent in *Patagonykus* than in the Asian taxa (Novas, 1997). Distally, this phalanx has a well-developed trochlea, with fossae for the collateral ligaments on both sides. The distal phalanx of digit I is robust and arched and forms a sharp claw (Perle et al., 1994). The flexor tubercle is essentially absent (Fig. 4.20). Instead, the ventral surface of the proximal end forms a flat platform, which in *Mononykus* and *Shuvuuia* is perforated on both sides by small foramina.

As shown in MPD 100/120, digits II and III of *Shuvuuia* are much smaller than digit I. Digit II bears three phalanges, and at least a similar number is present in digit III (Suzuki et al., 2002). Preservation, however, prevents deciding whether digit IV had a fourth phalanx. The claws of these digits are almost straight, gradually tapering toward their tips (Suzuki et al., 2002).

Pelvic Girdle

The pelvic girdle of the Asian alvarezsaurids exhibits strong differences from that of the Patagonian forms, which exhibit a less specialized morphology. The pelvic bones are best known for *Shuvuuia* and *Parvicursor,* but informative pelvic remains are available for all alvarezsaurid taxa. The bones of the pelvis are not fused in the only known specimen of *Patagonykus.* The condition is uncertain in *Alvarezsaurus* and *Shuvuuia,* but the pelvis is partially fused in *Mononykus* (holotype, MGI 107/6), in which the ischium and ilium are fused to each other. Karkhu and Rautian (1996) described a complete fusion in the acetabular region of *Parvicursor,* although their illustrations show that the pubis and ischium of the only known specimen are not connected to the iliac portion of the acetabulum.

Ilium. The ilium is low, especially in the postacetabular portion of the Asian taxa (Fig. 4.21). In *Alvarezsaurus,* pre- and postacetabular wings are subequal in length (Novas, 1996, 1997), and the former is slightly lower than the latter (Bonaparte, 1991). In contrast, the dimensions of the synsacrum of *Shuvuuia* (MGI 100/975; MGI N 100/99) suggest that the preacetabular wing of this taxon was somewhat shorter than the postacetabular one. Furthermore, in the Asian taxa, the preacetabular wing developed more vertically than the postacetabular wing, which is broad and remarkably expanded laterally (Figs. 4.21B, D).

In both *Alvarezsaurus* and the Asian taxa, the ventral surface of the postacetabular wing defines a shallow brevis fossa. This, however, is much narrower in *Alvarezsaurus* than in its Asian relatives.

The alvarezsaurid ilium possesses a supracetabular crest overhanging the acetabular fossa and restricted to the dorsal margin of the acetabulum. *Mononykus,* however, differs from both *Alvarezsaurus* and apparently *Patagonykus* in having this crest developed only in the cranial half of the acetabular fossa (Perle et al., 1994). The strong pubic peduncle is longer craniocaudally than transversely, and it is cranioventrally oriented (Perle et al., 1994; Novas, 1996, 1997). The ischiadic peduncle is smaller than the pubic one and much less distally projected. In *Patagonykus,* the ischiadic pedicel retains an articular facet for the ischium, but this is absent in the holotype specimen of *Mononykus* (MGI 107/6), in which the ilium and ischium are completely fused to each other. The antitrochanter of alvarezsaurids is stout and horizontally projected (Perle et al., 1994). Its articular surface is convex, and its main axis is oriented caudolaterally. The morphology of *Patagonykus* shows that the ilium formed most of the prominent antitrochanter (Novas, 1997).

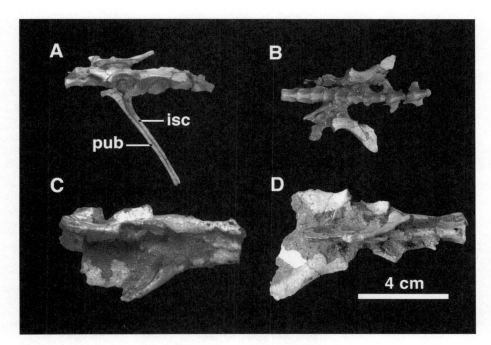

Figure 4.21. Pelvis and synsacrum of *S. deserti*. A, B, MGI N 100/99 in left lateral (A) and ventral (B) views. C, D, MGI 100/975 in right lateral (C) and ventral (D) views. Abbreviations: isc, ischium; pub, pubis.

Ischium. The ischium of *Shuvuuia* and *Parvicursor* (Karkhu and Rautian, 1996) is long, slender, and laterally compressed, abutting the pubis throughout its entire length (Figs. 4.21, 4.22G, H). Only the antitrochanteric portion of this bone is preserved in *Mononykus* (Perle et al., 1994). The compression of the proximal end (the only portion preserved) of the ischium of *Patagonykus* indicates that in the Patagonian alvarezsaurid the shaft of this bone was also strongly compressed (Novas, 1997). *Patagonykus* also shows that the ischium of alvarezsaurids contributed to the formation of the strong antitrochanter.

The ischium forms only a small portion of the acetabular fossa (Perle et al., 1994; Novas, 1997). The ischiadic shaft of *Shuvuuia* shows that the obturator process is absent and that the distal ends of the ischia do not form a terminal symphysis (*contra* Sereno, 2001).

Pubis. The alvarezsaurid pubis is also slender and long (Figs. 4.21, 4.22; Novas, 1996, 1997; Hutchinson and Chiappe, 1998). In proximal view, its proximal end is laterally compressed and kidney-shaped, although the articular facet of the Asian taxa is clearly longer than that of *Patagonykus*. Immediately distal to the ischiadic articulation, the caudal margin is notched, forming the cranial margin of the obturator foramen (Fig. 4.21A).

Alvarezsaurids have an opisthopubic pelvis (Perle et al., 1993a; Chiappe et al., 1996; Karkhu and Rautian, 1996; Novas, 1996). In *Shuvuuia*, for example, the shaft of the pubis forms an angle of approximately 45° with respect to the horizontal plane of the postacetabular wing of the ilium (MGI

100/975 and MGI N 100/99) (Fig. 4.21). In *Patagonykus,* the pubis was oriented caudoventrally (Novas, 1996, 1997), although because of the incompleteness of the ilium it is hard to estimate an accurate angle of orientation. The pubic shaft has a subtriangular section in *Mononykus* and a subcircular one in *Shuvuuia* and *Patagonykus*. The shaft of the last taxon exhibits a weak, medial, longitudinal ridge and a distal, caudally expanded foot (Novas, 1997). These structures are absent in the Asian taxa. Despite the presence of a distal foot, *Patagonykus* probably lacked a distal pubic symphysis (Novas, 1996, 1997), as in its Asian counterparts. The absence of a pubic symphysis is unique to birds among theropod dinosaurs (Weishampel et al., 1990). Alvarezsaurids, however, must have evolved this condition independently, since a distinct symphysis is present in *Confuciusornis* (Peters, 1996; Chiappe et al., 1999) and enantiornithine birds (Zhou and Hou, Chapter 7 in this volume; Chiappe and Walker, Chapter 11 in this volume).

Pelvic Limb

The anatomical information on the hindlimb is very complete for the Asian alvarezsaurids. Several portions of the hindlimb are also known for both *Patagonykus* and *Alvarezsaurus*.

Femur. The femur is slender and cranially bowed (Figs. 4.23, 4.24A–C; Perle et al., 1994; Karkhu and Rautian, 1996). Proximally, it has a stout, medially directed head that is subcircular in cross section. *Mononykus* and *Parvicursor* (Karkhu and Rautian, 1996) have well-developed, uninterrupted

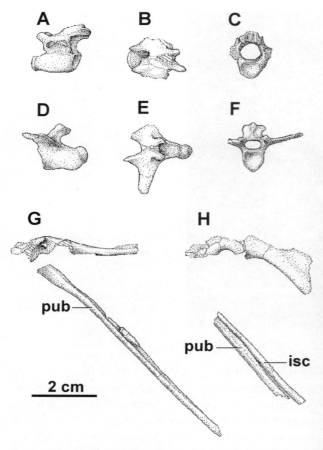

although these projections are not developed in either *Patagonykus* (Novas, 1997) or *Parvicursor* (Fig. 4.24H; Karkhu and Rautian, 1996). The robust medial condyle is subcircular in distal view and virtually flat; transversely, this condyle is less expanded in *Patagonykus*. The lateral condyle exhibits a prominent, cone-shaped distal surface (Fig.

Figure 4.22. Vertebral and pelvic elements of *P. remotus* (PIN 4487/25) (after Karhu and Rautian, 1996). A–C, last presacral vertebra in left lateral (A), dorsal (B), and caudal (C) view. D–F, first caudal vertebra in left lateral (D), dorsal (E), and cranial (F) view. G, H, acetabular and postacetabular portions of the left ilium in lateral (G) and dorsal (H) view. G, H, left pubis and ischium in lateral view (G) and right pubis and ischium in medial (H) view. Abbreviations as in Figure 4.21.

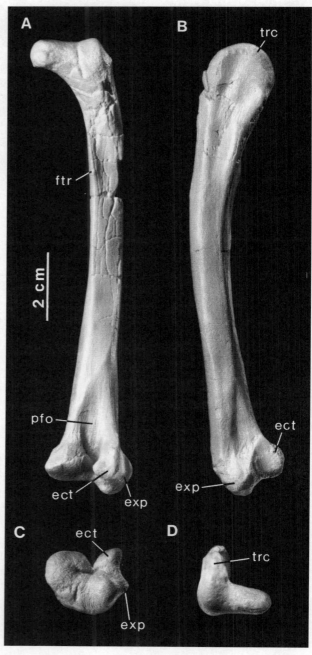

Figure 4.23. Right (A, C) and left (B, D) femora of *M. olecranus* (MGI 107/6) in caudal (A), lateral (B), distal (C), and proximal (D) views (after Perle et al., 1994). Abbreviations: ect, ectocondylar tuber; exp, external projection of lateral condyle; ftr, fourth trochanter; pfo, popliteal fossa; trc, trochanteric crest.

trochanteric crests, but in *Patagonykus* (Novas, 1997) and *Alvarezsaurus* (Bonaparte, 1991), the anterior (lesser) trochanter is separated from the greater trochanter by a low, narrow notch (Fig. 4.25A). In *Mononykus*, *Patagonykus*, and *Alvarezsaurus*, at the midpoint of the proximal half of the shaft, there is a weak, ridgelike fourth trochanter on the medial margin of the caudal surface (Fig. 4.23A). This appears to be absent in *Parvicursor* (Figs. 4.24B, C; Karkhu and Rautian, 1996), and it is only incipient in the juvenile specimen MGI N 100/99 of *Shuvuuia*. Unfortunately, this femoral portion is not preserved in the larger specimens of *Shuvuuia* to assess whether this feature is size-related.

Distally, the cranial surface of the femur lacks a distinct patellar groove (Figs. 4.23C, 4.24A). Caudally, the popliteal fossa of *Mononykus* is distally confined by projections of the lateral and medial condyles (Fig. 4.23A, C; Perle et al., 1994),

4.23A); a similar structure has been reported in the non-avian theropod *Bagaraatan* (Osmólska, 1996). Caudally, the lateral condyle forms a robust ectocondylar tuber. This tuber is proximally continuous with the lateral border of the popliteal fossa but distally separated from the distal articular surface by a depressed, concave area (Fig. 4.23B). The external surface of the lateral condyle is laterally projected.

Fibula. The proximal end is laterally compressed and projects caudally. The lateral surface of the shaft is convex, while the medial surface is flat. The tubercle for M. iliofibularis is prominent and laterally oriented (Fig. 4.26). In the Asian alvarezsaurids, the fibular shaft forms a very thin, short spine distal to this tubercle. In these taxa, the fibula does not articulate with the calcaneum. Nevertheless, the presence of a distinct fibular facet in the calcaneum of *Patagonykus* and a fragment of the distal end of the fibula of *Alvarezsaurus* articulated to the calcaneum (Figs. 4.25D, 4.27) indicate that the Patagonian taxa display the primitive condition, unlike in birds other than *Archaeopteryx*.

Tibiotarsus. In *Mononykus, Patagonykus,* and *Parvicursor,* the tibia and proximal tarsals are strongly united and partially co-ossified. The proximal tarsals are also tightly joined to the tibia in *Alvarezsaurus* (Figs. 4.25D, 4.27A, B) and *Shuvuuia,* although not co-ossified to it.

The proximal end of the tibia of alvarezsaurids bears a strong, short cnemial crest (Figs. 4.24D, E, 4.28; Perle et al., 1994). In *Mononykus,* on the medial side of this crest, there is a smooth, second crest (Perle et al., 1994), although this second crest is absent in *Patagonykus* (Novas, 1997) and apparently in *Parvicursor* (Karkhu and Rautian, 1996). The articular surface of the proximal end is laterally inclined (Fig. 4.28; Perle et al., 1994). In *Mononykus* and *Parvicursor,* the medial border of the medial articular facet strongly projects proximally, but this projection is absent in *Patagonykus.* The medial articular facet is suboval in proximal view. The lateral articular facet forms a round, strongly convex area that slopes caudolaterally. In *Mononykus,* this surface is separated from the cnemial crest by a broad, lateral indentation (Perle et al., 1994). In *Patagonykus,* how-

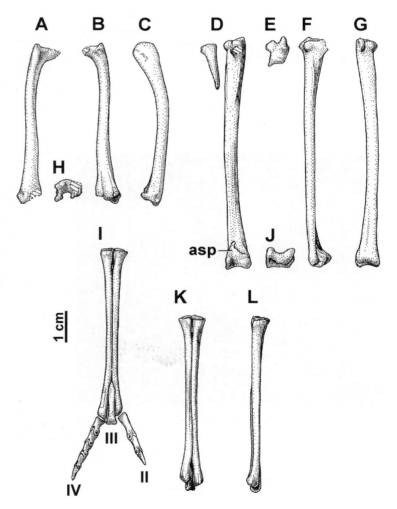

Figure 4.24. Hindlimb elements of *P. remotus* (PIN 4487/25) (after Karhu and Rautian, 1996). A–C, H, right femur in cranial (A), caudal (B), lateral (C), and distal (H) view. D–G, J, tibiotarsus and fibula in cranial (D), proximal (E), lateral (F), caudal (G), and distal (J) view. I, K, L, tarsometatarsus and pes in dorsal (I), plantar (K), and lateral (L) view. Abbreviations: asp, ascending process; II–IV, metatarsal and digit numbers.

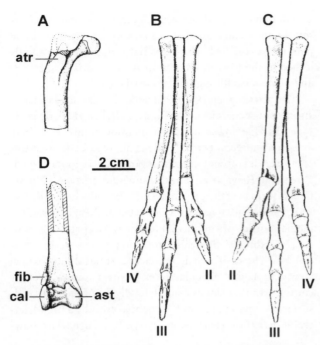

Figure 4.25. Hindlimb elements of *A. calvoi* (MUCPv 54) (after Bonaparte, 1991). A, proximal half of right femur in cranial view. B, C, dorsal and plantar views of the metatarsals and pes. D, distal end of tibia, fibula, and proximal tarsals. Abbreviations: ast, astragalus; atr, anterior (lesser) trochanter; cal, calcaneum; fib, fibula; other abbreviations as in Figure 4.24.

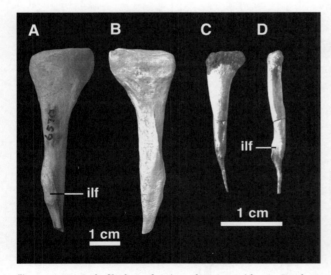

Figure 4.26. Left fibulae of Asian alvarezsaurids. A, B, alvarezsaurid fibula from Iren Dabasu (Inner Mongolia, China) in lateral (A) and medial (B) view (the incomplete nature of this specimen precludes a lower-level taxonomic identification). C, D, fibula of *S. deserti* (MGI N 100/99) in lateral (C) and caudal (D) view. Abbreviations: ilf, iliofibularis tubercle.

ever, there is a short, lateral projection in between these two areas (Novas, 1997). Medial and lateral articular facets are separated by a small proximal depression and a caudal notch in *Mononykus* and by a deep caudal notch in *Patagonykus*.

The shaft of the tibia is round to suboval. The fibular crest is well developed, running along the proximal third of the tibia. Caudal to its distal end there is a small nutrient foramen. The shaft is essentially straight in *Mononykus* (Perle et al., 1994) but somewhat bowed mediolaterally in *Shuvuuia* and *Parvicursor* (Figs. 4.24D, G). Distally, the tibial shaft is craniocaudally compressed. In *Shuvuuia*, the medial border of the distal end forms a sharp ridge. This is absent in *Mononykus*, *Patagonykus*, and *Alvarezsaurus*.

In *Mononykus*, *Patagonykus*, *Parvicursor*, and *Shuvuuia*, the astragalus and calcaneum are fused to each other (Figs. 4.24D, 4.27C, 4.28C). These elements, however, are not fused to each other in *Alvarezsaurus* (Fig. 4.27A; Bonaparte, 1991; Perle et al., 1994). The medial condyle is larger and broader than the lateral one. The ascending process of the astragalus is laminar. In the Asian alvarezsaurids, the medial border is indented. In contrast to *Shuvuuia* and *Parvicursor*, in *Mononykus* this medial indentation deeply excavates the base of the ascending process (Fig. 4.28C). *Mononykus* and *Parvicursor* differ from *Shuvuuia* in that their ascending processes rise from the medial margin of the astragalar (medial) condyle instead of from the lateral margin of this condyle. In *Patagonykus*, the ascending process is broad, and its medial margin is not indented (Novas, 1997). In contrast to the Asian alvarezsaurids, in both *Patagonykus* and *Alvarezsaurus* the calcaneum has a distinct articular facet for the distal end of the fibula (Fig. 4.27).

Distal Tarsals. A single, incomplete distal tarsal is preserved in a juvenile specimen of *Shuvuuia* (MGI N 100/99). It is a nearly flat bone that articulates primarily with metatarsal IV. A similar bone was reported by Karkhu and Rautian (1996) for *Parvicursor* (Fig. 4.24I, K, L). This taxon also preserves an individualized, smaller distal tarsal capping the plantar border of metatarsal II. A flat and disk-shaped distal tarsal, co-ossified to the proximal surface of metatarsals II and III, is preserved in *Patagonykus* (Novas, 1997).

Metatarsals. The metatarsals are elongated and slender; none of them is fused to another (Figs. 4.24I, K, 4.25B, C, 4.28D, 4.29; Perle et al., 1994). In the Asian alvarezsaurids, metatarsals II and IV have the same length, but metatarsal III is much shorter and does not contribute to the proximal articular surface (arctometatarsalian condition; see Holtz, 1994, 2001). This condition is absent in both *Alvarezsaurus* (Bonaparte, 1991) and *Patagonykus* (Novas, 1997), in which metatarsal III forms part of the proximal articular area, although its articular surface is smaller than that of meta-

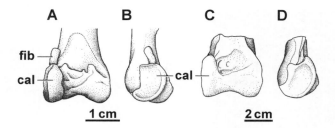

Figure 4.27. Distal end of crus of *A. calvoi* (MUCPv 54; after Novas, 1996) (A, B) and *P. puertai* (PVPH 37; modified from Novas, 1996) (C, D) in cranial (A, C) and lateral (B, D) view. Abbreviations as in Figure 4.25.

tarsals II and IV (Fig. 4.25B, C). The presence of a non-arctometatarsalian foot in these two taxa indicates an independent development of the arctometatarsalian condition in the Asian alvarezsaurids—a conclusion already advanced by Perle et al. (1994). *Alvarezsaurus* also differs from the Asian forms in that its metatarsal IV is significantly longer than metatarsal II and in that its metatarsal III is the longest (Bonaparte, 1991).

In the Asian taxa, the proximal articular surfaces of metatarsals II and IV are subtriangular and subquadrangular, respectively (Perle et al., 1994). The proximal articular surface of metatarsal II of *Patagonykus* is not subtriangular but subrectangular (Novas, 1997). The hypotarsus is absent in all known alvarezsaurid taxa. In the juvenile specimen MGI N 100/99 of *Shuvuuia*, a tiny, thin metatarsal V attaches to the plantar surface of the proximal end of metatarsal IV. Metatarsal I is known only for *Mononykus* and *Shuvuuia*. It is small and straight and tapers proximally (Perle et al., 1994). In the Asian taxa, the shafts of metatarsals II and IV are remarkably similar. They are straight and slightly compressed transversely (Fig. 4.29). These metatarsals define a median, axial furrow that runs through most of their length, both dorsally and plantarly. The medial and lateral margins of the plantar surfaces of metatarsals II and IV, respectively, form a sharp ridge. In these taxa, the shaft of metatarsal III is subtriangular in cross section and dorsally flat.

At the distal end, well-developed collateral foveae are present on both sides of the trochlea of metatarsal III. The trochlea for metatarsal II has a strong, plantarly projected lateral rim and a much thinner and less plantarly projected medial one (Perle et al., 1994). Laterally, this trochlea exhibits a large collateral fossa, while a small one is present in the medial face. The trochlea of metatarsal IV has a stout medial rim, and the much thinner lateral rim is twisted somewhat laterally. Well-developed collateral fossae are present on both sides of this trochlea.

Pedal Phalanges. The pedal phalanges are best known for *Mononykus* (Perle et al., 1994) and *Parvicursor* (Karkhu and Rautian, 1996). The phalangeal formula in these taxa is 2-3-4-5-*x* (phalangeal formula designation follows Padian [1992]). The hallux is not reversed (Fig. 4.28D; Suzuki et al.,

2002). This digit bears a short, asymmetrical proximal phalanx with a large, dorsally extended articular trochlea (Perle et al., 1994). Its ungual phalanx lacks a flexor tubercle, a condition comparable to that of the remaining claws. In the second digit, the proximal phalanx is much longer than the intermediate phalanx. Its ungual is the largest of the claws. As in the remaining unguals, it is sharp and weakly arched and bears deep, L-shaped axial grooves on both sides. The ungual phalanges of *Mononykus* and *Parvicursor* resemble the single pedal claw known for *Patagonykus* (Novas, 1997). The proximal and intermediate phalanges of the third digit distally decrease in length. The ungual phalanx of this digit is similar to, although slightly smaller than, that of the second digit. In the fourth digit, the proximal and intermediate phalanges are short (Fig. 4.28D).

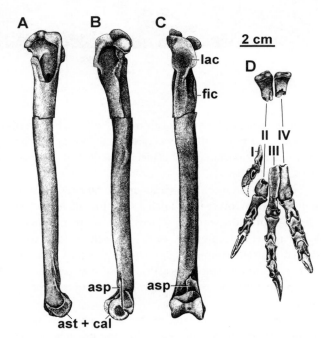

Figure 4.28. Left tibiotarsus (A, B, C) and pes (D) of *M. olecranus* (MGI 107/6) in medial (A), lateral (B), cranial (C), and dorsal (D) view (after Perle et al., 1994). Abbreviations: asp, ascending process of astragalus; fic, fibular crest; lac, lateral cnemial crest; others as in Figure 4.25.

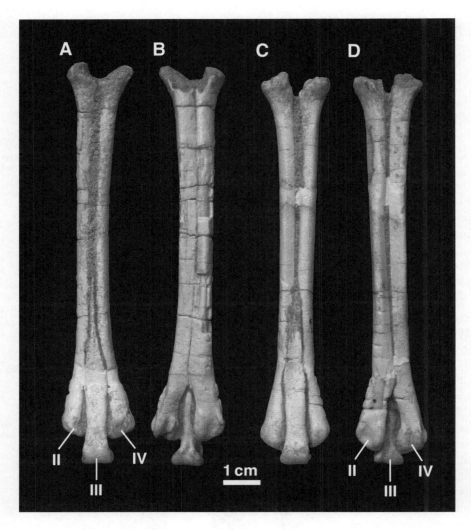

Figure 4.29. Metatarsals of *S. deserti* (MGI 100/975) in dorsal (A, C) and plantar (B, D) view. Abbreviations as in Figure 4.24.

Plumage

The phylogenetic placement of alvarezsaurids by Chiappe et al. (1998a), as well as all other cladistic analyses supporting the avian relationship of this clade, implies that these creatures were feathered.

Several portions of a specimen of *Shuvuuia* (MGI 100/977) were covered by small (~200 μm in diameter) fibrous structures with a preferred spatial arrangement. Schweitzer et al. (1997, 1999; see also Schweitzer, 2001) conducted multiple studies (e.g., microscopic, mass spectrometric, immunohistochemical) on these structures, concluding that they were epidermally derived and most likely feathers.

Although congruent with the hypothesis proposing an avian relationship for alvarezsaurids, the possible occurrence of feathers in alvarezsaurids should not be taken as the "ultimate proof" of their avian nature. Recent reports have documented the presence of feathers in several nonavian maniraptoriforms (Chen et al., 1998; Ji et al., 1998, 2001; Xu et al., 1999a,b, 2000, 2001; Zhou and Wang, 2000), thus showing that the presence of feathers is plesiomorphic for birds—an expected conclusion when considering that the feathered *Archaeopteryx* is by definition (phylogenetic) the most basal bird.

Phylogenetic Relationships

Although the placement of *Mononykus* by Perle et al. (1993a) as closer to neornithine birds than *Archaeopteryx* raised a great deal of controversy, this same hypothesis—later expanded to include all alvarezsaurids—was supported by the subsequent cladistic analyses of Chatterjee (1995), Chiappe et al. (1996, 1998a), Forster et al. (1996, 1998a), Novas (1996), and Ji et al. (1998) (Fig. 4.30). The better understanding of the cranial morphology of alvarezsaurids also added support to the proposed avian relationship, with the skull of *Shuvuuia* documenting several characters (e.g., absence of a jugal-postorbital bar, quadratojugal not sutured to the quadrate and without abutting the squamosal) otherwise unknown among nonavian archosaurs.

Early criticisms of the avian hypothesis focused on both empirical and methodological issues (e.g., Feduccia, 1994; Martin and Rinaldi, 1994; Ostrom, 1994; Wellnhofer, 1994; Zhou, 1995; Feduccia and Martin, 1996). Morphological objections, however, were shown to be misguided. As pointed out numerous times (e.g., Chiappe et al., 1995, 1996, 1997, 1998a; Novas and Pol, Chapter 5 in this volume), (1) only part of the supporting evidence was appraised (e.g., Ostrom, 1994), (2) the proposed synapomorphies were a priori deemed as convergent adaptations between *Mononykus* and birds (e.g., Feduccia, 1994; Zhou, 1995), and (3) the supporting evidence was disregarded because it was in conflict with assumptions about the evolutionary process (e.g., Wellnhofer, 1994; Feduccia and Martin, 1996). A good example of this last conflict can be found in Karkhu and Rautian's (1996) discussion of *Parvicursor*. Although these authors supported a close relationship between *Parvicursor* and *Mononykus,* these two taxa were regarded as nonavian maniraptoriforms. Karkhu and Rautian (1996) argued that the arctometatarsalian pes of *Mononykus* and *Parvicursor* (shared also by *Shuvuuia*), a condition different from the pedal design of other birds, prevents placement of these taxa within Aves. These authors stated that *Mononykus* and *Parvicursor* "have the metatarsal structure basically unsuitable for deducing the construction, which is characteristic of the Ornithurae . . . the general direction of the tarsometatarsus morphogenesis in the Ornithurae excludes the opportunity for evolving the highly specialized arctometatarsalian type" and continued that "among the Sauriurae . . . no examples of arctometatarsalian specializations are known. Hence . . . a close relationship between Sauriurae and *Mononykus* can hardly be supposed" (Karkhu and Rautian, 1996:591). However, their rationale stemmed more from assumptions about the evolution of developmental pathways in basal avian lineages than from a critical examination of the morphological evidence in support of or against the hypothesis placing *Mononykus* and its allies within birds. As pointed out by Perle et al. (1993a, 1994), the most parsimonious distribution of morphological characters indicates that, regardless of whether alvarezsaurids are considered to be birds, the arctometatarsalian condition of *Mononykus* evolved independently from that of other coelurosaurian dinosaurs (e.g., troodontids, ornithomimids). Under Karkhu and Rautian's (1996) interpretation, choice of a particular ontogenetic constraint (e.g., metatarsal fusion progressing proximally or distally) overthrows the morphological characters proposed in support of nesting *Mononykus* and its alvarezsaurid relatives within birds. Nevertheless, this assumption disregards the fact that developmental trajectories may themselves evolve (Rieppel, 1994; Shubin, 1994; Mabee, 2000).

Some other early disagreements with the avian hypothesis of alvarezsaurid relationships also stemmed from literal misreadings by some critics. For example, Martin and Rinaldi (1994:193) "disagreed" with parts of the first skeletal reconstruction of *Mononykus* (Perle et al., 1993a), despite the fact that it was clearly acknowledged that those parts were unknown and that their inclusion in the reconstruction was simply an educated guess. Similarly, Feduccia and Martin (1996:188) took issue with specific characters (e.g., biconvex thoracic vertebra) that were never even considered to provide support for the avian hypothesis. Certain critics also objected to the use of cladistic analysis as a way of inferring historical relationships. Martin and Rinaldi (1994:193), for example, stated that "clearly almost any outcome is possible in that sort of an analysis [cladistic analysis]." This is not the forum for a discussion of the effectiveness of cladistic methods in recovering the hierarchical structure of nature (see Clark, Norell, and Makovicky, Chapter 2 in this volume), but the reader should be aware that the methodological basis on which some of these early criticisms were constructed is profoundly different from that used by modern comparative biologists.

More constructive criticisms came from authors proposing specific hypotheses of relationships between Alvarezsauridae (or some of its taxa) and certain groups of nonavian theropods. For example, Martin and Rinaldi (1994; Martin, 1995) claimed *Mononykus* to be a specialized ornithomimid, although they provided little evidence to support it. Critical evaluation of this proposal indicates that evidence in support of their notion is at best weak (Chiappe et al., 1996; Novas and Pol, Chapter 5 in this volume). Work by Sereno (1997, 1999, 2001), Norell et al. (2001), and Novas and Pol (Chapter 5 in this volume) has also questioned the avian relationships of *Mononykus* and its alvarezsaurid relatives. Sereno (1997: Fig. 6) illustrated a generalized cladogram in which *Mononykus* was placed within basal Neotheropoda (Ceratosauria + Tetanurae) as an outgroup of Deinonychosauria plus Aves. Unfortunately, Sereno (1997) did not provide the specifics of his cladistic analysis. The more recent work by Sereno (1999, 2001) proposed a sister-taxon relationship between alvarezsaurids and ornithomimids. Character evidence in support of this proposal was evaluated by Suzuki et al. (2002), who concluded that the ornithomimid relationship of alvarezsaurids is weakly substantiated (see also Clark, Norell, and Makovicky, Chapter 2 in this volume, for a discussion of Sereno's 1999 study). Novas and Pol (Chapter 5 in this volume) have not published their data in support of their placement of alvarezsaurids as a distinct lineage of nonavian coelurosaurs, specifically as the sister group of a clade formed by oviraptorids, therizinosaurids, troodontids, dromaeosaurids, and birds. Interestingly, this solution is not the only one suggested by their analysis. When

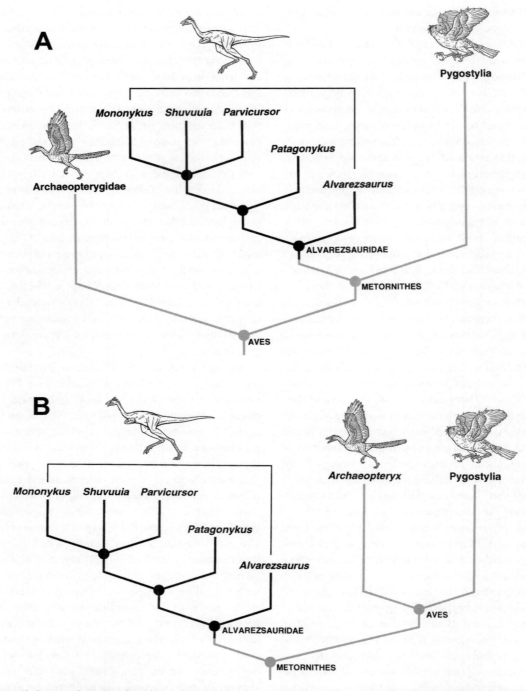

Figure 4.30. Cladogram depicting the relationships within Alvarezsauridae and between this clade and those of early birds. A, Avian hypothesis proposed by Perle et al. (1993a). B, Phylogenetic hypothesis proposed by Chiappe (Chapter 20 in this volume).

Goloboff's (1993) PeeWee computer program is used, the same data set supports the avian relationship of alvarezsaurids as the optimal solution (Novas and Pol, Chapter 5 in this volume). This suggests that support for their nonavian solution is not strong and that it may be based on highly homoplastic characters because PeeWee down-

weights homoplastic characters under the rationale that these are less likely to provide the correct solution (Goloboff, 1993). Perhaps the strongest proposals questioning the avian relationship of alvarezsaurids are those of Norell et al. (2001) and Chiappe (2001, Chapter 20 in this volume). While Norell et al. (2001) placed alvarezsaurids at the base of maniraptoriforms, Chiappe (2001, Chapter 20 in this volume) argued for a sister-group relationship between alvarezsaurids and birds. In the latter analysis, some of the characters previously used to support the inclusion of alvarezsaurids within birds are reinterpreted as synapomorphies of alvarezsaurids and birds (Chiappe, Chapter 20 in this volume). As in the case of the data set used by Novas and Pol (Chapter 5 in this volume), the shift between the avian hypothesis and this new proposal hinges on a few additional steps. Given Chiappe's data set, it would take only five extra steps to reroot the most parsimonious cladogram into the one proposed by Perle et al. (1993a) (Chiappe, Chapter 20 in this volume).

The highly apomorphic nature of the alvarezsaurid skeleton has complicated the phylogenetic placement of these dinosaurs. This is aggravated by the lack of consensus about the phylogenetic relationships of the different lineages of maniraptoriform dinosaurs. As pointed out by Novas and Pol (Chapter 5 in this volume), the addition of theropod taxa (both nonavian and avian) to the already conducted analyses may result in an optimal solution in which alvarezsaurids are placed outside birds. Thus far, however, the addition of more taxa has resulted in cladograms supporting both the avian (e.g., Forster et al., 1998a; Holtz, 1998, 2001; Ji et al., 1998; Chatterjee, 1999) and nonavian (Sereno, 1999; Norell et al., 2001; Novas and Pol, Chapter 5 in this volume; Chiappe, Chapter 20 in this volume) relationship of alvarezsaurids. The phylogenetic placement of Alvarezsauridae remains as controversial as those of several other lineages of early coelurosaurian theropods. Their crucial role in understanding the transition from nonavian theropods to birds is, however, undeniable.

Novas (1996, 1997) was first in recognizing the close relationship between the Patagonian *Alvarezsaurus* and *Patagonykus* and the Asian *Mononykus,* pointing out the more basal condition of the South American taxa. This proposal was strongly supported by Chiappe et al. (1998a) in their description of the Asian *Shuvuuia,* which together with *Mononykus* and *Parvicursor* were interpreted as forming a monophyletic taxon, Mononykinae. The discovery of a fragmentary alvarezsaurid from North America (Hutchinson and Chiappe, 1998) highlighted the common evolutionary origin of all Laurasian members of this bizarre group of theropods. As pointed out by Hutchinson and Chiappe (1998), the exclusive distribution of mononykines in Laurasia agrees with Novas's (1996) suggestion of a Gondwanan origin of the alvarezsaurids and the subsequent differentiation of mononykines.

Paleobiology

Alvarezsaurids were obligate cursorial animals with highly specialized forelimbs (Fig. 4.31). Their size varied significantly, as did their degree of forelimb specialization. Body masses of the different alvarezsaurid specimens were estimated by applying the standard equation of Campbell and Marcus (1992) ($W_{log} = 0.065 + 2.411C_{log}$; C = circumference) to the circumference of the midshaft of the femur of several specimens—these estimates rely on data derived from modern analogues and should be taken as gross approximations. Body mass estimates for the holotypes of *M. olecranus* (MGI 107/6), *A. calvoi* (MUCPv 54), and *P. puertai* (PVPH 37) were 2,944 g, 3,507 g, and 4,270 g, respectively. The mass of the juvenile specimen of *Shuvuuia deserti* (MGI 100/99) was estimated to be 659 g. The femoral shafts of larger specimens of *Shuvuuia* were either missing or not complete enough to provide an estimate. Yet given the fact that the holotype specimen of *S. deserti* (MGI 100/975) is roughly the size of the holotype of *M. olecranus* but displays a lesser degree of bone co-ossification between the tibia and the proximal tarsals, and between the metacarpals, it may be that fully grown specimens of *Shuvuuia* were larger than those of *Mononykus.* The small size of the holotype of *P. remotus* (PIN 4487/25) should not be taken to represent an early ontogenetic age. As pointed out by Karkhu and Rautian (1996), the complete obliteration of the neurocentral sutures of its vertebrae, along with the fusion of two of its synsacral elements, of the periacetabular bones, and of the proximal tarsals to the tibia, suggests that this specimen was not immature.

The elongate hindlimbs with short pedal digits and the wide pelvis, almost as wide as it is long, of alvarezsaurids strongly suggest that these creatures were predominantly terrestrial and most likely fast runners (Fig. 4.31; Storer, 1971; Mattison, 1997). The bizarre nature of their specialized forelimbs has puzzled researchers ever since the discovery of *Mononykus.* Although the forelimbs of alvarezsaurids were greatly abbreviated and were certainly not used in locomotion, the size and robustness of several of their tubercles and processes (e.g., olecranon, deltopectoral crest, ventral tubercle) as well as the massiveness of the wrist indicate that the enlarged digit I clearly must have been used for a powerful activity (Perle et al., 1994). Nevertheless, the specific function of this digit has been problematic (Perle et al., 1993a). The orientation and connection of the articular surfaces of the different elements suggest that forelimb movements were essentially restricted to abduction-adduction. The forelimbs of alvarezsaurids could have been used for burrowing, as was proposed by several authors who considered *Mononykus* as a fossorial organism (Ostrom, 1994; Zhou, 1995). At first glance, this notion may seem to be reasonable considering the fact that the robust forelimbs of alvarezsaurids are reminiscent of those of fossorial tetrapods, in particular mammals (Perle et al., 1993a, 1994; Zhou, 1995). Yet the rest of the body is that of a gracile cursor, and the forelimbs are too short to be consistent with a fossorial habit.

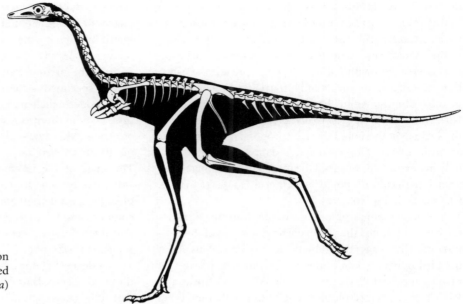

Figure 4.31. Skeletal reconstruction of a mononykine alvarezsaurid (based partially in *Mononykus* and *Shuvuuia*) (after Chiappe et al., 1996).

Perhaps the forelimbs were used for digging, but in the context of a predatory activity. Norell et al. (1993) speculated that the feeding habits of *Mononykus* may have been comparable to those of living aardvarks or anteaters, using its powerful forelimbs to tear apart insect nests or plants in search for food. The well-developed hyoid bones and the presence of numerous teeth in both the skull and mandible would agree with this insectivorous diet. An increase in the number of teeth has been documented in carnivorous mammals with a predominantly insectivorous diet (Ewer, 1973), and Varricchio (1997) argued for an insectivorous diet in the many-toothed troodontids. Although we acknowledge that inferring the behavior of extinct organisms is often a rather speculative exercise, it seems to us that this last alternative is less in conflict with the available morphological data.

Acknowledgments

We especially thank Perle A. for giving us the initial opportunity to study the holotype specimen. We also thank P. J. Currie, P. Makovicky, L. Marcus, F. Novas, and L. M. Witmer for discussions on different topics. S. Reuil, M. Ellison, S. Copeland, and L. Meeker prepared the illustrations, and S. Orell and D. Treon provided editorial assistance. This research has been funded by grants from the National Science Foundation (DEB-9407999), the John Simon Guggenheim Foundation, the Dinosaur Society, and the Chapman Memorial Fund (AMNH).

Literature Cited

Baumel, J. J., and L. M. Witmer. 1993. Osteologia; pp. 45–132 *in* J. J. Baumel, A. S. King, J. E. Breazile, H. E. Evans, and J. C. Van-den Berge (eds.), Handbook of Avian Anatomy: Nomina Anatomica Avium, 2nd edition. Nuttal Ornithological Club, Cambridge.

Berkey, C., and F. K. Morris. 1927. Geology of Mongolia. Natural History of Central Asia, American Museum of Natural History 2:1–475.

Bock, W. J. 1964. Kinetics of the avian skull. Journal of Morphology 114:1–42.

Bonaparte, J. F. 1991. Los vertebrados fósiles de la Formación Río Colorado de Neuquén y cercanias, Cretácico Superior, Argentina. Revista del Museo Argentino de Ciencias Naturales "Bernardino Rivadavia." Paleontología 4(3):17–123.

Brochu, C. A. 1996. Closure of neurocentral sutures during crocodilian ontogeny: implications for maturity assessment in fossil archosaurs. Journal of Vertebrate Paleontology 16(1): 49–62.

Calvo, J. O. 1991. Hullas de dinosaurios en la Formación Río Limay (Albiano-Cenomaniano), Picun Leufu, Provincia del Neuquén, República Argentina (Ornithischia-Saurischia: Sauropoda-Theropoda). Ameghiniana 2(3–4):241–258.

Campbell, K. E., Jr., and L. Marcus. 1992. The relationship of hind limb bone dimensions to body weight in birds; pp. 395–412 *in* K. E. Campbell (ed.), Papers in Avian Paleontology, Honoring Pierce Brodkorb. Science Series 36. Natural History Museum of Los Angeles County, Los Angeles.

Cazau, L. B., and M. A. Uliana. 1973. El Cretácico superior continental de la Cuenca Neuquina. V Congreso Geológico Argentino, Actas 3:131–163.

Chatterjee, S. 1991. Cranial anatomy and relationships of a new Triassic bird from Texas. Philosophical Transactions of the Royal Society London B 332:277–342.

———. 1995. The Triassic bird *Protoavis*. Archaeopteryx 13:15–31.

———. 1999. *Protoavis* and the early evolution of birds. Palaeontographica 254:1–100.

Chen P.-J., Dong Z.-M., and Zhen S.-N. 1998. An exceptionally well-preserved theropod dinosaur from the Yixian Formation of China. Nature 391:147–152.

Chiappe, L. M. 1996. Late Cretaceous birds of southern South America: anatomy and systematics of Enantiornithes and *Patagopteryx deferrariisi;* pp. 203–244 *in* G. Arratia (ed.), Contributions of Southern South America to Vertebrate Paleontology. Münchner Geowissenschaftliche Abhandlungen, Reihe A, Geologie und Paläontologie 30. Verlag Dr. Friedrich Pfeil, Munich.

Chiappe, L. M. 2001. Phylogenetic relationships among basal birds; pp. 125–139 *in* J. Gauthier and L. F. Gall (eds.), New Perspectives on the Origin and Early Evolution of Birds: Proceedings of the International Symposium in Honor of John H. Ostrom. Peabody Museum of Natural History, New Haven.

Chiappe, L. M., and J. O. Calvo. 1994. *Neuquenornis volans,* a new Enantiornithes (Aves) from the Upper Cretaceous of Patagonia (Argentina). Journal of Vertebrate Paleontology 14(2):230–246.

Chiappe, L. M., M. A. Norell, and J. M. Clark. 1995. Comment on Wellnhofer's new data on the origin and early evolution of birds. Comptes Rendus de l'Académie des Sciences, Paris 320 (série IIa):1031–1032.

———. 1996. Phylogenetic position of *Mononykus* from the Upper Cretaceous of the Gobi Desert. Memoirs of the Queensland Museum 39:557–582.

———. 1997. *Mononykus* and birds: methods and evidence. Auk 14(2):300–302.

———. 1998a. The skull of a new relative of the stem-group bird *Mononykus.* Nature 392:275–278.

Chiappe, L. M., R. A. Coria, L. Dingus, F. Jackson, A. Chinsamy, and M. Fox. 1998b. Sauropod dinosaur embryos from the Late Cretaceous of Patagonia. Nature 396: 258–261.

Chiappe, L. M., Ji S.-A., Ji Q., and M. A. Norell. 1999. Anatomy and systematics of the Confuciusornithidae (Aves) from the late Mesozoic of northeastern China. Bulletin of the American Museum of Natural History 242:1–89.

Chure, D. J., and J. H. Madsen. 1996. The presence of furculae in some non-maniraptoran theropods. Journal of Vertebrate Paleontology 16(3):573–577.

Clark, J. M., Perle A., and M. A. Norell. 1994. The skull of *Erlicosaurus andrewsi,* a Late Cretaceous "Segnosaur" (Theropoda: Therizinosauridae) from Mongolia. American Museum Novitates 3115:1–39.

Clarke, J., L. Dingus, and L. M. Chiappe. 1998. Review of Upper Cretaceous deposits at Neuquén, Argentina (Bajo de la Carpa Member, Río Colorado Formation). Journal of Vertebrate Paleontology 18(3):34A.

Cracraft, J. 1968. The lacrimal-ectethmoid bone complex in birds: a single character analysis. American Midland Naturalist 80(2):316–359.

Cruz, C. E., P. Condat, E. Kozlowski, and R. Manceda. 1989. Analisis estratigráfico secuencial del Grupo Neuquén (Cretácico superior) en el valle del Río Grande, Provincia de Mendoza. I Congreso Argentino de Hidrocarburos, Actas 2:689–714.

Currie, P. J. 1987. Bird-like characteristics of the jaws and teeth of trooodontid theropods (Dinosauria, Saurischia). Journal of Vertebrate Paleontology 7(1):72–81.

———. 1995. New information on the anatomy and relationships of *Dromaeosaurus albertensis* (Dinosauria, Theropoda). Journal of Vertebrate Paleontology 15(3):576–591.

Currie, P. J., and Zhao X.-J. 1993a. A new carnosaur (Dinosauria, Theropoda) from the Jurassic of Xinjiang, People's Republic of China. Canadian Journal of Earth Sciences 30(10–11): 2037–2081.

———. 1993b. A new troodontid (Dinosauria, Theropoda) braincase from Dinosaur Park Formation (Campanian) of Alberta. Canadian Journal of Earth Sciences 30(10–11): 2231–2247.

Dashzeveg D., M. J. Novacek, M. A. Norell, J. M. Clark, L. M. Chiappe, A. Davidson, M. C. McKenna, L. Dingus, C. Swisher, and Perle A. 1995. Extraordinary preservation in a new vertebrate assemblage from the Late Cretaceous of Mongolia. Nature 374:446–449.

Dingus, L., and D. Loope. 2000. Death in the dunes. Natural History (7–8):50–55.

Dingus, L., J. Clarke, G. R. Scott, C. C. Swisher III, L. M. Chiappe, and R. A. Coria. 2000. Stratigraphy and magnetostratigraphy/faunal constraints for the age of sauropod embryo-bearing rocks in the Neuquén Group (Late Cretaceous, Neuquén Province, Argentina). American Museum Novitates 3290: 1–11.

Eberth, D. A. 1993. Depositional environments and facies transitions of dinosaur-bearing Upper Cretaceous redbeds at Bayan Mandahu (Inner Mongolia, People's Republic of China). Canadian Journal of Earth Sciences 30:2196–2213.

Eldredge, N., and J. Cracraft. 1980. Phylogenetic Patterns and the Evolutionary Process. Columbia University Press, New York, 349 pp.

Elzanowski, A., and P. Wellnhofer. 1996. Cranial morphology of *Archaeopteryx:* evidence from the seventh skeleton. Journal of Vertebrate Paleontology 16(1):81–94.

Ewer, R. F. 1973. The Carnivores. Cornell University Press, Ithaca, N.Y., 494 pp.

Fastovsky, D. E., D. Badamgarav, H. Ishimoto, M. Watabe, and D. B. Weishampel. 1997. The paleoenvironments of Tugrikin-Shireh (Gobi Desert, Mongolia) and aspects of the taphonomy and paleoecology of *Protoceratops* (Dinosauria, Ornithischia). Palaios 12:59–70.

Feduccia, A. 1994. The great dinosaur debate. Living Bird 13(4):28–33.

Feduccia, A., and L. D. Martin. 1996. Jurassic Urvogels and the myth of the feathered dinosaurs; pp. 185–191 *in* M. Morales (ed.), The Continental Jurassic. Bulletin of the Museum of Northern Arizona 60, Flagstaff.

Forster, C. A., L. M. Chiappe, D. W. Krause, and S. D. Sampson. 1996. The first Cretaceous bird from Madagascar. Nature 382:532–534.

Forster, C. A., S. D. Sampson, L. M. Chiappe, and D. W. Krause. 1998a. The theropodan ancestry of birds: new evidence from the Late Cretaceous of Madagascar. Science 279:1915–1919.

———. 1998b. Genus correction. Science 280(5361):185.

Frost, D., and A. Kluge. 1994. A consideration of epistemology in systematic biology, with special reference to species. Cladistics 10:259–294.

Gao K. and M. A. Norell. 2000. Taxonomic composition and systematics of Late Cretaceous lizard assemblages from Ukhaa Tolgod and adjacent localities, Monolian Gobi Desert. Bulletin of the American Museum of Natural History 249:1–118.

Gauthier, J. 1986. Saurischian monophyly and the origin of birds; pp. 1–55 *in* K. Padian (ed.), The Origin of Birds and the Evolution of Flight. California Academy of Sciences, San Francisco.

Goloboff, P. A. 1993. Pee-Wee, version 2.8. Program and references. Privately published.

Gradzinski, R. 1970. Sedimentation of dinosaur-bearing Upper Cretaceous depostis of the Nemegt Basin, Gobi Desert. Palaeontologia Polonica 21:147–229.

Gradzinski, R., Z. Kielan-Jaworowska, and T. Maryanska. 1977. Upper Cretaceous Djadokhta, Barun Goyot and Nemegt Formations of Mongolia, including remarks of previous subdivisions. Acta Geologica Polonica 27:281–318.

Heredia, S., and J. O. Calvo. 1997. Sedimentitas eólicas en la Formación Río Colorado (Grupo Neuquén) y su relación con la fauna del Cretácico superior. Ameghiniana 34(1):120.

Holtz, T. R., Jr. 1994. The arctometatarsalian pes, an unusual structure of the metatarsus of Cretaceous Theropoda (Dinosauria: Saurischia). Journal of Vertebrate Paleontology 14(4):480–519.

———. 1998. A new phylogeny of the carnivorous dinosaurs. Gaia 15:5–61.

———. 2001. Arctometatarsalia revisited: the problem of homoplasy in reconstructing theropod phylogeny; pp. 99–122 *in* J. Gauthier and L. F. Gall (eds.), New Perspectives on the Origin and Early Evolution of Birds: Proceedings of the International Symposium in Honor of John H. Ostrom. Peabody Museum of Natural History, New Haven.

Hutchinson, J. R., and L. M. Chiappe. 1998. The first known alvarezsaurid (Theropoda: Aves) from North America. Journal of Vertebrate Paleontology 18(3):447–450.

Jerzykiewicz, T., and D. A. Russell. 1991. Late Mesozoic stratigraphy and vertebrates from the Gobi Basin. Cretaceous Research 12:345–377.

Jerzykiewicz, T., P. J. Currie, D. A. Eberth, P. A. Johnston, E. H. Koster, and Zheng J.-J. 1993. Djadokhta Formation correlative strata in Chinese Inner Mongolia: an overview of the stratigraphy, sedimentary geology, and paleontology and comparisons with the type locality in the pre-Altai Gobi. Canadian Journal of Earth Sciences 30:2180–2195.

Ji Q., P. J. Currie, M. A. Norell, and Ji S.-A. 1998. Two feathered dinosaurs from northeastern China. Nature 393:753–761.

Ji Q., M. A. Norell, Gao K.-Q., and Ren D. 2001. The distribution of integumentary structures in a feathered dinosaur. Nature 410:1084–1088.

Karkhu, A. A., and A. S. Rautian. 1996. A new family of Maniraptora (Dinosauria: Saurischia) from the Late Cretaceous of Mongolia. Paleontological Journal 30(5):583–592.

Kielan-Jaworowska, Z. 1969. Hunting for Dinosaurs. MIT Press, Cambridge, 177 pp.

Lavas, J. R. 1993. Dragons from the Dunes. Privately published, 138 pp.

Legarreta, L., and C. Gulisano. 1989. Análisis estratigráfico secuencial de la Cuenca Neuquina (Triásico Superior-Terciario Inferior); pp. 221–243 *in* G. Chebli and L. Spalletti (eds.), Cuencas Sedimentarias Argentinas. Serie Corre-

lación Geológica 6. Universidad Nacional de Tucumán, Tucumán.

Loope, D. B., L. Dingus, C. C. Swisher III, and C. Minjin. 1998. Life and death in a Late Cretaceous dune field, Nemegt Basin, Mongolia. Geology 26(1):27–30.

Mabee, P. M. 2000. The usefulness of ontogeny in interpreting morphological characters; pp. 84–114 *in* J. J. Wiens (ed.), Phylogenetic Analysis of Morphological Data. Smithsonian Institution Press, Washington.

Madsen, J. A. 1976. *Allosaurus fragilis*. A revised osteology. Bulletin of the Utah Geological and Mineral Survey 109: 1–163.

Makovicky, P. J., and P. J. Currie. 1998. The presence of a furcula in tyrannosaurid theropods, and its phylogenetic and functional implications. Journal of Vertebrate Paleontology 18(1):143–149.

Marsh, O. C. 1880. Odontornithes: A Monograph on the Extinct Toothed Birds of North America. United States Geological Exploration of the 40th Parallel. Government Printing Office, Washington, 201 pp.

———. 1892. Notice of new reptiles from the Laramie Formation. American Journal of Science 43:449–453.

Martin, L. D. 1995. The relationship of *Mononykus* to ornithomimid dinosaurs. Journal of Vertebrate Paleontology 15(3) (supp.):43A.

Martin, L. D., and C. Rinaldi. 1994. How to tell a bird from a dinosaur. Maps Digest 17(4):190–196.

Mattison, R. G. 1997. Quantifying the avian pelvis: statistical correlations of lifestyle to pelvic structure among non passeriform birds. Ph.D. dissertation, University of Massachusetts, Amherst, 301pp.

Mlikovsky, J. 1990. Foramen magnum area in birds. Acta Societatis Zoologicae Bohemoslovacae 54:97–106.

Nixon, K., and Q. Wheeler. 1990. An amplification of the phylogenetic species concept. Cladistics 6:211–223.

———. 1992. Extinction and the origin of species; pp. 119–143 *in* M. Novacek and Q. Wheeler (eds.), Extinction and Phylogeny. Columbia University Press, New York.

Norell, M. A., L. M. Chiappe, and J. Clark. 1993. New limb on the avian family tree. Natural History 102(9):38–43.

Norell, M. A., P. Makovicky, and J. M. Clark. 1997. A *Velociraptor* wishbone. Nature 389:447.

Norell, M. A., J. M. Clark, and P. Makovicky. 2001. Phylogenetic relationships among coelurosaurian theropods; pp. 69–98 *in* J. Gauthier and L. F. Gall (eds.), New Perspectives on the Origin and Early Evolution of Birds: Proceedings of the International Symposium in Honor of John H. Ostrom. Peabody Museum of Natural History, New Haven.

Novacek, M. J. 1996. Dinosaurs of the Flaming Cliffs. Anchor Books, New York, 367 pp.

Novas, F. E. 1996. Alvarezsauridae, Cretaceous maniraptorans from Patagonia and Mongolia. Memoirs of the Queensland Museum 39(3):675–702.

———. 1997. Anatomy of *Patagonykus puertai* (Theropoda, Maniraptora, Alvarezsauridae) from the Late Cretaceous of Patagonia. Journal of Vertebrate Paleontology 17(1):137–166.

———. 1998. Megaraptor namunhuaiquii, gen. et sp. nov., a large-clawed, Late Cretaceous theropod from Patagonia. Journal of Vertebrate Paleontology 18(1):4–9.

Novas, F. E., and P. Puerta. 1997. New evidence concerning avian origins from the Late Cretaceous of Patagonia. Nature 387: 390–392.

Osmólska, H. 1996. An unusual theropod dinosaur from the Late Cretaceous Nemegt Formation of Mongolia. Acta Palaeontologica Polonica 41(1):1–38.

Osmólska, H., E. Roniewicz, and Barsbold R. 1972. A new dinosaur, *Gallimimus bullatus* n. gen. n. sp. (Ornithomimidae) from the Upper Cretaceous of Mongolia. Palaeontologia Polonica 27:103–143.

Ostrom, J. H. 1969. Osteology of *Deinonychus antirrhopus,* an unusual theropod dinosaur from the Lower Cretaceous of Montana. Bulletin of the Peabody Museum of Natural History 30:1–165.

———. 1994. On the origin of birds and of avian flight; pp. 160–177 *in* D. R. Prothero and R. M. Schoch (convs.), Major Features of Vertebrate Evolution. Short Courses in Paleontology 7, The Paleontological Society, Knoxville.

Padian, K. 1992. A proposal for standardize tetrapod phalangeal formula designations. Journal of Vertebrate Paleontology 12(2):260–262.

Perle A., M. A. Norell, L. M. Chiappe, and J. M. Clark. 1993a. Flightless bird from the Cretaceous of Mongolia. Nature 362:623–626.

———. 1993b. Correction: flightless bird from the Cretaceous of Mongolia. Nature 362:623–626.

Perle A., L. M. Chiappe, Barsbold R., J. M. Clark, and M. A. Norell. 1994. Skeletal morphology of *Mononykus olecranus* (Theropoda: Avialae) from the Late Cretaceous of Mongolia. American Museum Novitates 3105:1–29.

Peters, S. 1996. Ein nahezu vollständiges Skelett eines urtümlichen Vogels aus China. Natur und Museum 126(9):298–302.

Powell, J. 1986. Revisión de los Titanosauroidea de América del Sur. Ph.D. dissertation. Facultad de Ciencias Naturales, Universidad Nacional de Tucumán, Tucumán, Argentina, 340 pp. and atlas.

Ramos, V. A. 1981. Descripción geológica de la Hoja 33c, Los Chihuidos Norte. Boletín del Servicio Geológico Nacional, Buenos Aires 182:1–103.

Rieppel, O. 1994. Homology, topology, and typology: the history of modern debates; pp. 64–100 *in* B. Hall (ed.), Homology: The Hierarchical Basis of Comparative Biology. Academic Press, San Diego.

Romer, A. S. 1956. Osteology of the Reptiles. University of Chicago Press, Chicago, 772 pp.

Sanz, J. L., L. M. Chiappe, B. P. Pérez-Moreno, J. J. Moratalla, F. Hernández-Carrasquilla, A. D. Buscalioni, F. Ortega, F. J. Poyato-Ariza, and X. Martínez-Delclòs. 1997. A nestling bird from the Early Cretaceous of Spain: implications for avian skull and neck evolution. Science 276:1543–1546.

Schweitzer, M. H. 2001. Evolutionary implications of possible protofeather structures associated with a specimen of *Shuvuuia deserti;* pp. 181–192 *in* J. Gauthier and L. F. Gall (eds.), New Perspectives on the Origin and Early Evolution of Birds: Proceedings of the International Symposium in Honor of John H. Ostrom. Peabody Museum of Natural History, New Haven.

Schweitzer, M. H., J. Watt, C. Forster, M. Norell, and L. M. Chiappe. 1997. Keratinous structures preserved with two Late Cretaceous avian theropods from Madagascar and Mongolia. Journal of Vertebrate Paleontology 17(Supp. 3):74A.

Schweitzer, M. H., J. Watt, R. Avci, L. Knapp, L. M. Chiappe, M. A. Norell, and M. Marshal. 1999. Beta-keratin specific immunological reactivity in feather-like structures of the Cretaceous alvarezsaurid, *Shuvuuia deserti.* Journal of Experimental Zoology 285:146–157.

Sereno, P. C. 1997. The origin and evolution of dinosaurs. Annual Review of Earth and Planetary Sciences 25:435–489.

———. 1999. The evolution of dinosaurs. Science 284:2137–2147.

———. 2001. Alvarezsaurids: birds or ornithomimosaurs?; pp. 69–98 *in* J. Gauthier and L. F. Gall (eds.), New Perspectives on the Origin and Early Evolution of Birds: Proceedings of the International Symposium in Honor of John H. Ostrom. Peabody Museum of Natural History, New Haven.

Shubin, N. 1994. History, ontogeny, and evolution of the archetype; pp. 249–271 *in* B. Hall (ed.), Homology: The Hierarchical Basis of Comparative Biology. Academic Press, San Diego.

Simonetta, A. M. 1960. On the mechanical implications of the avian skull and their bearing on the evolution and classification of birds. Quarterly Review of Biology 35:206–220.

Storer, R. W. 1971. Adaptive radiation of birds; pp. 150–186 *in* D. S. Farner and J. R. King (eds.), Avian Biology 1. Academic Press, New York.

Suzuki, S., L. M. Chiappe, G. J. Dyke, M. Watabe, Barsbold R., and Tsogtbaatar K. 2002. A new specimen of *Shuvuuia deserti* Chiappe et al., 1998 from the Mongolian Late Cretaceous with a discussion of the relationships of alvarezsaurids to other theropod dinosaurs. Contributions in Science 494:1–17.

Tumanova, T. A. 1987. Pantsiryne dinozavry Mongolii (the armored dinosaurs of Mongolia); pp. 1–80 *in* L. P. Tatarinov, Barsbold R., B. Luwsandansan, B. A. Trofimov, V. J. Reshetov, and M. A. Shishkin (eds.), Trudy. Sovmestnaya Sovetsko-Mongolskaya Paleontologicheskaya Ekspeditsya Nauka Moskva 32, Moscow.

Tverdokhlebov, V. P., and Y. I. Tsybin. 1974. Genezis verkhnemelovykh mestorozhdeniy dinozavrov Tugrikin-Us i Alag-teg; pp. 314–319 *in* N. N. Kramarenko et al. (eds.), Mesozoic i Kainozoik Faunas i Biostratigrafiya Mongolii. Sovmestnaya Sovetsko-Mongolskaya Paleontologicheskaya Ekspeditsia 1, Moscow.

Varricchio, D. J. 1997. Troodontidae; pp. 749–754 *in* P. J. Currie and K. Padian (eds.), The Encyclopedia of Dinosaurs. Academic Press, San Diego.

Weishampel, D. B., P. Dodson, and H. Osmólska. 1990. The Dinosauria. University of California Press, Berkeley, 733 pp.

Wellnhofer, P. 1974. Das fünfte Skelettexemplar von *Archaeopteryx. Palaeontographica* A 147:169–216.

———. 1994. New data on the origin and early evolution of birds. Comptes Rendus de l'Académie des Sciences, Paris 319(II):299–308.

Witmer, L. M. 1995. Homology of facial structures in extant archosaurs (birds and crocodilians), with special reference to paranasal pneumaticity and nasal conchae. Journal of Morphology 225:269–327.

———. 1997. The evolution of the antorbital cavity of archosaurs: a study in soft-tissue reconstruction in the fossil record with an analysis of the function of pneumaticity. Memoirs of the Society of Vertebrate Paleontology 3:1–73.

Xu X., Wang X.-L., and Wu X.-C. 1999a. A dromaeosaur dinosaur with filamentous integument from the Yixian Formation of China. Nature 401:262–266.

Xu X., Tang Z.-L., and Wang X.-L. 1999b. A therizinosauroid dinosaur with integumentary structures from China. Nature 399:350–354.

Xu X., Zhou Z., and Wang X.-L. 2000. The smallest known non-avian theropod dinosaur. Nature 408:705–708.

Xu X., Zhou Z., and R. O. Prum. 2001. Branched integumentary structures in *Sinornithosaurus* and the origin of feathers. Nature 410:200–204.

Zhou Z. 1995. Is *Mononykus* a bird? Auk 112:958–963.

Zhou Z. and Wang X.-L. 2000. A new species of *Caudipteryx* from the Yixian Formation of Liaoning, northeast China. Vertebrata PalAsiatica 38:111–127.

Zusi, R. L. 1985. A functional and evolutionary analysis of rhynchokinesis in birds. Smithsonian Contributions to Zoology 395:1–40.

———. 1993. Patterns of diversity in the avian skull; pp. 391–437 *in* J. Hanken and B. K. Hall (eds.), The Skull, Volume 2. University of Chicago Press, Chicago.

Zweers, G. A., and J. C. Vanden Berge. 1997. Birds at geological boundaries. Zoology 100(1997/98):183–202.

5

Alvarezsaurid Relationships Reconsidered

FERNANDO E. NOVAS AND DIEGO POL

 Mononykus, Shuvuuia, Patagonykus, Alvarez-saurus, and a few other more poorly known forms constitute Alvarezsauridae, a highly distinct group of theropods. Alvarezsaurids are easily distinguishable from other dinosaurs on the basis of an extensive list of apomorphic characters, which mainly correspond to their stout forelimbs and vertebral column. Alvarezsaurids have shaken our knowledge of early avian evolution because such very unusual morphological traits are found in association with clearly avian characters in the skull, vertebrae, pelvic girdle, and hindlimbs (Chiappe et al., 1996, 1998; Chiappe, Norell, and Clark, Chapter 4 in this volume). Such bizarre combinations of characters have generated disagreement among scientists as to their significance for the phylogenetic relationships of these taxa.

Variation in Hypotheses

Studies on alvarezsaurid morphology and systematics are divided between those supporting the avian relationships of alvarezsaurids and those regarding them as a different lineage of theropods, not related to birds. Proponents of the avian hypothesis (Perle et al., 1993, 1994; Chiappe et al., 1996, 1998; Novas, 1996, 1997; Forster et al., 1998; Chiappe, Norell, and Clark, Chapter 4 in this volume; Clark, Norell, and Makovicky, Chapter 2 in this volume) arrived at this conclusion through a cladistic approach, whereas most authors interpreting alvarezsaurids as nonavian theropods used a wide variety of arguments (Feduccia, 1994; Martin and Rinaldi, 1994; Ostrom, 1994; Wellnhofer, 1994; Zhou, 1995; Karkhu and Rautian, 1996; Martin, 1997; but see Sereno, 1997, 1999, who used strictly phylogenetic arguments). Divergence in opinion about the phylogenetic placement of alvarezsaurids is, however, wider than thought among supporters of nonavian hypotheses. For example, Karkhu and Rautian (1996) considered *Mononykus* and its kin as non-

avian maniraptorans, without specifying particular relationships among maniraptorans. Martin and Rinaldi (1994) and Martin (1997) interpreted alvarezsaurids as related to Ornithomimidae. Others simply concluded that they were nonavian theropods, not offering any particular hypothesis of relationships (e.g., Feduccia, 1994, 1996; Zhou, 1995).

Untestable Assumptions

Perhaps the main argument against the avian hypothesis is that of the "morphological difference," which is based on the presumed improbability that certain evolutionary transformations may occur (Chiappe et al., 1996, 1997). This approach is evident in the criticisms of Wellnhofer (1994) and Zhou (1995), who claimed that wings cannot be transformed into the specialized forelimbs of alvarezsaurids, which are obviously adapted for purposes other than flight, and in the work of Karkhu and Rautian (1996), who claimed that the arctometatarsalian metatarsus of *Mononykus* is unsuitable for giving rise to the metatarsal structure characteristic of the Ornithurae. The argument of the "morphological difference" of the forelimbs, however, may be used (with the same negative result) against hypotheses supporting ornithomimid, nonavian maniraptoran, coelurosaurian, theropodan, or even dinosaurian affinities of *Mononykus*.

Zhou (1995) constructed an intricate objection against the avian hypothesis on the assumption that similar characters (for example, those shared by *Mononykus* and ornithothoracine birds in the sternum) cannot be considered homologous simply because they pertain to different functions. In so doing, the phylogenetic informativeness of such morphological characters is dismissed a priori. This approach, however, neglects the possibility that the differentiation of a function may occur within a structurally homologous framework (Lauder, 1994; Chiappe et al., 1997).

From a cladistic point of view, phylogenetic relationships are justified on the basis of synapomorphies, and evolutionary processes must be tested and explained only after the pattern of character distribution has been analyzed. A priori assumptions about evolutionary processes cannot be used to overturn a particular phylogenetic hypothesis (see also Chiappe, Norell, and Clark, Chapter 4 in this volume). Thus, the criticisms enumerated earlier are irrelevant for phylogenetic inference.

Mononykus and Its Kin in the Context of Coelurosaurian Phylogeny

It has been shown that certain characters originally hypothesized to be synapomorphic of Metornithes (Perle et al., 1993, 1994) evolved independently in *Mononykus* and ornithurine birds (e.g., loss of pubic symphysis, greater and anterior trochanters fused, fibula not in contact with the proximal tarsals, presence of femoral popliteal fossa, and presence of an accessory cnemial crest on the tibiotarsus). This explanation emerged after the inclusion of two additional taxa, interpreted by Novas (1996, 1997) as basal alvarezsaurids. Even though the inclusion of the Patagonian *Alvarezsaurus* and *Patagonykus* in the character analysis somewhat reduced the support for the placement of *Mononykus* within birds, the avian hypothesis remained unchallenged.

In an unpublished cladistic analysis adding new characters to those used by previous cladistic studies (mainly Chiappe et al., 1996, 1998; Novas, 1996, 1997; but not yet including those characters recently advanced by Sereno, 1999) and including additional nonavian coelurosaurian taxa (e.g., Troodontidae, Oviraptorosauria, Therizinosauridae, *Ornitholestes*, *Allosaurus*, and *Unenlagia*), alvarezsaurids are placed outside birds, as a distinct group of nonavian coelurosaurians (Fig. 5.1). This new phylogeny (obtained using Hennig86; Farris, 1988) also proposes a new hierarchical arrangement of the other coelurosaur taxa. For example, the monophyly of Arctometatarsalia and Bullatosauria (Holtz, 1994) is not supported by this analysis; instead, troodontids are hypothesized as the sister taxon of Dromaeosauridae plus Avialae (*sensu* Padian, 1997) in one tree, and the monophyly of Deinonychosauria (i.e., Troodontidae plus Dromaeosauridae; Gauthier, 1986) is supported in the other. The results obtained here contrast with those of previous authors (e.g., Gauthier, 1986; Russell and Dong, 1993; Holtz, 1994). Bremmer support of the tree nodes is not high (Fig. 5.1), indicating that additional information (new characters or taxa) may easily change the proposed topology.

In this new hypothesis, Alvarezsauridae is the sister taxon of a group formed by Therizinosauridae, Oviraptorosauria, Troodontidae, Dromaeosauridae, and Aves (this group is labeled "Node X" in Figure 5.1, in which the strict consensus tree of the two most parsimonious trees is depicted). Alvarezsaurids lack characters synapomorphic of such a group (e.g., cervical centra with kidney-shaped cranial articular surface; pronounced proximodorsal lip on manual unguals; absence of supracetabular crest; iliac postacetabular blade subvertically oriented, and medial flange of ilium strongly reduced; pedal unguals of digits III and IV dorsoventrally deep and with enlarged flexor tubercles). Thus, in the context of this new hypothesis, all the characters previously considered synapomorphic of Metornithes (Chiappe et al., 1996, 1998; Novas, 1996, 1997) are reinterpreted as independently acquired in Alvarezsauridae and Ornithothoraces.

The clash between the new hypothesis and that supporting the monophyly of Metornithes relies not merely on the addition of new characters but, most important, on the inclusion of new coelurosaurian taxa. This is demonstrated by the fact that the results are congruent with the avian hypothesis if the new pool of characters is scored in the same set of theropods used in previous papers (e.g., Chiappe et al., 1996, 1998; Novas, 1996, 1997). Instead, the simple inclusion of Troodontidae, or both Therizinosauridae and Oviraptorosauria, is sufficient to challenge the avian hypothesis. Clearly, the phylogenetic position of Alvarezsauridae is contingent on the interrelationships among other coelurosaurian taxa.

Although these results place alvarezsaurids within nonavian theropods, the suggestion of Martin and Rinaldi (1994) and Martin (1997) concerning the ornithomimid relationships of *Mononykus* is difficult to sustain. First, ornithomimids lack the synapomorphies uniting Alvarezsauridae with Node X (neural spines confined to caudals 1–12, semilunate carpal, ilium with caudodorsal margin ventrally curved in lateral view, pubis caudoventrally oriented with respect to the pubic pedicle of ilium, proximal fibula with medial fossa reduced or absent, proximal ulna with a cranially oriented notch for reception of radius and reduced ulnar proximal articular processes, pubic pedicle of ilium craniocaudally narrow, midcaudal vertebrae with short prezygapophyses, absence of a dorsal fossa on caudal process of coracoid, prominent ventral processes on cervicodorsal vertebrae). Second, if alvarezsaurids and ornithomimids are constricted to form a monophyletic group, the most parsimonious hypothesis requires five additional steps. Incorporating the new data of Sereno (1999) might change this assessment somewhat, but it is likely that many of these impediments to ornithomimid relationships will remain (see also Suzuki et al., 2002; Clark, Norell, and Makovicky, Chapter 2 in this volume).

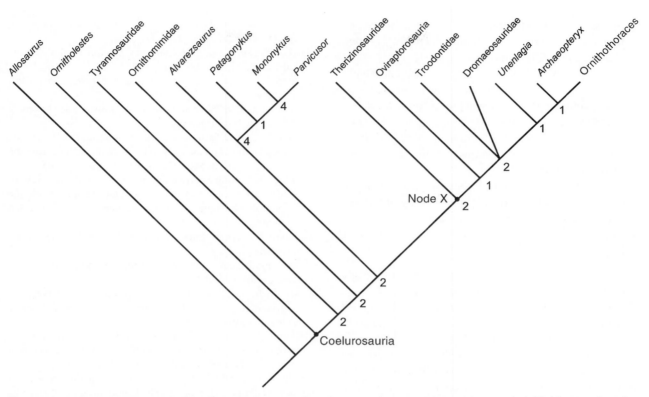

Figure 5.1. Strict consensus tree resulting from two most parsimonious trees (L = 309; CI = 0,48; RI = 0,54) yielded by Hennig86 (implicit enumeration command) (Farris, 1988). The information to construct the data matrix (Novas, unpublished data) was gathered from different sources: Gauthier, 1986; Russell and Dong, 1993; Holtz, 1994; Makovicky, 1995; Chiappe et. al., 1996, 1988; Martin, 1997; Novas, 1997; Novas and Puerta, 1997; the study of Sereno, 1999, was too recent to adequately incorporate into this analysis. The Bremmer support (Bremmer, 1994) is indicated in each node. This measure was calculated using the bs command in Nona (Goloboff, 1993a).

Although most authors devoted to theropod phylogeny currently use equally weighted parsimony as implemented in Hennig86 or PAUP (e.g., Gauthier, 1986; Russell and Dong, 1993; Holtz, 1994; Chiappe et al., 1996, 1998; Novas, 1996, 1997; Forster et al., 1998), a differential character weighting approach can also be used. For example, PeeWee (Goloboff, 1993b) is a program that sets differential weights for characters according to the amount of homoplasy they present on the tree. Under the PeeWee optimality criteria, the most parsimonious hypothesis is that in which the sum of all character weights (i.e., tree fit) is maximized. Our coelurosaurian data set was subjected to this kind of cladistic approach. Interestingly, under the exact solution commands ("wh" and "ms+"), the single most parsimonious tree supports the monophyly of Metornithes, placing alvarezsaurids within Aves. A possible explanation for this result is that most of the characters shared by alvarezsaurids and ornithothoracines are not present in other coelurosaurs, implying that such characters get the highest weights in trees supporting the monophyly of Metornithes. This is not the case for some of the characters supporting monophyly of Node X.

For example, cervical centra with kidney-shaped cranial articular surface are present in *Ornitholestes,* while a subvertically oriented postacetabular blade of ilium with a medial flange strongly reduced is absent in Troodontidae. In sum, PeeWee resolves the conflict in favor of the hypothesis supported by characters lacking homoplastic steps.

It is beyond the scope of this chapter to judge the validity of the different parsimony approaches mentioned previously. Our aim, instead, is to emphasize the highly conflictive pattern of character distribution among coelurosaurian taxa, which causes such disparity in the analyses using differentially or equally weighted characters.

Conclusions

This discussion emphasizes once again that alvarezsaurids are a problematic group of theropods, because, although they present obvious similarities with Ornithothoraces, they lack the following derived characters that in the present cladistic analysis are synapomorphies of different hierarchical groups within Coelurosauria:

1. *Archaeopteryx* + Ornithothoraces (= Aves): for example, forelimb exceeding 86% of hindlimb length; developed bicipital tubercle of coracoid
2. *Unenlagia* + Aves (= Avialae): for example, caudal end of ilium dorsoventrally low and sharply pointed; presence of a prominent proximodorsal process on ischium
3. Dromaeosauridae + Avialae (= Maniraptora): for example, mandibular fenestra reduced or absent; coracoid subrectangular in profile; elongate radius, 76% or more than humerus length
4. Troodontidae + Maniraptora: for example, distal haemal arches cranially bifurcated; ventral tubercle of humerus craniocaudally compressed and longitudinally elongated; pedal digit IV longer than II and closer to III in length; raptorial ungual in digit II

In sum, we are facing a group that presents homoplasies wherever it is located within Coelurosauria.

Criticisms of the avian hypothesis of alvarezsaurid relationships have relied largely on assumptions of the evolutionary process more than on cladistic analysis (Chiappe et al., 1997). This discussion shows that the issue is more complex than was previously thought. The results presented here, however, should not be considered an endorsement of the previous criticisms of the avian hypothesis (e.g., Feduccia, 1994; Martin and Rinaldi, 1994; Wellnhofer, 1994; Zhou, 1995; Karkhu and Rautian, 1996), since these were based on spurious assumptions.

The phylogenetic position of alvarezsaurids highlights a major problem in dinosaur systematics: coelurosaurian phylogeny. The resolution of this problem will certainly affect any hypothesis of alvarezsaurid relationships. Whatever the phylogenetic position of alvarezsaurids may be, alvarezsaurids are crucial to the knowledge of coelurosaurian evolution because they offer a mosaic of characters that illustrates the great amount of homoplasy existent within these theropod dinosaurs (including birds).

Acknowledgments

We thank L. M. Chiappe and L. M. Witmer for their invitation to participate in this book and for their thoughtful review of this chapter, and S. Orell for her editorial assistance.

Literature Cited

Bremmer, K. 1994. Branch support and tree stability. Cladistics 10(3):295–304.

Chiappe, L. M., M. A. Norell, and J. M. Clark. 1996. Phylogenetic position of *Mononykus* (Aves: Alvarezsauridae) from the Late Cretaceous of the Gobi Desert. Queensland Museum Memoirs 39:557–582.

———. 1997. *Mononykus* and birds: methods and evidence. Auk 114:300–302.

———. 1998. The skull of a relative of the stem-group bird *Mononykus*. Nature 392:275–278.

Farris, J. S. 1988. Hennig '86 references. Documentation for version 1.5. Privately published.

Feduccia, A. 1994. The great dinosaur debate. Living Bird 13(4):28–33.

———. 1996. The Origin and Evolution of Birds. Yale University Press, New Haven, 420 pp.

Forster, C. A., S. D. Sampson, L. M. Chiappe, and D. W. Krause. 1998. The theropod ancestry of birds: new evidence from the Late Cretaceous of Madagascar. Science 279:1915–1919.

Gauthier, J. A. 1986. Saurischian monophyly and the origin of birds; pp. 1–55 *in* K. Padian (ed.), The Origin of Birds and the Evolution of Flight. California Academy of Sciences, San Francisco.

Goloboff, P. A. 1993a. Nona, version 1.8. Program and references. Privately published.

———. 1993b. PeeWee, version 2.8. Program and references. Privately published.

Holtz, T., Jr. 1994. The phylogenetic position of the Tyrannosauridae: implications for theropod systematics. Journal of Paleontology 68(5):1100–1117.

Karkhu, A. A., and A. S. Rautian. 1996. A new family of Maniraptora (Dinosauria: Saurischia) from the Late Cretaceous of Mongolia. Paleontological Journal 30(5):583–592.

Lauder, G. V. 1994. Homology, structure and function; pp. 151–196 *in* B. Hall (ed.), Homology: The Hierarchical Basis of Comparative Biology. Academic Press, San Diego.

Makovicky, P. J. 1995. Phylogenetic aspects of the vertebral morphology of Coelurosauria (Dinosauria: Theropoda). M.S. dissertation, University of Copenhagen, 308 pp.

Martin, L. D. 1997. The differences between dinosaurs and birds as applied to *Mononykus*. Dinofest International, Proceedings 1997:337–343.

Martin, L. D., and C. Rinaldi. 1994. How to tell a bird from a dinosaur. Maps Digests 17(4):190–196.

Novas, F. E. 1996. Alvarezsauridae, Late Cretaceous maniraptorans from Patagonia and Mongolia. Queensland Museum Memoirs 39(3):675–702.

———. 1997. Anatomy of *Patagonykus puertai* (Theropoda, Maniraptora), from the Late Cretaceous of Patagonia. Journal of Vertebrate Paleontology 17(1):137–166.

Novas, F. E., and P. F. Puerta. 1997. New evidence concerning avian origins from the Late Cretaceous of Patagonia. Nature 387:390–392.

Ostrom, J. H. 1994. On the origin of birds and of avian flight; pp. 160–177 *in* D. R. Prothero and R. M. Schoch (eds.), Major Features of Vertebrate Evolution, Short Courses in Paleontology 7. The Paleontological Society, Knoxville.

Padian, K. 1997. Avialae; pp. 39–40 *in* P. Currie and K. Padian (eds.), The Encyclopedia of Dinosaurs. Academic Press, San Diego.

Perle A., M. A. Norell, L. M. Chiappe, and J. M. Clark. 1993. Flightless bird from the Cretaceous of Mongolia. Nature 362:623–626.

Perle A., L. M. Chiappe, Barsbold R., J. M. Clark, and M. A. Norell. 1994. Skeletal morphology of *Mononykus olecranus* (Theropoda: Avialae) from the Late Cretaceous of Mongolia. American Museum Novitates 3105:1–29.

Russell, D., and Dong Z. 1993. A nearly complete skeleton of a new troodontid dinosaur from the Early Cretaceous of the Ordos Basin, Inner Mongolia, People's Republic of China. Canadian Journal of Earth Sciences 30:2163–2173.

Sereno, P. C. 1997. The origin and evolution of dinosaurs. Annual Review of Earth and Planetary Sciences 25:435–489.

———. 1999. The evolution of dinosaurs. Science 284:2137–2147.

Suzuki, S., L. M. Chiappe, G. J. Dyke, M. Watabe, Barsbold R., and Tsogtbaatar K. 2002. A new specimen of *Shuvuuia deserti* Chiappe et al., 1998 from the Mongolian Late Cretaceous with a discussion of the relationships of alvarezsaurids to other theropod dinosaurs. Contributions in Science.

Wellnhofer, P. 1994. New data on the origin and early evolution of birds. Comptes Rendus de l'Académie des Sciences de Paris 319 (série II):299–308.

Zhou, Z. 1995. Is *Mononykus* a bird? Auk 112(4):958–963.

Part III

The Mesozoic Aviary
Anatomy and Systematics

6

Archaeopterygidae (Upper Jurassic of Germany)

 No other fossils have had more impact on the progress of biological thought than those of *Archaeopteryx* (Hecht, 1985). Just a year before the discovery of the first *Archaeopteryx* specimen, Louis Agassiz invoked "the definiteness of the characters of the class of birds" in support of his anti-evolutionary views (Mayr, 1970:302). Soon after its discovery in 1861, *Archaeopteryx* flew into the storm set off by Darwin's *Origin of Species* (1859). The importance of *Archaeopteryx* in filling the gap between reptiles and birds was immediately understood by Thomas Huxley (Rudwick, 1976) and Hugh Falconer, who wrote to Darwin in 1863: "Had the Solenhofen quarries been commissioned—by august command—to turn out a strange being à la Darwin—it could not have executed the behest more handsomely—than with the *Archaeopteryx*" (Kritsky, 1992:408). However, although privately delighted with the discovery, Darwin exercised remarkable restraint when commenting on it in *Descent of Man* (1871) and the 1872 edition of *Origin of Species* (Kritsky, 1992), probably because of Charles Lyell's insistence on the inherent incompleteness of the fossil record and thus its a priori minor significance as evidence of evolution (Rudwick, 1976). Darwin himself gave a considerable amount of thought to the origin of birds. He considered the bastard wing (alula) to be a remnant of a once major wing part and prophesied, in a 1858 letter to Lyell, the existence of an ancient bird with a double or bifurcate wing (Kritsky, 1992), a prophecy fulfilled, in a way, by *Archaeopteryx*.

Archaeopteryx inspired de Beer's (1954a) idea of mosaic evolution, which greatly contributed to the current understanding of evolutionary processes. Surprisingly and sadly enough, *Archaeopteryx* figures once again in the public debate on evolution (Cuffey, 1984; Halstead, 1984; Stephan, 1987; Ostrom, 1992) because of the recent surge of religious obscurantism. As a target of regular attacks by creationists, *Archaeopteryx* and other intermediate forms demonstrate their importance for the future of an enlightened worldview.

The Archaeopterygidae are currently represented by eight skeletal specimens (Fig. 6.1) and an isolated carbonized feather. Until recently, the names of skeletal specimens could be derived from the locations of museums where they were housed. Unfortunately, this system collapses because of the vicissitudes of privately owned specimens. Therefore, adopted here as a supplementary system (at least for interim use) is the numbering of the skeletal specimens in the order in which they were identified as pertaining to Archaeopterygidae: London (First) specimen (de Beer, 1954b) in the British Museum of Natural History (BMNH37001); Berlin (Second) specimen (Dames, 1884; Heilmann, 1926; Fischer and Krueger, 1979) in the Museum für Naturkunde (HMN1880/81); "Maxberg" (Third) specimen (Heller, 1959, 1960), lost from a private collection, referred to as "Solnhofen specimen" by Ostrom (1972); Haarlem (Fourth) specimen (Ostrom, 1972) in the Teyler Museum (6928/29), the Netherlands; Eichstätt (Fifth) specimen (Wellnhofer, 1974) in the Jura Museum (JM 2257); Solnhofen (Sixth) specimen (Wellnhofer, 1988, 1992) of contested ownership, at present in the Bürgermeister Müller Museum, Solnhofen; Munich (Seventh) specimen (Wellnhofer, 1993), in the Bayerische Staatssammlung für Paläontologie und historische Geologie; and an undescribed, privately owned eighth specimen, which consists of a compressed skull, shoulder girdle, and wing skeleton (Mäuser, 1997). First collected (in 1857) was the Haarlem specimen, which had been initially misidentified as a pterosaur and identified as *Archaeopteryx* more than a century later (Ostrom, 1972). A single feather impression was found in 1860 (von Meyer, 1862), and the London specimen was discovered a year after (in 1861).

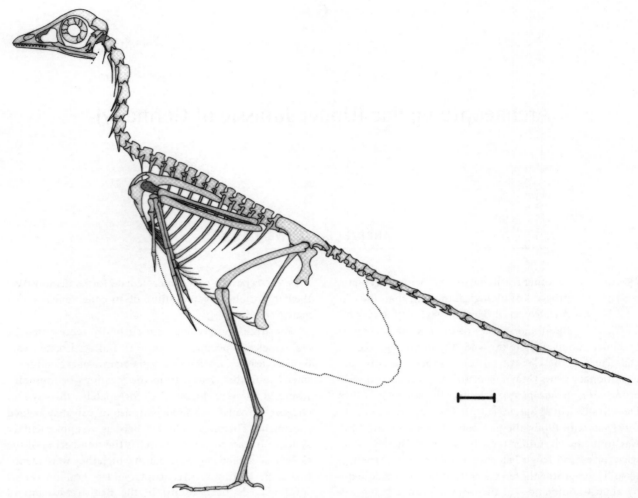

Figure 6.1. Skeletal reconstruction of *A. siemensii* with the thoracic girdle of the London specimen (*A. lithographica*). The wing is shown in a hypothetical, maximally folded position involving a distal shift of the radius relative to the ulna (Elzanowski and Pasko, 1999). The exact angular positions of the neck, tail, and hindlimb members are intuitively chosen within the likely ranges. Scale bar = 20 mm.

Geological Setting

The seven described specimens were found in the upper layers (i.e., above a horizon known as the *trennende krumme Lage*) of the Solnhofen Lithographic Limestone, which is of early Lower Tithonian, Upper Jurassic (Malm zeta 2) age. The depositional period of the Solnhofen Lithographic Limestone is maximally estimated as 500,000 years (Barthel et al., 1990), and thus the upper layers alone must have been deposited over a considerably shorter period. The eighth, undescribed specimen stems from the younger (Malm zeta 3) Mörnsheim layers (Mäuser, 1997).

The Solnhofen Lithographic Limestone was deposited in a lagoon that was closed off from the open sea by coral reefs (Barthel et al., 1990; Viohl, 1990). It is composed of the alternating layers of pure (*Flinz*) and marly (*Fäule*) limestone.

The pure limestone (which is exploited commercially) probably derives from the carbonate ooze that was first deposited in the open sea and periodically driven to the Solnhofen lagoon (possibly by storms). Because of high evaporation rates, the salinity in the lagoon was high, the water was stratified, and the bottom sediments were without macrobenthos and nearly lifeless because of the lack of oxygen. This accounts for the good preservation of Solnhofen fossils. The Solnhofen fauna derives from the open sea, the coral reef, the lagoon itself, the nearby German and Bohemian massifs, and possibly smaller islands, if they were present. In fact, the Jurassic massifs of what is now western and central Europe were of the size of large islands, and, in this sense, all the terrestrial elements of the Solnhofen fauna, including the archaeopterygids, were insular (Olson, 1987; Chatterjee, 1997).

The climate of the land masses was semiarid, with prolonged dry summers but rainy winters (Barthel et al., 1990; Viohl, 1990). The land masses were covered in part by the groves of araucarias, with some cycads, bennettites, and sparse gingko trees, and in part (probably mainly along the coast) by thickets of conifers (Hirmeriellaceae = Cheirolepidaceae) and seed ferns (Jung, 1995), but no logs have ever been recovered from the limestone (Barthel et al., 1990). The archaeopterygids, *Compsognathus,* lizards, rhynchocephalians, a palpigrade arachnid, and many insects, including Orthoptera, Blattodea, Hymenoptera, and some Coleoptera, are exclusively terrestrial throughout their entire life cycle and thus lived on the land masses (Barthel et al., 1990) or, possibly, on smaller islands (if present). The assemblage of insects is dominated by freshwater groups, some of which may have possibly lived in the coastal marine environment, and the same may be true of the Solnhofen crocodilians. Among flying vertebrates, finds of pterosaurs are much more frequent than those of archaeopterygids, which probably reflects differences in locomotor habits rather than original abundance.

Each of the archaeopterygid specimens has been found between two layers of limestone, but there is no regularity with respect to the alternation of pure and marly limestone layers (the Third and Eichstätt specimens were found between two pure layers, the Sixth and Munich specimens between a marly and a pure layer). The slab with most of the skeleton is referred to as the main slab (*Hauptplatte*) and the slab with the complementary skeletal impressions and bony fragments as a counterslab (*Gegenplatte*). For all specimens having stratigraphic documentation, the main slab is the upper slab (*Hangendplatte*), and the counterslab is the lower slab (*Liegendplatte*); this is probably true of all specimens (Rietschel, 1976). The Sixth specimen is represented by only one slab (probably the upper one) with nearly all postcranial bones; the counterslab was lost.

The preservation of all the specimens is remarkably similar (Rietschel, 1976). Each of them was buried upside down, as the main (upper) slab always shows the skeleton more or less in dorsal view. The head and neck are turned backward and to the side, and the tail is more or less raised, resulting in a lordotic curvature of the spine. In most specimens, both legs lie on the side opposite to the head, conveying a "biking" pose to the skeleton. Only the Berlin and Eichstätt specimens were complete (or nearly so) at the time of burial. The skull is the most damaged part in the London, Sixth, and Munich specimens, and the last two show a sharp dorsal bend in the tail just behind the pelvis. The postcranial skeleton alone is well articulated and almost complete in the Berlin, Eichstätt, Sixth, and Munich specimens and much less so in the London and Third specimens (Kemp and Unwin, 1997). However, the feather impressions are least distinct in the Eichstätt specimen (Rietschel, 1985a). None of

the specimens shows any evidence of predation or scavenging. The Berlin specimen, especially the skull, is somewhat worn by more than a century of use. The Sixth specimen suffered from unprofessional handling by the former owner, and most of its bones are obscured by an encrustation identified as iron oxide (Wellnhofer, 1988, 1992).

All that can be said about the cause of death is that the birds were blown in by wind or their carcasses transported by floods to the lagoon (Davis, 1996), but most probably they did not fly voluntarily above the sea. Because of the substantial differences from neornithines, especially in the structure of body cavities (Duncker, 1989), the archaeopterygid carcass may have behaved differently than the carcasses of neornithines (Helms, 1982), and its burial upside down does not necessarily indicate the same floating position (contra Helms, 1982). The uniform preservation of all specimens and the absence of predation signs indicate an underwater burial preceded by floating (Rietschel, 1976; Helms, 1982; de Buisonjé, 1985; Barthel et al., 1990; Elzanowski and Wellnhofer, 1996). Osmotic desiccation of the tendons in the hypersaline environment probably contributed to the lordotic bend of all archaeopterygid specimens and other Solnhofen vertebrates (de Buisonjé, 1985; Davis, 1996). The differences in preservation of the skull are accountable in terms of the avian pattern of disarticulation and/or specific preburial posture. In neornithine carcasses, the skulls and cervical vertebrae are the first to disarticulate (Davis and Briggs, 1998), and they did in the London specimen and possibly the Third, which are most damaged. In addition, drifting with a pendulant head and tail may have exposed specifically these parts to damage (Rietschel, 1976; Elzanowski and Wellnhofer, 1996). Regardless of the exact course of preburial events, all differences in the completeness of preservation of the known skeletons appear to be accountable in terms of the degree of decay and factors controlling it (the most evident of which are floating time and posture). In addition, the specimens differ in the extent of diagenetic damage by sediment pressure, which resulted in the collapsing of bone walls and crushing of superimposed bones.

Systematic Paleontology

Taxonomic Hierarchy

Aves Linnaeus, 1758
  Archaeopterygidae Huxley, 1871

The inclusion of *Wellnhoferia* (Elzanowski, 2001a) in this family is tentative (and hence so is the diagnosis of Archaeopterygidae) pending a phylogenetic analysis of *Archaeopteryx* and *Wellnhoferia* as separate taxa.

Tentative diagnosis—Four premaxillary, 8 to 9 maxillary, and 11–12 dentary teeth. Tooth crowns rounded in the basal part, with the tips of most teeth slightly curved caudally. Premaxilla projects beyond the tip of the mandible. Mandibula without fenestrae. Furcula without hypocleideum. Coracoid with a wing-shaped preglenoid process. Ischium with an intermediate process. Twenty-three presacral and 5 sacral vertebrae.

Archaeopteryx von Meyer, 1861

At present, there is not enough evidence to confirm a separation of the Eichstätt specimen as *Archaeopteryx recurva* Howgate, 1984, or to determine the precise taxonomic status of the fragmentary Third and Haarlem specimens. Each of these three specimens is provisionally classified as *Archaeopteryx* sp.

Diagnosis—Manual digit I with the ungual being approximately half the length of the basal phalanx. Manual digit III with phalanges 1 and 2 unfused (even if tightly connected). Metatarsal II not tapered proximally. Pedal digit IV with five phalanges, approximately 90% (more than 80%) of the length of digit III. Flexor tubercles of pedal claws incompletely differentiated or absent. Pedal digit IV with the ungual shorter than or subequal to the basal phalanx.

Archaeopteryx lithographica von Meyer, 1861

Holotype—The conserved name *A. lithographica* (International Commission on Zoological Nomenclature, 1977) was coined by von Meyer (1861) in his second preliminary note on the single Solnhofen feather for an animal species represented by the London specimen. Following comments on the feather, von Meyer (1861:678–679) wrote: "At the same time I am hearing from *Herrn Obergerichtsrath* Witte that a nearly complete skeleton of an animal covered with feathers was found in the lithographic slate. . . . *Archaeopteryx lithographica* is a name that I deem appropriate for the designation of *the* animal" (emphasis added). The London specimen (subsequently described as *Archaeopteryx macrura* Owen, 1864) should, therefore, be recognized as the holotype of *A. lithographica.*

Referred specimens—Provisionally none.

Locality and horizon (for the London specimen)—Langenaltheimer Haardt (former Ottmann's quarry), 3.5 m below the upper surface of the *Sieben-Lumpen-Schicht,* Upper Solnhofen Lithographic Limestone (Malm zeta 2b), lower Lower Tithonian.

Diagnosis—Larger than *Archaeopteryx siemensii.* Preacetabular ilium with iliofemoralis internus fossa and ventral process. Bases of pedal claws expanded into flexor tubercles.

A. siemensii Dames, 1897

Holotype—Berlin specimen (HMN1880/81).

Referred specimens—Provisionally none.

Locality and horizon—Dörr's quarry, Blumenberg near Eichstätt, Upper Solnhofen Lithographic Limestone (Malm zeta 2b), lower Lower Tithonian.

Diagnosis—Smaller than *A. lithographica,* close in size to *Archaeopteryx bavarica.* Preacetabular ilium without the iliofemoralis internus fossa and ventral process. Pedal claws without flexor tubercles. Tooth crowns consistently rounded in cross section. The humerus/ulna ratio above 110% and the femur/tibia ratio around 70% or more.

A. bavarica Wellnhofer, 1993

Holotype—Munich (Seventh) specimen.

Referred specimens—Provisionally none.

Locality and horizon—Langenaltheimer Haardt (quarry of the Solenhofer Aktien-Verein), 11 m above the upper surface of the *Sieben-Lumpen-Schicht,* Upper Solnhofen Lithographic Limestone (Malm zeta 2b), lower Lower Tithonian.

Diagnosis—Smaller than *A. lithographica,* close in size to *A. siemensii.* Preacetabular ilium without the iliofemoralis internus fossa and ventral process. Pedal claws without flexor tubercles. Tooth crowns compressed mesially (= rostrally). Spinous processes of the third and fourth cervicals high, contributing about one-third of the total vertebral height. The humerus/ulna ratio well below 110% and the femur/tibia ratio well below 70%.

Wellnhoferia Elzanowski, 2001

Diagnosis—Larger than *Archaeopteryx.* Manual digit I with ungual of approximately one-third the length of the basal phalanx. Manual digit III with phalanges 1 and 2 fused. Metatarsal II tapered proximally. Pedal digit IV with four phalanges, approximately 75% (less than 80%) of the length of digit III. Pedal digit IV with the ungual as the longest phalanx. Pedal claws with well-developed flexor tubercles.

Wellnhoferia grandis Elzanowski, 2001

Holotype—Sixth specimen.

Referred specimens—Provisionally none.

Locality and horizon—Exact locality and horizon unknown (in all probability Altmühl Valley), Upper Solnhofen Lithographic Limestone (Malm zeta 2b), lower Lower Tithonian.

Diagnosis—As for the genus.

Anatomy

Notwithstanding numerous studies of seven of the skeletal specimens, only superficial features of the archaeopterygid osteology are well known, a circumstance that is due to the essentially two-dimensional preservation of most specimens. The majority of articulations and the internal cranial structures remain poorly known.

Skull

Cranial bones are best preserved in the Munich, Eichstätt, and London specimens (Whetstone, 1983; Walker, 1985;

Elzanowski and Wellnhofer, 1995, 1996). Skull length is estimated as 39 mm in the Eichstätt, 45 mm in the Munich, and 52 mm in the Berlin specimens. In all known cases, the upper jaw protrudes for some 2–3 mm beyond the tip of the mandible. As far as it is known, the archaeopterygid skull shows a mosaic of nonavian theropodan and avian characters.

Braincase. A natural mold of the cranial cavity of the London specimen revealed that, as in other birds (as well as mammals, pterosaurs, and some small coelurosaurs), the brain completely filled the cranial cavity, the expanded hemispheres contacted the cerebellum dorsally in the midline, and the optic lobes (mesencephalic tectum) were placed laterally and thus separated (Jerison, 1973; Bühler, 1985). Accordingly, the hemispheric fossa in the Munich specimen extends onto the parietal for some 3 mm and ends caudally with a rounded ridge that separates it from the tectal fossa for the optic lobes, as in neornithines (Werner, 1963; Elzanowski and Galton, 1991). Contrary to Bühler's (1985) reconstruction and Stephan's (1987) assertion, it would not be possible to ascertain that the surface of the brain was smooth even if the entire cranial cavity did not reveal any indications of brain fissures. In fact, the cerebellar fossa is not known in the archaeopterygids, and, although the cerebellum of all extant birds is foliate, many of them do not have interfoliar ridges in the cerebellar fossa (Elzanowski and Galton, 1991). Hopson (1977) estimated the brain volume of the London specimen as 1.8 ml and its body mass as not greater than 400 g, which placed archaeopterygids in the lower range of the avian ratios. However, the London specimen probably weighed more than 400 g (see later).

As in neornithines, the frontal, parietal, and laterosphenoid contribute to the postorbital process, which is blunt (Fig. 6.2). The orbital margin of the frontal seems to be deflected into the orbit. The parietal has a caudolateral occipital process that contributes to a composite structure known as the paroccipital process. The laterosphenoid is tightly sutured to the parietal and prootic and tapers toward the postorbital process. The prootic has a large rostroventral wing that completely encloses the trigeminal foramen and has a slender preforaminal process that points rostrodorsally and contacted or approached the laterosphenoid (Elzanowski and Wellnhofer, 1996: Fig. 5).

The otic region, as exposed in the London specimen, has been interpreted (Walker, 1985; Witmer, 1990) under an assumption that its configuration in the archaeopterygids is similar in detail to that of the modern neognathous birds, although the braincases of *Archaeopteryx* and neornithines are otherwise dramatically different. A shallow depression in the dorsal part of the prootic (adjacent to the parietal and laterosphenoid) has been interpreted as the dorsal pneumatic recess. Similar prootic depressions are present in the same location in several nonavian theropods including *Syntarsus* (Raath, 1985), Ornithomimidae (Currie and Zhao, 1993), and Velociraptorinae (Barsbold and Osmólska, 1999). A small space caudoventral to the depression was identified as the caudal tympanic recess. A regular concavity that encloses this space ventrally was referred to by Walker (1985) as the "threshold" (based on comparison with the juveniles of modern neognaths) and a ridge that descends from it as the interfenestral bridge (crista interfenestralis).

There are several problems with these interpretations. The prootic depression does not look like a pneumatic recess (Walker, 1985; pers. obs.). More important, the dorsal recess in *Enaliornis* (Elzanowski and Galton, 1991) and all neornithines is enclosed by the squamosal, which is not incorporated in the braincase in *Archaeopteryx*. This depression may have possibly accommodated a part of one of the jaw adductors (pseudotemporalis or adductor mandibulae externus), similar to the way that a deep recess in this area accommodates the temporal head ("caput absconditum") of the pseudotemporalis muscle in the pelecaniforms and procellariiforms. The concavity called "threshold" was once identified as the quadrate cotyla (Whetstone, 1983) and in fact looks like a cotyla, as it is perfectly regular and evidently natural, not exaggerated by crushing (contra Walker, 1985). Since a cotyla in this location cannot be for anything but the quadrate, this alternative interpretation needs to be given serious consideration. If it were true, then the space immediately above the cotyla could represent the dorsal, not caudal, tympanic recess. The identification of the interfenestral bridge (Walker, 1985) is not compelling. The single piece composed of the "threshold" (or quadrate cotyla) and the putative interfenestral bridge is nowhere continuous with the main body of the prootic and may be dislocated. As a result, the putative dorsal recess and the angular positioning of the putative interfenestral bridge could be preservational artifacts.

The cranial base (Elzanowski and Wellnhofer, 1996: Fig. 2A) is poorly known. The parasphenoid rostrum is composed of two tightly superimposed bony laminae, suggesting an originally hollow structure. The pterygoid (basipterygoid) process is directed rostroventrolaterally. The articular facet for the pterygoid is slightly sinusoidal and oblique, its rostromedial end being dorsal to its caudolateral end. A prominent dorsoventral ridge at the base of the pterygoid process separates two small fossae, which correspond to the pneumatic recesses in *Syntarsus* (Raath, 1985). The basal tubercle ("mammillary process") seems to be traversed by the suture between the basisphenoid and basioccipital. The occipital condyle is traversed by the suture between the basioccipital and exoccipital.

Postorbital. The postorbital may be represented by a crescentic element in the Munich and Eichstätt specimens,

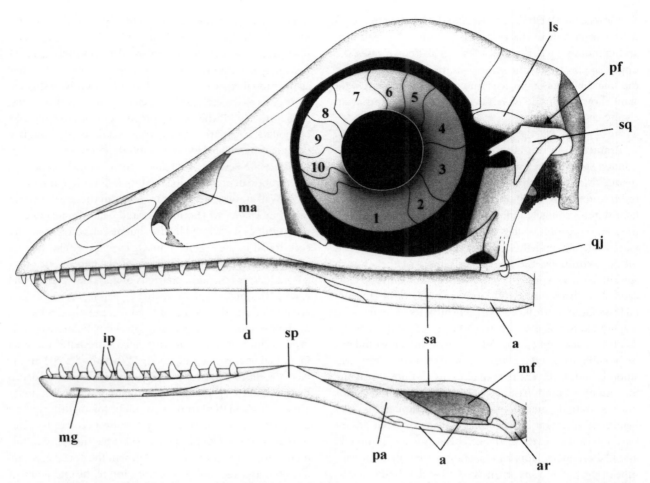

Figure 6.2. Scale reconstructions of the complete skull of *Archaeopteryx* (top, based on the Berlin, Eichstätt, and Munich specimens) and mandible in medial view (bottom, based on the Munich specimen). Structures in the nasal cavity (shown only in part) and the occipital area (broken lines) are barely known. The presence of the postfrontal and postorbital bones is uncertain. The caudal ends of the braincase and mandible and the exact angular position of the quadrate-squamosal complex cannot be reconstructed with any reliability. Abbreviations: a, angular; ar, articulare; d, dentary; ip, interdental plates; ls, laterosphenoid; ma, putative ascending ramus of the maxilla; mf, mandibular fossa; mg, meckelian groove; pa, prearticular; pf, prootic fossa; qj, quadratojugal; sa, surangular; sp, splenial; sq, squamosal.

in the latter just behind the postorbital process (Elzanowski and Wellnhofer, 1996: Figs. 1 and 7C), or by a small, poorly defined triradiate element in the Berlin specimen. If present, the postorbital is reduced, and its jugal process is lacking or vestigial. It did not contact the jugal. Chatterjee's (1991) reports of a large postorbital in the Berlin specimen and an identification of this element in the Eichstätt specimen are in error.

Squamosal. Although apparently still excluded from the braincase wall, the squamosal, when properly positioned, is not as similar to that of nonavian theropods as initially thought (Elzanowski and Wellnhofer, 1996). The quadrate cotyla of the squamosal appears to open rostroventrally rather than ventrally. The quadratojugal (descending)

process, which is probably homologous with that of nonavian theropods and, possibly, with the zygomatic process of the neornithines, probably did not contact the quadratojugal (Chatterjee, 1991, 1997). In front of the quadratojugal process, there is a slightly forked, rostroventral process. It has been previously interpreted as the postorbital process, but this is probably incorrect. In contrast with the nonavian theropods, it forms an acute angle with the quadratojugal process and thus must have been directed more or less rostroventrally (Fig. 6.2), which precludes the presence of a typical dorsotemporal ("supratemporal") arch. This rostroventral process corresponds in position to the rostral part of the zygomatic process of some birds (e.g., the gruiforms) and may have served as an attachment site for parts of the adductor mandibulae externus.

Quadrate. The quadrate is poorly understood. The bone identified as the quadrate in the London specimen (Walker, 1985; Chatterjee, 1991) does not seem to be similar in detail to what is exposed of the quadrate in the Munich specimen. The quadrate head in the Munich specimen (Elzanowski and Wellnhofer, 1996: Fig. 3) is damaged, as is the head of the putative quadrate of the London specimen (Walker, 1985). As preserved, the quadrate in the Munich specimen is incompatible with the presence of a broad head with two widely separated, approximately equivalent capitula, as seen in neornithines. However, it could be compatible with the configuration found in the oviraptorids, which show many other avian similarities (Elzanowski, 1999). In the oviraptorid quadrate, the apex of the otic process is formed by the prominent lateral (squamosal) capitulum, and the medial capitulum is attached below, on the medial slope of the lateral capitulum (Maryanska and Osmólska, 1997). The presence of a similar, small medial capitulum in *Archaeopteryx* is suggested by the presence of a small medial projection in the Munich specimen (Elzanowski and Wellnhofer, 1995: Fig. 1; 1996: Fig. 2B) and the putative cotyla in the tympanic fossa of the London specimen. The pterygoid articulation of the quadrate remains unclear but seems to be unlike that of other archosaurs, including neornithine birds.

The skulls of the Munich specimen and *Gobipteryx* share the total length of approximately 45 mm, and their quadrates are 11 mm long, which amounts to nearly one-fourth of total skull length, more than in either nonavian theropods or neornithine birds. A relatively longer quadrate is known in an alvarezsaurid *Shuvuuia* (Chiappe et al., 1998).

Quadratojugal. The quadratojugal is an L-shaped bone with jugal and squamosal (= ascending) processes. The jugal process is much stronger than the squamosal process. The latter probably did not contact the squamosal.

Jugal. The jugal is distinctly bowed and, based on its relative length, must have protruded rostrally far beyond the lacrimojugal articulation (Fig. 6.2). Caudally, it bifurcates into the quadratojugal and caudodorsal processes (contra Martin, 1991). The caudal margin of the caudodorsal process has two shallow embayments separated by an angular projection. The process itself narrows to a spike that apparently aligns with and thus may have been connected by a ligament to the quadratojugal process of the squamosal. The jugal lacks a typical postorbital process (contra Chatterjee, 1997), and the homology of the caudodorsal process is unclear. It strictly corresponds in position to a caudodorsal process that coexists with the postorbital process in therizinosaurid theropods (Elzanowski, 1999).

Lacrimal. The lacrimal is preserved in one piece in the London specimen and as fragments in the Eichstätt speci-

men. It has a prominent nasal process and a much smaller supraorbital process. The shaft descended in an oblique plane (intermediate between the sagittal and transverse) and straddled, at angles, the jugal bar, which was received in a broad ventral incisure. Witmer (1997:26) reported an opening for the nasolacrimal duct in the London specimen. As in all other birds, there is no evidence of a separate prefrontal.

Rostral to the lacrimal there are four ossicles that are connected and stacked one above the other. Although they could not be identified with any certainty, they seem to indicate some compartmentalization of the nasal cavity in its caudal part (Witmer, 1990:360; Elzanowski and Wellnhofer, 1996: Fig. 7). However, the most ventral of them may possibly represent the uncinatum (lacrimopalatinum).

Palate. The vomers (Fig. 6.3C) are partly or completely fused rostrally into a single bone. The premaxillary ramus of the vomer is about half as deep as, and distinctly more dorsal than, the pterygoid ramus. In this respect, the vomer is similar to that of nonavian theropods. However, at the

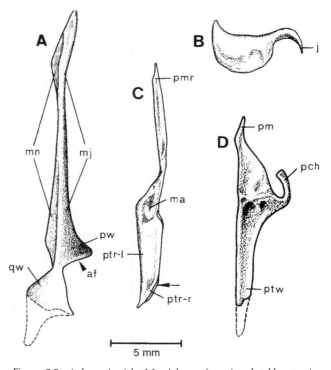

Figure 6.3. *A. bavarica* (the Munich specimen), palatal bones: A, pterygoid, probably in dorsolateral view; B, ectopterygoid in dorsal view; C, vomer in left lateral view; and D, palatine in dorsal view. Abbreviations: af, articular facet for the quadrate; j, jugal facet; ma, articular facet (possibly for the maxilla); mj, mn, two components of the pterygoid; pch, choanal (vomerine) process; pm, maxillary (= premaxillary) process; pmr, premaxillary ramus; pw, prequadrate wing; ptr-l, ptr-r, pterygoid rami (left and right); ptw, pterygoid wing; qw, quadrate wing (adapted from Elzanowski and Wellnhofer, 1996).

rostral end of the pterygoid ramus there is a distinct concave articular facet with a prominent caudal boss.

The palatine (Fig. 6.3D) is a bony slat with a medial process enclosing the choana. The maxillary (= premaxillary) process is short (approximately one-third of the total bone length) and narrows abruptly toward the rostral end, which turns laterally. The maxillary process of the palatine certainly overlapped the maxilla. The pterygoid wing is twice as long as the maxillary process. This wing certainly overlapped the pterygoid and may have contacted, or at least approached, the ectopterygoid. The choanal (vomerine) process is hook-shaped. At the junction of the maxillary process and pterygoid wing, there is a prominent transverse crest, which is higher medially than laterally. The crest passes into the caudal wall of the choana, whose medial rim is formed by the choanal process. A bone on the counterslab of the London specimen (some 15 mm from the skull) may represent the middle fragment of the palatine with the choanal process, but it differs in detail from the corresponding part in the Munich specimen.

The palatine of *Archaeopteryx* is distinctively avian and different from the tetraradiate palatine of nonavian theropods. As in *Hesperornis,* the choanal process is hook-shaped, and the pterygoid wing is long and slender. As in *Gobipteryx* (Elzanowski, 1995), the maxillary process is short and offset laterally. The maxillary process is also short in the nonstruthioniform palaeognaths, where it does not contact the palatal process of the premaxilla even though the latter extends far caudally. In the neognaths, the maxillary (= premaxillary) process reaches and often fuses to the premaxilla.

The ectopterygoid (Fig. 6.3B) is composed of the main plate and the hook-shaped jugal (lateral) process. The strong curvature of the process is comparable to that of nonavian theropods. However, the ectopterygoid of *Archaeopteryx* is positioned more caudally, and its main plate is much wider (mediolaterally) than it is long (rostrocaudally). A circular depression—possibly a pneumatic recess (Witmer, 1997)—is present in the Munich specimen. It probably marks the ventral surface of the bone, since no depression in the dorsal surface is present in the Eichstätt specimen.

The rostral end of the pterygoid (Fig. 6.3A) gently expands into an elongate, trapezoidal blade that presumably overlapped the pterygoid wing of the vomer ventrally. Caudally, the shaft flares out into a prominent, triangular quadrate wing with a distinct, flat articular surface along its caudal edge. The shaft is connected to the quadrate wing by a twist. The rostral blade corresponds to the spatulate ends of the hemipterygoid in *Hesperornis* (Elzanowski, 1991). In contrast, the caudal moiety substantially differs from the corresponding part in both birds and nonavian archosaurs, and so does the pterygoquadrate articulation.

The pterygoid reveals a puzzling, longitudinal division into a minor and a major component (Fig. 6.3A). The two components are connected very tightly and may be inseparable in the middle section of the bone, where the suture between the two components wanes. The minor component ends some 2 mm from the rostral end, whereby the rostralmost third of the blade is made of the major component only. Caudally, the major component contributes the quadrate wing and continues into the quadrate wing, while the minor component seems to end at the twist between the shaft and the quadrate wing.

Upper Jaw. The premaxilla distinctly protrudes beyond the tip of the mandible (Fig. 6.2). It is unclear whether the left and right premaxillae have their bodies co-ossified, as in neornithines, or separated by a median suture, as in reptiles. The nasal (= frontal) processes are separated by a suture. They are very thin, which suggests the possibility of a kinetic bending zone. The maxillary process, which overlies the maxilla, is short and extends for only one-third of the length of the nasal opening. The suture between the premaxilla and maxilla has a distinct twist, suggesting a complex interlocking of the two bones.

The maxilla has a slender nasal process and a long subfenestral part. The identification of structures in the antorbital fossa is uncertain. Wellnhofer (1974) and Witmer (1997) reconstructed a fenestrate "ascending ramus" of the maxilla, as found in nonavian theropods. However, in the Eichstätt specimen, caudal to the maxilla's nasal process and ventral to the nasal, there is a bony capsule (Elzanowski and Wellnhofer, 1996: Fig. 7: "Y"), which looks different from the "ascending ramus" of nonavian theropods. The capsule is ventrally supported by at least one strut, which is thought to separate the "subsidiary maxillary foramina" (Wellnhofer, 1974). However, it is not quite obvious that the nasal capsule is homologous to the "ascending ramus" of nonavian theropods. It may, in fact, represent a rostral ethmoid ossification, which was present in the Hesperornithidae (pers. obs.), as it is in several neornithines (e. g., Cathartidae), where it may be continuous with the nasal septum and form the floor of the nasal cavity (Hofer, 1949).

Each nasal has a rostral embayment for the nasal opening that is enclosed between two small processes (premaxillary and maxillary). Ventral to the nasals and caudal to the capsule, there is another uninterpreted, apparently median structure (Elzanowski and Wellnhofer, 1996: Fig. 7: "X") that may or may not represent the mesethmoid.

Cranial Kinesis. Several points of evidence suggest that the archaeopterygid skull was kinetic and had a flexion zone located in front of the braincase, as in other birds: (1) there is a sliding joint between the lacrimal and the jugal (Wellnhofer, 1974); (2) the palatine is slatlike, suggesting the presence of the pterygopalatine bar that transfers the force between the quadrate and the upper jaw (Elzanowski and Wellnhofer, 1996); (3) there is probably a propulsion joint between the quadrate and the pterygoid (Elzanowski and Wellnhofer, 1996); and (4) the cranial cavity was completely filled with the brain, which makes the presence of any flexion zone within the braincase extremely unlikely (Bühler, 1985). There is, however, very little evidence to establish the exact position of the kinetic flexion zone and no evidence whatsoever for a prokinetic flexion zone (contra Bühler, 1985). Wellnhofer (1974) suggested rhynchokinesis because the nasal processes of the premaxilla are remarkably thin. However, the archaeopterygid upper jaw differs from that of neornithines to such an extent (especially in the lack of bone fusions) that its mobility may not fall in any of the modern categories of cranial kinesis.

Mandible. Each mandibular ramus is rigid and solid, without any trace of intraramal mobility and fenestration, and very broad and robust caudally, especially in comparison with the corresponding region in neornithines. The dorsal surface of the surangular forms a remarkably wide platform in front of the quadratomandibular joint, and the prearticular is correspondingly broad under the mandibular fossa. Consequently, the mandibular fossa is very deep (Fig. 6.2). The lateral wall of the mandibular fossa is formed primarily by the surangular, and only its most ventral part is enclosed by the angular. Rostrally, the fossa narrows abruptly to the Meckelian groove in the dentary, which is deep and narrow, and extends up to the level of the second tooth. The groove divides the medial surface of the dentary into the dorsal, strongly convex, subdental part and the ventral ridge.

The surangular has a transverse descending edge, just rostral to the quadrate cotyla. The articular is wedged between the surangular and prearticular (Fig. 6.2). The articular forms the bottom of the medial quadrate cotyla and a distinct prominence just caudal to the cotyla. The articular overlies the prearticular, which extends far caudally. Rostrally, the prearticular expands into a lancet-shaped blade that fits exactly between the bulging dorsal border of the surangular and the caudal portion of the splenial. There is no indication of the coronoid, and the rostral blade of the prearticular does not leave any space for this bone.

The angular is exposed only marginally on the medial surface of the mandibular ramus, along and below the ventral edge of the prearticular. The angular extended at least up to the level of the dorsal apex of the splenial. The lateral surface of the angular is as described in the Eichstätt specimen (Elzanowski and Wellnhofer, 1996).

The splenial has the shape of an elongate triangle with the rostral arm longer than the caudal arm; the apex of the splenial reached and may have slightly protruded over the dorsal edge of the dentary (Fig. 6.2). The articulation between the splenial and prearticular appears to have been loose. The caudal end of the dentary and its articulation with the surangular remain unknown, although the preserved parts of the dentary suggest a caudal bifurcation.

The rostral ends of the mandibular rami were probably connected by a strong syndesmosis. In the dentigerous part, the buccal margin of the dentary is higher than the lingual margin. The latter bears unequivocal interdental plates (contra Martin and Stewart, 1999). They are distinctly separated from the lingual margin of the dentary by a narrow groove and widely separated from one another. The plates between the rostral seven to eight teeth are triangular and become rectangular farther caudally, where they seem to be nearly continuous with the buccal wall of the dentary.

Teeth. There are 4 teeth in the premaxilla, 8–9 in the maxilla, and 11–12 in the dentary. The premaxillary and all but the most caudal dentary teeth tend to be more procumbent than the maxillary teeth. Teeth are widely spaced, with the interdental spaces averaging one tooth diameter. All tooth crowns have rounded cross sections at their bases. While the front teeth tend to be peglike, most teeth have backward pointing tips, and thus their caudal profile is slightly concave, conveying the impression of a slight "waist" between the tip and the base of the crown (Howgate, 1984b). In addition, some of the teeth show a slight constriction at the base of the crown but no indication of a strongly expanded root (contra Martin and Stewart, 1999). There are no carinae and serrations on the crowns.

There is some variation in tooth structure between the specimens. The lower teeth of the Munich specimen have the crown tips compressed mesially, that is, with rostral edges sharp and caudal edges rounded (Elzanowski and Wellnhofer, 1996: Fig. 12A); the upper teeth were probably similar, since the opposite upper and lower teeth are similar in shape in the reptiles. The upper teeth of the London and Berlin specimens tend to be more peglike and stouter than those of the Eichstätt specimen (Howgate, 1984b), but otherwise these three specimens show similar teeth. The teeth of *Wellnhoferia* look similar to those of the London specimen.

Scleral Ring. The ring is composed of 11–12 plates (Fig. 6.2), fewer than in the majority of extant birds, except for the Columbidae, with 11 plates, and *Sula, Todus,* Cuculidae, *Opisthocomus,* Psittacidae, Trochilidae, and some Spheniscidae and Musophagidae, with 12 plates (Lemmrich, 1931; Curtis and Miller, 1938; de Queiroz and Good, 1988). Heilmann's (1926) figure of 14 plates for the Berlin specimen, in which the right side of the skull is exposed, evidently includes 2–3 plates (first, second, and last) of the left ring, preserved in the opening of the right ring.

Based on the ring in the Eichstätt specimen, the largest are plates 1 and 7, which occupy the most ventral and dorsal positions, respectively, and overlap both adjacent ossicles. Plate 4 and probably 8 (the rostrodorsal sector of the ring is disarrayed) are overlapped by both adjacent ossicles. This pattern of overlapping and overlapped ossicles (1,7;4,?8) is identical or similar to that in other taxa with 11–12 plates (Lemmrich, 1931; Curtis and Miller, 1938; de Queiroz and Good, 1988), including Columbidae and *Opisthocomus* (1,7;4,8), Musophagidae (1,7;4,9 and 1,7;4,8), Psittacidae (1,7;4,9), and Trochilidae (1,7;4,8). A sinuous boundary between the plates in the nasal canthus, especially between the 9th and 10th and between the 10th and 11th plates, suggests a mutual overlap called interlocking (Curtis and Miller, 1938) or *Verzahnung* (Lemmrich, 1931): one ossicle is overlapped on one side and overlaps on the other the same adjacent ossicle. As in other birds, the inner border of the scleral ring is reflected outward.

Vertebral Column

The vertebral column, which is still poorly known, includes 23 presacral, 5 sacral, and 21–22 caudal vertebrae in *Archaeopteryx* but only some 16–17 caudals in *Wellnhoferia*. The articulations between the vertebral bodies are platycoelous to slightly amphicoelous. The transition between the cervical and thoracic vertebrae is poorly preserved. There are 9 typical cervical vertebrae, 1–2 transitional cervicothoracic vertebrae (with moveable ribs), which are included in the neck in avian osteology, and 12–13 thoracic (= dorsal) vertebrae, including the 8 cranial thoracics with sternal ribs and the 4 caudal thoracics, which have short free ribs and probably correspond to the thoracosacrals, that is, segment 1 of the synsacrum of neornithines (Boas, 1933).

The atlas is poorly known. The identification of unfused arches of the atlas in the Eichstätt specimen (Wellnhofer, 1974) is uncertain. The cervicals first increase in length (from 5 mm of the third to 9.5 mm of the sixth vertebra in the Berlin specimen) and then decrease. Bonde (1996) reported that the longest cervical is the fifth in the Eichstätt and the ninth in the Berlin specimen, but the cervicals are difficult to count and measure, hence the differences between Wellnhofer (1974) and Bonde (1996), and the caudal cervicals are not measurable with any confidence in any specimen. The centra show lateral excavations and pneumatic foramina, which are present in cervical vertebrae (third through sixth or second through fifth) of at least the Berlin specimen and indicate the presence of cervical air sacs (Britt et al., 1998). The diapophyses are in the usual ventral position and seem to be knoblike and rounded. The spinous processes of the third and fourth cervicals in the Munich specimen are much higher than in other specimens and make up one-third of the total vertebral height. In contrast with neornithines, there is no indication of hypa-

pophyses on the cervicals. A pneumatic foramen is present behind the diapophysis in the cervicothoracic and perhaps the first thoracic vertebrae (Britt et al., 1998; pers. obs.).

The thoracic vertebrae have prominent spinous processes. The thoracic vertebrae proper (with the sternal ribs) are nowhere exposed, and their details are virtually unknown. In the caudal series of presumptive thoracosacrals, the transverse processes are very short, rounded knobs located near the cranial margins of the centra, straight above the parapophyses. This configuration corresponds to a narrow vertebral fork of the thoracic rib. The centra bear lateral excavations, which are commonly referred to as pleurocoels, although their relation to an air sac is not clear (Britt et al., 1998). However, a real pneumatic foramen seems to be present behind the diapophysis in at least one of the caudal thoracic vertebrae, 21st in the Berlin and 21st or 20th in the Munich specimen (pers. obs.).

The five sacral vertebrae are co-ossified into a synsacrum, but the boundaries between the vertebral bodies remain distinct (Wellnhofer, 1974: Fig. 7). The homology of single synsacral vertebrae remains unresolved. They may possibly correspond to a thoracosacral, two primary sacrals, and two caudosacrals of *Allosaurus* and *Tyrannosaurus* (Galton, 1999). However, a possibility that the three cranial sacrals correspond to the lumbosacrals, that is, segment 2 of the synsacrum of neornithines (Boas, 1933), is suggested by the cranial expansion of the ilium and the presence of presumptive thoracosacrals, in which case no caudal vertebrae would be included in the synsacrum.

The most proximal caudal vertebrae are compact and have long transverse processes directed ventrolaterally without the distal (caudal) slant seen in many neornithines. Farther distally from the synsacrum, the caudals gradually become more elongate, their zygapophyses slimmer, and the transverse processes become shorter and eventually disappear. The longest are caudals 10 through 14. Farther caudally the vertebrae become gradually shorter in *Archaeopteryx* but shorten abruptly just before a truncation by a break in the slab in *Wellnhoferia*, which suggests a tail much shorter than in *Archaeopteryx* (a suggestion obscured on the published photographs by the painting of the missing remainder of the tail, which misrepresents the relative tail length to be as in *Archaeopteryx*). The three to four proximal chevrons are broad and plate-shaped, the remaining are elongate and inverted T-shaped, and the last three to four are absent.

Ribs and Gastralia

Ribs are present on all cervical vertebrae including the epistropheus. The longest cervical ribs are on the 3rd through 5th vertebrae, at least in the Berlin specimen (pers. obs.); then the length decreases for 4 vertebrae and then increases in the first movable rib of the 10th (first cervicothoracic) vertebra. In the Eichstätt specimen, this rib seems to bear a pneumatic foramen between its capitulum and tuberculum, and the thoracic ribs are hollow (Britt et al., 1998), as in neornithines. In addition to the cervical ribs, attached to the cervical vertebrae, at least in the Munich specimen, are ossified tendons (pers. obs.), which also are present in the neck of some extant birds, for example, grebes (Podicipedidae) (James Vanden Berge, pers. comm.).

There are 11–12 pairs of thoracic ribs. Each rib consists of a single ossified piece, since there is no trace of separate ventral (sternal) parts, which are present in ornithurines. There are no uncinate processes. The first eight ribs increase in length (the longest being attached to the 19th vertebra), and, at least in the Munich specimen, their distal ends are distinctly spatulate (flared up), which suggests that they articulated with the cartilaginous sternum. The reconstruction (Fig. 6.1) also suggests that the first two, shortest thoracic ribs may have articulated with the bony sternum, but they are too poorly known to verify this possibility. Caudally, the four free ribs attached to the presumptive thoracosacral vertebrae are much shorter than the preceding eighth rib, and their tips are pointed. Their vertebral forks are narrow; that

is, the capitulum is close to the tuberculum, in accordance with the diapophysis being short and located straight above the parapophysis.

The series of 12–13 pairs of gastralia, some of them V-shaped, starts near the tips of the longest ribs and extends all the way to the pubis (Wellnhofer, 1993: Pl. 1). They decrease in length craniocaudally. Their exact life position must have been variable, depending on the extension of the abdominal wall.

Thoracic Girdle and Sternum

The thoracic girdle (Fig. 6.4A) is much smaller relative to the body and the wing (Fig. 6.1) than in ornithurine birds. The furcula is boomerang-shaped, that is, broadly open dorsally, almost twice as wide as it is deep, and its two rami meet at a right angle (Ostrom, 1976a: Fig. 23). Each ramus is much wider transversely than it is thick rostrocaudally, which makes the furcula of *Archaeopteryx* look sturdy in front view, as if it were designed to resist transverse compression, but less so in lateral view. In addition to being very shallow dorsoventrally, the furcula is small and clearly not "hypertrophied" (contra Olson and Feduccia, 1979). It

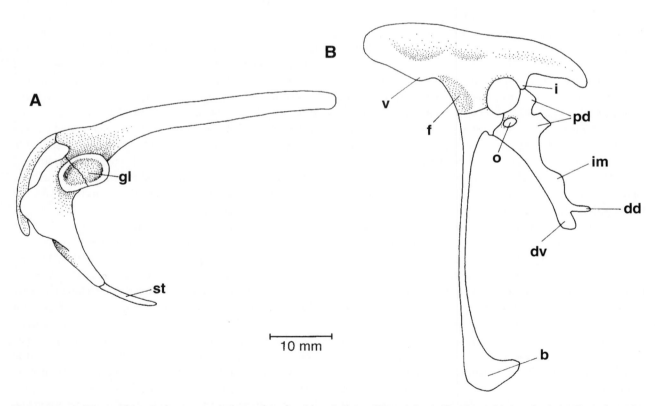

Figure 6.4. *A. lithographica.* Scale reconstructions of A, shoulder girdle, and B, pelvic girdle. The pubis is oriented as in *A. bavarica* (the Munich specimen). Abbreviations: b, calceus pubis (pubic boot); dd, dorsodistal process; dv, ventrodistal process; f, fossa for the origins of the femoralis internus ("cuppedicus") muscle; gl, glenoid; i, ischiadic process; im, intermediate process; o, foramen obturatum; pd, proximodorsal process; st, bony sternum (which may have been caudally extended by a cartilaginous part); v, ventral process.

resembles in shape the furcula of oviraptorids (Barsbold et al., 1990: Fig. 10.3), which, however, has a hypocleideum and a more rounded cross section. It is unclear which, if any, muscles attached to the furcula, but any major portion of the pectoralis responsible for the downstroke is out of the question (contra Olson and Feduccia, 1979) because very little of the furcula was located even slightly below the glenoid (Fig. 6.4A).

The scapula and coracoid form an acute angle of 60–65°. The two bones appear to articulate firmly in the Berlin, London, and Third but not in the Eichstätt and Munich specimens of *Archaeopteryx* and not in the *Wellnhoferia* specimen despite its large size. The glenoid (Fig. 6.4A) faces laterally (Ostrom, 1976b; Jenkins, 1993; Wellnhofer, 1993), as it does in some nonavian maniraptorans and other Mesozoic birds. The orientation of the archaeopterygid glenoid is intermediate between its dorsolateral position in neornithines and its ventrolateral position in the nonavian theropods. The glenoid is distinctly oval (Jenkins, 1993) and deeply concave along its long, rostrocaudal axis, the concavity being enclosed between two prominent labra, rostral (formed by the coracoid) and caudal (formed by the scapula). The shape of the glenoid surface along its short, dorsoventral axis is obscured by the irregularities of the coracoscapular articulation, which in life must have been smoothed out by the articular cartilage. The scapular and coracoid parts of the glenoid are nearly equal, at least in the London specimen (Jenkins, 1993), in which the scapular part is only marginally larger.

The scapula (Fig. 6.4A) is long, narrow, and straplike. Its blade maintains a nearly equal width and ends broadly rounded rather than pointed, as in neornithines. The acromion is prominent, more so than in neornithines. The coracoid (Fig. 6.4A) has a prominent process, here termed processus praeglenoidalis, which is rostroventral to the glenoid. The coracoid extends in two planes: the plane of the glenoid and preglenoid process is close to parasagittal, whereas the plane of the broad coracoid blade (that articulated with the sternum) is oblique, closest to the transverse plane but with a strong horizontal slant. If properly oriented (Ostrom, 1976a: Fig. 13, not Fig. 14), the bone is unlike the coracoid in either ornithurines or nonavian theropods (Tarsitano, 1991). Ostrom (1976a,b) exaggerated the nonavian theropod similarities of the archaeopterygid coracoid. He called the preglenoid process the biceps tubercle and, at the same time, proposed its homology with the acrocoracoid process of ornithurines. While the latter homology is indeed likely, Ostrom's terminology obscures a big difference between the coracoid of nonavian theropods and that of archaeopterygids. The preglenoid process of archaeopterygids has the form of an extensive wing or flange and most probably represents more than the

biceps tubercle, both structurally and functionally, as does the acrocoracoid process of ornithurines. The biceps tubercle is only a minor part of the acrocoracoid process, and this is probably true of the preglenoid process. At the very least, the preglenoid process (probably its rostromedial surface) provided attachment to the supracoracoid muscle and may have deflected its insertion tendon. The coracoid blade (Ostrom, 1976a: Fig. 14) is convex rostrally, concave caudally, and has a strongly concave lateral margin that ventrally ends with what seems to be a prominent sternocoracoid process (processus lateralis). As in nonavian theropods (Nicholls and Russell, 1985), the coracoid blade probably provided the origin for the coracobrachialis muscle(s), which probably served as the main depressor of the wing.

The thoracic girdle of archaeopterygids presents the mosaic of a birdlike scapula, which is close to the ornithurine condition, and a unique coracoid and sternum, which are very different from their homologues in either ornithurines or nonavian theropods. It lacks several distinctive features of ornithurine birds, including the triosseal foramen, the strutlike coracoid, and the large, keeled sternum. The last provides origins to those parts of the pectoralis muscle that are responsible for the pronation of the humerus (and the wing) during the downstroke (Jenkins, 1993), which is necessary for slow flight. This correlates with the absence of the acrocoracoid process on the coracoid, which in neornithines is dorsal to the shoulder joint and provides the attachment site for the coracohumeral ligament. Because of its dorsal attachment, this ligament prevents hyperpronation of the humerus (Sy, 1936). However, some pronating force is generated in any flight because of the postaxial attachment of the flight feathers, which raises the question of what prevented the archaeopterygid wing from being passively hyperpronated at the shoulder joint. It is possible that the horizontal position of the humeral head in the glenoid helped the collateral ligaments to prevent the humerus from pronating.

The discovery of a bony sternum in the Munich specimen (Wellnhofer, 1993) confirms the presence of its remnants, so far highly questionable, in the Berlin (Dames, 1897) and London (de Beer, 1954b) specimens. The sternum, as preserved in the Munich specimen, is a rectangular plate that apparently occupied a transverse position. The sternal plate is very thin, barely 0.5 mm thick in the Berlin specimen. One of the long sides, which has been interpreted as cranial, has a shallow embayment in the middle. With the series of gastralia starting only in the middle of the rib cage (near the tips of the longest ribs), there is some 25–30 mm left for a cartilaginous extension of the bony sternum (Fig. 6.1), whose presence is indicated by the club-shaped ends of the cranial thoracic ribs.

Thoracic Limb

The thoracic limb is approximately as long as the pelvic limb. The humerus/ulna/manus ratio for the Berlin specimen is 32% + 29% + 39%. The forelimb-to-hindlimb ratio changes from 0.86 in the smallest Eichstätt and 0.89 in the Munich specimens to 1.00 in the London and 1.02 in *Wellnhoferia* (Wellnhofer, 1993) (see Table 6.2 for shaft ratios between limb elements). Accordingly, the relative length of the manus slightly increases with size (Table 6.1).

The humerus differs from that of either nonavian theropods or ornithurines in the relative orientation of its ends. As revealed by the Berlin specimen, the angle between the head and the distal end is close to 90°, and the distal end of the right humerus lies flat in *Wellnhoferia* only because the bone is broken proximally (Wellnhofer, 1992: Fig. 10A). In nonavian theropods, the angle between the head and the distal end is much smaller (closer to 45°), and in ornithurines, the head and distal end are nearly parallel. The head is flattened dorsoventrally and thus appears to maintain its original tetrapod orientation and geometry. In contrast, the distal end has rotated 90° for horizontal abduction and adduction instead of parasagittal extension and flexion of the forearm. Consequently, the humerus has a structure intermediate between nonavian theropods and ornithurines despite dramatic differences from either group. In conjunction with osteological features unknown in living birds, a unique orientation of the humeral ends makes the manner and extent of wing folding in archaeopterygids extremely difficult to reconstruct (Fig. 6.1).

The dorsal surface of the proximal end of the humerus is flat (or nearly so), as exposed in the Berlin specimen. The ventral surface, visible in the London specimen (Jenkins, 1993: Fig. 5), shows a distinct concavity, which probably accommodated the attachments of biceps and coracobrachialis muscles. As in maniraptoran theropods and other primitive birds, the deltopectoral crest is continuous with the head; that is, there is no indication of either external or internal tuberosity. The shaft is slightly curved. The details of the distal end remain unknown.

Both the elbow and wrist joints are poorly understood. The olecranon is short and blunt, and there is no indication of the avian dorsal cotylar process. At the distal end, the ulna has a distinct labrum of the dorsal condyle and extensively articulates with the radius. The radius is shorter than the ulna (96% of its length, at least in the Eichstätt and London specimens).

The carpus is composed of at least three (Ostrom, 1976a) and probably four bones (Martin, 1991): a semilunate bone, which articulates with metacarpals I and II (but not metacarpal III); a roughly triangular radiale also known as scapholunar; a reniform pisiform also known as cuneiform or ulnare; and a minuscule carpal, probably distal carpal 3, preserved in the left wing of the Eichstätt specimen (Wellnhofer, 1974: Fig. 8A). The semilunate bone probably incorporates distal carpals I and II (Chatterjee, 1998; Wagner and Gauthier, 1999), although there is no trace of distal carpal I in neornithines, in which a single distal carpal II fuses with metacarpal 1 preaxially and an element X postaxially (Hinchliffe, 1985). Martin's (1991; Zhou and Martin, 1999) identification of the putative distal carpal 3 as element X seems to be rather inconsequential, because element X fuses with the semilunate and metacarpal 3 in the adult carpometacarpus and thus may in fact represent distal carpal 3.

The semilunate carpal and the expanded labrum of the dorsal ulnar condyle probably guided the sweeping movement of the manus, that is, its planar extension and flexion, in the upstroke-downstroke cycle (Rayner, 1991). The planar extension and flexion may have been accompanied by an automatic pronation and supination, respectively, if the semilunate carpal bore an asymmetrical ginglymus similar to that of maniraptorans (Ostrom, 1995). Such an automatic mechanism would lead to the supination of the manus upon folding of the wing.

In the wrist of neornithines, the interaction of an expanded labrum of the dorsal ulnar condyle with the U-shaped ulnare and trochlea blocks the pronation of the manus and guides the changes of its plane in slow flight; and the interaction of the polyhedral radiale and the

TABLE 6.1
Intramembral ratios (%) in *Archaeopteryx* in descending order
from smallest to largest specimen

Specimen	Hu + Ul + McII = 100%	Fe + Ti + MtIII = 100%	Hu/Ul	Fe/Ti
Eichstätt	43.3 + 38.1 + 18.6	30.8 + 44.1 + 25.1	113.7	69.8
Munich	(41.4 + 39.8 + 18.8)	(29.3 + 45.1 + 25.6)	(103.8)	(65.0)
Berlin	42.8 + 38.2 + 19.0	(32.6 + 44.3 + 23.1)	112.1	73.5
London	42.5 + 38.0 + 19.5	(32.6 + 43.9 + 23.5)	111.9	(74.4)

Note: Parentheses mark ratios based on at least one approximate measurement. Abbreviations: Fe, femur; Hu, humerus; McII, metacarpalII; MtIII, metatarsal III; Ti, tibia; Ul, ulna.

TABLE 6.2
Minimum shaft diameter-to-length ratios (in % of length) in *Archaeopteryx* in descending order from smallest to largest specimen

Specimen	Humerus	Femur	Ulna	Tibia*	Radius
Eichstätt	5.8	8.1	4.4	4.8	3.4
Munich	5.9	6.7	4.5	3.6	3.7
Berlin	5.7	6.5	4.8	4.1	3.5
London	5.3	7.0	4.3	4.0	4.4

*Ratios of the craniocaudal diameter (depth) to the length of the entire tibiotarsus.

trochlea controls supination during maneuvering (Sy, 1936; Vazquez, 1992). These functions must have been either lacking or performed with much less precision and more use of muscle power in archaeopterygids, which lack the fused carpometacarpus, the carpal trochlea, the U-shaped ulnare, and the polyhedral radial carpal. The modern avian features of the hand of *Archaeopteryx* as listed by Zhou and Martin (1999) could not be confirmed.

The manus must have been partly folded for the primaries to stay clear of the ground, but a complete folding as in neornithines seems to be impossible (Elzanowski and Pasko, 1999), and the left wing of *Wellnhoferia* is only partly folded (contra Stephan, 1994). The mechanism of hand folding remains unclear. At least some skeletal structures involved in the automatic coupling of the hand and forearm movements in neornithines are absent in archaeopterygids (Stegmann, 1937; Peters, 1985; Peters and Görgner, 1992). However, wing autofolding may have been present in *Deinonychus* (Gishlick, 2001), and bony structures are only in part responsible for the autofolding, which is helped by the muscles that actively displace the radius (Vazquez, 1994). Therefore, the lack of evident skeletal adaptations for elbow-wrist coupling implies only that the operation of the wing in the archaeopterygids was more expensive energetically than it is in neornithines, but not necessarily that the coupling did not exist. The autofolding of the wing involves longitudinal shifts between the radius and ulna, which seem necessary for the functionally required wing folding in *Archaeopteryx* (Elzanowski and Pasko, 1999).

The manus has three rays and may have been slightly arched with metacarpals I and III lying somewhat ventral to metacarpal II (Gishlick, 2001). Metacarpal I is much shorter and more robust than metacarpals II and III. The average length ratio for *Archaeopteryx* is 12.6% + 45.8% + 41.6%. Metacarpal I is concave dorsally. Metacarpal II is laterally compressed at the proximal end but flares out distally. Metacarpal III is deeper than wide throughout its length.

Metacarpal II ends flush with metacarpal III distally, but its proximal end lies distal to the other metacarpals, whereby metacarpal III is widely separated from the semilunate carpal. This configuration, as preserved in the intact wrist region of the Eichstätt specimen (Wellnhofer 1974: Fig. 22/3), seems to be the natural, in vivo condition, but Gishlick (2001) reconstructed the poorly preserved wrist of the Munich specimen and concluded otherwise.

The phalangeal formula is 2-3-4-*x*-*x*, and the evidence of reduction of digits IV and V in theropods generally (Ostrom, 1976a; Sereno and Novas, 1992; Padian and Chiappe, 1998) identifies the manual digits of archaeopterygids and other birds as I, II, and III. However, the position of carpal cartilages observed in the development of extant birds consistently points to the identification of neornithine digits as II, III, and IV (Hinchliffe, 1985; Burke and Feduccia, 1997). In view of the compelling evidence, both cranial and postcranial, for the theropod relationships of birds (Padian and Chiappe, 1998; Clark, Norell, and Makovicky, Chapter 2 in this volume), the observed pattern of carpal cartilages in the developing avian wing is much more likely to indicate a Haeckelian caenogenesis in the development of digits I, II, and III than homology with digits II, III, and IV. Wagner and Gauthier (1999) proposed that precartilaginous condensations for digits II, III, and IV were reassigned to develop into digits I, II, and III as a result of a genetic frameshift in the avian ancestors.

The digit proportions are highly conserved in all specimens: digit II is the longest, digit I is the shortest, and digit III is intermediate in length and more slender than digits I and II. The penultimate phalanx is the longest in each digit. Whereas the interphalangeal ratios of digit II are highly conserved, digit III has phalanx I relatively shorter and phalanx II longer in larger specimens. The nonungual phalanges of digit I are much deeper than wide, and those of digit II are much wider than deep. Phalanges 1 and 2 of digit III seem to be immovably connected and partly fused in *Wellnhoferia* (Wellnhofer, 1992: Fig. 10B) but clearly separate in *Archaeopteryx* (disarticulated in the Berlin specimen), although poor development of their ginglymoid articulation suggests limited mobility. The trochlea of phalanx 3 of digit III is rotated medially (Gishlick, 2001). The left acropodium of the Berlin specimen is partly pathological: a minuscule, rugose projection or flange is present on the basal phalanx of digit II (Heilmann, 1926: Fig. 19; Thulborn and Hamley, 1982: Fig. 4; Howgate, 1984a: Fig. 1; Wellnhofer, 1985: Fig. 1), and the neighboring phalanx 2 of digit III differs in shape from its counterpart in the right manus in that it is deeper distally.

The manual unguals of *Archaeopteryx*, at least of digits I and II, are more curved than are pedal unguals. The ungual of digit II is the longest (and most curved in *Archaeopteryx*) and digit III the shortest (and least curved in the

Berlin specimen) of the three unguals. The unguals of digit II and digit III are somewhat hyperextendable (Gishlick, 2001). The manual unguals bear distinct flexor tubercles, which, however, are relatively less prominent in the larger specimens of *Archaeopteryx* and seem to be poorly developed in the Third skeleton. In contrast, the proximal depth of the unguals increases with body size. In *Wellnhoferia*, the flexor tubercles are relatively smaller than in the small *Archaeopteryx* specimens but larger than in the large ones.

In all specimens digit I is at angles to the remaining two, and thus presumably moved largely independently, while digits II and III remain close to each other. In six out of the nine articulated hands, digit II is crossed *under* the joint between its proximal and the middle phalanx by digit III (Kemp and Unwin, 1997), and in the remaining three hands digits II and III lie more or less parallel. The observed association suggests a strong ligamentous connection between digits II and III, but the crossover between digits II and III is open to interpretation. Rau (1969) thought that digit III could provide a brace for the shafts of flight feathers attached to digit II, although, if anything, the flight feathers would need a dorsal, not a ventral, brace to withstand the air resistance. The crossing could conceivably be functional if some flight feathers attached to digit III, as proposed by Bohlin (1947), since then digit II would function as a dorsal brace of digit III, which is weaker than digit II, but this raises a question of why digit III should be weaker in the first place. Heilmann (1926), Bohlin (1947), Swinton (1975), Yalden (1984, but not 1985), and others accepted the crossing as a natural condition for their reconstructions, although it would be unprecedented, as no other vertebrate has two appendages crossed, and probably would have left some osteological evidence. In fact, there is no evidence of either a secondary articulation between the phalanges or a skin callosity, which would have formed and left traces on the bone if the two fingers were frequently pressed against each other. The crossing of digits II and III is most probably a taphonomic artifact resulting from their differential postmortem mobility (Kemp and Unwin, 1997; Gishlick, 2001).

The finger claws point forward in all wings of *Archaeopteryx* preserved in articulation and in the right wing of *Wellnhoferia*, as well as in *Protarchaeopteryx* (Ji et al., 1998), *Caudipteryx* (Ji et al., 1988; Zhou and Wang, 2000), and some specimens of *Confuciusornis* (Chiappe et al., 1999). In the left wing of *Wellnhoferia* the claws point backward. The forward orientation *as preserved* is probably in part artifactual as revealed by the postmortem twisting and luxation of phalanges (Heilmann, 1926:30; Thulborn and Hamley, 1982; Yalden, 1985; Stephan, 1994). In vivo, the claws were probably directed more or less ventrally, perpendicular to the wing surface (Yalden, 1985) and only upon burial turned

forward, together with some phalanges. However, the consistency of preservation in many related taxa from various deposits suggests an initial forward slant in vivo or an intrinsic anatomical proclivity for the forward slant, at least upon unfolding of the manus. Based on the orientation of vestigial claws in neornithines, Stephan (1994) proposed that the backward orientation in what seems to be a partly folded left wing of *Wellnhoferia* is close to the natural one, but this does not explain the forward orientation in the overwhelming majority of cases.

Pelvic Girdle

The pelvis (Fig. 6.4B) is unlike that of neornithines, and its component bones remain unfused. In many details it is similar to the pelvis of nonavian theropods but differs from it in proportions.

The preacetabular part of the ilium is 1.6–1.75 times longer than the postacetabular part, which is in contrast with the majority of nonavian theropods except for *Unenlagia*. The pubic peduncle is broad. The postacetabular part is much narrower than the preacetabular part and tapers toward the end, where it is slightly bent downward. It has an ischiadic process, which encloses the acetabulum caudally. The preacetabular ilium of the London but not the Berlin specimen has a well-marked fossa (in front of the pubic peduncle) for the origins of the iliofemoralis internus ("cuppedicus") muscle, which is also well developed in *Unenlagia;* a rounded but distinct ventral process (processus ventralis—nomen novum) on the ventral margin, caudal to the rostral tip of the ilium, as in *Caudipteryx, Protarchaeopteryx,* and *Unenlagia;* and two crescentic depressions, probably marking the origins of the iliotrochanteric muscles.

The ischium has four projections, which, in view of their uncertain homologies, are here called by morphologically neutral terms reflecting only their relative position: proximodorsal process (processus proximalis dorsalis—nomen novum), which is bipartite in the London specimen (as it seems to be in *Unenlagia*); processus intermedius (nomen novum) separating two embayments in the caudal (caudodorsal) margin of the bone; dorsodistal process (processus distalis dorsalis—nomen novum), which forms the dorsal prong of the terminal fork; and ventrodistal process (processus distalis ventralis—nomen novum), which forms the ventral prong of the terminal fork. Neornithines have two ventral processes of the ischium: the obturator process, which encloses the obturator fenestra, and the terminal process, which forms the distal end of the bone. Gauthier (1986:34) missed the terminal process of neornithines and pondered over the identity of the ventrodistal process in *Archaeopteryx,* but Novas and Puerta (1997) and Forster et al. (1998) simply assumed the homology of the ventrodistal process with the obturator process. In fact, it is possible that the ventrodistal process of the Archaeopterygidae, *Unenlagia,* and *Rahonavis* corresponds to the terminal process of neornithines. Near its suture with the pubis, the ischium of the London specimen has an oval foramen, most probably for the obturator nerve. At least an incisure is indicated in the same location in other specimens.

The pubis is very long, almost as long as the femur. The shaft seems to be cylindrical. The pubic symphysis was formed between the distal flat expansions of the shafts, known as pubic aprons. Each apron takes up 44% of the total length of the pubis in the London specimen. The distal end of each pubis is expanded into a "boot"—calceus pubis (nomen novum), with the tip directed caudally. In the London and Eichstätt specimens the pubic boots were replaced, at least in part, by a calcite mass, which suggests the former presence of cartilage.

The natural orientation of the pubis has been highly controversial. As preserved, the pubis is directed backward in the London and Berlin specimens, which look opisthopubic (Walker, 1980; Tarsitano, 1991). However, the pubis is nearly vertical in the Third, Eichstätt, and Munich specimens. The well-preserved pelvis in the Munich specimen (Wellnhofer, 1993) confirmed that the pubis was directed nearly vertically downward as in *Unenlagia* (Novas and Puerta, 1997) and *Rahonavis* (Forster et al., 1998). The pubis was only slightly retroverted, forming a cranial angle of 110° with the long axis of the ilium (Wellnhofer, 1985). The caudal orientation in the London and Berlin specimens is due to a dislocation in the acetabulum. Howgate (1985) realized that the pubis is particularly prone to dislocation but believed that only a limited extent of dislocation is possible and used the position of the pubis to support a taxonomic separation of the Eichstätt specimen. Wellnhofer (1993) provided evidence of a potent preservational factor that forced the pubis backward: while most of the pubis in the Munich specimen is nearly vertical, its distal end is broken and turned backward. However, Wellnhofer (1992) calculated a stronger backward slant of 128° in *Wellnhoferia,* which may add to the list of differences from *Archaeopteryx.* Whatever the exact angle, the orientation of the pubis is intermediate between neornithines and the majority of nonavian theropods, except for therizinosaurids and dromaeosaurids, which have the pubis strongly retroverted (Norell and Makovicky, 1997). Ruben et al.'s (1997) speculations about pulmonary ventilation in archaeopterygids being different from that of the nonavian theropods are based on a reconstruction that exaggerates the backward slant of the pubis in archaeopterygids.

Pelvic Limb

The average length ratio of femur/tibiotarsus/foot is 25% + 36% + 39% for *Archaeopteryx* and 27% + 37% + 36% for *Wellnhoferia.* The femur is distinctly bowed craniocaudally.

The proximal end bears three swellings, two of which may correspond to the greater and lesser trochanters in nonavian theropods (Ostrom, 1976a), and thus differs from that of neornithines, which have only a single trochanter. The femur differs from that of most nonavian theropods in lacking the fourth trochanter, which indicates the reduction of the caudofemoralis musculature and thus suggests a forward shift of the center of mass and a protracted, birdlike resting position of the bone (Gatesy, 1990, 1991). The medial side in the London specimen bears two small depressions, which were conjectured to mark the insertion of the caudofemoralis muscle (Raath, 1985) but may instead mark the insertion of the iliofemoralis internus. The femur is more robust in the Eichstätt specimen than it is in the Munich, Berlin, and London specimens, but no size-dependent trend is apparent among the last three specimens.

The crural bones are slender. The tibia bears a single cnemial crest, which, however, projects very little or not at all proximally. Wellnhofer (1974) identified it as the cranial (= internal or medial) crest, but Chiappe (1996) reinterpreted it as the lateral (= external) crest. In fact, it could be neither, as it is a proximal extension of the fibular crest (unless the lateral crest evolved from the proximal extension of the fibular crest), which also is present in *Vorona* (Forster et al., Chapter 12 in this volume). The tibial shaft is oval, approximately 1.5 times wider transversely than it is deep craniocaudally. There is no obvious directional variation in the proportions of the tibia between the specimens. The fibula extends from the knee to the tarsus, where it flares into a distal boot (apparently for articulation with the calcaneum), which has its nose directed caudally.

The tarsus is divided by a mesotarsal joint. The two proximal tarsals, astragalus and calcaneum, are tightly sutured to the tibia, forming a functional tibiotarsus. The astragalus has an ascending process as in nonavian theropods, *Vorona* (Forster et al., Chapter 12 in this volume), and embryonic palaeognaths (McGowan, 1985). The two or three distal tarsals are at least in part co-ossified with the metatarsals. Archaeopterygids have, therefore, an incipient tarsometatarsus.

The metatarsus is composed of five bones. As in most other birds and nonavian theropods, metatarsal III is the longest, and metatarsal I is short (it does not contact the tarsus) and appressed to the medioplantar surface of metatarsal II. Metatarsal II tapers proximally in *Wellnhoferia* but not in *Archaeopteryx.* Metatarsal V is vestigial and represented by a splint of bone attached behind the proximal end of metatarsal IV. It is possible that the metatarsals were superficially fused to some extent in all specimens, since, in contrast with nonavian theropods, no separation or displacement of any of the three main bones has ever been observed (Ostrom, 1976a). The three main metatarsals (II, III, and IV) were reported to be proximally fused in

Wellnhoferia (Wellnhofer, 1988, 1992) and the Third (now lost) specimen (Heller, 1959, 1960) but not in other specimens. The fusion in *Wellnhoferia* may indeed be more advanced than in *Archaeopteryx*, as it is evident on the surface (although a peculiar encrustation of this specimen may have obscured the sutures). In the Third specimen, the metatarsus was damaged and split longitudinally between the two slabs, which obscures comparisons with the intact bones of other specimens. The proximal ends of the metatarsals are distinctly separated, and their separation seems to fade away only subterminally, where Heller believed to see a complete dissolution of bony walls. However, X-ray stereophotographs of this specimen (courtesy of Prof. J. T. Groiss) reveal the walls separating the metatarsals throughout the length of the bone, which demonstrates that the fusion was at best incomplete and probably not very different from that observed in other specimens. There is very little evidence for the proximal-to-distal *ontogenetic* progression of fusion in the archaeopterygid metatarsus (contra Martin, 1991; contra Hou et al., 1996).

The foot of *Archaeopteryx* is of the unspecialized anisodactyl type. The phalangeal formula is 2-3-4-5-*x* in *Archaeopteryx* and 2-3-4-4-*x* in *Wellnhoferia*. Digit III is the longest, followed by digits IV, II, and I. The length ratios of the pedal digits (in percentage of the combined lengths of all four digits) are 9–11/22/36/32–33 in *Archaeopteryx* and 10/25/38/28 in *Wellnhoferia*. In digit IV of *Wellnhoferia* but not *Archaeopteryx*, the ungual is the longest of all phalanges.

In a single preserved foot of the London specimen, which is exposed in medial view, digit III covers digit IV except for the ungual and a part of the penultimate phalanx, making it impossible to count digit IV phalanges. However, there is no reason whatsoever to follow de Beer (1954b) in his expectation of an aberrant phalangeal formula in this specimen (Ostrom, 1972, 1992). Digit IV in the London as well as other *Archaeopteryx* specimens is close in length to digit III and thus much longer than digit II (de Beer, 1954b: Pl. II). In contrast, *Wellnhoferia* has digit IV close in length to digit II and much shorter than digit III, which is evidently correlated with the reduced number of phalanges in this species. The foot of *Wellnhoferia* is, therefore, more symmetrical, which suggests a more cursorial specialization than in *Archaeopteryx*.

The hallux is opposable but slightly elevated, which probably limited its grasping action (Bock and Miller, 1959). The basal phalanges tend to be flattened, and the distal pedal phalanges, including the unguals (Yalden, 1997), are mostly compressed laterally, although not as deep as the manual phalanges. Phalanges 1 and 2 of digit IV are flat, and phalanx 2 is proximally expanded into a lateral wing. The two phalanges seem to be connected by a nontrochlear, little movable joint.

The trochlea of the basal phalanx of digit II is slanted proximodorsally (and thus seemingly expanded proximally on the dorsal side), and phalanx 2 has a prominent extensor process and convex dorsal profile. This configuration is unlike that in dromaeosaurids and indicates a possibility of hyperflexion rather than hyperextension, as does a prominent extensor process of the ungual. The idea of a hyperextended pedal digit 2 recalls Gauthier's (1986) observation of what looks like a dorsal protrusion of the trochlea of the basal phalanx in the Eichstätt specimen (Wellnhofer, 1974: Fig.12). In fact, the phalanx is dislocated (as is the entire digit) and probably turned around, with the usual ventral convexity of the trochlea turned dorsally. In any case, there is no indication of the dorsal protrusion of this trochlea in the London and other specimens.

The pedal unguals of *Archaeopteryx* are less curved (81°–86° in digit III) than the manual unguals of digits I and II. The unguals of *Wellnhoferia* are strongly curved (126°–128°) dorsally and straight ventrally. Flexor tubercles separate from the ungual bases are present in *Wellnhoferia* and absent in *Archaeopteryx*, unless the expanded ungual bases in the London specimen are seen as containing the flexor tubercles.

Horny Sheaths

The horny claws of the hand and foot are similar (Peters and Görgner, 1992). Each horny sheath has a thick dorsal rib, which forms a needle point at the tip of the claw, and very thin ventral (medial and lateral) eaves. The dorsal and ventral curvatures differ, but each is uniform both dorsally and ventrally, without any threshold or abrupt tapering near the tip. The curvature of the horny sheaths alone is difficult to measure because of poorly preserved proximal ends, and the published measurements of the combined curvature of the unguals and horny sheaths in their preserved positions to a variable degree exaggerate the real curvature of the claws, because the horny claws slid away from the bases of their unguals and are more or less dislocated. Postmortem dislocations, so far noticed only in extreme cases such as that of the left third finger in the Eichstätt specimen (Peters and Görgner, 1992: Fig. 6B), explain striking inconsistencies in the measurements of preserved curvatures between the specimens and between the left and right sides of the same specimen as obtained by Peters and Görgner (1992), as well as the unusual gaps between the bases of the unguals and their horny sheaths (Peters and Görgner, 1992). The side walls (eaves), which in vivo covered and were supported by the ungual and appear as thin double blades after their postmortem separation, have been interpreted as preening devices (Wellnhofer, 1995).

With seven thoroughly studied specimens, which consistently show impressions of feathers and horny claws, the absence of impressions of rhamphotheca or podotheca suggests the absence of these structures in life. However, in rotting birds the rhamphotheca commonly falls off sooner than the claw sheaths, which, in conjunction with the likely selective exposure of the jaws to preburial damage, leaves marginal doubts about the in vivo absence of rhamphotheca in archaeopterygids.

Feathers, Wing, and Tail

The impressions of feathers associated with skeletal specimens of archaeopterygids demonstrate that most of the body, including the lower neck, wing, lower leg (but not tarsometatarsus), and tail, was covered by contour feathers (Dames, 1884; de Beer, 1954b; Heller, 1960; Stephan, 1987). None of the specimens shows any trace of feathers around the head and upper neck, which may or may not be due to preservation. The consistent evanescence of flight feather quills in all specimens is generally attributed to the presence of minor coverts, which did not leave identifiable impressions. A preservational bias against small feathers may also account for the conspicuous absence of semiplumaceous feathers that are frequent in Cretaceous deposits (Elzanowski, 1988; Martill and Filgueira, 1994). In addition, the dangling head and upper neck were exposed to damage when dragged over the bottom before the carcass subsided.

The wing span is estimated as 55 cm for the Berlin specimen, and the aspect ratio (of span to chord), as calculated from data in Yalden (1984), is in the range of 6.2–7.2. Yalden (1984) provided conservative estimates of 189 cm^2 for the single wing area, including the primaries and secondaries (i.e., distal to the humerus), and 479 cm^2 as the total wing area (two wings plus the area in between). The new reconstruction (Fig. 6.5) suggests that the total wing area was probably slightly higher, around 500 cm^2. Both estimates include a large part of the gap between the wing and the body and thus could be too high. With the calculated body mass of 276 g (see later), the minimum wing loading was around 0.55 g/cm^2. Shipman's (1998) estimates are based on an inaccurate reconstruction of the wing.

The wing looks modern (Savile, 1957), but the arrangement of feathers may prove substantially different from that in neornithines, as probably was their attachment to the wing bones (Fig. 6.6). There are 11 large primary flight feathers in the Berlin specimen, and a minute, most distal 12th primary may also have been present (Heilmann, 1926; Helms, 1982; Rietschel, 1985b). The distal primaries have asymmetric vanes, probably as asymmetric as the isolated feather, but the exact asymmetry ratios cannot be determined by direct measurements because the vanes are folded ventrally and their margins broken off (Helms, 1982; Rietschel, 1985b). Since the inner vanes (which are wider than the outer vanes) protruded ventrally, they were probably more damaged than the outer vanes. This makes the controversy over the asymmetry of flight feathers in

Figure 6.5. *A. siemensii*, restorations by Jerzy Desselberger in flight (from below) and on the ground, the latter based on the skeletal reconstruction in Figure 6.1. The flight silhouette assumes that the gap between the forearm and the body was largely filled out by the secondaries, as in the modern birds.

Archaeopteryx (Speakman and Thomson, 1994; Norberg, 1995) essentially groundless. The proximal primaries and the secondaries seem to be more symmetrical, and the feather identified as the first primary in the London specimen is symmetrical (de Beer, 1954b). The rachises bear ventral furrows extending over a half to three-fourths of their length. All the primaries of the London specimen and the digital primaries of the Berlin specimen are curved backward, as in neornithines, but the metacarpal primaries of the Berlin specimen are turned forward, which is probably a preservational artifact (Savile, 1957; R. A. Norberg, 1985) caused by the bending of feathers against the bottom when the carcass was subsiding. Rietschel (1985b) estimated the minimum lengths of the primaries in the Berlin specimen to be 130 mm for the 1st (the most proximal), 135 mm for the

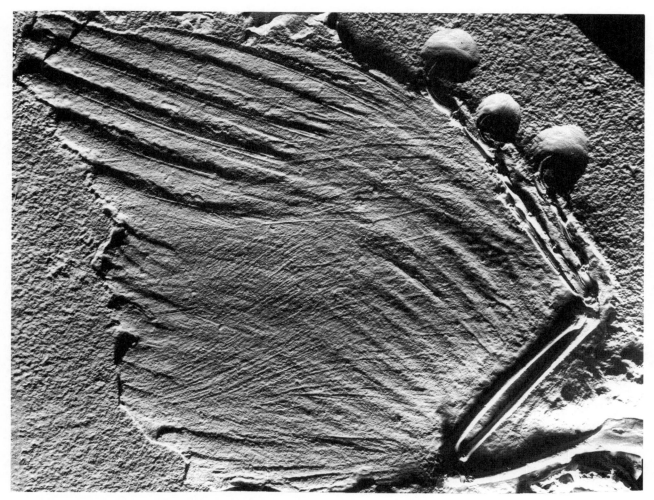

Figure 6.6. Forelimb and wing of *A. siemensii.*

3rd, 145 mm for the 5th, 140 mm for the 7th, 125 mm for the 9th, 95 mm for the 10th, and 55 mm for the 11th (the most distal) but reconstructed the 7th through 9th primaries as substantially longer than his own estimates. Yalden's (1971b, 1984) reconstruction of the wings assumes the inner 5–6 primaries to be shorter roughly by 20% than Rietschel's estimates, but their length probably cannot be exactly estimated because of their preservation. Moreover, their clear-cut differentiation from the secondaries in Rietschel's reconstruction is entirely hypothetical. The longest primaries in the London specimen are estimated to be more than 150 mm long (Savile, 1957).

At least 12 secondaries were present (Stephan 1987), although their exact number is of little importance, since it varies within the species of extant birds and the demarcation to the primaries is somewhat uncertain (Table 6.3; see also de Beer, 1954b). The secondaries show a backward curvature as in neornithines (R. A. Norberg, 1985). The long humerus and no evidence of the tertiaries (Stephan, 1987) suggest a gap between the body and the forearm, which is

plausible if flight feathers originated distally as steering rather than lift-producing devices (Garner et al., 1999). Rietschel's (1985b) reconstruction misrepresents the proportions of the wing parts by showing the forearm and the set of secondaries to be at least 1.5 times longer than permitted by the wing skeleton. Unfortunately, this reconstruction was used to draw sensational conclusions about the flight abilities of *Archaeopteryx* in a popular book (Shipman, 1998).

There is some evidence for the presence of all three tiers of undercoverts (major, middle, and minor) in the London specimen (Heilmann, 1926; Helms, 1982), of the major undercoverts in the Berlin specimen (de Beer, 1954b), and of the minor undercoverts in the Third specimen (Heller, 1959, 1960). The coverts of archaeopterygids may have been more different from those of extant birds than it is generally assumed, as they seem to be relatively longer than in extant birds and more plumaceous in comparison with the flight feathers (Helms, 1982). Their preserved, strongly diagonal position with respect to the primaries has yet to be ex-

TABLE 6.3

A comparison of the primary flight feather counts in the Berlin and London specimens

	Berlin		London
Heilmann	Heinroth	Savile	de Beer
1st secondary	10th	1st secondary	1st secondary
1st	9th	1st	1st
2nd : SS (R−)	—	—	SS
3rd	8th	2nd	2nd
4th : SS (R+)	7th (*)	—	—
5th	6th	3rd	3rd
6th : SS (R+)	—	—	—
7th	5th	4th	4th
8th : SS (R+)	4th (*)	—	—
9th	3rd	5th	5th
10th	2nd	6th	SS
11th	1st	7th	6th
12th	—	8th	—

Sources: Berlin specimen: Heinroth, 1923; Heilmann, 1926; Savile, 1957. London specimen: de Beer, 1954b.

Note: The shadow shafts are (SS) counted as separate feathers (Helms, 1982). The data are primarily for the left wing of the Berlin specimen and exclusively for the right wing in the London specimen, in which the left wing remains to be studied in detail. (R+), (R−) indicate the presence or absence of a shadow shaft or other evidence of a hidden feather in the right wing. (*) Presence inferred rather than observed by Heinroth (1923), who saw a growing feather in position 7 of the right wing.

plained. Overall, the fuzzy preservation of the coverts is in sharp contrast to the elaborate detail in Rietschel's (1985b) reconstruction of the wing.

While the upper slab exposes the dorsal side of the skeleton, the associated impressions of the wing feathers show their ventral surfaces, as revealed by the presence of furrows on the shafts and by each outer vane being overlapped by the inner vane of a more distal flight feather. Only the feather identified as the first primary in the London specimen breaks this pattern and overlaps, instead of being overlapped by, its distal neighbor (de Beer, 1954b). The lower slab shows the impressions of the upper slab impressions, that is, pseudomorphoses or natural casts. Superimposed on the complete traces (impressions and pseudomorphoses) of feathers, which show the detail of shafts and vanes, are traces of shafts without the accompanying vanes, the shadow shafts. The shadow shafts are preserved as pseudomorphoses on the upper slab and as impressions on the lower slab, which is the reverse of the preservation of feathers with vanes.

Helms (1982) developed an ingenious model that accounts for this peculiar preservation of wing feathers in the Berlin specimen. After the carcass subsided upside down on the bottom, the ventral surface of the wing was covered by the subsiding ooze and thereby imprinted itself on the lower surface of the overlying sediment. When the feather decayed away, the overlying layer imprinted the impression of the ventral surface onto the underlying layer of the former bottom sediment, thus forming a pseudomorphosis of the ventral wing surface on the lower slab. Apparently unaware of the orientation of *Archaeopteryx* skeletons on the slabs, Davis and Briggs (1995) assumed that the ventral surface of the wing contacted the bottom sediment and invoked bacterial lithification to explain the unlikely origin of detailed feather impressions in a bottom sediment.

Helms (1982) explained the shadow shafts as the impressions made in the bottom sediment by the shafts that were exposed on the dorsal surface of the wing but screened off from the ventral surface by other feathers. These (and possibly other) impressions of the dorsal surface were subsequently in part stamped out and in part flattened when the pseudomorphosis of the ventral wing surface was pressed over them by the overlying sediment. Rietschel (1985b) proposed that the shadow shafts were not impressed in the bottom sediments but molded by the sediment that precipitated between the feathers, that is, under the superficial layer of feathers of the ventral wing surface. Whether the sediment was pushed out or kept out, the shadow shafts are technically impressions of feather shafts hidden behind the ventral wing surface.

The preserved traces of flight feathers reveal substantial postmortem changes to the wing. The ventral furrows of the shafts are exaggerated by sediment pressure. The original overlap of the primaries, which in neornithines conceals the shafts on the ventral wing surface, was largely destroyed by pressing the wing into the sediment, which caused the vanes to be folded or deflected to the ventral side. If the wing were preserved intact, the shafts of all flight feathers should be preserved as shadow shafts. What remains to be resolved, however, is the question of why only even-numbered primaries are hidden in both wings of the Berlin and London specimens (Table 6.3). This regularity is unlikely to result from a haphazard disturbance by the sediment and suggests that the hidden feathers were situated and probably attached more dorsally than the odd-numbered feathers, which leads to a question of where the two tiers of attachments were placed. Bohlin (1947) proposed that only the even-numbered primaries formed a dorsal row and attached to digit II while the odd-numbered primaries formed a ventral row and attached to digit III. This proposal is consistent with the positioning of shadow shafts in relation to other primaries and the position of digit III on the trailing edge of the wing, which is where the flight feathers most probably originated. However, it is inconsistent with digit III being distinctly weaker than digit II. Another, never addressed possibility is that the shadow shafts represent the major dorsal coverts.

None of the specimens shows the attachment of flight feathers to the bones, which indicates that the minor undercoverts screened off the quills of flight feathers from the sediment (Dames, 1884; Heilmann, 1926:28; Heller, 1959, 1960). The mode of attachment is unknown. There are no quill knobs on the ulna, but these are also absent in many modern flying birds (Edington and Miller, 1941), and their absence does not necessarily imply a weak attachment of the secondaries (Yalden, 1985). However, the preservation of wing feathers, which stay together despite the disintegration of the hand skeleton in the London specimen (Yalden, 1985), and the absence of wing-shaped caudal expansions of the manual phalanges with depressions for the accommodation of the calami (Stephan, 1994) suggest that the feathers were attached more weakly than in neornithines. In the Berlin and Third specimens, the 1st through 6th proximal primaries are directed toward the metacarpals and the 7th through 11th distal primaries to the digital phalanges (Heinroth, 1923; Bohlin, 1947; Heller, 1959, 1960; Rau, 1969; Helms, 1982). The distal primaries most probably attached to the flattened phalanges of digit II, and the only free part of this digit was probably its claw, unless the controversial 12th primary attached to the ungual (Helms, 1982). There is no evidence of a bastard wing (alula), which in neornithines attaches to the first finger, but Rau (1969) observed that whatever feathers may have attached to the first finger, they would be on the dorsal wing surface and thus probably would not be marked on the preserved impressions of the ventral wing surface.

In contrast to the feather impressions associated with the skeletons, the isolated feather attributed to *Archaeopteryx* (von Meyer, 1862; de Beer, 1954b) is preserved as a carbonized trace (Helms, 1982; Davis and Briggs, 1995; Griffiths, 1996), and its vanes show traces of barbules. The calamus is poorly marked, and von Meyer (1862) thought it was soft. The trace of the calamus has become shortened, apparently as a result of wear of the specimen since its discovery: the feather was 69 mm long at von Meyer's (1862) time and now measures only 58 mm (Griffiths, 1996). It is 11–12 mm wide and strongly asymmetric: at 25% length from its tip, its proximal vane is 2.2 times wider than the distal one, which falls in the lower range of birds using flapping flight (Speakman and Thomson, 1994). Von Meyer (1862) described the end of the vane as rounded, but upon closer examination it turned out to be clipped at the angle of 110° (Griffiths, 1996). The proximal end of the vane has a tufted, downy appearance (Griffiths, 1996). The shaft is distinctly curved backward. The asymmetry and curvature of the isolated feather suggest a remex (de Beer, 1954b). It may represent a distal primary (Savile, 1957), since it is intermediate in size between the 11th and 10th primaries as reconstructed by Rietschel (1985b). Alternatively, it may possibly represent a secondary, since the secondaries of neornithines are also curved (R. A. Norberg, 1985), and the terminal clip suggests the airflow along rather than across the vane (Griffiths, 1996). However, if it was a secondary, it would have come from an individual much smaller than the Eichstätt specimen.

There are contour feathers associated with the tibia in the Berlin and Third specimens (see also Rietschel, 1985a: Fig. 1). In the Berlin specimen the lower leg feathers converge on the tibiotarsus and seem to form short feather breeches over the tarsometatarsus, as in some neornithines (such as raptors). They are relatively long, averaging 32 mm in the second specimen (Dames, 1884).

The tail, as preserved in the London specimen, is long and distinctly truncated at the end, with only slightly rounded corners (de Beer, 1954b: Pls. I and XIII). Each of the 16–17 pairs of symmetrical tail feathers (rectrices) apparently attached to a single vertebra, from the sixth caudal vertebra on (de Beer, 1954b; Stephan, 1987). Counting from this vertebra to the truncated end of the feathers, the tail is 28–29 cm long in the London specimen (de Beer, 1954b: Pl. I), approximately 20 cm in the Berlin specimen (Dames, 1884: Pl. 1), approximately 18 cm in the Munich specimen (Wellnhofer, 1993: Fig. 5), and probably slightly less in the Eichstätt specimen (the difference from the Munich is within the margin of error). The tail reaches its maximum span of 8 cm in the London, 7 cm in the second, 5.5 cm in the Munich, and 4 cm in the Eichstätt specimen at the level of the last caudal vertebra (except for the Munich specimen, in which the feathers seem to be displaced). Farther distally, the tail gradually tapers to the terminal truncation (ca. 6.5 cm wide in the London specimen). The tail area amounts to approximately 170 cm² in the London specimen (de Beer, 1954b: Pl. I), 110 cm² in the Berlin specimen (Dames, 1884: Pl. 1), and 44 cm² in the Eichstätt specimen (Wellnhofer, 1974: Pl. 20). Yalden's (1984) estimate for the Berlin specimen proved to be too high (D.W. Yalden, pers. comm.).

The shortest tail feathers (4.5 cm in the Berlin, 6 cm in the London specimen) are the proximal six pairs. Farther distally the length rapidly increases. The longest feathers (8.7 cm in the Berlin, 15.6 in the London specimen) are attached to the 15th through 17th caudal vertebrae. The feather length gradually decreases toward the end, and the terminal feathers are 9.8 cm long in the London specimen. As in neornithines, the tail feathers overlap. On the preserved ventral surface, the distal vane of each feather overlaps the following distal feather. The overlap is extensive between the proximal rectrices and only marginal between the distal ones (Dames, 1884). De Beer (1954b) made reference to the tail coverts in the London and Berlin specimens, but there seems to be no evidence of them.

As in the wing, the nature of feather attachments in the tail is unclear, since the quills do not contact the vertebrae. Dames (1884) explained the presence of a gap between the quills and caudal vertebrae by the former presence of small contour feathers, but this can be accounted for by the presence of other structures, such as muscles and skin. In most specimens, including the disarticulated London specimen, the tail feathers preserved their natural association with the caudal vertebrae, which seems to indicate the firm attachment of feathers to the vertebrae (Gatesy and Dial, 1996). However, in the Munich specimen the feathered tail seems to be out of phase by three vertebrae: it starts at the level of the ninth rather than the sixth caudal vertebra, and its widest point is far behind the last vertebra (Wellnhofer, 1993).

Phylogenetic Relationships

Archaeopterygidae is unquestionably a clade of birds by virtue of having functional flight feathers and avian features in both the skull (Elzanowski and Wellnhofer, 1996) and postcraneum (Feduccia, 1999) (see also discussion of definitions of Aves and birds in Witmer, Chapter 1 in this volume). The reptilian relationships of Archaeopterygidae are the same as those of all birds (Gauthier, 1986; Clark, Norell, and Makovicky, Chapter 2 in this volume). The osteology of Archaeopterygidae clearly supports a close relationship of theropods and birds, but not necessarily their current placement within theropod phylogenies (as reviewed by Zweers and Vanden Berge, 1997), in particular their repeatedly claimed closest relationship with Dromaeosauridae (Gauthier, 1986; Chiappe, 1997a; Padian and Chiappe, 1998). Cranial comparisons suggest an even closer relationship to other theropod groups (Elzanowski, 1999), and both the skull and thoracic girdle of archaeopterygids demonstrate a considerable evolutionary distance from any of the well-known theropod taxa, which makes the homology of detailed dromaeosaurid similarities of pelvic and vertebral morphology highly suspect. A strongly retroverted pelvis in dromaeosaurids (Norell and Makovicky, 1997) provides a case of convergence with ornithurine birds.

Close archaeopterygid relatives are likely to be found in the multitude of birdlike forms that have been recently identified as closest avian relatives. One candidate for a late survivor of the archaeopterygid lineage is *Unenlagia* (Forster et al., 1998), which has the archaeopterygid pelvis and scapula. *Unenlagia* was described as nonavian (Novas and Puerta, 1997) because of three characters, two of which (extensive pubic apron and the distal cranioventral process of the ischium) are present in the archaeopterygids and the third of which, hindlimb proportions, is irrelevant for establishing relationships between archaeopterygids and a large (2-m long) Late Cretaceous form. Another candidate is *Protarchaeopteryx* (Ji and Ji, 1997; Ji et al., 1998), which is larger than *Wellnhoferia* and was probably flightless (or nearly so), as its forelimb is 70% the length of the hindlimb and the pelvis is very robust. *Protarchaeopteryx* shares with

Archaeopterygidae a broad sternum, twice as broad as it is long; caudal dorsal vertebrae with lateral excavations, often referred to as pleurocoels (Britt et al., 1998); ilium with the ventral process on the preacetabular part (as in the London specimen); and metatarsals that keep their individual boundaries and yet seem to be tangentially co-ossified.

The Late Jurassic teeth from Guimarota, Portugal (Weigert, 1995), show only an overall similarity to archaeopterygid teeth but strongly differ in that they are pronouncedly sigmoid in side view and have serrated carinae, which are, curiously, deflected to the lingual side. While there is a possibility that they are derived from an archaeopterygid relative, their generic-level identification as "cf. *Archaeopteryx* sp." is unwarranted.

Archaeopterygids seem to have, in general, *the* most primitive postcranial skeleton and thus to represent the most basal lineage of all known unquestionably avian groups (see Padian and Chiappe, 1998, for a review of purported first birds). Because of the paucity of autapomorphies (Gauthier, 1986; Sereno, 1997), the archaeopterygid postcranial skeleton is intermediate between the archosaurian ancestors of birds and the most primitive of the Cretaceous birds including *Rahonavis* (Forster et al., 1998), Confuciusornithidae (Chiappe et al., 1999), *Iberomesornis* (Sereno, 2000; Sanz et al., Chapter 9 in this volume), and other enantiornithine birds (Chiappe and Walker, Chapter 11 in this volume). Archaeopterygids are more primitive than any of these birds in the structure of the thoracic girdle, with a platelike coracoid immovably sutured to the scapula. A possibility that the archaeopterygids gave rise to all more advanced birds cannot be ruled out, at least formally (Bonde, 1996).

A close relationship of archaeopterygids and *Rahonavis*, as suggested by one of the most parsimonious trees generated by Forster et al. (1998), is supported by proportions, which have little weight, and two simple details, the caudal projection of the pubic foot and the lack of a femoral neck, which may be primitive for birds. *Rahonavis* is more advanced toward neornithines than are archaeopterygids in at least four characters (heterocoelous vertebrae, six sacrals, movable coracoscapular articulation, and quill knobs). Both archaeopterygids and *Rahonavis* are more primitive than the remaining birds in lacking a pygostyle.

However, the basal position of archaeopterygids may necessitate several reversals in other, apparently more advanced Mesozoic birds. The jugal process of the postorbital has an extensive contact with the jugal in *Confuciusornis* (Peters and Ji, 1998; Chiappe et al., 1999; see also Zhou and Hou, Chapter 7 in this volume) and is relatively longer than that of *Archaeopteryx* and the El Montsec nestling LP-4450-IEI (Sanz et al., 1997). The mandibular ramus of both birds and many nonavian theropods is fenestrate but solid in *Archaeopteryx*. The prima facie derived absence of interdental plates in the El Montsec nestling is ambiguous because such secondary ossifications may appear late in ontogeny. In addition, the fifth metatarsal of *Confuciusornis* is relatively longer than that of archaeopterygids (Zhou, 1995).

Martin (1991) claimed that archaeopterygids are more closely related to the Enantiornithes than any other birds because they share (a) the proximal coalescence of the metatarsals and (b) fusions of thoracic vertebrae. Hou et al. (1996) redefined the first character as the proximal-to-distal ontogenetic fusion of the metatarsus and added (c) a caudolateral condylar ridge on the distal end of the femur, (d) a craniomedially directed "scapular process," and (e) the "anterodorsal" (= proximodorsal) ischiadic process. However, similarities (b) and (c) are nonexistent (thoracic vertebrae are not fused, and there is no ridge on the femur in archaeopterygids), and similarities (d) and (e) are in all probability symplesiomorphic, as they are present in *Rahonavis* and other primitive birds (Forster et al., 1998). There is little evidence of the proximodistal progression of fusion of the metatarsus in the ontogeny of *Archaeopteryx*, and the proximal coalescence of the metatarsals is evident only in *Wellnhoferia*, which is clearly more cursorial than *Archaeopteryx*. The fusion of metatarsals is a common locomotory adaptation, especially among bipeds. Whatever its extent in archaeopterygids, the proximal fusion of metatarsals among various taxa of primitive birds is a priori at least as likely to be convergent or symplesiomorphic as synapomorphic. It is present in an oviraptorosaur, *Elmisaurus* (Osmólska, 1981), and Avimimidae (Vickers-Rich, Chiappe, and Kurzanov, Chapter 3 in this volume).

Paleobiology

Although archaeopterygids certainly do not represent *the* most primitive stage of avian evolution and their biology is not *necessarily* representative of that of avian ancestors, they are likely to have retained some functional and ecological similarities to the avian ancestors because they *are* very primitive birds with a postcranial skeleton that is morphologically intermediate between nonavian theropods and more advanced birds. In addition, *Archaeopteryx* falls in the probable size range of avian ancestors. It is in consideration of the relatively short evolutionary distance to the avian ancestors that the life of archaeopterygids has been commonly categorized according to one's views on the origin birds. Such typological thinking commonly clashes with reality, especially with regard to locomotion and habitats of versatile animals of the size of *Archaeopteryx* (Regal, 1985; Rayner, 1991).

Body mass—For the Berlin specimen, Yalden (1984) arrived at the range of 200–300 g by using allometric extrapolations from the lengths and diameters of avian and mammalian long bones, and he calculated 271 g using his rigorous reconstruction of *Archaeopteryx* to determine the

approximate volumes of body parts and the carcasses of magpie (*Pica pica*) to estimate their specific masses.

By calculating the least circumferences of *Archaeopteryx* femora from their least diameters and using these in the regression of the least circumference of the femur on body mass in 69 nonpasserine families (Campbell and Marcus, 1992), body mass is here estimated as 174 g for the Eichstätt, 222 g for the Munich, 276 g for the Berlin, and 468 g for the London specimen. The estimate for the Berlin specimen is essentially identical to that obtained experimentally by Yalden (1984). The maximum figure of 400 g assumed by Hopson (1977) for the London specimen is probably too low and thus the ensuing relative brain size estimate too high. *Wellnhoferia* was larger than the London specimen, but no comparable estimate can be provided because the long bones of the sixth specimen are crushed.

Thermoregulation—A long discussion of the metabolic rates and thermoregulation of dinosaurs ended in agnostic conclusions (Farlow et al., 1995), which were extended to *Archaeopteryx* (Feduccia, 1999). However, the hard facts are that archaeopterygids had at least a partial thermocoel and that a thermocoel (whether made of feathers or other integumentary structures) was present in the ancestors of birds as evidenced by its presence in *Sinosauropteryx* (Chen et al., 1998) and the therizinosauroid *Beipiaosaurus* (Xu et al., 1999). Contrary to Ruben (1995), the presence of plumage or pelage is invariably correlated with and thus provides evidence for some level of homeothermy in a fossil terrestrial vertebrate even if some birds make use of external sources of heat and others manage to cool their bodies in flight through evaporation. The insulating layer of contour feathers with partly plumaceous bases (Griffiths, 1996) over most of the body almost certainly ensured some measure of homeothermy (Regal, 1985), which makes some level of endothermy probable despite the relatively small size of archaeopterygids. The second major piece of evidence in support of endothermy in archaeopterygids is the presence of an essentially modern avian brain that filled the cranial cavity and needed a fairly constant temperature and circulation for functioning, since relative brain size is known to be dependent on metabolic rate (Armstrong and Bergeron, 1985).

However, several lines of evidence indicate that archaeopterygids were poikilothermic and that their metabolic rate was lower than in neornithines. The absence in archaeopterygids of an extensive bony floor to the rib cage (i.e., a hypertrophied sternum), or the ossified sternal ribs that would hold it, and, possibly, of uncinate processes (although they are also absent in today's screamers—Anhimidae), suggests a low efficiency or absence of the volume control of the air sacs (primarily the thoracic sacs) and thus a low efficiency or absence of the unidirectional flow of respiratory air in the lungs (Randolph, 1994; Ruben et al., 1997). Even the hypothetical capability for active flight may not necessarily imply modern levels of endothermy if *Archaeopteryx* was able to sustain short (three-minute) bursts of active flight at the expense of anaerobic metabolism (Ruben, 1991). In addition, the bone microstructure in other Mesozoic birds shows the presence of lines of arrested growth (Chinsamy, Chapter 18 in this volume), which point to periodic metabolic slowdowns in the late growth phase of primitive birds. Finally, it is far from clear how a semiarid Jurassic ecosystem, based on the primary productivity of gymnosperms and a few pteridophytes, could have sustained a viable population of modern endothermic predators (Duncker, 1989; Farlow et al., 1995). The conflicting evidence suggests that thermoregulation in archaeopterygids was intermediate between today's poikilothermic and homeothermic vertebrates. Homeothermy is not an all-or-nothing phenomenon, as demonstrated by considerable temperature variation in some birds (such as kiwis) and mammals (such as monotremes).

Locomotion and habitat—One of the outcomes of the Eichstätt Conference (Hecht et al., 1985) was the realization that *Archaeopteryx* moved between the ground and the heights as "a bipedal cursor that was facultatively arboreal" (Dodson, 1985). Consequently, even the proponents of the arboreal theory of the origin of bird flight admitted that *Archaeopteryx* spent some of its time on the ground (Bock, 1986; Feduccia, 1999). Another corollary of the Eichstätt Conference was the refutation of the cursorial origin of avian flight on aerodynamic grounds (U. M. Norberg, 1985; Rayner 1985a,b). This notwithstanding, the false belief that the theropod origin of birds implies the cursorial origin of avian flight (Gauthier and Padian, 1985; Chiappe, 1997b; Padian and Chiappe, 1998) continues to inspire attempts to show that *Archaeopteryx* was purely cursorial and capable of a ground-up takeoff. All these attempts were unsuccessful and prove the contrary (Elzanowski, 2001b).

There is a remarkable congruence of evidence from several locomotor adaptations and their universal biological interdependencies that *Archaeopteryx* was a terrestrial forager and escape climber (Fig. 6.7). Because of the poor maneuverability of their wings and flight, *Archaeopteryx* was unable to take off from the ground up (save for special circumstances that cannot be reliably assumed) and to pursue its prey in the branches and thus must have foraged on the ground. Terrestrial foraging is indicated independently by the proportions of both limbs. The combination of aspect ratio and relative wing loading identifies *Archaeopteryx* as a ground forager in open spaces (Norberg, 1990). The hindlimb structure indicates that *Archaeopteryx* was a terrestrial, but not a cursorial, forager. The combination of maniraptoran hand, anisodactyl foot with laterally compressed

claws, and small size makes *Archaeopteryx* well adapted for climbing. In the presence of swift predators (such as *Compsognathus*), the combination of ground foraging and inability to take off from the ground makes running to climb to the safety of heights the most probable escape strategy (Fig. 6.7).

Archaeopteryx was capable of flight as evidenced by the size and aerodynamic asymmetries of the forelimb feathers (R. A. Norberg, 1985, 1995; Feduccia, 1999) and their similarity to those of neornithines; the size and arrangement of the tail feathers (Gatesy and Dial, 1996); and a combination of bipedalism and a long forelimb, which was as long as the hindlimb. These adaptations certainly allowed *Archaeopteryx* to glide once the body was set in motion by gravity and/or a thrust provided by the extension of the legs. The capability of *Archaeopteryx* for active (flapping) flight is sometimes questioned (Bock, 1986; Vazquez, 1992), but the majority of experts concluded that *Archaeopteryx* used flapping (Norberg, 1990; Rayner, 1991), albeit only weakly powered, that is, not sustained flight (Yalden, 1985; Speakman, 1993; Padian and Chiappe, 1998). The capability for flapping flight made *Archaeopteryx* largely independent of trees inasmuch as it could escape by using for takeoff only slightly elevated objects that by themselves were too low to offer protection against predators (Fig. 6.7). The flight capabilities of *Wellnhoferia* remain to be assessed.

Archaeopteryx appears to be equipped for relatively fast flight without sharp turns, which is simple aerodynamically and can be effected by simple oscillation (raising and lowering) of the wings combined with sweeping movements of the manus (i.e., flexion and extension in the plane of the wing). The minimum power speed (at which the least work has to be done) for a body mass of 250 g was estimated as 8.2 m/s (Yalden, 1971b) and below 8.0 m/s by Rayner (1985b: Fig. 4); approximately 8 m/s appears to be a safe value for the Berlin specimen, which was probably heavier than 250 g. Yalden (1971a) estimated the stalling speed below 6.3 m/s for a body mass of 200 g, and Padian (1985) concurred. The wingbeat frequency estimated for a wing span of 60 cm is 6.0–6.5 strokes/sec^{-1} (Norberg, 1990) and thus may have been slightly higher in the Berlin specimen with its wing span of 55 cm. The downstroke was effected by the coracobrachialis and the pectoralis muscles. The upstroke could have been effected automatically (Bock, 1986) or by the deltoideus muscle (Olson and Feduccia, 1979; Rayner, 1991). The supinatory rotation by the supracoracoideus muscle was apparently absent (Ostrom et al., 1999). The tail helped stabilize the body (Gatesy and Dial, 1996), especially its yaw (Norberg, 1990).

Archaeopteryx was poorly adapted (if at all) for flight at low speeds. The wings, as well as the tail (Gatesy and Dial, 1996), lacked the maneuverability used by neornithines for ground-up takeoffs, flight between obstacles (such as tree branches), and landings. Such low-speed maneuvers neces-

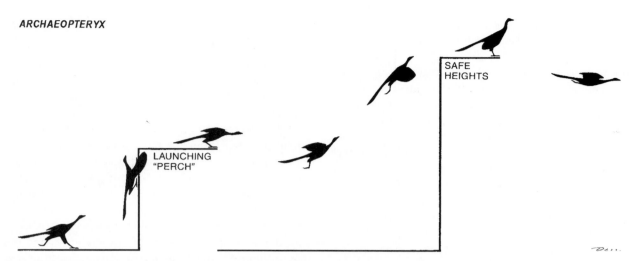

Figure 6.7. The escape behavior of *Archaeopteryx* as reconstructed from the locomotor adaptations and universal functional interdependencies (see text). The launching "perch" is assumed not to be predator-safe.

sitate controlled changes of the pitch of the entire wing relative to the body, in particular the effective upstroke supination (after the downstroke pronation) of the humerus (Ostrom et al., 1999) and flicking of the manus from the plane of the wing toward the body in the upstroke to avoid excessive drag (Rayner, 1991). Since a stalling speed of about 6 m/s calculated for *A. siemensii* is by 4 m/s higher than its estimated running speed of 2 m/s, it could possibly take off from the ground only by running into a wind (Rayner, 1991). However, a strong head wind would obviously decrease the maximum running speed, and *Archaeopteryx* lacked the maneuverability to control flight in a strong wind. Thus, it could not live in wind-swept areas that would otherwise enable them to take off whenever needed. In sum, even if it mastered the use of rare special circumstances (such as an appropriate wind from an appropriate direction), in most cases *Archaeopteryx* must have climbed in order to take off (Peters, 1985), and its flights may have ended with "sprawled crash landings" (Yalden, 1971b).

Archaeopterygids are well adapted for climbing by a combination of the maniraptoran hand, an unspecialized foot, laterally compressed claws with pointed tips, and small size, much smaller than that of nonavian maniraptorans. The forelimbs provided the upward propulsion, and their claws bear strong flexor tubercles, which provide leverage for the flexors, and conform in shape with those of modern tree and cliff climbers (Yalden, 1985, 1997; Peters and Görgner, 1992; Feduccia, 1993). The manual claws may have possibly had accessory functions such as preening, grasping, and fighting (Peters and Görgner, 1992). The hindlimb must have supported the body against the trunk, although the pedal claws are less curved than the manual claws (Yalden, 1985), which probably reflects an adaptive compromise between climbing and terrestrial locomotion. Having

climbed a tree, archaeopterygids were capable of perching, as evidenced by their opposable hallux and laterally compressed pedal claws (Yalden 1985, 1997). The preservation of the right foot of the Munich specimen in a grasping position, with the claws of the hallux and fourth toe superimposed, supports this conclusion (Wellnhofer, 1993).

Archaeopterygids were at ease on the ground, as demonstrated by their strong legs with the tibia longer than the femur; an unspecialized anisodactyl foot with trochlea III extending farther distally than trochleae II and IV; and a slightly elevated, short hallux that did not interfere with terrestrial locomotion. However, a full-length fibula and a pronounced asymmetry of the pes (Coombs, 1978), with digit IV much longer than digit II, are incompatible with truly cursorial specialization, and the hindlimb, pelvis, and vertebral column (especially synsacrum) of modern ground birds are much better adapted for cursorial locomotion than *Archaeopteryx* (Bock, 1986). Among birds of comparable size, Archaeopterygidae have intramembral hindlimb proportions that are nearly identical to those of some phasianid galliforms and tinamous (Gatesy and Middleton, 1997; Elzanowski, in press), which are slow-paced to multimode terrestrial foragers and run only in emergency, and widely different from those of truly cursorial foragers, such as roadrunners and some charadriiforms birds. Both the morphology and proportions of the hindlimb indicate that *Archaeopteryx* was a slow-paced or multimode forager but certainly could run in an emergency. The pes of *Wellnhoferia* is more symmetrical, which may possibly suggest more cursorial habits.

Feeding and food—Archaeopterygid teeth indicate broadly defined insectivory, that is, feeding on small animals, primarily arthropods. Archaeopterygid teeth do not seem adapted for either scavenging or crushing hard food items, such as some beetles with highly mineralized exoskeletons (Thulborn and Hamley, 1985). One can safely assume that relatively soft bodied arthropods made up at least a substantial part of the diet of smaller individuals of *Archaeopteryx*. Among the insects abundant in the Solnhofen assemblage, archaeopterygids almost certainly fed, at least at some stage of their life cycle, on the locusts and cockroaches, which are not very fast and agile in flight and, unlike most beetles, have leathery cuticles. However, next to nothing is known about the local diversity of flightless terrestrial arthropods, such as myriapods and arachnids, which may have been fed on by archaeopterygids.

The diets of the animals represented by the Eichstätt and London specimens may have differed considerably because of at least a twofold difference in body mass. Baby individuals must have fed exclusively on small, soft-bodied invertebrates (mainly arthropods), whereas large individuals probably had a broader dietary range. It is likely that small vertebrates contributed to the diet of *Wellnhoferia*, which is bigger and may have been better adapted for running than *Archaeopteryx*. The unique teeth of *A. bavarica* suggest a well-defined predatory adaptation, which has yet to be understood. A combination of pointed tips and sharp rostral edges seems to be appropriate for simultaneous piercing and cutting and thus propagating cracks in the arthropod cuticle.

Archaeopterygids may have exploited the floor of araucaria groves, the shores of inland pools and streams inhabited by numerous water insects, and beaches if they provided launching ramps for emergency takeoffs, but there is no evidence to postulate wading habits (contra Thulborn and Hamley, 1985), which, in addition, would be difficult to reconcile with the presence of feather breeches on the lower legs (Dyck, 1985). Archaeopterygids were clearly lacking the flight maneuverability or any other adaptation for the pursuit of arboreal prey, and there is no evidence of any abundant animal food that would be easily available for a clumsy flier on the scanty gymnosperm trees of Solnhofen.

The combined evidence from morphology and environment suggests that *Archaeopteryx* may have used trees or other elevated perches for taking off, especially to escape a predator, and possibly for other purposes (e.g., for roosting) but foraged primarily or exclusively on the ground (Peters, 1985) in a manner comparable to that of galliforms (Wellnhofer, 1995). *Wellnhoferia* may have possibly been more cursorial.

Garner et al. (1999) proposed that birds evolved from predators that specialized in ambush from elevated sites ("the pouncing proavis model"), which is in conflict with ecomorphological evidence from *Archaeopteryx* and fails to explain what a terrestrial avian ancestor would gain by becoming an ambush predator, given the considerable expense of both energy (for climbing) and time (for waiting) (Hedenström, 1999; Elzanowski, in press).

Development and reproduction—As in other primitive birds, the growth of archaeopterygids may have been biphasic, at first fast and then slow and extended (Chinsamy, Chapter 18 in this volume). This opens up the possibility that the observed size and other differences between the specimens result from the differences in individual age. Houck et al. (1990) followed this path and concluded that the six then known specimens (including *Wellnhoferia*) belong to a growth series of a single species, *A. lithographica*. However, this conclusion is based on a questionable interpretation of several osteological characters as being juvenile, the use of partly inaccurate measurements, and the conservation of growth allometry among closely related species, which is not unusual.

While different growth stages of the same species may possibly be represented by a pair or two among the known specimens of *Archaeopteryx* (especially including the Eichstätt and/or two fragmentary skeletons), the single growth series interpretation is untenable for all specimens or even for the set of five fairly complete specimens. First, despite great differences in size, there is not a single unequivocal age-related difference, especially in the texture and ossification of single bones (Howgate, 1985) between the known specimens. Second, some evidently size-independent differences between the specimens demonstrate the presence of more than one species, making it extremely difficult to decide whether slight differences in the fusion of some bones are age- or taxon-specific.

The extant phylogenetic bracket (Witmer, 1997) of archaeopterygids (i.e., birds and crocodiles) suggests that archaeopterygids defended breeding territories, as probably did some dinosaurs (Coombs, 1990), and provided some care (at least protection) to their eggs and young. Brooding behavior has been recently demonstrated in oviraptorids (Norell et al., 1995), which may be among the closest known relatives of birds (Elzanowski, 1999). Because of limited flying ability, the feeding and breeding habitats of archaeopterygids must have been close to each other. Very little can be said about breeding habitat except that nesting in the tree crowns seems unlikely because it requires behavioral adaptations of both parents and young that are probably derived among neornithines.

Acknowledgments

I thank P. Wellnhofer (Bayerische Staatssammlung für Paläontologie und Historische Geologie, Munich), H. Osmólska (Institute of Paleobiology, Warsaw), J. Desselberger (Warsaw), H. James (Smithsonian Institution, Washington, D.C.), D. S. Peters (Forschungsinstitut Senckenberg, Frankfurt), and D. W. Yalden (Victoria University, Manchester) for discussion; P. Wellnhofer,

D. S. Peters, J. T. Groiss (Universität Erlangen-Nürnberg), and A. D. Gishlick (Yale University, New Haven) for literature and unpublished data; Sandra Chapman and Angela Milner (British Museum, London), W. Heinrich (Museum für Naturkunde, Berlin), M. Röper and Bürgermeister Nürnberger (Bürgermeister-Müller-Museum, Solnhofen), G. Viohl (Jura-Museum, Eichstätt), and P. Wellnhofer for access to the specimens; and Poland's State Committee for Scientific Research (KBN) for support through grant 6 PO4C 059 17.

Literature Cited

Armstrong, E., and R. Bergeron. 1985. Relative brain size and metabolism in birds. Brain, Behavior and Evolution 26:141–153.

Barsbold R. and H. Osmólska. 1999. The skull of *Velociraptor* (Theropoda) from the Late Cretaceous of Mongolia. Acta Paleontologica Polonica 44:189–219.

Barsbold R., T. Maryanska, and H. Osmólska. 1990. Oviraptorosauria; pp. 249–258 *in* D. B. Weishampel, P. Dodson, and H. Osmólska (eds.), The Dinosauria. Berkeley, University of California Press.

Barthel, K. W., N. H. M. Swinburne, and S. Conway Morris. 1990. Solnhofen. A study in Mesozoic Palaeontology. Cambridge University Press, Cambridge, 236 pp.

Boas, J. E. V. 1933. Kreuzbein, Becken und Plexus lumbosacralis der Vögel. Kongelige Danske Vidensk. Selsk. Skrifter, Naturv. og Math. Afd., Ser. 9, 5:3–74.

Bock, W. J. 1986. The arboreal origin of avian flight. Memoirs of the California Academy of Sciences 8:57–72.

Bock, W. J., and W. D. Miller. 1959. The scansorial foot of the woodpeckers, with comments on the evolution of perching and climbing feet in birds. American Museum Novitates 1931:1–45.

Bohlin, B. 1947. The wing of Archaeornithes. Zoologiska Bidrag 25:328–334.

Bonde, N. 1996. The systematic and classificatory status of *Archaeopteryx*. Museum of Northern Arizona Bulletin 60: 193–199.

Britt, B. B., P. J. Makovicky, J. Gauthier, and N. Bonde. 1998. Postcranial pneumatization in *Archaeopteryx*. Nature 395:374–376.

Bühler, P. 1985. On the morphology of the skull of *Archaeopteryx*; pp. 135–140 in M. K. Hecht, J. H. Ostrom, G. Viohl, and P. Wellnhofer (eds.), The Beginnings of Birds. Freunde des Jura-Museums, Eichstätt.

Burke, A. C., and A. Feduccia. 1997. Developmental patterns and the identification of homologies in the avian hand. Science 278:666–668.

Campbell, K. E., and L. Marcus. 1992. The relationship of hindlimb bone dimensions to body weight in birds. Science Series Natural History Museum of Los Angeles County 36:395–412.

Chatterjee, S. 1991. Cranial anatomy and relationships of a new Triassic bird from Texas. Philosophical Transactions of the Royal Society of London B332:277–342.

———. 1997. The Rise of Birds. Johns Hopkins University Press, Baltimore, 312 pp.

———. 1998. Counting the fingers of birds and dinosaurs. Science 280:355–356.

Chen P., Dong Z., and Zhen S. 1998. An exceptionally well-preserved theropod dinosaur from the Yixian Formation of China. Nature 391:147–152.

Chiappe, L. M. 1996. Late Cretaceous birds of southern South America: anatomy and systematics of Enantiornithes and *Patagopteryx deferrariisi*; pp. 203–244 *in* G. Arratia (ed.), Contributions of Southern South America to Vertebrate Paleontology, Münchner Geowissenschaftliche Abhandlungen, Reihe A, Geologie und Paläontologie 30. Verlag Dr. Friedrich Pfeil, Munich.

———. 1997a. Aves; pp. 32–38 *in* P. J. Currie and K. Padian (eds.), Encyclopedia of Dinosaurs. Academic Press, San Diego.

———. 1997b. Climbing *Archaeopteryx*?: a response to Yalden. Archaeopteryx 15:109–112.

Chiappe, L. M., M. Norell, and J. M. Clark. 1998. The skull of a relative of the stem-group bird *Mononykus*. Nature 392:275–278.

Chiappe, L. M., Ji S., Ji Q., and M. A. Norell. 1999. Anatomy and systematics of the Confuciusornithidae (Theropoda: Aves) from the Late Mesozoic of Northeastern China. Bulletin of the American Museum of Natural History 242:1–89.

Coombs, W. P. 1978. Theoretical aspects of cursorial adaptations in dinosaurs. Quarterly Review of Biology 53:393–418.

———. 1990. Behavior patterns of dinosaurs; pp. 32–42 *in* D. B. Weishampel, P. Dodson, and H. Osmólska (eds.), The Dinosauria. University of California Press, Berkeley.

Cuffey, R. J. 1984. Paleontologic evidence and organic evolution; pp. 255–281 *in* A. Montagu (ed.), Science and Creationism. Oxford University Press, Oxford.

Currie, P. J., and Zhao X. 1993. A new troodontid (Dinosauria, Theropoda) braincase from the Dinosaur Park Formation (Campanian) of Alberta. Canadian Journal of Earth Sciences 30:2231–2247.

Curtis, E. L., and R. C. Miller. 1938. The sclerotic ring in North American birds. Auk 55:225–243.

Dames, W. 1884. Über *Archaeopteryx*. Palaeontologische Abhandlungen 2:119–196.

———. 1897. Über Brustbein, Schulter- und Beckengürtel der *Archaeopteryx*. Sitzungsberichte der Königlich Preussischen Akademie der Wissenschaften zu Berlin (1897):818–834.

Davis, P. G. 1996. The taphonomy of *Archaeopteryx*. Bulletin of the National Science Museum, Tokyo, Series C, 22:91–106.

Davis, P. G., and D. E. G. Briggs. 1995. Fossilization of feathers. Geology 23:783–786.

———. 1998. The impact of decay and disarticulation on the preservation of fossil birds. Palaios 13:3–13.

de Beer, G. R. 1954a. *Archaeopteryx* and evolution. Advancement of Science 11:160–170.

———. 1954b. *Archaeopteryx lithographica*. British Museum (Natural History), London, 68 pp.

de Buisonjé, P. H. 1985. Climatological conditions during deposition of the Solnhofen Limestones; pp. 45–65 *in* M. K. Hecht, J. H. Ostrom, G. Viohl, and P. Wellnhofer (eds.), The Beginnings of Birds. Freunde des Jura-Museums, Eichstätt.

de Queiroz, K., and D. A. Good. 1988. The scleral ossicles of *Opisthocomus* and their phylogenetic significance. Auk 105: 29–35.

Dodson, P. 1985. International *Archaeopteryx* Conference. Journal of Vertebrate Paleontology 5:177–179.

Duncker, H.-R. 1989. Structural and functional integration across the reptile-bird transition: locomotor and respiratory systems; pp. 147–169 *in* D. B. Wake and G. Roth (eds.), Complex Organismal Functions: Integration and Evolution in Vertebrates. John Wiley and Sons, New York.

Dyck, J. 1985. The evolution of feathers. Zoologica Scripta 14:137–154.

Edington, G. H., and A. E. Miller. 1941. The avian ulna: its quill-knobs. Proceedings of the Royal Society of Edinburgh 61:138–148.

Elzanowski, A. 1988. Ontogeny and evolution of the ratites; pp. 2037–2046 in H. Ouellet (ed.), Acta XIX Congressus Internationalis Ornithologici. University of Ottawa Press, Ottawa.

———. 1991. New observations on the skull of *Hesperornis* with reconstructions of the bony palate and otic region. Postilla 207:1–20.

———. 1995. Cretaceous birds and avian phylogeny. Courier Forschungsinstitut Senckenberg 181:41–59.

———. 1999. A comparison of the jaw skeleton in theropods and birds, with a description of the palate in the Oviraptoridae. Smithsonian Contributions to Paleobiology 89:311–323.

———. 2001a. A new genus and species for the largest specimen of *Archaeopteryx*. Acta Palaeontologica Polonica 46(4):519–532.

———. 2001b. The life style of *Archaeopteryx* (Aves). Publicaciones Especiales de la Asociación Paleontológica Argentina 7:91–99.

———. In press. Biology of basal birds and the origins of avian flight. Proceedings of the 5th International Meeting of the Society of Avian Paleontology and Evolution, Beijing.

Elzanowski, A., and P. M. Galton. 1991. Braincase of the Early Cretaceous bird *Enaliornis*. Journal of Vertebrate Paleontology 11:90–107.

Elzanowski, A., and L. Pasko. 1999. A skeletal reconstruction of *Archaeopteryx*. Acta Ornithologica 34:123–129.

Elzanowski, A., and P. Wellnhofer 1995. The skull of *Archaeopteryx* and the origin of birds. Archaeopteryx 13:41–46.

———. 1996. The cranial morphology of *Archaeopteryx*: evidence from the seventh skeleton. Journal of Vertebrate Paleontology 16:81–94.

Farlow, J. O., P. Dodson, and A. Chinsamy. 1995. Dinosaur biology. Annual Reviews of Ecology and Systematics 26:445–471.

Feduccia, A. 1993. Evidence from claw geometry indicating arboreal habits of *Archaeopteryx*. Science 259:790–793.

———. 1999. The Origin and Evolution of Birds. 2nd edition. Yale University Press, New Haven, 466 pp.

Fischer, K., and H. H. Krueger. 1979. Neue Präparationen am Berliner Exemplar des Urvogels *Archaeopteryx lithographica* H. v. Meyer, 1861. Zeitschrift für geologische Wissenschaften 7:575–579.

Forster, C. A., S. D. Sampson, L. M. Chiappe, and D. W. Krause. 1998. The theropod ancestry of birds: new evidence from the Late Cretaceous of Madagascar. Science 279:1915–1919.

Galton, P. 1999. Sex, sacra and *Sellosaurus gracilis* (Saurischia, Sauropodomorpha, Upper Triassic, Germany)—or why the character "two sacral vertebrae" is plesiomorphic in Dinosauria. Neues Jahrbuch für Geologie und Paläontologie 195:253–266.

Garner, J. P., G. K. Taylor, and A. L. R. Thomas. 1999. On the origins of birds: the sequence of character acquisition in the evolution of avian flight. Proceedings of the Royal Society of London B 266:1259–1266.

Gatesy, S. M. 1990. Caudofemoral musculature and the evolution of theropod locomotion. Paleobiology 16:170–186.

———. 1991. Hind limb scaling in birds and other theropods: implications for terrestrial locomotion. Journal of Morphology 209:83–96.

Gatesy, S. M., and K. P. Dial. 1996. From frond to fan: *Archaeopteryx* and the evolution of short-tailed birds. Evolution 50:2037–2048.

Gatesy, S. M., and K. M. Middleton. 1997. Bipedalism, flight, and the evolution of theropod locomotor diversity. Journal of Vertebrate Paleontology 17:308–329.

Gauthier, J. 1986. Saurischian monophyly and the origin of birds. Memoirs of the California Academy of Sciences, 8:1–46.

Gauthier, J., and K. Padian. 1985. Phylogenetic, functional, and aerodynamic analyses of the origin of birds and their flight; pp. 185–197 in M. K. Hecht, J. H. Ostrom, G. Viohl, and P. Wellnhofer (eds.), The Beginnings of Birds. Freunde des Jura-Museums, Eichstätt.

Gishlick, A. D. 2001. The function of the manus and forelimb of *Deinonychus antirrhopus* and its importance for the origin of avian flight; pp. 301–318 in J. A. Gauthier and L. F. Gall (eds.), New Perspectives on the Origin and Evolution of Birds: Proceedings of the International Symposium in Honor of John H. Ostrom. Peabody Museum of Natural History, New Haven.

Griffiths, P. J. 1996. The isolated *Archaeopteryx* feather. Archaeopteryx 14:1–26.

Halstead, L. B. 1984. Evolution—the fossils say yes!; pp. 240–254 in A. Montagu (ed.), Science and Creationism. Oxford University Press, Oxford.

Hecht, M. K. 1985. The biological significance of *Archaeopteryx*; pp. 199–207 in M. K. Hecht, J. H. Ostrom, G. Viohl, and P. Wellnhofer (eds.), The Beginnings of Birds. Freunde des Jura-Museums, Eichstätt.

Hecht, M. K., J. H. Ostrom, G. Viohl, and P. Wellnhofer (eds.). 1985. The Beginnings of Birds. Freunde des Jura-Museums, Eichstätt, 382 pp.

Hedenström, A. 1999. How birds became airborne. Trends in Ecology and Evolution 14:375.

Heilmann, G. 1926. The Origin of Birds. Witherby, London, 210 pp.

Heinroth, O. 1923. Die Flügel von *Archaeopteryx*. Journal für Ornithologie 71:277–283.

Heller, F. 1959. Ein dritter *Archaeopteryx*-Fund aus den Solnhofener Plattenkalken von Langenaltheim (Mfr.). Erlanger geologische Abhandlungen 31:1–25.

———. 1960. Der dritte *Archaeopteryx*-Fund aus den Solnhofener Plattenkalken des oberen Malm Frankens. Journal für Ornithologie 101:7–28.

Helms, J. 1982. Zur Fossilisation der Federn des Urvogels (Berliner Exemplar). Wissenschaftliche Zeitschrift der Humboldt-Universität zu Berlin, Mathematisch-Naturwissenschaftliche Reihe, 31:185–199.

Hinchliffe, J. R. 1985. "One, two, three" or "two, three, four": an embryologist's view of the homologies of the digits and carpus of neornithines; pp. 141–147 in M. K. Hecht, J. H. Ostrom, G. Viohl, and P. Wellnhofer (eds.), The Beginnings of Birds. Freunde des Jura-Museums, Eichstätt.

Hofer, H. 1949. Die Gaumenlücken der Vögel. Acta Zoologica 30:209–248.

Hopson, J. A. 1977. Relative brain size and behavior in archosaurian reptiles. Annual Review of Ecology and Systematics 8:429–448.

Hou L., L. D. Martin, Zhou Z., and A. Feduccia. 1996. Early adaptive radiation of birds: evidence from fossils from northeastern China. Science 274:1164–1167.

Houck, M. A., J. A. Gauthier, and R. E. Strauss. 1990. Allometric scaling in the earliest fossil bird, *Archaeopteryx lithographica*. Science 247:195–198.

Howgate, M. E. 1984a. *Archaeopteryx*'s morphology. Nature 310:104.

———. 1984b. The teeth of *Archaeopteryx* and a reinterpretation of the Eichstätt specimen. Zoological Journal of the Linnean Society 82:159–175.

———. 1985. Problems of the osteology of *Archaeopteryx*. Is the Eichstätt specimen a distinct genus?; pp. 105–112 *in* M. K. Hecht, J. H. Ostrom, G. Viohl, and P. Wellnhofer (eds.), The Beginnings of Birds. Freunde des Jura-Museums, Eichstätt.

International Commission on Zoological Nomenclature. 1977. Opinion 1070. Conservation of *Archaeopteryx lithographica* von Meyer 1861 (Aves). Bulletin of Zoological Nomenclature 33:165–166.

Jenkins, F. A. 1993. The evolution of the avian shoulder joint. American Journal of Science 293-A:253–267.

Jerison, H. J. 1973. Evolution of the Brain and Intelligence. Academic Press, New York, 482 pp.

Ji Q. and Ji S. 1997. *Protarchaeopteryx* gen. nov.—a Chinese archaeopterygid. Chinese Geology 238:38–41.

Ji Q., P. J. Currie, M. A. Norell, and Ji S. 1998. Two feathered dinosaurs from northeastern China. Nature 393:753–761.

Jung, W. 1995. Araukarienwälder oder *Brachyphyllum*-Dickichte —die Heimat des *Archaeopteryx*?; pp. 136–151 *in* Messekatalog der Mineralientage München 1995. Freunde der Bayerischen Staatssammlung für Paläontologie und historische Geologie, Munich.

Kemp, R. A., and D. M. Unwin. 1997. The skeletal taphonomy of *Archaeopteryx*: a quantitative approach. Lethaia 30:229–238.

Kritsky, G. 1992. Darwin's *Archaeopteryx* prophecy. Archives of Natural History 19:407–410.

Lemmrich, W. 1931. Der Skleralring der Vögel. Jenaische Zeitschrift für Naturwissenschaften 65:513–586.

Martill, D. M., and J. B. M. Filgueira. 1994. A new feather from the Lower Cretaceous of Brazil. Palaeontology 37:483–487.

Martin, L. D. 1991. Mesozoic birds and the origin of birds; pp. 485–540 *in* H.-P. Schultze and L. Trueb (eds.), Origins of the Higher Groups of Tetrapods. Comstock Publishing Associates, Ithaca, N.Y.

Martin, L. D., and J. D. Stewart. 1999. Implantation and replacement of bird teeth. Smithsonian Contributions to Paleobiology 89:295–300.

Maryanska, T., and H. Osmólska. 1997. The quadrate of oviraptorid dinosaurs. Acta Palaeontologica Polonica 42:377–387.

Mäuser, M. 1997. Der achte *Archaeopteryx*. Fossilien 3:156–157.

Mayr, E. 1970. Agassiz, Darwin, and evolution; pp. 299–307 *in* P. Appleman (ed.), Darwin. Norton, New York.

McGowan, C. 1985. Tarsal development in birds: evidence for homology with the theropod condition. Journal of Zoology 206:53–67.

Nicholls, E., and A. Russell. 1985. Structure and function of the pectoral girdle and forelimb of *Struthiomimus altus* (Theropoda: Ornithomimidae). Palaeontology 28:643–677.

Norberg, R. A. 1985. Function of vane asymmetry and shaft curvature in bird flight feathers; inferences on flight ability of *Archaeopteryx*; pp. 303–318 *in* M. K. Hecht, J. H. Ostrom, G. Viohl, and P. Wellnhofer (eds.), The Beginnings of Birds. Freunde des Jura-Museums, Eichstätt.

———. 1995. Feather asymmetry in *Archaeopteryx*. Nature 374:221.

Norberg, U. M. 1985. Evolution of flight in birds: aerodynamic, mechanical, and ecological aspects; pp. 293–302 *in* M. K. Hecht, J. H. Ostrom, G. Viohl, and P. Wellnhofer (eds.), The Beginnings of Birds. Freunde des Jura-Museums, Eichstätt.

———. 1990. Vertebrate Flight. Springer, Berlin, 291 pp.

Norell, M. A., and P. J. Makovicky. 1997. Important features of the dromaeosaur skeleton: information from a new specimen. American Museum Novitates 3215:1–28.

Norell, M. A., J. M. Clark, L. M. Chiappe, and Dashzeveg D. 1995. A nesting dinosaur. Nature 378:774–776.

Novas, F. E., and P. F. Puerta. 1997. New evidence concerning avian origins from the Late Cretaceous of Patagonia. Nature 387:390–392.

Olson, S. L. 1987. The beginnings of birds [a review]. American Scientist 75:74–75.

Olson, S. L., and A. Feduccia. 1979. Flight capability and the pectoral girdle of *Archaeopteryx*. Nature 278:247–248.

Osmólska, H. 1981. Coossified tarsometatarsi in theropod dinosaurs and their bearing on the problem of bird origins. Palaeontologia Polonica 42:79–95.

Ostrom, J. H. 1972. Description of the *Archaeopteryx* specimen in the Teyler Museum, Haarlem. Koninklige Nederlandse Akademie van Wetenschappen, Proceedings, Series B, 75:289–305.

———. 1974. *Archaeopteryx* and the origin of flight. Quarterly Review of Biology 49:27–47.

———. 1976a. *Archaeopteryx* and the origin of birds. Biological Journal of the Linnean Society 8:91–182.

———. 1976b. Some hypothetical anatomical stages in the evolution of avian flight. Smithsonian Contributions to Paleobiology 27:1–21.

———. 1992. Comments on the new (Solnhofen) specimen of *Archaeopteryx*. Science Series Natural History Museum of Los Angeles County 36:25–27.

———. 1995. Wing biomechanics and the origin of bird flight. Neues Jahrbuch für Geologie und Paläontologie 195:253–266.

Ostrom, J. H., S. O. Poore, and G. E. Goslow. 1999. Humeral rotation and wrist supination: important functional complex for the evolution of powered flight in birds? Smithsonian Contributions to Paleobiology 89:301–309.

Padian, K. 1995. The origins and aerodynamics of flight in extinct vertebrates. Palaeontology 28:413–433.

Padian, K., and L. M. Chiappe. 1998. The origin and early evolution of birds. Biological Reviews 73:1–42.

Peters, D. S. 1985. Functional and constructive limitations in the early evolution of birds; pp. 243–249 *in* M. K. Hecht, J. H. Ostrom, G. Viohl, and P. Wellnhofer (eds.), The Beginnings of Birds. Freunde des Jura-Museums, Eichstätt.

Peters, D. S., and E. Görgner. 1992. A comparative study on the claws of *Archaeopteryx*. Science Series Natural History Museum of Los Angeles County 36:29–37.

Peters, D. S., and Ji Q. 1998. The diapsid temporal construction of the Chinese fossil bird *Confuciusornis*. Senckenbergiana lethaea 78:153–155.

Poore, S. O., A. Sanchez-Haiman, and G. E. Goslow. 1997. Wing up-stroke and the evolution of flapping flight. Nature 387:799–802.

Raath, M. A. 1985. The theropod *Syntarsus* and its bearing on the origin of birds; pp. 219–227 *in* M. K. Hecht, J. H. Ostrom, G. Viohl, and P. Wellnhofer (eds.), The Beginnings of Birds. Freunde des Jura-Museums, Eichstätt.

Randolph, S. E. 1994. The relative timing of the origin of flight and endothermy: evidence from the comparative biology of birds and mammals. Zoological Journal of the Linnean Society 112:389–397.

Rau, R. 1969. Über den Flügel von *Archaeopteryx*. Natur und Museum 99:1–8.

Rayner, J. M. V. 1985a. Cursorial gliding in proto-birds; pp. 289–292 *in* M. K. Hecht, J. H. Ostrom, G. Viohl, and P. Wellnhofer (eds.), The Beginnings of Birds. Freunde des Jura-Museums, Eichstätt.

———. 1985b. Mechanical and ecological constraints on flight evolution; pp. 279–288 *in* M. K. Hecht, J. H. Ostrom, G. Viohl, and P. Wellnhofer (eds.), The Beginnings of Birds. Freunde des Jura-Museums, Eichstätt.

———. 1991. Avian flight evolution and the problem of *Archaeopteryx*; pp. 183–212 *in* J. M. V. Rayner and R. J. Wootton (eds.), Biomechanics in Evolution. Cambridge University Press, Cambridge.

Regal, P. J. 1985. Common sense and reconstruction of the biology of fossils: *Archaeopteryx* and feathers; pp. 67–74 *in* M. K. Hecht, J. H. Ostrom, G. Viohl, and P. Wellnhofer (eds.), The Beginnings of Birds. Freunde des Jura-Museums, Eichstätt.

Rietschel, S. 1976. *Archaeopteryx*—Tod und Einbettung. Natur und Museum 106:280–286.

———. 1985a. False forgery; pp. 371–376 *in* M. K. Hecht, J. H. Ostrom, G. Viohl, and P. Wellnhofer (eds.), The Beginnings of Birds. Freunde des Jura-Museums, Eichstätt.

———. 1985b. Feathers and wings of *Archaeopteryx,* and the question of her flight ability; pp. 251–260 *in* M. K. Hecht, J. H. Ostrom, G. Viohl, and P. Wellnhofer (eds.), The Beginnings of Birds. Freunde des Jura-Museums, Eichstätt.

Ruben, J. 1991. Reptilian physiology and the flight capacity of *Archaeopteryx*. Evolution 45:1–17.

———. 1995. The evolution of endothermy in mammals and birds: from physiology to fossils. Annual Review of Physiology 57:69–95.

Ruben, J., T. D. Jones, N. R. Geist, and W. J. Hillenius. 1997. Lung structure and ventilation in theropod dinosaurs and early birds. Science 278:1267–1270.

Rudwick, M. J. S. 1976. The Meaning of Fossils. Neale Watson Academic Publications, New York, 287 pp.

Sanz, J. L., L. M. Chiappe, B. Perez-Moreno, J. J. Moratalla, F. Hernandez-Carrasquilla, A. D. Buscalioni, F. Ortega, F. J. Poyato-Ariza, D. Rasskin-Gutman, and X. Martínez-Delclòs. 1997. A nestling bird from the Lower Cretaceous of Spain: implications for avian skull and neck evolution. Nature 276:1543–1546.

Savile, D. B. O. 1957. The primaries of *Archaeopteryx*. Auk 74:99–101.

Sereno, P. C. 1997. The origin and evolution of dinosaurs. Annual Review of Earth and Planetary Science 25:435–489.

———. 2000. *Iberomesornis* romerali (Aves, Ornthothoraces) reevaluated as an Early Cretaceous enanantyirnthine. Neues Jahrbuch für Geologie und Paläontologie 215:365–395.

Sereno, P. C., and F. E. Novas. 1992. The complete skull and skeleton of an early dinosaur. Science 258:1137–1140.

Shipman, P. 1998. Taking Wing: *Archaeopteryx* and the Evolution of Bird Flight. Simon and Schuster, New York, 336 pp.

Speakman, J. R. 1993. Flight capabilities in *Archaeopteryx*. Evolution 47:336–340.

Speakman, J. R., and S. C. Thomson. 1994. Flight capabilities of *Archaeopteryx*. Nature 370:514.

Stegmann, B. 1937. Über die Flügelhaltung von *Archaeornis* in der Ruhestellung. Ornithologische Monatsberichte 45:192–195.

Stephan, B. 1987. Urvögel. Ziemsen Verlag, Wittenberg Lutherstadt, 216 pp.

———. 1994. Die Orientierung der Fingerkrallen der Vögel. Journal für Ornithologie 135:1–16.

Swinton, W. E. 1975. Fossil Birds. British Museum (Natural History), London, 81 pp.

Sy, M. 1936. Funktionell-anatomische Untersuchungen am Vogelflügel. Journal für Ornithologie 84:199–296.

Tarsitano, S. 1991. *Archaeopteryx:* quo vadis?; pp. 541–576 *in* H. P. Schultze and L. Trueb (eds.), Origins of the Higher Groups of Tetrapods. Comstock, Ithaca, N.Y.

Thulborn, R. A., and T. L. Hamley. 1982. The reptilian relationships of *Archaeopteryx*. Australian Journal of Zoology 30:611–634.

———. 1985. A new palaeoecological role for *Archaeopteryx*; pp. 81–89 *in* M. K. Hecht, J. H. Ostrom, G. Viohl, and P. Wellnhofer (eds.), The Beginnings of Birds. Freunde des Jura-Museums, Eichstätt.

Vazquez, R. J. 1992. Functional osteology of the avian wrist and the evolution of flapping flight. Journal of Morphology 211:259–268.

———. 1994. The automating skeletal and muscular mechanisms of the avian wing (Aves). Zoomorphology 114:59–71.

Viohl, G. 1990. Solnhofen Lithographic Limestones; pp. 285–289 *in* D. E. G. Briggs and P. R. Crowther (eds.), Palaeobiology/A Synthesis. Blackwell, Oxford.

von Meyer, H. 1861. *Archaeopteryx lithographica* (Vogel-Feder) und *Pterodactylus* von Solnhofen. Neues Jahrbuch für Mineralogie, Geologie und Palaeontologie 1861:678–679.

———. 1862. *Archaeopteryx lithographica* from the Lithographic Slate of Solnhofen. Annals and Magazine of Natural History 9 (ser. 3):366–370.

Wagner, G. P., and J. A. Gauthier. 1999. 1,2,3 = 2,3,4: a solution to the problem of the homology of the digits in the avian hand. Proceedings of the National Academy of Sciences USA 96:5111–5116.

Walker, A. D. 1980. The pelvis of *Archaeopteryx*. Geological Magazine 117:595–600.

———. 1985. The braincase of *Archaeopteryx*; pp. 123–134 *in* M. K. Hecht, J. H. Ostrom, G. Viohl, and P. Wellnhofer (eds.), The Beginnings of Birds. Freunde des Jura-Museums, Eichstätt.

Weigert, A. 1995. Isolierte Zähne von cf. *Archaeopteryx* sp. aus dem Oberen Jura der Kohlengrube Guimarota (Portugal). Neues Jahrbuch für Geologie und Paläontologie 1995:562–576.

Wellnhofer, P. 1974. Das fünfte Skelettexemplar von *Archaeopteryx*. Palaeontographica A147:169–216.

———. 1985. Remarks on the digit and pubis problems of *Archaeopteryx*; pp. 113–122 *in* M. K. Hecht, J. H. Ostrom, G.

Viohl, and P. Wellnhofer (eds.), The Beginnings of Birds. Freunde des Jura-Museums, Eichstätt.

————. 1988. Ein neues Exemplar von *Archaeopteryx*. Archaeopteryx 6:1–30.

————. 1992. A new specimen of *Archaeopteryx* from the Solnhofen Limestone; pp. 3–23 *in* K. E. Campbell (ed.), Papers in Avian Paleontology, Honoring Pierce Brodkorb. Science Series 36. Natural History Museum of Los Angeles County, Los Angeles.

————. 1993. Das siebte Exemplar von *Archaeopteryx* aus den Solnhofener Schichten. Archaeopteryx 11:1–47.

————. 1995. *Archaeopteryx*/Zur Lebensweise Solnhofener Urvögel. Fossilien 5:296–307.

Werner, C. F. 1963. Schädel-, Gehirn-, und Labyrinthtypen bei den Vögeln. Morphologisches Jahrbuch 104:54–87.

Whetstone, K. N. 1983. Braincase of Mesozoic birds: I. New preparation of the "London" *Archaeopteryx*. Journal of Vertebrate Paleontology 2:439–452.

Witmer, L. M. 1990. The craniofacial air sac system of Mesozoic birds (Aves). Zoological Journal of the Linnean Society 100:327–378.

————. 1997. The evolution of the antorbital cavity of archosaurs: a study in soft tissue reconstruction in the fossil record with an analysis of the function of pneumaticity. Journal of Vertebrate Paleontology 17(suppl. to #1):1–73.

Xu X., Tang Z., and Wang X. 1999. A therizinosauroid dinosaur with integumentary structures from China. Nature 399:350–354.

Yalden, D. W. 1971a. *Archaeopteryx* again. Nature 234:478–479.

————. 1971b. The flying ability of *Archaeopteryx*. Ibis 113:349–356.

————. 1984. What size was *Archaeopteryx?* Zoological Journal of the Linnean Society 82:177–188.

————. 1985. Forelimb function in *Archaeopteryx*; pp. 91–97 *in* M. K. Hecht, J. H. Ostrom, G. Viohl, and P. Wellnhofer (eds.), The Beginnings of Birds. Freunde des Jura-Museums, Eichstätt.

————. 1997. Climbing *Archaeopteryx*. Archaeopteryx 15:107–108.

Zhou Z. 1995. New understanding of the evolution of the limb and girdle elements in early birds—evidences from Chinese fossils; pp. 209–214 *in* Sun A. and Wang Y. (eds.), Sixth Symposium on Mesozoic Terrestrial Ecosystems and Biota. China Ocean Press, Beijing.

Zhou Z., and L. D. Martin. 1999. Feathered dinosaur or bird?: a new look at the hand of *Archaeopteryx*. Smithsonian Contributions to Paleobiology 89:289–293.

Zhou Z., and Wang X. 2000. A new species of *Caudipteryx* from the Yixian Formation of Liaoning, Northeastern China. Vertebrata Palasiatica 38:111–127.

Zweers, G. A., and J. C. Vanden Berge. 1997. Birds at geological boundaries. Zoology 100:183–202.

7

The Discovery and Study of Mesozoic Birds in China

ZHOU ZHONGHE AND HOU LIANHAI

 The discovery and study of Mesozoic birds in China are relatively recent events, with the first find reported only in 1984 (Hou and Liu, 1984). Other areas, such as Spain and Argentina, have also recently yielded important new discoveries, but since 1990 more Mesozoic birds have been unearthed in China than anywhere else. The findings from China excel not only in abundance, diversity, and completeness but also in the controversies they have inspired.

The extensive distribution of continental Mesozoic deposits is characteristic of the distinctive geological history of China. Chinese continental Jurassic and Cretaceous sequences are probably more complete than anywhere else. The Late Jurassic–Early Cretaceous sediments from which all the Chinese Mesozoic birds are found are all deposited in lacustrine environments (Hou et al., 1995a). Diverse vertebrates, including fishes, amphibians, reptiles, and mammals, have also been recovered from these deposits (Zhou, 1992; Jin et al., 1993, 1995a; Lu, 1994; Zhang et al., 1994; Wang et al., 1995; Hu et al., 1997; Chen et al., 1998; Ji et al., 1998; Xu et al., 1999). Many of the fishes (e.g., *Lycoptera*, *Peipiaosteus*) and reptiles (e.g., *Psittacosaurus*) are endemic to the East and Central Asia region, although others may have a wider distribution (Chen, 1988).

One long-standing geological problem has been the age of these freshwater deposits (Wang, 1990; Swisher et al., 1999). This is largely due to a lack of direct correlation with corresponding marine deposits in Europe. Though some progress has been made in recent years through the study of fossils from continental sediments interbedded with marine sedimentary layers, no consensus has been reached among Chinese geologists. As a result, because of the complex geological phenomena and insufficient dating information, we are presented with a situation of having discovered very important fossils but with incomplete knowledge of their precise geological age.

Historical Review

The first discovered Mesozoic bird from China, *Gansus yumenensis*, was collected by the late paleoichthyologist Liu Zhichen from Early Cretaceous mudstone of Gansu Province, northwest China, in August 1981. The material of this bird includes only a distal tibiotarsus articulated with a complete tarsometatarsus and pedal phalanges. Associated with this bird are abundant teleost fishes. Hou and Liu (1984) reported this finding and regarded it as ancestral to shorebirds (i.e., Charadriiformes).

The second Mesozoic bird from China came from the Early Cretaceous shale of the Jiufotang Formation. This bird was discovered by a farmer and amateur fossil collector, Yuan Zhiyou, from an outcrop near his house in Liaoning Province, northeast China, in 1987. The specimen is nearly complete and is housed in the Beijing Natural History Museum. Sereno and Rao (1992) described this material under the name *Sinornis santensis* (see also Sereno, Rao, and Li, Chapter 8 in this volume) and placed it in a phylogenetic position between *Archaeopteryx* and all later birds. Despite efforts to find more specimens of *Sinornis* from the same site, none has yet been obtained, and the poor outcrops prohibit further efficient excavation. Associated with *Sinornis* are abundant teleost fishes.

Another important discovery of Mesozoic birds in China was made by the senior author in September 1990 at the Boluochi site, in Liaoning Province. He recovered three partial skeletons at that time; two of them were subsequently named *Cathayornis yandica* (Zhou et al., 1992) and *Boluochia zhengi* (Zhou, 1995a). The exposure (shales and mudstone) is also of the lacustrine Early Cretaceous Jiufotang Formation. This site is only a few kilometers from the locality of *Sinornis*. Since its discovery, a number of productive excavations on various scales have been made, yielding nearly 30 more bird specimens. One of the most significant results was the discovery of *Chaoyangia beishanensis* (Hou and Zhang, 1993), a bird of more modern appearance that is clearly distinguishable from all other birds from the same site, birds that were subsequently referred to Enantiornithes. The holotype of *Chaoyangia* was luckily saved by the second author from the rock debris that was going to be buried. This site has now proven to be one of the most prolific Early Cretaceous bird localities in the world. Associated with these birds are various teleosts, *Peipiaosteus, Protopsephurus,* and *Sinamia* (Zhou, 1992; Lu, 1994; Jin et al., 1995b).

In November 1993, paleomammologist Hu Yaoming and the senior author were shown a new fossil bird by an amateur fossil collector, Zhang He, at his home in Jinzhou, Liaoning Province. Zhang had bought this specimen in a local fossil and antique market. This fossil later became the holotype of *Confuciusornis sanctus* (Hou et al., 1995b,c). In December 1993, the senior author was notified by a farmer, Yang Yushan, of a bird with feathers attached to the leg bones, which turned out to be another specimen of *Confuciusornis.* These two specimens came from the same locality of the Late Jurassic–Early Cretaceous lacustrine Yixian Formation in Beipiao, Liaoning Province (Hou et al., 1995c). Since then, several hundred specimens of *Confuciusornis* have been recovered from virtually one locality of the Yixian Formation at Sihetun (Chiappe et al., 1999). Owing to the richness of this fossil locality and the commercial value of these excellently preserved fossils (insects, fishes, and birds), this region produced an unusually large number of specimens. Associated with *Confuciusornis* are abundant plants, insects, conchostracheans, ostracods, fishes (including *Peipiaosteus* and *Lycoptera*), amphibians, various reptiles (turtles, lizards, theropod dinosaurs, etc.), and symmetrodontan mammals (Hu et al., 1997). In 1996, another equally important avian find was discovered at Sihetun. This bird, named *Liaoningornis longidigitris* (Hou et al., 1996; Hou, 1997a), is based on a nearly complete pelvic limb and a keeled sternum. In 1996, a larger feathered animal from the same site was briefly described by Ji Qiang and Ji Shuan (1997). They named this animal *Protarchaeopteryx robusta* and suggested a close relationship with *Archaeopteryx.* Subsequent research on this and other feathered specimens from the Yixian Formation has suggested that some of these specimens may not be birds but feathered theropod dinosaurs (Xu et al., 1999a,b, 2000; Zhou and Wang, 2000a,b). The compsognathid theropod *Sinosauropteryx prima*was also collected from this site (Ji and Ji, 1996; Chen et al., 1998).

Another important Early Cretaceous bird was discovered in Otoge of Inner Mongolia by the Sino-Canadian dinosaur expedition in 1990. Although it was first misidentified as a pterosaur, it was described as an enantiornithine bird by Dong (1993) and named *Otogornis genghisi* by Hou (1994). Though incomplete, the preservation is one of the best of all known Chinese Early Cretaceous birds. The material has been carefully prepared from both sides, exposing a nearly three-dimensional skeleton.

The Early Cretaceous bird locality in Fenglin County, Hebei Province, was discovered in 1993 by farmers. At least three nearly complete and excellently preserved specimens have been recovered. None of them has been described in detail. Together with the birds are abundant sturgeon fossils as well as teleosts (Jin et al., 1995a). To our knowledge no extensive excavation has been made at this site, mainly because of the limited outcrop and the relatively low commercial value deriving from the fish and insect fossils. The age of these fossil-bearing lacustrine deposits is also controversial. According to some geologists (Wang Shien and Jin Fan, pers. comm.) the strata could be Late Jurassic or the lowest part of the Early Cretaceous.

Two Early Cretaceous sites from Shandong Province and Ningxia Autonomous Region yielded feather impressions. The site in Shandong Province was found by Zhang Junfeng, a paleoentomologist from the Shandong Museum, in 1992. The feathers were collected from the lacustrine shale associated with teleosts and abundant insects. Though tremendous efforts were made by Zhang and us after the feather was discovered, no avian skeletal fossils have been collected from this site. The site in Ningxia was discovered by the senior author when he was collecting teleosts with his colleague Zhang Jiangyong, a paleoichthyologist. The deposits are hard black lacustrine shale. Because of extremely difficult transportation and excavation conditions, the feasibility of further collection is doubtful.

Since preparation of this chapter, which summarizes the Mesozoic record of Chinese birds available up to the late 1990s, many other Early Cretaceous birds have been found in China. Although some of them have already been preliminarily described (e.g., *Changchengornis hengdaoziensis* [Ji et al., 1999], *Eoenantiornis buhleri* [Hou et al., 1999], *Protopteryx fengningensis* [Zhang and Zhou, 2000], *Longipteryx chaoyangensis* [Zhang et al., 2001], *Yanornis martini* and *Yixianornis grabaui* [Zhou and Zhang, 2001]), limitations in time prevented their inclusion in this review—the reader is referred to the original literature cited earlier. At least 11 unquestionable genera of birds have been described from the Early Cretaceous of China, 7 of which are enantiornithine birds (Chiappe and Walker, Chapter 11 in this volume). They are *Sinornis, Cathayornis, Boluochia, Otogornis, Eoenantiornis, Protopteryx,* and *Longipteryx.* The enantiornithine birds are dominant not only in diversity but also in the absolute number of specimens known from Early Cretaceous deposits in China. Furthermore, some of the undescribed birds from Hebei and Liaoning Provinces may also be enantiornithines.

In summary, discoveries of Late Jurassic–Early Cretaceous birds in China have been overwhelming in recent

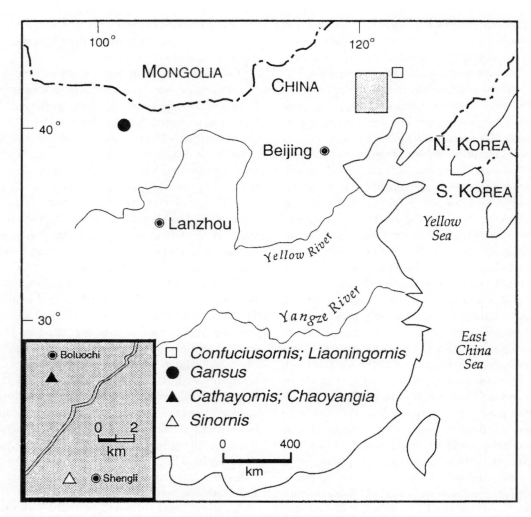

Figure 7.1. The distribution of some of the most important Mesozoic birds in China.

years (Fig. 7.1). Though most of them have been named, many have yet to be described in detail. Many of the specimens are not yet properly prepared. All Mesozoic bird fossils in China were recovered from lacustrine deposits (shale or mudstone), and their preservation is highly variable. Most birds were found with abundant fishes, invertebrates, and plants.

Most of the sites are believed to be Early Cretaceous, though there is disagreement on their precise age. The sites that produced *Confuciusornis, Liaoningornis,* and *Protarchaeopteryx* are regarded by some geologists as Late Jurassic, largely based on the K-Ar and Rb-Sr dating (Wang and Diao, 1984); this age is supported by conchostracheans and bivalves (Chen, 1988; Wang, 1990). Other geologists proposed an Early Cretaceous age (Pu and Wu, 1985; Li et al., 1994) for these sites. Their conclusions were drawn mainly from the study of pollen, ostracods, and plants as well as from the occurrence of the dinosaur *Psittacosaurus,* which is otherwise known only from Early Cretaceous rocks

of Mongolia. Recent Ar-Ar dating of the Sihetun basalts by a joint Canadian-Chinese team indicated an even younger date (Smith et al., 1995). This team proposed an age of about 120 million years for the Yixian Formation. Although the validity of these dates depend on whether it can be demonstrated that the samples used are nonintrusive (intrusion is common in the area), comparable ages were obtained by Swisher et al. (1999), who dated the tuffaceous sediments. Although the age of the Yixian Formation is still controversial, cumulative evidence is leaning more toward an Early Cretaceous age.

Finally, we should point out that, although trading and exporting of important fossil vertebrates are forbidden by Chinese law, hundreds of specimens of *Confuciusornis* and other birds from the Yixian Formation in Liaoning Province have been smuggled out of China in recent years. The upside is that many excellently preserved birds and other important vertebrates have been discovered at an unprecedented rate. The downside is that the locality where these birds came from has suffered an unbridled exploitation in an unscientific way. Much stratigraphic information unfortunately has been lost during the excavation and smuggling process. Many specimens have been artistically assembled and coupled with faked feathers and bones. Worse, most of these illegally exported specimens went to private collections with no assured access by the scientific community. At the Fourth International Meeting of the Society of Avian Paleontology and Evolution (SAPE) in Washington, D.C., in 1996, the participants signed a letter to Professor Zhou Guangzhao, then president of the Chinese Academy of Sciences, appealing to the Chinese government for more effective efforts to protect the *Confuciusornis* locality, arguably one of the most important paleontological localities in the world. Fortunately, this locality is now largely under scientific control.

Systematic Paleontology

Taxonomic Hierarchy

Aves Linnaeus, 1758
  Confuciusornithidae Hou et al., 1995
    Confuciusornis Hou et al., 1995
      C. sanctus Hou et al., 1995

Holotype—Skull, humerus, ulna, radius, and manus. Institute of Vertebrate Paleontology and Paleoanthropology (IVPP) V10918, Beijing.

Referred specimens—IVPP V10895, 10919–10925, and many other specimens housed in various Chinese and foreign institutions (see Chiappe et al., 1999).

Locality and horizon—Huanghuagou and Sihetun, Shangyuan, Beipiao, Liaoning Province, northeast China; Yixian Formation (Late Jurassic–Early Cretaceous).

Diagnosis—Teeth totally lost. Presence of horny beak. Postorbital robust, Y-shaped, and in contact with jugal bone and squamosal. Orbit large. Dorsal process of maxilla and nasal separating naris and small antorbital fenestra. Quadratojugal Y-shaped, with slender ascending process. Quadrate high with no orbital process developed. Dentary branched caudally. Presence of a large rostral mandibular fenestra and a small, rounded caudal mandibular fenestra. Centrum of dorsal vertebrae with deep lateral depressions. Synsacrum composed of seven vertebrae. Pygostyle composed of eight or nine vertebrae. Sternum without keel. Gastralia present. Coracoid short and fused with scapula. Furcula robust, lacking hypocleideum. Humerus large at the proximal end with a developed deltopectoral crest and an elliptical fenestra or depression. Humerus slightly longer than ulna. Semilunate bone fused with major metacarpal. Ulnare slender. Major and minor metacarpals of nearly equal length. Minor metacarpal reduced and narrow proximally. Manual phalangeal formula 2-3-4-*x*-*x*. First phalanx

of minor digit short. Ungual phalanges large and curved in alular and minor digits but small in major digit. Major digit and minor digit of nearly equal length. Ischium with a high, strutlike proximodorsal process. Fibula about three-fourths as long as tibiotarsus. Metatarsal V about one-third the length of metatarsal III.

Enantiornithes Walker, 1981
  Boluochia Zhou, 1995
    B. zhengi Zhou, 1995

Holotype—A pair of nearly complete pelvic limbs, a pygostyle, incomplete skull, sternum, vertebrae, and pelvic girdle. IVPP V9770, Beijing.

Locality and horizon—Xidagou, Boluochi, Chaoyang, Liaoning Province, northeast China; Jiufotang Formation (Early Cretaceous).

Diagnosis—Teeth absent on premaxilla but present on dentary. Premaxilla hook-shaped. Sternum caudally notched with a low keel distributed caudally. Pygostyle longer than tarsometatarsus. Pubis retroverted and strongly curved. Intercondylar groove narrow at the distal tibiotarsus. Medial condyle of tibiotarsus nearly as wide as lateral condyle, and cranial margin of the medial condyle flat in distal view. Metatarsals II–IV of nearly equal length. Trochlea of metatarsal II wider than those of metatarsals III and IV. Trochlea of metatarsals II–IV on approximately the same plane. Pedal claws long and curved.

Cathayornithidae Zhou et al., 1992

Diagnosis—Teeth on premaxilla inclined cranially. Dentary about half the length of skull. Presence of elongate median groove on dorsal major and minor metacarpals. First phalanx of alular digit extremely slender. Pubis slender and curved. Pubic foot small, triangular, and pointed dorsally. Ischium thin, distally pointed, and curved dorsally. Trochlea of metatarsal II slightly higher than that of III but slightly lower than that of IV. Presence of a triangular dorsal ridge at the proximal half of pygostyle.

Sinornis Sereno and Rao, 1992
  S. santensis Sereno and Rao, 1992

Holotype—A nearly complete individual. Beijing Natural History Museum (BPV) 538A (part) and 538B (counterpart).

Locality and horizon—Huanghuagou, Shengli, Chaoyang, Liaoning Province, northeast China; Jiufotang Formation (Early Cretaceous).

Diagnosis—Slightly smaller than *Cathayornis*. Pygostyle short (pygostyle 0.55 of femoral length). Wing claws on alular and major digits of equal length. Phalanx of minor digit curved and long relative to *Cathayornis*. Antitrochanter present. Ulnare heart-shaped with a small metacarpal incision.

Cathayornis Zhou et al., 1992

C. yandica Zhou et al., 1992

Holotype—A nearly complete individual. V9769A, B (part and counterpart) (IVPP), Beijing.

Referred specimens—IVPP V9936, 10896.

Locality and horizon—Xidagou, Boluochi, Chaoyang, Liaoning Province, northeast China; Jiufotang Formation (Early Cretaceous).

Diagnosis—Humerus nearly as long as ulna. Wing claw on alular digit slightly longer than on major digit. Phalanx of minor digit expanded at proximal end and about half as long as first phalanx of major digit. Antitrochanter absent. Acetabular foramen craniodorsally bordered by a high ridge. Ulnare with a large, shallow metacarpal incision. Sereno, Rao, and Li (Chapter 8 in this volume) argue for the synonymy of *C. yandica* and *S. santensis*.

Otogornis Hou, 1994

O. genghisi Hou, 1994

Holotype—A pair of scapulae and coracoids, forelimb, and feather impression. V9607 (IVPP), Beijing.

Locality and horizon—Chaibu-Sumi, Otog-qi, Yikezhao-meng, Inner Mongolia, China; Yijingholuo Formation (Early Cretaceous).

Diagnosis—Ulna about 120% as long as humerus. Ulna nearly twice as wide as radius. Medial margin of scapula straight. Coracoid with a long neck. Kurochkin (2000) supported a close relationship between *Otogornis* and *Ambiortus dementjevi* from the Early Cretaceous of Mongolia, placing both of them within palaeognaths.

Ornithurae Haeckel, 1866

Liaoningornis Hou, 1996

L. longidigitris Hou, 1996

Holotype—incomplete right forelimb, sternum, coracoid, complete right hindlimb, left tarsometatarsus, and pedal phalanges. IVPP V11303, Beijing.

Locality and horizon—Sihetun, Shangyuan, Beipiao, Liaoning Province, northeast China; Yixian Formation (Late Jurassic–Early Cretaceous).

Diagnosis—Smaller than *Cathayornis*. Sternum elongated with a keel extending along its full length. Craniolateral process of sternum long. Xiphoid end of sternum expanded laterally. Sternum broad cranially and progressively narrow caudally, with a pair of large deep notches on both sides. Tarsometatarsus short and about half as long as tibiotarsus. Tarsometatarsus fused at both ends but not along majority of shaft. Claws curved.

Chaoyangia Hou and Zhang, 1993

C. beishanensis Hou and Zhang, 1993

Holotype—Pelvis, femora, tibiotarsus, vertebrae, and ribs with uncinate processes. V9934 (IVPP), Beijing.

Referred specimens—IVPP V9937, V10913 (this specimen has also been used as holotype of *Songlingornis linghensis* [Hou, 1997b], also from the Jiufotang Formation of Boluochi).

Locality and horizon—Xidagou, Boluochi, Chaoyang, Liaoning Province, northeast China; Jiufotang Formation (Early Cretaceous).

Diagnosis—Teeth present on premaxilla and dentary. Sternum elongated, with a keel extending along its full length. Craniolateral process, costal process, and deep facet for articulation with coracoid developed on sternum. A pair of small medial fenestra and short, expanded lateral trabecula present caudally on sternum. Coracoid with procoracoidal process and deep, round scapular facets developed. Furcula U-shaped with hypocleideum lacking. Synsacrum composed of more than eight vertebrae. Uncinate processes on ribs long and slender and ventrally distributed. Caudal process of ilium slightly rounded. Pubic symphysis about one-third as long as pubis. Crest of trochanter high on femur. Cnemial crest present on proximal tibiotarsus. Tarsometatarsus partially fused distally. External rim of trochlea of pedal digit IV well developed. Ungual phalanx of digit I smaller than others of the foot.

Gansus Hou and Liu, 1984

G. yumenensis Hou and Liu, 1984

Holotype—Distal tibiotarsus, complete left tarsometatarsus and foot. IVPP V6862, Beijing.

Locality and horizon—Shenjiawan, Changma, Yumen, Gansu Province, northwest China; Xiagou Formation (Early Cretaceous).

Diagnosis—Lateral condyle of tibiotarsus larger than medial condyle. Trochlea of metatarsal II remarkably higher and projected more plantarly than those of metatarsals III and IV. Tarsometatarsus nearly completely fused. Pedal digit IV longer than III, and both longer than tarsometatarsus. Pedal phalanges long with ungual phalanges relatively short. Ungual phalanges slightly curved. Extensor tubercles on ungual phalanges well developed.

Anatomy

C. sanctus

A detailed description of this bird, based on specimens of the National Geological Museum of China (Beijing) and of several institutions outside China, was given by Chiappe et al. (1999). The present description is based on specimens housed at the Institute of Vertebrate Paleontology and Paleoanthropology (Beijing).

Skull. *Confuciusornis* lacks teeth but has a horny bill (Figs. 7.2, 7.3A). Nasal processes of the premaxillae are long and meet nearly along their full length. The nasal forms a rounded caudal margin of the narial opening. The orbit is large. The frontal is domed. The lacrimal is large and C-shaped. The postorbital bone is robust and Y-shaped. Ventrally, the postorbital is in broad contact with the jugal bone. Caudodorsally, it articulates with the squamosal, and both of them form a complete ventral bar for the upper temporal fenestra. Rostrodorsally, it overlaps the frontal and defines the caudal margin of the orbit. The jugal bone lacks a dorsal process, a process that is characteristic of most diapsid archosaurs and articulates with the postorbital. The jugal bifurcates caudally into the postorbital process and quadratojugal process. This resembles *Archaeopteryx* (Elzanowski and Wellnhofer, 1996). The jugal bone overlaps the maxilla rostrally. The maxilla has a pointed dorsal process that contacts the nasal and forms a septum between the narial opening and a small triangular antorbital fenestra. The quadratojugal is similar to that of *Archaeopteryx*. It is Y-shaped and has a long slender ascending process. It is not clear how this bone articulated with the quadrate. The quadrate also is similar to that of *Archaeopteryx*. It is high and single-headed and lacks the orbital process. The dentary is forked caudally into dorsal and ventral branches. The ventral branch extends caudally nearly to the end of the lower jaw. There is a large rostral mandibular fenestra and a small, rounded caudal mandibular fenestra in the lower jaw. The splenial resembles that of a chicken.

Vertebral Column. There are about eight cervical vertebrae. The thoracic vertebrae bear a high spinal crest; the centra are pleurocoelous with deep excavations on both sides. None of the ribs have uncinate processes. The synsacrum is well fused and composed of seven vertebrae. The

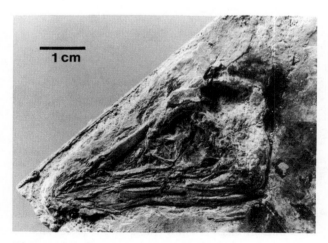

Figure 7.2. Skull of *C. sanctus*, part of the holotype from Liaoning, northeast China. IVPP Collection V 10918.

transverse processes of the vertebrae become longer and more robust caudally. The pygostyle is composed of eight or nine fused vertebrae. In adult specimens, the pygostyle is so well fused that the sutures between vertebrae are hardly observable. The pygostyle has a long, high lamina with a long base. It also has a long transverse ridge (Fig. 7.3D). There are about 10 pairs of gastralia caudal to the sternum.

Thoracic Girdle. The thoracic girdle is relatively primitive. The furcula is similar to that of *Archaeopteryx*. It is robust and nearly U-shaped; it has no hypocleideum (Fig. 7.3C). The furcula has a pair of facets for articulation with the acromion of the scapula. The scapula and the coracoid seem to be fused with each other. The angle between the scapula and coracoid is about 90°. The scapula is much longer than the coracoid. The scapular blade is rodlike. The coracoid is probably different from that of other birds in that it has a very expanded proximal end. The glenoid facet for the articulation of the humerus faces dorsolaterally. The sternum (Fig. 7.3B) is flat and longitudinally short, and its main body is nearly round, a condition similar to that of *Cathayornis, Concornis,* and ratites (Zhou, 1995c). *Confuciusornis* also has caudal notches on the sternum, a significant difference from that of *Archaeopteryx* (Wellnhofer, 1993). The keel is absent.

Thoracic Limb. The humerus has an extraordinarily expanded proximal end with a rounded proximal margin. It has a large deltopectoral crest and a unique ellipsoid fenestra or depression (Figs. 7.3F, 7.4). The capital incision is shallow. The humerus is slightly longer than the ulna and the radius, which is primitive, as in *Archaeopteryx* (Wellnhofer, 1993; Zhou, 1995b). The ulna is significantly wider than the radius. The semilunate carpal is centered on and exclusively fused with the major metacarpal. The ulnare is slender and rodlike and does not conform to the V shape typically seen in neornithines. The metacarpal incision of the ulnare is shallower than in both *Cathayornis* and *Sinornis*. The radiale is about as large as the ulnare. There is an additional small carpal bone lying proximal to the minor metacarpal, as in *Archaeopteryx*. The major and minor metacarpals are fused proximally, although the suture between them can be recognized. The major metacarpal is more robust than the minor metacarpal. The minor metacarpal is curved and narrow throughout the proximal half, a feature that is unknown in other birds. The major metacarpal extends past the minor metacarpal distally in contrast to the condition in enantiornithines. The alular metacarpal is short and robust. The number of digits of the hand of *Confuciusornis* is unreduced. The phalangeal formula is 2-3-4-*x*-*x*. The major digit is the most robust among the three of the hand. The first phalanx of the major digit is the most robust and has a curved, caudally expanded margin. The second phalanx of

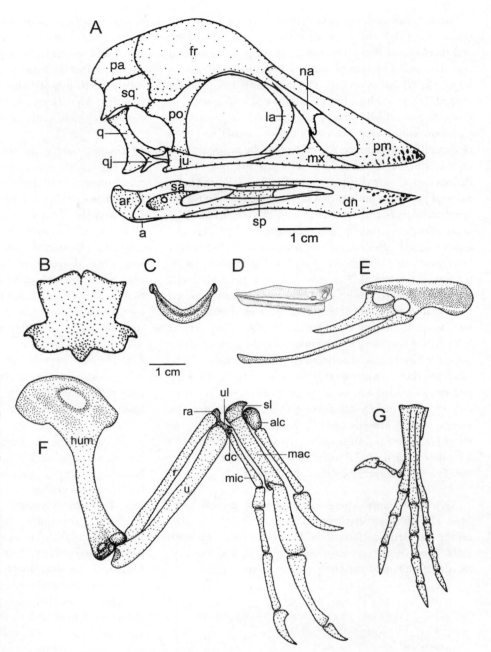

Figure 7.3. *C. sanctus.* Reconstruction of the skull in right lateral view (A), sternum in ventral view (B), furcula in ventral view (C), pygostyle in lateral view (D), pelvis in right lateral view (E), left thoracic limb in ventral view (F), and left foot in dorsal view (G). Abbreviations: a, angular; alc, alular carpal; ar, articular; dc, distal carpal; dn, dentary; fr, frontal; hum, humerus; ju, jugal; la, lachrymal; mac, major metacarpal; mic, minor metacarpal; mx, maxilla; na, nasal; pa, parietal; pm, premaxilla; po, postorbital; q, quadrate; qj, quadratojugal; r, radius, ra, radiale; sa, surangular; sl, semilunate bone; sp, splenial; sq, squamosal; u, ulna; ul, ulnare.

the major digit is also curved and with a convex caudal margin. The first phalanx of the minor digit is significantly shorter than the others. The first phalanx of the alular digit is long and robust compared with those of *Cathayornis* and *Sinornis*. Another distinctive character of *Confuciusornis* is that the ungual phalanges on the alular and major digits are usually extremely large, sharp, and strongly curved, while the one on the minor digit is significantly smaller. This probably indicates that the major digit was exclusively used for flight while the other two digits were still important for climbing.

Pelvic Girdle. The pelvis is primitive in retaining a pubic foot, which is attached to its mate medially but probably not fused (Fig. 7.3E). The pubis is remarkably retroverted, similar to the condition in *Cathayornis*. The ischium is robust and about half as long as the pubis. It has a dorsal process that contacts the ilium nearly at its caudal end. The ilium is blade-shaped and tapers caudally. The ilium is similar to those of *Archaeopteryx* and enantiornithines in that the ilium cranial to the acetabulum is large and expanded whereas the ilium caudal to the acetabulum is short and reduced.

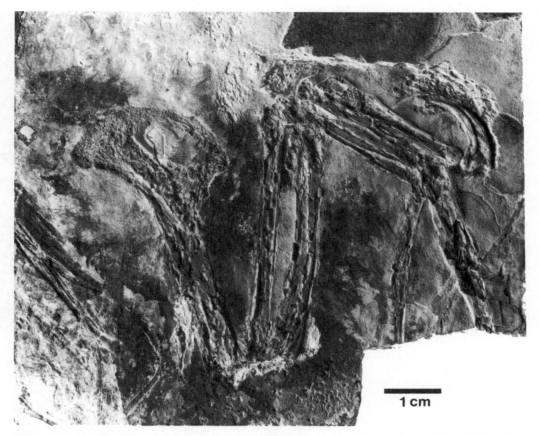

Figure 7.4. Thoracic limb of *C. sanctus,* part of the holotype from Liaoning, northeast China. IVPP Collection V 10918.

Pelvic Limb. The femur is slightly curved and is slightly shorter than the humerus. The tibiotarsus is slender and straight. The fibula does not extend to the distal end of the tibiotarsus and becomes significantly reduced distally. The tarsometatarsus is relatively short and fused only proximally. Metatarsal V (Fig. 7.3G) is present and is about one-third the length of metatarsal III. It is relatively short in *Archaeopteryx,* where it is about one-fourth as long as metatarsal III (Zhou, 1995b). The first metatarsal articulates with the second at a higher position than in *Sinornis* or *Cathayornis.* Metatarsals II–IV are similar in width. The trochlea of metatarsal II is higher than that of metatarsal IV, and the trochlea of metatarsal III is the lowest. The phalanges and claws are typical of perching birds (Fig. 7.4), though the position of the hallux is not.

Feathers. Finally, many feathers have been excellently preserved. The flight feathers are highly asymmetrical. No alula or bastard wing has been recognized. It was probably not developed in *Confuciusornis* considering that abundant specimens with delicate preservation of feathers have been discovered and that the alular digit is unreduced with a large and curved claw. Tail feathers vary significantly, probably

because of sexual dimorphism. Those with a pair of long tail feathers are probably males. Most of the specimens do not have long paired tail feathers and probably are females.

Enantiornithes

A detailed description of *Sinornis* is provided by Sereno, Rao, and Li (Chapter 8 in this volume); again, they regarded *Cathayornis* as synonymous with *Sinornis,* but we prefer to retain them, at least provisionally, as separate taxa, partly because they come from different localities.

Cathayornis. *Cathayornis* may be the only indisputably Early Cretaceous bird that is represented by more than a dozen individuals, many of which are preserved as nearly complete skeletons (Figs. 7.5–7.7). Like nearly all known Early Cretaceous birds, *Cathayornis* is smaller than *Archaeopteryx.*

Skull. The skull is similar to that of *Archaeopteryx* in that it retains teeth in the premaxilla, maxilla, and dentary. Four teeth are on the premaxilla, and they incline slightly rostrally. The teeth closely resemble those of *Ichthyornis* (Martin and Zhou, 1997). The premaxillae are fused rostrally. The nasal process of the premaxilla is dorsoventrally com-

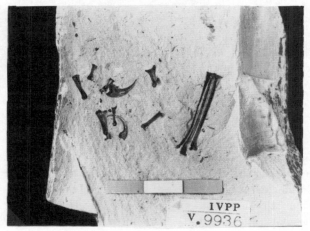

Figure 7.5. *C. yandica,* an Early Cretaceous enantiornithine bird. Upper: holotype, IVPP Collection V 9769; Lower: tarsometatarsus and toes. IVPP Collection V 9936. Scale bar = 1 cm.

pressed, especially at the caudal end. The nasals meet in the midline along their full length and are primitive in having a very short maxillary process. In dorsal view, a depression is present between the rostral parts of the two nasals. We suggest that cranial kinesis (e.g., rhynchokinesis) was present in *Cathayornis.* The frontal is large, the parietal small. It is unclear whether the postorbital is present. The dentary is about half as long as the skull. The splenial is triangular (Fig. 7.6).

Vertebral Column. The number of cervical vertebrae in *Cathayornis* is uncertain, but probably there were no more than 10. Cervical centra are neither amphicoelous nor heterocoelous and probably represent a transitional form be-

tween these two conditions (Zhou, 1995c). The thoracic vertebrae have well-developed pleurocoels. The parapophysis is positioned in the middle of each centrum. The synsacrum is composed of eight vertebrae. The transverse processes of the last three vertebrae are more robust and longer than those of the rest of the synsacrum; the transverse processes of the last two vertebrae are united distally (Fig. 7.6). There are probably fewer than eight caudal vertebrae. The pygostyle is long, with a long, low lamina. Uncinate processes and gastralia are absent.

Thoracic Girdle. The thoracic girdle is more advanced than those of *Archaeopteryx* and *Confuciusornis.* Among the derived characters in this structure, many are also charac-

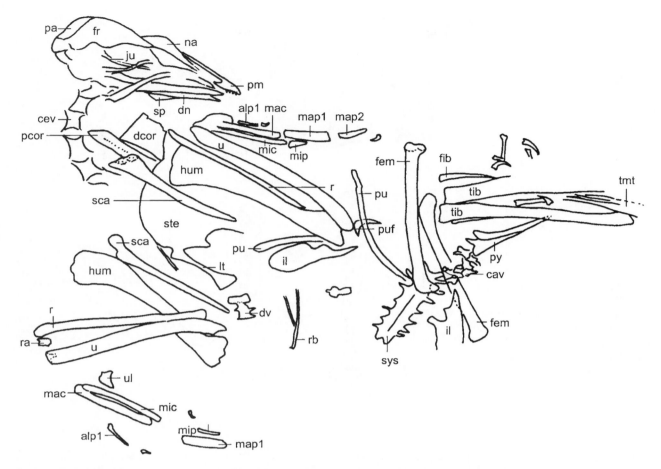

Figure 7.6. *C. yandica*. Line drawing of the holotype. Abbreviations: alp1, first phalanx of alular digit; cav, caudal vertebra; cev, cervical vertebra; dcor, distal coracoid; dv, dorsal vertebra; fem, femur; fib, fibula; il, ilium; ju, jugal bar; lt, lateral trabecula of sternum; map1, first phalanx of major digit; map2, second phalanx of major digit; mip, phalanx of minor digit; pcor, proximal coracoid; pu, pubis; puf, pubic foot; py, pygostyle; rb, rib; sca, scapula; ste, sternum; sys, synsacrum; tib, tibiotarsus; tmt, tarsometatarsus; others as in Figure 7.3. Scale refers to Figure 7.5.

teristic of other enantiornithine birds (Chiappe and Walker, Chapter 11 in this volume). The furcula is peculiar in possessing a V-shaped main body that contains a small U-shaped base near the symphysis of the two clavicles (Fig. 7.7C). It also has a distinct, long, and slender hypocleideum. Obviously, the furcula was laterally compressible during flight, though the degree of movement may be limited compared with ornithurine birds (Jenkins et al., 1988). The scapula has a straplike blade with a straight lateral margin. The articulating facet for the humerus faces dorsolaterally. The coracoid is strutlike but lacks a procoracoid process. It bears a protruding process instead of a cotyla for articulation with the scapula (Fig. 7.7A). The dorsal side of the coracoid is concave, especially near the sternal end. The sternal half of the coracoid has a thin and convex lateral margin (Figs. 7.7A, B).

The sternum has two pairs of caudal notches and a pair of long, slender lateral trabeculae with expanded triangular distal extremities (Fig. 7.7D). Lateral trabeculae are parallel to the keel. A pair of intermediate trabeculae are present and are short. The sternum has a pair of rounded cranial margins instead of facets for the articulation of the coracoid. The keel is low and restricted to the caudal portion of the sternum. The keel has V-shaped cranial extensions that form low ridges (Fig. 7.7D). The sternum is primitive in retaining a rounded main body, somewhat similar to those of ratite birds. The dorsal side of the sternum is concave.

Thoracic Limb. The thoracic limb is basically avian, though many of the features (e.g., carpometacarpus not fused completely, retention of claws) are still primitive compared with extant birds. The humerus is twisted and

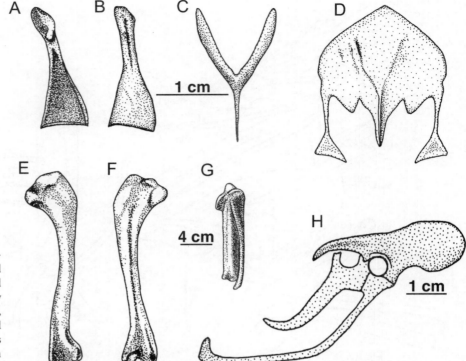

Figure 7.7. *C. yandica.* Reconstruction of the right coracoid in dorsal view (A), left coracoid in ventral view (B), furcula in ventral view (C), sternum in ventral view (D), right humerus in ventral (E) and dorsal (F) views, carpometacarpus in ventral view (G), and pelvis in lateral view (H). A–F same scale.

nearly as long as the ulna. The bicipital crest is bulbous (Fig. 7.7F). The ventral tuberculum is well developed. The dorsal tuberculum is not developed. The capital incision is wide and deep (Fig. 7.7E). Distally, the tricipital groove is present, the flexor process is slender, the dorsal condyle is developed, and the ventral condyle is small. The dorsal epicondyle is well developed, whereas the ventral epicondyle is not.

The radius is about two-thirds as wide as the ulna (Figs. 7.5, 7.6). The ulna is curved, especially at the proximal end. The caudal papillae for remiges are absent. The olecranon is low. The distal end of the ulna is nearly as wide as the midshaft. The dorsal condyle is semilunate in shape, whereas the ventral condyle is not well developed. At the distal end of the ulna there is a longitudinal groove on the dorsal side. The radius is straight and slightly expanded at its proximal end. The bicipital tuberculum is slightly developed.

The manus is shorter than the ulna. The ulnare has a wide and shallow metacarpal incision. The radiale is smaller than the ulnare and is nearly square. The carpometacarpus is fused only at the proximal end (Fig. 7.7G), and the carpal trochlea is well developed and ball-shaped. The extensor process is absent. The minor metacarpal extends distally past the major one, as is typical of enantiornithine birds. The minor metacarpal is curved and extends farther ventrally at the proximal end. This is a typical avian

feature initially developed in *Archaeopteryx* (Zhou and Martin, 1999). The first phalanx of the alular digit is short and very slender. The first phalanx of the major digit is craniocaudally expanded, especially distally. The second phalanx of the major digit is shorter and more slender than the first phalanx. There is only one phalanx left on the minor digit. It is short, expanded proximally, and closely attached to the first phalanx of the major digit (Figs. 7.5, 7.6). Ungual phalanges remain on both alular and major digits even though they are small. This also occurs in *Sinornis* (Sereno and Rao, 1992). The ungual phalanx on the alular digit is slightly longer than that on the major digit.

Pelvic Girdle. The pubis of *Cathayornis* is slender, curved, and remarkably retroverted (Fig. 7.7H). Two pubic feet are united but not fused. The foot is small, triangular, and dorsally pointed as in *Sinornis*. The ischium has a distinct proximodorsal process extending toward the ilium. The ischium forms a thin blade and tapers caudally with the caudal end curved dorsally. The cranial portion of the ilium is significantly larger than the postacetabular process, which tapers toward the caudal end. The three pelvic bones are not fused to each other. An antitrochanter is not present, and the acetabular foramen is craniodorsally bordered by a semicircular high ridge.

Pelvic Limb. The femur lacks the well-developed trochanter that is present in neornithines. The tibiotarsus is long. A cnemial crest and supratendinal bridge are lacking on the tibiotarsus. The intercondylar incision is narrow, and the medial condyle is large. The fibula is reduced and short. The tarsometatarsus is fused only proximally. The fifth metatarsal is lost. Metatarsal I is slender and J-shaped. The articulation of metatarsal I with metatarsal II is relatively lower (i.e., more distal) than in *Liaoningornis* and *Confuciusornis.* Metatarsal IV is slender. The number of distal tarsals fused into the tarsometatarsus is uncertain. The trochlea of metatarsal II is slightly higher than that of III but slightly lower than that of IV. The pedal phalanges are typical of perching birds, with sharp and strongly curved claws.

Boluochia. *Boluochia* (Zhou, 1995a) is an enantiornithine bird that differs remarkably from *Sinornis* and *Cathayornis.* Like other Early Cretaceous enantiornithines, it is small (Fig. 7.8).

Skull. Teeth are absent from the premaxilla, which is strongly curved with a convex cranial margin, probably indicating the presence of a hooked beak (Fig. 7.9). The nasal process of the premaxilla is straight, slender, and slightly dorsoventrally compressed. There is a concave dorsal surface between the nasal process and the cranial portion of the premaxilla. The premaxilla has a maxillary process that tapers caudally. The dentary is incomplete, with a single tooth preserved. The tooth is conical, and the crown has a constricted base.

Vertebral Column. The free caudal vertebrae have a simple structure. The pygostyle is well fused and longer than the tarsometatarsus. It has a long, low lamina.

Sternum. The sternum is incomplete. The low keel is restricted to the caudal portion, and marked caudal notches are recognizable. A pair of lateral trabeculae are slender and parallel to each other.

Pelvic Girdle. The pubis is articulated with the ischium at the proximal end. It is slender, retroverted, and strongly curved. The pubic foot is small, triangular, and dorsally pointed. The ischium is long and blade-shaped and tapers distally. There is a proximodorsal process near the proximal end on the ischium, as in *Cathayornis, Sinornis,* and *Confuciusornis.* The ilium caudal to the acetabulum is reduced as in other enantiornithine birds (Fig. 7.9).

Pelvic Limb. The pelvic limb of *Boluochia* is nearly completely preserved. Many of the features are distinctive from those of other known enantiornithine birds. The head of the femur is rounded and well developed, and the great

trochanter is relatively well developed with a convex lateral margin. The tibiotarsus lacks a cnemial crest. Distally, the tibiotarsus has a narrow intercondylar fossa. The medial condyle is only slightly wider than the lateral one, probably suggesting a more primitive condition than in advanced enantiornithine birds (Zhou, 1995a). The cranial margin of the medial condyle appears fairly flat in distal view, which is characteristic of enantiornithine birds. The tarsometatarsus is fused only proximally. Metatarsal II is relatively straight. Metatarsal IV is slightly curved and narrower, and the midshaft is mediolaterally compressed and strongly constricted. Metatarsal I is J-shaped and articulates with metatarsal II at a low position. The three trochlea of digits II–IV are nearly equal in height. The trochlea of digit II is slightly wider than that of digit III, and both are much wider than that of digit IV. The structures of the three trochlea are all simple, with no significant caudal projections. They lie nearly in the same plane. The ungual phalanges are longer than other phalanges, and the horny claws are strongly curved and sharply pointed distally.

Otogornis. *Otogornis* is another enantiornithine bird from the Early Cretaceous of China (Dong, 1993; Hou, 1994). Although only the pectoral girdle and forelimbs are preserved, they are represented by three-dimensional bones prepared from both sides of the specimen.

Thoracic Girdle. The scapula and coracoid are articulated with each other at an angle less than 90°. The scapula is basically rod-shaped with a straight medial margin. The articulating facet for the humerus is large and elliptical and faces dorsolaterally (Figs. 7.10, 7.11). The coracoid is long and strutlike. It has a long and slender neck. The sternal part of the coracoid has a convex lateral margin and a concave dorsal surface. The acrocoracoidal process is cone-shaped. The procoracoidal process is absent, as in other enantiornithines.

Thoracic Limb. The humerus has an internally slanting head and a flat deltoid crest. The ventral tubercle is well developed. The capital incision is deep and large. The distal end of the humerus is flattened and extends dorsally. Both ventral and dorsal condyles are small. The bicipital crest is bulbous. The ulna is about 120% as long as the humerus. It is nearly twice as wide as the radius. The carpometacarpus is not well preserved, and one wing claw, probably belonging to the alular digit, is as small as in *Cathayornis* and *Sinornis.*

Ornithurae

Liaoningornis. *Liaoningornis* is a small bird and the oldest known ornithurine (Hou et al., 1996).

Thoracic Girdle. The coracoid is not completely preserved. The sternal end is extended medially. The lateral mar-

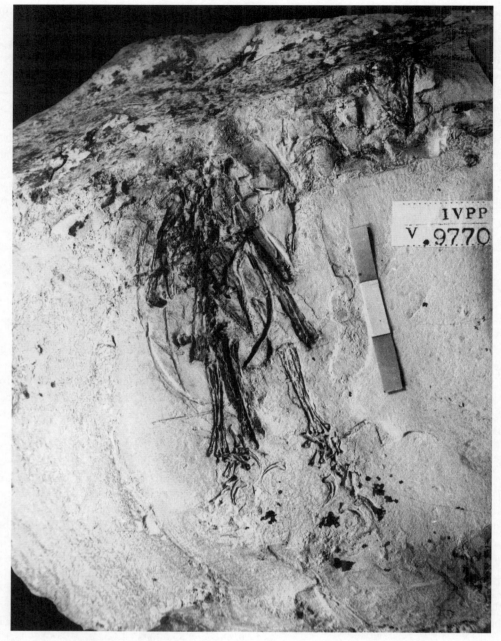

Figure 7.8. *B. zhengi*, an Early Cretaceous enantiornithine bird from Liaoning, China. Holotype, IVPP Collection V 9770. Scale bar = 1 cm.

gin is slightly concave at the sternal end. The lateral process of the coracoid is weakly developed. The sternum is long and has a keel that extends nearly to the front of the sternum (Fig. 7.12). Coracoidal sulci are present on the sternum. A pair of craniolateral processes are long and well developed. The sternum is broad cranially and narrows caudally, where it forms a pair of large lateral notches. The xiphoid end of the sternum is expanded laterally, and the sternal notch is indented.

Thoracic Limb. The humerus is twisted. The ulna is curved and about twice as wide as the radius. The radius is straight.

Pelvic Girdle. The distal pubis is preserved with a small pubic foot. The bone initially described by Hou (1997a) as the scapula may be another incomplete pubis.

Pelvic Limb. The femur is long and slender. The greater trochanter is well developed. The fibula is reduced. The tar-

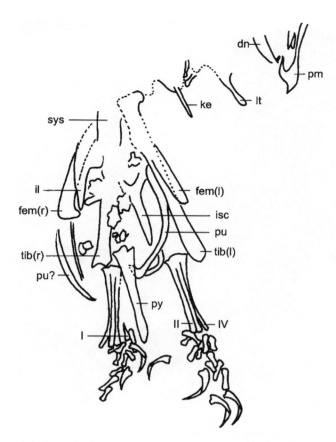

Figure 7.9. *B. zhengi.* Line drawing of the holotype. Abbreviations: dn, dentary; fem(l), left femur; fem(r), right femur; il, ilium; isc, ischium; ke, keel; lt, lateral traberculum of sternum; pm, premaxilla; pu, pubis; pu?, pubis?; py, pygostyle; sys, synsacrum; tib(l), left tibiotarsus; tib(r), right tibiotarsus; I, II, and IV, metatarsals I, II, and IV. Scale bar refers to Figure 7.8.

Thoracic Girdle. The thoracic girdle is generally modern in appearance (Fig. 7.15). The coracoid has a deep, round scapular cotyla and a procoracoidal process. The furcula is U-shaped and lacks a hypocleideum. The sternum is elongated, with a keel extending its full length. The craniolateral process, costal process, and deep facet for articulation with the coracoid are well developed on the sternum. A pair of small medial fenestrae and short, laterally expanded trabeculae are present caudally on the sternum.

Pelvic Girdle. It is uncertain whether the ilium, ischium, and pubis were fused with each other. The pubis is retroverted, and the pubic symphysis is about one-third as long as the pubis. The ischium has a dorsal expansion in the middle part. This expansion differs from the strutlike proximodorsal process found in enantiornithines and *Confuciusornis.* The pubes are expanded and united distally. The ilium caudal to the acetabulum is reduced, and the ilium cranial to the acetabulum is large and slightly rounded cranially.

Pelvic Limb. The crest of the greater trochanter is high and well developed on the femur. The tibiotarsus is not completely known, but the cnemial crest seems to be present. The proximal tarsometatarsus is not preserved, but the distal end is partly fused, although the rest of the preserved three main metatarsals are not. Metatarsal I is long and J-shaped. The trochlea of digit II is slightly higher than that of digit IV, and both of them are higher than that of digit III. The trochlea of digit II and III are about the same width, and they are slightly wider than that of digit IV. The external rim of the trochlea of digit IV is well developed and extends plantarly. Ungual phalanges are slightly curved. The ungual phalanx on digit I is relative small.

sometatarsus is short and broad; it is about half as long as the tibiotarsus. It appears to be fused at both ends but remains unfused along most of its shaft. Metatarsal I is long and J-shaped. Its articulation with metatarsal II is relatively high. The ungual phalanges are sharp and strongly curved.

Chaoyangia. *Chaoyangia* is another small ornithurine bird. It is slightly larger than *Liaoningornis,* which was collected from a lower horizon.

Skull. The skull bones are poorly known. The premaxilla and dentary are toothed. Teeth are typical of those found in other toothed birds; they are conical and constricted at the base of the crown.

Vertebral Column. The synsacrum is composed of more than eight vertebrae. Uncinate processes on the ribs are long and slender and ventrally distributed (Figs. 7.13, 7.14).

Gansus. The only known specimen of *Gansus* includes only a complete left tarsometatarsus and phalanges as well as a distal tibiotarsus articulated with the tarsometatarsus (Fig. 7.16).

Pelvic Limb. The supratendinal bridge is absent. The lateral condyle of the tibiotarsus is significantly larger than the medial condyle. The intercondylar fossa is shallow and somewhat V-shaped. The tarsometatarsus is completely fused. The trochlea of digit II is remarkably higher than those of digits III and IV. The phalangeal formula of this bird is 2-3-4-5-*x*. The phalanges are slender and elongated. The fourth digit is the longest of the four. Both the third and fourth digits are longer than the tarsometatarsus. The ungual phalanges are relatively short and slightly curved. The ungual phalanges, especially those of the second and fourth digits, are distinctive in possessing well-developed extensor tubercles.

Figure 7.10. *O. genghisi*, an Early Cretaceous enanti-ornithine bird from Inner Mongolia, northern China. Holotype, IVPP Collection V 9706.

1.4 cm

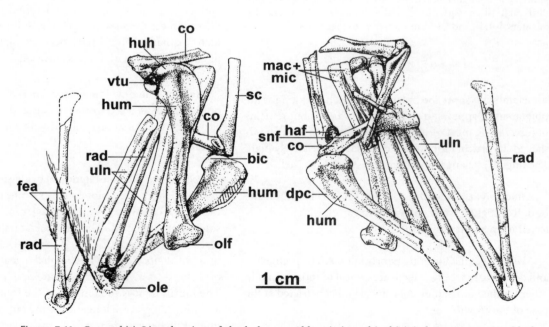

Figure 7.11. *O. genghisi*. Line drawing of the holotype. Abbreviations: bic, bicipital crest; co, coracoid; dpc, deltopectoral crest; fea, feathers; haf, humeral articular facet; huh, humeral head; hum, humerus; mac, major metacarpal; mic, minor metacarpal; ole, olecranon; olf, olecranon fossa; rad, radius; sc, scapula; snf, supra-coracoid nerve foramen; uln, ulna; vtu, ventral tubercle.

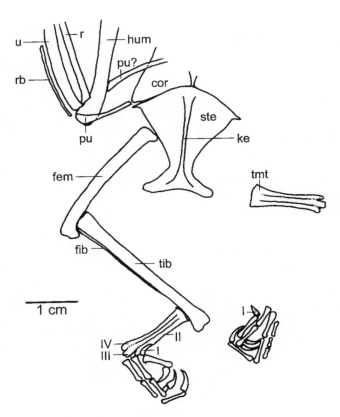

Figure 7.12. *L. longidigitris.* Line drawing of the holotype (IVPP V11303). Abbreviations: cor, coracoid; fem, femur; fib, fibula; hum, humerus; ke, keel; pu, pubis; pu?, pubis?; r, radius; rb, rib; ste, sternum; tib, tibiotarsus; tmt, tarsometatarsus; u, ulna; I–IV, metatarsals I–IV.

Evidence of the Earliest Diversification of Birds

Chinese Mesozoic birds provide evidence for at least two steps in early avian evolution. The first is represented by the Late Jurassic–Early Cretaceous *Confuciusornis* and *Liaoningornis* and one undescribed genus from the Yixian Formation. The second is represented by the indisputably Early Cretaceous birds *Gansus, Chaoyangia, Sinornis, Cathayornis, Otogornis,* and *Boluochia* from the Jiufotang Formation and other formations of similar age. Evidence of a remarkable diversification of birds has been found in both of these stages.

Confuciusornis is probably the next most basal bird after *Archaeopteryx* (*Protarchaeopteryx, Caudipteryx,* and *Sinosauropteryx* are discussed later in this chapter; see also Witmer, Chapter 1 in this volume), even though *Confuciusornis* is also specialized in many aspects and had lost its teeth in favor of a horny beak (for discussion of the avian status of *Mononykus* and other alvarezsaurids, see Zhou, 1995d; Feduccia, 1996; Chiappe et al., 1997; Chiappe, Norell, and Clark, Chapter 4 in this volume; Novas and Pol, Chapter 5 in this volume). It is also close to *Archaeopteryx* in size.

The postorbital was still present in the skull, and it has a wide contact with the jugal bone, which is unknown in other birds (Sanz et al., 1997). The jugal bifurcates caudally into postorbital and quadratojugal processes; in this it resembles *Archaeopteryx.* The postorbital and squamosal are in contact and complete the ventral bar of the supratemporal opening (Fig. 7.3A). The quadratojugal is similar to that of *Archaeopteryx*; it is unfused with the jugal bone and has an ascending process. The sternum is short and lacks a carina (Fig. 7.3B). The scapula and coracoid are fused, a more primitive condition than in other birds. The scapular blade is rodlike. The furcula is robust and probably laterally incompressible. Gastralia are present, as in *Archaeopteryx,* but probably not in any of the other known birds (see also Sereno, Rao, and Li, Chapter 8 in this volume). The wing has unreduced phalanges and claws (Fig. 7.3F). The ischium is robust. The fifth metatarsal is present (Fig. 7.3G). Nearly all these characters are more primitive than in enantiornithines or ornithurines from the Early Cretaceous.

Liaoningornis, which is from the same site and horizon as *Confuciusornis,* has a significantly different appearance. It is much smaller than *Confuciusornis.* The sternum is elongated, with a modern articulation with the coracoid. It also has a pair of craniolateral processes such as are usually seen in neornithines. A keel is well developed and, more important, extends along the full length of the sternum, which indicates a powerful capacity for flight (Fig. 7.12). The ulna is nearly twice as wide as the radius, providing a firm attachment for secondary flight feathers. The tarsometatarsus is short and fused at both ends, indicating improved perching capacity.

Another bird that was also collected from the Yixian Formation is probably the oldest known enantiornithine. This bird is toothed and is different from both *Confuciusornis* and *Liaoningornis,* indicating the presence of a remarkable diversification of birds at this early stage.

By the Early Cretaceous birds began to diversify over all the continents. Enantiornithine birds were probably the most important group of birds at that time (Walker, 1981; Martin, 1983, 1991; Chiappe, 1993, 1995; Hou et al., 1996; Chiappe and Walker, Chapter 11 in this volume). They were dominant not only in number but also in diversity. For instance, among a total of 30 specimens collected from Boluochi, an Early Cretaceous site in Chaoyang, Liaoning Province, only 3 can be referred to the ornithurine bird *Chaoyangia.* All but one of the other specimens belong to either *Cathayornis* or *Boluochia,* both of which are enantiornithine birds. *Chaoyangia* and *Gansus* are the only two ornithurine birds known from the Early Cretaceous of China, but five taxa have been recognized as belonging to the Enantiornithes (*Cathayornis, Sinornis, Boluochia, Otogornis,* and one unnamed genus).

Significant diversification also exists among these enantiornithines. *Cathayornis* has a half-ring-shaped process

Figure 7.13. *C. beishanensis,* an Early Cretaceous ornithurine bird form the Early Cretaceous of Liaoning, China. Holotype, IVPP Collection V 9934.

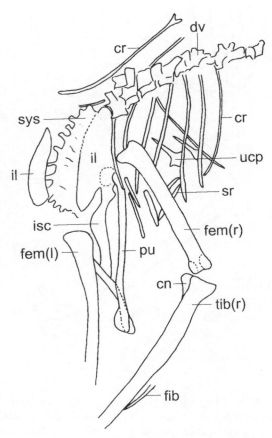

Figure 7.14. *C. beishanensis.* Line drawing of the holotype. Abbreviations: cn, cnemial crest; cr, costal rib; dv, dorsal vertebra; fem(l), left femur; fem(r), right femur; fib, fibula; il, ilium; isc, ischium; pu, pubis; sr, sternal rib; sys, synsacrum; tib(r), right tibiotarsus; ucp, uncinate process. Scale refers to Figure 7.13.

bordering the craniodorsal margin of the acetabulum. The ulnare has a wide metacarpal incision. *Sinornis* has a relatively short pygostyle, a well-developed antitrochanter, and a heart-shaped ulnare with a small metacarpal incision. *Boluochia* has a long pygostyle and lacks teeth on the hooked premaxilla. Trochlea of metatarsals II–IV are nearly of equal length. *Otogornis* has a relatively long ulna and radius compared with the humerus. Its ulna is nearly twice as wide as the radius. The undescribed genus is similar to *Cathayornis* but has a more primitive sternum and coracoids and retains three wing claws.

These enantiornithine birds were probably capable of sustained flight. However, the sternum is relatively short, and the keel is low and restricted to the caudal part of the sternum. There probably were no uncinate processes. The pectoral girdle is primitive compared with that of neornithines. The coracoid lacks the procoracoidal process.

The bones of the pelvic girdle are unfused. The tarsometatarsus is fused only proximally, though in the Late Cretaceous enantiornithine, *Avisaurus gloriae,* the tarsometatarsus has a distal fusion (Varricchio and Chiappe, 1995).

In the ornithurine bird *Chaoyangia,* the thoracic girdle is basically of modern avian appearance. The sternum is typical of that of modern volant birds. The sternum is long, with a pair of coracoidal facets. The keel is deep and extends along the full length of the sternum (Figs. 7.13–7.15). Both the craniolateral process and costal process of the sternum are well developed. The coracoid has a well-developed procoracoidal process as well as a deep, round scapular cotyla. Distally, the coracoid has a well-developed lateral process. Uncinate processes are developed on the ribs. A cnemial crest is present on the tibiotarsus. The tarsometatarsus is partially fused distally. None of these

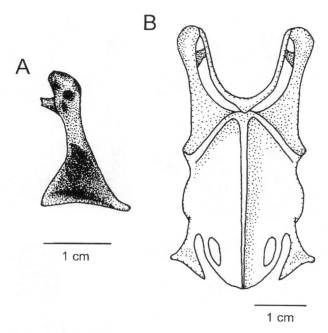

Figure 7.15. *C. beishanensis*. Composite reconstruction of right coracoid in dorsal view (A), sternum, furcula, and coracoids in ventral and cranial view (B).

features is present in enantiornithines of the same age. They also indicate a significantly improved flight capability. In *Gansus*, the tarsometatarsus is nearly completely fused, a condition unknown in contemporaneous enantiornithine birds. Ungual phalanges are equipped with distinctive extensor tubercles that may further indicate water adaptation. *Gansus* is probably the oldest bird known to have aquatic adaptations.

Chaoyangia probably is closely related to the ornithurine bird *Ambiortus* from the Early Cretaceous of Mongolia (Kurochkin, 1985). This relationship is supported by similarities in the structure of the coracoid, furcula (Kurochkin, pers. comm.), and sternum. These birds are members of the endemic Jehol Fauna, which is distributed only in East and Central Asia from the Late Jurassic to Early Cretaceous. According to Kurochkin (pers. comm.), some enantiornithines have also been found from the Early Cretaceous of Mongolia that are similar to *Cathayornis*.

In summary, the major diversification of birds in the Late Jurassic–Cretaceous or Early Cretaceous of China is crucial to our understanding of avian evolution and unexpected only a decade ago. It underscores the complexity of the early evolution of birds, and it may further indicate that *Archaeopteryx* is only one side branch of early avian evolution. Such early avian diversity suggests that bird origin occurred earlier than the Late Jurassic.

Paleobiology

Early Evolution of Body Size and Flight

The discovery of abundant and diverse Mesozoic birds in China also provides clues for exploring the correlation between body size and flight in the early evolution of birds. Size is a characteristic that influences almost all functional processes (Peters, 1983). It is interesting to note that the two most primitive birds known (*Archaeopteryx* and *Confuciusornis*) are close in size. The primitive nature of their sternum, thoracic girdle, and wing demonstrates that neither structure could enable sustained, powerful flight. Both of these birds have strongly curved and well-developed pedal claws. These most likely were used for perching, although the equally developed and curved wing claws might have been adapted to climbing trees (Zhou and Hou, 1998). We have no idea how big the immediate ancestor of birds could have been. However, it seems reasonable to believe that it was larger than the earliest known birds because it would probably have been more efficient to reduce its size to get off the ground whether it initially took off from the ground or jumped from trees. Therefore, we suppose that with the development of the initial flight capacity the oldest known birds had probably already achieved smaller size compared with their immediate ancestors .

The ornithurine bird *Liaoningornis* is contemporaneous with *Confuciusornis*. This bird is, however, much smaller than *Confuciusornis* (Fig. 7.12). *Liaoningornis* possessed an improved flight capacity, as is indicated by features such as a keeled sternum. The small size of *Liaoningornis* is probably the earliest known fossil evidence for the coevolution of improved flight capacity and reduced body size (Zhou and Hou, 1997). One negative impact that accompanied the reduction of body size is the requirement of relatively more energy for the maintenance of body temperature. The study of fossil plants indicates that from the Late Jurassic to Early Cretaceous, the dominant climate in northeastern China was temperate to subtropical. This environment and the abundance of food, including flowering plants, provided the necessary conditions for the transition from large and primitive birds to small and more powerful volant birds.

In the Early Cretaceous, nearly all birds, including the enantiornithines and ornithurines, became relatively small compared with *Archaeopteryx* and *Confuciusornis* (Zhou and Hou, 1998). This is most likely correlated with the improvement of flight capacity in both these groups. This is evidenced by the findings from China and supported by fossil birds from Mongolia, Australia, and Spain (Kurochkin, 1985; Molnar, 1986; Sanz et al., 1988, 1996, 1997; Sanz and Bonaparte, 1992; Sanz and Buscalioni, 1992; see also Sanz et al., Chapter 9 in this volume).

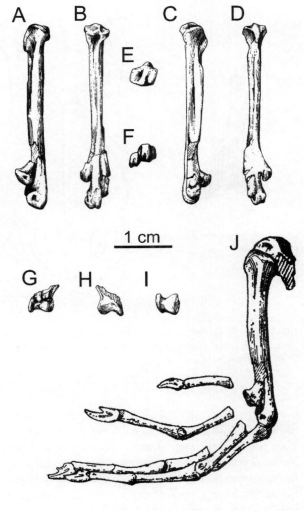

Figure 7.16. *G. yumenensis,* an Early Cretaceous ornithurine bird from Gansu, northwest China. Holotype, IVPP Collection V 6862. Left tarsometatarsus in medial view (A); dorsal view (B); lateral view (C); plantar view (D); proximal view (E); distal view (F). Left tibiotarsus in cranial view (G); caudal view (H); distal view (I). J, specimen in articulation.

1 cm

As the flight mechanism became well established and the modern breathing system evolved, the selection for small size was no longer a factor affecting the evolution of early birds. The Early Cretaceous ornithurine *Chaoyangia* is actually slightly larger than *Liaoningornis* and *Cathayornis* (Figs. 7.5, 7.12, 7.13). *Chaoyangia* has a thoracic girdle of modern appearance. The sternum is long and broad so that it could act as a pump for the air sacs. A modern avian-style lung, permitting a high rate of oxygen consumption during flight, probably already existed (Hou et al., 1996). Uncinate processes and hinged ribs as well as an expansive sternum maintained a proper ventilatory airflow, as in the neornithine lung (Ruben et al., 1997). The procoracoidal process of the coracoid is essential for the formation of the triosseal canal, through which the supracoracoidal tendon passes to elevate the wing (Fig. 7.15). By the Late Cretaceous, a variety of birds, including the ornithurines *Ichthyornis* and *Hesperornis* and a variety of enantiornithines, had achieved large size (Martin, 1991; Chiappe, 1995). There are many reasons that can explain

this increase of body size, but probably all these birds, except those that secondarily became flightless, were better fliers than their ancestors.

It seems that reduction of body size played an important role in the early stages of avian evolution. The enhancement of flight capability was accompanied by the reduction of body size, at least in the very early stages of avian history. The relatively large sizes of *Archaeopteryx* and *Confuciusornis* probably can be regarded as one of the primitive characters of birds. The selection for large size in volant birds would probably have been impossible if the power of flight was still in its initial stages.

Early Ecological Differentiation

It is notable that all the oldest known birds—*Archaeopteryx, Confuciusornis,* and *Liaoningornis*—have often been interpreted as perching birds (Feduccia, 1993, 1996; Hou et al., 1995c, 1996; but see Chiappe, 1995; Hopson, 2001). All these birds are characterized by sharp and strongly curved pedal claws. Digit I is opposed to the rest of the pedal digits, indicating grasping capability. *Archaeopteryx* is most likely the oldest known side branch of avian evolution. Regardless of how and why birds first left the ground, the innovation of perching, combined with the power of flight in these early birds, provided them with a better means of escape and avoidance from the attack of terrestrial predators. Among these potential predators were mammals, lizards, amphibians, and dinosaurs. In addition, food resources available to these birds increased with the extended scope of activity. The discovery of abundant fossil plants (including trees and seeds) associated with early Chinese birds provides another clue to the whole story of the early evolution of flight and perching. It should be pointed out that the perching capability of both *Confuciusornis* and *Liaoningornis* was limited compared with those of many neornithines. The articulation of metatarsal I with metatarsal II in these two birds is as high as in *Archaeopteryx*.

In the Early Cretaceous, enantiornithine birds inherited and perfected the perching capability of their ancestors. In these birds, metatarsal I articulates at the distal end of metatarsal II (Sereno and Rao, 1992). The hooked beak, strongly curved pedal claws, and relatively wide trochlea of digit II in *Boluochia* are probably indicative of the raptorial nature of *Boluochia*. It may also have had a more powerful perching capacity than *Cathayornis* and other early enantiornithines. Among the ornithurine birds, *Chaoyangia* possessed a much more sophisticated power of flight and could probably have spent more time on the ground with little risk. It has relatively short pedal claws and long pelvic limbs. This bird probably lived near the shores of lakes, thus avoiding the competition from the then dominant enantiornithine birds. *Gansus* is another ornithurine bird that was found in northwestern China. This bird had a well-fused tarsometatarsus, reduced pedal claws, and remarkably elongated phalanges. The claws (especially those on the second and fourth digits) had well-developed extensor tubercles. The elongate phalanges offer evidence of aquatic adaptation (Fig. 7.16). We believe *Gansus* represents the oldest bird known that was specifically well adapted to life in or around water.

Comments on *Sinosauropteryx, Protarchaeopteryx,* and *Caudipteryx*

Sinosauropteryx was initially described as a primitive bird linking dinosaurs and later birds (Ji and Ji, 1996). The main reason this fossil was referred to as a bird is that it was asserted to have possessed the most primitive feathers, even though it has a dinosaurian skeleton (Ji and Ji, 1996). Chen et al. (1998) have shown that it is a theropod dinosaur most closely related to *Compsognathus*. Like *Confuciusornis* and *Liaoningornis*, this animal was discovered in the Yixian For-

mation. Many people have been able to look at the putative feathers, and no consensus yet exists on their identity. Some have flatly dismissed their identity as feathers, interpreting these structures as collagenous fibers underneath the skin (Geist et al., 1997; Martin and Ruben, pers. comm.). Others argued that they are integumentary structures rather than internal fibers and agreed that they are not modern feathers. They suggest that these structures indicate an evolutionary stage prior to the origin of modern feathers (Gibbons, 1997; Padian, 1997; Chen et al., 1998; Unwin, 1998; Xu et al., 2001). Branching of the filaments was reported as evidence of further similarity to feathers (Unwin, 1998).

We have examined most of the specimens of *Sinosauropteryx*. Although few conclusions can be drawn from known evidence, we would like to provide a summary of the controversies and our opinions. (1) We believe that whether the featherlike filaments are internal or external structures is still an unsettled question. Unless we can provide convincing evidence that they are skin or integumentary structures, it is probably premature to discuss their significance with regard to the origin of feathers or flight. (2) The filaments are mostly distributed along the dorsal midline of the body as well as the ventral midline of the tail. Although dubious evidence of the distribution of the filaments on the other parts of the body has been reported, we are not convinced that these structures are as extensively present as has been suggested (Chen et al., 1998). (3) As Chen et al. (1998) correctly pointed out, the filaments were so delicate and piled up that a single one has not been isolated for examination. This fact alone raises the question of how a branching featherlike structure could be positively distinguished from complex overlapping of filaments. We have seen more than one example of "branching" appearance but failed to confirm any actual branching.

Protarchaeopteryx is another animal recently reported from the Yixian Formation (Ji and Ji, 1997). Feathers were preserved with the skeleton. It is larger than both *Archaeopteryx* and *Confuciusornis*. *Protarchaeopteryx* was referred by its authors to the Archaeopterygidae based on some shared primitive characters. Although the avian status of this animal was later rejected (Ji et al., 1998), *Protarchaeopteryx* does seem to possess several avian characteristics: retroverted pubis with a pubic foot similar to that of *Confuciusornis*; opposing hallux of the foot; reduced fibula; ulna remarkably wider than the radius; major metacarpal and digit more robust than the other two; and, most important, the presence of real feathers. *Protarchaeopteryx* was reported to have about 30 caudal vertebrae (Ji and Ji, 1997). This cannot be confirmed from its published figure. The humerus is longer than the ulna and the radius; this character is also present in *Archaeopteryx* and *Confuciusornis*.

The thoracic limb of *Protarchaeopteryx* appears to be small and short compared with the robust pelvic limb. The ratio of the length of the forelimb to the pelvic limb is about 0.7 (Ji and Ji, 1997). This is smaller than in *Archaeopteryx* and *Confuciusornis*. The ratio of the length of the hand to the ulna in *Protarchaeopteryx* is larger than those in *Archaeopteryx* and *Confuciusornis*. Compared with the major metacarpal, the alular metacarpal is relatively longer in *Protarchaeopteryx* than in *Archaeopteryx* and *Confuciusornis*. *Protarchaeopteryx* appears to have retained more primitive characters (e.g., the wing) than *Archaeopteryx*.

Caudipteryx was described as another feathered animal from the Early Cretaceous of Liaoning (Ji et al., 1998). Although no one doubts that it has feathers, it was, however, regarded as a flightless bird by some workers (Feduccia, 1999; Jones et al., 2000). The description of a new species, *Caudipteryx dongi* (Zhou and Wang, 2000a), and the discoveries of several more additional specimens of *Caudipteryx* (Zhou and Wang, 2000b) indicate that it is probably more similar to oviraptorosaur dinosaurs than to birds, although it has some unexpected bird characters. For instance, it is similar to advanced birds in having a hand with a digital formula of "2-2-1" rather than "2-3-4" (Zhou and Wang, 2000b). In addition, it possesses uncinate processes and an antorbital fenestra smaller than the nasal opening. On the other hand, *Caudipteryx* is basically similar to other small nonavian theropod dinosaurs. It has a nearly vertical pubis, its ischium has a large obturator process, and metatarsal II is pinched in the middle, which is different from that in birds. Although we cannot absolutely conclude that *Caudipteryx* could not be a secondarily flightless bird, we believe that current evidence does support its close relationship with oviraptorosaur dinosaurs.

New Advances and Future Prospects

Several other birds have recently been recognized from the Mesozoic of Liaoning, China. One of them comes from the Yixian Formation. It is a toothed bird intermediate in size between *Archaeopteryx* and *Cathayornis*. It has been recently named *Eoenantiornis* (Hou et al., 1999). It probably represents the most primitive known enantiornithine. It is more primitive than the Early Cretaceous enantiornithines, such as *Cathayornis*, in having short coracoids and a short sternum without caudal notches. However, this bird possessed an alula (bastard wing), indicating the presence of a more sophisticated flight capacity. The alula was also reported from an Early Cretaceous enantiornithine, *Eoalulavis* (Sanz et al., 1996, Chapter 9 in this volume). We suppose that this flight structure was common in Early Cretaceous enantiornithines as well as Early Cretaceous ornithurines.

The second recently described bird is *Liaoxiornis* (Hou and Chen, 1999). The type specimen of this genus was un-

doubtedly a juvenile; however, several additional specimens are nearly of equal size. The skeletal characters indicate that it was an enantiornithine as advanced as *Cathayornis*. The full development of the feathers in *Liaoxiornis* probably indicates that its adult size was still very small and hence it was probably still the smallest bird in the Mesozoic (Hou and Chen, 1999).

The third new bird is from the Early Cretaceous Jiufotang Formation. It is from the same site (Boluochi) that produced *Cathayornis*, *Boluochia*, and *Chaoyangia*. The senior author is currently describing this bird as a cathayornithid enantiornithine. This bird is as small as *Cathayornis* but relatively more primitive, although it was collected from a slightly higher horizon than the other birds from the same site.

More birds from the Early Cretaceous of Liaoning are currently under study. It is difficult to predict what and how many new birds will be discovered in the next decade from the Mesozoic of China. However, many of the known birds are represented by incomplete specimens. Some of the taxa that we believe are important and will likely be further collected include the following. (1) *Liaoningornis*, the oldest known ornithurine bird, is known by only one partial skeleton. However, collecting at the type locality continues, and we expect additional findings of this bird. (2) *Chaoyangia* is an Early Cretaceous ornithurine bird. Much is left to complete our knowledge of this bird, particularly its skull. (3) *Gansus*, the only Early Cretaceous bird to have aquatic adaptations, is known only from incomplete feet. New findings of this bird may provide evidence concerning its precise life habit. (4) *Boluochia*, the Early Cretaceous enantiornithine, lacks teeth in the premaxilla and differs from other enantiornithines in its foot structure. This bird is important for our understanding of the early diversification among enantiornithines. (5) Many questions remain about *Confuciusornis*. How many species were present? What is the nature of sexual dimorphism between males and females? What are the individual differences and ontogenetic variation among species? We will focus on the answers to these questions. (6) The controversy on *Caudipteryx* and *Protarchaeopteryx*: Are they feathered nonavian dinosaurs or secondarily flightless birds? No matter what we conclude today, the debate on its phylogenetic position will continue for years. We believe that the debate can be finally settled only with a better understanding of its anatomy in the future.

Acknowledgments

We owe a great deal to those of our colleagues who have provided support or advice, all of whom cannot be mentioned here. We list a few of them: L. D. Martin, A. Feduccia, S. Olson, H. James, L. M. Chiappe, E. Kurochkin, P. Wellnhofer, J. H. Ostrom, D. S. Peters, L. M. Witmer, P. C. Sereno, J. Cracraft, M. A. Norell,

A. Elzanowski, R. L. Zusi, S. L. Cumbaa, J. Chorn, Sun A., Chang M., Miao D., Jin F., Zhang J., Hu Y., Wang Y., You H., Gu Y., Chen P., and Zheng G. We are indebted to Zhang H., Yang Y., Wang F., Sun Y., Xie S., Li Y., Hou J., and many others who have either contributed fossils or been of great help to us in the field excavations. Zhang J. has taken many pictures for us over the years. The senior author also thanks R. O. Prum and M. Robbins for their support and suggestions in comparing the skeletons of neornithines that are in their care. J. Chorn, P. J. Currie, L. M. Witmer, and L. M. Chiappe critically reviewed the manuscript of this chapter and provided many valuable suggestions and assistance. Finally, we thank the following institutions for their hospitality during visits made by one or both of us: the Natural History Museum of the University of Kansas, the Forschungsinstitut Senckenberg (Frankfurt, Germany), the National Museum of Natural History, and the American Museum of Natural History. The research was supported by the Chinese Academy of Sciences (CAS, KZ951-B1-410 and KZCX3-J-03), the "Hundred Talent Project" of the CAS, the Chinese National Science Foundation, the National Geographic Society, and the Special Funds for Major State Basic Research Projects of China (G2000077700).

Literature Cited

Chen P. 1988. Distribution and migration of Jehol fauna with a note on the continental Jurassic-Cretaceous boundary in China. Acta Paleontologica Sinica 27:659–683.

Chen P., Dong Z., and Zhen S. 1998. An exceptionally well-preserved theropod dinosaur from the Yixian Formation of China. Nature 391:147–152.

Chiappe, L. M. 1993. Enantiornithine (Aves) tarsometatarsi from the Cretaceous Lecho Formation of northwestern Argentina. American Museum Novitates 3083:1–27.

———. 1995. The first 85 million years of avian evolution. Nature 378:349–355.

Chiappe, L. M., M. A. Norell, and J. M. Clark. 1997. *Mononykus* and birds: methods and evidence. Auk 114:300–302.

Chiappe, L. M., Ji S.-A., Ji Q., and M. Norell. 1999. Anatomy and systematics of the Confuciusornithidae (Aves) from the Mesozoic of Northeastern China. Bulletin of the American Museum of Natural History 242:1–89.

Dong Z. 1993. A lower Cretaceous enantiornithine bird from the Ordos Basin of Inner Mongolia, People's Republic of China. Canadian Journal of Earth Science 30:2177–2179.

Elzanowski, A., and P. Wellnhofer. 1996. Cranial morphology of *Archaeopteryx*: evidence from the seventh skeleton. Journal of Vertebrate Paleontology 16(1):81–94.

Feduccia, A. 1993. Evidence from claw geometry indicating arboreal habits of *Archaeopteryx*. Science 259:790–793.

———. 1996. The Origin and Evolution of Birds. Yale University Press, New Haven, 420 pp.

———. 1999. The Origin and Evolution of Birds. 2nd edition. Yale University Press, New Haven, 466 pp.

Geist, N. R., T. D. Jones, and J. A. Ruben. 1997. Implications of soft tissue preservation in the compsognathid dinosaur, *Sinosauropteryx*. Journal of Vertebrate Paleontology 17(3):48A.

Gibbons, A. 1997. Plucking the feathered dinosaur. Science 278:1229.

Hopson, J. A. 2001. Ecomorphology of avian and nonavian theropod phalangeal proportions: implications for the arboreal versus terrestrial origin of bird flight; pp. 211–235 *in* J. Gauthier and L. F. Gall (eds.), New Perspectives on the Origin and Early Evolution of Birds: Proceedings of the International Symposium in Honor of John H. Ostrom. Peabody Museum of Natural History, New Haven.

Hou L. 1994. A Late Mesozoic bird from Inner Mongolia. Vertebrata PalAsiatica 32(4):258–266.

———. 1997a. A carinate bird from the Upper Jurassic of western Liaoning, China. Chinese Science Bulletin 42(5):413–416.

———. 1997b. Mesozoic Birds of China. Taiwan Provincial Feng Huang Ku Bird Park. Nan Tou, Taiwan, 228 pp.

Hou L. and Chen P. 1999. *Liaoxiornis delicates* gen. et sp. nov., the smallest Mesozoic bird. Chinese Science Bulletin 44(9):834–838.

Hou L. and Liu Z. 1984. A new fossil bird from Lower Cretaceous of Gansu and early evolution of birds. Scientific Sinica (Series B) 27(12):1296–1302.

Hou L. and Zhang J. 1993. A new fossil bird from Lower Cretaceous of China. Vertebrata PalAsiatica 31(3):217–224.

Hou L., Zhou Z., Gu Y., and Sun Y. 1995a. Introduction to Mesozoic birds from Liaoning, China. Vertebrata PalAsiatica 33(4):261–271.

Hou L., Zhou Z., Gu Y., and Zhang H. 1995b. *Confuciusornis sanctus*, a new Late Jurassic sauriurine bird from China. Chinese Science Bulletin 40(18):1545–1551.

Hou L., Zhou Z., L. D. Martin, and A. Feduccia. 1995c. A beaked bird from the Jurassic of China. Nature 277:616–618.

Hou L., L. D. Martin, Zhou Z., and A. Feduccia. 1996. Early adaptive radiation of birds: evidence from fossils from northeastern China. Science 274:1164–1167.

———. 1999. *Archaeopteryx* to opposite birds—missing link from the Mesozoic of China. Vertebrata PalAsiatica 37(2):88–95.

Hu Y., Wang Y., Luo Z., and Li C. 1997. A new symmetrodont mammal from China and its implications for mammalian evolution. Nature 390:137–142.

Jenkins, F. A., Jr., K. P. Dial, and G. E. Goslow Jr. 1988. A cineradiographic analysis of bird flight: the wishbone in starlings is a spring. Science 241:1495–1498.

Ji Q. and Ji S.-A. 1996. The finding of the oldest Chinese bird fossil and the origin of birds. Chinese Geology 233:30–33.

———. 1997. *Protarchaeopteryx* gen. nov.—a Chinese *Archaeopteryx* fossil. Chinese Geology 238:38–41.

Ji Q., P. J. Currie, M. Norell, and Ji S.-A. 1998. Two feathered theropods from the Upper Jurassic/Lower Cretaceous strata of northeastern China. Nature 393:753–761.

Ji Q., L. M. Chiappe, and Ji S.-A. 1999. A new Late Mesozoic confuciusornithid bird from China. Journal of Vertebrate Paleontology 19(1):1–7.

Jin F., Zhang J., and Zhou Z. 1993. A review of *Longdeichthys* (Teleostei: ?Clupeocephala) from northern China. Vertebrata PalAsiatica 31(4):241–256.

Jin F., Tian Y., Yang Y., and Dong S. 1995a. An early fossil sturgeon (Aciperseriformes, Peipiaosteidae) from Fengning of Hebei, China. Vertebrata PalAsiatica 33(1):1–16.

Jin F., Zhang J., and Zhou Z. 1995b. Late Mesozoic fish fauna from western Liaoning, China. Vertebrata PalAsiatica 33(3): 169–193.

Jones, T. D., J. O. Farlow, J. A. Ruben, D. M. Henderson, and W. J. Hillenius. 2000. Cursoriality in bipedal archosaurs. Nature 406:716–718.

Kurochkin, E. 1985. A true carinate bird from Lower Cretaceous deposits in Mongolia and other evidence of Early Cretaceous birds in Asia. Cretaceous Research 6:271–278.

———. 2000. Mesozoic birds of Mongolia and the former USSR; pp. 533–559 in M. J. Benton, M. A. Shishkin, D. M. Unwin, and E. N. Kurochkin (eds.), The Age of Dinosaurs in Russia and Mongolia. Cambridge University Press, Cambridge.

Li P., Su D., Li Y., and Yu J. 1994. The age of the *Lycoptera*-bearing strata. Geologica Sinica. 68(1):87–100.

Lu L. 1994. A new paddlefish from the Upper Jurassic of Northeast China. Vertebrata PalAsiatica 32(2):134–142.

Martin, L. D. 1983. The origin and early radiation of birds; pp. 291–338 in A. H. Brush and G. A. Clark (eds.), Perspectives in Ornithology. Cambridge University Press, London.

———. 1991. Mesozoic birds and the origin of birds; pp. 485–540 in H.-P. Schultze and L. Trueb (eds.), Origins of the Higher Groups of Tetrapods: Controversy and Consensus. Cornell University Press, Ithaca, N.Y.

Martin, L. D., and Zhou Z. 1997. *Archaeopteryx*-like skull in enantiornithine bird. Nature 389:556.

Molnar, R. E. 1986. An enantiornithine bird from the Lower Cretaceous of Queensland, Australia. Nature 322:736–738.

Padian, K. 1997. How did bird flight begin?: an integrative approach. Journal of Vertebrate Paleontology 17(3):68A.

Peters, R. H. 1983. The ecological implications of body size. Cambridge University Press, Cambridge, 329 pp.

Pu R. and Wu H. 1985. Spore and pollen assemblage in the Mesozoic of western Liaoning and their stratigraphic significance; pp. 121–212 in Mesozoic Stratigraphy and Paleontology of West Liaoning, Volume 2. Geological Publishing House, Beijing.

Ruben, J. A., T. D. Jones, N. R. Geist, and W. J. Hillenius. 1997. Lung structure and ventilation in theropod dinosaurs and early birds. Science 278:1267–1270.

Sanz, J. L., and J. F. Bonaparte. 1992. A new order of birds (Class Aves) from the Lower Cretaceous of Spain; pp. 39–49 in K. E. Campbell (ed.), Papers in Avian Paleontology, Honoring Pierce Brodkorb. Science Series 36, Natural History Museum of Los Angeles County, Los Angeles.

Sanz, J. L., and A. D. Buscalioni. 1992. A new bird from the Early Cretaceous of Las Hoyas, Spain. Paleontology 35:829–845.

Sanz, J. L., J. F. Bonaparte, and L. Lacasa. 1988. Unusual Early Cretaceous birds from Spain. Nature 331:433–435.

Sanz, J. L., L. M. Chiappe, B. P. Pérez-Moreno, A. D. Buscalioni, J. Moratalla, F. Ortega, and F. J. Poyato-Ariza. 1996. A new Lower Cretaceous bird from Spain: implications for the evolution of flight. Nature 382:442–445.

Sanz, J. L., L. M. Chiappe, B. P. Pérez-Moreno, J. Moratalla, F. Hernández-Carrasquilla, A. D. Buscalioni, F. Ortega, F. J. Poyato-Ariza, D. Rasskin-Gutman, and X. Martínez-Delclòs. 1997. A nestling bird from the Early Cretaceous of Spain: implications for avian skull and neck evolution. Science 276: 1543–1546.

Sereno, P. C., and Rao C. 1992. Early evolution of avian flight and perching: new evidence from the Lower Cretaceous of China. Science 255:845–848.

Smith, P. E., N. M. Evensen, D. York, Chang M.-M., Jin F., Li J.-L., S. Cumbaa, and D. Russell. 1995. Dates and rates in an ancient lake: ^{40}Ar-^{39}Ar evidence for an Early Cretaceous age of the Jehol Group, northeast China. Canadian Journal of Earth Science 32:1426–1431.

Swisher, C. C., III, Wang Y.-Q., Wang X.-L., Xu X., and Wang Y. 1999. Cretaceous age for the feathered dinosaurs of Liaoning, China. Nature 400:58–61.

Unwin, D. M. 1998. Feathers, filaments and theropod dinosaurs. Nature 391:119–120.

Varricchio, D. J., and L. M. Chiappe. 1995. A new enantiornithine bird from the Upper Cretaceous Two Medicine Formation of Montana. Journal of Vertebrate Paleontology 15: 201–204.

Walker, C. A. 1981. New subclass of birds from the Cretaceous of South America. Nature 292(2):51–53.

Wang D. and Diao N. 1984. Geochronology of Jura-Cretaceous volcanic in west Liaoning, China; pp. 1–12 in Scientific Papers on Geology from International Exchange 27th International Geological Congress, Volume 1. Geological Publishing House, Beijing.

Wang S. 1990. The origin and evolutionary mechanism of the Jehol Fauna. Acta Geologica Sinica 64(4):350–360.

Wang Y., Hu Y., Zhou M., and Li C. 1995. Mesozoic mammal localities in western Liaoning, northeastern China; pp. 221–227 in Sun A. and Wang Y. (eds.), Sixth Symposium on Mesozoic Terrestrial Ecosystems and Biota, Short Papers. China Ocean Press, Beijing.

Wellnhofer, P. 1993. Das siebte Exemplar von *Archaeopteryx* aus den Solnhofener Schichten. *Archaeopteryx* 11:1–47.

Xu X., Wang X.-L., and Wu X.-C. 1999a. A dromaeosaur dinosaur with filamentous integument from the Yixian Formation of China. Nature 401:262–266.

Xu X., Tang Z.-L., and Wang X.-L. 1999b. A therizinosauroid dinosaur with integumentary structures from China. Nature 399:350–354.

Xu X., Zhou Z., and Wang X.-L. 2000. The smallest known non-avian theropod dinosaur. Nature 408:705–708.

Xu X., Zhou Z., and R. O. Prum. 2001. Branched integumentary structures in *Sinornithosaurus* and the origin of feathers. Nature 410:200–204.

Zhang F. and Zhou Z. 2000. A primitive enantiornithine bird and the origin of feathers. Science 290:1955–1959.

Zhang F., Zhou Z., Hou L., and Gu G. 2001. Early diversification of birds: evidence from a new opposite bird. Chinese Science Bulletin 46(11):945–949.

Zhang J., Jin F., and Zhou Z. 1994. A review of Mesozoic osteoglossomorph fish *Lycoptera longicephalus*. Vertebrata PalAsiatica 32(1):41–59.

Zhou Z. 1992. Review on *Peipiaosteus pani* based on new materials of *P. pani*. Vertebrata PalAsiatica 30(2):85–101.

———. 1995a. Discovery of a new enantiornithine bird from the Early Cretaceous of Liaoning, China. Vertebrata PalAsiatica 33(2):99–113.

———. 1995b. New understanding of the evolution of the limb and girdle elements in early birds—evidences from Chinese fossils; pp. 209–214 in Sun A. and Wang Y. (eds.), Sixth Symposium on Mesozoic Terrestrial Ecosystems and Biota, Short Papers. China Ocean Press, Beijing.

———. 1995c. The discovery of Early Cretaceous birds in China. Courier Forschungsinstitut Senckenberg 181:9–22.

———. 1995d. Is *Mononykus* a bird? Auk 112(4):958–963.

Zhou Z. and Hou L. 1997. Diversification of birds from the "Late Jurassic" of China. Journal of Vertebrate Paleontology 17(3):86A.

———. 1998. *Confuciusornis* and the early evolution of birds. Vertebrata PalAsiatica 36(2):136–146.

Zhou Z. and L. D. Martin. 1999. Feathered dinosaur or bird?—a new look at the hand of *Archaeopteryx*. Smithsonian Contributions to Paleobiology 89:289–293.

Zhou Z. and Wang X. L. 2000a. A new species of *Caudipteryx* from the Yixian Formation of Liaoning, northeast China. Vertebrata PalAsiatica 38(2):111–127.

———. 2000b. *Caudipteryx* and the origin of the flight of birds. Vertebrata PalAsiatica Supplement to 38:38.

Zhou Z. and Zhang F. 2001. Two new ornithurine birds from the Early Cretaceous of western Liaoning, China. Chinese Science Bulletin 46(15):1258–1264.

Zhou Z., Jin F., and Zhang J. 1992. Preliminary report on a Mesozoic bird from Liaoning, China. Chinese Science Bulletin 37(16):1365–1368.

8

Sinornis santensis (Aves: Enantiornithes) from the Early Cretaceous of Northeastern China

PAUL C. SERENO, RAO CHENGGANG, AND LI JIANJUN

Our current understanding of the origins and early evolution of birds (Chiappe, 1995; Norell and Makovicky, 1997; Sereno, 1997, 1999b; Forster et al., 1998; Padian and Chiappe, 1998) has been strongly influenced by remarkable skeletons of basal birds and other coelurosaurs discovered recently in lacustrine deposits of Early Cretaceous age in Liaoning Province, China. The first Mesozoic avian remains reported from this region consisted of a partial skeleton of a sparrow-sized bird named *Sinornis santensis* (Sereno and Rao, 1992). *S. santensis* was discovered in a richly fossiliferous lacustrine facies of the Jiufotang Formation (Chen, 1988). Shortly after its discovery, several additional avian skeletons of similar size were discovered in the same local region and horizons and described as *Cathayornis yandica*, *Boluochia zhengi*, and others (Zhou et al., 1992; Zhou, 1995a,b; Hou, 1997; Zhou and Hou, Chapter 7 in this volume).

In this chapter, we provide a brief description of the skeletal anatomy of *S. santensis*, discuss its relationship with other avian remains from the same levels, and consider its phylogenetic position among early avians.

The following institutional abbreviations are used in this chapter: BVP, Beijing Natural History Museum, Paleovertebrate Collection, Beijing, China; IEI, Institut d'Estudis Illerdencs, Castile–La Mancha, Spain; IVPP, Institute of Vertebrate Paleontology and Paleoanthropology, Beijing, China; UAM, Universidad Autónoma de Madrid, Madrid, Spain.

Geological Setting

In July 1987, a young boy living in the countryside near Chaoyang city in Liaoning Province (Fig. 8.1) noticed some fossil bone on a small mudstone slab. He brought pieces of the slab to Yan Zhiyou, a local farmer known for his interest in fossils. Yan Zhiyou, in turn, reported the possible discovery of a small fossil bird to the Beijing Natural History Museum in 1988. Two of us (Li J. and Rao C.) visited Yan Zhiyou and confirmed the avian affinity of the specimen and site of discovery. The specimen underwent acid preparation, and an initial description was published (Rao and Sereno, 1990; Sereno and Rao, 1992). Additional specimens were discovered in the same area in 1990 by Zhou Zhonghe of the Institute of Vertebrate Paleontology and Paleoanthropology and described as *C. yandica* (Zhou et al., 1992; Zhou, 1995a; Zhou and Hou, Chapter 7 in this volume) and *B. zhengi* (Zhou, 1995b).

All these specimens were discovered in lacustrine facies of the Jiufotang Formation in association with the Jehol fauna, characterized by abundant remains of the teleost *Lycoptera*, the insect *Ephemeropsis*, and the conchostracan *Eosestheria* (Hong, 1988). Two species of the ornithischian dinosaur *Psittacosaurus* (*P. meileyingensis*, *P. mongoliensis*) are recorded in overlying nonmarine facies of the Jiufotang Formation (Sereno et al., 1988).

The Jiufotang Formation lies stratigraphically above the Yixian and Chaomidianzi Formations (Wang et al., 1998; Ji et al., 1999). The Chaomidianzi Formation, which has yielded many skeletons of *Confuciusornis sanctus* (Hou et al., 1995, 1996; Chiappe et al., 1999) and other taxa (Barrett, 2000), has recently been radiometrically dated as Barremian in age (Swisher et al., 1999), which suggests that the Jiufotang Formation may be as young as Aptian or Albian in age (100 million–120 million years old).

Systematic Paleontology

The phylogenetic definitions of taxa in the hierarchy listed subsequently and shown in Figure 8.9A are briefly reviewed.

Aves—Chiappe (1992:348) defined Aves as "the common ancestor of *Archaeopteryx* and modern birds [Neornithes] plus all its descendants." Sereno (1997:459) used *Archaeop-*

Figure 8.1. Map of Liaoning Province in northeastern China showing Shengli (type locality of *S. santensis*), Boluochi (type locality for *C. yandica*), and the Sihetun-Jianshangou region (type locality of *C. sanctus*).

teryx and Passeriformes for the same. Later, in a more formal consideration of taxa, Sereno used *Archaeopteryx* and Neornithes as reference taxa (1998:65), with Neornithes, in turn, defined as *Struthio, Passer*, their common ancestor, and all descendants (1998:59). The use of vernacular terms as reference taxa, such as "extinct birds," is best avoided for clarity (Sereno, 1999b). All these definitions, nonetheless, correspond well with the traditional taxonomic content of Aves.

Gauthier (1986:36), on the other hand, suggested that Aves be restricted to crown birds alone and defined as "the most recent common ancestor of Ratitae, Tinami, and Neognathae, and all of its descendants." This crown group definition has priority over those mentioned previously (Sereno, 1998: table 2) but excludes from Aves many extinct sister taxa that have been regarded as birds for more than a century. Substituting a new taxon, Avialae, to identify a clade long recognized as Aves (Gauthier, 1986) has not gained wide acceptance. Invoking entrenched historical usage, however, may be the only reason to set aside Gauthier's phylogenetic definition of Aves (see also Witmer, Chapter 1 in this volume; Clark, Norell, and Makovicky, Chapter 2 in this volume).

Ornithurae—Gauthier (1986:36) provided the first phylogenetic definition of Ornithurae. He coined a stem-based taxon that included "Aves [i.e., living birds] plus all extinct maniraptorans that are closer to Aves than is *Archaeopteryx*." Cracraft (1986:385) used the same definition, stating that "*Archaeopteryx* is the sister-group of all other birds; this latter taxon is here termed the Ornithurae (following Martin 1983a [Martin 1983 in this chapter]; and Gauthier, 1986)."

Sereno (1997:459) followed this definition, recognizing *Archaeopteryx* and Neornithes as formal reference taxa (1998:65).

Chiappe (1996:205), on the other hand, redefined Ornithurae as a node-based taxon including only "Hesperornithiformes and Neornithes plus all taxa descended from it." Two reasons were given: convenience and historical usage. Chiappe (1991:337) stated that Ornithurae was "defined by previous authors as a taxon including extant birds and all extinct birds that are closer to them than is *Archaeopteryx* (Cracraft, 1986; Gauthier, 1986). These authors diagnosed Ornithurae on the basis of several derived features, most of which are accepted here. However, the recent discovery of several new fossil birds which lack the synapomorphies noted earlier (e.g., the Las Hoyas bird (Sanz et al., 1988), Enantiornithes (Walker, 1981; Martin, 1983)) makes it convenient to adopt a stricter definition of Ornithurae." This reasoning is in conflict with one of the principal aims of phylogenetic taxonomy—to replace taxa based on suites of characters with node- or stem-based definitions that reduce ambiguity in taxonomic content and eliminate arbitrary definitional revisions of this kind. Traditional usage (Chiappe, 1996:205), likewise, cannot be invoked to alter the original node-based definition of Ornithurae. Historical use of Ornithurae does not correspond to that preferred by Chiappe (Sereno, 1999b), and recent use of Ornithurae, when associated with informal phylogenetic definitions (e.g., Martin, 1983; Cracraft, 1986), is consistent with Gauthier's original definition (Sereno, 1999c). The editors of this volume, nevertheless, have instructed contributing authors to adopt Chiappe's more restrictive definition for Ornithurae. To avoid confusion, the taxon Ornithurae will not be used in the text of this chapter.

Ornithothoraces—"Ornithopectae" was defined as "the common ancestor of the Las Hoyas bird and extant birds, plus all their descendants" (Chiappe, 1991:337). Chiappe and Calvo (1994: Fig. 9) and Chiappe (1996:205) substituted Ornithothoraces for "Ornithopectae" and *Iberomesornis romerali* for the "Las Hoyas bird." Sereno (1998:65) defined Ornithothoraces in a similar manner as a node-based taxon but used *S. santensis*—a more completely known member of Enantiornithines—and Neornithes as reference taxa.

Enantiornithes—Sereno (1998:65) provided a stem-based definition for Enantiornithes—"all ornithothoracines closer to *Sinornis* than to Neornithes."

Taxonomic Hierarchy

    Aves Linnaeus, 1758
        Ornithothoraces Chiappe and Calvo, 1994
           Enantiornithes Walker, 1981
              Sinornis Sereno and Rao, 1992
              (= *Cathayornis* Zhou, Jin, and Zhang, 1992)

Diagnosis—See type species.

S. santensis Sereno and Rao, 1992
(= C. yandica Zhou, Jin, and Zhang, 1992)

Holotype—BPV 538a and 538b constitute part and counterpart, respectively, of a single articulated skeleton in the collections of the Beijing Natural History Museum. A triangle in the center of the skeleton is missing from both part and counterpart, and the smaller counterpart slab covers only the caudal one-half of the skeleton. The counterpart (BPV 538b) is preserved as discovered, with the majority of the bones exposed in cross section. The part (BPV 538a) was prepared in acid and now is preserved as a natural mold, from which an epoxy cast was made. All the descriptive detail is based on the epoxy cast derived from the part (BPV 538a), drawn with a camera lucida (Figs. 8.2–8.7).

An earlier drawing was made prior to acid preparation and was based on tracings from part and counterpart slabs (Sereno and Rao, 1992: Fig. 2). Right and left hindlimbs were misidentified, and the view chosen (right lateral) is the reverse of that now available from the epoxy cast of the part slab (left lateral) (Figs. 8.2–8.7).

Referred specimens—Referred material includes IVPP V9769A, B (holotypic specimen of C. yandica) and several other specimens in the collections of the IVPP from the same region (just south of Boluochi, Liaoning Province; Zhou, 1995a) (Fig. 8.1).

Synonomy—Like S. santensis, the holotypic specimen of C. yandica (IVPP V9769A, B; Zhou et al., 1992) comes from lacustrine facies in the lower portion of the Jiufotang Formation near the city of Chaoyang, Liaoning Province, in northeastern China, approximately 15 km from the type locality of S. santensis (Fig. 8.1; Zhou, 1995a: Fig. 1). S. santensis and C. yandica were named and described independently (the former shortly before the latter) without comparison. Now that both holotypic skeletons have undergone acid preparation, a detailed comparison is possible. The holotypic skeleton of C. yandica is extremely similar to that of S. santensis.

Cranial comparisons are limited to the maxilla and nasal. Each specimen preserves the caudal portion of the antorbital fossa and the flat external surface of the maxilla below the fossa in lateral view. Unlike Archaeopteryx (Elzanowski and Wellnhofer, 1996: Fig. 7), the external surface of the maxilla has parallel dorsal and ventral margins and does not taper in depth caudally until very near its articular end for the jugal. Both specimens also preserve articulated nasals in dorsal view. In contrast to Archaeopteryx (Elzanowski and Wellnhofer, 1996: Fig. 7), the caudal half of the nasal is more transversely expanded, flaring abruptly caudal to the external naris. In both specimens the nasal terminates caudally as a broad triangular sheet separated in the

midline from its opposite by an equally broad V-shaped frontal embayment. The caudal end of the nasal and the median frontal embayment appear to be narrower in Archaeopteryx (Elzanowski and Wellnhofer, 1996: Fig. 7).

Comparisons of the axial column are limited to thoracic, sacral, and caudal vertebrae and the pygostyle. The vertebrae are very similar to each other and to those of other enantiornithines (Chiappe and Walker, Chapter 11 in this volume). Midthoracic vertebrae have amphicoelous, spool-shaped centra, and the cupped articular surface of the parapophysis is caudally displaced on the neural arch. There are eight sacral vertebrae, the centra of which are broad and flat ventrally and the ribs of which increase in length caudally (Zhou, 1995a: Fig. 5). The relatively short caudal centra are broader than deep and have broad, bulbous chevron facets ventrally. The pygostyle is characteristic in C. yandica and S. santensis. Unlike Iberomesornis (Sanz and Bonaparte, 1992; Sereno, 2000), dorsal and ventral processes flare beyond the dorsal and ventral margins of the base of the pygostyle, and the lateral crest is located at midheight along the pygostyle (Fig. 8.2). The basal processes are smaller in Iberomesornis, and the lateral crest is situated along the ventral margin (Sereno, 2000: Fig. 7). In C. sanctus, the basal processes of the pygostyle are not as pronounced as in S. santensis and C. yandica, but the lateral crest is located in a similar position at midheight along the pygostyle (Martin et al., 1998: Fig. 2E; Chiappe et al., 1999: Fig. 25).

The long bones of the postcranial skeleton are approximately 10% longer in the holotypic specimen of C. yandica than in S. santensis (Tables 8.1, 8.2). Ratios between and within the forelimb and hindlimb, however, are nearly identical and similar to those in two other enantiornithines (Table 8.3). In the forelimb, the ungual of manual digit II is larger than that of digit I in S. santensis (BPV 538a); in the holotypic specimen of C. yandica, the unguals appear to be subequal in size and length. Phalanx 1 of manual digit III curves away from digit II to a greater degree in S. santensis (BPV 538a), but the form of this reduced phalanx appears to be somewhat variable among ornithothoracines. Zhou (1995a: Figs. 8, 9) and Hou (1997:107) have shown metacarpals II and III in contact, or fused, along their shafts, lacking the intermetacarpal space that is present in S. santensis and most other avians. The lack of an obvious intermetacarpal space in C. yandica, however, is the result of postmortem flattening of the metacarpus, which has deflected the narrow shaft of metacarpal III against the more robust shaft of metacarpal II. An intermetacarpal space is present in the right metacarpus of IVPP V9769B, as seen in ventral view.

The pelvic girdle is very similar in S. santensis and C. yandica. The ilium and pubis are identical in shape and proportions, although the dorsal margin of the ilium in S. santensis and the pubic foot in C. yandica are not preserved and

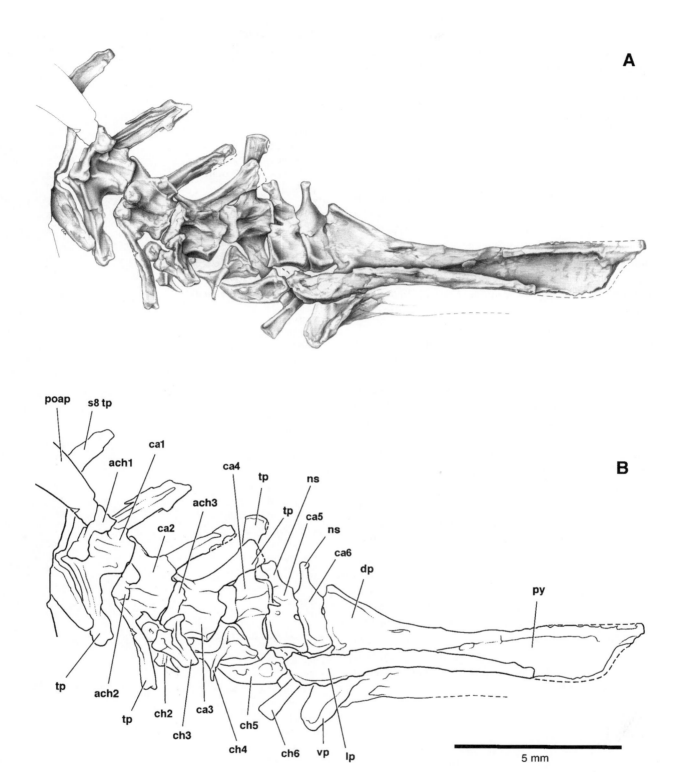

Figure 8.2. Caudal vertebrae and pygostyle of *S. santensis* (BPV 538a, epoxy cast) in left lateral view (A, B). Broken lines indicate bone margins visible as impressions. Abbreviations: ach1–3, articular surface for chevrons 1–3; ca1–6, caudal vertebrae 1–6; ch2–6, chevrons 2–6; dp, dorsal process; lp, lateral process; ns, neural spine; poap, postacetabular process; py, pygostyle; s8, eighth sacral vertebra; tp, transverse process; vp, ventral process.

cannot be compared. The elongate, scimitar-shaped is-chium in *S. santensis* has an unusual reflected lamina on its lateral side, and the elongate, curving profile and lamina are exposed on the right ischium of *C. yandica* (IVPP V9769B). No differences are apparent in the hindlimb.

Although clearly very similar to each other, *S. santensis* and *C. yandica* are also very similar to *Concornis lacustris* (Sanz et al., 1995) and *I. romerali* (Sanz and Bonaparte, 1992; Sereno, 2000; Sanz et al., Chapter 9 in this volume). Ulti-mately, synonymy must be based on the presence of au-tapomorphies, of which there are only a few. In both *S. san-tensis* and *C. yandica*, the base of the pygostyle has prominent and flaring dorsal and ventral processes, and a lateral crest is present at midheight along the length of the pygostyle. In both of these species, the ischium has a thin lamina in its proximal half that is reflected ventrally. The is-chiadic lamina and the elongate, curved shape of the blade are probably more diagnostic and less variable than pygo-style form. As these features have not been observed in other enantiornithines, *C. yandica* is regarded as a junior syn-onym of *S. santensis*, a synonymy acknowledged previously as a possibility by Zhou (1995a:6; see also Zhou and Hou, Chapter 7 in this volume).

The systematic status of other basal avians described re-cently from the same beds will remain unclear until they are described in more detail. *B. zhengi* appears to be distinct (Zhou, 1995b), but *Cathayornis caudatus* and *Longchengor-nis sanyanensis* (Hou, 1997: Figs. 51, 59) may eventually be re-garded as junior synonyms of *S. santensis*.

Locality and horizon—Chaoyang County, Liaoning Province, People's Republic of China; lacustrine facies in the lower part of the Jiufotang Formation; Early Cretaceous (? Aptian-Albian) (Fig. 8.1; Sereno and Rao, 1992:848; Zhou, 1995a; Swisher et al., 1999).

Revised diagnosis—Sparrow-sized enantiornithine very close in form to *C. lacustris* and characterized by an elon-gate, scimitar-shaped ischiadic blade (narrowest width ap-proximately one-tenth ischiadic length), a ventrally re-flected lamina on the lateral aspect of the proximal half of the ischium, flaring dorsal and ventral processes at the base of the pygostyle, and a lateral crest situated at midheight along the pygostyle.

Anatomy

The following description is based primarily on an epoxy cast made from the natural mold of the part (BPV 538a) of the holotypic skeleton. The smaller counterpart was not prepared in acid. Additional information and compar-isons are based on epoxy casts of the part and counterpart of a referred specimen, IVPP V9769 (*C. yandica*), also pre-served as a natural mold. Epoxy casts of these two speci-

TABLE 8.1

Measurements of the holotypic skeleton of *Sinornis santensis* (BPV 538a, epoxy cast) in mm

Skull and Axial Column

Skull length			(26.5)
Dorsal vertebral column, length			(30.5)
Sacrum length			12.9
CA1–6 length			8.4
Pygostyle length			11.5
Furcula, dorsolateral ramus, length			(8.0)
Hypocleideum length			6.0
?C12	2.4	CA1	1.6
?D1	2.4	CA2	1.5
?D8	2.3	CA3	1.5
S1	(2.5)	CA4	1.3
S8	(1.5)	CA5	1.0
		CA6	0.6

Pectoral Girdle and Forelimb

Scapula (left)	
Length (left)	(18.0)
Width of distal blade	1.5
Humerus (right)	
Length	24.0
Ulna (left)	
Length	(19.2)
Midshaft diameter	2.0
Radius (left)	
Length	(22.2)
Midshaft diameter	1.0
Carpometacarpus	
Metacarpal I (right)	(1.9)
Metacarpal II (left)	10.3
Metacarpal III (right)	(10.8)
Manual phalanges[1] (right)	
I-1 (left)	4.0
I-ungual	1.6
II-1	5.4
II-2	3.8
II-ungual	2.1
III-1	3.9

Pelvic Girdle and Hindlimb

Ilium (left)	
Length	(13.0)
Acetabulum, craniocaudal diameter	1.8
Femur (right, left)	
Length	(21.0)
Tibiotarsus	
Length (right)	26.4
Length (left)	25.7
Tarsometatarsus (right)	
Metatarsal I	3.2
Metatarsal II	14.1
Metatarsal III	14.6
Metatarsal IV (left)	14.1

(continued)

TABLE 8.1 (continued)

Pelvic Girdle and Hindlimb

Pedal phalanges[1] (right)	
I-1	3.7
I-ungual[2]	5.6
II-1	3.2
II-2	4.1
II-ungual[2]	5.8
III-1	4.3
III-2	3.6
III-3	3.9
III-ungual[2]	6.7
IV-1	2.1
IV-2	2.1
IV-3	2.1
IV-4	3.0
IV-ungual[2]	5.3

Note: Parentheses indicate estimated measurement. Abbreviations: C, cervical; CA, caudal; D, dorsal; S, sacral.
[1]Nonungual phalangeal length is measured from the deepest part of the proximal articular cotyla to the apex of the farthest distal condyle.
[2]Pedal ungual length measured as a chord from the extensor process of the base to the tip of the ungual sheath.

mens are identified hereafter as the "holotypic" and "referred" specimens.

Skull

Dermal Skull Roof. The skull of the holotypic specimen is partially disarticulated and rotated so that the snout is pointing caudally. The premaxillary portion of the snout extends beyond the preserved edge of the slab. The nasals are flattened in the plane of the slab and are exposed in dorsal view. The orbital region of the skull is visible in right lateral view, and several sclerotic plates from the left sclerotic ring are preserved within the orbit. The caudal skull roof and occipital condyle are preserved in dorsal view. Portions of the dermal skull roof (maxilla, nasal, lacrimal), palate, occiput, and portions of the lower jaws (surangular, angular, splenial) are exposed.

The skull of the referred specimen is more articulated and is preserved in left lateral view on the part (IVPP V9769A; Zhou, 1995a: Fig. 3). Only the tip of the snout is preserved on the counterpart (IVPP V9769B). The oval external naris is large and retracted from the rostral end of the snout as in *Archaeopteryx* and nearly all other birds. The antorbital cavity is invaginated along its ventral margin as best seen in the holotypic specimen. An oval maxillary fenestra may have been present on the medial wall of the antorbital cavity; the region is best preserved in the right maxilla of the referred specimen, which is exposed in medial view. Another enantiornithine skull has been shown as lacking the

TABLE 8.2

Comparative measurements of the holotypic skeleton of *Cathayornis yandica* (IVPP V9769A, B, epoxy cast) in mm

Skull and Axial Column

Skull length	28.2
Sacrum length	13.1
Pygostyle length	14.4
Furcula, dorsolateral ramus length	9.6
Hypocleideum length	4.5
?C9	2.7
?C10	2.5
?D4	2.7
CA1	2.0
CA5	1.6

Pectoral Girdle and Forelimb

Scapula	
Length (right)	19.6
Width of distal blade (left)	1.3
Humerus (right)	
Length	27.0
Ulna (right)	
Length	26.4
Midshaft diameter	2.3
Radius (right)	
Length	25.3
Midshaft diameter	1.2
Carpometacarpus	
Metacarpal I (right)	2.2
Metacarpal II (left)	11.5
Metacarpal III (right)	13.0
Manual phalanges[1] (left)	
I-1 (right)	4.7
I-ungual	1.8
II-1	6.9
II-2	4.2
II-ungual	1.8
III-1 (right)	3.3

Pelvic Girdle and Hindlimb

Ilium (left)	
Length	13.9
Femur (left)	
Length	23.1
Tibiotarsus (right)	
Length	29.0
Tarsometatarsus (right)	
Metatarsal I	3.0
Metatarsal II	14.8

Note: Abbreviations: C, cervical; CA, caudal; D, dorsal; S, sacral.
[1]Nonungual phalangeal length is measured from the deepest part of the proximal articular cotyla to the apex of the farthest distal condyle.

TABLE 8.3

Comparative ratios of long bones in the holotypic specimens of *Iberomesornis romerali*,
Concornis lacustris, *Sinornis santensis*, and *Cathayornis yandica*

Ratios	Iberomesornis	Concornis	Sinornis BPV 538a	Sinornis (= Cathayornis) IVPP V9769
h/r	(0.97)	1.13	(1.08)	1.07
f/tt	0.82	0.69	(0.81)	0.80
tt/mtIII	1.70	1.74	1.79	(1.88)
h/f	1.06	1.31	1.14	1.17
h+r/f+tt	0.98	1.00	0.98	1.00

Source: Measurements for *Iberomesornis* and *Concornis* are from Sereno, 2000.

Note: Parentheses indicate ratios that are based on one estimated length measurement. Abbreviations: f, femur; h, humerus; mtIII, metatarsal III; r, radius; tt, tibiotarsus.

maxillary fenestra (Sanz et al., 1997), but that part of the bone is not well preserved.

The elongate caudodorsal process of the premaxilla extends to the caudal end of the external naris (Zhou, 1995a: Fig. 3) as in *Archaeopteryx*. In *Confuciusornis* and more advanced ornithurines, in contrast, this process extends farther caudally to contact the frontals (Martin et al., 1998; Chiappe et al., 1999). The external surface of the caudal ramus of the maxilla is strap-shaped with parallel dorsal and ventral margins, as preserved in the holotypic and referred specimens. Rostrally, the nasal is transversely narrow and arched. Caudally, it expands as a broad sheet, and there is a V-shaped embayment in the midline between opposing nasals. The elongate jugal is strap-shaped (transversely compressed), rather than rod-shaped, below the orbit, and it has a well-developed, caudodorsally inclined postorbital process. The slender postorbital process of the jugal is very similar to that in an enantiornithine skull from Spain (IEI LH-4450), although this process was not described in the latter (Sanz et al., 1997: Fig. 3). A postorbital may be preserved in the holotypic specimen but cannot be identified with certainty. The postorbital, however, is preserved in other enantiornithines (Sanz et al., 1997; Chiappe and Walker, Chapter 11 in this volume). The frontal and parietal form the domed caudal portion of the cranium. A portion of the occipital condyle may be exposed in the holotypic specimen.

Palate. The quadrate head is partially exposed in the holotypic specimen and appears to be developed as a single condyle. Zhou (1995a: Fig. 3) identified a quadrate in the referred skull, but we cannot confirm this identification. The area of bone in question in the referred skull is very broken. In the holotype, the right pterygoid is exposed in articulation with the basisphenoid, caudally, and ectopterygoid, laterally. The quadrate wing of the pterygoid expands caudally as a sheet.

Lower Jaw. The lower jaw is deeper caudally than rostrally. The dentary ramus, in particular, is very slender. Its pointed rostral end is slightly upturned in lateral view, as seen in the referred skull. The splenial has a broad triangular shape and is pierced by a foramen in both specimens. The left angular and prearticular are exposed in medial view in the holotypic specimen and are similar in shape to that in *Archaeopteryx* (Elzanowski and Wellnhofer, 1996).

Teeth. Preserved only in the referred specimen, the teeth are fewer in number than in *Archaeopteryx*. There are four premaxillary teeth that increase in size caudally (Zhou, 1995a). The crowns are subconical with a subtle basal constriction as in *Archaeopteryx*. One crown is preserved toward the rostral end of the left maxilla in the referred specimen. At least the caudal two-thirds of the maxilla, however, is edentulous. Similarly, the slender dentary clearly lacks teeth along most of its length. One dentary tooth, preserved on both sides in the referred specimen, projects into the gap between premaxillary and maxillary teeth.

Axial Column

Presacral Vertebrae. None of the cervicals are preserved in the holotypic specimen. A series of five midcervical vertebrae are preserved in the referred specimen, which are exposed in dorsal view (IVPP V9769A). The caudal two of these are also exposed in ventral view (IVPP V9769B). The elongate, spool-shaped centra have a low ventral keel that spans the length of the centrum. Partially exposed articular faces show some development of heterocoely; rostral surfaces are transversely concave and dorsoventrally convex, similar to that in a better exposed enantiornithine cervical series (Sanz et al., 1997). In dorsal view, the neural arches are proportionately broad, and the vertebral canal is spacious. The postzygapophyses arch dorsally, and the neural spines are low.

Best exposed in the holotypic specimen, the thoracic vertebrae have a broad, poorly defined pleurocoel. An especially robust ventral keel is present on the centrum of the most cranial thoracic vertebra. The parapophysis projects over the pleurocoel and is deeply cupped, as in other enantiornithines (Chiappe and Walker, Chapter 11 in this volume). Middorsal vertebrae have spool-shaped amphicoelous centra without ventral keels. The neural canal is large, and the flat zygapophyseal articulations are set at about 45° from the horizontal (BPV 538a). The broad neural spines are craniocaudally longer than tall and angle caudodorsally (IVPP V9769A).

Sacral Vertebrae. The sacrum consists of eight fused vertebrae, as best seen in the referred specimen, and is disarticulated from the ilium in both specimens. The cranial-most sacral, exposed in ventral view in the holotypic specimen, has a spool-shaped centrum cranially but flattens caudally. The cranial articular face is gently concave. Sacrals 1–5 have broad centra and plate-shaped transverse processes. In *Iberomesornis*, sacrals 1–3 and 6–8 are associated with pre- and postacetabular processes, respectively, and sacrals 4 and 5 attach to the ilium just above the acetabulum (Sereno, 2000). The shape of the sacral series in *Sinornis* suggests that it attached to the ilium in a similar manner. The transverse processes that attach to the postacetabular process, for example, angle backward and decrease in length and width, as is well exposed in the referred specimen (Zhou, 1995a).

Caudal Vertebrae. There are six free caudal vertebrae, as in *Iberomesornis* (Sereno, 2000; contra Sanz and Bonaparte, 1992). The number of caudal vertebrae is most easily ascertained in the holotypic specimen (Fig. 8.2), although the same number is present between the sacrum and pygostyle in the referred specimen (contra Zhou, 1995a:11–12). The transverse processes decrease in length and width and shift from a caudolateral to a lateral orientation near the pygostyle. The transverse processes are long on the fourth caudal but appear to be absent on the fifth and sixth caudal vertebrae, lateral to which projects the lateral process of the pygostyle. The form of the caudal neural arches is not well known because the proximal four caudal vertebrae are exposed in ventral view (Fig. 8.2). Neural spines are present on the fifth and sixth caudal vertebrae. Robust paired chevron facets are present, and it is noteworthy that these are located on the proximal, rather than distal, margin of the centra.

Pygostyle. As in other enantiornithines and *Confuciusornis*, the pygostyle is large, exceeding the collective length of the free caudals (Fig. 8.2). Best preserved in the holotypic specimen, the pygostyle is deepest proximally and narrowest at midlength. At its proximal end, sagittal, dorsal, and ventral processes are well developed. A strong lateral process extends proximally lateral to the sixth, and the distal half of

the fifth, free caudal. The lateral process extends distally as a lateral shelf along most of the length of the pygostyle.

Ribs. The short cervical ribs are fused to the transverse processes, as preserved in the referred specimen and in other enantiornithines (Sanz et al., 1996, 1997). They extend parallel to the centra and stop short of its caudal articular surface. The first thoracic rib is transitional in form between the short cervical ribs and the broad and long cranial dorsal ribs (BPV 538a). About half the length and width of the other cranial thoracic ribs, the first thoracic rib tapers to a slender pointed distal end. The rib shafts of the second and more caudal cranial thoracic ribs are very broad and strap-shaped. The second through the fifth ribs probably articulated distally with the ossified sternal ribs preserved in the referred specimen. The tuberculum of the second rib is long and set at nearly a right angle to the rib shaft and capitulum. The mid- and caudal thoracic rib shafts decrease in width and length but retain an oval cross section.

Portions of the shafts of the caudal thoracic ribs were previously interpreted as gastralia (Fig. 8.4; Sereno and Rao, 1992). These apparently short and tapered elements are best identified as caudal thoracic rib shafts that are passing into, and out of, the part slab. Their shafts are comparable to those of nearby ribs, rather than being much more slender, as are the gastralia in *Archaeopteryx* and *Confuciusornis* (Chiappe et al., 1999). The absence of gastralia is now demonstrated in the articulated skeletons of other very well-preserved enantiornithines, such as *Concornis* (Sanz et al., 1995) and *Eoalulavis* (Sanz et al., 1996), and we believe that *Sinornis* is no exception.

Chevrons. All the free caudal vertebrae are associated with chevrons (Fig. 8.2). The first chevron articulates between the eighth sacral and first caudal, as shown by the pair of large articular facets along the cranial margin of the first caudal vertebra. The last chevron is as long as, or longer than, the others and articulates between the fifth and sixth caudal vertebrae.

Pectoral Girdle

Scapula. The scapula closely resembles that of *Concornis* and is preserved in both specimens. Cranially, a robust, sub-rectangular acromial process projects away from the axis of the scapular blade at an angle of approximately 45°. The scapular glenoid has a gently concave, kidney-shaped surface. When the scapular blade is held in a horizontal plane, the glenoid faces primarily laterally and slightly dorsally, as reported (Zhou, 1995a:12). It also is canted to face slightly cranially. The costal (dorsal) and lateral (ventral) margins of the blade are gently arched and straight, respectively. The blade is broadest about two-thirds of the way down its length and tapers to a rounded distal end (IVPP V9769A).

Coracoid. Very little of the coracoid is exposed in the holotypic specimen. The coracoids of the referred specimen, however, are well exposed and resemble those in other enantiornithines (Chiappe and Walker, Chapter 11 in this volume). A rounded and very prominent acrocoracoid process projects dorsally from the proximal end of the coracoid. The shaft broadens distally into a curved sheet of bone with convex cranial, and concave caudal, surfaces. The lateral margin is convex. The straight distal articular edge for the sternum is offset slightly from a perpendicular to the axis of the coracoid.

Furcula. The robust Y-shaped furcula is exposed in both holotypic and referred specimens. The clavicular rami join ventrally, with an intrafurcular angle of approximately 50°. The long hypocleideum is blade-shaped with sharp cranial and caudal margins. In both specimens, the edges of the hypocleideum join a crest along the edge of each dorsal ramus. The shaft is flattened near the distal end of the dorsal ramus and curves laterally toward the acromion, as in *Concornis* (UAM LH-2814). Given the prominence of both the scapular acromion and coracoidal acrocoracoid process, there can be little doubt that the curved articular end of the dorsal process of the furcula completed a triosseal canal that would have accommodated the tendon of the supracoracoideus muscle, as in living birds.

Sternum. The broad plate-shaped sternum is partially preserved in ventral view in the holotypic specimen and nearly fully exposed in dorsal and ventral views in the referred specimen. The cranial margin is thickened with broad, trough-shaped articular surfaces for the coracoids (IVPP V9769B). A short median slit accommodates the dorsal edge of the hypocleideum of the furcula as in *Eoalulavis* (Sanz et al., 1996). The slender caudolateral processes flare at their distal ends, and the caudal margin of the sternum is W-shaped (Zhou, 1995a: Fig. 5a), as in *Concornis* (Sanz et al., 1995).

Forelimb

Humerus. The holotypic specimen preserves the proximal and distal ends of the right humerus. Both humeri are preserved in the referred specimen. At the proximal end, both specimens show the hypertrophied bicipital tubercle that characterizes enantiornithines (Walker, 1981; Chiappe, 1996). In the holotypic specimen, this tubercle has a curved columnar form with a concave distal end, as in *Eoalulavis* (UAM LH 13500; Sanz et al., Chapter 9 in this volume). The distal condyles of the humerus are similar to those in other enantiornithines. In both specimens, the radial condyle is transversely narrower and more prominent than the ulnar condyle, but the radial condyle is not shifted significantly proximally as in Neornithes. The ulnar condyle is broad and

has a distinctly flattened articular surface. The saddle-shaped form of the humeral head and other features are characteristic of enantiornithines and are well described elsewhere (Chiappe, 1996).

Radius and Ulna. The radius and ulna are well exposed in the referred specimen; in the holotypic specimen, the left forearm is folded against the humerus, and the right forearm is preserved only proximally. The radius and ulna are very similar to those in *Concornis, Eoalulavis,* and other enantiornithines (Chiappe, 1996; Chiappe and Walker, Chapter 11 in this volume). The slightly expanded proximal end of the radius has an oval proximal articular surface, a subtriangular facet for the ulna, and a low biceps tuberosity (IVPP V9769A). The radial shaft is marked by a shallow groove (Zhou, 1995a:13) as in other enantiornithines and *Ichthyornis.* The distal end is curved and strongly flattened, with a narrow distal articular surface.

The ulnar shaft has a minimum width approximately twice that of the radius. The proximal one-third of the shaft curves toward the elbow joint. A biceps tubercle is located just before this articulation. The distal end has an arcuate external condyle and a trough-shaped intercondylar sulcus, as noted by Zhou (1995a:18). There are no papillae for flight feathers on the shaft of the ulna in either specimen.

Proximal Carpals. The radiale and ulnare have yet to be described in detail among enantiornithines. Both ossifications are preserved in the holotypic specimen, although only the form of the radiale is well shown. Initially identified as the ulnare (Sereno and Rao, 1992: Fig. 3E), the left radiale of the holotypic specimen is preserved disarticulated a short distance from the wrist joint. Its concave, heart-shaped distal articular surface is exposed. In the referred specimen, the right radiale is also disarticulated a short distance from the wrist joint and is exposed in proximal and distal views. The left radiale, however, is in place at the distal end of the ulna, with its concave, heart-shaped distal articular surface exposed. This surface articulates against the convex trochlea of the semilunate carpal.

The left ulnare is present in the holotypic specimen but not well exposed. It is preserved in place on the external side of the left wrist joint, articulating in the angle between the forearm and manus. It appears to have been similar to the large V-shaped ulnare in the enantiornithines *Eoalulavis* (UAM LH 13500; Sanz et al., 1996) and *Neuquenornis* (Chiappe and Calvo, 1994: Fig. 1), in the euornithine *Ichthyornis* (Marsh, 1880: Pl. 30), and in Neornithes. The internal surface of this V-shaped bone forms a deep articular notch for the metacarpus, an unusual configuration that is absent in *Archaeopteryx* and only poorly formed in *Confuciusornis.* In the latter, the ulnare has a broader trough shape and is only about one-half the size of the radiale. An enlarged V-shaped ulnare, therefore, appears to have evolved in basal ornithothoracines (Sereno, 1999a: Fig. 4).

Carpometacarpus. The carpometacarpus is preserved in both specimens and is at least partially co-ossified. Prior to acid preparation of the holotypic specimen, the fragmented base of the right carpometacarpus gave the appearance that metacarpal II and the semilunate carpal had yet to co-ossify (Fig. 8.3; Sereno and Rao, 1992:845). This portion of the carpometacarpus is better preserved on both sides of the referred specimen, in which metacarpal II and the semilunate carpal are clearly co-ossified. The same is not uniformly true of metacarpal I, which in the left manus of the holotypic specimen is disarticulated, lying near the base of metacarpal III (Fig. 8.3). The length of the disarticulated left metacarpal I, however, suggests that it may include part of the semilunate carpal, which may have broken away from the remainder of the carpometacarpus. In the right carpometacarpus of the holotypic specimen, the distal condyles of right metacarpal I are in their proper position alongside metacarpal II, but the base of the bone has broken away. In the referred specimen, metacarpal I is preserved in place in both manus and appears to be co-ossified with the medial edge of the semilunate carpal. The trochlea of the semilunate carpal is situated largely proximal to metacarpal II.

Metacarpal I is very short, although longer than that shown in an enantiornithine from Argentina (Walker, 1981: Fig. 2F). The medial margin of the proximal half of the shaft is convex and may constitute a rudimentary extensor process (IVPP V9769A), similar to that seen in *Confuciusornis, Archaeopteryx,* and many nonavian maniraptorans. The distal condyles are best exposed in the holotypic specimen and closely resemble those in *Archaeopteryx;* the deeper medial distal condyle projects farther distally and is raised above the level of the medial condyle. These features, and the well-developed dorsal extensor depression between the condyles, indicate that the phalanges of the first digit were deflected medially and were capable of substantial rotation against the metacarpal.

Metacarpal II is the most robust. The shaft appears to be subquadrate in cross section but is dorsoventrally flattened in sections by postmortem crushing. There is no dorsal extensor depression distally. Distal condyles are not apparent in dorsal view, because they are beveled proximoventrally. The distal condylar surface is developed as a broad saddle, with the condyles canted medially.

An interosseous space separates metacarpal III from metacarpal II, and, although fused proximally, these bones remain separate distally. As in most neornithines, the shaft of metacarpal III is transversely flattened, particularly in its proximal one-half. The shaft, therefore, is readily deflected by postmortem crushing, a situation that has given rise to

spurious interpretations, such as the absence of an interosseous space in *Sinornis* (Zhou, 1995a: Fig. 8) or the notion that the shaft is as robust as that of metacarpal II in *Neuquenornis* (Chiappe and Calvo, 1994:236). As in other enantiornithines (Chiappe and Walker, Chapter 11 in this volume) and some neornithines, metacarpal III is longer than metacarpal II, and the medially curving shaft terminates in subequal distal condyles, as seen in both specimens.

Phalanges. The phalangeal formula is 2-3-1, with digits I and II terminating in recurved unguals (Fig. 8.3; Zhou, 1995a). Although the manual phalanges are completely preserved in *Sinornis,* evidence from other enantiornithines is consistent with this pattern (e.g., *Concornis, Eoalulavis*).

The first phalanx of digit I is long, slender, and gently arched dorsoventrally. The distal end has well-developed condyles, extensor and flexor pits, and medial and lateral collateral ligament pits. The ungual is slightly shorter and more recurved than that on digit II (Fig. 8.3). The tip of the ungual of digit I falls short of the length of metacarpal II, suggesting that digit I may be relatively shorter in *Sinornis* than in some other enantiornithines (e.g., *Eoalulavis*).

The first phalanx of digit II has a concave proximal articular surface, weakly divided into two unequal articular facets (IVPP V9769). A proximally projecting heel is present on the ventral aspect of the base, distal to which is located a large foramen and associated canal (Fig. 8.3). This unusual foramen, which has never been reported before among avians, is present in both the holotypic and referred specimens and in the enantiornithine *Eoalulavis* (UAM LH 13500). It may have accommodated a flexor tendon on the ventral aspect of the digit in some as yet poorly understood configuration; the foramen and canal seem too large to have transmitted neurovascular structures and are difficult to envision as a pneumatic invasion of the phalanx (Sereno, 2000). A flange is present on the lateral side of the distal two-thirds of the shaft, and, as a result, the phalanx appears to broaden in width distally until just before the condyles. The distal condyles are weakly separated, and there are no extensor and flexor depressions or collateral ligament pits.

The second phalanx of digit II tapers distally, although, as in the preceding phalanx, a lateral flange is present. The distal trochlea is well developed and symmetrical, with extensor and flexor depressions and collateral ligament pits (Fig. 8.3). The ungual is larger and less recurved than that of the first digit, although both unguals are relatively shorter and less recurved than those in *Archaeopteryx* and *Confuciusornis.*

The sole phalanx of digit III is anchored laterally against the shaft of the first phalanx of digit II (Fig. 8.3). In the holotypic specimen, only the proximal third of the shaft articulates against the adjacent phalanx. The remainder of the shaft curves away (Fig. 8.3). In the referred specimen, left

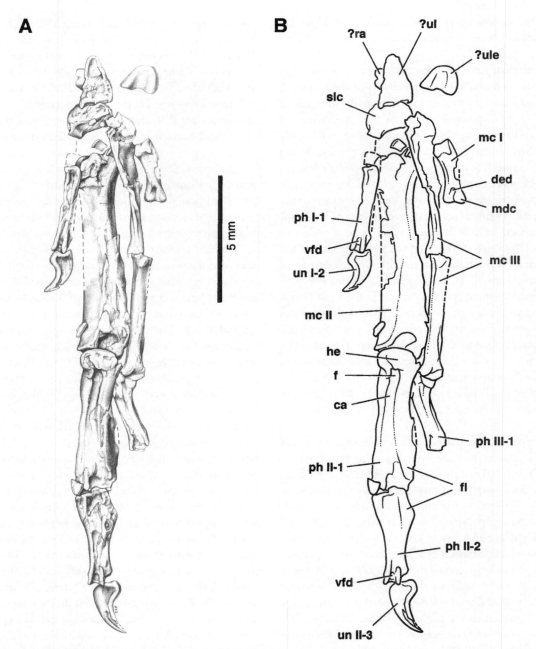

Figure 8.3. Right carpus and manus of *S. santensis* (BPV 538a, epoxy cast), mostly in ventral view (A, B). Metacarpal I is in dorsal view, digit I ungual is in medial view, and digit II ungual is in lateral view. Broken lines indicate bone margins visible as impressions. Abbreviations: I–III, metacarpals or digits I–III; 1–3, phalangeal number; ca, canal; ded, dorsal extensor depression; f, foramen; fl, flange; he, heel; mc, metacarpal; mdc, medial distal condyle; ph, phalanx; ra, radius; slc, semilunate carpal; ul, ulna; ule, ulnare; un, ungual; vfd, ventral flexor depression.

and right sides show a different pattern. There is a narrow medial flange that contacts the adjacent phalanx along most of its straighter shaft. Only the narrow distal end is free and, despite the absence of an additional ossified phalanx, appears to be developed as a rudimentary condyle.

Pelvic Girdle

In the holotypic specimen, the right half of the pelvic girdle has been displaced ventrally relative to the left side, and the sacrum has been rotated to expose its ventral side (Fig. 8.4). In the referred specimen, the pelvic girdle is completely disarticulated. It is clear that the pelvic girdle and sacrum were not fully co-ossified in either specimen. The natural configuration of the pelvic girdle is best preserved on the left side of the holotypic specimen, where two of the three main articulations (iliopubic and ilioischiadic) appear to be co-ossified. The ischiadic and pubic shafts parallel each other, and the pubic shaft is rotated significantly caudal to a vertical orientation. The acetabulum is broadly open medially.

Ilium. The ilium has a deep preacetabular process and a narrower, arched postacetabular process (Fig. 8.4) as in other enantiornithines (Walker, 1981; Sanz et al., 1996; Sereno, 2000; Chiappe and Walker, Chapter 11 in this volume). The pubic peduncle is longer than the ischiadic peduncle, as in *Archaeopteryx* and many other paravians. The pubic peduncle is narrower craniocaudally than in *Archaeopteryx* (Wellnhofer, 1974), *Rahonavis* (Forster et al., 1998), and dromaeosaurids but is quite similar to that in *Confuciusornis*. Well-developed dorsal and ventral antitrochanters are present on the dorsal margin and ischiadic peduncle of the ilium, respectively. The ventral antitrochanter is prominent (Fig. 8.4); it is partially broken away in the exposed left ilium of the referred specimen, which is why its presence was formerly questioned (Zhou, 1995a:13). As in other enantiornithines, the dorsal rim of the postacetabular process caudal to the dorsal antitrochanter projects laterally (Fig. 8.4; Walker, 1981: Fig. 2C).

Ischium. The iliac peduncle of the ischium is articulated with the ilium in the holotypic specimen (Fig. 8.4). The ventral antitrochanter is shared between these bones, the ventral one-quarter of which forms a raised platform on the acetabular margin of the ischium. The remainder of the acetabular margin of ischium is rounded and tapers in width toward the end of the pubic peduncle, which is developed as a broad, thin sheet of bone.

The proximodorsal process of the ischium is developed as a tongue-shaped flange, the tip of which articulates on the medial aspect of the postacetabular process of the ilium. The elongate ischiadic blade is scimitar-shaped, gradually tapering toward its gently curved tip. At its narrowest width, the blade is approximately one-tenth of total ischiadic length. The distal portion of the blade is exposed on the right side of the holotypic specimen (Fig. 8.4). A reflected lamina is present on the lateral side of the blade, an unusual feature that can also be observed on the left ischium of the referred specimen (IVPP V9769B). At its proximal end, the reflected lamina is about one-half the width of the blade. Distally, the lamina decreases in width to form the laterally protruding dorsal margin of the distal one-half of the blade. The shorter length of the ischiadic blade in a previous reconstruction (Zhou, 1995a: Fig. 10B) was based on the incomplete distal ends of the ischia of the referred specimen.

Pubis. The pubis is longer than the ischium. Its cranial margin, from the iliac peduncle to the foot, is gently curved caudally. The acetabular margin of the pubis is broad and joins the prominent supracetabular crest of the ilium (Fig. 8.4). The plate-shaped ischiadic peduncle is very short; its height exceeds its length. The proximal one-half of the pubic shaft appears to have a subtriangular cross section; the medial side is flattened, and a rounded edge is present laterally. The distal two-thirds of the shaft is blade-shaped, with a rounded lateral edge and sharp medial edge. Proximal to the foot, the blade expands in width, as is best exposed in the right pubis of the holotypic specimen (Fig. 8.4). This well-preserved distal end suggests that the pubic symphysis, which had yet to co-ossify in either specimen, was limited to the distal one-quarter of the pubic blade. The subtriangular pubic foot curves caudally.

Hindlimb

Femur. The femoral head is approximately 60% the maximum craniocaudal width of the femur; it is not particularly small in this regard. It is separated from the greater trochanter by a shallow articular saddle. A prominent caudal trochanter, preserved in both femora of the referred specimen, is present on the caudolateral side of the proximal end, distal to the greater trochanter. A trochanter of similar position and size is present in *Archaeopteryx* and other basal paravians. The shaft is very gently caudally curved. There is no trace of a fourth trochanter. The distal condyles are subequal in size and separated caudally by the popliteal fossa. The lateral distal condyle has a distinct caudal extension, the crista tibiofibularis (IVPP V9769A).

Tibiotarsus. The proximal articular surface is oval, only slightly longer craniocaudally than broad. The proximal end of the tibiotarsus appears to have a very low, laterally projecting cnemial crest, separated proximally from the remainder of the proximal articular surface by a shallow groove (contra Zhou, 1995a:14). This area is only preserved in the tibiae of the referred specimen and has suffered some crushing (IVPP V9769A, B). The fibular crest projects from the lateral aspect of the shaft as a thin flange, reaching its

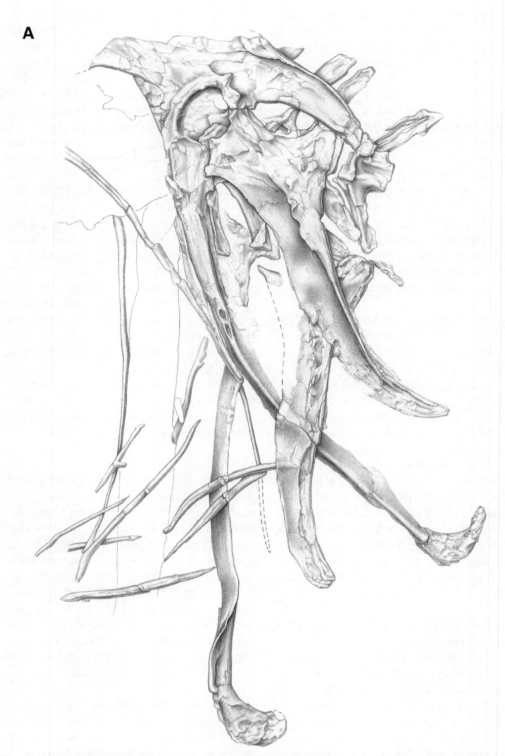

A

Figure 8.4. Sacral and cranial caudal vertebrae, caudal dorsal ribs, and pelvic girdle of *S. santensis* (BPV 538a, epoxy cast), as shown in full-tone (A) and outline (B) drawings. Vertebrae are in ventral view, and pelvic elements are in left lateral view. Broken lines indicate bone margins visible as impressions. Abbreviations: acet, acetabulum; ach, articular surface for chevron; c, centrum; ca1, first caudal vertebra; dant, dorsal antitrochanter; dpr, proximodorsal process; fe, femur; fl, flange; il, ilium; is, ischium; pfo, pubic foot; poap, postacetabular process; pped, pubic peduncle; prap, preacetabular process; pu, pubis; r, rib (also right); s1–8, sacral vertebrae 1–8; sym, symphyseal surface; tp, transverse process; vant, ventral antitrochanter.

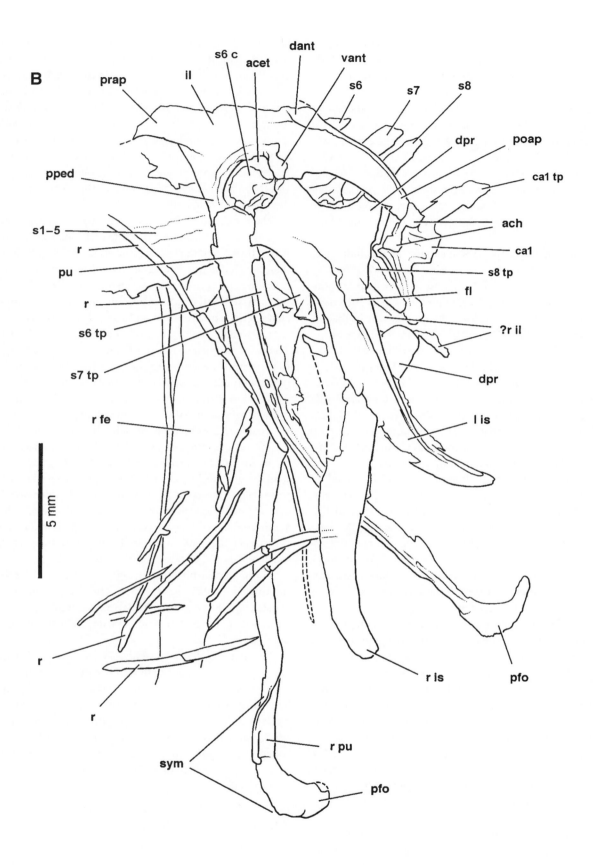

broadest width near its distal end. The distal condyles are large; in side view, the lateral condyle is nearly circular, whereas the medial condyle projects more strongly caudally.

Fibula. The fibula, disarticulated in both specimens, extends along only the upper one-third of the tibiotarsus. The flattened proximal end is triangular, the caudal corner projecting farther from the axis of the shaft (IVPP V9769B). The iliofibularis tubercle is developed as a distinct swelling on the craniolateral margin of the shaft (BPV 538a). Distal to this tubercle, the shaft tapers into a narrow rod that extends a short distance beyond the fibular crest on the tibia.

Tarsometatarsus. The tarsometatarsi of the holotypic specimen are well preserved and exposed, showing the distal, fully retroverted (anisodactyl) position of digit I. Zhou (1995a:14) described the absence of a hypotarsus and commented on the length and form of metatarsal II in the referred specimen (IVPP V9769B), which shows the expanded, medially deflected distal condyles that characterize the holotypic specimen and most other enantiornithines (Chiappe, 1993). Co-ossified distal tarsals appear to form a slightly expanded cap fused to the proximal ends of metatarsals II–IV, although no sutures remain. A very short section of the suture between the proximal ends of metatarsals II and III is also co-ossified. This co-ossification has maintained the integrity of the tarsometatarsi in both specimens.

Metatarsal I has a splint-shaped shaft and is located at the distal end of metatarsal II (Figs. 8.5–8.7). On the right side of the holotypic specimen, digit I has shifted caudally, so that metatarsal I is in contact with metatarsal IV (Fig. 8.6). In side view, metatarsal I is J-shaped as a result of its bulbous distal condyle, which is single cranially but divided by a groove caudally. A broad, asymmetrical dorsal extensor depression is present just behind the condyle, and the collateral ligament pit is deeper on the medial side.

Metatarsals II–IV are very similar to those in other basal enantiornithines, such as *Iberomesornis* (Sanz and Bonaparte, 1992; Sereno, 2000; Sanz et al., Chapter 9 in this volume) and *Concornis* (Sanz et al., 1995). A suite of derived enantiornithine features include the very broad distal condyles of metatarsal II (Fig. 8.6), the caudally prominent medial distal condyle of metatarsal III (Fig. 8.7), extremely narrow shaft of metatarsal IV (Fig. 8.5), and C-shaped distal condyles of metatarsal IV (Chiappe, 1993, 1996). The derived form of metatarsal I and the reduced size of metatarsal IV are also visible in the referred specimen (IVPP V9769B). Metatarsal V is absent.

Phalanges. The pedal phalanges are articulated and well exposed in the holotypic specimen. The phalangeal formula is 2-3-4-5-*x* (Figs. 8.5, 8.6). Horny claw sheaths are partially preserved, especially the hard outer rim (Figs. 8.5–8.7). This keratinized outer rim extends as a slender arcuate extension from the tip of each ungual and is preserved in several enantiornithine specimens encased in lacustrine sediments. The juncture between the tip of the ungual and the outer rim of the claw sheath is visible under high magnification in the holotypic specimens of *Iberomesornis* and *Concornis,* which have not undergone acid preparation. After acid preparation, such a distinction is not possible; Sereno and Rao (1992) misinterpreted the slender outer rim of the claw sheaths in the holotypic specimen as the particularly slender tips of the pedal unguals.

The proximal phalanx of digit I is robust and almost as long as the proximal phalanx of digit III (Fig. 8.7). The base is very broad, and the lateral distal condyle is slightly longer than the medial distal condyle, as in *Concornis.* The ungual is as long and recurved as that in digit II. *Sinornis,* like the other basal enantiornithines *Iberomesornis* and *Concornis,* is characterized by a robust, long, and fully retroverted digit I.

The phalanges of digit II are distinctive in several regards. The proximal phalanx has a particularly broad base (Fig. 8.6) to accommodate the transversely expanded distal condyles of metatarsal II. The penultimate phalanx is longer than that proximal to it, as in digits III and IV. Its distal condyles are more expanded and the intercondylar depression is deeper than comparable phalanges of digits III or IV (either the second or penultimate phalanges). Likewise, the ungual is more recurved and the flexor tubercle significantly more robust than the unguals in digits III or IV (Figs. 8.5, 8.6). These features are also present in *Archaeopteryx* (Wellnhofer, 1974: Fig. 13) and *Confuciusornis* and may well be homologous with the similar, but hypertrophied, condition of digit II in *Rahonavis* (Forster et al., 1998) and basal paravians, such as dromaeosaurids and troodontids (Sereno, 1999a).

The proximal phalanx of digit III is the most robust nonungual phalanx in the pes. The second phalanx is shorter, and the penultimate phalanx is as long as, but more slender than, the first phalanx. The ungual is the longest in the pes.

The first two phalanges of digit IV are subequal in length (Figs. 8.5, 8.7), whereas the second is significantly shorter in *Archaeopteryx* and *Confuciusornis* (Chiappe et al., 1999). The third phalanx is somewhat shorter, and the penultimate phalanx is much longer. These proportional differences are readily understood as adaptations that enhance perching function (Hopson and Chiappe, 1998; Hopson, 2001). The ungual is shorter than that of digits I–III and is the least recurved in the pes.

Phylogenetic Relationships

Basal Avian Relationships

Recent years have witnessed a profound transformation in the quality of the early avian fossil record, as this volume at-

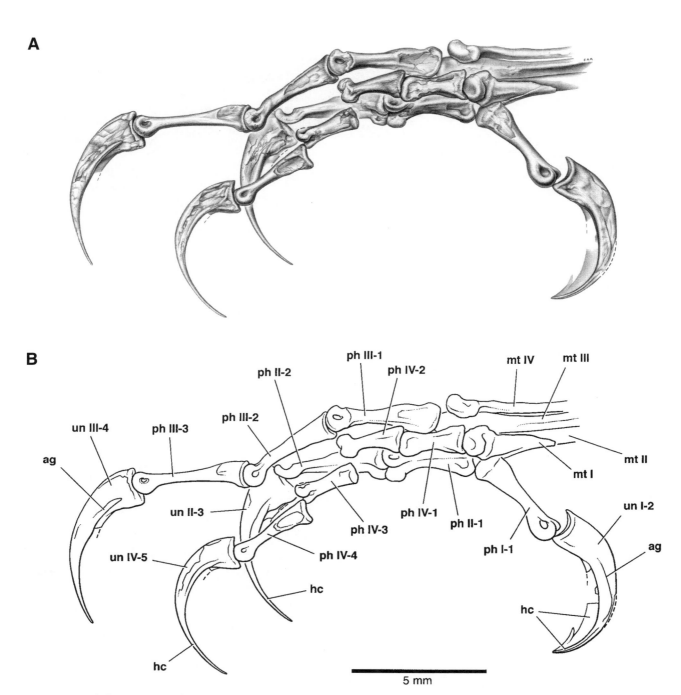

Figure 8.5. Left distal metatarsals (most in dorsal view) and pedal phalanges (most in medial view) of *S. santensis* (BPV 538a, epoxy cast) (A, B). Broken lines indicate the margins of bone or horny claw that are visible as impressions. Abbreviations: I–IV, metatarsals or digits I–IV; 1–5, phalangeal number; ag, attachment groove; hc, horny claw; mt, metatarsal; ph, phalanx; un, ungual.

tests. Despite the presence of several taxa of uncertain association or controversial affinity, such as *Protoavis* (Chatterjee, 1995) and the alvarezsaurids (Chiappe et al., 1996; Novas, 1997; Sereno, 1997, 1999a, 2001; Chiappe et al., 1998), a phylogenetic framework has emerged (Chiappe, 1995, 1996; Sereno 1997, 1999a; Chiappe et al., 1998). This framework

(Fig. 8.9C) constitutes a consensus of recent cladistic analyses with explicit outgroups that attempt to incorporate the increasing quantity of skeletal data now available. In a recent analysis (Sereno, 1999a), for example, approximately 100 synapomorphies have been identified in support of three nodes at the base of Aves (Fig. 8.9C).

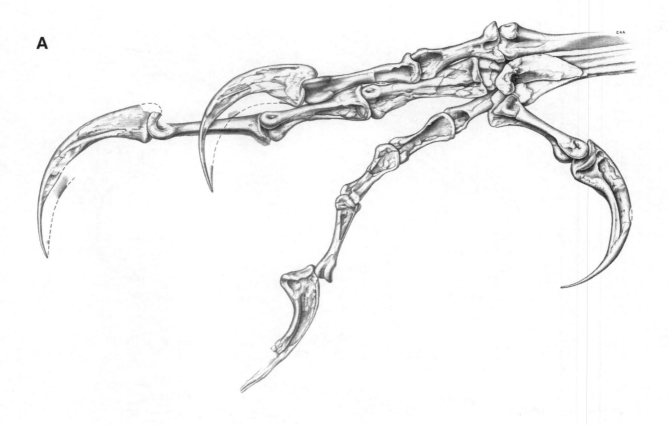

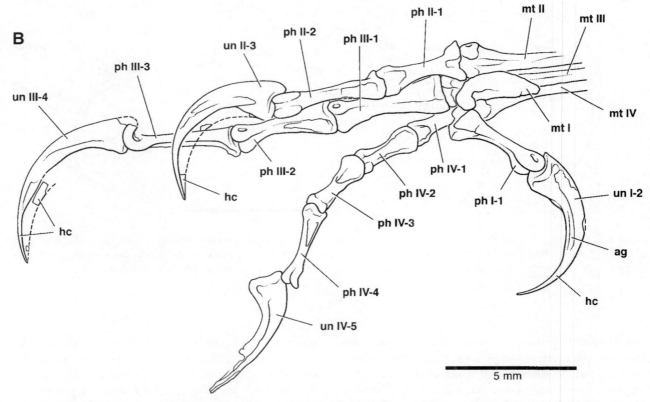

Figure 8.6. Right distal metatarsals and pedal phalanges of *S. santensis* (BPV 538a, epoxy cast), most in ventral view (A, B). Broken lines indicate the margins of bone or horny claw that are visible as impressions. See Figure 8.5 for abbreviations.

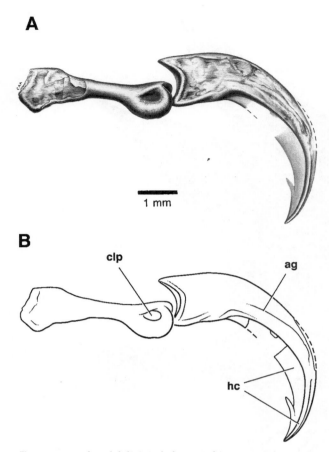

A

1 mm

B

clp

ag

hc

Figure 8.7. Left pedal digit I phalanges of *S. santensis* (BPV 538a, epoxy cast) in lateral view (A, B), showing the side of digit I that is farthest from digit II. Broken lines indicate the margins of bone or horny claw that are visible as impressions. Abbreviations: clp, collateral ligament pit; ag, attachment groove; hc, horny claw.

Enantiornithine Status of *Sinornis*

Sereno and Rao (1992) initially described the holotypic specimen of *S. santensis* as the most basal avian after *Archaeopteryx* (Fig. 8.10A). At that time, articulated remains of other basal avians were rare and included only *Ambiortus* (Kurochkin, 1985), a partial postcranial skeleton from Mongolia, and *Iberomesornis* (Sanz et al., 1988; Sanz and Bonaparte, 1992), a more complete postcranial skeleton from Spain. It is now clear that the two features used by Sereno and Rao (1992) to establish the basal position of *Sinornis* relative to *Ambiortus* and *Iberomesornis*—the presence of gastralia and a plate-shaped coracoid—were based on misidentifications (of rib tips and a portion of the left humerus, respectively).

Sereno and Rao (1992) did not compare *Sinornis* with the available disarticulated material of enantiornithines (Walker, 1981; Molnar, 1986). As more complete enantiornithine remains came to light (Chiappe, 1991, 1993; Sanz and Buscalioni, 1992; Zhou et al., 1992, 1995a; Chiappe and

Calvo, 1994; Sanz et al., 1995), the status of *Sinornis* as an enantiornithine was proposed (Chiappe, 1995; Zhou, 1995a). *Ambiortus* and *Iberomesornis*, in turn, also appear to belong within Enantiornithes. Initially, these genera were allied with Enantiornithes based on overall similarity rather than synapomorphy (Martin, 1995; Kurochkin, 1996). More recently, several enantiornithine synapomorphies were highlighted in a revision of the skeletal anatomy of *Iberomesornis* (Sereno, 2000).

Enantiornithine synapomorphies were first outlined by Walker (1981) and Chiappe (1991, 1993) based largely on disarticulated material. Chiappe and Calvo (1994) and Chiappe (1996) listed 21 and 22 synapomorphies, respectively. Kurochkin (1996:32–33) listed 34 characters in his diagnosis of Enantiornithes, but these constitute a mixture of enantiornithine synapomorphies and symplesiomorphies that characterize many other avians (e.g., "digit I reversed").

Sereno (2000) accepted 12 of the 22 synapomorphies listed by Chiappe (1996:213) and added another 15 for a total of 27 enantiornithine synapomorphies. Most of these 27 synapomorphies are present in *S. santensis*, but the following 19 are the easiest to observe in the postcranial skeleton (Fig. 8.8; BPV 538a; IVPP V9769A, B). In the axial column, (1) the parapophyses of cranial dorsal vertebrae are positioned in the center of the centrum. In the pectoral girdle, (2) the dorsolateral rami of the furcula are L- or V-shaped in cross section; (3) the hypocleideum is 40% or more of the length of the dorsolateral ramus of the furcula; (4) the scapular acromial process is deflected medially at about 45° from the long axis of the scapula; and (5) the ventral ramus of the coracoid has a convex lateral margin and is transversely arched. In the forelimb, (6) the proximal apex of the humeral head is gently saddle-shaped; (7) the internal tuberosity (located on the proximal end of the humerus ventral to the head) is associated with a pneumatic opening; (8) a humeral dorsal trochanter is present; (9) the humeral ventral trochanter is developed as a pendant process with a subtriangular cross section and terminal fossa; (10) humeral distal condyles with transverse axis angled ventromedially at approximately 25° from the perpendicular to the shaft axis; (11) humeral distal end with maximum width across the epicondyles three times maximum craniocaudal depth; and (12) foramen present on the ventral aspect of the base of manual phalanx II-1. In the pelvic girdle, (13) the iliac postacetabular process has a maximum depth one-third the maximum depth of the preacetabular process. In the hindlimb, (14) the distal condyles of metatarsal I are reflected caudally; (15) the distal condyles of metatarsal I are joined along their dorsal margin; (16) the distal condyles of metatarsal II and the proximal articular surface of opposing phalanx II-1 are broader in width than those of metatarsal III and phalanx III-1; (17) the medial distal condyle of metatarsal III extends significantly caudal to the lateral

Figure 8.8. Silhouette reconstruction of
S. santensis showing the preserved bones,
based on BPV 538 and IVPP V9769.

condyle; (18) the distal shaft of metatarsal IV is much narrower than that of metatarsals II and III; and (19) the distal
condyles of metatarsal IV are **C**-shaped in distal view. These
synapomorphies were adapted from Chiappe, 1991 (synapomorphy 5); Chiappe, 1993 (synapomorphies 14, 16–19); Chiappe and Calvo, 1994 (synapomorphies 6, 11); Chiappe, 1996
(synapomorphies 1, 2); and Sereno, 2000 (synapomorphies
3, 4, 7–10, 12, 13, 15).

Relationships within Enantiornithes

What are the relationships of *S. santensis* within Enantiornithes? Enantiornithine relationships have been considered by Chiappe (1993), Sanz et al. (1995), and Kurochkin
(1996). These studies highlight just how little is known
about the internal structure of this major avian radiation.

Chiappe (1993) described 10 characters in six enantiornithine genera, using *Mononykus* and *Patagopteryx* as out-
groups (Fig. 8.10B). Three of these characters (1, 4, 10) are
enantiornithine synapomorphies and, as scored by Chiappe
(1993: Fig. 13), are uninformative regarding relationships
among enantiornithines. Three of the remaining characters
(2, 5, 6; metatarsal I distal condyles strongly reflected or "J-
shaped"; caudally extended medial distal condyle on
metatarsal III; and metatarsal IV distal condyles **C**-shaped)

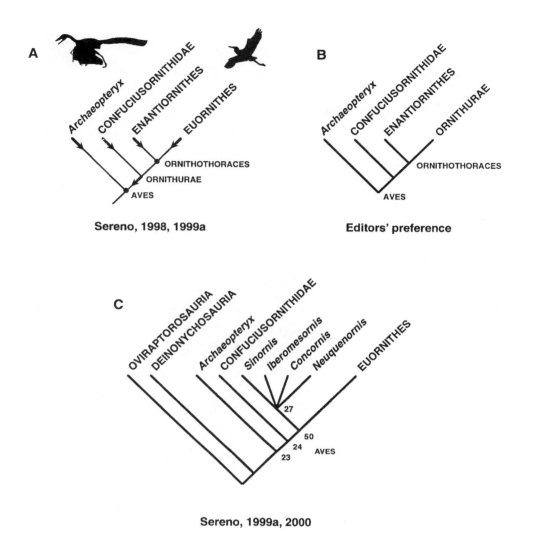

Figure 8.9. Basal avian taxonomy and phylogeny. (A) Higher-level avian taxonomy (from Sereno, 1998, 1999a). Dots and arrows indicate taxa with node-based and stem-based definitions, respectively. Euornithes Dementjev 1940 was inadvertently listed as a new taxon (Sereno, 1998). (B) Editors' preferred higher taxonomy for the same basal phylogeny as in (A), differing in use of the taxon Ornithurae. (C) Higher-level avian phylogeny, with number of synapomorphies shown at each node (under delayed character-state optimization). Alvarezsaurids are regarded as ornithomimosaurs (and so are not included on the cladogram; Sereno, 1999a, 2000), and *Iberomesornis* is regarded as an enantiornithine (Sereno, 2000).

we also regard as enantiornithine synapomorphies; these features, for example, are present in *Sinornis*. One of the remaining characters (4; tubercle for the tibialis cranialis muscle on metatarsal II) is probably plesiomorphic within Enantiornithes, given its presence in the outgroups *Confuciusornis* and dromaeosaurids (Norell and Makovicky, 1997:10–11, Fig. 6). Although Chiappe correctly reported three minimum-length (12-step) trees with a consistency index of 0.83, the robustness of this result was not considered. If the data as presented are accepted, all structure among enantiornithines breaks down with one additional step (13 steps, 17 trees; Fig. 8.10B).

Sanz et al. (1995) added *C. lacustris* to the matrix in Chiappe (1993), using the same 10 characters and pair of outgroups (Fig. 8.10C). Their results were the same: three minimum-length trees with the same consistency index. After examining the holotypic skeleton of *Concornis*, we would score four characters (2, 4, 6, 7) with the derived state (as opposed to primitive or missing states). Nonetheless, if the data as originally presented are accepted, all structure within Enantiornithes breaks down with one additional step (13 steps; 32 trees; Fig. 8.10C). This is not the result of a single incomplete taxon. If *Lectavis*, the least complete taxon in the matrix (50% complete), is removed, all struc-

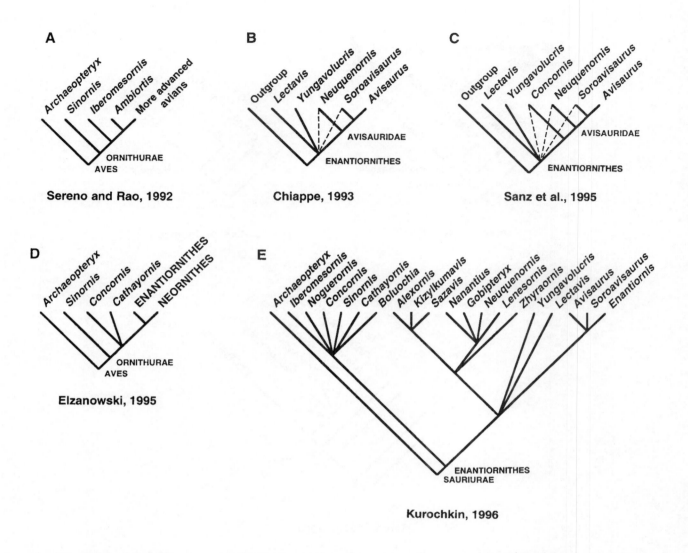

Figure 8.10. Enantiornithine relationships. (A) Phylogeny presented by Sereno and Rao (1992). (B) Relationships proposed by Chiappe (1993), with broken lines showing loss of all resolution for trees one step longer than the minimum. (C) Minimum-length tree presented by Sanz et al. (1995), with broken lines showing loss of all resolution for trees one step longer than the minimum. (D) Phylogenetic diagram presented by Elzanowski (1995). (E) Phylogenetic diagram presented by Kurochkin (1996) showing phylogenetic resolution that is impossible to generate from its associated data matrix.

ture still breaks down with one additional step (13 steps, 13 trees). The tarsometatarsus alone, we conclude, is insufficient to clarify much about higher-level relationships within Enantiornithes.

Elzanowski (1995: Fig. 4) presented a cladogram showing a paraphyletic arrangement of taxa here considered enantiornithines (Fig. 8.10D). Eight synapomorphies were listed to demonstrate that some enantiornithines were more advanced than *Concornis*, *Sinornis*, and *Cathayornis*. As argued previously, *Cathayornis* is a junior synonym of *Sinornis*. The five synapomorphies Elzanowski listed to unite *Concornis*, *Cathayornis*, and other avians to the exclusion of *Sinornis* are flawed, such as the absence of a pubic foot (which is clearly present in the holotypic specimens

of *Cathayornis* and *Concornis*), absence of gastralia (also absent in *Sinornis*, as described earlier), and a prominent antitrochanter on the ilium (which is clearly present and prominent in *Sinornis*). The information available to Elzanowski for several of these genera was limited. The holotypic skeletons of *Sinornis* and *Concornis* clearly how many features that establish their inclusion within a monophyletic Enantiornithes (Chiappe, 1991, 1996; Sereno, 2000).

Kurochkin (1996: Fig.13) presented a phylogeny for 19 enantiornithine species (Fig. 8.10E). Although accepting "the principal thesis of a [*sic*] Hennig's taxonomy," Kurochkin (1996:38) suggested that use of parsimony would lead to "an obviously false phylogeny" for basal avians. Perhaps not surprisingly, his "eclectic" phylogeny bears no discernible relation to an extensive matrix of character data tabulated in the same paper. The matrix includes 122 features scored across 40 terminal entities, which include 27 named taxa, several unnamed specimens, one embryo, and some isolated bones. The data are not a list of characters but rather a mixture of primitive and derived character states, which were then scored, with few exceptions, as present or absent. Missing entries abound; 13 terminal entities are scored for 6 or fewer of the 122 features in the matrix. One terminal unit, an isolated femur, is scored for only one state—the absence of a particular feature. Twenty-four features are constant or otherwise uninformative.

These data give rise to an incalculably enormous number of minimum-length trees of 158 steps. If one omits all terminal units except the outgroups (*Archaeopteryx,* Ichthyornithiformes, Neornithes) and the seven enantiornithines considered by Sanz et al. (1995), 80 features are phylogenetically uninformative. The remainder give rise to 72 minimum-length trees of 51 steps, the strict consensus of which yields a basal polytomy encompassing all included enantiornithines. Although we disagree with many aspects of Kurochkin's data, this result seems to be an accurate gauge of current understanding of enantiornithine phylogeny (see Chiappe and Walker, Chapter 11 in this volume, for additional discussion).

S. santensis, I. romerali, C. lacustris, and *Eoalulavis hoyasi* are known from well-preserved skeletons of Early Cretaceous age. Their postcranial remains are remarkably uniform and bear few autapomorphies. Reasonably complete cranial remains are known only for *Sinornis.* The Late Cretaceous *Neuquenornis volans* is less complete but also very similar in structure. It is clear, then, that sparrow- to starling-sized enantiornithines with this conservative skeletal design achieved worldwide distribution during the Early Cretaceous. Their interrelationships, nonetheless, are currently unknown. More derived Late Cretaceous enantiornithines include the long-limbed genera *Nanantius* and *Lectavis* and the asymmetrically footed genus *Yungavolucris* (Chiappe, 1993; Kurochkin, 1996). More complete material of these genera may yield character data that will distinguish subgroups within Enantiornithes.

Paleobiology

S. santensis, the most completely known enantiornithine, provides a detailed view of a basal avian that, judging from its skeleton, would have been similar in flight performance and perching capability to sparrow-sized birds living today in arboreal habitats (Fig. 8.8).

The thorax is strengthened to resist forces generated by an increase in pectoral muscle mass. The coracoids expand distally to form broad, lengthened struts to the sternum, which is enlarged and partially keeled. The cranial ribs are particularly robust and are associated with ossified sternal ribs.

The scapula, coracoid, and furcula articulate to form a triosseal canal adjacent to the shoulder joint. Martin (1995) and Ostrom et al. (1999), to the contrary, have stated that a triosseal canal is absent in Enantiornithes. The form of the three bones that would compose the canal, however, suggests otherwise. In enantiornithines the scapula has a long, robust, and craniomedially deflected acromial process; the coracoid, similarly, has a well-developed, dorsally prominent acrocoracoid process; and the furcula has dorsolateral rami that curve laterally and terminate in an expanded articular head (preserved in *Sinornis, Concornis,* and *Eoalulavis*). The form of these bones and their configuration as preserved in partial articulation (Sanz et al., 1995: Figs. 6, 7) and in restoration strongly suggest that a triosseal canal of modern avian design was present and would have provided passage for the tendon of the supracoracoideus muscle. The tendon would have inserted on the prominent dorsal tuberosity of the humerus. In enantiornithines, the supracoracoideus muscle with its tendinous attachment to the humerus thus appears to have achieved its modern configuration and function—as an important elevator and, especially, rotator of the wing, functions that are critical for takeoff and landing (Poore et al., 1997; Ostrom et al., 1999). The triosseal canal, in contrast, is not present in either *Archaeopteryx* or *Confuciusornis.*

In the wing, the wrist joint is characterized by a laterally displaced, V-shaped ulnare. Unlike in *Archaeopteryx,* in which the ulnare is positioned distal to the ulna, in enantiornithines and other ornithothoracines the ulnare is displaced lateral to the long axis of the ulna. The V-shaped incisure on the ulnare for the metacarpus allowed greater flexion at the wrist during upstroke, which is important in small-bodied fliers for decreasing drag (Jenkins et al., 1988). At the same time, the enlarged ulnare forms a rigid interlocking wrist joint during downstroke (Vazquez, 1992, 1994).

The first digit of the manus bears alular feathers in the enantiornithine *Eoalulavis* (Sanz et al., 1997). The similar slender form of the first digit in *Sinornis* and other enantiornithines suggests that it also may have borne alular feathers. Alular feathers, which are clearly absent in *Confuciusornis* and *Archaeopteryx,* help to prevent turbulent airflow over the wing, enhancing maneuverability at slow speeds.

Several features in the pes suggest that *Sinornis* and other well-known enantiornithines (e.g., *Neuquenornis, Concornis, Iberomesornis*) had achieved advanced perching function and lived primarily in an arboreal habitat. Advanced perching function is indicated by the presence of a fully opposable hallux with a particularly large ungual (Sereno and Rao, 1992). In *Sinornis* the three principal digits of the pes are long compared with the tibia or femur, the penultimate pedal phalanges are lengthened, and the pedal claws are strongly recurved (Peters and Görgner, 1992; Feduccia, 1993). The lengthening of the penultimate phalanges of the pedal digits is probably the best single indicator of habitat preference (Hopson and Chiappe, 1998; Hopson, 2001).

Conclusions

We have drawn the following conclusions based on our study of *S. santensis* and other basal avians:

1. *C. yandica* is a junior synonym of *S. santensis*.
2. *S. santensis* is currently the most completely known enantiornithine and is very close in skeletal form and body size to *I. romerali, C. lacustris,* and *E. hoyasi.*
3. Enantiornithes, defined as all ornithothoracines more closely related to *Sinornis* than to Neornithes, is a monophyletic clade united by many skeletal features.
4. Phylogenetic relationships among enantiornithines remain poorly established, because several genera (*Sinornis, Iberomesornis, Concornis, Eoalulavis, Neuquenornis*) differ in only minor ways and because taxa with more derived features (*Nanantius, Lectavis, Yungavolucris, Avisaurus, Soroavisaurus*) are, to date, very incomplete.
5. The increase in robustness of the ribcage and sternum, a triosseal canal, the advanced design of the wrist bones, the presence of alular feathers, and the design of the pes suggest that *Sinornis* and other similar-sized enantiornithines were arboreal birds that had achieved a level of performance in flight and perching that would be indistinguishable from sparrow-sized birds living today.

Acknowledgments

We thank C. Abraczinskas for execution of the camera lucida drawings and matching labeled figures. We also thank J. L. Sanz, A. D. Buscalioni, L. M. Chiappe, L. D. Martin, L. M. Witmer, and Zhou Z. for discussing aspects of anatomy and for access to specimens in their care.

Literature Cited

Barrett, P. M. 2000. Evolutionary consequences of dating the Yixian Formation. Trends in Ecology and Evolution 15:99–103.

Chatterjee, S. 1995. The Triassic bird *Protoavis. Archaeopteryx* 13:15–31.

Chen P. 1988. Distribution and migration of Jehol fauna with reference to nonmarine Jurassic-Cretaceous boundary in China. Acta Palaeontologica Sinica 27:681–683.

Chiappe, L. M. 1991. Cretaceous avian remains from Patagonia shed new light on the early radiation of birds. Alcheringa 15:333–338.

———. 1992. Enantiornithine (Aves) tarsometatarsi and the avian affinities of the Late Cretaceous Avisauridae. Journal of Vertebrate Paleontology 12:344–350.

———. 1993. Enantiornithine (Aves) tarsometatarsi from the Cretaceous Lecho Formation of northwestern Argentina. American Museum Novitates 3083:1–27.

———. 1995. The first 85 million years of avian evolution. Nature 378:349–355.

———. 1996. Late Cretaceous birds of southern South America: anatomy and systematics of Enantiornithes and *Patagopteryx deferrariisi;* pp. 203–244 *in* G. Arratia (ed.), Contributions of Southern South America to Vertebrate Paleontology, Münchner Geowissenschaftliche Abhandlungen, Reihe A, Geologie und Paläontologie 30. Verlag Dr. Friedrich Pfeil, Munich.

Chiappe, L. M., and J. O. Calvo. 1994. *Neuquenornis volans,* a new Late Cretaceous bird (Enantiornithes: Avisauridae) from Patagonia, Argentina. Journal of Vertebrate Paleontology 14:230–246.

Chiappe, L. M., M. A. Norell, and J. M. Clark. 1996. Phylogenetic position of *Mononykus* (Aves: Alvarezsauridae) from the Late Cretaceous of the Gobi Desert. Memoirs of the Queensland Museum 39:557–582.

———. 1998. The skull of a relative of the stem-group bird *Mononykus.* Nature 392:275–278.

Chiappe, L. M., Ji S., Ji Q., and M. A. Norell. 1999. Anatomy and systematics of the Confuciusornithidae (Theropoda:Aves) from the Late Mesozoic of northeastern China. Bulletin of the American Museum of Natural History 242:1–89.

Cracraft, J. 1986. The origin and early diversification of birds. Paleobiology 12:383–399.

Dementjev, G. P. 1940. Handbook on Zoology. Volume 6, Vertebrates. Birds. Academy of Sciences, Moscow-Leningrad, 856 pp. [Russian].

Elzanowski, A. 1995. Cretaceous birds and avian phylogeny. Courier Forschungsinstitut Senckenberg 181:37–53.

Elzanowski, A., and P. Wellnhofer. 1996. Cranial morphology of *Archaeopteryx:* evidence from the seventh skeleton. Journal of Vertebrate Paleontology 16:81–94.

Feduccia, A. 1993. Evidence from claw geometry indicating arboreal habits of *Archaeopteryx.* Science 259:790–793.

Forster, C. A., S. D. Sampson, L. M. Chiappe, and D. W. Krause. 1998. The theropod ancestry of birds: new evidence from the Late Cretaceous of Madagascar. Science 279:1915–1919.

Gauthier, J. 1986. Saurischian monophyly and the origin of birds; pp. 1–55 *in* K. Padian (ed.), The Origin of Birds and the Evolution of Flight. California Academy of Sciences, San Francisco.

Hong Y. 1988. Early Cretaceous Orthoptera, Neuroptera, Hymenoptera (Insecta) of Kezuo in west Liaoning Province. Entomotaxonomia 10:128–130.

Hopson, J. A. 2001. Ecomorphology of avian and nonavian theropod phalangeal proportions: implications for the arboreal versus terrestrial origin of bird flight; pp. 211–235 *in* J. Gauthier and L. F. Gall (eds.), New Perspectives on the Origin and Early Evolution of Birds: Proceedings of the International Symposium in Honor of John H. Ostrom. Peabody Museum of Natural History, New Haven.

Hopson, J. A., and L. M. Chiappe. 1998. Pedal proportions of living and fossil birds indicate arboreal or terrestrial specialization. Journal of Vertebrate Paleontology 18:52A.

Hou L. 1997. Mesozoic Birds of China. Taiwan Provincial Feng Huang Ku Bird Park, Nan Tou, Taiwan, 228 pp.

Hou L., Zhou Z., L. D. Martin, and A. Feduccia. 1995. A beaked bird from the Jurassic of China. Nature 377:616–618.

Hou L., L. D. Martin, Zhou Z., and A. Feduccia. 1996. Early adaptive radiation of birds: evidence from fossils from northeastern China. Science 274:1164–1167.

Jenkins, F. A., Jr., K. P. Dial, and G. E. Goslow Jr. 1988. A cineradiographic analysis of bird flight: the wishbone in starlings is a spring. Science 241:1495–1498.

Ji Q., Ji S., Ren D., Lu L., Fang X., and Gou Z. 1999. On the sequence and age of the proto-bird bearing deposits in the Sihetun-Jianshangou area, Beipiao, western Liaoning. Professional Papers in Stratigraphy and Paleontology 27:75–80. [Chinese]

Kurochkin, E. N. 1985. A true carinate bird from Lower Cretaceous deposits in Mongolia and other evidence of Early Cretaceous birds in Asia. Cretaceous Research 6:271–278.

———. 1996. A New Enantiornithid of the Mongolian Late Cretaceous, and a General Appraisal of the Infraclass Enantiornithes (Aves). Palaeontological Institute, Moscow, 60 pp.

Marsh, O. C. 1880. Odontornithes: A Monograph on the Extinct Toothed Birds of North America. Report of Geological Exploration of the 40th Parallel. Government Printing Office, Washington, D.C., 201 pp.

Martin, L. D. 1983. The origin and early radiation of birds; pp. 291–338 in A. H. Bush and G. A. Clark Jr. (eds.), Perspectives in Ornithology. Cambridge University Press, Cambridge.

———. 1995. The Enantiornithes: terrestrial birds of the Cretaceous. Courier Forschungsinstitut Senckenberg 181:23–36.

Martin, L. D., Zhou Z., Hou L., and A. Feduccia. 1998. *Confuciusornis sanctus* compared to *Archaeopteryx lithographica*. Naturwissenschaften 85:286–289.

Molnar, R. E. 1986. An enantiornithine bird from the Lower Cretaceous of Queensland, Australia. Nature 322:736–738.

Norell, M. A., and P. J. Makovicky. 1997. Important features of the dromaeosaur skeleton: information from a new specimen. American Museum Novitates 3215:1–28.

Novas, F. E. 1997. Anatomy of *Patagonykus puertai* (Theropoda, Maniraptora, Alvarezsauridae) from the Late Cretaceous of Patagonia. Journal of Vertebrate Paleontology 17:137–166.

Ostrom, J. H., S. O. Poore, and G. E. Goslow Jr. 1999. Humeral rotation and wrist supination: important functional complex for the evolution of powered flight in birds?; pp. 301–309 in S. L. Olson (ed.), Paleontology at the Close of the 20th Century: Proceedings of the 4th International Meeting of the Society of Avian Paleonotology and Evolution, Washington, D.C., 4–7 June 1996. Smithsonian Contributions to Paleobiology 89. Smithsonian Institution Press, Washington.

Padian, K., and L. M. Chiappe. 1998. The origin and early evolution of birds. Biological Reviews 73:1–42.

Peters, D. S., and E. Görgner. 1992. A comparative study on the claws of *Archaeopteryx*; pp. 29–37 in K. E. Campbell Jr. (ed.), Papers in Avian Paleontology, Honoring Pierce Brodkorb. Science Series 36. Natural History Museum of Los Angeles County, Los Angeles.

Poore, S. O., A. Sanchéz-Haiman, and G. E. Goslow Jr. 1997. Wing upstroke and the evolution of flapping flight. Nature 387:799–802.

Rao C. and P. C. Sereno. 1990. Early evolution of the avian skeleton: new evidence from the Lower Cretaceous of China. Journal of Vertebrate Paleontology 10:38–39A.

Sanz, J. L., and J. F. Bonaparte. 1992. A new order of birds (Class Aves) from the Lower Cretaceous of Spain; pp. 39–49 in K. E. Campbell Jr. (ed.), Papers in Avian Paleontology, Honoring Pierce Brodkorb. Science Series 36. Natural History Museum of Los Angeles County, Los Angeles.

Sanz, J. L., and A. D. Buscalioni. 1992. A new bird from the Early Cretaceous of Las Hoyas, Spain, and the early radiation of birds. Palaeontology 35:829–845.

Sanz, J. L., J. F. Bonaparte, and A. Lacasa. 1988. Unusual Early Cretaceous birds from Spain. Nature 331:433–435.

Sanz, J. L., L. M. Chiappe, and A. D. Buscalioni. 1995. The osteology of *Concornis lacustris* (Aves: Enantiornithes) from the Lower Cretaceous of Spain and a reexamination of its phylogenetic relationships. American Museum Novitates 3133:1–23.

Sanz, J. L., L. M. Chiappe, B. P. Pérez-Moreno, A. D. Buscalioni, J. J. Moratalla, F. Ortega, and F. J. Poyato-Ariza. 1996. An Early Cretaceous bird from Spain and its implications for the evolution of avian flight. Nature 382:442–445.

Sanz, J. L., L. M. Chiappe, B. P. Pérez-Moreno, J. Moratalla, F. Hernández-Carrasquilla, A. D. Buscalioni, F. Ortega, F. J. Poyato-Ariza, D. Rasskin-Gutman, and X. Martínez-Delclòs. 1997. A nestling bird from the Lower Cretaceous of Spain: implications for avian neck and skull evolution. Science 276:1543–1546.

Sereno, P. C. 1997. The origin and evolution of dinosaurs. Annual Review of Earth and Planetary Science 25:435–489.

———. 1998. A rationale for phylogenetic definitions, with application to the higher-level taxonomy of Dinosauria. Neues Jahrbuch für Geologie und Paläontologie 210:41–83.

———. 1999a. The evolution of dinosaurs. Science 284:2137–2147.

———. 1999b. Definitions in phylogenetic taxonomy: critique and rationale. Systematic Biology 48:329–351.

———. 1999c. A rationale for dinosaurian taxonomy. Journal of Vertebrate Paleontology 19:788–790.

———. 2000. *Iberomesornis romerali* (Aves, Ornithothoraces) reevaluated as an Early Cretaceous enantiornithine. Neues Jarhbuch für Geologie und Paläontologie Abhandlungen 215:365–395.

———. 2001. Alvarezsauridae (Dinosauria, Coelurosauria): birds or ornithomimosaurs?; pp. 69–98 in J. Gauthier and L. F. Gall (eds.), New Perspectives on the Origin and Early Evolution of Birds: Proceedings of the International Symposium in Honor of John H. Ostrom. Peabody Museum of Natural History, New Haven.

Sereno, P. C., and Rao C. 1992. Early evolution of avian flight and perching: new evidence from the Lower Cretaceous of China. Science 55:845–848.

Sereno, P. C., Chao S., Cheng Z., and Rao C. 1988. *Psittacosaurus meileyingensis* (Ornithischia: Ceratopsia), a new psittacosaur from the Lower Cretaceous of northeastern China. Journal of Vertebrate Paleontology 4:366–377.

Swisher, C. C., III, Wang Y.-Q., Wang X.-L., Xu X., and Wang Y. 1999. Cretaceous age for the feathered dinosaurs of Liaoning, China. Nature 400:58–61.

Vazquez, R. J. 1992. *Archaeopteryx* and powered flight. Research and Exploration 8:387–388.

———. 1994. The automating skeletal and muscular mechanisms of the avian wing (Aves). Zoomorphology 114:59–71.

Walker, C. A. 1981. A new subclass of birds from the Cretaceous of South America. Nature 292:51–53.

Wang X.-L., Wang Y.-Q., Xu X., Tang Z., Zhang F., and Hu Y. 1998. Stratigraphic sequence and vertebrate-bearing beds of the lower part of the Yixian Formation in Sihetun and neighboring area, western Liaoning, China. Vertebrata PalAsiatica 36:96–101.

Wellnhofer, P. 1974. Das fünfte Skelettexemplar von *Archaeopteryx*. Palaeontographica Abhandlungen A 147:169–216.

Zhou Z. 1995a. Discovery of a new enantiornithine bird from the Early Cretaceous of Liaoning, China. Vertebrata PalAsiatica 1994:99–113.

———. 1995b. The discovery of Early Cretaceous birds in China. Courier Forschungsinstitut Senckenberg 181:9–22.

Zhou Z., Fan F. J., and Zhang J. 1992. Preliminary report on a Mesozoic bird from Liaoning, China. Chinese Science Bulletin 37:1365–1368.

9

The Birds from the Lower Cretaceous of Las Hoyas
(Province of Cuenca, Spain)

JOSÉ L. SANZ, BERNARDINO P. PÉREZ-MORENO, LUIS M. CHIAPPE, AND ANGELA D. BUSCALIONI

 Our knowledge of the early evolutionary history of birds has changed dramatically since the early 1980s (Chiappe, 1995a). The increasing information about the Mesozoic diversification of birds is generating insightful hypotheses on their phylogenetic structure and evolutionary history. Since 1986, the Konservat-Lagerstätte of Las Hoyas (province of Cuenca, Spain) has provided exceptional evidence contributing to the flourishing revision on this paleobiological issue.

The first report on *Iberomesornis romerali* (referred to as "the Las Hoyas bird" in literature prior to 1992) appeared in 1988 (Sanz et al., 1988). This preliminary paper stressed the significance of this primitive bird in regard to early avian diversification. *Iberomesornis,* lacking the skull, the cranial portion of the neck, and most of the hands (see Anatomy), presents apomorphic characters (e.g., pygostyle, strutlike coracoid) with respect to *Archaeopteryx* and plesiomorphic characters (e.g., primitive sacropelvic elements, unfused metatarsus) with respect to neornithine birds. Therefore, *Iberomesornis* was proposed to be the closest known sister group of Ornithurae, a suggestion promptly accepted by several authors (Cracraft, 1988; Chiappe, 1991). This conclusion was presented in the Second International Symposium of the Society of Avian Paleontology and Evolution, held in Los Angeles, California, in September 1988. Four years later, a formal denomination of the Las Hoyas bird, along with a further description and discussion, was published in the Proceedings of the above-mentioned Symposium (Sanz and Bonaparte, 1992).

At the end of the 1980s, a new bird skeleton was discovered at Las Hoyas. This new specimen was published by Sanz and Buscalioni (1992) with the name *Concornis lacustris. Concornis,* roughly twice the size of *Iberomesornis,* is known by an almost complete skeleton with some feather evidence, although it lacks the skull and neck. Sanz et al. (1995) fully described the holotype (and only known) spec-

imen and provided strong evidence in support of its placement within Enantiornithes. One of the most striking features of *Concornis* is the presence of a broad, notched sternum with a caudal carina, similar to that described for *Cathayornis* (Zhou, 1995a; see also Zhou and Hou, Chapter 7 in this volume).

In 1994, a third avian skeleton was found at Las Hoyas. This new bird specimen is better preserved than the other two and consists of much of the skeleton but lacks the skull. Wing feathers are preserved in position. It was interpreted as a new enantiornithine taxon and named *Eoalulavis hoyasi* (Sanz et al., 1996). A peculiar assemblage of four juvenile specimens was also discovered that same year. This rare finding was interpreted as a regurgitated pellet on the basis of taphonomic considerations (Sanz et al., 2001a). In this chapter we describe *I. romerali, C. lacustris,* and *E. hoyasi,* placing more emphasis on this last taxon, which to date has been only preliminarily described. We review how these taxa have been interpreted and discussed according to recent phylogenetic hypotheses. We also discuss the contribution made by the birds from Las Hoyas to our current understanding of the early phases of avian flight.

Geological Setting

The Las Hoyas fossil site is located in the southern part of the Serranía de Cuenca, in the Castilla–La Mancha region of Spain (Fig. 9.1). The outcropping is stratigraphically placed within the Calizas de la Huérguina Formation and is located about 30 km east of the city of Cuenca. It is Early Cretaceous (Late Barremian) in age, as dated by means of palinomorphs, ostracods, and charophytes (Diéguez et al., 1995). The fossiliferous facies of Las Hoyas are rhythmically laminated limestones derived from a wide lacustrine carbonate system. Several phases of deepening and shallowing occurred in this lake. The fossil-bearing limestones were deposited in the deepening periods (Fregenal-Martínez and Meléndez, 1995).

Most of the faunal assemblage and facies from Las Hoyas indicate that the basin was a freshwater lacustrine environment. Yet the vicinity of contemporaneous marine deposits (a few tens of kilometers to the southeast) and the presence of coelacanth and pycnodontiform fishes may suggest some sort of marine influence. Nevertheless, a strontium isotope analysis of the bones of these fishes and carbon and oxygen analyses of the carbonate rocks from the Las Hoyas basin are consistent with a freshwater lacustrine environment (Talbot et al., 1995). Thus, the limestone isotopic analyses clearly indicate that the carbonates were precipitated in a hydrologically closed basin. This type of restricted basin is consistent with the warm, seasonally dry climate proposed for the Las Hoyas area during Early Cretaceous times (Gómez-Fernández and Meléndez, 1991).

Las Hoyas is a Konservat-Lagerstätte that has yielded the fossil remains of a high diversity of organisms (Sanz et al., 2001b). The floral assemblage includes charophytes, bryophytes, filicales, cicadophytes, gnetales, conifers, and angiosperms. A large part of the floral remains belongs to the enigmatic species *Montsechia vidali* (Diéguez et al., 1995). Most of the invertebrates are arthropods: 16 forms of ostracods (Rodríguez-Lázaro, 1995); 2 decapods (Martínez-Delclòs and Nell, 1995a); and 31 different forms of insects, mainly belonging to Odonata, Blattodea, Heteroptera, Orthoptera, Coleoptera, Diptera, Neuroptera, and Ephemeroptera (Martínez-Delclòs and Nell, 1995b).

The ichthyofauna includes Coelacanthiformes, Semionotiformes, Amiiformes, Pycnodontiformes, and primitive Teleostei (Pholidophoriformes and Gonorynchiformes) (Poyato-Ariza and Wenz, 1995). Amphibians are represented by the salamanderlike Albanerpetontidae, salamanders, and frogs (Evans et al., 1995). Lizards (Barbadillo and Evans, 1995) and turtles (Jiménez-Fuentes, 1995) were also present in the Las Hoyas biota. The crocodilian fauna is peculiar, as it is represented by four small-sized taxa belonging to Gobiosuchidae and Neosuchia (Buscalioni and Ortega, 1995; Ortega et al., 2000). The dinosaur fauna includes fragmentary evidence of Sauropoda, Ornithopoda, and Theropoda. Along with the birds, the most complete remains belong to the ornithomimosaur *Pelecanimimus polyodon* (Pérez-Moreno et al., 1994; Sanz and Pérez-Moreno, 1995). Ichnofossils include invertebrate traces, a large amount of unstudied coprolites, probably from fishes, and a trackway from a medium-sized crocodile (Fregenal-Martínez and Moratalla, 1995).

The exceptional quality of preservation of the Las Hoyas Konservat-Lagerstätte is illustrated by the preservation of soft tissues in several fossils. An outstanding example is the flight feathers of the enantiornithine bird *E. hoyasi,* which are preserved in situ, attached to their respective bones (Sanz et al., 1996). Other examples include both isolated and adhered feathers (such as in *Concornis*), as well as other epidermal remains (such as the ornithomimosaur *Pelecanimimus* (Briggs et al., 1997) or the albanerpetontid *Celtedens* (McGowan and Evans, 1995).

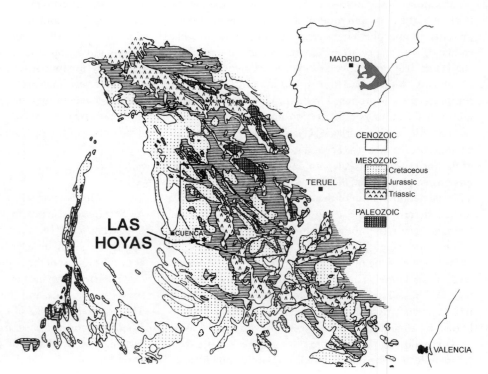

Figure 9.1. Locality and general stratigraphy of south-central Spain, where the fossil site of Las Hoyas is located.

Systematic Paleontology

Taxonomic Hierarchy

> Aves Linnaeus, 1758
>> Ornithothoraces Chiappe, 1995
>>> Enantiornithes Walker, 1981
>>>> *I. romerali* Sanz and Bonaparte, 1992

Holotype—LH-22; Las Hoyas Collection, Museo de Cuenca, Cuenca, Spain (provisionally housed in the Unidad de Paleontología of the Universidad Autónoma de Madrid, Spain). Nearly complete articulated specimen, lacking the skull, the cranial portion of the neck, and most of the hands (Figs. 9.2 and 9.3).

Referred specimens—LH-8200; Las Hoyas Collection, Museo de Cuenca, Cuenca, Spain (provisionally housed in the Unidad de Paleontología of the Universidad Autónoma de Madrid, Spain). Left metatarsals I–IV and their respective articulated digits (Fig. 9.4). Poorly preserved specimen comparable in size to that of the holotype. Sanz and Buscalioni (1994) regarded it as cf. *Iberomesornis* sp.

Horizon and locality—Las Hoyas fossil site, province of Cuenca, Spain. Calizas de La Huérguina Formation, Late Barremian (Lower Cretaceous) (Fregenal-Martínez and Meléndez, 1995).

Diagnosis—Autapomorphic characters include cervicals with a lateral shelf linking cranial and caudal zygapophyses; high vertebral arches of dorsal vertebrae, with relatively high neural arches and laminar spines; a very large and laminar pygostyle; and the absence of a cnemial crest on the tibiotarsus.

> Euenantiornithes Chiappe, Chapter 20 in this volume
>> *C. lacustris* Sanz and Buscalioni, 1992

Holotype—LH-2814; Las Hoyas Collection, Museo de Cuenca, Cuenca, Spain (provisionally housed in the Unidad de Paleontología of the Universidad Autónoma de Madrid, Spain). Articulated specimen, including the right forelimb and thoracic girdle, sternum, some dorsal, synsacral, and caudal vertebrae, pubis, ischium, and hindlimbs (Fig. 9.5).

Figure 9.2. *I. romerali* (LH-22), Las Hoyas, Upper Barremian (Lower Cretaceous), Cuenca, Spain. Holotype. Scale bar = 1 cm.

Figure 9.3. *I. romerali* (LH-22), ultraviolet-induced fluorescence photograph.

Horizon and locality—Same as in *I. romerali* (see previously).

Diagnosis—Autapomorphic characters include a ribbonlike ischium, a transverse ginglymoid articulation of trochlea of metatarsal I, and a strongly excavated distal end of metatarsal IV.

E. hoyasi Sanz et al., 1996

Holotype—LH-13500 a/b; Las Hoyas Collection, Museo de Cuenca, Cuenca, Spain (provisionally housed in the Unidad de Paleontología of the Universidad Autónoma de Madrid, Spain). Articulated specimen, including feathers, the last 5 cervical and the first 10 dorsal vertebrae, the thoracic girdle, sternum and ribs, both wings, and remains of the pelvic girdle and femora (Fig. 9.6–9.9).

Horizon and locality—Same as in *I. romerali* (see previously).

Diagnosis—Autapomorphic characters include the presence of laminar, keel-like cervical and dorsal centra; an undulating ventral surface of the furcula; a distal end of the humerus with a thick, caudally projected ventral margin; the presence of several small tubercles on the distal, caudal surface of the minor metacarpal; a depressed, spear-shaped sternum, with a footlike caudal expansion and a faint carina; and a deep, rostral cleft in the sternum.

A

B

Figure 9.4. *I. romerali* (LH-8200). Las Hoyas, Upper Barremian (Lower Cretaceous), Cuenca, Spain. Referred specimen; part (A) and counterpart (B) of an isolated foot. Scale bar = 1 cm.

Figure 9.5. *C. lacustris* (LH-2814), Las Hoyas, Upper Barremian (Lower Cretaceous), Cuenca, Spain. Scale bar = 1 cm.

Figure 9.6. *E. hoyasi* (LH-13500a), Las Hoyas, Upper Barremian (Lower Cretaceous), Cuenca, Spain. Specimen before preparation. Scale bar = 1 cm.

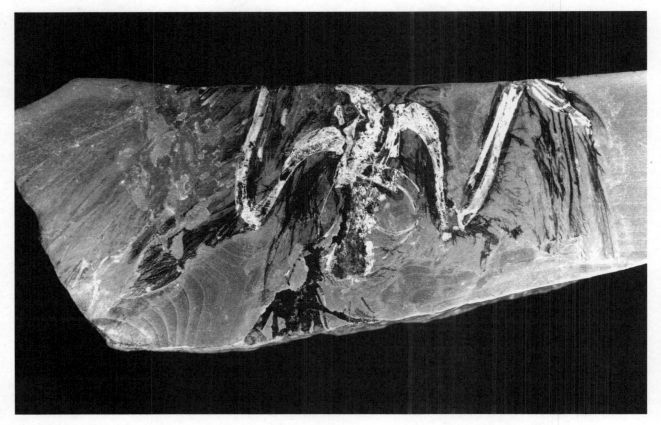

Figure 9.7. *E. hoyasi* (LH-13500a). Ultraviolet-induced fluorescence photograph of the specimen before preparation.

Figure 9.8. *E. hoyasi* (LH-13500a). Dorsal view of the specimen after preparation.

Figure 9.9. *E. hoyasi* (LH-13500b). Ventral view of the specimen after preparation.

Anatomy

I. romerali Sanz and Bonaparte, 1992

New preparation of the holotype specimen has revealed new details of its anatomy (Figs. 9.2–9.4, 9.10, and 9.11). Thus, we include here this new information that revises the comprehensive description of Sanz and Bonaparte (1992). Sereno (2000) also discussed the newly prepared holotype

specimen and provided an account of its anatomy and systematic significance, but it has not been possible to adequately address any issues raised in that paper.

Ontogenetic Stage. As is discussed in detail subsequently, the holotype of *I. romerali* is probably an adult individual (contra Martin, 1995; Feduccia, 1996). This assessment is supported by the fusion between the pubis and ischium and the absence of neurocentral sutures in the pre-sacral vertebral series.

Vertebral Column. The vertebral column is nearly complete, lacking only the cranial portion of the cervical series. The fifth preserved vertebra, starting from the front, is probably the first dorsal. The two foremost cervical elements are elongate and have a lateral shelf linking cranial and caudal zygapophyses, while the two subsequent cervicals are shorter, lacking the lateral ridge. The first two preserved cervicals lack a spinous process; this process is also very reduced in the third preserved cervical, but it appears to increase in height in the last cervical vertebra. Pneumatic foramina (pleurocoels) are not visible in any of these vertebrae. Although exposed in lateral view, the cervical centra were almost certainly not heterocoelous.

There are 11 thoracic vertebrae; the first three have a transitional morphology with the caudalmost cervicals. The cranial thoracic vertebrae, like the last cervical, bear prominent ventral crests (Chiappe, 1996a). The thoracic centra are slightly longer than those of the last cervicals, becoming progressively longer and lower caudal to the fourth dorsal. They have small, lateral depressions and appear to be platycoelous. The vertebral arch is relatively high, with well-developed zygapophyses and with cranial zygapophyses projecting beyond the cranial articular surface. From the fifth thoracic, the costal fovea seems to be situated on the vertebral arch. The spinous processes are laminar and relatively high.

The sacrum is composed of five elements (Fig. 9.10). Sanz and Bonaparte (1992) considered the sacral series to be

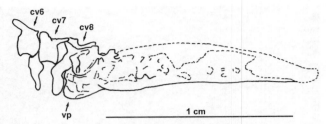

Figure 9.11. *I. romerali* (LH-22). Camera lucida drawing of the last free caudal vertebrae and pygostyle. Abbreviations: cv, caudal vertebrae; vp, proximal ventral process of the pygostyle.

exposed in lateral view and proposed the presence of keel-like ventral processes. Nevertheless, the sacral vertebrae are most likely exposed in ventral view, and no ventral processes are visible. This is consistent with the position of the iliac blades and the disposition of the three cranialmost free caudal vertebrae.

There are eight free caudal vertebrae and a large pygostyle (Fig. 9.11). The distalmost free centra appear to be slightly procoelous. In the cranial part of the pygostyle, the first three elements can be individualized, while in the caudal part there is no differentiation between the elements. Sanz and Bonaparte (1992) suggested that the pygostyle was composed of vertebral elements 10–15. Chiappe (1996a) emphasized the primitiveness of the caudal vertebral count of *Iberomesornis*, which, when pygostyle elements are added to the more proximal, free vertebrae, approaches the number of elements present in *Archaeopteryx*. Cranially, the pygostyle seems to have a ventral expansion, with an inverted-T cross section, becoming distally laminar. As a whole, the caudal series is subequal in length to the dorsal series.

There is no evidence of either uncinate processes or gastralia.

Thoracic Girdle and Limb

Sternum. A fragment of the sternum is visible. Five stout sternal ribs are attached to it. It is probable that a portion of the sternum is concealed by the ulna. The presence or absence of a carina is uncertain.

Furcula. The furcula is almost completely preserved. The rami are robust, becoming thinner toward the hypocleideum. The interclavicular angle is approximately 60° at the base of the clavicular rami and approximately 45° if it is measured at the shoulder end of the rami. The hypocleideum is partially hidden by the radius, but it seems to be long.

Scapula. A fragment of the scapula is preserved. The scapular blade seems to be rather straight, long, and narrow. It articulates with the coracoid at a right angle, although it

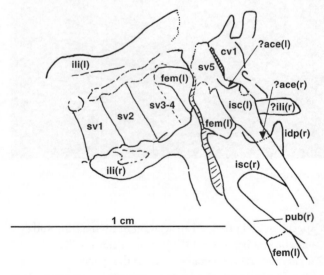

Figure 9.10. *I. romerali* (LH-22). Camera lucida drawing of the sacropelvic region. Abbreviations: ace, acetabulum; cv, caudal vertebra; fem, femur; idp, dorsal process of ischium; ili, ilium; isc, ischium; pub, pubis; sv, synsacral vertebrae; (l, r), left or right element.

is uncertain whether this condition was also present at the level of the humeral articular facet (Chiappe, 1996a).

Coracoid. The avian, strutlike coracoids of *Iberomesornis* are not well preserved. The left coracoid, exposed in ventral view, is nearly complete, but its shaft and head lack any anatomical details. The sternal margin is broad and straight, and the lateral margin lacks the typical enantiornithine convexity.

Humerus. The left humerus is exposed craniodorsally, and its shaft is fractured in two points. It is distinctly shorter than the ulna. The deltopectoral crest is well developed, and the bicipital crest is low. There is no visible transverse ligamental groove. Neither condyles nor an intercondylar depression is observed. Nevertheless, the outline of the ventral epicondyle seems to be placed more distally than the dorsal epicondyle. The absence of visible distal condyles is not necessarily related to the suggested juvenile condition of the specimen but rather to the preservation of this area.

Ulna and Radius. Both are nearly articulated with the humerus. The ulnar shaft is almost twice the radial shaft in width. The former is curved caudally in its proximal area. The ends of both the ulna and radius are badly preserved.

Manus. Some bones placed close to the midshaft of the radius could be the distal end of the metacarpus and some phalanges, although no clear features can be observed.

Pelvic Girdle and Limb

Ilium. The interpretation of the pelvic girdle is still problematic, even after the new preparation of the specimen. Both ilia are preserved in ventral view at both sides of the sacrum (Fig. 9.10). These bones are so poorly preserved that nothing can be said about their morphology.

Pubis. The new preparation revealed that Sanz and Bonaparte (1992) confused the right pubis with the proximal half of the femur. The pubis is slightly retroverted, with a robust shaft of suboval cross section. It is not possible to verify whether there is a pubic foot, as the distal ends of both pubes are missing. No suture is visible between the pubis and ischium, and it seems that these pelvic elements are fused to each other. This condition again suggests that this specimen was not a juvenile.

Ischium. Both ischia are exposed. The left one is displaced dorsally with respect to the right one. Behind the left ischiadic shaft, a process of the right ischium can be observed (Fig. 9.10). The base of this process is visible in the proximal area of the left ischium, as well. As pointed out by Zhou (1995b), this projection may compare to the caudodorsal

process of the ischium of Enantiornithes (see Chiappe and Walker, Chapter 11 in this volume). Yet it can be alternatively interpreted as the iliac process of the ischium (contra Zhou, 1995b) that forms the caudal margin of the acetabulum. If this latter interpretation is correct, the concave outline dorsal to this process is the ventral margin of the acetabulum. Both interpretations are plausible, although its proximal position supports its interpretation as part of the acetabulum. The ischiadic shaft lacks the characteristic caudal process of *Archaeopteryx* (Wellnhofer, 1974).

Femur. The right femur is clearly visible in medial view. The left is badly preserved, and its proximal half is virtually missing (Fig. 9.10). The femur is relatively long but shorter than the tibia. Distally, it bears a well-developed popliteal fossa.

Tibia. The right tibia is almost complete, exposed in medial view. The left tibia is broken into two, displaced pieces. There is no visible cranial cnemial crest. A round proximal articular surface is present.

Tarsus. Both astragali are in their original positions. They are not fused to the tibia. The ascending process is broad and low. Two small, free distal tarsals are also visible.

Metatarsals. Left and right metatarsals II–IV are preserved, although the right set is difficult to interpret because of its bad preservation. The left metatarsals are visible in dorsal view. The first metatarsal is slightly displaced from its original position. Metatarsals II–IV are subequal in length, although metatarsal III is somewhat longer. There is no evidence of either proximal or distal fusion between metatarsals.

Foot. The phalangeal formula is 2-3-4-5-*x*. The pes is anisodactyl, with three main digits (II–IV) directed dorsally, and a reversed hallux.

C. lacustris Sanz and Buscalioni, 1992

A detailed description of *C. lacustris* after the definitive preparation of the holotype was given elsewhere (Sanz et al., 1995). Thus, only its most significant features are mentioned here (Figs. 9.5, 9.12, and 9.13).

Ontogenetic Stage. The only known specimen of *Concornis* is an adult individual, possessing a well-ossified sternum, tibiotarsus, and tarsometatarsus.

Vertebral Column. Eight vertebrae are preserved. The four preserved thoracic centra are amphicoelous, compressed, and excavated laterally by a longitudinal depression, as occurs in other enantiornithine birds (Chiappe and

Calvo, 1994; Chiappe and Walker, Chapter 11 in this volume). The last two synsacral vertebrae have elongated transverse processes. The first two caudals are amphicoelous, with caudally directed transverse processes (as are those of the synsacrum). A haemal arch is preserved at the proximal border of the second caudal vertebra.

Several dorsal ribs are well preserved, but none of them shows evidence of uncinate processes. This suggests that these processes, if present, did not ossify.

Thoracic Girdle and Limb

Sternum. The sternum has a rounded cranial margin. The carina is developed only in its caudal half, with lateral crests diverging from its cranial end. The caudal portion of the sternum is deeply notched, with a stout and distally expanded lateral process and a less robust medial one (Fig. 9.12). On the lateral margin of the sternum, there is a short notch, delimited caudally by a cranially directed process, which is interpreted as the articular area for the ribs.

Furcula. The furcula is very robust and excavated laterally, with the cranial margin projected externally. The interclavicular angle is 60°. The hypocleideum is long and tapers distally, with a weak crest on its cranial face.

Scapula. The scapular blade is straight. The shoulder end has a medial, subrectangular area interpreted as the articular facet for the furcula.

Coracoid. The strutlike coracoid is long, lacking lateral and procoracoid processes. It has a typical enantiornithine morphology: a dorsal fossa, a sternal half with a convex lateral margin and a concave medial one, and a supracoracoidal nerve foramen opening into a medial groove and separated from the dorsal surface by a thick bar (see Chiappe and Calvo, 1994; Chiappe and Walker, Chapter 11 in this volume). However, the sternal margin is only slightly concave, lacking the remarkable concavity typical of several Enantiornithes.

Humerus. The humerus also presents the characteristic morphology of euenantiornithine birds (Chiappe and Walker, Chapter 11 in this volume). The humerus has a distinct torsion between proximal and distal ends, a cranially concave and caudally convex head, a strong development and cranioventral projection of the bicipital area, and a craniocaudally compressed distal end.

Ulna and Radius. The ulna is nearly twice the width of the radius. The latter is badly preserved, so the presence of the longitudinal groove typical of other Euenantiornithes (Chiappe and Calvo, 1994; Chiappe and Walker, Chapter 11 in this volume) cannot be confirmed. The ulna has a prox-imal end with a convex and strongly caudodorsally projected dorsal cotyla, which is separated from the concave ventral cotyla and the olecranon by a shallow depression. The bicipital tubercle is distally placed to the proximal articular surface.

Hand. The manus is slightly shorter than the forearm. The robust and ventrally convex major metacarpal is subequal in length to the minor one, and these two metacarpals appear to be fused distally. The intermetacarpal space is narrow. The alular metacarpal must have been short, although it is not preserved.

The alular digit has two phalanges: the proximal one is slender and elongate, and the distal one is a claw. The major digit has three phalanges, with the proximal one being the largest and the distal one being a claw. The minor digit has only one preserved phalanx.

Pelvic Girdle and Limb

Pubis. The pubis is slender, elongate, and suboval in cross section and has a short distal symphysis (Fig. 9.13). Based on its preservation, the pubis was probably oriented caudally.

Ischium. The ischium is about 25% shorter than the pubis. It lacks a distal contact and is narrow in its distal two-thirds, having a ribbonlike appearance (Figs. 9.12 and 9.13). It has a prominent obturator process and a dorsal process similar to those of other enantiornithines (see Chiappe and Walker, Chapter 11 in this volume).

Femur. The femur is notably shorter than the tibiotarsus. It is robust and slightly curved, and the cranial surface is convex. The femoral head lacks the depression for the insertion of the round ligament (fovea lig. capitis). The medially inclined lateral margin of the proximal end suggests the presence of a well-developed posterior trochanter. There is no patellar groove in the distal end. The lateral margin of the lateral condyle projects caudally, as in other euenantiornithines.

Tibiotarsus. The tibia is gracile, with the proximal tarsals fused entirely to its distal end. The fibular crest is relatively short and well developed. The proximal articular surface of the tibiotarsus is round, and there is a single cnemial crest—two characters that are also present in most Enantiornithes (see Chiappe, 1996a; Chiappe and Walker, Chapter 11 in this volume). The cross section of the shaft is subcircular, unlike the craniocaudally compressed condition of *Lectavis bretincola* (Chiappe, 1993). The medial condyle of the distal end of the tibiotarsus is transversely broad, the same condition present in other enantiornithines.

Tarsometatarsus. The tarsometatarsus is shorter than the tibiotarsus. The third metatarsal is the longest, with the

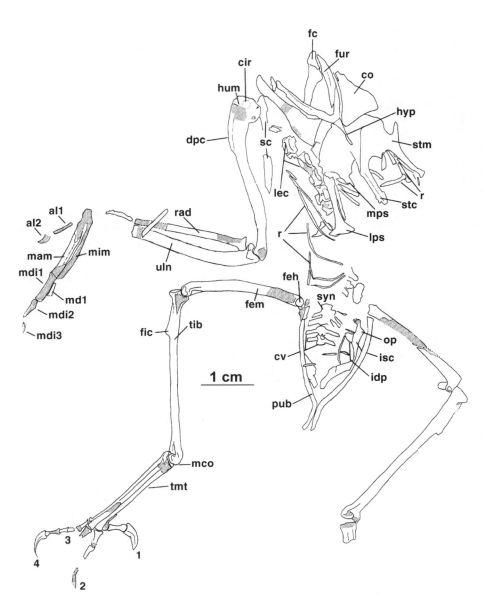

Figure 9.12. *C. lacustris* (LH-2814). Camera lucida drawing. Abbreviations: al, alular phalanx; cir, circular fossa on proximocranial end of humerus; co, coracoid; dpc, deltopectoral crest; fc, facet; feh, femoral head; fic, fibular crest; fur, furcula; hum, humerus; hyp, hypocleideum; lec, lateral excavation of centra; lps, lateral process of sternum; mam, major metacarpal; mco, medial condylus; md, phalanges of minor digit; mdi, phalanges of major digit; mim; minor metacarpal; mps, medial process of sternum; op, obturator process; r, rib; rad, radius; sc, scapula; stc, sternal carina; stm, sternum; syn, synsacrum; tib, tibiotarsus; tmt, tarsometatarsus; uln, ulna; 1–4, digits 1–4; others as in Figure 9.10.

fourth one being the most slender. Metatarsals II–IV are straight and co-planar, although the shaft of metatarsal III is strongly convex in its central portion, being projected somewhat forward from metatarsals II and IV. Metatarsals II–IV are fused proximally and closely connected to each other, lacking the distal divergence of metatarsal II that appears in some avisaurid enantiornithines (Chiappe, 1993). The proximal end lacks an intercotylar eminence as well as the dorsal sloping present in several avisaurids (Varricchio and Chiappe, 1995). The trochlea of metatarsal IV is the most reduced, and that of metatarsal II the most developed. Metatarsal I articulates with the medial surface of metatarsal II. This metatarsal is not J-shaped in mediolateral view as in other enantiornithines (Chiappe, 1992, 1993; Chiappe and Calvo, 1994) but straight.

Pedal Phalanges. Digit I is robust and reverted, with a large claw. The unguals lack well-developed flexor tubercles.

E. hoyasi Sanz et al., 1996

The specimen (Figs. 9.6–9.9, 9.14, and 9.15) is preserved as a slab and a counterslab. The slab (LH-13500a) exhibits the specimen in dorsal view (Figs. 9.8 and 9.14), and contains 15 vertebrae (of which 5 are cervicals), the two scapulae, the left coracoid and the proximal half of the right coracoid, the distal portion of the furcula including the hypocleideum, several ribs, almost all of the left wing skeleton and the right wing lacking only the hand, part of the left side of the pelvic girdle, the proximal end of the left femur, and the synsacrum. The counterslab (LH-13500b) exhibits the specimen in ventral view (Figs. 9.9 and 9.15), including seven ver-

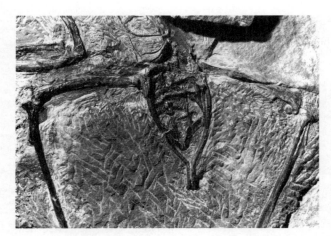

Figure 9.13. *C. lacustris* (LH-2814). Detail of the sacropelvic region.

tebrae (five of which are cervicals), both coracoids and the proximal end of the left scapula, the complete furcula, the sternum, several ribs, and the left wing and the humerus and ulna-radius of the right wing.

From the ninth preserved vertebra, the caudal part of the specimen is preserved as a cast (in plastic resin) of the diagenetic dissolution of the bones. Thus, the available data from this area of the specimen are poor.

Ontogenetic Stage. The well-ossified skeleton of *Eoalulavis*, lacking the pattern of foramina in the periosteal surface that appears in the humerus and other bones in the nestling from El Montsec (Sanz et al., 1997), indicates that it

pertains to an adult individual (note: *Concornis* and *Iberomesornis* also lack this punctate periosteal texture).

Vertebral Column. The last five cervicals are preserved, of which the cranialmost preserved cervical is longer than the others. All the cervicals have very compressed spinous processes. These are relatively small in the first three vertebrae and somewhat taller in the fourth and fifth. The epipophyses are small. In ventral view, the cervical centra are very compressed, forming a ventral keel. From the second to the fifth preserved cervicals, this ventral keel projects into a cranial ventral crest. The cervical centra are roughly amphiplatyan. There are robust cervical ribs, of about the same length as the centrum. The fourth and fifth have a transitional morphology with the thoracic vertebrae.

Ten thoracic vertebrae are preserved in articulation between the last cervical and the cranial margin of the ilium. In the second and the subsequent vertebrae, the spinous processes are tall and compressed. In the second thoracic vertebra, the tip of the neural spine forms a rectangular platform. This appears to be absent in the third thoracic vertebra (although it could be broken off), but it appears again in the fourth and subsequent vertebrae. In the fourth thoracic vertebra this platform is forked caudally, and in the fifth vertebra it is forked both caudally and cranially. This cranial and caudal bifurcation appears to be present in all the remaining thoracic vertebrae, because a bifurcation is visible in all the preserved portions. In the platform of the seventh thoracic vertebra, the caudal bifurcation is doubled, with one fork situated on top of the other. The spinous process of the sixth thoracic vertebra is missing, so it is un-

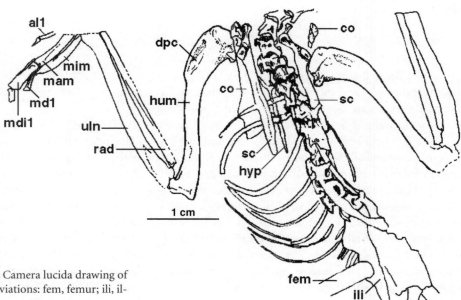

Figure 9.14. *E. hoyasi* (LH-13500a). Camera lucida drawing of the specimen in dorsal view. Abbreviations: fem, femur; ili, ilium; others as in Figure 9.12.

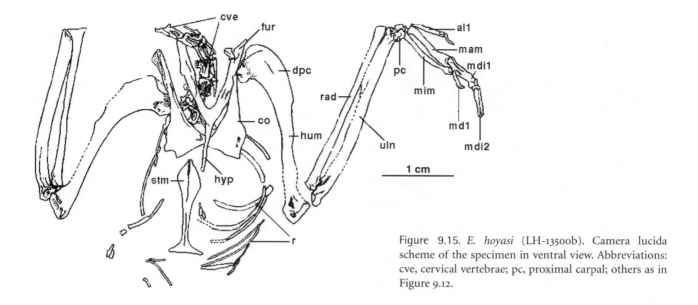

Figure 9.15. *E. hoyasi* (LH-13500b). Camera lucida scheme of the specimen in ventral view. Abbreviations: cve, cervical vertebrae; pc, proximal carpal; others as in Figure 9.12.

certain whether this feature was present in this vertebra. This condition is also present in the eighth and ninth dorsals, but the area is not preserved in the tenth vertebra.

The centra of the 7th to 10th thoracic vertebrae are copied by the resin. These show very clearly that the bodies are excavated laterally by a large fossa, a condition common to several other enantiornithine birds (Chiappe, 1996a; Chiappe and Walker, Chapter 11 in this volume). The transverse processes are short and virtually centered. Only the first and second vertebrae are exposed ventrally; the hypapophyses of the third and fourth are seen through the right coracoid. These centra bear a ventral keel that is larger than that of the cervicals. In the cranial half of the centrum, this keel projects into a well-developed ventral process. In the second, third, and fourth thoracic vertebrae, this process is deep and very robust. In the second thoracic vertebra, the cranial articular surface is flat; yet the condition for the caudal articular surface of this vertebra is uncertain.

There are several preserved thoracic ribs, some of them nearly complete. In none of these is there evidence of uncinate processes, which suggests their true absence as ossified elements. No evidence of gastralia has been found.

Thoracic Girdle and Limb

Sternum. The morphology of the sternum is singular within dinosaurs (Fig. 9.16), although a similarly shaped sternum has been recently found for the Early Cretaceous *Liaoningornis* of China (Hou et al., 1996). The sternum is a depressed, spear-shaped element, with a footlike caudal expansion. On its ventral surface there is a faint carina. The rostral third of the sternum is cleft by a narrow indentation, presumably an area interlocking with the extremely long hypocleideum of the furcula. The margins of the sternum

are smooth and well defined, lacking either facets for the articulation of the ribs or the coracoids. Regardless of the peculiar morphology of the element, its median carina and position with respect to the ribs indicate that it constitutes the central portion of the sternum. Conceivably, the sternum had cartilaginous extensions that formed the articular areas for both the ribs and coracoids (Sanz et al., 1996).

Furcula. The furcula is very robust and V-shaped, with an angle of approximately 45° between the two rami. It is compressed mediolaterally, and its ventral margin undulates in lateral view, with two "hills" and three "valleys." The latter are placed at the midpoint of each third of the rami. The dorsal margin is thinner than the ventral margin, as in *Concornis* (Sanz et al., 1995) and *Neuquenornis* (Chiappe and Calvo, 1994), although it is not possible to discern whether there is a lateral excavation. The hypocleideum is hypertrophied and strongly compressed. It is roughly 77% of the ramal length. In ventral view, there is a strong buttress in the middle of the hypocleideum, although it may be a pathological condition.

Scapula. The scapular blade is straight with a sharp distal end. The well-developed acromion is roughly depressed and placed somewhat perpendicular to the scapular blade. The humeral articular facet is concave and subtriangular in outline.

Coracoid. The coracoid is very long and thin. The ventral surface is convex. Dorsally, there is a subtriangular fossa. The shoulder end of this fossa converges toward the foramen for the supracoracoideus nerve, which opens medially into a longitudinal groove. The shoulder end of the coracoid is very compressed, and it is oriented caudodorsally. In dorsal view,

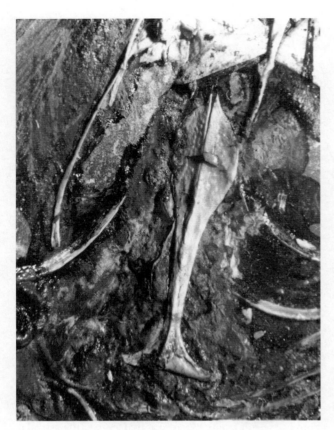

Figure 9.16. *E. hoyasi* (LH-13500b). Detail of the spear-shaped sternum, in ventral view.

this end is sigmoidal in shape, with a laterally convex humeral articular facet and a laterally concave articular facet for the scapula. The glenoid is flat and becomes thinner dorsally to form a convex surface for the scapula. In the middle of the two articular surfaces (glenoid and sternal) there is a small depression in the medial margin. As in the El Brete enantiornithine birds, there is no tubercle between the glenoidal and the scapular articular surfaces. The sternal third of the coracoid becomes wider abruptly, with the lateral margin strongly convex, although this becomes straight distally. In ventral view, the sternal margin is straight.

Humerus. The humerus is sigmoidal in shape, with a length of about 2.7 cm. It presents an evident torsion between the two ends. The humeral head is concave cranially and strongly convex caudally. The deltopectoral crest runs along the proximal third of the humerus. This crest is cranially flat without a cranial torsion. Just distal to the head, there is a small transverse fossa. Dorsodistally to this fossa, a rounded scar appears on a small elevation. The bicipital area is prominent and projected cranioventrally. On the cranial face of this area, distally displaced, there is a small fossa for a muscular insertion. This fossa is also found in other enantiornithines (Chiappe and Walker, Chapter 11 in this

volume). Proximal to the bicipital area, a distinct ligamental furrow can be observed. In caudal view, the ventral tubercle is well developed and projected caudally, separated from the humeral head by the capital incision. The caudal end of the ventral tubercle bears a distinct subtriangular depression. Unlike in *Enantiornis leali,* the ventral tubercle is not perforated proximodistally by a fossa (Chiappe, 1996b). There is no evidence of either pneumatic foramen or fossa.

Distally, the humerus is craniocaudally compressed. The dorsal epicondyle is projected dorsally and proximally, and it possesses a shallow depression over its dorsal face. In caudal view, the ventral margin presents a thick ridge projected caudally, which defines the ventral margin of a deep olecranon fossa. The ventral epicondyle is projected somewhat distally. On the ventral margin of this epicondyle two small suboval fossae appear. The condyles are situated on the humeral cranial surface. The ventral condyle is oriented transversely, and its distal surface is planar. The dorsal condyle is subspherical, oriented transversely, and situated more proximally than the ventral condyle.

Ulna and Radius. The antebrachium is longer than the humerus (ulnar length: 3.1 cm). It is straight, although the ulna is curved somewhat caudally in its proximal area. Thus, the interosseous space is wider proximally than distally. The radial shaft is notably thinner than the ulnar shaft, which is roughly twice the radial shaft in width.

The radius has a subcircular proximal end. Its bicipital tubercle is conspicuous. The radial shaft has a longitudinal groove, similar to that of several Enantiornithes (Chiappe and Calvo, 1994; Chiappe and Walker, Chapter 11 in this volume).

The ulna has a weak olecranon. The ventral cotyla is somewhat planar. There is an elongated and strong scar for the bicipital muscle.

Hand. The hand is clearly shorter than the ulna and humerus, with a length of about 1.8 cm (length of major digit, including the metacarpal). The ratio between the different elements of the forelimb is 0.88:1:0.58 (humerus: ulna:major digit).

Only the left hand is preserved. In ventral view, there is a small bone in the wrist, situated in the angle formed by the distal end of the ulna and the proximal end of the minor metacarpal. It could be interpreted as a displaced carpal bone, probably the ulnare.

The alular metacarpal is not preserved. The straight major metacarpal is clearly more robust and somewhat shorter than the minor metacarpal, and it becomes wider distally. The distal ends of the major and minor metacarpals are not fused to each other; it is uncertain whether these metacarpals are fused one to another proximally. The intermetacarpal space is very narrow, even thinner than in

Neuquenornis (Chiappe and Calvo, 1994). The minor metacarpal is curved slightly caudally, and its proximal end is compressed craniocaudally, with the shaft becoming rounded gradually. On the caudal face of the distal half of this metacarpal there are several small tubercles.

The alular digit has two phalanges, and it is slightly shorter than the major metacarpal. The first phalanx is thin and relatively long, with the distal end having a ginglymoid articulation and collateral foveae. The distal phalanx is an ungual, having a strong flexor tubercle. The major digit has three phalanges. The proximal one is very robust, and it is the longest of the hand. It has a subquadrangular section unlike the depressed, comparable element of extant birds. This phalanx presents collateral foveae in the distal end. The second phalanx of the major digit is smaller than the first one, with a ginglymoid distal articulation and collateral foveae. The last phalanx is a small ungual. Only one phalanx of the minor digit has been preserved.

Pelvic Girdle and Limb

Ilium. Most of the left ilium is copied by the resin. It has a postacetabular wing that is shorter and much lower than the preacetabular one. The postacetabular wing tapers caudally.

Femur. The proximal portion of the left femur has been copied by the resin. It possesses a well-developed posterior trochanter, a condition typical of euenantiornithine birds (Chiappe and Walker, Chapter 11 in this volume).

Feathers. There is visible evidence of feathers in both slabs of the specimen. Most of the isolated feathers from the Las Hoyas fossil site are preserved as carbonized remains, which is the most common way of feather preservation in the fossil record (Davis and Briggs, 1995). Nevertheless, the feathers of *Eoalulavis* are usually covered, like most of its bones, by limonite and, in some cases, by piroluxite. Most of the oxide is removed during the transference process and the acid preparation, but, fortunately, some remains. In any case, the plastic resin copies the structure of the feathers, so all the information is maintained after the preparation of the specimen.

Remiges are preserved in both wings, and body feathers are visible around the humeri and pectoral girdle. There are eight preserved primary remiges, although the proximal two-thirds of the major metacarpal lacks any feather evidence. As in neornithine birds, some feathers are attached to the phalanges (five of the preserved ones) and others to the major metacarpal (only three of these are preserved). Eight secondary remiges are preserved. Seven of these are in the proximal portion of the ulna, and the eighth attaches to its distal end. Although only 16 primary and secondary remiges have been preserved, their number was probably comparable to that of extant flying birds.

The alular digit of the left wing presents at least one feather adhered to the first phalanx (although it is possible that two existed). This feather is the most ancient evidence of an alula known in the fossil record, and *Eoalulavis* is the most primitive bird preserving an alula. A clearly defined rachis seems to be lacking from the alular feather, which is very long. Although the existence of a rachis is not evident, the disposition of barbs appears to be symmetrical.

Phylogenetic Relationships

Two alternative phylogenetic hypotheses have been proposed for *I. romerali*, the first bird from Las Hoyas. Sanz et al. (1988) placed the Las Hoyas bird in an intermediate position between *Archaeopteryx* and Neornithes, a hypothesis highly corroborated by a number of studies (e.g., Chiappe, 1991, 1995a,b, 1996a; Sereno and Rao, 1992; Chiappe and Calvo, 1994; Chatterjee, 1995; Sanz et al., 1995). Specifically, this hypothesis supported the sister-group relationship between *Iberomesornis* and a clade formed by Enantiornithes and Ornithurae. Chiappe (2001, Chapter 20 in this volume) proposed a sister-group relationship between *Iberomesornis* and Euenantiornithes, including *Iberomesornis* within Sereno's (1998) new phylogenetic definition of Enantiornithes (all taxa closer to *Sinornis* than to Neornithes). Enantiornithes (including *Iberomesornis* and Euenantiornithes) and its sister group, Ornithuromorpha (*Patagopteryx*, *Vorona*, plus Ornithurae), constitutes Ornithothoraces (Fig. 9.17). This clade is phylogenetically defined as the common ancestor of *Iberomesornis* and neornithine birds plus all its descendants (Chiappe, 1996a). Recently, Sereno (2000) also included *Iberomesornis* within his phylogenetically defined Enantiornithes. Several authors (e.g., Feduccia, 1995, 1996; Hou et al., 1995; Kurochkin, 1995; Martin, 1995) had placed *Iberomesornis* within Enantiornithes but followed Martin's (1983) "Sauriurae," endorsing a sister-group relationship between *Archaeopteryx* and the latter clade. This notion is dramatically different from the one supported by us because it implies that the closest ancestor shared by *Iberomesornis* and neornithine birds is the ancestor of *all* birds. The nonmonophyletic status of "Sauriurae" has been pointed out many times (e.g., Olson, 1985; Chiappe, 1991, 1995a,b, 1996a; Forster et al., 1996; Sanz et al., 1996; see Chiappe [1995b, Chapter 20 in this volume] for a detailed discussion of this view) and does not need to be repeated here. Unfortunately, none of those placing *Iberomesornis* within "Sauriurae" have structured their views under a rigorous cladistic framework; nor have they tried to explain the enormous amount of convergence (i.e., the independent origin of a pygostyle, a strutlike coracoid, a low interclavicular angle, and many other characters shared by

Iberomesornis, Euenantiornithes, and neornithine birds) implied in such a hypothesis.

Several synapomorphies support placement of *Iberomesornis* within Ornithothoraces (Chiappe and Calvo, 1994; Chiappe, 1995a,b, 1996a, 2001, Chapter 20 in this volume; Sanz et al., 1995, 1996). Some of these are the presence of prominent ventral processes on the cervicothoracic vertebrae, 11 thoracic vertebral elements, a pygostyle, the presence of a strutlike coracoid and a sharp caudal end of the scapula, a humerus shorter than or nearly equivalent to the ulna, and a radial shaft that is considerably thinner than that of the ulna. Although an unquestionable ornithothoracine, *Iberomesornis* lacks (or fails to evidence) most euenantiornithine synapomorphies, such as the convex lateral margin of the coracoid, the strong lateral depressions on the centra of thoracic vertebrae, or metatarsal IV being significantly smaller than metatarsals II and III, for example (see appendix II in Chiappe and Calvo [1994] for a list of the synapomorphies of this clade). Additional data may support placement of *Iberomesornis* as the sister taxon of Euenantiornithes—with this clade outside "Sauriurae" (see Chiappe, 2001, Chapter 20 in this volume).

Of importance for establishing its phylogenetic relationships is the fact that some of the authors placing *Iberomesornis* within Enantiornithes have regarded its holotype as a juvenile. For example, Kurochkin (1995) points out that the unfused metatarsals, incomplete ossification of the sacrum, absence of co-ossification between astragalus and tibia, and poor definition of the articular surfaces of the humerus, femur, and tibia are suggestive of its immature status. Nevertheless, these features can alternatively be regarded as a combination of its general poor preservation and its more plesiomorphic condition (see Anatomy). The partial fusion of the pelvic elements and the absence of neurocentral sutures in the presacral vertebral series (see Brochu, 1996) do not support Kurochkin's point of view.

In the original description of *C. lacustris*—the second bird from Las Hoyas—Sanz and Buscalioni (1992) regarded this taxon as more derived than *Iberomesornis*. Specifically, these authors placed *Concornis* as the most basal taxon of a node situated between *Iberomesornis* and the neornithine birds. This preliminary description of *Concornis* used the available information on several small slabs containing the specimen. After its complete preparation and restudy, *Concornis* was placed within Enantiornithes (Sanz et al., 1993, 1995). This hypothesis has been accepted by most authors (e.g., Hou et al., 1995; Kurochkin, 1995; Martin, 1995), except for Elzanowski (1995), who regards both *Concornis* and *Cathayornis* as nonenantiornithine birds, placing them in a more basal position within an expanded Carinatae. Like *Concornis*, the recently described *E. hoyasi*—Las Hoyas's third bird—has also been identified as an Enantiornithes (Sanz et al., 1996) (Fig. 9.17). A large number of synapo-

morphies support placement of these two taxa within Enantiornithes, and in particular within Euenantiornithes. These include the presence of strong lateral depressions and of centered costal foveae (parapophyses) on the bodies of the dorsal vertebrae, a coracoid with convex lateral margin and a supracoracoidal nerve foramen opening into a medial furrow, a laterally excavated furcula, a prominent and cranioventrally projected bicipital crest of the humerus, a convex external cotyla of the ulna that is separated from the olecranon by a groove, and several others.

Paleobiology

The birds from Las Hoyas have made a significant contribution to our current understanding of the early evolutionary phases of avian flight. The fossil record indicates that several evolutionary novelties essential for the acquisition of a modern flight apparatus can be traced back to Early Cretaceous times (Chiappe, 1995a).

Three fundamental evolutionary transformations are noticeable in the birds from Las Hoyas: a strutlike coracoid, a modern, V-shaped furcula, and a pygostyle. These features, and presumably the functions correlated to them, are definitively absent in *Archaeopteryx* (Wellnhofer, 1974, 1992, 1993; Ostrom, 1976; Hecht et al., 1985).

The elongated, strutlike coracoid of *Iberomesornis, Concornis,* and *Eoalulavis* is an important modification from the relatively short coracoid of *Archaeopteryx.* While a triosseal foramen has not been directly documented in any of the birds from Las Hoyas, the morphology of the shoulder end of the coracoid, scapula, and furcula of *Eoalulavis* and *Concornis* strongly supports its presence. This foramen, which in extant birds allows a pulley system of M. supracoracoideus (Raikow, 1985), the most important muscle in the upstroke phase of the wingbeat cycle, was absent in *Archaeopteryx.* The actual function of this muscle in *Archaeopteryx* is not well understood. Despite traditional views regarding this muscle as a wing elevator (Norberg, 1989; Rayner, 1991), new data from nerve stimulation and electromyography suggest that its main action is to rotate the humerus on its longitudinal axis (Poore et al., 1997a,b). This action is considered essential for positioning the wing in place for the subsequent downstroke (Poore et al., 1997a,b). The elongation of the coracoid may have increased the efficiency of M. supracoracoideus as well as its area of origin on a larger coracoclavicular membrane. The presence of a basically modern design in the birds from Las Hoyas suggests that the mechanisms involved in the recovery stroke of these birds may have been comparable to those of their modern counterparts.

The boomerang-shaped furcula of *Archaeopteryx* has been interpreted as an important flight element (Olson and Feduccia, 1979). This may be the case, but very similar fur-

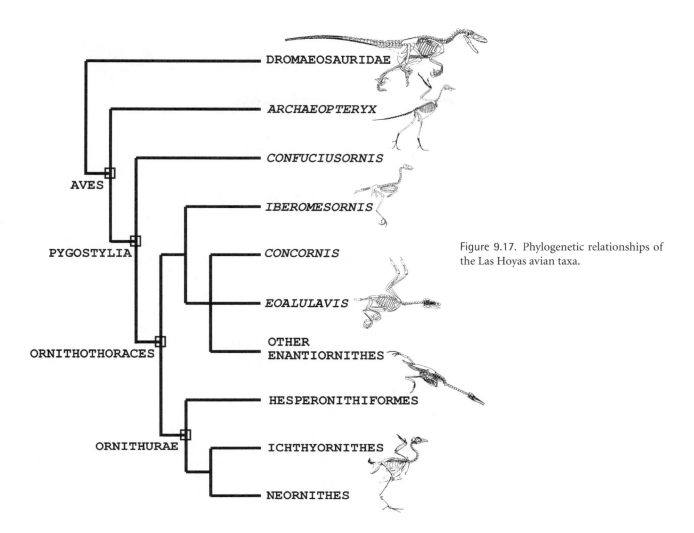

DROMAEOSAURIDAE

ARCHAEOPTERYX

CONFUCIUSORNIS

IBEROMESORNIS

CONCORNIS

EOALULAVIS

OTHER ENANTIORNITHES

HESPERONITHIFORMES

ICHTHYORNITHES

NEORNITHES

AVES

PYGOSTYLIA

ORNITHOTHORACES

ORNITHURAE

Figure 9.17. Phylogenetic relationships of the Las Hoyas avian taxa.

culae are present in cursorial, nonavian theropods (Barsbold et al., 1990; Bryant and Russell, 1993; Chure and Madsen, 1996; Makovicky and Currie, 1996). Thus, it is conceivable that the derived (flight) function of the furcula, or at least a more important role in this novel activity, was attained with its transformation into a V-shaped element similar to that of *Iberomesornis* and the other birds from Las Hoyas, a structure equivalent to that of extant birds. Then, as now, a V-shaped furcula may have played a significant role in the aeration of the lungs and air sacs during flight (Bailey and DeMont, 1991).

The presence of a pygostyle in *Iberomesornis* (the tail is not complete in the remaining birds from Las Hoyas) implies the presence of a large, fleshy rectricial bulb (Parson's nose). Although no feather evidence has been found in *Iberomesornis* (contra Chatterjee, 1997), it is reasonable to suppose that this structure was flanked by a number of tail feathers. Therefore, as compared with *Archaeopteryx*, the flight capabilities of the Las Hoyas bird probably included greater mobility of the rectrices and thus a larger capacity for steering and braking. The long, frond-shaped tail of

Archaeopteryx can be better interpreted as a device increasing lift, with a low contribution to flight maneuverability.

Our understanding of the evolution of bird flight has been greatly clarified by the concept of "locomotor modules" (Gatesy and Dial, 1996; see also Gatesy, Chapter 19 in this volume). The anatomy of the birds from Las Hoyas has been instrumental in clarifying the temporal pattern of locomotor modular evolution (Fig. 9.18). Gatesy and Dial (1996) recognized three locomotor modules in extant birds—pectoral, hindlimb, and tail—derived from an ancestral theropod locomotor module composed by the hindlimb and tail. The early evolution of flight can be viewed as a transition involving three sequential stages (Gatesy and Dial, 1996): (1) formation of the pectoral module; (2) decoupling of the hindlimb and tail; and (3) novel allegiance of the pectoral and tail modules to form the flight apparatus. The similarities between *Archaeopteryx* and other nonavian theropods indicate that their main locomotor module retains the coupling of the tail and hindlimb for running. The dimensions of the bony tail of *Archaeopteryx* and other skeletal features suggest the presence of a signifi-

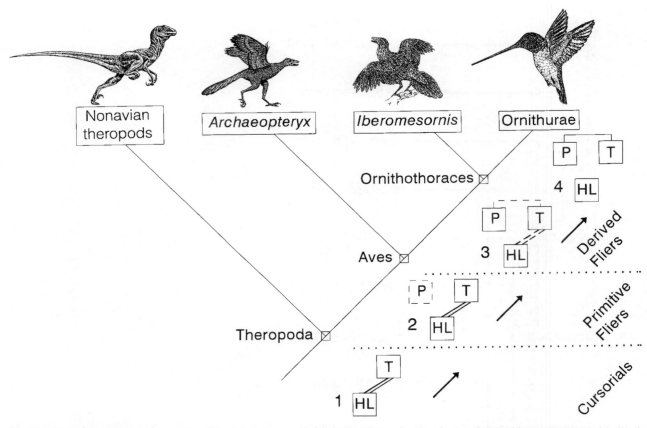

Figure 9.18. Schematic transformation of the locomotor modules during the evolution of avian flight. Abbreviations: HL: hindlimb module; P: pectoral module; T: tail module. 1: primitive, single locomotor module (nonavian theropods). 2: dominance of the primitive single theropod locomotor module; appearance of the pectoral module (*Archaeopteryx*). 3: Pectoral module fully developed; partial decoupling between tail and hindlimb modules; beginning of the allegiance between pectoral and tail modules (*Iberomesornis*). 4: Complete decoupling between tail and hindlimb locomotor module; allegiance between pectoral and tail modules fully accomplished (Ornithurine birds). *Deinonychus, Archaeopteryx, Iberomesornis,* and hummingbird, from Ostrom (1969), Charig (1979), Sanz and Bonaparte (1992), and Proctor and Lynch (1993), respectively.

cant caudofemoral musculature to retract the hindlimbs, the caudal appendage counterbalancing the cranial part of the body (Gatesy, 1990; Gatesy and Dial, 1996). This implies that its tail and hindlimb were still combined in a single unit. On the other hand, the avian pectoral module is present in *Archaeopteryx*, although lacking the fine-tuning that appears in *Iberomesornis, Concornis, Eoalulavis,* and other ornithothoracine birds as well as any allegiance between this module and the tail module (Fig. 9.18). Thus, it is possible to conclude that, from the conceptual point of view of the "locomotor module," *Archaeopteryx* is not a flying bird in the modern sense.

Although the impressive rate of new discoveries steadily changes our understanding of early character evolution, the known Early Cretaceous avian record (e.g., *Iberomesornis* and the other birds from Las Hoyas, *Sinornis, Cathayornis*) appears to indicate a correlation between the evolution of the modern avian pectoral and tail modules. The evolution

of derived wing proportions and strutlike coracoids, along with the furcula, correlates with a shortened tail and the presence of a pygostyle. Thus, the evidence of *Iberomesornis* (confirmed later by other taxa) shows a simultaneous, more than a sequential, development of stages (2) and (3). Yet this scenario gets more complicated when considering that decoupling of tail and hindlimb modules may not have been fully accomplished in *Iberomesornis*. The number of total caudal elements of the Las Hoyas bird could be comparable to that of *Archaeopteryx* (20–23). Furthermore, the large pygostyle and the presence of a primitive pelvis and synsacrum in *Iberomesornis,* which indicates the absence of a hiatus in the epaxial musculature between the trunk and tail, suggest the persistence of a significant amount of caudofemoral musculature and consequently a greater participation of the tail during terrestrial locomotion.

The birds from Las Hoyas have also rendered crucial data on the early development of nonskeletal structures correlated to the fine-tuning of avian flight. The presence of an alula in *Eoalulavis* is of great significance in understanding the early steps toward the evolution of enhanced flight capabilities (Sanz et al., 1996). This structure plays a fundamental role in modern avian aerodynamics, allowing low-speed flights and high maneuverability (Graham, 1932; Brown, 1963; Nachtigall and Kempf, 1971). *Eoalulavis* documents the oldest known alula. It also represents the most primitive bird for which an alula is known. The alula appears to be absent in the feathered specimens of *Archaeopteryx*. This suggests some sort of flight constraints, mainly related to low-speed flight and maneuverability (Rayner, 1991). This interpretation is consistent with the skeletal differences and inferred functional aptitudes pointed out earlier.

The discovery of the birds from Las Hoyas revealed that an advanced perching capacity, then known only for early Tertiary birds (Feduccia, 1980), was an early avian specialization. The pedal anatomy of *Iberomesornis* and *Concornis* hints at improved perching capabilities. These birds have an anisodactyl foot with an opposing hallux that is proportionally longer than that of *Archaeopteryx*. For example, in *Iberomesornis*, the hallux is almost as long as digit II, whereas in *Archaeopteryx* the hallux is about half the length of digit II. More important is that the hallux of the birds from Las Hoyas is lower than that of *Archaeopteryx*. As pointed out by Bock and Miller (1959), among others (e.g., Ostrom, 1976; Sereno and Rao, 1992; Chiappe, 1995a), the size and position of the hallux of *Archaeopteryx* would reduce its effectiveness as an opposable toe. Conversely, the pedal morphology of the birds from Las Hoyas supports a significant improvement in perching and grip effectiveness.

The birds from Las Hoyas have also provided important evidence on the feeding specializations of early birds. Sanz et al. (1996) reported scattered organic particles preserved within a limonitic mass in the thoracic cavity of *Eoalulavis*. These particles, interpreted as stomach contents, appear to be exoskeletal elements of crustaceans. This evidence provides the oldest direct manifestation of trophic habits in Mesozoic birds. The fact that *Eoalulavis* fed on crustaceans hints at aquatic feeding habits for this particular bird, and it reinforces previous interpretations of the Las Hoyas birds as residing in nearshore or aquatic habitats (Sanz and Buscalioni, 1992).

In conclusion, the fossil birds from Las Hoyas have provided evidence indicating that enhanced flight capabilities similar to those of neornithine birds were already present at the outset of the Cretaceous. These aptitudes were probably present in basal ornithothoracines and almost certainly in enantiornithine birds.

Acknowledgments

This work was supported by Junta de Comunidades de Castilla–La Mancha, DGICYT (Promoción General del Conocimiento, project number PB93-0284), the European Union (Human Capital and Mobility Program, network contract number ERBCHRXCT 930164), and grants from the Guggenheim Foundation and the Dinosaur Society to L. M. Chiappe.

Literature Cited

Bailey, J. P., and M. E. DeMont. 1991. The function of the wishbone. Canadian Journal of Zoology 69:2751–2758.

Barbadillo, J., and S. E. Evans. 1995. Lacertilians; pp. 57–58 in N. Meléndez (ed.), Las Hoyas. A Lacustrine Konservat-Lagerstätte (Cuenca, Spain). Field Trip Guide Book. II International Symposium on Lithographic Limestones. Universidad Complutense, Madrid.

Barsbold R., T. Maryanska, and H. Osmólska. 1990. Oviraptorosauria; pp. 249–258 in D. B. Weishampel, P. Dodson, and H. Osmólska (eds.), The Dinosauria. University of California Press, Berkeley.

Bock, W., and W. D. Miller. 1959. The scansorial foot of the woodpeckers, with comments on the evolution of perching and climbing feet in birds. American Museum Novitates 1931:1–45.

Briggs, D. E. G., P. R. Wilby, B. P. Pérez-Moreno, J. L. Sanz, and M. A. Fregenal-Martínez. 1997. The mineralization of dinosaur soft tissue in the Lower Cretaceous of Las Hoyas, Spain. Journal of the Geological Society of London 154:587–588.

Brochu, C. A. 1996. Closure of neurocentral sutures during crocodilian ontogeny: implications for maturity assessment in fossil archosaurs. Journal of Vertebrate Paleontology 16(1): 49–62.

Brown, R. H. J. 1963. The flight of birds. Biological Reviews of the Cambridge Philosophical Society 36:460–489.

Bryant, H. N., and A. P. Russell. 1993. The occurrence of clavicles within Dinosauria: implications for the homology of the avian furcula and the utility of negative evidence. Journal of Vertebrate Paleontology 12:171–184.

Buscalioni, A. D., and F. Ortega. 1995. Crocodylomorphs; pp. 59–61 in N. Meléndez (ed.), Las Hoyas. A Lacustrine Konservat-Lagerstätte (Cuenca, Spain). Field Trip Guide Book. II International Symposium on Lithographic Limestones. Universidad Complutense, Madrid.

Charig, A. 1979. A New Look at the Dinosaurs. Heinemann/British Museum (Natural History), London. 160 pp.

Chatterjee, S. 1995. The Triassic bird *Protoavis*. Archaeopteryx 13:15–31.

———. 1997. The Rise of Birds. Johns Hopkins University Press, Baltimore, 312 pp.

Chiappe, L. M. 1991. Cretaceous avian remains from Patagonia shed new light on the early radiation of birds. Alcheringa 15(3–4):333–338.

———. 1992. Enantiornithine tarsometatarsi and the avian affinity of the Late Cretaceous Avisauridae. Journal of Vertebrate Paleontology 12(3):344–350.

———. 1993. Enantiornithine (Aves) tarsometatarsi from the Cretaceous Lecho Formation of Northwestern Argentina. American Museum Novitates 3079:1–27.

———. 1995a. The first 85 million years of avian evolution. Nature 378:349–355.

———. 1995b. The phylogenetic position of the Cretaceous birds of Argentina: Enantiornithes and *Patagopteryx deferrariisi*; pp. 55–63 in D. S. Peters (ed.), Proceedings of the 3rd

Symposium of the Society of Avian Paleontology and Evolution. Courier Forschungsinstitut Senckenberg 181, Frankfurt am Main.

———. 1996a. Late Cretaceous birds of southern South America: anatomy and systematics of Enantiornithes and *Patagopteryx deferrariisi;* pp. 203–244 *in* G. Arratia (ed.), Contributions of Southern South America to Vertebrate Paleontology, Münchner Geowissenschaftliche Abhandlungen, Reihe A, Geologie und Paläontologie 30. Verlag Dr. Friedrich Pfeil, Munich.

———. 1996b. Early avian evolution in the southern hemisphere: fossil record of birds in the Mesozoic of Gondwana. Memoirs of the Queensland Museum 39(3):533–555.

———. 2001. Phylogenetic relationships among basal birds; pp. 125–139 *in* J. Gauthier and L. F. Gall (eds.), New Perspectives on the Origin and Early Evolution of Birds: Proceedings of the International Symposium in Honor of John H. Ostrom. Peabody Museum of Natural History, New Haven.

Chiappe, L. M., and J. O. Calvo. 1994. Neuquenornis volans, a new Upper Cretaceous bird (Enantiornithes: Avisauridae) from Patagonia, Argentina. Journal of Vertebrate Paleontology 14(2):230–246.

Chure, D. J., and J. H. Madsen Jr. 1996. The furcula in allosaurid theropods and its implication for determining bird origins. Journal of Vertebrate Paleontology 16(3):28A.

Cracraft, J. 1988. Early evolution of birds. Nature 331:389–390.

Davis, P. G., and D. E. G. Briggs. 1995. Fossilization of feathers. Geology 23(9):783–786.

Diéguez, M. C., C. Martín-Closas, P. Trincao, and N. López-Morón. 1995. Flora; pp. 29–32 *in* N. Meléndez (ed.), Las Hoyas. A Lacustrine Konservat-Lagerstätte (Cuenca, Spain). Field Trip Guide Book. II International Symposium on Lithographic Limestones. Universidad Complutense, Madrid.

Elzanowski, A. 1995. Cretaceous birds and avian phylogeny. Courier Forschungsinstitut Senckenberg 181:37–53.

Evans, S. E., G. McGowan, A. R. Milner, and B. Sanchiz. 1995. Amphibians; pp. 51–53 *in* N. Meléndez (ed.), Las Hoyas. A Lacustrine Konservat-Lagerstätte (Cuenca, Spain). Field Trip Guide Book. II International Symposium on Lithographic Limestones. Universidad Complutense, Madrid.

Feduccia, A. 1980. The Age of Birds. Harvard University Press, Cambridge, 196 pp.

———. 1995. Explosive evolution in Tertiary birds and mammals. Science 267:637–638.

———. 1996. The Origin and Early Evolution of Birds. Yale University Press, New Haven, 420 pp.

Forster, C. A., L. M. Chiappe, D. W. Krause, and S. D. Sampson. 1996. The first Cretaceous bird from Madagascar. Nature 382:532–534.

Fregenal-Martínez, M. A., and N. Meléndez. 1995. Geological setting; pp. 1–10. *in* N. Meléndez (ed.), Las Hoyas. A Lacustrine Konservat-Lagerstätte (Cuenca, Spain). Field Trip Guide Book. II International Symposium on Lithographic Limestones. Universidad Complutense, Madrid.

Fregenal-Martínez, M. A., and J. J. Moratalla. 1995. Paleoichnology; pp. 71–75. *in* N. Meléndez (ed.), Las Hoyas. A Lacustrine Konservat-Lagerstätte (Cuenca, Spain). Field Trip Guide Book. II International Symposium on Lithographic Limestones. Universidad Complutense, Madrid.

Gatesy, S. M. 1990. Caudofemoral musculature and the evolution of theropod locomotion. Paleobiology 16(2):170–186.

Gatesy, S. M., and K. P. Dial. 1996. Locomotor modules and the evolution of avian flight. Evolution 50(1):331–340.

Gómez-Fernández, J. C., and N. Meléndez. 1991. Rhythmically laminated lacustrine deposits in the Lower Cretaceous of La Serranía de Cuenca basin (Iberian Ranges, Spain); pp. 247–258 *in* P. Anadón, L. Cabrera, and K. Kelts (eds.), Lacustrine Facies Analysis. IAS Special Publication 13. Blackwell Scientific Publications, Oxford.

Graham, R. R. 1932. Slots in the wings of birds. Journal of the Royal Aeronautical Society 36:598–600.

Hecht, M. K., J. H. Ostrom, G. Viohl, and P. Wellnhofer. 1985. The Beginnings of Birds. Freunde des Jura-Museums, Eichstätt, 382 pp.

Hou L., Zhou Z.-H., Gu Y., and Zhang. 1995. *Confuciusornis sanctus,* a new Late Jurassic sauriurine bird from China. Chinese Science Bulletin 40(18):1545–1551.

Hou L., L. D. Martin, Zhou Z., and A. Feduccia. 1996. Early adaptive radiation of birds: evidence from fossils from northeastern China. Science 274:1164–1167.

Jiménez-Fuentes, E. 1995. Turtles; pp. 55–56 *in* N. Meléndez (ed.), Las Hoyas. A Lacustrine Konservat-Lagerstätte (Cuenca, Spain). Field Trip Guide Book. II International Symposium on Lithographic Limestones. Universidad Complutense, Madrid.

Kurochkin, E. N. 1995. Synopsis of Mesozoic birds and early evolution of Class Aves. *Archaeopteryx* 13:47–66.

Makovicky, P. J., and P. J. Currie. 1996. Discovery of a furcula in tyrannosaurid theropods. Journal of Vertebrate Paleontology 16(3):50A.

Martin, L. D. 1983. The origin and early radiation of birds; pp. 291–338 *in* A. H. Brush and F. A. Clark Jr. (eds.), Perspectives in Ornithology. Cambridge University Press, Cambridge.

———. 1995. The Enantiornithes: terrestrial birds of the Cretaceous. Courier Forschungsinstitut Senckenberg 181:23–36.

Martínez-Delclòs, X., and A. Nell. 1995a. Decapods and Mysidaceans; p. 35 *in* N. Meléndez (ed.), Las Hoyas. A Lacustrine Konservat-Lagerstätte (Cuenca, Spain). Field Trip Guide Book. II International Symposium on Lithographic Limestones. Universidad Complutense, Madrid.

———. 1995b. Insects; pp. 37–41 *in* N. Meléndez (ed.), Las Hoyas. A Lacustrine Konservat-Lagerstätte (Cuenca, Spain). Field Trip Guide Book. II International Symposium on Lithographic Limestones. Universidad Complutense, Madrid.

McGowan, G., and S. E. Evans. 1995. Albanerpetontid amphibians from the Cretaceous of Spain. Nature 373:143–145.

Nachtigall, W., and B. Kempf. 1971. Vergleichende Untersuchungen zur flugbiologischen Funktion des Daumenfittichs (Alula spuria) bei Vögeln. I. Der Daumenfittich als Hochauftriebserzeuger. Zeitschrift für Vergleichende Physiologie 71:326–341.

Norberg, U. M. 1989. Vertebrate Flight. Springer-Verlag, Berlin, 291 pp.

Olson, S. L. 1985. The fossil record of birds; pp. 79–238 *in* D. S. Farner, J. King, and K. C. Parkes (eds.), Avian Biology, Vol. 8. Academic Press, New York.

Olson, S. L., and A. Feduccia. 1979. Flight capability and the pectoral girdle of *Archaeopteryx.* Nature 278:247–248.

Ortega, F., Z. Gasparini, A. D. Buscalioni, and J. O. Calvo. 2000. A new species of *Araripesuchus* (Crocodylomorpha, Mesoeucrocodylia) from the Lower Cretaceous of Patagonia (Argentina). Journal of Vertebrate Paleontology 20 (1):57–76.

Ostrom, J. H. 1969. Osteology of *Deinonychus antirrhopus,* an unusual theropod from the Lower Cretaceous of Montana. Bulletin of the Peabody Museum of Natural History 30:1–165.

———. 1976. *Archaeopteryx* and the origin of birds. Biological Journal of the Linnean Society 8:91–182.

Pérez-Moreno, B. P., J. L. Sanz, A. D. Buscalioni, J. J. Moratalla, F. Ortega, and D. Rasskin-Gutman. 1994. A unique multi-toothed ornithomimosaur from the Lower Cretaceous of Spain. Nature 370:363–367.

Poore, S. O., A. Ashcroft, A. Sánchez-Haiman, and G. E. Goslow Jr. 1997a. The contractile properties of the M. supracoracoideus in the pigeon and starling: a case for long-axis rotation of the humerus. Journal of Experimental Biology 200:2987–3002.

Poore, S. O., A. Sánchez-Heiman, and G. E. Goslow Jr. 1997b. Wing upstroke and the evolution of flapping flight. Nature 387:799–802.

Poyato-Ariza, F. J., and S. Wenz. 1995. Ichthyofauna; pp. 43–49 *in* N. Meléndez (ed.), Las Hoyas. A Lacustrine Konservat-Lagerstätte (Cuenca, Spain). Field Trip Guide Book. II International Symposium on Lithographic Limestones. Universidad Complutense, Madrid.

Proctor, N. S., and P. J. Lynch. 1993. Manual of Ornithology. Avian Structure and Function. Yale University Press, New Haven, 340 pp.

Raikow, R. J. 1985. Locomotor system; pp. 57–147 *in* A. S. King and J. McLelland (eds.), Form and Function in Birds, Vol. 3. Academic Press, New York.

Rayner, J. M. V. 1991. Avian flight evolution and the problem of *Archaeopteryx;* pp. 183–212 *in* J. M. V. Rayner and R. J. Wooton (eds.), Biomechanics in Evolution. Cambridge University Press, Cambridge.

Rodríguez-Lázaro, J. 1995. Ostracods; pp. 33–34 *in* N. Meléndez (ed.), Las Hoyas. A Lacustrine Konservat-Lagerstätte (Cuenca, Spain). Field Trip Guide Book. II International Symposium on Lithographic Limestones. Universidad Complutense, Madrid.

Sanz, J. L., and J. F. Bonaparte. 1992. A new order of birds (Class Aves) from Early Cretaceous of Spain; pp. 39–49 *in* K. E. Campbell (ed.), Papers in Avian Paleontology. Honoring Pierce Brodkorb. Science Series 36, Natural History Museum of Los Angeles County, Los Angeles.

Sanz, J. L., and A. D. Buscalioni. 1992. A new bird from the Early Cretaceous of Las Hoyas, Spain, and the early radiation of birds. Paleontology 35(4):829–845.

———. 1994. An isolated bird foot from the Barremian (Lower Cretaceous) of Las Hoyas (Cuenca, Spain). Geobios 16:213–217.

Sanz, J. L., and B. P. Pérez-Moreno. 1995. Dinosaurs; pp. 63–65 *in* N. Meléndez (ed.), Las Hoyas. A Lacustrine Konservat-Lagerstätte (Cuenca, Spain). Field Trip Guide Book. II International Symposium on Lithographic Limestones. Universidad Complutense, Madrid.

Sanz, J. L., J. F. Bonaparte, and A. Lacasa. 1988. Unusual Early Cretaceous birds from Spain. Nature 331:433–435.

Sanz, J. L., L. M. Chiappe, J. F. Bonaparte, and A. D. Buscalioni. 1993. The Spanish Lower Cretaceous bird *Concornis lacustris* reevaluated. Journal of Vertebrate Paleontology 13(3):56A.

Sanz, J. L., L. M. Chiappe, and A. D. Buscalioni. 1995. The osteology of *Concornis lacustris* (Aves: Enantiornithes) from the Lower Cretaceous of Spain and a reexamination of its phylogenetic significance. American Museum Novitates 3133: 1–23.

Sanz, J. L., L. M. Chiappe, B. P. Pérez-Moreno, A. D. Buscalioni, J. J. Moratalla, F. Ortega, and F. J. Poyato-Ariza. 1996. A new Lower Cretaceous bird from Spain: implications for the evolution of flight. Nature 382:442–445.

Sanz, J. L., L. M. Chiappe, B. P. Pérez-Moreno, J. J. Moratalla, F. Hernández-Carrasquilla, A. D. Buscalioni, F. Ortega, F. J. Poyato-Ariza, D. Rasskin-Gutman, and X. Martínez-Delclòs. 1997. A nestling bird from the Early Cretaceous of Spain: implications for avian skull and neck evolution. Science 276:1543–1546.

Sanz, J. L., C. Diéguez, F. J. Poyato-Ariza. 1999. Il Cretacico inferiore di Las Hoyas; pp. 155–160 *in* G. Pinna (ed.), Ale radici della Storia Naturale d'Europa. Associazione Paleontologica Europea. Jaca Book, Milan, Italy.

Sanz, J. L., L. M. Chiappe, Y. Fernández-Jalvo, F. Ortega, B. Sánchez-Chillón, F. J. Pyato-Ariza, and B. P. Pérez-Moreno. 2001a. An Early Cretaceous pellet. Nature 409:998–1000.

Sanz, J. L., M. A. Fregenal-Martínez, N. Meléndez, and F. Ortega. 2001b. Las Hoyas; pp. 356–359 *in* D. E. G Briggs and P. R. Crowther (eds.), Paleobiology II. Blackwell Science, Oxford.

Sereno, P. C. 1998. A rationale for phylogenetic definitions, with application to the higher-level taxonomy of Dinosauria. Neues Jahrbuch für Geologie und Paläontologie 210(1):41–83.

———. 2000. *Iberomesornis romerali* (Aves, Ornithothoraces) reevaluated as an Early Cretaceous enantiornithine. Neues Jahrbuch für Geologie und Paläontologie 215(3):365–395.

Sereno, P. C., and Rao C. 1992. Early evolution of avian flight and perching: new evidence from Lower Cretaceous of China. Science 225:845–848.

Talbot, M. R., N. Meléndez, and M. A. Fregenal-Martínez. 1995. The waters of the Las Hoyas lake: sources and limnological characteristics; pp. 11–16 *in* N. Meléndez (ed.), Las Hoyas. A Lacustrine Konservat-Lagerstätte (Cuenca, Spain). Field Trip Guide Book. II International Symposium on Lithographic Limestones. Universidad Complutense, Madrid.

Varricchio, D. J., and L. M. Chiappe. 1995. A new bird from the Cretaceous Two Medicine Formation of Montana. Journal of Vertebrate Paleontology 15(1):201–204.

Walker, C. A. 1981. New subclass of birds from the Cretaceous of South America. Nature 292:51–53.

Wellnhofer, P. 1974. Das fünfte Skelettexemplar von *Archaeopteryx.* Palaeontographica A 147:169–216.

———. 1992. A new specimen of *Archaeopteryx* from the Solnhofen limestone; pp. 3–23 *in* K. E. Campbell (ed.), Papers in Avian Paleontology, Honoring Pierce Brodkorb. Science Series 36, Natural History Museum of Los Angeles County, Los Angeles.

———. 1993. Das siebte Exemplar von *Archaeopteryx* aus den Solnhofener Schichten. Archaeopteryx 11:1–48.

Zhou Z. 1995a. The discovery of Early Cretaceous birds in China. Courier Forschungsinstitut Senckenberg 181:9–22.

———. 1995b. New understanding of the evolution of the limb and girdle elements in early birds—evidences from Chinese fossils; pp. 209–214 *in* Sixth Symposium on Mesozoic Terrestrial Ecosystems and Biota, Short Papers. China Ocean Press, Beijing.

10

Noguerornis gonzalezi (Aves: Ornithothoraces) from the Early Cretaceous of Spain

LUIS M. CHIAPPE AND ANTONIO LACASA-RUIZ

 The first Mesozoic avian remains from the Iberian Peninsula were reported by Vidal (1902), who mentioned the accidental destruction of a fossil bird from the Lower Cretaceous quarry of "La Pedrera de Meiá" (42°01′ N; 4°35′ W) in the Spanish province of Lleida, Catalonia (Lacasa-Ruiz, 1985). In subsequent years, this site yielded a large number of isolated feathers (Lacasa-Ruiz, 1985; Gómez Pallerola, 1986; Kellner, Chapter 16 in this volume). However, the first significant skeletal discovery—the skeletal remains of *Noguerornis gonzalezi*—was not made until the 1980s (Lacasa-Ruiz, 1986, 1989a,b, 1991a).

The discovery of Early Cretaceous avians from Spain provided relevant new information about the early evolution of birds, clarifying the early pattern of morphological transformation (see Sanz et al., 1988, Chapter 9 in this volume; Chiappe, 1991, 1995a) and establishing some of the earliest records of volant birds with enhanced aerodynamic specializations (Sanz et al., 1988, 1996, Chapter 9 in this volume; Chiappe, 1995b). The significance of the spectacular specimens of *Iberomesornis romerali* (Sanz and Bonaparte, 1992; Sanz et al., Chapter 9 in this volume), *Concornis lacustris* (Sanz and Buscalioni, 1992; Sanz et al., 1995, Chapter 9 in this volume), and *Eoalulavis hoyasi* (Sanz et al., 1996, Chapter 9 in this volume) was immediately recognized (see Cracraft, 1988; Milner, 1993; Padian, 1996); this was not the case, however, for the less complete *N. gonzalezi*. Although several papers by Lacasa-Ruiz (e.g., Lacasa-Ruiz, 1986, 1989a,b, 1991a) remarked on the peculiarities and significance of *Noguerornis* (one of the earliest known avian members), this information has remained virtually unnoticed. In this chapter, after further mechanical preparation, we provide a detailed description of the holotype and still unique specimen of *Noguerornis* and discuss both its relationships to other basal birds and its contribution to our understanding of early avian evolution.

Geological Setting

The Sierra del Montsec is roughly 65 km north of the town of Lleida (Catalonia), extending approximately 50 km from east to west (Fig. 10.1). The rivers Noguera Pallaresa and Noguera Ribagorzana cross the Montsec range at right angles. In the eastern region of the Montsec range on the eastern slope of the valley of the Noguera Pallaresa, there is a famous quarry of lithographic limestones that since 1902 has yielded numerous well-preserved fossils of continental facies (Lacasa-Ruiz, 1991b; Martínez-Delclòs, 1991a, 1995). This quarry gained prominence as an important paleontological site when, at the end of the nineteenth century, a small mining company started to process the fine stone for lithographic purposes. Mining engineer L. M. Vidal (1902) was pivotal in recognizing the importance of this fossil site and reporting, along with other workers, the first fossil findings. The quarry reported by Vidal (1902) is known as "Las Calizas Litográficas de Santa María de Meyá" (also known as "La Pedrera de Meiá," "La Pedrera de Rúbies," or simply "La Pedrera") in reference to its proximity to the town of Santa María de Meyá. In more recent years, a second quarry was opened nearby; this second quarry has been named "La Cabrúa."

Vidal (1902) proposed a Kimmeridgian (Late Jurassic) age for the lithographic limestones of Montsec. Later studies, however, determined the age of these deposits as Early Cretaceous but did not arrive at a consensus regarding their precise age (Fig. 10.2). Whereas paleofloristic (Barale, 1973, 1991, 1995) and ostracod (Peybernés and Oertli, 1972; Brenner et al., 1974) data placed these beds within the Late Berriasian–Early Valanginian, recent charophyte studies point to a younger, Late Hauterivian–Early Barremian age (Martín-Closas and López-Morón, 1995).

Peybernés (1976:199) defined this 50-meter thick deposit of lithographic limestones as "calcaires lithographiques à

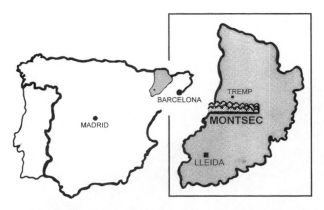

Figure 10.1. Map indicating the area of the Montsec range within the province of Lleida, Catalonia (Spain), where the holotype of *N. gonzalezi* was found.

plantes et vertébrés de La Pedrera de Rúbies" (Unit N2) and included it within a unit he named as "Calcaires à Charophites du Montsec." This unit has also been referred to as "La Pedrera de Rúbies Lithographic Limestones" (see Fregenal-Martínez and Meléndez, 1995).

Rocks at "La Pedrera" are rhythmically laminated lithographic limestones deposited in the distal areas of a lake that developed under a warm, subtropical-semiarid climate (Fregenal-Martínez and Meléndez, 1995). During the Early Cretaceous, the Iberian Plate was close to the 30° N parallel, and this lake was in the proximity of the seashore. However, the presence of a marine connection with the predominantly freshwater environment has not been definitively established (Fregenal-Martínez and Meléndez, 1995).

The fossil biota yielded by the "La Pedrera de Rúbies Lithographic Limestones" is highly diverse (Martínez-Delclòs, 1991a, 1995). The record of plants (Barale, 1991, 1995) is composed of pteridophytes (Ferns and Equisetales), prespermatophytes (e.g., Ginkgoales, Cycadales, Bennettitales), and spermatophytes (Coniferales and Angiospermae), along with spores of algae and bryophytes. The fossil fauna is composed of mollusks (Martinell and Doménech, 1991), crustaceans (Lacasa-Ruiz and Via, 1991), arachnids (Selden, 1991), abundant insects (Martínez-Delclòs, 1991b), and vertebrates. Vertebrate fossils are known primarily by the abundant "holosteans" and teleostean fishes (Wenz, 1991a), anurans (Wenz, 1991b), nonavian reptiles (Buscalioni and Sanz, 1991), and birds (Lacasa-Ruiz, 1989a,b, 1991b). The avian fossil remains comprise approximately 25 specimens of isolated feathers (see Lacasa-Ruiz, 1985; Kellner, Chapter 16 in this volume), the holotype of *N. gonzalezi* (Lacasa-Ruiz, 1989a), and the enantiornithine hatchling reported by Sanz et al. (1997).

Systematic Paleontology

Taxonomic Hierarchy

Aves Linnaeus, 1758
  Ornithothoraces Chiappe, 1995
    Enantiornithes Walker, 1981
      N. gonzalezi Lacasa-Ruiz, 1989

Holotype—Slab numbered LP.1702 P (collection from "La Pedrera," Institut d'Estudis Ilerdencs, Lleida, Spain), including portions of both humeri, radii, and carpometacarpi, left ulna, furcula, tibia, three trunk vertebrae, pelvis, other osseous remains, and feathers (Fig. 10.2).

Locality and horizon—La Pedrera de Meiá, Sierra del Montsec, province of Lleida, Spain. "La Pedrera de Rúbies Lithographic Limestones," Lower Cretaceous, Upper Berriasian–Lower Barremian.

Diagnosis—*N. gonzalezi* is diagnosed by the presence of a strongly curved humerus, a clear autapomorphy of this taxon.

Anatomy

N. gonzalezi is preserved on a single slab (Fig. 10.2). Although most bones are incomplete, significant anatomical details are visible on both their preserved portions and the impressions left by them on the rock. Based on furcular and humeral comparisons, *Noguerornis* is intermediate in size between the smaller *Iberomesornis* (Sanz and Bonaparte, 1992) and the larger *Concornis* (Sanz et al., 1995) and *Eoalulavis* (Sanz et al., 1996).

Vertebral Column

The most caudal thoracic vertebra and the cranial two synsacral vertebrae are preserved. These three elements lay on their right lateral side within the pelvic remains, exposing their left lateral sides (Figs. 10.2, 10.3, 10.4C).

The caudalmost thoracic vertebra shows a broad depression excavating its centrum, although it is not clear whether this is an artifact of its preservation. A broad lateral fossa is present in the thoracic centra of several enantiornithine taxa (Chiappe and Walker, Chapter 11 in this volume), *Confuciusornis* (Chiappe et al., 1999), *Ichthyornis* (Marsh, 1880), and some neornithine birds (e.g., charadriiforms, procellariiforms). The articular surfaces of the thoracic and synsacral vertebrae are apparently amphyplatian. The spinous processes are broad and laminar, with their craniodorsal apex projected cranially (Figs. 10.3, 10.4C). In the caudal thoracic vertebra, a small cranial zygapophysis is preserved. The two synsacral vertebrae are fused to each other, although the boundary between them is still visible. The spinous processes of these two elements form a continuous spinous crest. Zygapophyses are not preserved in these synsacral vertebrae.

Figure 10.2. Holotype of *N. gonzalezi* (LP.1702.P) under ultraviolet light.

Thoracic Girdle and Limb

Furcula. The furcula of *Noguerornis* is preserved nearly complete and exposed apparently in ventral view. It is robust and U-shaped, having ventral and medial faces that are nearly flat and perpendicular to each other. The interclavicular angle is approximately 46°. The furcula bears a hypertrophied hypocleideum, which, as in the enantiornithine *Sinornis* (Zhou, 1995; see Sereno, Rao, and Li, Chapter 8 in this volume, for a discussion of the synonymy of *Sinornis* and *Cathayornis*) and *Eoalulavis* (Sanz et al., 1996), exceeds half the length of the clavicular rami. The hypocleideum is thin and relatively compressed and tapers distally (Figs. 10.2, 10.3, 10.4B).

Humerus. The left humerus is articulated with the ulna and radius, while the right element is disarticulated. The humerus of *Noguerornis* is shorter than the ulna and radius, a condition shared by most ornithothoracines (Chiappe, 1996). It is robust and strongly curved, with a convex dorsal margin and a concave ventral one (Figs. 10.2, 10.3, 10.4C). The proximal end is better preserved in the right, disarticulated element, being exposed in its cranial aspect. In this view, the proximal border of the head is slightly inclined dorsally (Fig. 10.4C). A slightly depressed area located just distal to the proximal margin is dorsally and ventrally bordered by smooth elevated areas. As in other basal birds (e.g., Enantiornithes, *Patagopteryx*) and nonavian theropods (e.g., *Deinonychus*), the humeral head is concave cranially and presumably convex caudally. There is a short bicipital area ventral to the head (Fig. 10.4C). The distal end is badly preserved in both humeri; in the left element, which is exposed caudally, the olecranon fossa is poorly developed.

Ulna. Only the left ulna is preserved, exposed in dorsal aspect. Both distal and proximal ends are damaged. The shaft is nearly straight, round in cross section, and uniformly broad (Figs. 10.2, 10.3). The midshaft width appears to be only slightly broader than the radial midshaft (Table 10.1). Papillae for the remigial feathers are absent.

Radius. The left radius is preserved in articulation with the humerus, ulna, and carpometacarpus; the right element is disarticulated and mostly preserved as an impression on the slab. The radius of *Noguerornis* is straight, uniformly wide, and subcylindrical in cross section (Figs. 10.2, 10.3). The distalmost end curves slightly caudalward. The proximal end is not preserved in either radii. In contrast to the

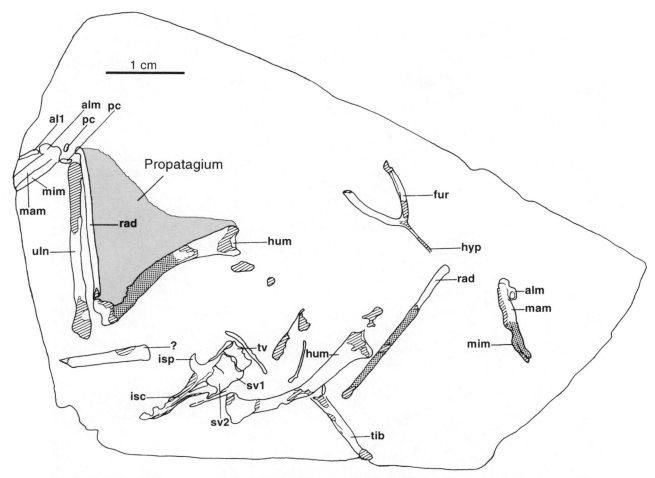

Figure 10.3. Holotype of *N. gonzalezi* (LP.1702.P). Abbreviations: al1, proximal phalanx of alular digit; alm, alular metacarpal; fur, furcula; hum, humerus; hyp, hypocleideum; isc, ischium; isp, ischiadic process (see text for interpretation); mam, major metacarpal; mim, minor metacarpal; pc, proximal carpal; rad, radius; sv1, first synsacral vertebra; sv2, second synsacral vertebra; tib, tibia; tv, thoracic vertebra; uln, ulna.

condition of some Enantiornithes (Chiappe and Calvo, 1994; Chiappe and Walker, Chapter 11 in this volume), the radial shaft of *Noguerornis* lacks a caudal, axial groove.

Carpometacarpus. Only the shaft and proximal portion of the carpometacarpi are preserved. Despite previous statements (Lacasa-Ruiz, 1989a), the three metacarpals of the hand of *Noguerornis* are not unfused; they are indeed fused, at least proximally (Figs. 10.2, 10.3, 10.4C). A semilunate carpal—corresponding to the semilunate bone of nonavian theropods—is co-ossified with the metacarpals (mostly the major and alular ones), a condition more clearly visible under ultraviolet light. The alular metacarpal is small, forming a semicircular medial margin without development of an extensor process (Fig. 10.4A). The central area of this metacarpal possesses a round depression on its dorsal face. Articulated to the alular metacarpal of the left element is a spinelike proximal phalanx, with the distal end

missing. The major metacarpal is broader and more robust than the minor one. These two metacarpals are firmly united throughout their length, leaving no intermetacarpal space (Fig. 10.4A). The distal end is missing in both carpometacarpi. The impression left by the most distal portion of the right element, however, indicates that the major and minor metacarpals curve cranially (Fig. 10.4A) and that, as in Enantiornithes (Zhou, 1995; Chiappe and Walker, Chapter 11 in this volume), the latter was the longer.

Pelvic Girdle and Limb

Ischium. Among the presumed pelvic remains, only the ischium, which is exposed dorsally, can be confidently identified (Figs. 10.2, 10.3, 10.4C). The ischium of *Noguerornis* is a laminar, straplike bone that gradually narrows distally. On its proximal half, this bone possesses a large, hook-shaped process (Fig. 10.4A). Pending the interpretation of its position as either cranially or caudally oriented, this process can

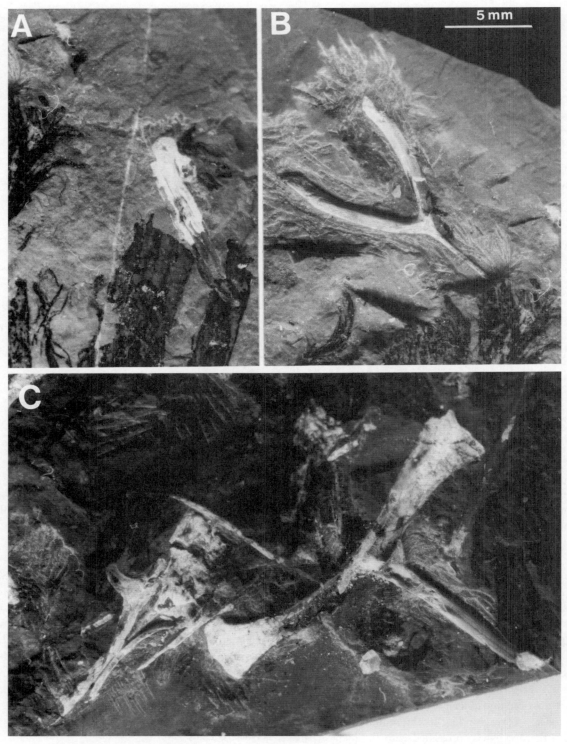

Figure 10.4. Detail of carpometacarpus (A), furcula (B), and ischium, vertebrae, and right humerus (C) of *N. gonza-lezi* (LP.1702.P). Scale bar same for A–C.

TABLE 10.1
Measurements (mm) of *Noguerornis gonzalezi*

	Left	Right
Maximum width of proximal end of humerus	4.6	4.6
Maximum width of distal end of humerus	4.2	4.7
Maximum width of midshaft of humerus	1.8	2
Maximum length of ulna	24.3	—
Maximum width of midshaft of ulna	1.3	—
Maximum length of radius	23.1	23.2
Maximum width of midshaft of radius	0.9	0.9
Maximum width of proximal end of carpometacarpus	2.9	2.9
Maximum width of proximal end of alular metacarpal	1.1	1.1
Maximum width of proximal end of major metacarpal	1.2	1
Maximum width of proximal end of minor metacarpal	0.7	0.7
Maximum length of clavicular branch	9.3	10
Maximum length of hypocleideum	5.3	—
Maximum length of trunk vertebra	1.6	—
Maximum length of first synsacral vertebra	2	—
Maximum length of second synsacral vertebra	2.4	—
Length of ischium from the obturator process to distal end	10.5	—

be alternatively regarded as a large obturator process (Chiappe, 1995b) or the proximodorsal ischiadic process of several other basal birds and nonavian theropods (e.g., Enantiornithes, *Confuciusornis, Unenlagia, Rahonavis*) (see Hou et al., 1995; Martin, 1995; Novas and Puerta, 1997; Forster et al., 1998a,b; Chiappe et al., 1999; Chiappe and Walker, Chapter 11 in this volume). Although the poor preservation of the ischium prevents clarification of this structure, the fact that a proximodorsal process is widespread among basal birds and a prominent obturator process is not makes more probable the identification of the ischiadic process of *Noguerornis* as the proximodorsal ischiadic process. It is clear, on the other hand, that both ischia contact each other on their distal halves, forming a long symphysis. This is of particular interest, as no other Mesozoic bird has been shown to have an ischiadic symphysis. An ischiadic symphysis has not been definitively reported for *Archaeopteryx* (Wellnhofer, 1974, 1993), *Rahonavis* (Forster et al., 1998a,b), or *Iberomesornis* (Sanz and Bonaparte, 1992), and the ischia are separated from each other in *Confuciusornis* (Chiappe et al., 1999), *Patagopteryx* (Chiappe, 1996), Ornithurae (Marsh, 1880; Baumel and Witmer, 1993), and those enantiornithines for which data are available (e.g., *Concornis;* Sanz et al., 1995). An ischiadic symphysis is also absent in dromaeosaurid theropods (Norell and Makovicky, 1997).

Tibia. The straight limb bone perpendicular to the right humerus is interpreted as the tibia. This interpretation is based on the fact that the bone exhibits a crest (considered the fibular crest) and that the most distal portion is craniocaudally compressed, as is typical of avian tibiotarsi (Figs.

10.2, 10.3). It is not possible to discern if it is a right or left element.

Feathers

Two main aggregations of feathers, corresponding to wing feathers, are preserved. Isolated coverts are also visible on the slab. The secondary remiges are directed at an angle of 70° with respect to the ulnar shaft on the left ulna. Portions of coverts are preserved cranial to the left radius and proximal end of the left humerus, documenting the development of a propatagium. Unfortunately, the detailed structure of these feathers is not preserved.

Phylogenetic Relationships

The avian relationship of *N. gonzalezi* is unquestionably supported by the combination of feathers with derived characters that are either exclusive to birds (e.g., U-shaped furcula with hypertrophied hypocleideum, ulna longer than humerus) or uncommon among nonavian theropods (e.g., carpometacarpus). Yet the fragmentary nature of the holotype and only known specimen complicates any attempt to understand the relationships of this species to other basal birds.

Martin (1995) placed *Noguerornis* within Enantiornithes, allying it to both *Iberomesornis* and *Concornis*. This claim was based on the purported similarity of the humeri and furcula of these taxa and the alleged presence of a broad distal humeral end bearing a short articular surface (supposed to be found in Enantiornithes). The remarkable curvature of the humerus of *Noguerornis,* however, is unique, and it makes this bone quite unlike that of either *Iberomesornis* or *Concornis*. In addition, the hypocleideum of the furcula of the latter two taxa (based on what is known) is definitely shorter than that of *Noguerornis*. Whereas the furcula of *Iberomesornis* differs from that of *Concornis* in lacking the euenantiornithine lateral excavation (Chiappe and Calvo, 1994; Sanz et al., 1995, 1996; Chiappe and Walker, Chapter 11 in this volume), the presence or absence of this condition remains uncertain in the available material of *Noguerornis*. Furthermore, Martin's observations of the distal end of the humerus of *Noguerornis* are equivocal because the distal ends of the humeri of the only known specimen are completely abraded, preserving no other details besides a weak olecranon fossa on the left element.

Kurochkin (1995:51) also supported the inclusion of the Spanish bird within Enantiornithes. This claim was based on the presence of a "sharp-angled furcula with hypocleideum, pronounced external and internal tuberosities of the proximal humerus, oblique-shaped distal humerus with the distal protrusion of the internal condyle, and metacarpals fused only proximally." Apparently, the condition of the furcula simply refers to the presence of both a low interclavicular angle and a hypocleideum, attributes that are clearly not exclusive to Enantiornithes within birds (Chiappe, 1996). Furthermore, Kurochkin's (1995) observations of the humerus are not manifest in the poorly preserved material, and whether minor (III) and major (II) metacarpals fuse to each other only proximally is also uncertain because the distal ends of these metacarpals are not preserved but known only from the impressions left by the right elements (Fig. 10.4A).

Thus, the allocation of *Noguerornis* within Enantiornithes, as expressed by Martin (1995) and Kurochkin (1995), does not seem to have any evidential support. This, however, does not mean that *Noguerornis* and at least certain Enantiornithes do not share some derived characteristics. As pointed out by Zhou (1995), the furcula of the enantiornithine *Sinornis* has a long hypocleideum, which exceeds half the length of the clavicular rami. This condition is comparable to that of *Noguerornis* and the enantiornithine *Eoalulavis* (Sanz et al., 1996). Likewise, metacarpal I of *Sinornis*, *Enantiornis*, *Neuquenornis*, and *Noguerornis* has a rounded medial margin lacking an extensor process, and the minor metacarpal appears to be longer than the major metacarpal and slightly curved distally. These two conditions are comparable to that of other enantiornithines (see Chiappe and Walker, Chapter 11 in this volume).

In any case, the relationships of *Noguerornis* to other birds should be examined on the basis of the distribution among taxa of all the characters available for study in the Spanish bird. In this cladistic framework (Chiappe, Chapter 20 in this volume), *Noguerornis* is clustered with all Ornithothoraces by the fusion of the semilunate carpal and the metacarpals and a U-shaped furcula with an interclavicular angle smaller than 90°. Within Ornithothoraces, the well-developed hypocleideum and distal projection of the minor metacarpal, which extends distally more than the major metacarpal, group *Noguerornis* together with Euenantiornithes and *Iberomesornis* (Fig. 10.5), although this last character is missing in the latter taxon. *Noguerornis*, however, lacks several derived characters of Euenantiornithes, such as the presence of a concave central portion of the

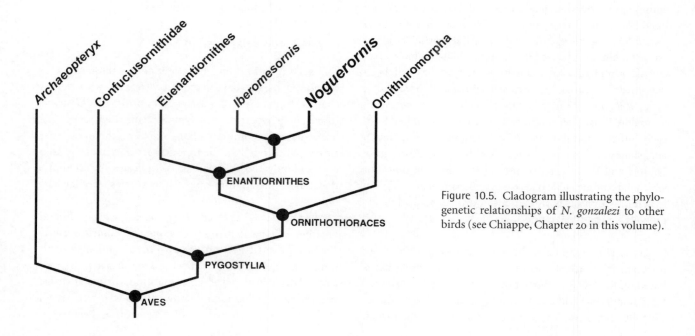

Figure 10.5. Cladogram illustrating the phylogenetic relationships of *N. gonzalezi* to other birds (see Chiappe, Chapter 20 in this volume).

proximal margin of the humeral head, a prominent bicipital crest of the humerus, and a longitudinal groove on the shaft of the radius (Chiappe and Walker, Chapter 11 in this volume). The absence of a procoelous synsacrum clusters *Noguerornis* with *Iberomesornis* (Chiappe, Chapter 20 in this volume) (Fig. 10.5). The sister-taxon relationship between these taxa and between them and Euenantiornithes (Fig. 10.5) is weakly supported. Nevertheless, the result of Chiappe's analysis (Chapter 20 in this volume) suggests that together with *Iberomesornis, Noguerornis* is likely an Enantiornithes and a close outgroup of Euenantiornithes. Characters that appear to be plesiomorphic for Ornithothoraces, such as its ischiadic symphysis and the subequal diameter of the ulna and radius, may indeed be homoplastic. This is very likely the case for the ischiadic symphysis, which is definitively absent in *Confuciusornis* (Chiappe et al., 1999) and dromaeosaurid theropods (Norell and Makovicky, 1997).

Paleobiology

The fragmentary nature of the single known specimen of *N. gonzalezi* prevents inferring much about its lifestyle. Nevertheless, the modern proportions and general aspect of its wing (e.g., ulna-radius longer than humerus, distal carpals fused to metacarpals) suggest that this bird was capable of some sort of flight. In agreement with this is the presence of a U-shaped furcula with an enlarged hypocleideum. The U shape of the furcula of *Noguerornis*—which is in contrast to the boomerang shape of that of *Archaeopteryx* and nonavian theropods—suggests that this element may have functioned as a spring, bending laterally during downstroke and recoiling during upstroke, a mechanical property observed in several neornithine clades (Jenkins et al., 1988; Bailey and DeMont, 1991). Although the functional correlation of this mechanical characteristic is not fully understood (Bailey and DeMont, 1991), this property of the avian furcula is essential for a postulated secondary respiratory cycle in connection with the increased metabolic demands of flight (Jenkins et al., 1988). Likewise, the long hypocleideum of *Noguerornis* may have provided a larger area for the attachment of the coracoclavicular membrane, enlarging the area for the most cranial origin of the M. pectoralis (Olson and Feduccia, 1979), the main flight muscle (Raikow, 1985; Dial et al., 1987, 1988), as well as providing additional strength to the thoracic girdle as suggested by Sanz et al. (1996) for the enantiornithine *Eoalulavis*. Perhaps the strongest evidence in support of the flying capabilities of *Noguerornis* rests on the presence of a propatagium (Fig. 10.2), the largest of the wing patagia. This feathered skinfold has been regarded as essential for flight (Brown et al., 1994, 1995), although it is likely that most basal birds flew without it. A propatagium has never been reported for *Archaeopteryx.* Interestingly, a nineteenth-century sketch of the

Berlin specimen shows a fold of feathers in between the humerus and the ulna-radius (see Ostrom, 1985), suggesting that a propatagium may indeed have been present in *Archaeopteryx.* Nevertheless, given the fact that this area of the Berlin specimen has not been prepared and also that examination of this and other specimens fails to document the presence of a propatagium, it seems quite likely that the "propatagium" of this early sketch is simply inaccurate. Likewise, available evidence fails to document the presence of a propatagium in any feathered nonavian theropod. Definitive propatagia are preserved in multiple specimens of *Confuciusornis* (Chiappe et al., 1999), a taxon that, albeit more primitive than *Noguerornis*, is known from deposits that are several million years younger (see Zhou and Hou, Chapter 7 in this volume). Therefore, although *Noguerornis* is not the most basal avian exhibiting a propatagium, it is the earliest bird for which this structure is known.

The occurrence of a propatagium in such an ancient bird adds to our knowledge of the early evolution of flight. The presence of this airfoil implies that a ligamentum propatagiale was already developed on the leading edge of the wing, between the shoulder and the carpus. The neornithine avian ligamentum propatagiale fulfills a crucial role in the wing's extension-flexion mechanism and is essential for avoiding distortion of this surface, allowing the wing's proper function as an airfoil. As hypothesized by Brown et al. (1995), the ligamentum propatagiale may also support the distal portion of the wing against drag. This modern design of the wing's airfoil and control appears to have been present in *Noguerornis*, one of the earliest known birds after *Archaeopteryx.*

Acknowledgments

We are grateful to N. Frankfurt and S. Copeland for preparing some illustrations and to D. Treon and S. Orell for editorial assistance. Support for this research was provided by grants to L. M. Chiappe from the Dinosaur Society and the Guggenheim Foundation.

Literature Cited

Bailey, J. P., and M. E. DeMont. 1991. The function of the wishbone. Canadian Journal of Zoology 69:2751–2758.

Barale, G. 1973. Présence du genre *Frenelopsis* Schenk dans les calcaires lithographiques du Montsec, province de Lérida (Espagne). Comptes Rendus de l'Académie des Sciences 227: 1349–1340.

———. 1991. The fossil flora of the Lower Cretaceous (Berriasian-Valanginian) lithographical limestones of Montsec (Lleida Province, Spain); pp. 39–47 *in* X. Martínez-Delclòs (ed.), The Lower Cretaceous Lithographic Limestones of Montsec. Ten Years of Paleontological Expeditions. Institut Estudis Ilerdencs, Lleida, Spain.

———. 1995. The fossil flora megarests and microrests; pp. 31–38 *in* X. Martínez-Delclòs (ed.), Montsec and Mont-Ral-Alcover, Two Konservat-Lagerstätten. Catalonia, Spain. Field

Trip Guide Book, II International Symposium on Lithographic Limestones. Institut Estudis Ilerdencs, Lleida, Spain.

Baumel, J. J., and L. M. Witmer. 1993. Osteologia; pp. 45–132 *in* J. J. Baumel, A. S. King, J. E. Breazile, H. E. Evans, and J. C. Vanden Berge (eds.), Handbook of Avian Anatomy: Nomina Anatomica Avium, 2nd edition. Nuttal Ornithological Club, Cambridge.

Brenner, P., W. Geldmacher, and R. Schroeder. 1974. Ostrakoden und Alter der Plattenkalke von Rubies (Sierra del Montsech, Prov. Lérida, NE Spanien). Neues Jahrbuch für Geologie und Paläontologie 9:513–524.

Brown, R. E., J. J. Baumel, and R. D. Klemm. 1994. Anatomy of the propatagium: the great horned owl (*Bubo virginianus*). Journal of Morphology 219:205–224.

———. 1995. Mechanics of the avian propatagium: flexion-extension mechanism of the avian wing. Journal of Morphology 225:91–105.

Buscalioni, A. D., and J. L. Sanz. 1991. Diapsid reptiles from "La pedrera de Rúbies." Early Cretaceous, Lleida, Spain; pp. 89–93 *in* X. Martínez-Delclòs (ed.), The Lower Cretaceous Lithographic Limestones of Montsec. Ten Years of Paleontological Expeditions. Institut Estudis Ilerdencs, Lleida, Spain.

Chiappe, L. M. 1991. Cretaceous avian remains from Patagonia shed new light on the early radiation of birds. Alcheringa 15(3–4):333–338.

———. 1995a. The phylogenetic position of the Cretaceous birds of Argentina: Enantiornithes and *Patagopteryx deferrariisi*; pp. 55–63 *in* D. S. Peters (ed.), Proceedings of the 3rd Symposium of the Society of Avian Paleontology and Evolution. Courier Forschungsinstitut Senckenberg 181, Frankfurt am Main.

———. 1995b. The first 85 million years of avian evolution. Nature 378:349–355.

———. 1996. Late Cretaceous birds of southern South America: anatomy and systematics of Enantiornithes and *Patagopteryx deferrariisi*; pp. 203–244 *in* G. Arratia (ed.), Contributions of Southern South America to Vertebrate Paleontology, Münchner Geowissenschaftliche Abhandlungen, Reihe A, Geologie und Paläontologie 30. Verlag Dr. Friedrich Pfeil, Munich.

Chiappe, L. M., and J. O. Calvo. 1994. *Neuquenornis volans*, a new Enantiornithes (Aves) from the Upper Cretaceous of Patagonia (Argentina). Journal of Vertebrate Paleontology 14(2):230–246.

Chiappe, L. M., Ji S.-A., Ji Q., and M. A. Norell. 1999. Anatomy and systematics of the Confuciusornithidae (Aves) from the late Mesozoic of northeastern China. Bulletin of the American Museum of Natural History 242:1–89.

Cracraft, J. 1988. Early evolution of birds. Nature 331:389–390.

Dial, K. P., S. R. Kaplan, G. E. Goslow Jr., and F. A. Jenkins Jr. 1987. Structure and neural control of the pectoralis in pigeons: implications for flight mechanics. Anatomical Record 218:284–287.

———. 1988. A functional analysis of the primary upstroke and downstroke muscles in the domestic pigeon (*Columba livia*) during flight. Journal of Experimental Biology 134:1–16.

Forster, C. A., S. D. Sampson, L. M. Chiappe., and D. W. Krause. 1998a. The theropodan ancestry of birds: new evidence from the Late Cretaceous of Madagascar. Science 279:1915–1919.

———. 1998b. Genus correction. Science 280:185.

Fregenal-Martínez, M., and N. Meléndez. 1995. Geological setting; pp. 12–24 *in* X. Martínez-Delclòs (ed.), Montsec and Mont-Ral-Alcover, Two Konservat-Lagerstätten. Catalonia, Spain. Field Trip Guide Book, II International Symposium on Lithographic Limestones. Institut Estudis Ilerdencs, Lleida, Spain.

Gómez Pallerola, E. 1986. Nota preliminar sobre una pluma penna del yacimiento eocretácico de la Pedrera de Meiá (Lérida). Boletín Geológico Minero de España 97:22–24.

Hou L., Zhou Z., L. D. Martin, and A. Feduccia. 1995. A beaked bird from the Jurassic of China. Nature 377:616–618.

Jenkins, F. A., Jr., K. P. Dial, and G. E. Goslow Jr. 1988. A cineradiographic analysis of bird flight: the wishbone in starlings is a spring. Science 241:1495–1498.

Kurochkin, E. 1995. Synopsis of Mesozoic birds and early evolution of Class Aves. Archaeopteryx 13:47–66.

Lacasa-Ruiz, A. 1985. Nota sobre las plumas fósiles del yacimiento eocretácico de "La Pedrera–La Cabrua" en la Sierra del Montsec (Prov. Lleida, España). Ilerda 46:227–238.

———. 1986. Nota preliminar sobre el hallazgo de restos óseos de un ave fósil en el yacimiento neocomiense del Montsec. Prov. De Lérida. España. Ilerda 47:203–206.

———. 1989a. Nuevo género de ave fósil del yacimiento neocomiense del Montsec (Provincia de Lérida, España). Estudios Geológicos 45:417–125.

———. 1989b. An Early Cretaceous fossil bird from Montsec Mountain (Lleida, Spain). Terra Nova 1(1):45–46.

———. 1991a. The fossil birds from the lithographical limestones of Montsec; pp. 95–97 *in* X. Martínez-Delclòs (ed.), The Lower Cretaceous Lithographic Limestones of Montsec. Ten Years of Paleontological Expeditions. Institut Estudis Ilerdencs, Lleida, Spain.

———. 1991b. The fossiliferous outcrops of the Montsec lithographic limestones; pp. 11–14 *in* X. Martínez-Delclòs (ed.), The Lower Cretaceous Lithographic Limestones of Montsec. Ten Years of Paleontological Expeditions. Institut Estudis Ilerdencs, Lleida, Spain.

Lacasa-Ruiz, A., and L. Via. 1991. The crustacean fauna from the lithographical limestones of Montsec; pp. 59–60 *in* X. Martínez-Delclòs (ed.), The Lower Cretaceous Lithographic Limestones of Montsec. Ten Years of Paleontological Expeditions. Institut Estudis Ilerdencs, Lleida, Spain.

Marsh, O. C. 1880. Odontornithes: A Monograph on the Extinct Toothed Birds of North America. Report of Geological Exploration of the 40th Parallel. Government Printing Office, Washington, D.C., 201 pp.

Martin, L. D. 1995. The Enantiornithes: terrestrial birds of the Cretaceous. Courier Forschungsinstitut Senckenberg 181:23–36.

Martín-Closas, C., and N. López-Morón. 1995. The charophyte flora; pp. 29–31 *in* X. Martínez-Delclòs (ed.), Montsec and Mont-Ral-Alcover, Two Konservat-Lagerstätten. Catalonia, Spain. Field Trip Guide Book, II International Symposium on Lithographic Limestones. Institut Estudis Ilerdencs, Lleida, Spain.

Martinell, J., and R. Doménech. 1991. Malacological contents of the Montsec lithographical limestones (Lower Cretaceous); pp. 51–52 *in* X. Martínez-Delclòs (ed.), The Lower Cretaceous Lithographic Limestones of Montsec. Ten Years of

Paleontological Expeditions. Institut Estudis Ilerdencs, Lleida, Spain.

Martínez-Delclòs, X. 1991a. The Lower Cretaceous Lithographic Limestones of Montsec. Ten Years of Paleontological Expeditions. Institut Estudis Ilerdencs, Lleida, Spain, 105 pp.

———. 1991b. Insects from the lithographical limestones of the Serra del Montsec. Lower Cretaceous of Catalonia, Spain; pp. 61–71 in X. Martínez-Delclòs (ed.), The Lower Cretaceous Lithographic Limestones of Montsec. Ten Years of Paleontological Expeditions. Institut Estudis Ilerdencs, Lleida, Spain.

———. 1995. Montsec and Mont-Ral-Alcover, Two Konservat-Lagerstätten. Catalonia, Spain. Field Trip Guide Book, II International Symposium on Lithographic Limestones. Institut Estudis Ilerdencs, Lleida, Spain.

Milner, A. 1993. Ground rules for early birds. Nature 362:589.

Norell, M. A., and P. J. Makovicky. 1997. Important features of the dromaeosaur skeleton: information from a new specimen. American Museum Novitates 3215:1–28.

Novas, F. E., and P. Puerta. 1997. New evidence concerning avian origins from the Late Cretaceous of Patagonia. Nature 387: 390–392.

Olson, S. L., and A. Feduccia. 1979. Flight capability and the pectoral girdle of *Archaeopteryx*. Nature 278:247–248.

Ostrom, J. H. 1985. The Yale *Archaeopteryx*: the one that flew the coop; pp. 359–367 in M. K. Hecht, J. H. Ostrom, G. Viohl, and P. Wellnhofer (eds.), The Beginnings of Birds. Freunde des Jura-Museums, Eichstätt.

Padian, K. 1996. Early bird in slow motion. Nature 382:400–401.

Peybernés, B. 1976. Le Jurassique et le Crétacée inférieur des Pyrénées Franco-Espagnoles entre la Garonne et la Méditerranée. Thèse Doctorat de Sciences Naturelles Université Paul-Sabatier, Toulouse, Imp. CRDP. Toulouse, 459pp.

Peybernés, B., and H. Oertli. 1972. La série de passage du Jurassique au Crétacé dans le Bassin sud-pyrénéen (Espagne). Comptes Rendus de l'Académie des Sciences 276:2501–2504.

Raikow, R. J. 1985. Locomotor system; pp. 57–147 in A. S. King and J. McLelland (eds.), Form and Function in Birds. Volume 3. Academic Press, London.

Sanz, J. L., and J. F. Bonaparte. 1992. A new order of birds (Class Aves) from the Lower Cretaceous of Spain; pp. 39–49 in K. E. Campbell (ed.), Papers in Avian Paleontology, Honoring Pierce Brodkorb. Science Series 36. Natural History Museum of Los Angeles County, Los Angeles.

Sanz, J. L., and A. D. Buscalioni. 1992. A new bird from the Early Cretaceous of Las Hoyas, Spain, and the Early radiation of birds. Paleontology 35(4):829–845.

Sanz, J. L., J. F. Bonaparte, and A. Lacasa. 1988. Unusual Early Cretaceous birds from Spain. Nature 331:433–435.

Sanz, J. L., L. M. Chiappe, and A. Buscalioni. 1995. The osteology of *Concornis lacustris* (Aves: Enantiornithes) from the Lower Cretaceous of Spain and a re-examination of its phylogenetic relationships. American Museum Novitates 3133:1–23.

Sanz, J. L., L. M. Chiappe, B. P. Pérez-Moreno, A. D. Buscalioni, J. Moratalla, F. Ortega, and F. J. Poyato-Ariza. 1996. A new Lower Cretaceous bird from Spain: implications for the evolution of flight. Nature 382:442–445.

Sanz, J. L., L. M. Chiappe, B. P. Pérez-Moreno, J. Moratalla, F. Hernández-Carrasquilla, A. D. Buscalioni, F. Ortega, F. J. Poyato-Ariza, D. Rasskin-Gutman, and X. Martínez-Delclòs. 1997. A nestling bird from the Early Cretaceous of Spain: implications for avian skull and neck evolution. Science 276: 1543–1546.

Selden, P. 1991. Lower Cretaceous spiders from the Serra del Montsec (Spain); pp. 53–58 in X. Martínez-Delclòs (ed.), The Lower Cretaceous Lithographic Limestones of Montsec. Ten Years of Paleontological Expeditions. Institut Estudis Ilerdencs, Lleida, Spain.

Vidal, L. M. 1902. Sobre la presencia del tramo Kimeridgense del Montsech y hallazgo de un batracio en sus hiladas. Memórias de la Real Academia de Ciencias y Artes de Barcelona 4(18):263–267.

Wellnhofer, P. 1974. Das fünfte Skelettexemplar von *Archaeopteryx*. Palaeontographica A 147: 169–216.

———. 1993. Das siebte Exemplar von *Archaeopteryx* aus den Solnhofener Schichten. Archaeopteryx 11:1–48.

Wenz, S. 1991a. Lower Cretaceous fishes from the Serra del Montsec (Spain); pp. 73–84 in X. Martínez-Delclòs (ed.), The Lower Cretaceous Lithographic Limestones of Montsec. Ten Years of Paleontological Expeditions. Institut Estudis Ilerdencs, Lleida, Spain.

———. 1991b. Amphibia (Anura) from the lithographical limestones of Montsec; pp. 85–87 in X. Martínez-Delclòs (ed.), The Lower Cretaceous Lithographic Limestones of Montsec. Ten Years of Paleontological Expeditions. Institut Estudis Ilerdencs, Lleida, Spain.

Zhou Z. 1995. The discovery of Early Cretaceous birds in China. Courier Forschungsinstitut Senckenberg 181:9–22.

11

Skeletal Morphology and Systematics of the Cretaceous Euenantiornithes (Ornithothoraces: Enantiornithes)

LUIS M. CHIAPPE AND CYRIL A. WALKER

 Between 1974 and 1976, J. F. Bonaparte (then of the Universidad Nacional de Tucumán, Argentina) quarried a site within the continental deposits of the Maastrichtian Lecho Formation (Bonaparte et al., 1977; Bonaparte and Powell, 1980), on lands of the Estancia El Brete, in the northwestern Argentine province of Salta. These excavations resulted in a collection of different dinosaurian taxa, including the armored titanosaurid sauropod *Saltasaurus loricatus* (Bonaparte and Powell, 1980). As the remains of *Saltasaurus* were being collected and prepared, a series of small and mostly disarticulated bones was found. These bones, originally identified as avian by J. F. Bonaparte (Bonaparte et al., 1977; Bonaparte and Powell, 1980), formed the collection that led Walker (1981) to recognize a monophyletic ensemble of early avians: Enantiornithes.

Subsequent studies of Mesozoic birds demonstrated that other Cretaceous taxa discovered earlier could be allocated within the supraspecific taxon that Walker (1981) had recognized. Walker (1981) pointed out characters shared by the bones from El Brete and those of *Alexornis antecedens* from the Late Cretaceous of Mexico (Brodkorb, 1976). *Alexornis* was consequently assigned to Enantiornithes by Martin (1983), who also included within this taxon the Late Cretaceous *Gobipteryx minuta* (Elzanowski, 1974) from the Mongolian Gobi Desert. Later discoveries, primarily in the 1990s, showed that Enantiornithes embodied a diverse monophyletic group of birds evolving throughout the Cretaceous (Chiappe, 1991, 1993, 1995a, 1997; Zhou et al., 1992; Chiappe and Calvo, 1994; Martin, 1995; Sanz et al., 1995, 1996, 1997; Zhou, 1995a; Feduccia, 1996; Chatterjee, 1997; Hou, 1997; Padian and Chiappe, 1997, 1998) (Fig. 11.1).

Although the taxon name Enantiornithes was erected in 1981, Sereno (1998) was the first to publish a phylogenetic definition for this speciose clade. Sereno (1998) defined Enantiornithes as all taxa closer to *Sinornis* than to Ne-

ornithes. Under this phylogenetic definition, most attributes previously used to diagnose Enantiornithes (e.g., Chiappe, 1992, 1996a; Chiappe and Calvo, 1994; Sanz et al., 1995, 1996, 1997; Zhou, 1995a,b), were left to support the monophyly of a clade that is only a subset of Sereno's newly defined Enantiornithes. As discussed elsewhere (Sereno, 2000; Sereno, Rao, and Li, Chapter 8 in this volume; Chiappe and Lacasa-Ruiz, Chapter 10 in this volume; Chiappe, Chapter 20 in this volume), the Spanish Early Cretaceous *Iberomesornis* and *Noguerornis* may be primitive members of the *Sinornis* lineage and thus basal Enantiornithes. All the remaining and more advanced enantiornithines form a monophyletic taxon, Euenantiornithes, defined phylogenetically as all taxa closer to *Sinornis* than to *Iberomesornis* (Chiappe, Chapter 20 in this volume).

Several chapters in this book (e.g., Zhou and Hou, Chapter 7; Sereno, Rao, and Li, Chapter 8; Sanz et al., Chapter 9; Chiappe and Lacasa-Ruiz, Chapter 10; Chinsamy, Chapter 18) provide in-depth coverage of anatomical, ecological, and physiological aspects of Enantiornithes (see the Preface of this volume for a justification for so many chapters on this one avian clade). The anatomical treatment of these, however, is specific to a few taxa (e.g., *Iberomesornis*, *Concornis*, and *Eoalulavis* by J. Sanz et al.; *Noguerornis* by L. Chiappe and A. Lacasa-Ruiz; *Sinornis* by P. Sereno et al.). In this chapter we provide a summary of the osteology and historical relationships of Euenantiornithes, the clade that contains most enantiornithine species, followed by a brief discussion of the biological attributes that can be inferred from their morphology and fossil record.

The following institutional abbreviations are used in this chapter: IVPP, Institute of Vertebrate Paleontology and Paleoanthropology, Beijing, China; MUCPv, Museo de Ciencias Naturales, Universidad del Comahue, Neuquén, Argentina; PVL, Instituto Miguel Lillo, Tucuman, Argentina; PU, Princeton University (currently at Yale

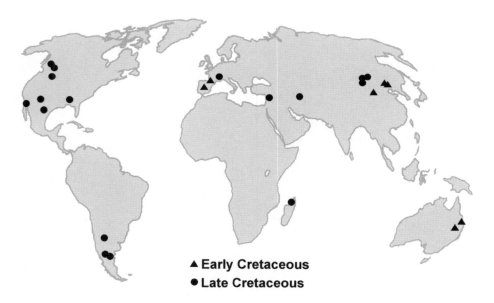

▲ **Early Cretaceous**
● **Late Cretaceous**

Figure 11.1. Early and Late Cretaceous occurrences of euenantiornithine birds.

Peabody Museum, New Haven, Connecticut, United States), Princeton, New Jersey, United States; YPM, Yale Peabody Museum, New Haven, Connecticut, United States.

Geological Setting

Euenantiornithine birds have been found in a variety of Cretaceous strata worldwide (Martin, 1983; Molnar, 1986; Chiappe, 1993, 1996a; Chiappe and Calvo, 1994; Dashzeveg et al., 1995; Kurochkin, 1995, 1996, 2000; Varricchio and Chiappe, 1995; Zhou and Hou, Chapter 7 in this volume; Sanz et al., Chapter 9 in this volume) (Fig. 11.1; Table 11.1). Although the available space precludes a discussion of the paleoenvironment, associated fauna, and chronology of each site, a brief summary is given (see Table 11.1 for specifics on the chronology).

Most euenantiornithine birds are from nonmarine deposits. Early Cretaceous euenantiornithines are known exclusively from Europe, Asia, and Australia (Fig. 11.1). The European (e.g., *Concornis lacustris, Eoalulavis hoyasi*) and Asian (e.g., *Sinornis santensis, Otogornis genghisi, Boluochia zenghi, Longipteryx chaoyangensis*) species come from lacustrine deposits (e.g., Las Hoyas and the Montsec range in Spain, Liaoning Province and Inner Mongolia in China) indicating a subtropical environment with a variety of plants, arthropods, fishes, amphibians, and reptiles (Fregenal-Martínez, 1991; Martínez-Delclòs, 1995; Meléndez, 1995; Zhou and Hou, Chapter 7 in this volume; Sanz et al., Chapter 9 in this volume). The Australian *Nanantius eos* constitutes the only Early Cretaceous species known from marine deposits (Molnar, 1986; Kurochkin and Molnar, 1997), although the occurrence of abundant fossil logs within these beds suggests that they were deposited not far from the coast; this evidence also points to a forested shoreline

(Dettmann et al., 1992). Late Cretaceous enantiornithines are known from North and South America, Europe, Asia, and Madagascar (Fig. 11.1). Most taxa (e.g., *Soroavisaurus australis, Neuquenornis volans, Avisaurus archibaldi, A. antecedens*) occur in fluvial deposits (e.g., western North American, Argentine, and Malagasy localities; see Table 11.1) indicating fluvial plains or systems of braided rivers. These species are usually associated with the typical dinosaurs, mammals, lizards, and other reptiles of the Late Cretaceous of western North America (Lillegraven and McKenna, 1986; Bryant, 1989; Currie and Padian, 1997) and Gondwana (Bonaparte, 1996; Krause et al., 1997). Some of them (e.g., *G. minuta*), however, come from deposits indicating arid or semiarid environments sometimes containing vast fields of dunes (e.g., Khulsan; Loope et al., 1998). These same deposits entombed much of the rich Late Cretaceous fauna of central Asia (Jerzykiewicz and Russell, 1991; Dashzeveg et al., 1995). The only known marine Late Cretaceous enantiornithine is *Halimornis thompsoni* from the Mooreville Formation of western Alabama (Lamb et al., 1993; Chiappe et al., 2002). Chiappe et al. (2002) estimated that the depositional site of *Halimornis* was as far as 50 km offshore in the North American Western Interior.

Systematic Paleontology

Taxonomic Hierarchy

    Aves Linnaeus, 1758
        Ornithothoraces Chiappe, 1996
            Enantiornithes Walker, 1981

Definition—Following Sereno's (1998) phylogenetic definition, Enantiornithes is defined as all taxa closer to *S. santensis* than to Neornithes.

TABLE 11.1

Geographical and stratigraphical distribution of euenantiornithine birds

Taxon	Distribution	Horizon/Age	Reference
Early Cretaceous			
Euenantiornithes indet.	La Pedrera, Lleida, Spain	La Pedrera de Rúbies Fm., Berriasian–Barremian	Sanz et al. (1997)
	Las Hoyas, Cuenca, Spain	"Calizas de La Huerguina" Fm., Barremian	Sanz et al. (2001)
C. lacustris	Las Hoyas, Cuenca, Spain	"Calizas de La Huerguina" Fm., Barremian	Sanz and Buscalioni (1992); Sanz et al. (1995)
E. hoyasi	Las Hoyas, Cuenca, Spain	"Calizas de La Huerguina" Fm., Barremian	Sanz et al. (1996)
E. buhleri	Sihetun, Liaoning, China	Yixian Fm., Early Cretaceous	Hou et al. (1999)
Liaoxiornis delicatus[*1]	Dawangzhangzi, Lingyuan, Liaoning, China	Yixian Fm., Early Cretaceous	Hou and Chen (1999); Ji and Ji (1999)
Lonchengornis sanyanensis[*]	Chaoyang, Liaoning, China	Jiufotang Fm., Early Cretaceous	Hou (1997)
Cuspirostrisornis houi[*]	Chaoyang, Liaoning, China	Jiufotang Fm., Early Cretaceous	Hou (1997)
Largirostornis sexdentoris[*]	Chaoyang, Liaoning, China	Jiufotang Fm., Early Cretaceous	Hou (1997)
S. santensis[2]	Chaoyang, Liaoning, China	Jiufotang Fm., Early Cretaceous	Sereno and Rao (1992); Zhou et al. (1992)
B. zhengi	Chaoyang, Liaoning, China	Jiufotang Fm., Early Cretaceous	Zhou (1995b)
Longipteryx chaoyangensis	Chaoyang, Liaoning, China	Jiufotang Fm., Early Cretaceous	Zhang et al. (2000)
O. genghisi	Chaibu-Sumi, Inner Mongolia, China	Yijinhuoluo Fm., Early Cretaceous	Hou (1994)
N. eos	Warra Station and Canary, Queensland	Toolebuc Fm., Albian	Molnar (1986); Chiappe (1996b); Kurochkin and Molnar (1997)
Late Cretaceous			
K. cretacea[*]	Dzhyrakuduk, Kizylkum Desert, Uzbekistan	Bissekty Fm., Coniacian	Nessov (1984)
S. prisca[*]	Dzhyrakuduk, Kizylkum Desert, Uzbekistan	Bissekty Fm., Coniacian	Nessov in Nessov and Jarkov (1989)
E. martini[*]	Dzhyrakuduk, Kizylkum Desert, Uzbekistan	Bissekty Fm., Coniacian	Nessov and Panteleyev (1993)

Taxon	Locality	Formation, Age	References
*E. walkeri**	Dzhyrakuduk, Kizylkum Desert, Uzbekistan	Bissekty Fm., Coniacian	Nessov and Panteleyev (1993)
*"Ichthyornis" minusculus**	Dzhyrakuduk, Kizylkum Desert, Uzbekistan	Bissekty Fm., Coniacian	Nessov (1990)
N. volans	Neuquén City, Neuquén; Puesto Tripailao, Río Negro, Argentina	Río Colorado Fm., Campanian	Chiappe and Calvo (1994); Chiappe (1996b)
A. antecedens	El Rosario, Baja California, Mexico	Bocana Roja Fm., Campanian	Brodkorb (1976); Martin (1983)
G. minuta[3]	Khulsan, Khermeen Tsav, and Ukhaa Tolgod, South Gobi Aimak, Mongolia	Barun Goyot Fm. and Djadokhta Fm., Campanian	Elzanowski (1974); Martin (1983); Kurochkin (1996); Chiappe et al. (2001)
Euenantiornithes indet.	Tugrugeen Shireh, South Gobi Aimak, Mongolia	Djadokhta Fm. Campanian	Suzuki et al. (1999)
Euenantiornithes indet.	New Mexico, United States	Indet. Fm., Campanian	Martin (1983)
Avisauridae, nov. sp.	southern Utah, United States	Kaiparowits Fm., Campanian	Hutchison (1993)
H. thompsoni	Greene County, Alabama, United States	Mooreville Fm., Campanian	Lamb et al. (1993); Chiappe et al. (2002)
A. gloriae	Glacier County, Montana, United States	Two Medicine Fm., Campanian	Varricchio and Chiappe (1995)
Euenantiornithes indet.	Mahajanga, Madagascar	Maevarano Fm., Campanian–Maastrichtian	Forster et al. (1996, 1998a)
Euenantiornithes indet.	Massecaps, Hérault, France	Indet. Fm., Campanian–Maastrichtian	Buffetaut (1998)
*Gurilynia nessovi**	Gurilyn Tsav, South Gobi Aimak, Mongolia	Nemegt Fm., Campanian–Maastrichtian	Kurochkin (1999)
A. archibaldi	Garfield County, Montana, United States	Hell Creek Fm., Maastrichtian	Brett-Surman and Paul (1985); Chiappe (1993)
Euenantiornithes indet.	Niobrara County, Wyoming, United States	Lance Fm., Maastrichtian	This chapter
E. leali	El Brete, Salta, Argentina	Lecho Fm., Maastrichtian	Walker (1981); Chiappe (1996b)
S. australis	El Brete, Salta, Argentina	Lecho Fm., Maastrichtian	Walker (1981); Chiappe (1993)
L. bretincola	El Brete, Salta, Argentina	Lecho Fm., Maastrichtian	Walker (1981); Chiappe (1993)
Y. brevipedalis	El Brete, Salta, Argentina	Lecho Fm., Maastrichtian	Walker (1981); Chiappe (1993)
Euenantiornithes indet.	El Brete, Salta, Argentina	Lecho Fm., Maastrichtian	Walker (1981); Chiappe (1996a)

* Species based on very fragmentary material; its validity can only be established through further examination.

[1] Including its junior synonym, *Lingyuanornis parvus* (Ji and Ji, 1999); this species is based on a juvenile specimen and its validity is therefore dubious.

[2] Including its junior synonym, *C. yandica* (Zhou et al., 1992).

[3] Including its junior synonym, *N. valifanovi* (Kurochkin, 1986).

Included taxa—Enantiornithes includes *Iberomesornis romerali* (Sanz and Bonaparte, 1992; Sanz et al., Chapter 9 in this volume), *Noguerornis gonzalezi* (Lacasa-Ruiz, 1989; Chiappe and Lacasa-Ruiz, Chapter 10 in this volume), and a large variety of Euenantiornithes. One of us (C. W.) thinks that *Wyleyia valdensis* (Harrison and Walker, 1973) may be an Enantiornithes.

Euenantiornithes Chiappe, this volume

Definition—Following the phylogenetic analysis of Chiappe (Chapter 20 in this volume), Euenantiornithes is stem-based, phylogenetically defined as all taxa closer to *S. santensis* than to *I. romerali*.

Included species—Eighteen valid species of euenantiornithine birds have been described from Cretaceous beds worldwide (Table 11.1). These are (1) *Enantiornis leali* (this species was listed on a table legend by Walker [1981] and properly diagnosed by Chiappe [1996b]), (2) *G. minuta* (Elzanowski, 1974; "*Nanantius valifanovi*" [Kurochkin, 1996] is here regarded as a junior synonym of *G. minuta* [see Chiappe et al., 2001]), (3) *A. antecedens* (Brodkorb, 1976), (4) *A. archibaldi* (Brett-Surman and Paul, 1985), (5) *N. eos* (Molnar, 1986), (6) *S. santensis* (Sereno and Rao, 1992; following Sereno, Rao, and Li, Chapter 8 in this volume; "*Cathayornis yandica*" [Zhou et al., 1992] is regarded as a junior synonym of *S. santensis*), (7) *C. lacustris* (Sanz and Buscalioni, 1992), (8) *Lectavis bretincola* (Chiappe, 1993), (9) *Yungavolucris brevipedalis* (Chiappe, 1993), (10) *S. australis* (Chiappe, 1993), (11) *Neuquenornis volans* (Chiappe and Calvo, 1994), (12) *Otogornis genghisi* (Hou, 1994), (13) *Avisaurus gloriae* (Varricchio and Chiappe, 1995), (14) *B. zenghi* (Zhou, 1995b), (15) *E. hoyasi* (Sanz et al., 1996), (16) *Eoenantiornis buhleri* (Hou et al., 1999), (17) *L. chaoyangensis* (Zhang et al., 2000), and (18) *H. thompsoni* (Chiappe et al., 2002). Although they are unquestionably euenantiornithines, the validity of the fragmentary *Kizylkumavis cretacea* (Nessov, 1984), *Sazavis prisca* (Nessov and Jarkov, 1989), *Enantiornis martini* (Nessov and Panteleyev, 1993), *Enantiornis walkeri* (Nessov and Panteleyev, 1993), *Lonchengornis sanyanensis* (Hou, 1997), *Cuspirostrisornis houi* (Hou, 1997), *Largirostornis sexdentoris* (Hou, 1997), and *Gurilynia nessovi* (Kurochkin, 1999) from the Cretaceous of Asia remains to be confirmed. Likewise, the early ontogenetic age of *Liaoxiornis delicatus* (Hou and Chen, 1999) and its junior synonym, *Lingyuanornis parvus* (Ji and Ji, 1999), posits a caveat to the validity of this species, and whether the primitive enantiornithine *Protopteryx fengningensis* (Zhang and Zhou, 2000) is a member of Euenantiornithes needs yet to be examined.

Diagnosis—Based on only partial phylogenetic hypotheses (e.g., Chiappe, 1993; Sanz et al., 1995; Varricchio and Chiappe, 1995) and derived similarities present in several taxa, Euenantiornithes (Enantiornithes of many previous authors) was variously diagnosed by different lists of characters (e.g., Chiappe and Calvo, 1994; Chiappe, 1996a). The present phylogenetic definition of Euenantiornithes has not altered the traditional concept of this clade (cf. Chiappe, 1993, 1995b, 1996a; Sanz et al., 1995, 1996, 1997), but a better understanding of their cladistic relationships has logically divided the list of characters previously thought to diagnose Euenantiornithes into synapomorphies of the whole clade and synapomorphies that diagnose subclades within Euenantiornithes. The relationships among euenantiornithine taxa, however, are still largely unresolved, and, consequently, it is likely that future studies will substantially modify this diagnosis.

A cladistic analysis presented in Appendix 11.1 provides a diagnosis of Euenantiornithes based on unambiguous synapomorphies (see Sereno, Rao, and Li, Chapter 8 in this volume, for other putative diagnostic characters): parapophyses (costal foveae) located in the central part of the bodies of dorsal vertebrae (character 1); distinctly convex lateral margin of coracoid (character 2); supracoracoid nerve foramen opening into an elongate furrow medially and separated from the medial margin of the coracoid by a thick bony bar (character 3); broad, deep fossa on the dorsal surface of the coracoid (character 4); costal surface of scapular blade with a prominent, longitudinal furrow (character 6); ventral margin of furcula distinctly wider than the dorsal margin (character 7); well-developed hypocleideum (character 8); dorsal (superior) margin of the humeral head concave in its central portion, rising ventrally and dorsally (character 13); prominent bicipital crest of humerus (character 14); ventral face of the humeral bicipital crest with a small fossa for muscular attachment (character 15); strongly convex dorsal cotyla of ulna (character 17), separated from the olecranon by a groove (character 18); shaft of radius with a long axial groove on its interosseous surface (character 19); minor metacarpal (III) projecting distally more than major metacarpal (II) (character 20); hypertrophied femoral posterior trochanter (character 21); very narrow, deep intercondylar sulcus on tibiotarsus that proximally undercuts condyles (character 23); metatarsal IV significantly thinner than metatarsals II and III (character 25). Several other characters with an ambiguous optimization are also diagnostic of Euenantiornithes: lateral process of sternum (character 10); medial process of sternum (character 11); distal end of humerus very compressed craniocaudally (character 16).

Monophyletic status of the El Brete specimens—When erecting Enantiornithes, Walker (1981) assumed the monophyly of all bird specimens from El Brete, even though very few had skeletal elements in common. Subsequent discoveries of articulated skeletons from other fossil sites, sharing derived characters with the El Brete specimens, supported the inclusion of several isolated bones within a monophyletic group (e.g., Chiappe, 1992). The large number of articulated enantiornithine skeletons known today has confirmed that nearly all the isolated bones and incomplete specimens from El Brete belong to a monophyletic group of birds. Perhaps the only noticeable exception is an isolated mandibular right ramus (PVL-4698; Fig. 11.3). Considering that more than 50 other specimens collected from El Brete have all been found to be Euenantiornithes, we tentatively regard this isolated ramus as an euenantiornithine element. Nevertheless, the reader should be aware that enantiornithine or euenantiornithine synapomorphies are unknown for mandibular elements.

Anatomy

Skull and Mandible

Despite the abundance of euenantiornithine material in the Cretaceous deposits around the world, informative cranial remains are known for only a handful of taxa. Most of our knowledge of the cranial anatomy of these birds is based on the skulls of the Early Cretaceous *S. santensis* (Zhou et al., 1992; Martin and Zhou, 1997) and *E. buhleri* (Hou et al., 1999) from northwestern China (see Zhou and Hou, Chapter 7 in this volume), an unamed hatchling from the Early Cretaceous of Catalonia (Spain), and the Late Cretaceous *G. minuta* (Elzanowski, 1974, 1977, 1995; Kurochkin, 1996; Chiappe et al., 2001) from the Mongolian Gobi Desert. In addition, the Chinese Early Cretaceous *B. zhengi* (Zhou, 1995b) and *L. chaoyangensis* (Zhang et al., 2000), and the Patagonian Late Cretaceous *N. volans* (Chiappe and Calvo, 1994) preserve fragmentary skull material. Also, an isolated mandibular ramus was collected among the enantiornithines from the El Brete (Argentina).

The rostral morphology of Euenantiornithes is quite diverse. Most taxa, such as *Sinornis, Eoenantiornis, Boluochia, Longipteryx,* and the Catalan hatchling, are toothed (Fig. 11.2), whereas teeth are completely absent in *Gobipteryx. Boluochia* (Zhou and Hou, Chapter 7 in this volume) bears a prominent, toothless hook at the end of its upper beak—only mandibular teeth are known (Zhou, 1995b)—but in *Gobipteryx,* the beak is broader and without a hook (Chiappe et al., 2001). Enantiornithine teeth (Fig. 11.2) are conical and unserrated, with a constriction between their crown and their base (Martin and Zhou, 1997). The premaxillae are fused to each other. Their maxillary processes are short (Fig. 11.2); most of the rostrum is typically formed by the maxilla (Martin and Zhou, 1997; Sanz et al., 1997). The frontal process of the premaxilla extends near the caudal margin of the naris in the Catalan hatchling, *Eoenantiornis,* and *Sinornis* (Fig. 11.2), but in *Gobipteryx* it appears to reach as far as the level of the lacrimals. In *Sinornis* and

apparently in *Eoenantionis* (Hou et al., 1999), paired nasals dorsally line the caudal half of the snout (Fig. 11.2D) (Martin and Zhou, 1997).

The nares are subelliptical in *Sinornis* and the hatchling from Catalonia (Fig. 11.2) and more subtriangular in *Eoenantiornis* and *Gobipteryx.* They are typically smaller than the antorbital fenestra, which is subtriangular (Fig. 11.2A–C). Martin and Zhou (1997) reported the presence of maxillary and promaxillary fenestrae within the antorbital fossa of *Sinornis,* a conclusion Sereno, Rao, and Li (Chapter 8 in this volume) appear to endorse for at least the maxillary fenestra. These fenestrae are absent in *Gobipteryx,* in which the nasal process (ascending ramus) of the maxilla is reduced (Chiappe et al., 2001). Their absence was also described for *Eoenantiornis* (Hou et al., 1999), although this region is poorly preserved in the only known specimen of this taxon. The jugal forms the caudoventral angle of the antorbital fenestra (Fig. 11.2B) and the ventral margin of the orbit. The shaft of this bone lacks a dorsal, postorbital process, but caudally it ends in a fork with a caudodorsally directed process. This process is comparable to that of the jugal of *Archaeopteryx lithographica,* which some authors (e.g., Wellnhofer, 1974; Elzanowski and Wellnhofer, 1996; but see Elzanowski, Chapter 6 in this volume) reconstructed in direct articulation with the postorbital. As evidenced by the skull of the Catalan hatchling (Fig. 11.2A, B), however, this jugal process does not join the postorbital but abuts the lateral surface of the quadrate's orbital process (Sanz et al., 1997; Sereno, Rao, and Li, Chapter 8 in this volume). This condition is most likely the one present in *Archaeopteryx,* although these bones are not in articulation in the available specimens. The enantiornithine orbit is round and large—a minimum of nine ossicles have been found to form the sclerotic ring of *Longipteryx* (Zhang et al., 2000) and *Gobipteryx* (Chiappe et al., 2001). A postorbital bone is present in at least some, if not all, Euenantiornithes, as the Catalan hatchling and other taxa have unquestionably documented (Sanz et al., 1997; Zhang and Zhou, 2000). In the Catalan hatchling, this splint-shaped bone does not reach the jugal, leaving a gap between the orbit and the laterotemporal fenestra. The postorbital articulates with the squamosal, forming the lateral margin of a small dorsotemporal fenestra (Sanz et al., 1997). The squamosal is thus not incorporated into the braincase. Martin and Zhou's (1997) reconstruction of the suspensorial region of *Sinornis* without either a postorbital or a "free" squamosal is most likely incorrect. The configuration of the euenantiornithine temporal region was not far removed from that of a typical diapsid reptile; the skull of these birds bears a fully enclosed dorsotemporal fenestra and a laterotemporal fenestra only partially connected to the orbit (Fig. 11.2A). The euenantiornithine quadrate bears a broad orbital process (contra Martin and Zhou, 1997). This is comparable to the orbital process

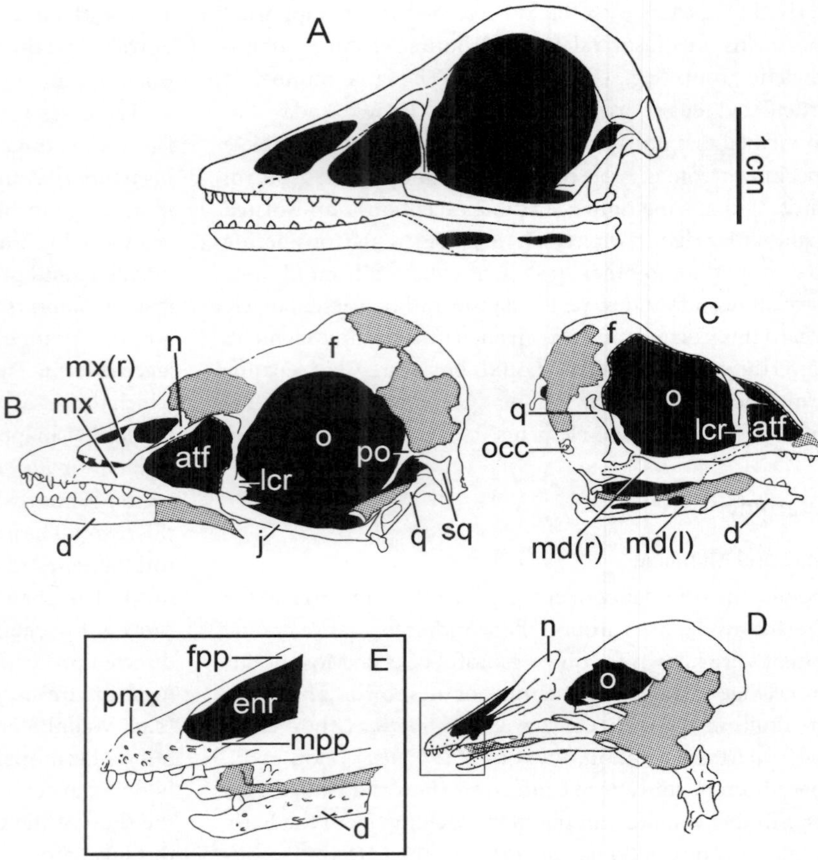

Figure 11.2. Skull and mandible of euenantiornithine birds. A, reconstruction of the skull and mandible of the Early Cretaceous hatchling from Catalonia (Spain) (from Sanz et al., 1997). B, C, skull and mandible of the Catalan hatchling in left (B) and right (C) lateral view (from Sanz et al., 1997). D, E, skull and mandible of *S. santensis* from the Early Cretaceous of Liaoning (China) (E, augmented view of box in D). Abbreviations: atf, antorbital fenestra; d, dentary; enr, external nares; f, frontal; fpp, frontal process of premaxilla; j, jugal; lcr, lacrimal; md, mandible; mpp, maxillary process of premaxilla; mx, maxilla; n, nasal; o, orbit; occ, occipital condyle; pmx, premaxilla; po, postorbital; q, quadrate; sq, squamosal; (l, r), left or right element. Scale bar applies for A–D. Gray areas represent poorly preserved or broken bone.

(pterygoid ramus) of nonavian theropods, *Archaeopteryx,* and the alvarezsaurid *Shuvuuia* (Chiappe et al., 1996, 1998; Sanz et al., 1997). Distally, the quadrate has two transversally oriented condyles (Elzanowski, 1977).

The palatal region of Euenantiornithes is only well known for *Gobipteryx,* for which four partial skulls have been discovered (Elzanowski, 1977, 1995; Chiappe et al., 2001). The palate of *Gobipteryx* is characterized by small and rostrally located choanae and a more rostral horseshoe-shaped opening that connects the oral cavity with a large chamber of the palatal vault (Chiappe et al., 2001). The vomers, fused throughout their rostral half, form a rodlike, central element that medially bounds the choanae. The rostral portion of this element does not contact the premaxilla, and an exquisite new skull suggests that in contrast to previous claims (Elzanowski, 1977; Witmer and Martin, 1987) the caudal ends of the vomers were not articulated to the pterygoids (Chiappe et al., 2001). The palatines are long and paddlelike, articulating rostrally with the vomers. The pterygoids are robust, rostrally forked elements, very different from the rodlike bones that characterized the palate of living birds (Chiappe et al., 2001). An ossification believed to be the ectethmoid was described for *Gobipteryx* by Chiappe et al. (2001), and an ectopterygoid was reported by

Elzanowski (1995). This latter bone compares well with the hooked ectopterygoid of *Archaeopteryx* (Wellnhofer, 1974) and nonavian theropods (Weishampel et al., 1990).

Euenantiornithines have a vaulted cranial roof (Figs. 11.2, 11.3). The parietals are much shorter than the frontals, but they are also vaulted (Fig. 11.3A). In *Neuquenornis,* the cranial roof is separated from the occiput by a shallow transverse nuchal crest (Chiappe and Calvo, 1994). *Neuquenornis* has a small occipital condyle and a proportionally much larger foramen magnum (Fig. 11.3B). Coupled with its short parietals, the presence of a large foramen magnum suggests that *Neuquenornis* may have had a fairly "modern" brain organization, although data from either the inner surface of the braincase or endocranial casts are not available for any enantiornithine.

The mandibles of the Catalan hatchling (Sanz et al., 1997), *Sinornis* (Martin and Zhou, 1997), *Eoenantiornis* (Hou et al., 1999), and *Gobipteryx* (Elzanowski, 1977; Chiappe et al., 2001) are low and straight (Fig. 11.2). While in *Gobipteryx* the two rami fuse to each other at the symphysis, they remain separated in *Sinornis* and the Catalan hatchling, although the presence of this condition in the latter may be related to its early ontogenetic age. The caudal half of the mandible of the Catalan hatchling is perforated by

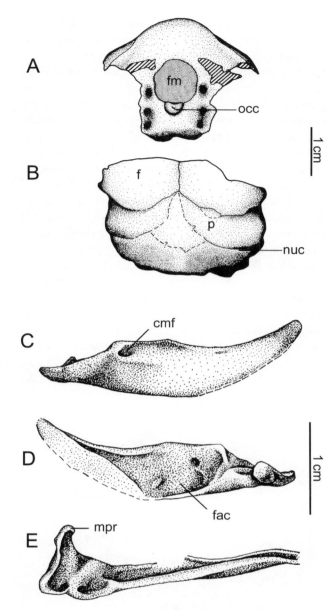

A

B

C

D

E

Figure 11.3. Skull and mandible of euenantiornithine birds. Dorsal (A) and occipital (B) views of the skull of the Late Cretaceous *N. volans* from Patagonia (Argentina) (modified from Chiappe and Calvo, 1994). C–E, Right mandibular ramus (PVL-4698) from the Late Cretaceous of El Brete (Argentina) in lateral (C), medial (D), and dorsal (E) views. This mandible is tentatively regarded as an euenantiornithine element, although euenantiornithine mandibular synapomorphies have not yet been discovered (see section "Systematic Paleontology"). Abbreviations: cmf, caudal mandibular fenestra; f, frontal; fac, fossa aditus canalis; fm, foramen magnum; mpr, medial process of mandible; nuc, nuchal crest; occ, occipital condyle; p, parietal.

two elongate mandibular fenestrae (Fig. 11.2A, C). In contrast to the condition in neornithine birds, the rostral mandibular fenestra of the hatchling is not ventrally lined by the dentary but rather by the angular, an ancestral condition present in nonavian theropods (Sanz et al., 1997). Elzanowski (1977) described a lateral depression where the dentary of *Gobipteryx* bifurcates into caudodorsal and caudoventral prongs; he regarded this depression as the rostral mandibular fenestra (even though the mandible is not perforated). If this interpretation is correct, *Gobipteryx* resembles more derived avians in this respect, a condition that must have evolved independently. The El Brete mandible, the caudal portion of a right ramus, markedly differs from those of the Catalan hatchling, *Sinornis*, and *Gobipteryx* in that it is heavily upcurved in lateral view, with a strongly concave dorsal margin (Fig. 11.3C). The articular bone of this mandible shows a depressed retroarticular area with a short, rounded retroarticular process (Fig. 11.3E). A similar but larger retroarticular process is present in *Gobipteryx* (Elzanowski, 1977). As in *Gobipteryx*, the mandibular medial process is elongate, but in the El Brete mandible it ends in a hook that projects slightly rostrally. In the mandibles of the El Brete specimen and *Gobipteryx*, there is a small surangular foramen (i.e., caudal mandibular fenestra) laterally, just rostral to the rim of the articular facets (Fig. 11.3C). Medially, the El Brete mandible is excavated by a large, elongate, and rostrally pointed fossa auditus (Fig. 11.3D). The rostral margin of this fossa is bounded by a long prearticular (angular of Elzanowski, 1977: Fig. 1b).

Axial Skeleton

Several cervical vertebrae are preserved in articulation in the Early Cretaceous *Eoalulavis* (Sanz et al., 1996, Chapter 9 in this volume), *Sinornis* (Zhou et al., 1992), *Longipteryx* (Zhang et al., 2000), and the Catalan hatchling (Sanz et al., 1997). Two isolated midcervicals (PVL-4050 and PVL-4057) are part of the El Brete collection. The neck of the hatchling from Catalonia is composed of 9 elements; this number is concordant with estimates of 9–10 elements in the cervical series of *Sinornis* (Zhou, 1995a) and *Longipteryx* (Zhang et al., 2000). Hou et al. (1999) reported 11 cervicals for the neck of *Eoenantiornis*. The atlas is known only from the Catalan hatchling. As in *Archaeopteryx* (Wellnhofer, 1974) and the alvarezsaurid *Shuvuuia* (Chiappe, Norell, and Clark, Chapter 4 in this volume), it is formed by three elements—two atlantal arches and the centrum. This condition may be a reflection of the early ontogenetic age of this specimen, although the presence of unfused atlantal hemiarches is also the primitive condition for birds (Chiappe, 1992). The remaining cervicals are elongate (in particular the midcervicals from El Brete) and strongly compressed laterally. The spinous processes are low to completely reduced, although they become somewhat taller in the most caudal cervicals.

In the axis and the following two vertebrae of the Catalan hatchling, prominent epipophyses project farther caudally than the postzygapophyseal facets (Sanz et al., 1997). This condition is reminiscent of that of the nonavian theropod *Deinonychus* (Ostrom, 1969). In the mid- and caudal cervicals, as evidenced by the condition in the Catalan hatchling, the El Brete vertebrae, and *Eoalulavis*, the epipophyses are less developed, forming sagittal ridges of varying degree that do not extend beyond the postzygapophyseal facets. Tapering costal processes fuse to the centra; in the caudal cervicals of *Eoalulavis* these are as long as their respective centra (Sanz et al., Chapter 9 in this volume). The centra have sharp, bladelike ventral borders. In none of the known cervicals is the portion of the centrum caudal to the transverse foramen perforated by pneumatic foramina. In the El Brete cervicals and in the axis and first four postaxial elements of the Catalan hatchling, the cranial articular surfaces are heterocoelous, being wider than tall. The caudal articular surfaces of these vertebrae, however, are taller than wide, and their degree of heterocoely is incipient, with only a slight convexity transversely and a slight concavity dorsoventrally. In *Eoalulavis*, however, both articular surfaces of the caudal cervicals are not heterocoelous but amphiplatyan (Sanz et al., Chapter 9 in this volume). Because the El Brete cervicals are midcervicals and heterocoely progresses from the most cranial vertebrae toward the last presacrals (Chiappe, 1996a), the most caudal cervicals of the taxa represented by the isolated vertebrae from El Brete may have also been nonheterocoelous. Future findings, however, will have to clarify this issue.

Thoracic vertebrae are preserved in articulation in the El Brete collection, *Neuquenornis* (Chiappe and Calvo, 1994), *Concornis* (Sanz et al., 1995), *Eoalulavis* (Sanz et al., 1996), *Longipteryx* (Zhang et al., 2000), and the Catalan hatchling. Disarticulated thoracic vertebrae are known for *Sinornis* (Sereno and Rao, 1992; Zhou et al., 1992) and the Late Cretaceous *Halimornis* from Alabama (Chiappe et al., 2002). The spinous processes of the thoracic vertebrae are laminar and broad (Fig. 11.4). The caudodorsal tip of the mid- and caudal thoracic vertebrae form a distinct fork; a double fork is present in the caudalmost thoracic vertebrae of *Eoalulavis* (Sanz et al., Chapter 9 in this volume). The articular surfaces of the thoracic centra are typically amphiplatyan. Nevertheless, an isolated cranial thoracic vertebrae from El Brete has an opisthocoelous centrum. This recalls the condition in the Late Cretaceous Patagonian *Patagopteryx*, in which opisthocoelous cranial thoracic centra make the transition to incipiently heterocoelous caudal cervicals (Chiappe, 1996a; Chapter 13 in this volume). The cranial thoracic vertebrae bear ventral processes (Sereno, Rao, and Li, Chapter 8 in this volume); these are in some instances very prominent (e.g., *Eoalulavis;* see Sanz et al., Chapter 9 in this volume). In several articulated thoracic series from El Brete (e.g., PVL-4051,

PVL-4041-2, PVL-4047) and in the cranial thoracic vertebrae of *Eoalulavis,* the centra are remarkably compressed. This, however, is not true either for more recently collected specimens from El Brete or for *Halimornis* (Chiappe et al., 2002). All euenantiornithine thoracic vertebrae exhibit prominent lateral excavations of their centra (Fig. 11.4), the development of which increases progressively caudally (Chiappe, 1996a). These can be simple grooves as in *Neuquenornis* (Chiappe and Calvo, 1994), suboval fossae (e.g., *Halimornis, Sinornis, Concornis,* and the El Brete specimens), or remarkable excavations that turn the centra into compressed keels as in the cranial thoracic vertebrae of *Eoalulavis.* Euenantiornithines are unique among avians in that the parapophyses (costal fovea) of the thoracic vertebrae are situated in a midcentral position, ventral to the transverse process and dorsal to the lateral excavation (Fig. 11.4).

Synsacral vertebrae are known from several specimens from El Brete, *Sinornis* (Zhou, 1995a), *Concornis* (Sanz et al., 1995), *Longipteryx* (Zhang et al., 2000), and *Gobipteryx* ("*N. valifanovi*" [Kurochkin, 1996]). The euenantiornithine synsacrum is formed by at least eight fused vertebrae. The cranial articular surface is concave, a condition shared by alvarezsaurids and *Patagopteryx*. The spinous processes of the

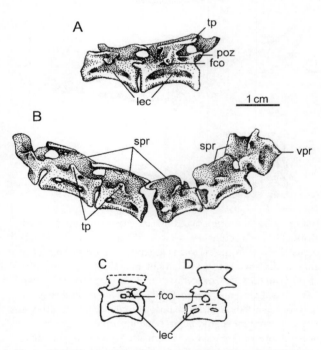

Figure 11.4. Euenantiornithine thoracic vertebrae. A, PVL-4041-2 from the Late Cretaceous of El Brete (Argentina), right lateral view. B, PVL-4051 from the Late Cretaceous of El Brete (Argentina), left lateral view. C, *H. thompsoni* from the Late Cretaceous of Alabama (United States), left lateral view. D, *S. santensis* from the Early Cretaceous of Liaoning (China), left lateral view. Abbreviations: fco, fovea costalis; lec, lateral excavation of centra; poz, postzygapophysis; spr, spinous process; tp, transverse process; vpr, ventral process. C and D not in scale.

synsacral vertebrae are fused to each other, forming a dorsal crest that decreases in height caudally. Long, individualized transverse processes project laterally from both the cranial and caudal sections of the synsacrum. In specimen PVL-4041-4, those of the first two vertebrae fused together, forming left and right bars parallel to the spinous crest. The following caudal ones coalesce with the ilium. The ventral surface of the synsacrum bears a shallow, axial furrow in *Gobipteryx* (Kurochkin, 1996) and some El Brete specimens (e.g., PVL-4045-2).

Free caudal vertebrae and/or a pygostyle are known for *Sinornis* (Sereno and Rao, 1992; Zhou et al., 1992; Sereno, Rao, and Li, Chapter 8 in this volume), *Boluochia* (Zhou, 1995b), *Concornis* (Sanz et al., 1995), *Longipteryx* (Zhang et al., 2000), and *Halimornis* (Chiappe et al., 2002). Zhou et al. (1992) estimated the presence of eight free caudals in *Sinornis*. The caudal vertebrae are amphiplatyan and typically have elongate transverse processes. The long, euenantiornithine pygostyle is formed of at least eight vertebrae (Lamb et al., 1993; Chiappe et al., 2002), although its remarkable elongation in certain taxa (e.g., *Boluochia*, in which it is longer than the metatarsals; see Zhou, 1995b; Zhou and Hou, Chapter 7 in this volume) suggests that many more elements may have been incorporated in some euenantiornithines. In *Halimornis*, the pygostyle's centrum is laterally compressed with paired laminar processes projecting ventrally; similar ventral processes are present in *Sinornis* (Sereno, Rao, and Li, Chapter 8 in this volume). A proximally forked, axial platform forms the dorsal roof of the pygostyle in *Halimornis* (Chiappe et al., 2002). Zhou (1995a,b) mentions the presence of a dorsal crest in the pygostyle of both *Boluochia* and *Sinornis*. In *Sinornis* (IVPP-V-9769A), however, the roof of the pygostyle bears an overhanging axial platform comparable to that of *Halimornis*, and there seems to be no evidence of a dorsal crest.

The long thoracic ribs articulate to short sternal ribs in several taxa (e.g., *Sinornis, Concornis, Neuquenornis, Longipteryx*), and uncinate processes have been reported for *Longipteryx* (Zhang et al., 2000). Although several authors pointed out the absence of gastralia among euenantiornithines (Sereno, Rao, and Li, Chapter 8 in this volume), a typically reptilian basket of gastralia is present in *Eoenantiornis*, and evidence of these elements is also available for other euenantiornithines (e.g., *Longipteryx*). The fact that evidence of gastralia is missing from many well-articulated specimens of *Confuciusornis*, whereas gastralia are definitively present in this basal bird (Chiappe et al., 1999), further cautions against assuming that these elements have been reduced in a particular taxon when the sample of specimens is small.

Thoracic Girdle and Sternum

Elements of the thoracic girdle and sternum are known for a variety of euenantiornithine taxa, including the Early Cretaceous *Sinornis* (Sereno and Rao, 1992; Zhou et al., 1992; Zhou, 1995a), *Boluochia* (Zhou, 1995b), *Concornis* (Sanz et al., 1995), *Eoalulavis* (Sanz et al., 1996), *Eoenantiornis* (Hou et al., 1999) and *Longipteryx* (Zhang et al., 2000), and the Late Cretaceous *Neuquenornis* (Chiappe and Calvo, 1994), *Alexornis* (Brodkorb, 1976), *Gobipteryx* ("*N. valifanovi,*" Kurochkin [1996]), and *Halimornis* (Chiappe et al., 2002). A framentary sternum, several coracoids, and scapulae are included within the El Brete collection, and portions of the thoracic girdle are known for the Early Cretaceous hatchling from Catalonia (Sanz et al., 1997) and a yet undescribed avisaurid from the Late Cretaceous Kaiparowits Formation of Utah (Hutchison, 1993).

The enantiornithine coracoid is long and slender; this condition is extreme in *Eoalulavis* and *Gobipteryx*. Its lateral and medial borders are convex and concave, respectively, and a broad triangular fossa excavates its dorsal surface (Fig. 11.5). Its shoulder end is laterally compressed; the acrocoracoid process, the humeral articular facet, and the scapular articulation are more or less aligned (Fig. 11.5A, E). The articular facet for the scapula is located medial and perpendicular to the humeral articular facet, but the procoracoid process is absent. The scapular articular facet is characteristically convex (Fig. 11.5C, D). The acrocoracoid process is poorly developed; the humeral articular facet essentially converges on it. In *Enantiornis* and other taxa from El Brete (Fig. 11.5), as well as in a recently discovered euenantiornithine from the Late Cretaceous of France (Buffetaut, 1998), there is a peglike medial process between the acrocoracoid process and the humeral articular facet (Walker, 1981). Martin (1995) suggested that this peglike process would have articulated with the furcula, but this seems unlikely given its position in the coracoid and its topological relationship to the scapula. This process may have been connected by a ligament to a pit on the shoulder end of the scapula (Chiappe, 1996a), although a better understanding of its connections must await the discovery of additional, articulated specimens. This peg-pit complex is absent in *Gobipteryx* and *Halimornis*.

The coracoidal shaft expands to varying degrees below the shoulder end. A large supracoracoid nerve foramen perforates the shoulder half of the shaft. Its relative position is variable, but it consistently opens into a medial, longitudinal groove (Fig. 11.5D). In most taxa (e.g., *Enantiornis, Sinornis, Gobipteryx*, the Kaiparowits avisaurid, French enantiornithine), this foramen opens above the dorsal fossa, but in *Neuquenornis*, as in PVL-4034 from El Brete, it opens inside this fossa. The coracoidal sternal end is transversely straight to concave (Fig. 11.5). It lacks a lateral process. As seen in *Neuquenornis* and *Concornis*, the coracoids articulate close to each other on the sternum (Chiappe and Calvo, 1994; Sanz et al., 1995).

The euenantiornithine scapula is robust and straight, with a well-developed acromion (Fig. 11.5). The acromion of

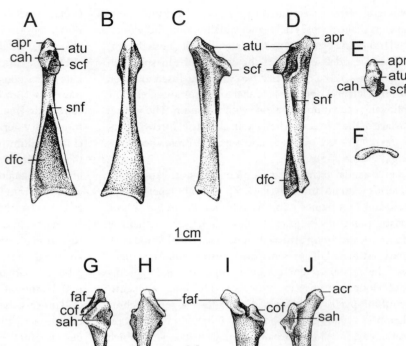

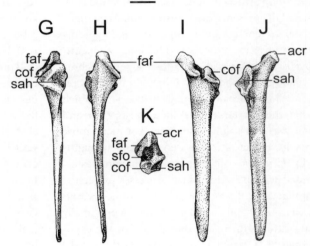

Figure 11.5. Coracoid (PVL-4035) and scapula (PVL-4055) of *E. leali* from the Late Cretaceous of El Brete (Argentina). A, dorsal view. B, ventral view. C, lateral view. D, medial view. E, shoulder view. F, sternal view. G, ventral view. H, dorsal view. I, costal view. J, lateral view. K, shoulder view. Abbreviations: acr, acromion; apr, acrocoracoidal process; atu, acrocoracoidal tubercle; cah, coracoidal articular humeral facet; cof, coracoidal facet of scapula; dfc, dorsal fossa of coracoid; faf, furcular articular facet; sah, scapular articular humeral facet; scf, scapular facet of coracoid; sfo, scapular fossa; snf, supracoracoid nerve foramen.

Eoalulavis and *Eoenantiornis* is remarkably long (Sanz et al., Chapter 9 in this volume). The scapula of *Enantiornis* and *Halimornis* has an expanded shoulder end. The acromion is wide, bearing a crescentic to semilunate, flat articular surface (Fig. 11.5). Martin (1995) argued that this facet articulated with the cranial dorsals. Although possible, this unusual condition has not been observed in any of the euenantiornithine specimens (e.g., *Eoalulavis*, *Neuquenornis*) for which the preserved thoracic girdles and vertebrae retain their topological relationships. Furthermore, the vertebral facet of this alleged articulation is absent in known enantiornithine thoracic vertebrae. Alternatively, the proximity between the acromion and the shoulder end of the furcula in *Eoalulavis* (Sanz et al., Chapter 9 in this volume) suggests that this acromial facet may have articulated with the furcula. The euenantiornithine acromion is separated from the humeral articular facet by a thick neck. In *Enantiornis* this neck is excavated by a deep, circular pit (Chiappe, 1996a,b; Fig. 11.5K); this pit is absent in *Halimornis*. The humeral articular facet of the scapula is distinctly concave (Fig. 11.5). A subtriangular coracoidal facet develops

craniomedially and perpendicular to the latter. The scapular blade is relatively short. Costally, it is scarred by a distinct axial furrow (Fig. 11.6).

The enantiornithine furcula has a narrow interclavicular angle (approximately 60° in *Concornis*, 50° in *Longipteryx*, 45° in *Eoalulavis* and *Eoenantiornis*, and 48° in *Sinornis*). The ascending rami are compressed, with laterally projected ventral borders. As Martin (1995) pointed out, this lateral convexity of the furcula may have been a place for the origin of flight muscles. The furcula of *Neuquenornis* has a short hypocleideum (Fig. 11.6), whereas those of *Concornis*, *Sinornis*, *Eoenantiornis*, and *Eoalulavis* bear a much longer one. In *Eoalulavis*, the hypocleideum is three-fourths the length of the clavicular rami.

Although a definitive, articulated triosseal foramen has never been reported in any enantiornithine bird, the connection between the scapula and the coracoid, and the position of the furcula in some articulated specimens (e.g., *Eoalulavis*), strongly suggest that this important avian structure was present in Euenantiornithes. As is typical of neornithine birds, the triosseal foramen of Euenanti-

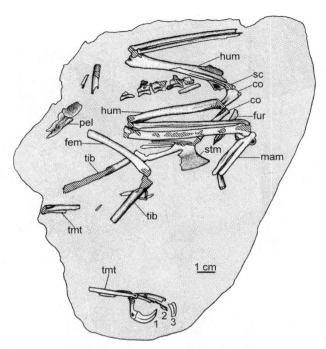

Figure 11.6. The Late Cretaceous *N. volans* from Patagonia (Argentina) (modified from Chiappe and Calvo, 1994). Abbreviations: co, coracoid; fem, femur; fur, furcula; hum, humerus; mam; major metacarpal; pel, pelvis; sc, scapula; stm, sternum; tib, tibiotarsus; tmt, tarsometatarsus; 1, 2, 3, digits 1–3.

ornithes was most likely formed by the coracoid, scapula, and furcula.

The morphology of the sternum of Euenantiornithes is highly variable (Fig. 11.7). Its ossified portions can be relatively broad as in *Concornis, Longipteryx,* or *Sinornis,* fan-shaped as in *Eoenantiornis,* or slender and spear-shaped as in *Eoalulavis.* The sternum can have a prominent carina projecting from its cranial margin (e.g., *Neuquenornis*) (Fig. 11.6), a low keel restricted to its caudal half (e.g., *Concornis, Sinornis, Longipteryx*), or only a faint ridge (e.g., *Eoalulavis*) (Fig. 11.7). In *Concornis, Longipteryx,* and *Sinornis,* the cra-

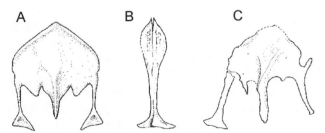

Figure 11.7. Enantiornithine sterna in ventral view (from Sanz et al.,1996). A, *S. santensis* from the Early Cretaceous of Liaoning (China); B and C, *E. hoyasi* and *C. lacustris,* respectively, from the Early Cretaceous of Las Hoyas (Spain). Drawings not to scale.

nial margin is distinctly parabolic. With the exception of *Eoalulavis,* the length of the hypocleideum appears to be related to the extension of the sternal carina. Several euenantiornithines have two sternal notches (e.g., *Concornis, Sinornis, Boluochia*). In these cases, the lateral process is robust, more than the medial process, and bears a distinct caudal expansion (e.g., *Neuquenornis, Concornis, Sinornis;* Fig. 11.7A, C). *Eoalulavis* lacks any evidence of sternal processes, although Sanz et al. (1996) argued that cartilaginous portions of the sternum may have surrounded this central ossified element. The sternum of *Eoalulavis* (Fig. 11.7B) differs from all other euenantiornithines by the presence of a narrow, cranial notch and an inverted T-shaped caudal foot; the latter is reminiscent of the condition present in the ornithurine *Liaoningornis* (Hou et al., 1996). Sanz et al. (1996) argued that the long hypocleideum of *Eoalulavis* could have articulated with this cranial sternal notch, strengthening the thorax. The sternum of *Eoenantiornis* also displays a distinct morphology (Hou et al.,1999). It has an overall round shape with a long, slender caudomedial process. The caudolateral margins of the sternum of *Eoenantiornis* bear short caudal processes that lack the distal expansions of other euenantiornithines. Although Hou et al. (1999) mentioned the presence of a carina in the sternum of *Eoenantiornis,* this bone is exposed in dorsal view, making it impossible to visualize the alleged ventral keel.

Thoracic Limb

Nearly complete wings are known for the Early Cretaceous *Concornis* (Sanz et al., 1995), *Eoalulavis* (Sanz et al., 1996), *Sinornis* (Zhou et al., 1992; Zhou, 1995a), *Eoenantiornis* (Hou et al., 1999), *Otogornis* (Hou, 1994), and *Longipteryx* (Zhang et al., 2000), and the Late Cretaceous *Enantiornis* (Walker, 1981; Chiappe, 1996b) and *Neuquenornis* (Chiappe and Calvo, 1994). Additional forelimb material is available for most of the remaining euenantiornithine taxa (Brodkorb, 1976; Sereno and Rao, 1992; Hutchison, 1993; Kurochkin, 1996; Sanz et al., 1997). The El Brete collection includes several disarticulated wing elements; at least four morphotypes of humerus other than that of *Enantiornis* are known from this collection.

The euenantiornithine humerus retains to some extent the primitive torsion of nonavian theropods. In El Brete specimens PVL-4022 and PVL-4054, the major axes of the proximal and distal ends are at an angle of 42–44°. The deltopectoral crest is flat, lacking any cranial curvature (Fig. 11.8; Zhou and Hou, Chapter 7 in this volume). The bicipital crest is robust and prominently projected cranioventrally. Proximally, this crest is excavated by the transverse ligamental groove; this is a short and shallow groove in *Concornis* (Sanz et al., 1995) but longer and deeper in some of the El Brete humeri (Chiappe, 1996a). On the ventral margin of the bicipital crest's distal portion there is a small but

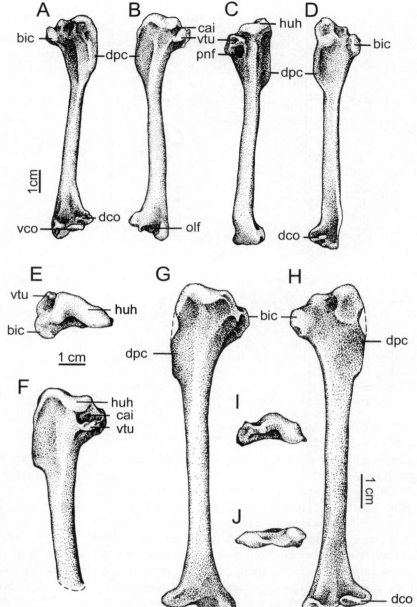

Figure 11.8. Euenantiornithine humeri from the Late Cretaceous of El Brete (Argentina). PVL-4054 in cranial (A) and caudal (B) views. PVL-4022 in caudal (C) and cranial (D) views. PVL-4035 (*E. leali*) in proximal (E) and caudal (F) views. PVL-4025 in caudal (G), cranial (H), proximal (I), and distal (J) views. Abbreviations: bic, bicipital crest; cai, capital incision; dco, dorsal condyle; dpc, deltopectoral crest; huh, humeral head; olf, olecranal crest; pnf, pneumotricipital formamen; vco, ventral condyle; vtu, ventral tubercle.

distinct fossa, presumably for muscle attachment (Fig. 11.8H). The humeral head is concave cranially and convex caudally (Fig. 11.8E, I). Its dorsal margin is concave in its middle area (Fig. 11.8). Caudally, the ventral tubercle is well projected (Fig. 11.8), and in *Enantiornis* and other forms from El Brete (PVL-4020, PVL-4043) it is perforated by a proximodistal canal (Walker, 1981). Although the humerus of several euenantiornithines bears a distinct pneumotricipital fossa, a pneumotricipital foramen is known only for an isolated humerus from El Brete (PVL-4022; Fig. 11.8C). This is important because the occurrence of such a foramen suggests the presence of a diverticulum of the clav-icular air sac pneumatizing the humerus as in neornithine birds. Pulmonary air sacs have been inferred for other basal birds (and nonavian theropods) on the basis of other skeletal features (Britt et al., 1998). The presence of a pneumotricipital foramen in PVL-4022 provides further support to the idea that at least some basal birds had already evolved some of the characteristics of the respiratory apparatus of neornithines. The distal end of the humerus is craniocaudally compressed and transversely expanded (Fig. 11.8; Zhou and Hou, Chapter 7 in this volume). In some forms, the ventral epicondyle projects distally. The dorsal condyle is rectangular, and it is commonly horizontally oriented

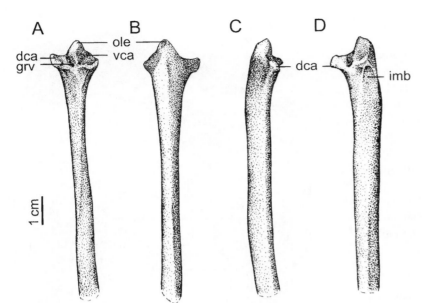

Figure 11.9. Ulna of *E. leali* (PVL-4023) from the Late Cretaceous of El Brete (Argentina). A, interosseous view. B, caudal view. C, dorsal view. D, ventral view. Abbreviations: dca, dorsal cotyla; grv, groove dorsal cotyla/olecranon; imb, impression for m. brachialis; ole, olecranon; vca, ventral cotyla.

(Fig. 11.8). The olecranon fossa is wide, and there are no scapulotricipital or humerotricipital grooves (contra Zhou, 1995a).

The ulna is longer than or nearly as long as the humerus (Zhou and Hou, Chapter 7 in this volume). The dorsal cotyla is elongated; in *Enantiornis* (Walker, 1981) and *Concornis* (Sanz et al., 1995), its surface is distinctively convex (Fig. 11.9). In *Enantiornis,* a deep groove separates both cotylae (Fig. 11.9A, D). This groove is shallower, but still distinct, in *Concornis.* In most forms, the ulnar shaft lacks any caudal papillae for the insertion of the secondary remigial feathers. These, however, have been reported in the avisaurid from the Kaiparowits Formation (Hutchison, 1993). Eight secondary feathers are preserved in their natural connection in the ulna of *Eoalulavis* (Sanz et al., 1996, Chapter 9 in this volume). At the distal end of the ulna, the caudal margin of the dorsal condyle is semilunar (Fig. 11.6). In *Enantiornis* a large foramen perforates the distal end of the ulna on its interosseous surface. This foramen is not present in *Sinornis,* although this taxon has a fossa in the same position. The radial shaft is much more slender than the ulna; the ratio between the diameters of the two is approximately 1:2. On its interosseous surface, the radial shaft is scarred by a longitudinal groove (Fig. 11.6; Zhou and Hou, Chapter 7 in this volume; Sereno, Rao, and Li, Chapter 8 in this volume). Its distal end is spoon-shaped in *Neuquenornis* and *Enantiornis.*

The euenantiornithine proximal ends of the metacarpals are typically fused to the semilunate carpal, forming a carpometacarpus. An exception was reported by Zhang et al. (2000), who described the semilunate carpal of *Longipteryx* as incompletely fused to its metacarpals. The distal ends of metacarpals II and III are typically not fused to one another (Figs. 11.6, 11.10). Metacarpal I is subcircular in *Neuquenornis* and other euenantiornithines and subrectangular in *Eoenantiornis* and apparently *Longipteryx.* This metacarpal lacks a distinct extensor process. Metacarpal III is longer than metacarpal II, projecting somewhat distally (Figs. 11.6, 11.10). The intermetacarpal space between these two metacarpals is very narrow, if not absent.

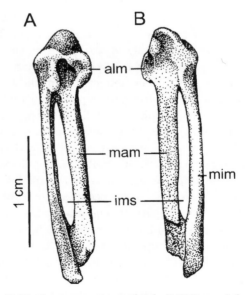

Figure 11.10. Carpometacarpus of *E. leali* (PVL-4049) from the Late Cretaceous of El Brete (Argentina). A, ventral view. B, dorsal view. Abbreviations: alm, alular metacarpal; ims, intermetacarpal space; mam, major metacarpal; mim, minor metacarpal.

Manual digit II is the longest, as in all saurischian dinosaurs. Its proximal phalanx is longer than the intermediate phalanx. This condition differs from that of other early avians (e.g., *Archaeopteryx, Confuciusornis, Patagopteryx*) and is comparable to that of ornithurine birds (Chiappe, 1996a). Digit I is well developed, but its full length is shorter than, or subequal to, metarsals II–III as opposed to more primitive birds (e.g., *Archaeopteryx, Confuciusornis*) and nonavian theropods (Weishampel et al., 1990). In *Eoalulavis*, a well-developed alula attaches to the proximal phalanx of this digit (Sanz et al., 1996). Digit III is typically reduced to a single phalanx; a tiny second phalanx appears to be present in *Sinornis* and *Longipteryx*. Manual digits I and II bear distinct claws (e.g., *Concornis, Eoenantiornis, Sinornis, Longipteryx*).

Pelvic Girdle

Pelvic remains are known for the Early Cretaceous *Sinornis* (Sereno and Rao, 1992; Zhou, 1995a), *Concornis* (Sanz et al., 1995), *Boluochia* (Zhou, 1995b), *Eoenantiornis* (Hou et al., 1999), and *Eoalulavis* (Sanz et al., Chapter 9 in this volume) and the Late Cretaceous *Gobipteryx* ("*N. valifanovi*" [Kurochkin, 1996]) and two isolated pelvises from El Brete (PVL-4041-4, PVL-4032-3).

The pelvic bones of most euenantiornithines are fused to each other (Fig. 11.11), although Zhou et al. (1992) reported that these bones are not fused in *Sinornis*. The ilium is subtriangular, with a short, pronglike postacetabular wing (Fig. 11.11A). The slightly longer preacetabular wing is laterally concave. In PVL-4041-4 the two ilia contact each other in the midline, although this may be a preservational artifact (Fig. 11.11B). In *Gobipteryx*, the ilia are separated from each other (Kurochkin, 1996). The acetabulum is round, and its medial face is completely perforated (Fig. 11.11), a different condition than that in other Mesozoic birds in which the floor of the acetabulum is partially occluded (Martin, 1983, 1987; Chiappe, 1996a). Above the acetabulum, the dorsal iliac crest expands into a triangular supracetabular tubercle, the apex of which projects laterally. A prominent antitrochanter is developed on the caudodorsal corner of the acetabulum. The pubic peduncle is robust and caudoventrally directed (Fig. 11.11A). In PVL-4032-3, the ventral margin of the ilium is perforated by a small fossa just cranial to the base of the pubic peduncle. This fossa, presumably part of the origin of the cuppedicus muscle (Rowe, 1986), is also present in *Gobipteryx* (Kurochkin, 1996).

The pubis is opisthopubic, although not to the degree seen in ornithurine birds (Chiappe, 1996a). It is a slender bone with a suboval to straplike cross section. In most euenantiornithines, the distal ends of the pubes contact each other, forming a short symphysis (Sanz et al., 1995, Chapter 9 in this volume; Kurochkin, 1996; Zhou and Hou,

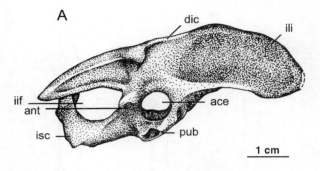

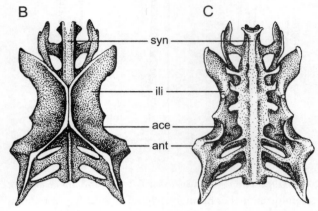

Figure 11.11. Euenantiornithine pelvises from the Late Cretaceous of El Brete (Argentina). A, right lateral view (PVL-4032-3). B, dorsal view (PVL-4041-4). C, ventral view (PVL-4041-4). Abbreviations: ace, acetabulum; ant, antitrochanter; dic, dorsal iliac crest; iif, ilioischiadic foramen; ili, ilium; isc, ischium; pub, pubis; syn, synsacrum.

Chapter 7 in this volume). The distal ends of the pubes are not in contact in *Sinornis* and *Eoenantiornis*, however, even though their tips expand into caudally projected feet (Fig. 11.16). The distal pubic foot is, however, absent in *Gobipteryx* (Kurochkin, 1996) and presumably in *Concornis* (Sanz et al., 1995).

The euenantiornithine ischium is roughly two-thirds to three-fourths the length of the pubis. It is laterally compressed and much longer than the postacetabular wing of the ilium. The morphology of the ischium ranges from beltlike in *Concornis* to knifelike in *Sinornis* and *Eoenantiornis*. In its proximal half, the caudodorsal margin of the ischiadic shaft projects into a proximodorsal process (Fig. 11.11A; Zhou and Hou, Chapter 7 in this volume), which in certain cases (e.g., PVL-4032-3, *Sinornis*) abuts against the medial margin of the ilium, near the end of the postacetabular wing. This contact caudally encloses an ilioischiadic fenestra (Fig. 11.11A) comparable, although not homologous, to that of neognathine birds. Ischiadic proximodorsal processes are present in a variety of early birds [e.g., *Archaeopteryx* (Wellnhofer, 1985), *Rahonavis* (Forster et al.,

1998a,b), *Confuciusornis* (Chiappe et al., 1999)] and non-avian theropods (e.g., *Unenlagia* [Novas and Puerta, 1997]), and it probably represents a primitive condition for birds. The two ischia approach each other but do not form a distal symphysis in *Concornis* (Sanz et al., 1995, Chapter 9 in this volume). It is unclear whether this is also the case for other euenantiornithine taxa.

Pelvic Limb

Elements of the pelvic limb are known for almost all species of euenantiornithines. Several euenantiornithine species (e.g., *A. archibaldi, S. australis, Y. brevipedalis*) are exclusively known from hindlimb bones.

The femur bears a small, cylindrical head separated from the shaft by a distinct neck (Fig. 11.12). A clear fossa for the capital ligament is present in the El Brete femora (e.g., PVL-4060, PVL-4273, PVL-4037). The cranial surface of the proximal end bears a low trochanteric crest. Laterally, the proximal end is excavated by a prominent posterior trochanter (Fig. 11.12). This structure is present in varying degrees among enantiornithine species (Fig. 11.12). Its consistent hypertrophy, however, is characteristic of Euenantiornithes among all other birds, suggesting that important

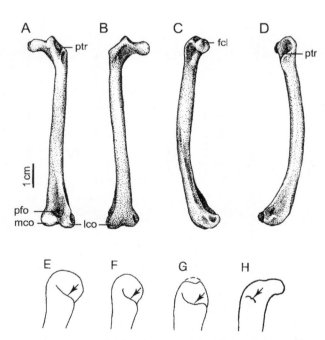

Figure 11.12. A–D, euenantiornithine femur (PVL-4037) from the Late Cretaceous of El Brete (Argentina) in caudal (A), cranial (B), medial (C), and lateral (D) views. E–H, euenantiornithine femoral posterior trochanter (arrow). E, PVL-4037. F, PVL-4060. G, *N. volans*. H, *S. santensis*. Abbreviations: fcl, fossa for capital ligament; lco, lateral condylus; mco, medial condylus; pfo, popliteal fossa; ptr, posterior trochanter. Scale applies to Figures A–D only.

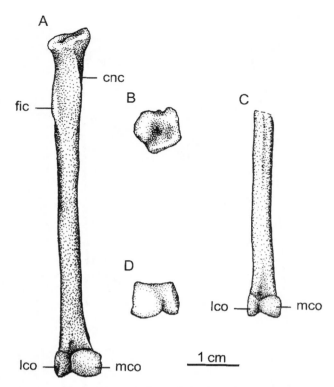

Figure 11.13. Tibiotarsus of *S. australis* from the Late Cretaceous of El Brete (Argentina). A, C, cranial view [PVL-4033 (A) and PVL-4030 (C)]. B, proximal view (PVL-4033). D, distal view (PVL-4033). Abbreviations: cnc, cranial cnemial crest; fic, fibular crest; lco, lateral condylus; mco, medial condylus.

postural or locomotor muscles must have inserted there. Currie and Peng (1993) postulated that the posterior trochanter received the insertion of the ischiofemoralis muscle, although it would be conceivable that other muscles also attached to this area. The femoral shaft is slightly bowed in *Concornis, Sinornis,* and some of the El Brete femora (Fig. 11.12); in certain specimens from El Brete (PVL-4038), however, it is completely straight. Distally, the femur lacks a patellar groove. In *Neuquenornis, Concornis,* and certain taxa from El Brete (e.g., PVL-4037, PVL-4038, MACN-S-01), the distal lateral margin of the femur projects caudally, forming a ridge continuous with the lateral condyle (Fig. 11.12). This ridge is absent in *Halimornis* and the Kaiparowits avisaurid.

The tibia is typically fused to the proximal tarsals, forming a true tibiotarsus (Figs. 11.13, 11.14); a faint suture defining a tall ascending process of the astragalus is preserved in the holotype of *N. eos* (Molnar, 1986) and the specimen of *Gobipteryx* (i.e., "*N. valifanovi*") described by Kurochkin (1996). The proximal articular end is subcircular in *Nanantius, Concornis,* and *Soroavisaurus* (Fig. 11.13B), but it has a transversely wider axis in *Lectavis* (Fig. 11.14C), in which the

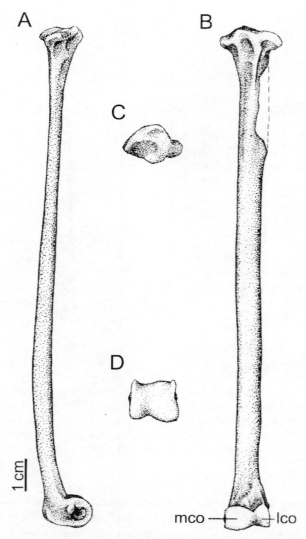

aspect in cranial view (Figs. 11.13, 11.14). In *Gobipteryx* ("*N. valifanovi*" [Kurochkin, 1996]), however, the medial condyle is more suboval, and a much wider groove separates the two condyles.

No free tarsals have ever been reported in any adult euenantiornithine, although Stidham and Hutchison (2001) identified the outlines of three distal tarsals by analysis of x-rays of *A. archibaldi*. In consequence, it seems reasonable to assume that these were fused to the proximal end of the metatarsals (contra Martin, 1983, 1995). In general, metatarsals II–IV are fused only proximally (Figs. 11.6, 11.15; Zhou and Hou, Chapter 7 in this volume), but distal fusion of the metatarsals has been reported for *A. gloriae* (Varricchio and Chiappe, 1995). Euenantiornithines retain the ancestral coplanar orientation of metatarsals II–IV. Metatarsal IV is less prominent than metatarsals II–III (Fig. 11.15; Zhou and Hou, Chapter 7 in this volume). Metatarsal II bears a distinct tubercle on its dorsal surface, presumably the insertion of the tibialis cranialis muscle (Chiappe, 1996a). This tubercle was previously considered a synapomorphy of enantior-

Figure 11.14. Tibiotarsus of *L. bretincola* (PVL-4021-1) from the Late Cretaceous of El Brete (Argentina). A, medial view. B, cranial view. C, proximal view. D, distal view. Abbreviations: lco, lateral condylus; mco, medial condylus.

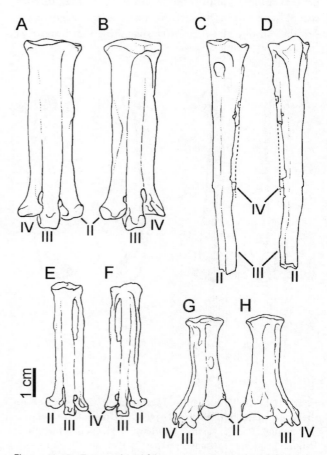

Figure 11.15. Euenantiornithine tarsometatarsi (modified from Chiappe, 1993). A, B, *A. archibaldi.* C, D, *L. bretincola.* E, F, *S. australis.* G, H, *Y. brevipedalis.* A, C, E, G, dorsal view. B, D, F, H, plantar view. Abbreviations: I–IV, metatarsals I–IV.

articular end is more ellipsoidal (Chiappe, 1993). There is only a single weak cnemial crest on the craniolateral border of the tibiotarsus (Figs. 11.13, 11.14). This is thought to correspond to the ancestral cnemial crest of nonavian theropods (Chiappe, 1996a). Distally, both the extensor canal and the supratendinal bridge are absent (Figs. 11.13, 11.14). As in other basal birds (e.g., *Confuciusornis, Patagopteryx, Vorona*), the medial condyle is wider than the lateral one. In most taxa, these condyles are separated by a narrow intercondylar notch that proximally undercuts them. In *Lectavis, Concornis, Nanantius,* and *Soroavisaurus,* the medial condyle has a subcylindrical shape, which, along with the lateral condyle, gives the distal end of the tibiotarsus an hourglass

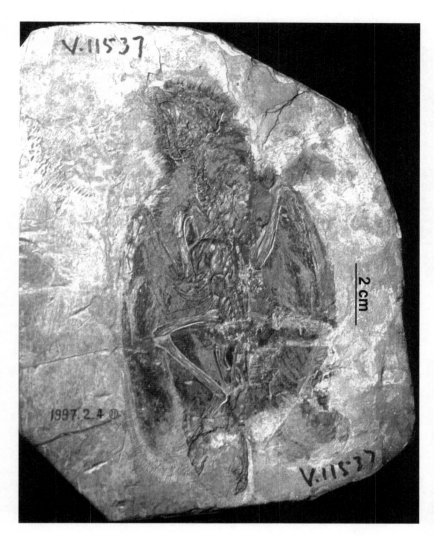

Figure 11.16. *E. buhleri* from the Early Cretaceous of Liaoning, China.

nithines, but its occurrence in *Confuciusornis* (Chiappe et al., 1999) and dromaeosaurid theropods (Norell and Makovicky, 1997) suggests that it is a primitive feature. Distally, the trochlea of metatarsal II is broader than the trochleae of metatarsals III and IV. Despite these common morphological features, the morphological range of the euenantiornithine tarsometatarsi is striking (Fig. 11.15). While *Lectavis* is characterized by a long and slender tarsometatarsus, with a keel-like hypotarsus, *Yungavolucris*'s tarsometatarsus is short and broad, with a stongly ginglymoid trochlea on metatarsal II and laterally divergent trochleae of metatarsals III and IV (Chiappe, 1993). Other euenantiornithine species (e.g., avisaurids, *Concornis*, *Boluochia*, *Sinornis*) have less specialized tarsometatarsi. All euenantiornithines for which the foot is known have a retroverted hallux (Fig. 11.8). Metatarsal I, however, is not reversed, and it retains the primitive articulation on the medial surface of metatarsal II of nonavian theropods (e.g., Norell and Mako-

vicky, 1997). Unfortunately, the feet of the highly specialized *Lectavis* and *Yungavolucris* are not known.

Phylogenetic Relationships

Euenantiornithes is universally accepted as an important basal chapter in the early evolution of birds, and with rare exceptions (e.g., Elzanowski [1995], who regarded *Sinornis*, *Concornis*, and the remaining enantiornithines as successive outgroups of ornithurine birds), they are considered monophyletic. This was not always the case, however. The history of the study of this Cretaceous avian radiation illustrates a general shift in the perception of the early evolution of birds. *Gobipteryx* was the first enantiornithine bird to be properly recognized as a bird at the time of its description; Elzanowski (1974) identified *Gobipteryx* as a Cretaceous palaeognath. This conclusion, however, was criticized by others (e.g., Brodkorb, 1976, 1978), who regarded *Gobipteryx*

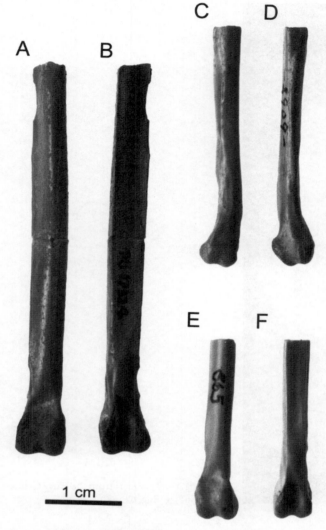

1 cm

Figure 11.17. Euenantiornithine remains collected in the nineteenth century. A, B, metatarsal III (PU-17324) of *A. archibaldi* in dorsal (A) and plantar (B) view. C, D, metatarsal II (YPM-2909) in dorsal (C) and plantar (D) view. E, F, metatarsal III (YPM-865) in dorsal (E) and plantar (F) view.

as a nonavian reptile. The discovery of the large assemblage of bones from El Brete and their recognition as a monophyletic ensemble of birds by Walker (1981) did little to change this view (J. F. Bonaparte, J. Cracraft, and L. D. Martin are exceptions); the avian status of Enantiornithes was still not fully accepted (e.g., Steadman, 1983; Brett-Surman and Paul, 1985; Kurochkin, 1985; Olson, 1985). Interestingly, enantiornithine bones had been found decades earlier, during explorations of the Cretaceous outcrops of the western United States in the nineteenth century (Chiappe, 1995a; Padian and Chiappe, 1998)—these specimens (Fig. 11.17), however, were also mistakenly identified as nonavian theropods. In 1892, for example, O. C. Marsh named a new species of ornithomimid theropod, *Ornithomimus minutus,* on the

basis of several portions of metatarsals (Marsh, 1892). The holotype is now lost, but one of these metatarsal fragments (USNM 2909) was of an avisaurid euenantiornithine (Fig. 11.17C, D). This is also the case for several other "old" specimens such as YPM 865 (Fig. 11.17E, F), which, although its original label reads "bird or *Ornithomimus,*" belongs to an euenantiornithine avisaurid bird. *A. archibaldi* is another euenantiornithine that was also initially regarded as a nonavian theropod (Brett-Surman and Paul, 1985), a taxonomic conclusion these authors extended to specimens later

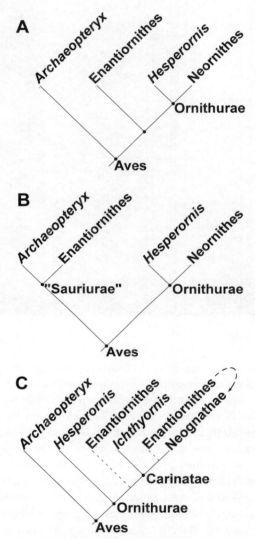

Figure 11.18. Enantiornithine relationships to other birds. A, Enantiornithes as phylogenetically intermediate between *Archaeopteryx* and Ornithurae (e.g., Walker, 1981; Chiappe, 1991, 1995b, 1996b; Sanz et al., 1995, 1996, 1997; Forster et al., 1996). B, Enantiornithes as the sister taxon of *Archaeopteryx* (e.g., Martin, 1983, 1991, 1995; Hou et al., 1995, 1996; Kurochkin, 1996). C, Enantiornithes as a member of the Ornithurae, either as the sister taxon of *Ichthyornis* or Neognathae or included within Neognathae (e.g., Cracraft, 1986).

named *S. australis* (Chiappe, 1993). These frequent misidentifications evoke the statement commonly used in discussions of avian origins, namely, that the theropod ancestry of birds is illustrated by the fact that specimens of *Archaeopteryx* for which feathers are not clearly visible (e.g., Eichstätt) were first misidentified as nonavian theropods. Undeniably, this has also been the case for enantiornithine birds. Only with the advent of more complete discoveries (e.g., Chiappe, 1992; Zhou et al., 1992) came the notion that these primitive and sometimes odd creatures were indeed birds and not nonavian theropods. Along with this notion came the realization that several early lineages of birds were vastly different from their extant counterparts.

The phylogenetic position of Enantiornithes and Euenantiornithes within basal birds has been a matter of intense debate (see Chiappe, 1995b, 1996a, 2001, Chapter 20 in this volume, for a more extensive discussion). These birds have variously been considered the sister group of *Archaeopteryx* (Martin, 1983), an ornithurine lineage (Cracraft, 1986), or an intermediate lineage between *Archaeopteryx* and Ornithurae (Walker, 1981; Thulborn, 1984) (Fig. 11.18). More recent cladistic analyses with varying data sets and included taxa (e.g., Chiappe, 1991, 1995b, 1996a; Sanz et al., 1995, 1996; Chiappe et al., 1996; Forster et al., 1996, 1998a) have consistently supported Walker's (1981) early view that Enantiornithes is phylogenetically intermediate between *Archaeopteryx* and Neornithes (Fig. 11.18A). These same analyses have rejected the proposal of Martin (1983, 1991, 1995) and his followers (e.g., Hou et al., 1995, 1996; Feduccia, 1996; Kurochkin, 1996) of a sister-taxon relationship between *Archaeopteryx* and Enantiornithes, a paraphyletic group that Martin (1983) called "Sauriurae" (Fig. 11.18B). Proponents of a close relationship between Enantiornithes and *Archaeopteryx* have substantiated their views using primarily noncladistic methods, and thus it is hard to evaluate their proposals in light of cladistic parsimony. The only cladistic analysis supporting the monophyly of "Sauriurae" is that of Hou et al. (1996; data matrix and character list available from *Science*'s Web site). This analysis included only 36 characters, approximately a third or a fifth of the information used by Chiappe et al. (1996) and Chiappe (Chapter 20 in this volume), respectively. Perhaps of greater significance is the exclusion of Neornithes as a terminal taxon (see Hou et al., 1996; supplementary information at www.sciencemag.org/feature/data/hou.shl)—a choice with profound consequences for the analysis of Hou et al. (1996), in which "Sauriurae" is supported by five unambiguous synapomorphies, the same ones proposed by Martin (1983) in his original paper. These characters, however, have very little (if any) phylogenetic information because they have been shown to be either equivocal or plesiomorphic (see Chiappe, 1995b). Even Zhou (1995a:18), one of the coauthors of the analysis, disregarded the

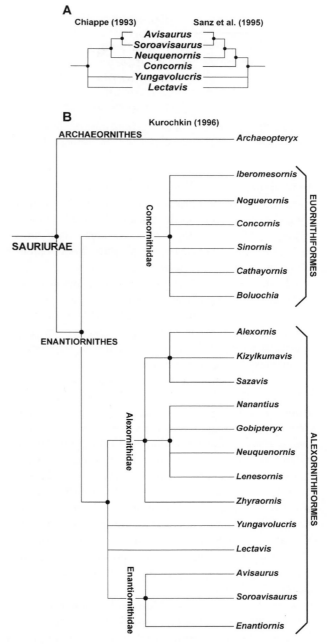

Figure 11.19. Interrelationships among enantiornithine species. A, cladistic analyses of Chiappe (1993) and Sanz et al. (1995) based on pedal morphology only. B, Kurochkin's (1996) diagram of relationships (note that this is not a consensus cladogram derived from any cladistic numerical analysis).

phylogenetic meaning of most of them, concluding that "*Cathayornis* [here regarded as a synonym of *Sinornis*] lies between *Archaeopteryx* and *Ichthyornis* in terms of both evolutionary grade and flight capability." In Chiappe's most recent analysis (2001; Chapter 20 in this volume), Enantiornithes (including Euenantiornithes plus *Iberomesornis* and *Noguerornis*) shows a sister-group relationship with

Eoenantiornis
Sinornis
Eoalulavis
Gobipteryx
Enantiornis
Concornis
Neuquenornis

Figure 11.20. Strict consensus trees derived from the analyses of all euenantiornithine genera with 70% or less missing data.

Ornithuromorpha—a relationship that is supported by 13 unambiguous synapomorphies. As pointed out by Chiappe (2001; Chapter 20 in this volume), any cladistic hypothesis supporting the monophyly of "Sauriurae" over those placing Enantiornithes closer to Neornithes would require a large number of additional steps. At the same time, it would require accepting that flight was independently fine-tuned in both Enantiornithes and Neornithes, because many of the characters that are shared by all these birds have been consistently correlated to flight enhancement (e.g., carinate sternum, strutlike coracoid, pygostyle, modern proportions of wing elements, alula).

Although the intermediate placement of Enantiornithes and Euenantiornithes between *Archaeopteryx* and Neornithes is well supported today, the phylogenetic inter-relationships among different euenantiornithine species have been substantially less explored. To date, only cladistic analyses including a handful of euenantiornithine species have been conducted (e.g., Chiappe, 1993; Sanz et al., 1995; Varricchio and Chiappe, 1995; Fig. 11.19A). Kurochkin (1996) provided a hypothesis of relationships including most known enantiornithine species (Fig. 11.19B). Unfortunately, this study was not framed under cladistic methods, making it harder to evaluate its merit over previous cladistic hypotheses (e.g., Chiappe, 1993; Sanz et al., 1995; see Sereno, Rao, and Li, Chapter 8 in this volume, for further discussions on Kurochkin's work). More problematic is the fact that the rationale for the proposed clusters of species depicted in his single tree, which is not a consensus cladogram, was not explained. This is particularly disturbing when we consider that some of Kurochkin's (1996) proposed groups cluster taxa for which overlapping skeletal elements are unknown. For example, the rationale underlying the grouping of *Avisaurus, Soroavisaurus,* and *Enantiornis* in Enanti-ornithidae (Fig. 11.19B) is unclear when we consider that while *Avisaurus* and *Soroavisaurus* are known by hindlimb elements (Chiappe, 1993), *Enantiornis* is known from elements of the forelimb and thoracic girdle (Chiappe, 1996b).

For this chapter, a list of 36 characters (one multistate) was scored for 16 valid genera of euenantiornithine birds (all genera listed in the section "Systematic Paleontology" except

for *Longipteryx,* which holotype had not been examined by the authors at the time this chapter was written) (Appendix 11.1). Scoring was based on direct examination of the holo-types and referred specimens. Because many of the taxa are highly incomplete, the character matrix includes a high percentage of missing data. The data matrix was analyzed using Goloboff's EST computer program (Goloboff and Farris, 1998, in Horovitz, 1999). This analysis resulted in a "bush-like" consensus cladogram lacking any kind of resolution. Subsequent analyses of the data matrix excluding taxa with 90% or more (i.e., *Otogornis, Alexornis,* and *Nanantius*) and 80% or more missing data (i.e., *Otogornis, Alexornis, Nanantius, Yungavolucris, Avisaurus,* and *Lectavis*) also pro-duced consensus trees without resolution. Only when the analyzed matrix was restricted to taxa with 70% or less miss-ing data (i.e., *Eoalulavis, Eoenantiornis, Neuquenornis, Enan-tiornis, Gobipteryx, Sinornis,* and *Concornis*) did EST produce a consensus tree with some resolution. In this consensus cladogram, *Concornis, Enantiornis,* and *Neuquenornis* formed a group that joined the remaining four taxa in an unresolved polychotomy (Fig. 11.20). Clearly, an understanding of the in-terrelationships within Euenantiornithes has been hampered by the incomplete material available for several of the species, which often are known by either one bone or another (e.g., *Yungavolucris, Avisaurus, Nanantius*), and the close similarity of many of these birds (Sereno, Rao, and Li, Chapter 8 in this volume). Although the results of this new cladistic analysis are largely incomplete and will certainly show substantial modifications with the addition of new data, it is interesting to note that this consensus cladogram recovers Sanz et al.'s (1995) proposed relationship between *Concornis* and *Neu-quenornis* (Sanz et al. [1995] also included *Soroavisaurus* and *Avisaurus* along with the former two taxa [Fig. 11.19A]). Deeper comparative studies of the known euenantiornithine species as well as the discovery of additional material are nec-essary for a better understanding of the cladistic relationships among enantiornithine birds, which remains one of the most challenging issues of basal avian systematics.

Paleobiology

Enantiornithes is probably the group of Mesozoic birds for which we are able to provide the most reliable inferences re-garding their general lifestyle, reproductive biology, and growth strategies. Most enantiornithine fossils occur in nonmarine deposits, indicating their preponderance in the terrestrial avifaunas of the Cretaceous. Nevertheless, as the discovery of the partial skeleton of *Halimornis* in marine beds formed as far as 50 km offshore suggests, these early birds may have played a more important role in the marine and littoral ecosystems than was earlier understood (Lamb et al., 1993; Chiappe et al., 2002).

The morphological disparity seen among the different euenantiornithine species suggests a diversity of lifestyles. The anisodactyl foot of *Sinornis, Concornis, Neuquenornis, Longipteryx,* and *Soroavisaurus* suggests perching capabilities (Sereno and Rao, 1992; Chiappe, 1993; Chiappe and Calvo, 1994; Sanz et al., 1995), but the pedal morphology of *Yungavolucris* and *Lectavis* has been correlated with aquatic and wading habits, respectively (Walker, 1981; Chiappe, 1993). Aquatic or nearshore habits have also been proposed for *Eoalulavis* (Sanz et al., 1996), for which the presence of exoskeletal remains of aquatic crustaceans inside its digestive tract indicates that, at the very least, it frequented nearshore environments. Although the presence of stomach contents in *Eoalulavis* is the only direct evidence of feeding habits available for enantiornithine birds, the hooked beak of *Boluochia* suggests more raptorial habits (Zhou, 1995b). Similar habits have been inferred for *Neuquenornis* and *Soroavisaurus,* in which the hallucial claw is hypertrophied (Chiappe, 1993; Chiappe and Calvo, 1994). Feduccia (1996) pointed out that the long hypocleideum and caudally restricted sternal keel of some euenantiornithines (e.g., *Sinornis, Concornis*) may indicate the existence of a digestive system with a large crop. He compared this design to that of the folivore hoatzin (*Opisthocomus*) and suggested that *Sinornis* may have had similar feeding habits. Indeed, these birds may have been folivores, as observed by Feduccia (1996), but the presence of small, sharp teeth in *Sinornis* suggests an insectivorous diet more than one based on leaves.

During their evolution throughout the Cretaceous there is a general increment in size and size range among the euenantiornithine species. All known Early Cretaceous euenantiornithines are small to medium-size birds (see comparative measurements in Bochenski, 1996). *Sinornis* and *Eoalulavis,* for example, were the size of a sparrow or finch, and the size of *Nanantius* and *Concornis* was comparable to that of a medium-sized thrush. Late Cretaceous euenantiornithines, however, are generally larger: *Gobipteryx* was the size of a jackdaw and *Neuquenornis* that of a small falcon. Some of the Late Cretaceous taxa were much larger. *Enantiornis* had a wingspan of about 1.2 meters, ranging in size between a skua and a turkey vulture (Chiappe, 1996b), and a recently discovered enantiornithine from France has been compared in size to a herring gull (Buffetaut, 1998). Some Late Cretaceous euenantiornithines, however, retained the small size of their Early Cretaceous relatives; *Alexornis* was sparrow-sized.

On the basis of the presence of a deep notch in the sternum, Walker (1981) argued that enantiornithines had a limited capacity for flight. This view, however, was challenged by Chiappe and Calvo (1994), who, based on the morphology of the wing, sternum, and thoracic girdle of *Neuquenornis,* proposed that these birds were probably good fliers. A decisive find was that of *Eoalulavis* (Sanz et al., 1996), in which a nearly intact alula was preserved attached to its manual digit I (Sanz et al., Chapter 9 in this volume). The presence of an alula in *Eoalulavis,* along with derived wing proportions and other flight-correlated structures (e.g., strutlike coracoid, fused wrist bones, pygostyle) in this and other euenantiornithines, strongly suggests that these birds had achieved an enhanced flying capacity and control of maneuverability, including in their repertoire the ability to perform low-speed flight (Sanz et al., 1996).

Studies of euenantiornithine embryos (Elzanowski, 1981), hatchlings (Sanz et al., 1997), and eggshell microstructure (Mikhailov, 1991, 1996; Schweitzer et al., 2002), as well as of the bone histology of adult individuals (Chinsamy et al., 1994, 1995), have provided important data on the reproductive biology and rates of growth of these early birds. Elzanowski (1981) reported on a series of embryos from the Late Cretaceous of Mongolia, which he tentatively referred to *Gobipteryx.* The forelimb bones of these embryos are proportionally longer than those of most neornithine hatchlings, and their degree of ossification is more advanced than that of hatchlings of extant precocial birds (e.g., chicken, skua) (Elzanowski, 1981). This precocial development of the forelimb, along with the differential degree of ossification between the fore- and hindlimbs of these embryos, led Elzanowski (1981, 1995) to argue that the developmental strategy of *Gobipteryx* was superprecocial, that is, juveniles were completely self-sufficient in terms of finding food and were able to fly soon after hatching. Elzanowski (1995) also argued for male incubation, because this strategy is generally associated with the high energetic demands expected of embryos with such a precocial degree of ossification. A precocial developmental mode is primitive for neornithine birds (Starck, 1993), since phylogenetically basal extant birds such as ratites and galliforms have this mode of development. Thus, it is likely to be the case for Euenantiornithes (Chiappe, 1995a). The degree of embryonic ossification, however, appears not to be necessarily correlated with one or another type of developmental mode (Starck, 1993). Phylogenetic inference also indicates that biparental care is actually primitive for neornithine birds (McKitrick, 1992). Given this and the fact that the extreme ossification of the *Gobipteryx* embryos may not be indicative of superprecociality, Elzanowski's (1995) argument for superprecociality and male incubation in Enantiornithes should be taken with caution. Elzanowski (1995) also expanded his developmental conclusions to the Patagonian *Neuquenornis,* which he regarded as a very young specimen. Feduccia (1996), following Elzanowski (1995), regarded *Neuquenornis* as late embryonic. These conclusions, however, are most likely incorrect. First, the holotype specimen of *Neuquenornis* (MUCPv-142; Chiappe and Calvo, 1994) is actually relatively large; the wingspan of this specimen exceeds 40 cm. Second, many of the long bone ends of MUCPv-142 are completely ossified, including metacarpal I, which does not ossify before hatching (Elzanowski, 1981). Third, the sternum of this specimen is fully ossified, including its prominent carina. As Elzanowski (1981) pointed out, most of the ossification of the sternum of neornithine birds occurs during postembryonic development and typically at a slow rate of ossification. Consequently, neither the size nor the bony structures of MUCPv-142 indicate an embryonic or near-hatching stage. The poor preservation of some of its bones is nothing more than a preservational artifact, and it compares well with the preservation shown by other fossils from the same site.

Data on the growth rates of euenantiornithines were derived from the histological studies of Chinsamy et al. (1994, 1995). These studies revealed the presence of growth rings (lines of arrested growth, or LAGs) in the cross sections of two euenantiornithine femora from El Brete. The presence of growth rings indicates a pause of postnatal growth, presumably a cyclical interruption of this growth. This pattern of bone deposition is strikingly different from that of neornithine birds and apparently of ornithurine birds in general (Chinsamy and Elzanowski, 2001; Padian et al., 2001; Chinsamy, Chapter 18 in this volume), in which uninterrupted growth leads to the development of adult size in the range of a single year (Starck, 1993). The presence of growth rings in the compacta of euenantiornithine limb bones suggests that these birds took more than one year to achieve mature size (Chinsamy et al., 1995). Furthermore, the microstructure of the euenantiornithine

bone was unique, exhibiting virtually no vascularization and a lamellar, slowly deposited type of bone. The slow rate of growth inferred from this pattern of bone deposition appears to be consistent with the idea that euenantiornithine birds were precocial (Chiappe, 1995a; Chinsamy and Elzanowski, 2001), since precocial birds have rates of growth that are slower than those of altricial birds (Starck, 1993). The presence of this characteristic type of bone tissue led Chinsamy et al. (1994, 1995) to argue that important physiological differences with respect to their modern counterparts may have been present in Euenantiornithes. Specifically, they argued that because growth is strongly correlated with body temperature in living birds (Prosser, 1973), these bone differences may be indicating that these early birds were not fully endothermic. Other authors incorrectly used this to claim that euenantiornithine birds were ectothermic (e.g., Feduccia and Martin, 1996), but Chinsamy et al. (1994, 1995) simply stated that their thermal metabolism was probably lower than that of neornithine birds. If so, the metabolic rates of Euenantiornithes probably fit an intermediate level within the spectrum of ectothermy-endothermy (Chiappe, 1995a). This argument, although somewhat imprecise and more conjectural than reading growth rates from the pattern of bone deposition (see Padian et al., 2001), is interesting because euenantiornithine birds were fully feathered, and feathers have consistently been correlated to endothermy. More sampling of euenantiornithine bones and other metabolic-indicator studies are necessary before the meaning of the peculiar pattern of euenantiornithine bone microstructure can be fully understood.

Characters, Character-States, and Character Matrix Used for Cladistic Analysis of Euenantiornithine Taxa

Key: 0 = primitive state; 1, 2 = derived states; ? = missing entries.

1. Premaxillary teeth: present (0), absent (1).
2. Parapophyses located in the cranial (0) or central (1) part of the bodies of dorsal vertebrae.
3. Cranial end of pygostyle dorsally forked and with a pair of laminar, ventrally projected processes: absent (0), present (1).
4. Laterally compressed shoulder end of coracoid, with nearly aligned acrocoracoid process, humeral articular surface, and scapular facet, in dorsal view: absent (0), present (1).
5. Distinctly convex lateral margin of coracoid: absent (0), present (1).
6. Supracoracoid nerve foramen opening into an elongate furrow medially and separated from the medial margin of the coracoid by a thick, bony bar: absent (0), present (1).
7. Broad, deep fossa on the dorsal surface of the coracoid: absent (0), present (1).
8. Supracoracoid nerve foramen opens above the coracoidal fossa (0) or inside of it (1).
9. Scapular acromion costolaterally wider than deeper: absent (0), present (1).
10. Costal surface of scapular blade with a prominent, longitudinal furrow: absent (0), present (1).
11. Dorsal and ventral margins of the furcula: subequal in width (0), ventral margin clearly wider than the dorsal margin (1).
12. Hypocleideum: absent or poorly developed (0) or well developed (1).
13. Ossified sternal keel: absent or incipient (0), present and near or projecting cranially from the cranial border of the sternum (1), present and not reaching the cranial border of the sternum (2).
14. Lateral process on sternum: absent (0), present (1).
15. Lateral process of sternum without (0) or with (1) a subtriangular distal expansion.
16. Medial process of sternum: absent (0), present (1).
17. Dorsal (superior) margin of the humeral head concave in its central portion, rising ventrally and dorsally: absent (0), present (1).
18. Prominent bicipital crest of humerus, cranioventrally projecting: absent (0), present (1).
19. Ventral face of the humeral bicipital crest with a small fossa for muscular attachment: absent (0), present (1).
20. Distal end of humerus very compressed craniocaudally: absent (0), present (1).
21. Dorsal cotyla of ulna separated from the olecranon by a groove: absent (0), present (1).
22. Shaft of radius with a long longitudinal groove on its caudoventral surface: absent (0), present (1).
23. Subcircular extensor process of metatarsal I: absent (0), present (1).
24. Minor metacarpal (III) projecting distally more than major metacarpal (II): absent (0), present (1).
25. Manual digit II: proximal phalanx shorter (0) or longer (1) than intermediate phalanx.
26. Pubic foot: present (0), absent (1).
27. Femoral posterior trochanter: absent or weakly developed (0), hypertrophied (1).
28. Caudal projection of the lateral border of the distal end of femur: absent (0), present (1).
29. Very narrow, deep intercondylar sulcus on tibiotarsus that proximally undercuts condyles: absent (0), present (1).
30. Trochlea of metatarsal II much broader than the trochlea of metatarsal III: absent (0), present (1).
31. Metatarsal IV significantly thinner than metatarsals II and III: absent (0), present (1).
32. Medial rim of the trochlea of metatarsal III with a strong plantar projection: absent (0), present (1).
33. Metatarsal I distinctly J-shaped in medial aspect: absent (0), present (1).
34. Proximal articular surface strongly inclined dorsally: absent (0), present (1).
35. Strong transverse convexity of the dorsal surface of the midshaft of metatarsal III: absent (0), present (1).
36. Distal end of metatarsal II strongly curved medially: absent (0), present (1).

Taxon	1	2	3	4	5	6	7	8	9	10	11	12	13	14	15	16	17	18	19	20	21	22	23	24	25	26	27	28	29	30	31	32	33	34	35	36
A. lithographica	0	0	0	0	0	0	n	0	0	0	0	0	0	0	n	0	0	0	0	0	0	0	0	0	0	0	0	0	0	0	0	0	0	0	0	0
H. thompsoni	?	1	1	?	?	?	?	?	?	1	1	?	?	?	?	?	?	1	1	1	?	?	?	?	?	?	0	?	?	?	?	?	?	?	?	?
C. lacustris	?	1	?	1	1	1	1	?	1	?	1	1	2	1	1	1	1	1	1	1	1	?	?	?	1	1	1	?	1	?	1	1	?	0	0	1
S. santensis	0	1	1	?	1	1	1	0	?	0	?	1	2	1	1	1	1	1	1	?	?	1	1	1	1	0	1	0	1	?	1	?	0	?	?	0
G. minuta	1	?	?	1	?	?	1	0	0	?	1	1	?	?	?	?	?	?	?	?	?	1	?	1	?	1	?	?	?	1	1	?	0	1	0	1
L. bretincola	?	?	?	?	?	?	?	?	?	?	?	?	?	?	?	?	?	?	?	?	?	?	?	?	?	?	?	0	?	1	?	?	0	0	?	?
S. australis	?	?	?	?	?	?	?	?	?	?	?	?	?	?	?	?	?	?	?	?	?	?	?	?	?	?	?	?	1	1	1	1	1	1	1	1
Y. brevipedalis	?	?	?	?	?	?	?	?	?	?	?	?	?	?	?	?	?	?	?	?	?	?	?	?	?	?	?	?	?	1	1	0	?	0	0	0
A. gloriae	?	?	?	?	?	?	?	?	?	?	?	?	?	?	?	?	?	?	?	?	?	?	?	?	?	?	?	?	?	1	1	1	?	1	1	1
E. leali	?	?	?	1	1	1	1	0	1	1	?	?	?	?	?	?	1	1	1	0	1	1	1	1	?	?	?	?	?	?	?	?	?	?	?	?
N. volans	?	1	?	?	1	1	1	1	?	1	1	0	1	1	1	?	?	?	?	1	?	1	1	1	?	1	1	1	?	1	1	1	1	?	1	0
B. zenghi	1	?	?	?	?	?	?	?	?	?	?	?	?	2	1	0	1	?	?	?	?	?	?	?	?	?	?	?	?	1	1	0	?	?	?	1
N. eos	?	?	?	?	?	?	?	?	?	?	?	?	?	?	?	?	?	?	?	?	?	?	?	?	?	?	?	1	?	?	?	?	?	?	?	?
O. genghisi	?	?	?	?	?	1	?	?	?	?	?	?	?	?	?	?	1	1	?	?	?	?	?	?	?	?	?	?	?	?	?	?	?	?	?	?
A. antecedens	?	?	?	1	?	?	?	?	?	?	?	?	?	?	?	?	?	?	?	?	1	?	?	?	?	?	?	?	?	?	?	?	?	?	?	?
E. hoyasi	?	?	?	1	1	1	1	1	0	?	1	1	0	0	n	0	1	1	1	1	?	1	?	1	1	1	?	1	?	?	?	?	?	?	?	?
E. buhleri	1	?	?	1	0	?	1	?	?	?	1	1	?	1	0	0	0	?	?	?	?	?	?	1	0	1	1	0	1	?	?	1	1	?	?	?

Acknowledgments

We are grateful to J. Bonaparte, J. Calvo, Chen Pei-Ji, Hou L.-H., Ji Q., Ji S.-A., E. Kurochkin, M. Norell, H. Osmólska, J. Powell, R. Prum, J. Sanz, D. Varricchio, and Zhou Z., who facilitated specimens for study and/or provided useful discussions, and to K. Padian and L. M. Witmer for their constructive reviews. We also thank S. Orell for her editorial work on the manuscript. S. Reuil provided assistance with the graphics and illustrations. Figure 11.20 was photographed by L. Meeker. Support for this research was provided by the Chapman Fund of the American Museum of Natural History, the Natural History Museum of Los Angeles County, and the National Science Foundation (DEB-9873705).

Literature Cited

Bochenski, Z. 1996. Enantiornithes—a dominant group of the Cretaceous terrestrial birds. Przeglad Zoologiczny 40(3–4): 175–184.

Bonaparte, J. F. 1986. History of the terrestrial Cretaceous vertebrates of Gondwana. IV Congreso Argentino de Paleontología y Bioestratigrafía, Actas 2:63–95.

Bonaparte, J. F., and J. E. Powell. 1980. A continental assemblage of tetrapods from the Upper Cretaceous beds of El Brete, northwestern Argentina (Sauropoda-Coelurosauria-Carnosauria-Aves). Mémoires de la Société Géologique de France (Nouvelle Série) 139:19–28.

Bonaparte, J. F., J. A. Salfitty, G. Bossi, and J. E. Powell. 1977. Hallazgos de dinosaurios y aves cretácicas en la Formación Lecho de El Brete (Salta), próximo al límite con Tucumán. Acta Geológica Lilloana 14:5–17.

Brenner, P., W. Geldmacher, and R. Schroeder. 1974. Ostrakoden und Alter der Plattenkalke von Rubies (Sra. Del Montsech) Prov. Lérida, NE-Spanien. Neues Jarbuch für Geologie und Paläontologie Monatshefte 9:513–524.

Brett-Surman, M. K., and G. Paul. 1985. A new family of bird-like dinosaurs linking Laurasia and Gondwanaland. Journal of Vertebrate Paleontology 5(2):133–138.

Britt, B. B., P. J. Makovicky, J. Gauthier, and N. Bonde. 1998. Postcranial pneumatization in Archaeopteryx. Nature 395: 374–376.

Brodkorb, P. 1976. Discovery of a Cretaceous bird, apparently ancestral to the Orders Coraciformes and Piciformes (Aves: Carinatae); pp. 67–73 in S. L. Olson (ed.), Collected Papers in Avian Paleontology Honoring the 90th Birthday of Alexander Wetmore, Smithsonian Contributions to Paleobiology 27. Smithsonian Institution Press, Washington, D.C.

———. 1978. Catalogue of fossil birds, Part 5. Bulletin of the Florida State Museum, Biological Sciences 15:163–266.

Bryant, L. J. 1989. Non-dinosaurian lower vertebrates across the Cretaceous-Tertiary boundary in Northeastern Montana. University of California Publications in Geological Sciences 134:1–107.

Buffetaut, E. 1998. First evidence of enantiornithine birds from the Upper Cretaceous of Europe: postcranial bones from Cruzy (Hérault, Southern France). Oryctos 1:131–136.

Chatterjee, S. 1997. The Rise of Birds. Johns Hopkins University Press, Baltimore, 312 pp.

Chiappe, L. M. 1991. Cretaceous avian remains from Patagonia shed new light on the early radiation of birds. Alcheringa 15(3–4):333–338.

———. 1992. Enantiornithine tarsometatarsi and the avian affinity of the Late Cretaceous Avisauridae. Journal of Vertebrate Paleontology 12(3):344–350.

———. 1993. Enantiornithine (Aves) tarsometatarsi from the Cretaceous Lecho Formation of northwestern Argentina. American Museum Novitates 3083:1–27.

———. 1995a. The first 85 million years of avian evolution. Nature 378:349–355.

———. 1995b. The phylogenetic position of the Cretaceous birds of Argentina: Enantiornithes and Patagopteryx deferrariisi; pp. 55–63 in D. S. Peters (ed.), Proceedings of the 3rd Symposium of the Society of Avian Paleontology and Evolution. Courier Forschungsinstitut Senckenberg 181, Frankfurt am Main.

———. 1996a. Late Cretaceous birds of southern South America: anatomy and systematics of Enantiornithes and Patagopteryx deferrariisi; pp. 203–244 in G. Arratia (ed.), Contributions of Southern South America to Vertebrate Paleontology, Münchner Geowissenschaftliche Abhandlungen, Reihe A, Geologie und Paläontologie 30. Verlag Dr. Friedrich Pfeil, Munich.

———. 1996b. Early avian evolution in the southern hemisphere: fossil record of birds in the Mesozoic of Gondwana. Memoirs of the Queensland Museum 39:533–556.

———. 1997. Aves; pp. 32–38 in P. J. Currie and K. Padian (eds.), The Encyclopedia of Dinosaurs. Academic Press, New York.

———. 1999. Early avian evolution: roundtable report. International Meeting of the Society of Avian Paleontology and Evolution. Washington, 1996. Smithsonian Contributions to Paleobiology 88:335–340.

———. 2001. Phylogenetic relationships among basal birds; pp. 125–139 in J. Gauthier and L. F. Gall (eds.), New Perspectives on the Origin and Early Evolution of Birds: Proceedings of the International Symposium in Honor of John H. Ostrom. Peabody Museum of Natural History, New Haven.

Chiappe, L. M., and J. O. Calvo. 1994. Neuquenornis volans, a new Enantiornithes (Aves) from the Upper Cretaceous of Patagonia (Argentina). Journal of Vertebrate Paleontology 14(2):230–246.

Chiappe, L. M., M. A. Norell, and J. M. Clark. 1996. Phylogenetic position of Mononykus from the Late Cretaceous of the Gobi Desert. Memoirs of the Queensland Museum 39(3):557–582.

———. 1998. The skull of a new relative of the stem-group bird Mononykus. Nature 392:275–278.

Chiappe, L. M., Ji S., Ji Q., and M. A. Norell. 1999. Anatomy and systematics of the Confuciusornithidae (Aves) from the Late Mesozoic of northeastern China. Bulletin of the American Museum Novitates 242:1–89.

Chiappe, L. M., M. A. Norell, and J. M. Clark. 2001. A new skull of Gobipteryx minuta (Aves: Enantionithes) from the Cretacious of the Gobi Desert. American Museum Novitates 3346:1–15.

Chiappe, L. M., J. P. Lamb Jr., and P. G. P. Ericson. 2002. An enantiornithine bird from the marine Upper Cretaceous of Alabama. Journal of Vertebrate Paleontology 22(1):169–173.

Chinsamy, A., L. M. Chiappe, and P. Dodson. 1994. Growth rings in Mesozoic birds. Nature 368:196–197.

———. 1995. Mesozoic avian bone microstructure: physiological implications. Paleobiology 21(4):561–574.

Chinsamy, A., and A. Elzanowski. 2001. Evolution of growth pattern in birds. Nature 412:402–403.

Cracraft, J. 1986. The origin and early diversification of birds. Paleobiology 12(4):383–399.

Currie, P., and K. Padian. 1997. Encyclopedia of Dinosaurs. Academic Press, San Diego, 869 pp.

Currie, P., and Peng J.-H. 1993. A juvenile specimen of Saurornithoides mongoliensis from the Upper Cretaceous of northern China. Canadian Journal of Earth Sciences 30(10–11):2224–2230.

Dashzeveg D., M. J. Novacek, M. A. Norell, J. M. Clark, L. M. Chiappe, A. Davidson, M. C. McKenna, L. Dingus, C. Swisher, and Perle A. 1995. Extraordinary preservation in a new vertebrate assemblage from the Late Cretaceous of Mongolia. Nature 374:446–449.

Dettmann, M. E., R. E. Molnar, J. G. Douglas, D. Burger, C. Fielding, H. T. Clifford, J. Francis, J. Jell, T. Rich, M. Wade, P. Vickers-Rich, N. Pledge, A. Kemp, and A. Rozefelds. 1992. Australian Cretaceous terrestrial faunas and floras: biostratigraphic and biogeographic implications. Cretaceous Research 13:207–262.

Elzanowski, A. 1974. Preliminary note on the paleognathous birds from the Upper Cretaceous of Mongolia. Paleontologica Polonica 30:103–109.

———. 1977. Skulls of Gobipteryx (Aves) from the Upper Cretaceous of Mongolia. Palaeontologica Polonica 37:153–165.

———. 1981. Embryonic bird skeletons from the Late Cretaceous of Mongolia. Palaeontologica Polonica 42:147–176.

———. 1995. Cretaceous birds and avian phylogeny. Courier Forschungsinstitut Senckenberg 181:37–53.

Elzanowski, A., and P. Wellnhofer. 1996. Cranial morphology of Archaeopteryx: evidence from the seventh skeleton. Journal of Vertebrate Paleontology 16(1):81–94.

Feduccia, A. 1996. The Origin and Evolution of Birds. Yale University Press, New Haven, 420 pp.

Feduccia, A., and L. Martin. 1996. Jurassic Urvogels and the myth of the feathered dinosaurs; pp. 185–191 in M. Morales (ed.), The Continental Jurassic. Bulletin of the Museum of Northern Arizona 60. Museum of Northern Arizona, Flagstaff.

Forster, C. A., L. M. Chiappe, D. W. Krause, and S. D. Sampson. 1996. The first Cretaceous bird from Madagascar. Nature 382:532–534.

Forster, C. A., S. D. Sampson, L. M. Chiappe, and D. W. Krause. 1998a. The theropodan ancestry of birds: new evidence from the Late Cretaceous of Madagascar. Science 279:1915–1919.

———. 1998b. Correction. Science 280:185.

Fregenal-Martínez, M. A. 1991. El sistema lacustre de Las Hoyas (Cretácico inferior, Serranía de Cuenca): estratigrafía y sedimentología. Publicaciones de la Excelentísima Diputación Provincial de Cuenca, Paleontología 1:1–181.

Harrison, C. J. O., and C. A. Walker. 1973. Wyleyia: a new bird humerus from the Lower Cretaceous of England. Palaeontology 16:721–728.

Horovitz, I. 1999. A report on "One Day Symposium on Numerical Cladistics." Cladistics 15:177–182.

Hotton, C. L. 1988. Palynology of the Cretaceous-Tertiary boundary in central Montana, U.S.A., and its implications for extraterrestrial impact, Part I. Ph.D. dissertation, University of California, Davis, 269 pp.

Hou L. 1994. A late Mesozoic bird from Inner Mongolia. Vertebrata PalAsiatica 32(4):258–266.

———. 1997. Mesozoic Birds of China. Taiwan Provincial Feng Huang Ku Bird Park. Nan Tou, Taiwan, 228 pp. [Chinese]

Hou L. and Chen P.-J. 1999. Liaoxiornis delicatus gen. et sp. nov., the smallest Mesozoic bird. Chinese Science Bulletin 44(9):834–838.

Hou L., Zhou Z., L. D. Martin, and A. Feduccia. 1995. A beaked bird from the Jurassic of China. Nature 377:616–618.

Hou L., L. D. Martin, Zhou Z., and A. Feduccia. 1996. Early adaptive radiation of birds: evidence from fossils from northeastern China. Science 274:1164–1167.

———. 1999. Archaeopteryx to opposite birds—missing link from the Mesozoic of China. Vertebrata PalAsiatica 4:88–95.

Hutchison, J. H. 1993. Avisaurus: a "dinosaur" grows wings. Journal of Vertebrate Paleontology 13(3):43A.

Jerzykiewicz, T., and D. Russell. 1991. Late Mesozoic stratigraphy and vertebrates of the Gobi Basin. Cretaceous Research 12:345–377.

Ji Q. and Ji S.-A. 1999. A new genus of the Mesozoic birds from Lingyuan, Liaoning, China. Chinese Geology 262:45–48. [Chinese]

Krause, D. W., J. H. Hartman, and N. A. Wells. 1997. Late Cretaceous vertebrates from Madagascar: implications for biotic change in deep time; pp. 3–43 in B. Patterson and S. Goodman (eds.), Natural and Human-Induced Change in Madagascar. Smithsonian Institution Press, Washington, D.C.

Kurochkin, E. N. 1985. A true carinate bird from the Lower Cretaceous deposits in Mongolia and other evidence of early Cretaceous birds in Asia. Cretaceous Research 6:271–278.

———. 1995. Synopsis of Mesozoic birds and early evolution of Class Aves. Archaeopteryx 13:47–66.

———. 1996. A new enantiornithid of the Mongolian Late Cretaceous, and a general appraisal of the infraclass Enantiornithes (Aves). Russian Academy of Sciences, Special Issue:1–50.

———. 1999. A new large enantiornithid from the Upper Cretaceous of Mongolia (Aves, Enantiornithes). Proceedings of the Zoological Institute, Russian Academy of Sciences, St. Petersburg, 277:130–141. [Russian]

———. 2000. Mesozoic birds of Mongolia and the former USSR; pp. 533–559 in M. J. Benton, M. A. Shishkin, D. M. Unwin, and E. N. Kurochkin (eds.), The Age of Dinosaurs in Russia and Mongolia. Cambridge University Press, Cambridge.

Kurochkin, E. N., and R. E. Molnar. 1997. New material of enantiornithine birds from the Early Cretaceous of Australia. Alcheringa 21:291–297.

Lacasa-Ruiz, A. 1989. Nuevo género de ave fósil del yacimiento neocomiense del Montsec (provincia de Lérida, España). Estudios Geológicos 45:417–425.

Lamb, J. P., Jr., L. M. Chiappe, and P. G. P. Ericson. 1993. A marine enantiornithine from the Cretaceous of Alabama. Journal of Vertebrate Paleontology 13(3):45A.

Lillegraven, J. A., and M. C. McKenna. 1986. Fossil mammals from the "Mesaverde" Formation (Late Cretaceous, Judithian) of the Bighorn and Wind River Basins, Wyoming, with definitions of Late Cretaceous North American land-mammal "ages." American Museum Novitates 2840:1–68.

Loope, D. B., L. Dingus, C. C. Swisher III, and C. Minjin. 1998. Life and death in a Late Cretaceous dune field, Nemegt Basin, Mongolia. Geology 26(1):27–30.

Marsh, O. C. 1892. Notice of new reptiles from the Laramie Formation. American Journal of Science (ser. 3) 43:449–453.

Martin, L. D. 1983. The origin and early radiation of birds; pp. 291–338 in A. H. Brush and G. A. Clark Jr. (eds.), Perspectives in Ornithology. Cambridge University Press, New York.

———. 1987. The beginning of the modern avian radiation; pp. 9–19 in C. Mourer-Chauviré (ed.), L'évolution des Oiseaux d'Après le Témoignage des Fossiles. Documents des Laboratoires de Géologie 99, Lyon. Université Claude-Bernard, Villeurbanne Cedex.

———. 1991. Mesozoic birds and the origin of birds; pp. 485–539 in H.-P. Schultze and L. Trueb (eds.), Origins of the Higher Groups of Tetrapods: Controversy and Consensus. Comstock Publishing Associates, Ithaca, N.Y.

———. 1995. The Enantiornithes: terrestrial birds of the Cretaceous. Courier Forschungsinstitut Senckenberg 181:23–36.

Martin, L. D., and Zhou Z. 1997. Archaeopteryx-like skull in enantiornithine bird. Nature 389:556.

Martínez-Delclòs, X. 1995. Montsec and Montral-Alcover, Two Konservat-Lagerstätten, Catalonia, Spain. Pre-Symposium field-trip. International Symposium on Lithographic Limestones. Institut d'Estudis Ilerdencs, Lleida, Catalonia, 97 pp.

McKitrick, M. C. 1992. Phylogenetic analysis of avian parental care. Auk 109(4):828–846.

Meléndez, M. N. 1995. Las Hoyas, a lacustrine Konservat-Lagerstätte, Cuenca, Spain. International Symposium on Lithographic Limestones. Field trip guide book. Institut d'Estudis Ilerdencs, Lleida, Catalonia, 89 pp.

Mikhailov, K. E. 1991. Classification of fossil eggshells of amniotic vertebrates. Palaeontologica Polonica 36(2):193–238.

———. 1996. The eggs of birds in the Upper Cretaceous of Mongolia. Paleontologichesky Zhurnal 1:19–121. [Russian]

Molnar, R. E. 1986. An enantiornithine bird from the Lower Cretaceous of Queensland, Australia. Nature 322:736–738.

Nessov, L. A. 1984. Pterosaurs and birds of the Late Cretaceous of Central Asia. Paleontologicheskii Zhournal 1:47–57. [Russian]

———. 1990. Small ichthyornithiform bird and other bird remains from Bissekty Formation (Upper Cretaceous) of central Kizylkum Desert. USSR Academy of Sciences. Proceedings of the Zoological Institute 210:59–62. [Russian]

Nessov, L. A., and A. A. Jarkov. 1989. New Cretaceous-Paleogene birds of the USSR and some remarks on the origin and evolution of the Class Aves. USSR Academy of Sciences. Proceedings of the Zoological Institute 197:78–97. [Russian]

Nessov, L. A., and A. V. Panteleyev. 1993. About similarities of the Late Cretaceous ornithofaunas of South America and western Asia. Proceedings of the Zoological Institute, Russian Academy of Sciences, St. Petersburg 252:84–94. [Russian]

Norell, M. A., and P. J. Makovicky. 1997. Important features of the dromaeosaur skeleton: information from a new specimen. American Museum Novitates 3215:1–28.

Novas, F. E., and P. Puerta. 1997. New evidence concerning avian origins from the Late Cretaceous of Patagonia. Nature 387:390–392.

Olson, S. L. 1985. The fossil record of birds; pp. 79–238 in D. S. Farner, J. R. King, and K. C. Parkes (eds.), Avian Biology, Vol. 8. Academic Press, New York.

Ostrom, J. H. 1969. Osteology of Deinonychus antirrhopus, an unusual theropod dinosaur from the Lower Cretaceous of Montana. Bulletin of the Peabody Museum of Natural History 30:1–165.

Padian, K., and L. M. Chiappe. 1997. Bird origins; pp. 71–79 in P. Currie and K. Padian (eds.), The Encyclopedia of Dinosaurs. Academic Press, New York.

———. 1998. The origin and early evolution of birds. Biological Reviews 73:1–42.

Padian, K., A. J. de Ricqulès, and J. R. Horner. 2001. Dinosaurian growth rates and bird origins. Nature 412:405–408.

Prosser, C. L. 1973. Comparative Animal Physiology. Saunders, New York, 1011 pp.

Rowe, T. 1986. Homology and evolution of the deep dorsal thigh musculature in birds and other Reptilia. Journal of Morphology 189:327–346.

Sanz, J. L., and J. F. Bonaparte. 1992. A new order of birds (Class Aves) from the Lower Cretaceous of Spain; pp. 39–49 in K. E. Campbell (ed.), Papers in Avian Paleontology, Honoring Pierce Brodkorb. Science Series 36. Natural History Museum of Los Angeles County, Los Angeles.

Sanz, J. L., and A. D. Buscalioni. 1992. A new bird from the Early Cretaceous of Las Hoyas, Spain, and the early radiation of birds. Paleontology 35(4):829–845.

Sanz, J. L., L. M. Chiappe, and A. D. Buscalioni. 1995. The osteology of Concornis lacustris (Aves: Enantiornithes) from the Lower Cretaceous of Spain, and a reexamination of its phylogenetic relationships. American Museum Novitates 3133:1–23.

Sanz, J. L., L. M. Chiappe, B. P. Pérez-Moreno, A. D. Buscalioni, J. Moratalla, F. Ortega, and F. J. Poyato-Ariza. 1996. A new Lower Cretaceous bird from Spain: implications for the evolution of flight. Nature 382:442–445.

Sanz, J. L., L. M. Chiappe, B. P. Pérez-Moreno, J. Moratalla, F. Hernández-Carrasquilla, A. D. Buscalioni, F. Ortega, F. J. Poyato-Ariza, D. Rasskin-Gutman, and X. Martínez-Delclòs. 1997. A nestling bird from the Early Cretaceous of Spain: implications for avian skull and neck evolution. Science 276:1543–1546.

Sanz, J. L., L. M. Chiappe, Y. Fernández-Jalvo, F. Ortega, B. Sánchez-Chillón, F. J. Poyato-Ariza, and B. P. Pérez-Moreno. 2001. An Early Cretaceous pellet. Nature 409:998–1000.

Schweitzer, M. H., F. D. Jackson, L. M. Chiappe, J. G. Schmitt, J. O. Calvo, and D. E. Rubilar. 2002. Late cretaceous avian eggs with embryos from Argentina. Journal of Vertebrate Paleontology 22(1):191–195.

Sereno, P. 1998. A rationale for phylogenetic definitions, with application to the higher-level taxonomy of Dinosauria. Neues Jahrbuch für Geologie und Paläontologie 210(1):41–83.

———. 2000. Iberomesornis romerali (Aves, Ornithothoraces) reevaluated as an Early Cretaceous enantiornithine. Neues Jahrbuch für Geologie und Paläontologie 215(3):365–395.

Sereno, P., and Rao C. 1992. Early evolution of avian flight and perching: new evidence from Lower Cretaceous of China. Science 255:845–848.

Starck, J. M. 1993. Evolution of avian ontogenies; pp. 275–366; in D. M. Power (ed.), Current Ornithology, Vol. 10. Plenum, New York.

Steadman, D. W. 1983. Commentary [on Martin, 1983]; pp. 338–344 in A. H. Bush and G. A. Clark Jr. (eds.), Perspectives in Ornithology. Cambridge University Press, New York.

Stidham, T. A., and J. H. Hutchison. 2001. The North American avisaurids (Aves: Enantiornithes): new data on biostratigraphy and biogeography. Publicación Especial de la Asociación Paleontológica Argentina 7:175–177.

Suzuki, S., M. Watabe, and Tsogtbaatar K. 1999. A new enanti-ornithine bird from the Upper Cretaceous Djadokhta Formation of Gobi Desert, Mongolia. Journal of Vertebrate Paleontology 19(Supplement to 3):79A–80A.

Thulborn, R. 1984. The avian relationships of *Archaeopteryx* and the origin of birds. Zoological Journal of the Linnean Society, London 82:119–158.

Varricchio, D. J., and L. M. Chiappe. 1995. A new bird from the Cretaceous Two Medicine Formation of Montana. Journal of Vertebrate Paleontology 15(1):201–204.

Walker, C. A. 1981. New subclass of birds from the Cretaceous of South America. Nature 292:51–53.

Weishampel, D. B., P. Dodson, and H. Osmólska. 1990. The Dinosauria. University of California Press, Berkeley, 733 pp.

Wellnhofer, P. 1974. Das fünfte Skelettexemplar von *Archaeopteryx*. Palaeontographica A 147:169–216.

———. 1985. Remarks on the digit and pubis problem of *Archaeopteryx*; pp. 113–122 *in* M. K. Hecht, J. H. Ostrom, G. Viohl, and P. Wellnhofer (eds.), The Beginnings of Birds. Freunde des Jura-Museums, Eichstätt.

Witmer, L. M., and L. D. Martin. 1987. The primitive features of the avian palate, with special reference to Mesozoic birds; pp. 21–40 *in* C. Mourer-Chauviré (coord.), L'évolution des oiseaux d'après le témoignage des fossiles. Documents des Laboratoires de Géologie 99, Lyon. Université Claude-Bernard, Villeurbanne Cedex.

Zhang F. and Zhou Z. 2000. A primitive enantiornithine bird and the origin of feathers. Science 290:1955–1959.

Zhang F., Zhou Z., Hou L., and Gu G. 2000. Early diversification of birds—evidence from a new opposite bird. Chinese Science Bulletin 45(24):2650–2657.

Zhou Z. 1995a. The discovery of Early Cretaceous birds in China. Courier Forschungsinstitut Senckenberg 181:9–22.

———. 1995b. Discovery of a new enantiornithine bird from the Early Cretaceous of Liaoning, China. Vertebrata PalAsiatica 33(2):99–113.

Zhou Z., Jin F., and Zhang J. 1992. Preliminary report on a Mesozoic bird from Liaoning, China. Chinese Science Bulletin 37(16):1365–1368.

12

Vorona berivotrensis, a Primitive Bird
from the Late Cretaceous of Madagascar

CATHERINE A. FORSTER, LUIS M. CHIAPPE, DAVID W. KRAUSE, AND SCOTT D. SAMPSON

Vorona berivotrensis was discovered in the Upper Cretaceous (Maastrichtian) Maevarano Formation, Mahajanga Basin, northwestern Madagascar (Rogers et al., 2000) in the austral winter of 1995 by a joint State University of New York at Stony Brook–Université d'Antananarivo expedition. Three specimens assigned to *Vorona* and previously described by Forster et al. (1996a) were recovered from a productive quarry (site MAD 93-18) near the village of Berivotra. This quarry has also produced a partial skeleton of the primitive bird *Rahonavis ostromi* (Forster et al., 1998), teeth of the large abelisaurid theropod *Majungatholus atopus,* and a nearly complete skeleton and partial skull of the titanosaurid *Rapetosaurus krausei,* as well as isolated elements of the small abelisaurid *Masiakasaurus knopfleri,* crocodilians, fish, frogs, snakes, turtles, and at least three additional taxa of birds (Forster et al., 1996a,b; Krause et al., 1997a, 1999; Sampson et al., 2001; Curry Rogers and Forster, 2001; Fig. 12.1). Interestingly, the faunal composition at site MAD 93-18 is nearly identical to that of the Late Cretaceous El Brete site (Lecho Formation), Salta Province, Argentina, where a quarry in fluvial sandstone produced four taxa of birds, associated titanosaurid sauropod remains, teeth of a large theropod, and scattered elements of a small theropod (Bonaparte and Powell, 1980; Chiappe, 1993).

Vorona was the first pre–Late Pleistocene avian described from Madagascar. The record of Mesozoic birds in Gondwana is very poor, and no Jurassic birds have yet been discovered (Chiappe, 1995, 1996b; Fig. 12.2). For the Cretaceous there is a growing number of taxa known from South America, particularly Argentina (e.g., Walker, 1981; Chiappe, 1991, 1993; Alvarenga and Bonaparte, 1992; Chiappe and Calvo, 1994; Novas, 1996). However, there are only three occurrences of skeletal remains of Cretaceous birds outside South America: two from the Antarctic Peninsula (Chatterjee,

1989; Noriega and Tambussi, 1995; see also Hope, Chapter 15 in this volume) and one from eastern Australia (Molnar, 1986). Several footprints from a single locality in the Cretaceous of Morocco (Ambroggi and Lapparent, 1954; see also Lockley and Rainforth, Chapter 17 in this volume) are the only record of Mesozoic birds from a large part of Gondwana that includes Africa, Madagascar, the Indian subcontinent, and mainland Antarctica. *Vorona* thus marks the first Mesozoic record of avian skeletal remains from this large area of the world.

The extant Malagasy avifauna is striking in its low diversity (only 256 species) and high endemism (53% of species) (Langrand 1990), which are odd characteristics in view of the high habitat diversity on the island and the high vagility of birds relative to most other vertebrates. Whether or not Madagascar obtained some of its modern and recently extinct avifauna prior to separation from other major landmasses has been the subject of some considerable speculation and controversy (Andrews, 1894; Lamberton, 1934; Rand, 1936; Darlington, 1957; Battistini, 1965; Moreau, 1966; Dorst, 1972; Mahé, 1972; Sauer, 1972; Cracraft, 1973, 1974; Feduccia, 1980; Keith, 1980; Briggs, 1995; Quammen, 1996; Chatterjee, 1997). Keith (1980:106), for instance, posed the question in the following manner: "Are any of the birds found on Madagascar today descendants of those that remained behind after its separation, or did all of these become extinct, leaving Madagascar to be populated by birds that later flew in from overseas?" He concluded that "the fossil record is so poor that unless further discoveries are made we may never have a proper answer to this question."

The published record of fossil birds from Madagascar is indeed poor. Outside of the avifauna recovered from the Maevarano Formation, it is restricted to the Holocene (MacPhee et al., 1985). This, in turn, is owing in no small measure to the overwhelming predominance of marine

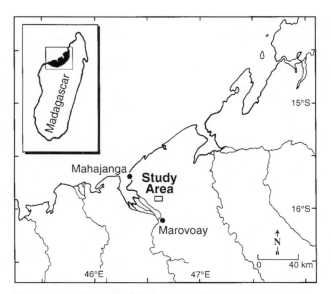

Figure 12.1. Map showing the general location of quarry site MAD 93-18 (marked "Study Area") in the Upper Cretaceous (Maastrichtian) Maevarano Formation. Inset map shows the location of the Mahajanga Basin (in black) in northwestern Madagascar. At least four taxa of fossil birds, including *V. berivotrensis* (described here) and *R. ostromi* (Forster et al., 1998), were recovered from this site. Exact locality information is on file in the Department of Geology, Field Museum of Natural History, Chicago.

strata throughout the Cenozoic (Besairie, 1972). Included among extinct forms are the giant, flightless elephant-birds (Aepyornithidae), which according to some workers (e.g., Rothschild, 1911; Lambrecht, 1929, 1933; Sauer and Rothe, 1972; Sauer, 1976) are represented outside Madagascar as well (Egypt, Algeria, India, Mongolia, Canary Islands, Turkey). Others (e.g., Feduccia, 1980, 1996; Olson, 1985; Rasmussen et al., 1987), however, are skeptical about these records. In any case, the biogeographic origins of the Malagasy avifauna remain enigmatic and have previously been interpreted largely through the relationships of modern taxa. The affinities of most species are generally regarded to be with Africa, but a few species indicate Asian or even Australasian connections (e.g., Rand, 1936; Dorst, 1972; Cracraft, 1973; Keith, 1980). As stressed by Walker (1978:208) with regard to the origin of the Malagasy mammalian fauna, the "one major piece of evidence missing is . . . fossils of Cretaceous, Paleocene or Eocene age." The discovery of *Vorona, Rahonavis,* and other taxa in the Late Cretaceous of Madagascar provides the first pre–Late Pleistocene fossil evidence with which to address questions concerning the biogeographic origins of the extant and recently extinct Malagasy avifauna.

The following institutional abbreviations are used in this chapter: FMNH, Field Museum of Natural History, Chicago, Illinois, United States; UA, Université d'Antananarivo, Service de Paléontologie, Antananarivo, Madagascar.

Geological and Paleogeographic Setting

The Maevarano Formation is the uppermost nonmarine unit of Cretaceous age in the Mahajanga Basin and has recently been considered Maastrichtian in age (Rogers and Hartman, 1998; Rogers at al., 2000). The Maevarano Formation is divided into three distinct members: Masarobe, Anembalemba, and Miadana (Rogers et al., 2000). The quarry at MAD 93-18, like most of the fossil-bearing sites in the Maevarano Formation, occurs in the Anembalemba Member. This member averages 10–15 meters in thickness and is composed of two distinctive, alternating sandstone facies (facies 1 and facies 2). Facies 1 consists of light gray to white, fine- to coarse-grained, poorly sorted sandstone with an abundant clay matrix that displays small- to medium-scale tabular and trough cross stratification. Facies 2 consists of a massive, fine- to coarse-grained, poorly sorted, clay-rich sandstone that is olive green in color (Rogers et al., 2000). Specimens of *Vorona* described here were recovered from both the top of a facies 1 layer and the overlying facies 2 layer.

The importance of *Vorona* to the Mesozoic avian record is enhanced by its paleogeographic location in eastern Gondwana. As discerned from a wealth of geophysical data, the Cretaceous represents the most active period in the breakup of Gondwana. The sequence and pattern of breakup, however, are poorly tested paleontologically. The current geophysical model for plate tectonic evolution of the western Indian Ocean holds that Madagascar has long been an island (summarized in Storey, 1995; Krause and Hartman, 1996; Krause et al., 1997b, 1999). With the Indian subcontinent attached to its eastern coast, Madagascar separated from Africa in the Late Jurassic, approximately 160–150 Ma, considerably earlier than generally regarded in previous biogeographic scenarios involving the island's avifauna (e.g., Sauer, 1972; Cracraft, 1973; Keith, 1980). Indo-Madagascar attained its current position relative to the east coast of Africa in the Early Cretaceous, approximately 130–125 Ma, and, according to most geophysical reconstructions, rifted from Antarcto-Australia at about the same time. Hay et al. (1999), however, posited a lingering connection between Indo-Madagascar and western Antarctica across the Kerguelen Plateau that lasted well into the Late Cretaceous, possibly as late as 80 Ma. Separation of Madagascar from the Indian subcontinent did not occur until late in the Late Cretaceous, at approximately 88 Ma (Storey et al., 1995). Thus, it is only since the Late Cretaceous that Africa has been the nearest major landmass to Madagascar, and it appears more likely that the Late Cretaceous terrestrial and freshwater vertebrate fauna of Madagascar was more similar to that of the Indian subcontinent, and, more distantly, that of Antarctica, Australia, and South America, than that of Africa (Krause et al., 1997b, 1999; Sampson et al., 1998).

TABLE 12.1
Measurements (in mm) of specimens of *Vorona berivotrensis*

	UA 8651	FMNH PA 715	FMNH PA 717
Femur			
Total length	93.7	—	94.1
Craniocaudal width trochanteric crest	10.9	—	—
Greatest diameter at midshaft	8.4	—	8.1
Mediolateral width distal condyles	15.3	—	15.5
Fibula			
Length of preserved portion	—	76.0	—
Craniocaudal length proximal end	—	9.3	—
Mediolateral width proximal end	—	3.9	—
Tibiotarsus			
Total length	—	165.8	
Height astragalar ascending process	33.3	32.5	—
Mediolateral width proximal end	—	11.8	—
Craniocaudal length proximal end	—	16.0	—
Mediolateral diameter at midshaft	5.9	6.1	—
Craniocaudal diameter at midshaft	7.1	7.0	—
Mediolateral width distal condyles	12.3	12.4	—
Craniocaudal width distal condyles	11.8	12.0	—
Tarsometatarsus			
Total length	60.9	—	—
Mediolateral width proximal end	12.9	—	—
Dorsoplantar width proximal end	8.0	—	—
Mediolateral width at midshaft	9.0	—	—
Dorsoplantar depth at midshaft	4.2	—	—
Mediolateral width across trochleae	13.7	—	—
MT II: total length	3.8	—	—
Mediolateral width trochlea	4.2	—	—
Dorsoplantar depth trochlea	4.7	—	—
MT III: total length	61.0	—	—
Mediolateral width trochlea	5.6	—	—
Dorsoplantar depth trochlea	6.3	—	—
MT IV: total length	58.4	—	—
Mediolateral width trochlea	3.9	—	—
Dorsoplantar depth trochlea	5.7	—	—
MT V: total length	16.4	—	—
Length of tarsometatarsus/tibiotarsus (UA 8651/FMNH PA 715)	0.37		

Systematic Paleontology

Taxonomic Hierarchy

Aves Linnaeus, 1758
  Ornithothoraces Chiappe and Calvo, 1994
    Vorona Forster et al., 1996a
     V. berivotrensis Forster et al., 1996a

Holotype—UA 8651, a distal left tibiotarsus found in articulation with a complete tarsometatarsus.

Referred specimens—FMNH PA 715, associated right tibiotarsus, fibula, and femur; FMNH PA 717, left femur.

Locality and horizon—Upper Cretaceous (Maastrichtian) Maevarano Formation, Mahajanga Basin, near the village of Berivotra, northwestern Madagascar. Specimens were collected from quarry site MAD 93-18. Locality coordinates are on file at FMNH and UA.

Diagnosis—Possesses short, blunt cnemial crest on tibiotarsus; irregular low ridge on medial surface of proximal tibiotarsus; narrow but deep notch on proximodorsal tarsometatarsus between metatarsals (MTs) II and III; expanded vascular groove proximal to distal foramen on tarsometatarsus.

Anatomy

The three specimens assigned to *Vorona* together comprise nearly complete left and right hindlimbs. The tibiotarsus

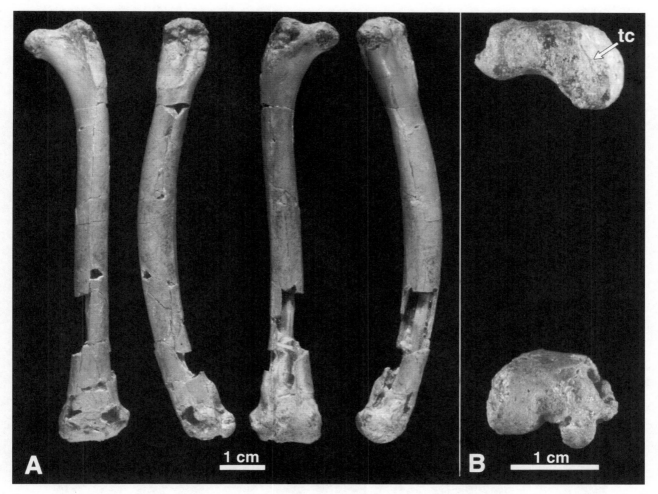

Figure 12.2. Right femur of referred specimen FMNH PA 715: A, cranial, lateral, caudal, and medial views (left to right); B, proximal view, femoral head to the left (top), and distal view, cranial surface facing up (bottom). Abbreviation: tc, trochanteric crest.

and tarsometatarsus of UA 8651 were found in direct articulation in the lower facies-1 level. The tibiotarsus, fibula, and femur of FMNH PA 715 were loosely articulated in the base of the overlying facies-2 level; the fibula was lying next to and paralleling the tibia, and the proximal portion of the femur was lying next to the proximal tibiotarsus. The shaft and distal portion of the femur were recovered a few centimeters south of the rest of the specimen.

At the time of its original description (Forster et al., 1996a), only a small portion of the proximal right femur (FMNH PA 715) could be assigned with confidence to *Vorona* (Fig. 12.3). Since then, additional preparation revealed contacts between this femoral fragment and the shaft and distal right femur of a then unidentified avian. This discovery has allowed us to reconstruct the now nearly complete right femur and to assign another specimen from the quarry, a nearly complete left femur (FMNH PA 717), to *Vorona*. Additionally, a more proximal segment of the holotype tibiotarsus (UA 8651) was recovered during prepara-

tion of quarry material. Measurements of all limb elements are given in Table 12.1.

Pelvic Limb

Femur. The neck of the femur is short, robust, and weakly constricted (Figs. 12.2, 12.3). The outline of the femoral head is nearly round in proximomedial view. The flattened proximomedial surface of the femoral head bears a broad, shallow depression for the attachment of the capital ligament. A fossa for the capital ligament is also present in Enantiornithes and Ornithurae but is absent in nonavian maniraptorans, *Unenlagia* (Novas and Puerta, 1997), *Archaeopteryx*, and *Mononykus* (Perle et al., 1994). There is a distinct trochanteric crest projecting slightly above the level of the femoral head. The lateral surface of the proximal femur is slightly depressed, but, unlike the condition seen in Enantiornithes, *Archaeopteryx*, and some nonavian maniraptorans, it bears no evidence of a posterior trochanter.

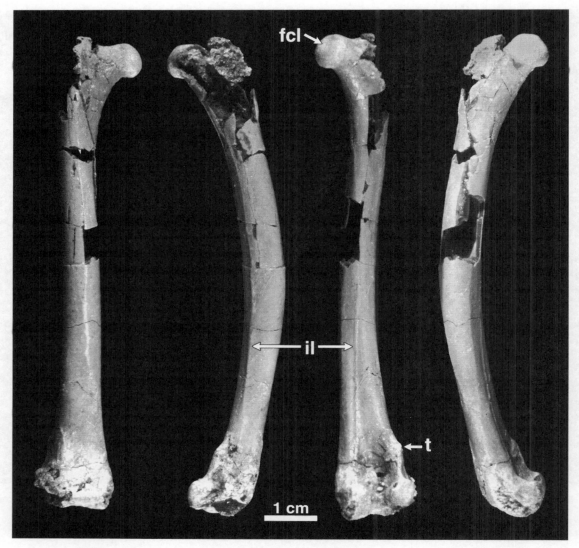

Figure 12.3. Left femur of referred specimen FMNH PA 717 in (left to right) cranial, lateral, caudal, and medial views. Abbreviations: fcl, fossa for capital ligament; il, intermuscular line; t, tubercle on lateral supracondylar ridge.

The shaft of the femur is cranially bowed and lacks a fourth trochanter. An intermuscular line courses from the trochanteric crest down to the medial side of the distal femur, crossing the cranial aspect of the femoral shaft. A caudal intermuscular line runs from the medial condyle proximally across approximately three-fourths of the femur. A third intermuscular line runs the length of the entire caudolateral edge of the shaft.

The distal femur is flat cranially and lacks a patellar groove. A well-developed patellar groove is a synapomorphy of ornithurine birds (Chiappe, 1996a). On the caudal aspect, the triangular popliteal fossa is bounded distally by an intercondylar bridge. This condition is comparable to that of Enantiornithes, *Patagopteryx,* and Ornithurae. The lateral margin of the popliteal fossa is developed into a supracondylar ridge that projects caudally beyond the level of the medial margin, as in Enantiornithes (Chiappe and Calvo, 1994; Sanz et al., 1995; Chiappe, 1996a). Approximately 0.5 cm above the lateral condyle this ridge swells into a tubercle. This tubercle does not occur in any known enantiornithine bird. The lateral supracondylar ridge, along with the lateral condyle, defines the medial boundary of a weak fibular trochlea. This trochlea, and the tibiofibular crest of the lateral condyle, appear less developed than in neornithines.

Tibiotarsus. The tibiotarsus is long, slender, and straight (Figs. 12.4, 12.5). The shaft is piriform in cross section, with an evenly rounded medial surface that narrows to the lateral margin. The tibiotarsus has a very blunt, broad, and robust cnemial crest. This single cnemial crest is homologous to the lateral cnemial crest of neornithines (Chiappe, 1996a). A single cnemial crest is also present in nonavian maniraptorans, *Archaeopteryx*, Alvarezsauridae, and *Patagopteryx*, while two distinct crests are typical of Ornithurae (Chiappe, 1996a). *Iberomesornis* (Sanz and Bonaparte, 1992) is specialized in having no distinct cnemial crest, whereas Enantiornithes (Walker, 1981) have a faint, craniomedially placed crest. The proximal articular surface of the tibiotarsus slopes slightly caudolaterally. Except for the rather bulbous, subcircular lateral articular facet, the tilted proximal articular surface is nearly planar. The lateral and medial articular facets are well developed and separated caudally by a broad notch; the medial articular facet projects slightly more caudally than the lateral one. Running down the medial surface of the proximal one-fifth of the tibiotarsus is a low, irregular ridge not found on any other known Mesozoic bird. For most of its length, this ridge defines the cranial boundary of an elongate rugose area, almost certainly an area of muscle attachment. The lateral surface of the proximal one-fourth of the tibiotarsus bears a prominent fibular crest, which projects more laterally at its distal end. A small vascular foramen pierces the shaft caudal to the distal fibular crest, as in neornithines.

The calcaneum and astragalus are firmly co-ossified but are only partially fused to the distal tibia. A well-defined line of contact on the caudal, medial, and lateral aspects of the tibia delimit the edges of the proximal tarsals. Although patterns of fusion clearly have an ontogenetic component, there nevertheless appears to be a phylogenetic signal present in the co-ossification of the proximal tarsals, as well as their incorporation into the tibia (e.g., McGowan, 1984, 1985). The two proximal tarsals are apparently fused to each other in some nonavian maniraptorans. For example, in juvenile specimens of the troodontid *Saurornithoides* there is no sign of a separate calcaneum, implying that fusion may have occurred early in ontogeny (Currie and Peng, 1993). These elements are also fused in the troodontid *Sinornithoides* (Russell and Dong, 1993). The calcaneum and astragalus are co-ossified in all known birds except *Archaeopteryx* (Wellnhofer, 1992), *Iberomesornis*, and *Alvarezsaurus* (Bonaparte, 1991; Novas, 1997). In nonavian maniraptorans, *Archaeopteryx*, Alvarezsauridae, and *Iberomesornis*, the proximal tarsals are not co-ossified to the tibia. However, the proximal tarsals are completely fused to the tibia in Enantiornithes, *Patagopteryx*, and Ornithurae (Chiappe, 1996a).

The astragalus, including the ascending process, is one-fifth the length of the tibia in *Vorona*, a proportion similar to that in other birds and nonavian maniraptorans. The tall, triangular ascending process is laterally placed and nearly fused to the tibia, although a faint suture can still be seen around its perimeter on the cranial aspect.

Martin et al. (1980) disputed the homology of the ascending process of nonavian theropods with that of *Archaeopteryx* on the basis of the lateral position of the ascending process of the latter and its interpretation as a non-

homologous structure called the pretibial bone. Additional embryological and paleontological data have shown the proposal of Martin et al. (1980) to be incorrect. McGowan (1985) demonstrated that the pretibial bone is indeed a modification of the astragalar ascending process and pointed out the occurrence of a lateralized ascending process in several nonavian theropods (e.g., *Allosaurus*), as well as in ratites and tinamous. Additionally, the morphology and position of the ascending process of *Vorona*, along with those of other recently found basal birds (e.g., *Rahonavis*), strongly support the hypothesis that the ascending processes of nonavian theropods and birds are homologous (contra Martin et al., 1980; Tarsitano and Hecht, 1980; Martin, 1983).

The medial condyle on the distal tibiotarsus is twice the width of the lateral one in cranial view. This condition resembles that of *Patagopteryx* and Enantiornithes but contrasts with that of Ornithurae, in which the condyles are subequal in width in cranial view. A narrow, deep intercondylar sulcus separates the condyles on their cranial aspect. This sulcus opens proximally into a small, deep fossa; the fossa and most of the sulcus are overhung by the median edges of both condyles. This condition also occurs in Enantiornithes and *Patagopteryx*. The lateral margin of the distal tibiotarsus (calcaneum) lacks any trace of a fibular articulation.

Fibula. The proximal end of the gracile fibula is transversely compressed with a slightly concave medial face and a convex lateral face (Fig. 12.4A, B). Although medially concave, it does not possess the prominent medial fossa typical of nonavian maniraptorans and other theropods (e.g., *Gallimimus, Deinonychus, Allosaurus*). Distally, the fibula narrows rapidly to a slender splint. Twenty-eight millimeters down the fibula (approximately two-thirds of the way down the fibular crest), and just proximal to the fibular spine, is a small, laterally directed tubercle for the insertion of M. iliofibularis. The lateral orientation of this tubercle, shared with *Mononykus* and *Patagopteryx*, has been interpreted as an intermediate condition between the craniolateral position of the tubercle in nonavian maniraptorans and the caudally or caudolaterally oriented tubercle of Ornithurae (Chiappe, 1996a).

Although incomplete distally, the fibula clearly did not reach the distal end of the tibiotarsus as it does in *Archaeopteryx, Alvarezsaurus*, and nonavian maniraptorans. This derived condition is confirmed by the lack of a fibular facet on the lateral (calcaneal) condyle of the distal tibiotarsus.

Tarsometatarsus. The tarsometatarsus is approximately one-third the length of the tibiotarsus (Fig. 12.6). MTs II, III, IV, and V are preserved, and MTs II, III, and IV are completely fused except along the midshaft of MTs III and IV

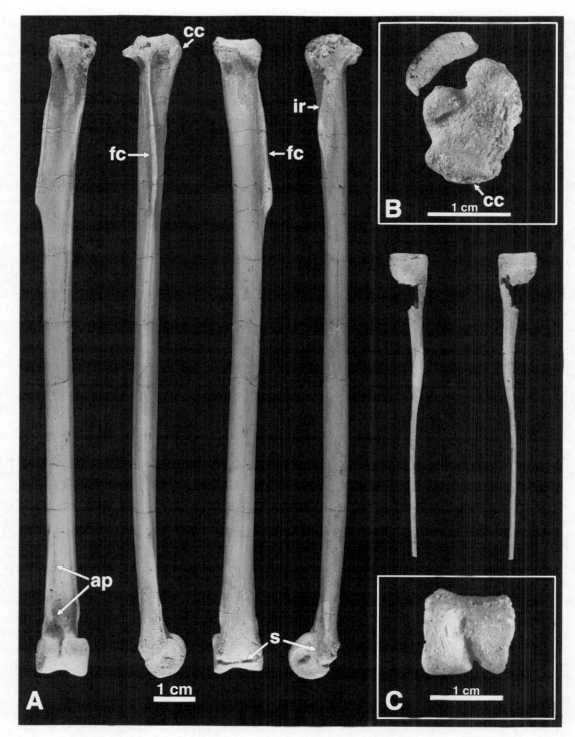

Figure 12.4. Referred specimen FMNH PA 715: A, right tibiotarsus in (left to right, on left) cranial, lateral, caudal, and medial views; and partial right fibula in (left to right, on right) lateral and medial views; B, articulated fibula and tibiotarsus in proximal view, cranial end facing down; C, tibiotarsus in distal view, caudal end facing up. Abbreviations: ap, ascending process of the astragalus; cc, cnemial crest; fc, fibular crest; ir, irregular ridge; s, suture between tibia and proximal tarsals.

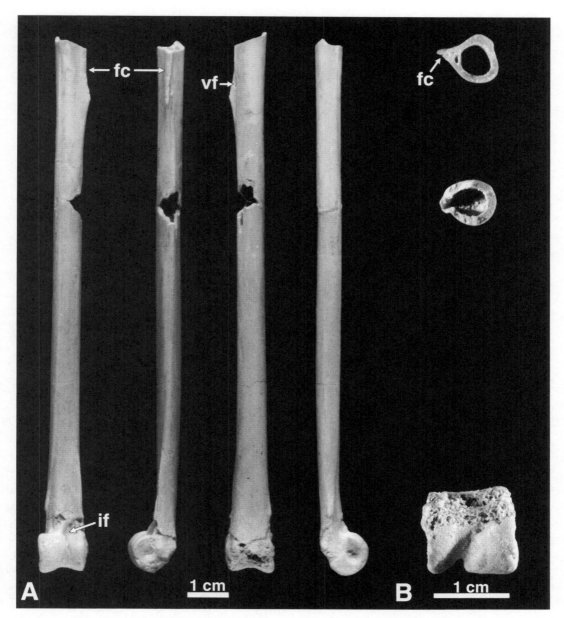

Figure 12.5. Left tibiotarsus of holotype UA 8651: A, cranial, lateral, caudal, and medial views (left to right); B, cross section at proximal broken end, lateral margin to the left (top); cross section at break below fibular crest, lateral margin to left (middle); distal view, caudal margin at top (bottom). Abbreviations: fc, fibular crest; if, intercondylar fossa; vf, vascular foramen.

distal to the proximolateral foramen. This part of the tarsometatarsus is only partially fused. A lesser degree of fusion is seen in the middle portion of their shafts: MTs II and III are fused along their entire length, whereas a small gap between MTs III and IV is present proximally. MTs II–IV are completely co-ossified in adults of *Patagopteryx* and Ornithurae but exhibit a wide range of co-ossification within other Mesozoic birds and nonavian maniraptorans (Chiappe, 1996a). The metatarsals are fused only proximally in Enantiornithes (e.g., *Concornis, Yungavolucris, Soroavi-*

saurus), some specimens of *Archaeopteryx* (e.g., London specimen, Solnhofen specimen), and some nonavian theropods (e.g., *Ceratosaurus, Elmisaurus, Avimimus*). One enantiornithine, *Avisaurus gloriae,* shows MTs III and IV fused distally (Varricchio and Chiappe, 1995).

The proximal portions of MTs II–IV are coplanar, a primitive condition shared by nonavian maniraptorans and basal birds such as *Archaeopteryx, Rahonavis,* Alvarezsauridae, *Iberomesornis,* Enantiornithes, and *Patagopteryx* (Chiappe, 1996a). In proximal view, the articular surface of the

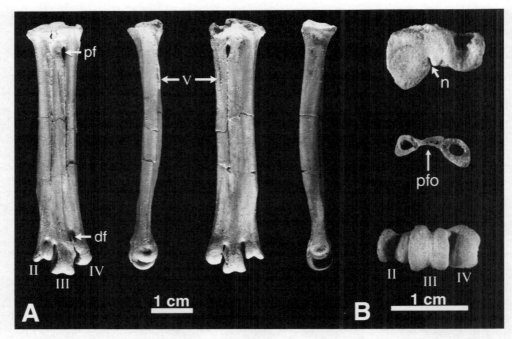

Figure 12.6. Left tarsometatarsus of holotype UA 8651: A, dorsal, lateral, plantar, and medial views (left to right); B, proximal view, plantar margin up (top); cross section at midshaft, dorsal margin up (middle); distal view, dorsal margin up. Abbreviations: df, distal foramen; n, notch between MTs III and IV; pf, proximolateral foramen; pfo, palmar fossa; Roman numerals refer to metatarsal number.

tarsometatarsus is kidney-shaped (concave dorsally), and the plantar margin of the articular area is slightly elevated above the dorsal one. No free distal tarsals are present despite the fact that the distal tibiotarsus and tarsometatarsus of UA 8651 were found in articulation. We hypothesize that the distal tarsals are fused to the metatarsus and contribute to the proximal articular surface. MT II is dorsoplantarly expanded and projects dorsally with respect to MTs III and IV. A deep, very narrow notch is present on the dorsal margin between MTs III and IV. Like most primitive birds (e.g., *Archaeopteryx*, Alvarezsauridae, most Enantiornithes), the tarsometatarsus of *Vorona* lacks a hypotarsus and an intercondylar eminence. The medial cotyla is transversely expanded and larger than the lateral one. It extends over the expanded proximal surface of MT II and the small, restricted proximal surface of MT III. The lateral cotyla is not well preserved but appears to have been circular and restricted to the proximal surface of MT IV.

MT V is a very narrow, splintlike element that is sutured to the lateroplantar margin of the proximal end of MT IV. MT V is slightly less than one-third the length of MT IV. MT IV is not reduced to a noticeable degree relative to MTs II and III, unlike the condition seen in Enantiornithes (Chiappe, 1993). Aside from the grooves separating each individual metatarsal, the shaft of the tarsometatarsus has a nearly flat dorsal surface. Plantarly, however, it is strongly concave with a deep longitudinal fossa along most of the shaft. This plantar fossa has MT III as its roof and is bounded by MTs II and IV. A similar plantar fossa is present in *Patagopteryx* and some Enantiornithes (e.g., *Soroavisaurus*), as well as in some nonavian maniraptorans (e.g., *Elmisaurus*). The shaft of MT II is dorsoplantarly expanded, and its proximal plantar surface bears a blunt edge reminiscent of the ridge seen in *Soroavisaurus*. In medial view, MT II is plantarly bowed, although less so than in *Soroavisaurus* (Chiappe, 1993). There is no proximal, dorsolaterally placed tubercle on MT II (an Enantiornithes synapomorphy; Chiappe, 1993). On the proximal end of MT III is a small, dorsally projecting tubercle. This tubercle is situated at the level of a very weak muscular scar on the dorsolateral margin of the proximal end of MT II. Most likely, these two features indicate the attachment area for M. tibialis cranialis. Between the proximal ends of MTs III and IV is a small, elongate proximolateral foramen. This foramen also occurs in other primitive birds (e.g., *Patagopteryx*) and in at least one nonavian maniraptoran (*Elmisaurus*).

MT II is slightly shorter than MT IV, which, in turn, is shorter than MT III. MT III is more slender than MT II and especially MT IV proximally, but it expands distally to exceed the others in width. The dorsal surface of the distal one-third of MT III is laterally excavated by a broad vascular groove. This groove ends in a broad distal foramen, which opens distally between the trochleae of MTs III and IV rather than onto the plantar surface. This distal foramen is dorsally roofed by a bony bridge formed by projections of these two trochleae that fail to fuse together. An elongate, shallow facet for the articulation of MT I is present on the medial surface of MT II just proximal to its trochlea. The distal location of MT I in *Vorona* is consistent with the capability for perching.

The equally spaced trochleae are coplanar. The trochleae of MTs II and III are square in distal view and bear well-formed ginglymi. The trochlea of MT II is significantly smaller than that of MT III, the reverse of the condition in Enantiornithes. Proximal to each of these trochleae, on the dorsal surface, there are small, circular fossae. MT III bears the largest trochlea and lacks the medial plantar projection of avisaurid enantiornithines. The trochlea of MT IV is subrectangular in distal view, with its main axis dorsoplantarly oriented. This trochlea is nonginglymoid, and its lateral face has a well-developed collateral fossa. No phalanges are preserved.

Phylogenetic Relationships

Although *V. berivotrensis* is represented by incomplete specimens, its avian nature was clearly outlined by Forster et al. (1996a). In this initial article, a preliminary phylogenetic placement was established for *Vorona* by scoring its preserved characters into the data matrix of Sanz et al. (1995: app. 2) with minor modifications and additions. This analysis placed *Vorona* in a trichotomy with Enantiornithes and a clade that consists of *Patagopteryx deferrariisi* plus Ornithurae (Forster et al., 1996a).

Because of the incompleteness of the known specimens, the position of *Vorona* within primitive birds has proved difficult to refine. A reanalysis of *Vorona* among primitive birds is presented elsewhere in this volume (Chiappe, Chapter 20), in a new phylogenetic analysis based on many more characters and taxa than that of Sanz et al. (1995). Importantly, femoral characters for *Vorona* not available for the initial analysis were included in the reanalysis. In this new analysis, *Vorona* forms a trichotomy with *Patagopteryx* and Ornithurae.

Although the relationship of *Vorona* to other basal avians is still not fully resolved, its separation from the Enantiornithes is supported by several derived characters shared among *Vorona, Patagopteryx,* and Ornithurae. These characters include the presence of a fossa for the capital ligament in the head of the femur, the near complete fusion of MTs II–IV, and the complete enclosure of the vascular distal foramen by MTs III and IV. The superficial resemblance of *Vorona* and certain Enantiornithes (e.g., *Soroavisaurus australis;* Chiappe, 1993) is shown to be plesiomorphic, since these characters (e.g., bulbous medial condyle of tibiotarsus, excavated plantar surface of the tarsometatarsus) are also present in the far more primitive bird *Confuciusornis sanctus* (see Chiappe, Chapter 20 in this volume).

The hypothesis that *Vorona* lies outside Enantiornithes as a basal member of the Ornithuromorpha (see Chiappe, Chapter 20 in this volume) reminds us that the phylogenetic diversity of Mesozoic Gondwanan avifaunas is likely far greater than that sampled thus far. Of particular note is the consistent phylogenetic proximity of *Vorona* and *Patagopteryx,* another nonenantiornithine basal bird from Gondwana, in both cladistic analyses (Forster et al., 1996a; Chiappe, Chapter 20 in this volume). Nevertheless, characters supporting a sister-taxon relationship between *Vorona* and *Patagopteryx* have not been found yet, although future findings may support such a relationship. Clearly, whatever the outcome, more complete material of *Vorona* is necessary for a full understanding of its phylogenetic relationships.

Paleobiogeography

The discovery of *Vorona* demonstrates conclusively that birds were present on Madagascar during the Late Cretaceous. This occurrence represents a significant geographic range extension for Gondwanan birds, since the only previous Late Cretaceous records of avian skeletal material are known from the western part of Gondwana (Argentina and the Antarctic Peninsula) and from eastern Australia. The virtual absence of Mesozoic birds from other landmasses surrounding the western Indian Ocean, coupled with our incomplete knowledge of morphology for *Vorona* (and thus its unresolved phylogenetic placement), precludes the derivation of a strong biogeographic signal from this new record.

The current biogeographic significance of *Vorona,* however, derives in large part from what it is not. It is not a sister taxon to, or a basal representative of, any of the modern or recently extinct groups of Malagasy birds (including the Aepyornithidae), all of which belong within Neornithes. The occurrence of *Vorona,* as well as that of the other four non-neornithine birds from quarry MAD93-18, hardly constitutes strong and persuasive evidence that the ancestors of modern and recently extinct groups were not present on Madagascar in the Late Cretaceous. It is, nevertheless, the only fossil evidence currently available that bears directly on the question. Speculation that ancestors of these groups might have been isolated on Madagascar prior to the island's separation from Africa is not supported by any fossil evidence.

Acknowledgments

None of this work could have been accomplished without the effort and endurance of the 1995 Madagascar field crew: R. Asher, G. Buckley, Prosper, A. Rabarison, L. Rahantarisoa, L. Randriamiarimanana, F. Ravoavy, and C. Wall. We also thank B. Rakotosamimanana, B. Andriamihaja, the staff of the Institute for the

Conservation of Tropical Environments, P. Wright, and S. Goodman for crucial logistical help in Madagascar. We especially thank R. Fox for an early review of this work and L. M. Witmer for a later review. The specimens were prepared by V. Heisey and photographed by M. Stewart; L. Betti-Nash drafted all the figures. We also thank the Field Museum of Natural History, Chicago, for additional preparation and specimen assistance. This work was supported by National Science Foundation grants EAR-9418816 and EAR-9706302 and the Dinosaur Society.

Literature Cited

Alvarenga, H. M. F., and J. F. Bonaparte. 1992. A new flightless land bird from the Cretaceous of Patagonia; pp. 51–64 *in* K. E. Campbell (ed.), Papers in Avian Paleontology, Honoring Pierce Brodkorb. Science Series 36. Natural History Museum of Los Angeles County, Los Angeles.

Ambroggi, R., and A. F. de Lapparent. 1954. Les empreintes des pas fossiles du Maestrichtien d'Agadir. Notes du Service Géologique du Maroc 10:43–57.

Andrews, C. W. 1894. On some remains of *Aepyornis* in the British Museum (Natural History). Proceedings of the Zoological Society of London 1894:108–123.

Battistini, R. 1965. Sur le découverte de l'*Aepyornis* dans le Quaternaire de l'Extrême-Nord de Madagascar. Compte Rendu Sommaire des Séances de la Société Géologique de France 2:171–175.

Besairie, H. 1972. Géologie de Madagascar. I. Terrains sedimentaires. Annales Géologiques de Madagascar 35:1–463.

Bonaparte, J. F. 1991. Los vertebrados fósiles de la Formación Río Colorado de Neuquén y cercanias, Cretácico Superior, Argentina. Revista del Museo Argentina de Ciencias Naturales "Bernardino Rivadavia" (Paleontología) 4:17–123.

Bonaparte, J. F., and J. E. Powell. 1980. A continental assemblage of tetrapods from the Upper Cretaceous beds of El Brete, northwestern Argentina (Sauropoda-Coelurosauria-Carnosauria-Aves). Mémoires de la Société Géologique de France, N. S. 139:19–28.

Briggs, J. C. 1995. Global Biogeography. Developments in Paleontology and Stratigraphy 14. Elsevier Science B. V., Amsterdam, 452 pp.

Chatterjee, S. 1989. The oldest Antarctic bird. Journal of Vertebrate Paleontology 9(supplement):16A.

———. 1997. The Rise of Birds. Johns Hopkins University Press, Baltimore, 312 pp.

Chiappe, L. M. 1991. Cretaceous birds of Latin America. Cretaceous Research 12:55–63.

———. 1993. Enantiornithine (Aves) tarsometatarsi from the Cretaceous Lecho Formation of northwestern Argentina. American Museum Novitates 3083:1–27.

———. 1995. The first 85 million years of avian evolution. Nature 378:349–355.

———. 1996a. Late Cretaceous birds of southern South America: anatomy and systematics of Enantiornithes and *Patagopteryx deferrariisi*; pp. 203–244 *in* G. Arratia (ed.), Contributions of Southern South America to Vertebrate Paleontology, Münchner Geowissenschaftliche Abhandlungen, Reihe A, Geologie und Paläontologie 30. Verlag Dr. Friedrich Pfeil, Munich.

———. 1996b. Early avian evolution in the Southern Hemisphere: the fossil record of birds in the Mesozoic of Gondwana. Memoirs of the Queensland Museum 39:533–556.

Chiappe, L. M., and J. O. Calvo. 1994. *Neuquenornis volans,* a new Late Cretaceous bird (Enantiornithes: Avisauridae) from Patagonia, Argentina. Journal of Vertebrate Paleontology 14:230–246.

Cracraft, J. 1973. Continental drift, paleoclimatology, and the evolution and biogeography of birds. Journal of Zoology 169:455–545.

———. 1974. Phylogeny and evolution of the ratite birds. Ibis 116:494–521.

Currie, P. J., and Peng J.-H. 1993. A juvenile specimen of *Saurornithoides mongoliensis* from the Upper Cretaceous of northern China. Canadian Journal of Earth Sciences 30:2224–2230.

Curry Rogers, K., and C. A. Forster. 2001. The last of the dinosaur titans: a new sauropod from Madagascar. Nature 412:530–534.

Darlington, P. J., Jr. 1957. Zoogeography: The Geographical Distribution of Animals. John Wiley and Sons, New York, 675 pp.

Dorst, J. 1972. The evolution and affinities of the birds of Madagascar; pp. 615–627 *in* R. Battistini and G. Richard-Vindard (eds.), Biogeography and Ecology in Madagascar, Dr. W. Junk B. V., The Hague.

Feduccia, A. 1980. The Age of Birds. Harvard University Press, Cambridge, 196 pp.

———. 1996. The Origin and Evolution of Birds. Yale University Press, New Haven, 420 pp.

Forster, C. A., L. M. Chiappe, D. W. Krause, and S. D. Sampson. 1996a. The first Cretaceous bird from Madagascar. Nature 382:532–534.

———. 1996b. The first Mesozoic bird from Madagascar. Society of Avian Paleontology and Evolution Programs and Abstracts, p. 5.

Forster, C. A., S. D. Sampson, L. M. Chiappe, and D. W. Krause. 1998. The theropod ancestry of birds: new evidence from the Late Cretaceous of Madagascar. Science 279:1915–1919.

Hay, W. W., R. M. DeConto, C. N. Wold, K. M. Wilson, S. Voigt, M. Schulz, A. Wold-Rossby, W.-Chr. Dullo, A. B. Ronov, A. N. Balukhovsky, and E. Soeding. 1999. An alternative global Cretaceous paleogeography; pp. 1–48 *in* E. Barrera and C. Johnson (eds.), The Evolution of the Cretaceous Ocean/Climate Systems, Geological Society of America Special Paper 332. Geological Society of America, Boulder, Colo.

Hoffstetter, R. 1961. Nouveaux restes d'un serpent Boïde (*Madtsoïa madagascariensis* nov. sp.) dans le Crétacé supérieur de Madagascar. Bulletin du Muséum National d'Histoire Naturelle 33:152–160.

Keith, S. 1980. Origins of the avifauna of the Malagasy region; pp. 99–108 *in* D. N. Johnson (ed.), Proceedings of the 4th Pan-African Ornithological Congress. Southern African Ornithological Society.

Krause, D. W., and J. H. Hartman. 1996. Late Cretaceous fossils from Madagascar and their implications for biogeographic relationships with the Indian subcontinent. Memoir of the Geological Society of India 37:135–154.

Krause, D. W., J. H. Hartman, and N. A. Wells. 1997a. Late Cretaceous vertebrates from Madagascar: implications for biotic change in deep time; pp. 3–43 *in* S. M. Goodman and B. D. Patterson (eds.), Natural Change and Human Impact in

Madagascar, Smithsonian Institution Press, Washington, D.C.

Krause, D. W., J. V. R. Prasad, W. von Koeningswald, A. Sahni, and F. E. Grine. 1997b. Cosmopolitanism among Late Cretaceous Gondwanan mammals. Nature 390:504–507.

Krause, D. W., R. R. Rogers, C. A. Forster, J. H. Hartman, G. A. Buckley, and S. D. Sampson. 1999. The Late Cretaceous vertebrate fauna of Madagascar: implications for Gondwana biogeography. GSA Today 9:1–7.

Lamberton, C. 1934. Contribution à la connaisance de la faune subfossile de Madagascar. Lémuriens et Ratites. Mémoires de l'Academie Malgache 17:1–168.

Lambrecht, K. 1929. Ergebnisse der Forschungsreisen Prof. E. Stromers in den Wüsten Ägyptens. V. Tertiäre Wirbeltiere. 4. *Stromeria fajumensis* n. g., n. sp., die kontinentale Stammform der Aepyornithidae, mit einer Übersicht über die fossilen Vögel Madagaskars und Afrikas. Abhandlungen der Bayerischen Akademie der Wissenschaften Mathematisch-Naturwissenschaftliche Abteilung 4:1–18.

———. 1933. Handbuch der Palaeornithologie. Gebrüder Borntraeger, Berlin, 1024 pp.

Langrand, O. 1990. Guide to the Birds of Madagascar. Yale University Press, New Haven, 364 pp.

MacPhee, R. D. E., D. A. Burney, and N. A. Wells. 1985. Early Holocene chronology and environment of Ampasambazimba, a Malagasy subfossil lemur site. International Journal of Primatology 6:463–489.

Mahé, J. 1972. The Malagasy subfossils; pp. 339–365 in R. Battistina and G. Richard-Vindard (eds.), Biogeography and Ecology in Madagascar. Dr. W. Junk B.V., The Hague.

Martin, L. D. 1983. The origin and early radiation of birds; pp. 291–338 in A. H. Bush and G. A. Clark Jr. (eds.), Perspectives in Ornithology, Cambridge University Press, New York.

Martin, L. D., J. D. Stewart, and K. N. Whetstone. 1980. The origin of birds: structures of the tarsus and teeth. Auk 97:86–93.

McGowan, C. 1984. Evolutionary relationships of ratites and carinates: evidence from ontogeny of the tarsus. Nature 307:733–735.

———. 1985. Tarsal development in birds: evidence for homology with the theropod condition. Journal of Zoology, London (A) 206:53–67.

Molnar, R. E. 1986. An enantiornithine bird from the Lower Cretaceous of Queensland, Australia. Nature 322:736–738.

Moreau, R. E. 1966. Bird Faunas of Africa and Its Islands. Academic Press, New York, 424 pp.

Noriega, J. I., and C. P. Tambussi. 1995. A Late Cretaceous Presbyornithidae (Aves: Anseriformes) from Vega Island, Antarctic Peninsula: paleobiogeographic implications. Ameghiniana 32:57–61.

Novas, F. E. 1996. Alvarezsauridae, basal birds from Patagonia and Mongolia. Memoirs of the Queensland Museum 39:675–702.

———. 1997. Anatomy of *Patagonykus puertae* (Theropoda, Avialae, Alvarezsauridae) from the Late Cretaceous of Patagonia. Journal of Vertebrate Paleontology 17:137–166.

Novas, F. E., and P. F. Puerta. 1997. New evidence concerning avian origins from the Late Cretaceous of Patagonia. Nature 387:390–392.

Olson, S. L. 1985. The fossil record of birds; pp. 79–238 in D. S. Farner, J. R. King, and K. C. Parkes (eds.), Avian Biology, Vol. 8. Academic Press, New York.

Perle A., L. M. Chiappe, Barsbold R., J. M. Clark, and M. A. Norell. 1994. Skeletal morphology of *Mononykus olecranus* (Theropoda: Avialae) from the Late Cretaceous of Mongolia. American Museum Novitates 3105:1–29.

Quammen, D. 1996. The Song of the Dodo: Island Biogeography in an Age of Extinctions. Scribner, New York, 702 pp.

Rand, A. L. 1936. The distribution and habits of Madagascar birds. Bulletin of the American Museum 72:143–499.

Rasmussen, D. T., S. L. Olson, and E. L. Simons. 1987. Fossil birds from the Oligocene Jebel Qatrani Formation, Fayum Province, Egypt. Smithsonian Contributions to Paleobiology 62:1–20.

Rogers, R. R., and J. H. Hartman. 1998. Revised age of the dinosaur-bearing Maevarano Formation (Upper Cretaceous), Mahajanga Basin, Madagascar. Journal of African Earth Sciences 27:160–162.

Rogers, R. R., J. H. Hartman, and D. W. Krause. 2000. Stratigraphic analysis of Upper Cretaceous rocks in the Mahajanga Basin, northwestern Madagascar: implications for ancient and modern faunas. Journal of Geology 108:275–301.

Rothschild, W. 1911. On the former and present distribution of the so-called Ratitae or ostrich-like birds with certain deductions and a description of a new form by C. W. Andrews; pp. 144–174 in Verhandlungen des V. Internationalen Ornithologen-Kongresses, Berlin.

Russell, D. A., and Dong Z.-M. 1993. A nearly complete skeleton of a new troodontid dinosaur from the Early Cretaceous of the Ordos Basin, Inner Mongolia, People's Republic of China. Canadian Journal of Earth Science 30:2163–2173.

Sampson, S. D., L. M. Witmer, C. A. Forster, D. W. Krause, P. M. O'Connor, P. Dodson, and F. Ravoavy. 1998. Predatory dinosaur remains from Madagascar: implications for the Cretaceous biogeography of Gondwana. Science 280:1048–1051.

Sampson, S. D., M. T. Carrano, and C. A. Forster. 2001. A bizarre new carnivorous dinosaur from Madagascar. Nature 409:504–506.

Sanz, J. L., and J. F. Bonaparte 1992. A new order of birds (Class Aves) from the Lower Cretaceous of Spain; pp. 39–49 in K. E. Campbell Jr. (ed.), Papers in Avian Paleontology, Honoring Pierce Brodkorb. Science Series 36. Natural History Museum of Los Angeles County, Los Angeles.

Sanz, J. L., L. M. Chiappe, and A. D. Buscalioni. 1995. The osteology of *Concornis lacustris* (Aves: Enantiornithes) from the Lower Cretaceous of Spain and a reexamination of its phylogenetic relationships. American Museum Novitates 3133:1–23.

Sauer, E. G. F. 1972. Ratite eggshells and phylogenetic questions. Bonner Zoologische Beiträge 23:3–48.

———. 1976. Aepyornithoide Eierschalen aus dem Miozän und Pliozän von Anatolien, Turkei. Palaeontographica A 153:62–115.

Sauer, E. G. F., and P. Rothe. 1972. Ratite eggshells from Lanzarote, Canary Islands. Science 176:43–45.

Storey, B. C. 1995. The role of mantle plumes in continental breakup: case histories from Gondwanaland. Nature 377:301–308.

Storey, M., J. J. Mahoney, A. D. Saunders, R. A. Duncan, S. P. Kelly, and M. F. Coffin. 1995. Timing of hot spot-related volcanism and the break-up of Madagascar and India. Science 267: 852–855.

Tarsitano, S., and M. K. Hecht. 1980. A reconsideration of the reptilian relationships of *Archaeopteryx*. Zoological Journal of the Linnean Society 69:149–182.

Varricchio, D. J., and L. M. Chiappe. 1995. A new enantiornithine bird from the Upper Cretaceous Two Medicine Formation of Montana. Journal of Vertebrate Paleontology 15:201–204.

Walker, A. 1978. Prosimian primates; pp. 90–99 *in* V. J. Maglio and H. B. S. Cooke (eds.), Evolution of African Mammals. Harvard University Press, Cambridge.

Walker, C. A. 1981. New subclass of birds from the Cretaceous of South America. Nature 292:51–53.

Wellnhofer, P. 1992. A new specimen of *Archaeopteryx* from the Solnhofen limestone; pp. 3–23 *in* K. E. Campbell (ed.), Papers in Avian Paleontology, Honoring Pierce Brodkorb. Science Series 36. Natural History Museum of Los Angeles County, Los Angeles.

13

Osteology of the Flightless *Patagopteryx deferrariisi* from the Late Cretaceous of Patagonia (Argentina)

LUIS M. CHIAPPE

The evolution of Mesozoic Gondwanan birds is poorly known. This book amply attests to the fact that regardless of the significant contribution of Gondwanan birds to our understanding of early avian history (Chiappe, 1991, 1996a; Forster et al., 1996, 1998, Chapter 12 in this volume; Clarke and Chiappe, 2001), the majority of the available data come from the Northern Hemisphere.

Patagopteryx deferrariisi (Alvarenga and Bonaparte, 1992) is a hen-sized, flightless bird (Fig. 13.1) known exclusively from Late Cretaceous beds in the northwestern limits of the Patagonian city of Neuquén, Argentina (Fig. 13.2). *Patagopteryx* is important not only because it documents a distinct lineage of basal birds but also because it is the best-represented Mesozoic avian from the Southern Hemisphere, being known from several specimens.

The discovery of the first specimens of *Patagopteryx* in 1984–1985 was made in connection with the expansion of the campus of the Universidad Nacional del Comahue (UNC, Neuquén) and the resultant clearing and leveling of the hills that form the southwestern margin of the Neuquén River, near its confluence with the Limay River (Fig. 13.2). Oscar de Ferrariis, the director of the UNC's Museo de Ciencias Naturales at the time, recovered the first specimens and passed them on to José Bonaparte for study (Alvarenga and Bonaparte, 1992). Additional specimens were later collected by Jorge Calvo and Leonardo Salgado from the UNC.

The first reference to *Patagopteryx* was made by Bonaparte (1986:78), who mentioned the presence of an "indeterminate family of ratite-like birds" in the Late Cretaceous of Patagonia. The idea of *Patagopteryx* as a basal ratite was retained during its formal description (Alvarenga and Bonaparte, 1992; see also Alvarenga, 1993). Although originally subscribing to the ratite hypothesis (Chiappe, 1987), I challenged its allocation within Neornithes, and even Ornithurae, in subsequent studies (Chiappe, 1989, 1991). This

conclusion was strongly supported by later phylogenetic analyses of basal avians (Chiappe, 1992, 1995a,b, 1996b, see Chapter 20 in this volume; Chiappe and Calvo, 1994).

The following institutional abbreviations are used in this chapter: MACN, Sección Paleontología de Vertebrados, Museo Argentino de Ciencias Naturales, Buenos Aires, Argentina; MUCPv, Museo de Ciencias Naturales, UNC, Neuquén, Argentina.

Geological Setting

P. deferrariisi is known only from the continental beds of the Bajo de la Carpa Member of the Río Colorado Formation (Neuquén Group), exposed at Boca del Sapo in the northeast corner of the city of Neuquén (see Chiappe and Calvo, 1994, and references cited therein; Fig. 13.2).

At Boca del Sapo, the Bajo de la Carpa Member is predominantly composed of medium to fine, friable, whitish gray to pink sandstones, generally made of clean quartz with very little clay (Chiappe and Calvo, 1994).

The fauna from Boca del Sapo is composed of a diverse ensemble of small tetrapods (Bonaparte, 1991). Particularly abundant are the crocodilian *Notosuchus terrestris* (Smith Woodward, 1896; Gasparini, 1971; Bonaparte, 1991) and the snake *Dinilysia patagonica* (Smith Woodward, 1901; Bonaparte, 1991). This fauna also includes the crocodilian *Comahuesuchus brachibuccalis* (Bonaparte, 1991), the nonavian theropod *Velocisaurus unicus* (Bonaparte, 1991), the enantiornithine *Neuquenornis volans* (Chiappe and Calvo, 1994; Chiappe and Walker, Chapter 11 in this volume), and *Alvarezsaurus calvoi*, a close relative of the Mongolian *Mononykus olecranus* (Chiappe, 1995a; Novas, 1996; Chiappe, Norell, and Clark, Chapter 4 in this volume).

Cazau and Uliana (1973) regarded the depositional environment of the Bajo de la Carpa Member as a system of braided streams. Opalized trees and large, disarticulated di-

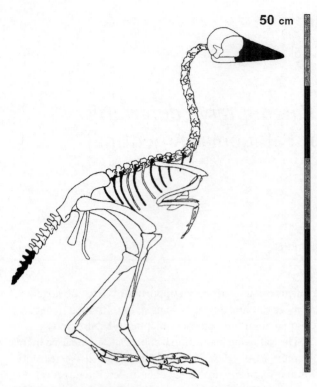

50 cm

Figure 13.1. Skeletal reconstruction of *P. deferrariisi* (modified from Alvarenga and Bonaparte, 1992). Black areas are not preserved in any available specimen.

nosaurian remains were found in other localities of this member, suggesting transport. In contrast, at Boca del Sapo the fauna is composed of small and usually articulated tetrapods, and large dinosaurian remains are very rare (Bonaparte, 1991; Chiappe and Calvo, 1994). The deposits at Boca del Sapo apparently represent a different depositional environment, one with lower energy than that generally envisioned for the Bajo de la Carpa Member (Chiappe and Calvo, 1994). Heredia and Calvo (1997) have recently suggested that the sandstones of Boca del Sapo may represent interdune deposits in an eolian depositional regime, although Clarke et al. (1998) found little basis to support such a claim.

The Río Colorado Formation, along with the entire Neuquén Group, has traditionally been considered as Late Cretaceous, generally as Campanian (e.g., Cazau and Uliana, 1973; Uliana and Dellapé, 1981; Danderfer and Vera, 1992; see also Bonaparte, 1991). A Campanian-Maastrichtian charophyte assemblage was reported for the Anacleto Member (Musacchio, 1993), yet sequential stratigraphic analyses in the northern area of the Neuquenian Basin hinted at an older age for the entire Río Colorado Formation. Cruz et al. (1989) proposed a Coniacian–Lower Campanian age, while Legarreta and Gulisano (1989) suggested a Santonian-Campanian age. The pre-Campanian age of the Río Colorado Formation also received indirect support from the

placement of the basal unit of the Neuquén Group (Río Limay Formation) as either Cenomanian (Ramos, 1981; Cruz et al., 1989) or Albian-Cenomanian (Calvo, 1991), as well as from the suggestion that the whole Neuquén Group may be older than was traditionally thought (Ramos, 1981). Recent magnetostratigraphic analyses of rocks from the Río Colorado Formation at Auca Mahuevo, a sauropod nesting site some 150 km northwest of the city of Neuquén (Chiappe et al., 1998), have shed important light on the age of this fossil-rich unit. Dingus et al. (2000) have provided the first reliable age for the Anacleto Member of the Río Colorado Formation, constraining it between 83.5 and 79.5 million years ago. This age can reasonably be extrapolated to the Bajo de la Carpa sandstones of Boca del Sapo, which are therefore considered to be early to middle Campanian.

Systematic Paleontology

Taxonomic Hierarchy

Aves Linnaeus, 1758

Ornithothoraces Chiappe, 1995

P. deferrariisi Alvarenga and Bonaparte, 1992

Holotype—MACN-N-03, partial skeleton including 5 cervical vertebrae, 11 thoracic vertebrae, a complete synsacrum, 2 caudal vertebrae, both humeri and the proximal portion of the radius and ulna, the shoulder ends of both scapulae and coracoids, the acetabular and post-acetabular portions of both ilia, and portions of the femora and tibiotarsi.

Referred specimens—MACN-N-10, five fragments of tarsometatarsi (of at least three individuals) and pedal phalanges; MACN-N-11, nearly complete skeleton including skull and jaws; MACN-N-14, four articulated thoracic vertebrae and other vertebral fragments; MUCPv-48, complete hindlimb, several vertebrae, and fragments of the pelvis and skull; MUCPv-207, portions of hindlimb, synsacrum, and several other vertebrae.

Locality and horizon—Boca del Sapo, city of Neuquén, province of Neuquén, Argentina (Fig. 13.2). Bajo de la Carpa Member, Río Colorado Formation, Late Cretaceous (Campanian).

Diagnosis—*P. deferrariisi* is diagnosed by the following autapomorphies: quadrate fused to the pterygoid; quadrate pneumatic foramen located laterally; biconvex fifth thoracic vertebra; thoracic vertebrae 6–11 procoelous, with wide, kidney-shaped centra; synsacrum procoelous; shoulder end of coracoid formed by a broad, tonguelike caudolateral surface that includes the humeral articular facet and the area of scapular articulation; acromion of scapula dorsoventrally expanded and keyhole-shaped in shoulder (proximal) view; strong muscular lines in the shaft of the humerus; distal half of shaft of ulna strongly compressed craniocaudally; minor

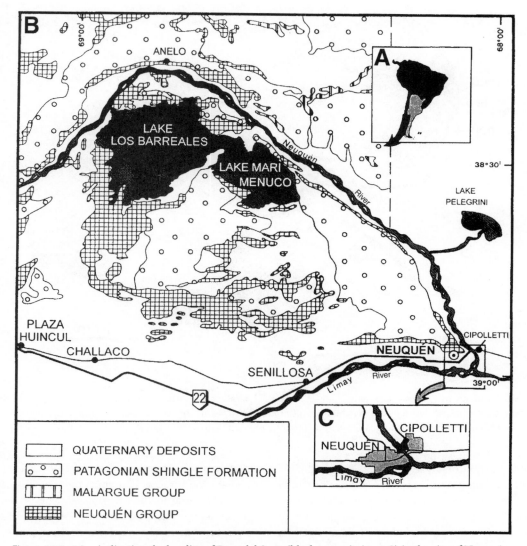

Figure 13.2. Map indicating the locality of Boca del Sapo (black arrow in inset C) in the city of Neuquén, northwestern Patagonia, Argentina (modified from Gasparini et al., 1991).

metacarpal more robust than the major metacarpal; major metacarpal with a cranioventral, laminar projection; ilium with a prominent iliac crest and a well-developed caudolateral spine; craniocaudally compressed, straplike pubis, with its caudal third curved cranioventrally; paddle-shaped ischium; prominent M. iliofibularis tubercle on the fibula; fibular spine distally fused to the cranial surface of tibiotarsus; transversely wide tarsometatarsus; pamprodactyl foot.

Anatomy

Skull and Mandible

The skull of *P. deferrariisi* is known only for MACN-N-11 and MUCPv-48. Only the braincase, caudal portion of the orbits, quadrates, and caudal end of the right pterygoid are preserved (Figs. 13.3, 13.4).

The skull roof is formed primarily by the frontals, which are separated from one another by a straight suture. The frontals are flat and slope rostrally. In MUCPv-48, in the rostralmost preserved portion of the right frontal, there is a median depression that may have received the caudal end of the nasal process of the premaxilla. Both MACN-N-11 and MUCPv-48 preserve portions of the endocranial cavity as a natural cast in some of the areas where the frontals are missing.

The orbits are large and are bound by the frontals caudodorsally and the laterosphenoids caudoventrally. The laterosphenoid appears to form a short postorbital process, suggesting that *Patagopteryx* lacks a postorbital bone.

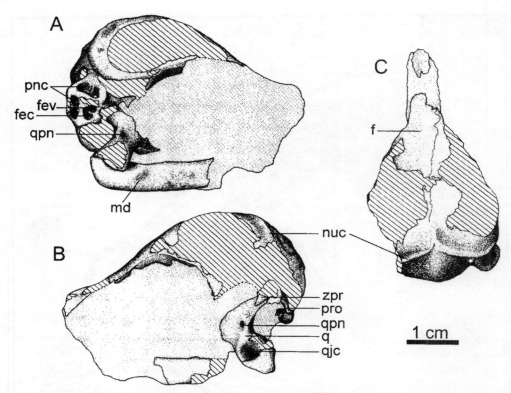

Figure 13.3. Skull and jaw of *P. deferrariisi* (MACN-N-11). A, right lateral view; B, left lateral view; C, dorsal view. Abbreviations: f, frontal; fec, fenestra cochlearis; fev, fenestra vestibularis; md, mandible; nuc, nuchal crest; pnc, pneumatic cavities; pro, prootic; q, quadrate; qjc, quadratojugal cotyla of the quadrate; qpn, quadrate pneumatic foramen; zpr, zygomatic process.

The caudal portion of the skull roof is formed by the parietals. These are significantly shorter than the frontals. In MACN-N-11, the frontoparietal contact is not well defined, but in MUCPv-48 there is a clear, transverse suture between these two bones. The caudal boundary of the parietals is marked by the strong transverse nuchal crest, which has an inverted V-shape in caudal view that wedges between the parietals (Figs. 13.3, 13.4).

The squamosal is preserved only in the area of its articulation with the quadrate and prootic, and, along with the latter bone, it caps the otic process of the quadrate (Figs. 13.3B, 13.4A, 13.5). The articulation of the squamosal to the prootic and its position with respect to other elements indicate that this bone was incorporated into the braincase, a derived condition contrasting with that of other basal avians (e.g., *Archaeopteryx, Confuciusornis,* basal enantiornithines). The squamosoprootic articulation leaves no space for the entrance of the dorsal tympanic recess (Fig. 13.5B). This suggests that the otic process of the quadrate was not differentiated into two separate condyles (Baumel and Witmer, 1993), although this cannot be established confidently. Laterally, the squamosal possesses a short zygomatic process (Figs. 13.3B, 13.4A, 13.5B) that is firmly adhered to the lateral surface of the quadrate's otic process. This latter condition recalls the condition of modern ratites

(Starck, 1995). Interestingly, the squamosal also possesses a medial, stout process that abuts the medial surface of the quadrate, distal to the otic process. The presence of this process along with the closely appressed zygomatic process must have seriously constrained the quadrate's kinesis (Starck, 1995).

The prootic is represented by the pila otica (the opisthotic may be taking part in it; see Baumel and Witmer, 1993) and portions surrounding the columellar recess. The pila otica articulates with the otic process of the quadrate and connects it with the dorsal margin of the columellar recess (Figs. 13.4D, 13.5B). The fact that the pila otica is quite long suggests that the otic process of the quadrate was not close to the columellar recess. The latter is a heart-shaped recess with its apex pointing caudodorsally (Fig. 13.4A). This recess houses the cochlear and vestibular fenestrae and the entrance to the caudal tympanic recess. The latter opens dorsal to the vestibular fenestra, and it probably connected to a highly pneumatized area caudodorsal to the columellar recess.

The quadrate is laterally compressed, with both dorsal and ventral ends lying in the same vertical plane. In lateral view, its caudal margin is strongly notched (Figs. 13.3, 13.4). The rostral margin extends into a broad, laminar orbital process. The caudal surface is deeply excavated by a dorso-

The quadrate of *Patagopteryx* is pneumatic; its hollowed body is exposed in the right element of MACN-N-11. In contrast to other birds, the entrance of the quadrate diverticulum of the tympanic air sac is centered on the lateral face of the quadrate (Figs. 13.3, 13.4A). In the left quadrate of MACN-N-11, immediately dorsal to the pneumatic foramen, there is a small fossa perforating the quadrate. This fossa is not present on the right element and most likely represents an artifact.

The mandibular process of the left quadrate of MACN-N-11 preserves the rostral margin of the quadratojugal cotyla (Fig. 13.3B). Judging from the shape of this portion, the quadratojugal cotyla must have been large and round. The mandibular articulation is formed by three condyles placed on the vertices of a triangular surface (Fig. 13.4C). These condyles are ventrally projected, defining a deep, central depression. The larger, rostral condyle is transversely oriented and slopes dorsomedially. A smaller and pointed medial condyle forms the medial vertex of the triangle. The caudal condyle is flat and less ventrally projected than the others.

The caudal end of the right pterygoid of MACN-N-11 is the only known portion of the palate (Fig. 13.5A). The pterygoid, at least at its caudal end, is broad and robust, and it becomes broader rostrally (Figs. 13.4C, 13.5A). Interestingly, close examination of the pterygoid-quadrate complex does not reveal even a suture between these two bones. The only conclusion to be drawn from this condition is that the pterygoid is completely fused to the quadrate, a very unusual condition among birds (Simonetta, 1960). This condition combined with the presence of a robust, medial process of the squamosal restricting the quadrate's movements suggests that the skull of *Patagopteryx* had limited, if any, capacity for streptostylic movement (Chiappe, 1996b).

The occiput is nearly vertical, and there are no sutures demarcating its component bones. The cerebellar prominence is feeble (Fig. 13.4D). The foramen magnum and the occipital condyle are covered by the articulated atlas. The angle formed by the intersection of the planes of the foramen magnum and the basitemporal plane is slightly larger than 90°. This suggests that the longitudinal axis of the skull was continuous with the main axis of the neck (Saiff, 1983).

To the right of the occipital condyle of MACN-N-11 there are two very close foramina (joined by a narrow slit on the left side) forming the exit of cranial nerve XII (n. hypoglossi) (Figs. 13.4D, 13.5). Ventrolaterally from these foramina there is another foramen (best observed on the right side of MACN-N-11) that was probably an additional exit for this nerve. The number of hypoglossal foramina in birds is variable, but most have three (Webb, 1957). Lateral to the hypoglossal foramina is a large foramen forming the exit of cranial nerve X (n. vagi) and presumably cranial nerve XI (n. glossopharyngealis) (Fig. 13.5). On the left side of

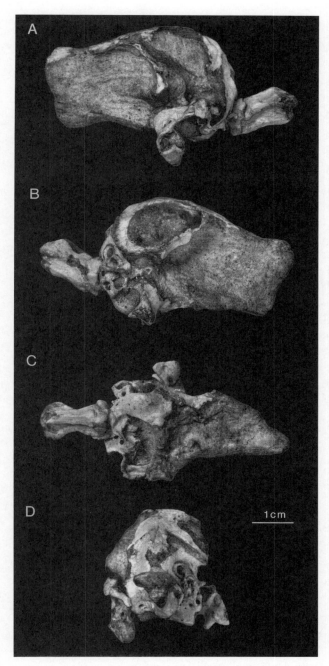

Figure 13.4. Skull, proatlas, atlas, and axis of *P. deferrariisi* (MACN-N-11). A, left lateral view; B, right lateral view; C, ventral view; D, occipital view.

ventral furrow (Figs. 13.4C, 13.5B). The latter starts right at the level of the apex of the lateral notch, and it extends up to the end of the otic process. This furrow is flanked by dorsoventral bars, the lateral of which is thinner than the medial. Dorsal to this furrow is the tied contact between the squamosal and prootic bones, both articulating with the otic process of the quadrate.

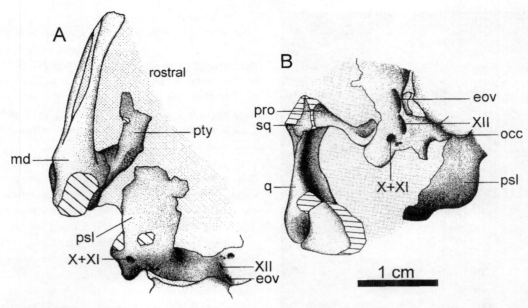

Figure 13.5. Skull and jaw of *P. deferrariisi* (MACN-N-11). A, ventral view; B, left caudolateral view. Abbreviations: eov, exit for the external occipital vein; md, mandible; occ, occipital condyle; pro, prootic; psl, parasphenoidal lamina; pty, pterygoid; q, quadrate; sq, squamosal; X–XII, cranial nerves.

MACN-N-11, immediately medial to the latter foramen, are three tiny foramina laid in a triangle. The most medial one appears to be the counterpart of the one interpreted as an additional exit of cranial nerve XII. The correspondence of the remaining ones is unclear. Over the left margin of the foramen magnum of MACN-N-11 there is a slitlike foramen that most likely corresponds to the exit of the external occipital vein (Fig. 13.5); this area has not been preserved on the right side.

Rostral to the occipital condyle there is a large, deep subcondylar fossa (Figs. 13.4C, 13.5A). Rostral to this fossa there is an ample parasphenoidal lamina that is depressed centrally, with its lateral borders projecting ventrally. Because only the right half of this lamina is known (preserved in MACN-N-11), it is unclear whether the parasphenoidal lamina was ventrally perforated by a recess (i.e., basisphenoidal recess of Currie, 1995), as in the alvarezsaurids *Shuvuuia* (Chiappe et al., 1998) and several nonavian theropods (e.g., Currie, 1995). In contrast to many Neornithes (except ratites), the parasphenoidal lamina projects ventrally well below the level of the occipital condyle. Even though the basipterygoid processes are not preserved, the presence of a scar on the right pterygoid of MACN-N-11 indicates that these processes were prominent and located caudally, articulating with the pterygoid near its juncture to the quadrate—such a condition is comparable to that of palaeognaths (Bock, 1963).

The mandible of *Patagopteryx* is known only from its caudal portion (Fig. 13.6). It is slender and transversely compressed, except for its robust articular area. The retro-articular process is absent. Individual bones are not discernible except for a ventral, longitudinal cleft separating the surangular from the angular (Figs. 13.5A, 13.6C). It is probable that this cleft was housed in a long, caudal projection of the dentary. These two portions, however, fuse caudal to the beginning of the articular area.

The lateral surface of the mandible is flat. Just rostral to the articular area, the surangular is perforated by a small, subcircular caudal fenestra (Fig. 13.6A), an apparently primitive condition known for a variety of nonavian theropods (Weishampel et al., 1990). This surangular fenestra is the only opening perforating the mandible; the presence of an extensive, imperforate area of the surangular rostral to this fenestra suggests that a rostral mandibular fenestra was absent in *Patagopteryx*. The medial surface of the mandible is strongly excavated by an extensive aditus fossa (Baumel and Witmer, 1993). Caudally, this fossa is bound by a small, medial process.

The quadrate articular fossa bears three cotylae for the articulation with the three condyles of the quadrate (Fig. 13.6B). The lateral cotyla is large, shallow, and elliptical. Its main axis is oriented rostromedially. This cotyla is separated from the medial cotyla by a thick, prominent intercotylar crest. The medial cotyla is large and forms a deep basin between the intercotylar crest, the margin of the medial process, and the dorsally projected caudal cotyla. Like the lateral cotyla, the main axis of the medial cotyla is oriented rostromedially. Interestingly, the caudal cotyla crowns a strong, dorsally projected process that forms the caudal end of the mandible.

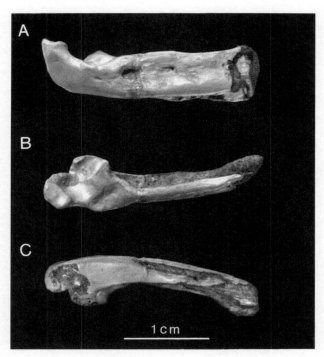

Figure 13.6. Left jaw of *P. deferrariisi* (MACN-N-11). A, lateral view; B, dorsal view; C, medial view.

Vertebral Column

Cervical Vertebrae. The cervical vertebrae are elongate and heterocoelous (Figs. 13.7, 13.8). *Patagopteryx* is the most basal bird with fully heterocoelous cervical centra. Pre- and postzygapophyses are strongly separated from one another. The postzygapophyses bear prominent epipophyses. The development of the spinous process is variable. Fairly high spinous processes are present in the cranial region of the neck, whereas they are almost absent in the last cervicals. The cervical ribs fuse to the para- and diapophyses (Fig. 13.7B, D) and all together enclose an ample, round transverse foramen. The cervical count of *Patagopteryx* is unknown. Nevertheless, the difference between the epipophysis of the eighth cervical of MACN-N-11, the last one preserved in articulation, and the first of the five last cervicals preserved in MACN-N-03 suggests that *Patagopteryx* had at least 13 cervical vertebrae. Such a number is comparable to that of many neornithine groups (Verheyen, 1960) and larger than the 9–10 cervicals of enantiornithines (Chiappe and Walker, Chapter 11 in this volume) and other basal avians (Chiappe et al., 1999).

Despite the retention of some primitive features (e.g., free proatlas, absence of fusion between the atlantal vertebral arch and centrum), the cervical series of *Patagopteryx* is essentially comparable to that of neornithine birds. This is presumably the case from a functional point of view. As in neornithine birds (see Zusi and Storer, 1969; Zusi and Bentz, 1984; Zusi, 1985), the cervical series of *Patagopteryx*

can be divided into three functional sections that most likely performed similar movements as those of its living relatives. Cervicals 3–5 (along with the atlas-axis) and 6–8 constitute sections I and II, respectively, while the last four cervicals, and possibly the first preserved element of MACN-N-03, constitute section III. The range of movements of sections I and III of *Patagopteryx* would have been ventral to the normal position of the neck, while that of section II would have been dorsal with respect to the neck's prevalent position. This functional complex was also present in Enantiornithes (Chiappe and Walker, Chapter 11 in this volume) and in basal Ornithurae but absent in *Archaeopteryx* and nonavian theropods.

Proatlas. In MACN-N-11, between the atlas and the skull, there is a tiny, bony element that is interpreted here as the proatlas (Figs. 13.4A, 13.7F). This element is preserved only in the left side, abutting against the lateral surface of the atlantal vertebral arch. In lateral view it has an inverted V-shape, with its cranial arm projecting ventromedially. Its caudal arm is shorter and rounder than the cranial one, recalling the shape of a postzygapophysis.

The presence of an individualized proatlas in *P. deferrariisi* is startling. In adult extant birds this element is fused to the skull, and to the best of my knowledge it has not been reported in any extinct birds. Yet an individualized proatlas of comparable shape is known for the nonavian theropod *Herrerasaurus* (Sereno and Novas, 1993).

Atlas. The atlas has a ringlike shape. Its centrum is small, and the neural arch is not fused to it (Figs. 13.4A–C, 13.7F). The dorsal and dorsolateral portions of the neural arch are wide and laminar. The neural arch narrows ventrally and ends in a round border that articulates with the centrum.

Axis. The axis is elongate, as are those cervical vertebrae that follow it (Fig. 13.4A–C). Its centrum is formed by a single element (Fig. 13.7F). It is laterally compressed, and its sides are excavated. The axis possesses the primitive type of heterocoelous condition present in the last vertebrae of the cervical series. Unfortunately, both spinous processes and zygapophyses are not adequately preserved.

Cervicals 3–5. The centra of these vertebrae are heterocoelous. Cranioventrally, between the parapophysis, the vertebral body has a pronounced depression that receives a ventral projection of the caudal articular surface of the previous vertebra. Caudal to this fossa, the centrum of the third cervical narrows ventrally. In the two following vertebrae (C_4 and C_5), the ventral margin of the centrum becomes a broader surface with an axial depression bound by slender ridges. Dorsal to these ridges, the vertebral body of these vertebrae widens remarkably to form a horizontal plane that connects pre- and postzygapophyses. The zygapophy-

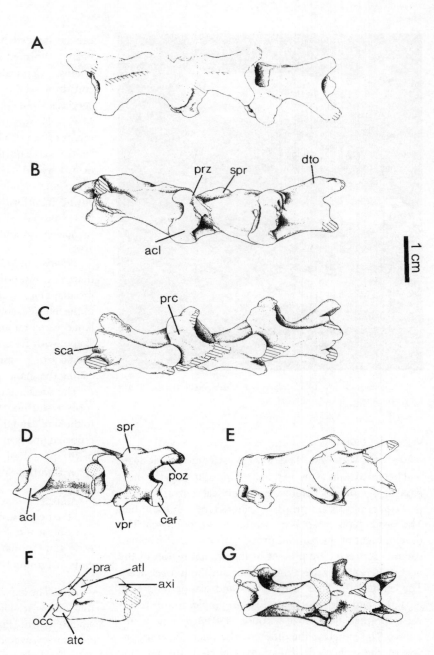

Figure 13.7. Cervical vertebrae of *P. de-ferrariisi*. A, B, C, first three preserved vertebrae of MACN-N-03 in dorsal (A), left lateral (B), and ventral (C) view. D, E, G, last two preserved vertebrae of MACN-N-03 in left lateral (D), dorsal (E), and ventral (G) view. F, left lateral view of the proatlas, atlas, and axis of MACN-N-11. Abbreviations: acl, arco-costal lamina; atc, atlantal centrum; atl, atlantal arch; axi, axis; caf, cranial articular facet; dto, dorsal torus (epi-pophysis); occ, occipital condyle; poz, postzygapophysis; pra, proatlas; prc, carotic process; prz, prezygapophysis; sca, carotic sulcus; spr, spinal process; vpr, ventral process.

seal surfaces are ample, elliptic, and broadly separated one from the other. Of these three vertebrae, only the spinous process of C_5 is preserved. It is broad and well developed.

In C_3 and apparently in C_4 and C_5, the cranial articular surface faces cranioventrally. This orientation of the cranial articular surfaces combined with the strong development of the cranioventral fossa of their centra suggests that these three elements belong functionally to section I of the cervical series of neornithine birds (Zusi and Storer, 1969).

Cervicals 6–8. These vertebrae are poorly preserved in MACN-N-11, the only specimen for which the middle and

cranial sections of the neck are known. The spinous process of C_6, as in C_5, is broad and well developed. Spinous processes are missing in C_7 and C_8. In C_7, ventrally, the body is flat—lacking the slender ridges of C_4 and C_5. The epiphyses of these vertebrae are well developed. They are laterally compressed and occupy a central position above the postzygapophyses. The fact that the cranial articular surfaces of C_7 and C_8 face craniodorsally and that the cranio-ventral fossa of their centra is poorly developed suggests that these vertebrae—and presumably C_6, as well—compare functionally to section II of the neck of neornithine birds (Zusi and Storer, 1969).

Caudal Cervicals. The caudal cervicals are known for MACN-N-03, in which the last five vertebrae are preserved. Because the actual cervical count of *Patagopteryx* is unknown, the precise position of these elements within the cervical series remains uncertain. The first three vertebrae behind the first preserved of these are slightly smaller than the latter, but the last cervical is considerably smaller (Table 13.1; Fig. 13.7).

In the last four cervicals, the spinous processes are reduced (Fig. 13.7A, D, E). In the first vertebra preserved in MACN-N-03, the spinous process is broken in a way that suggests that it might have been more developed. The epipophyses of these five vertebrae are moderately developed (Figs. 13.7B, 13.8A, C), though considerably less than in the middle cervicals. In none of these vertebrae is there evidence of an area for the elastic ligament.

The vertebral bodies of the last cervicals are also heterocoelous. The heterocoely of these vertebrae, however, differs from the typical neornithine condition in that it possesses a lesser degree of transverse convexity and concavity of the caudal and cranial articular surfaces, respectively.

The bodies are ventrally flat (Figs. 13.7C, G, 13.8B, E), with the exception of the last cervical, in which the ventral face is sharp and forms a small ventral process in the cranial portion (Figs. 13.7G, 13.8E). The morphology of this vertebra is transitional between the cervical and thoracic series.

It possesses a ventral process of the centrum and postzygapophyses that are less separated from one another, with their articular surfaces facing more laterally.

Thoracic Vertebrae. The 11 thoracic vertebrae of *Patagopteryx* are shorter than the cervicals. The transverse-processes are long. In the cranial portion of the series, the distance between both prezygapophyses and postzygapophyses decreases caudalward. This distance reaches a minimum value in the midtrunk region, and it increases again in the caudal thoracic vertebrae (Table 13.1). The vertebral centra lack pleurocoels (Figs. 13.9, 13.10), and the first seven vertebrae (cranial and midthoracic vertebrae) have ventral processes (Fig. 13.10H). The articular surfaces of the centra are procoelous, biconvex, opisthocoelous, and heterocoelous depending on the position in the thoracic series. The thoracic vertebrae are free, and there is no formation of a notarium.

The large number of thoracic vertebrae of *Patagopteryx* indicates an early stage in the trend toward cervicalization of the thoracic series and the incorporation of its elements into the synsacrum.

Thoracic Vertebrae (D_1–D_3). The first thoracic vertebra has a broad cranioventral depression in the body and smooth epipophyses over the postzygapophyses; both fea-

TABLE 13.1
Measurements (in mm) of the vertebral column of *Patagopteryx deferrariisi*

	MACN-N-03	MACN-N-14	MUCPv-48
Cervical vertebrae[1]			
Total length of centrum of Cp_2	20.0	—	—
Maximum width at the level of postzygapophyses of Cp_2	>14.1	—	—
Total length of centrum of Cp_2	16.0	—	—
Maximum width at the level of postzygapophyses of Cp_2	12.5	—	—
Thoracic vertebrae			
Total length of centrum of D_2	13.0	11.6	—
Maximum width at the level of postzygapophyses of D_2	10.3	9.0	—
Total length of centrum of D_5	14.5	>11.5	—
Maximum width at the level of postzygapophyses of D_5	7.0	—	—
Total length of centrum of D_{10}	>12.1	—	—
Maximum width at the level of postzygapophyses of D_{10}	10.1	—	—
Synsacrum			
Total length	52.6	—	—
Width at the level of acetabulum	12.3	—	—
Caudal vertebrae			
Maximum length of proximal caudal[2]	9.0	—	—
Maximum length of midcaudal[3]	—	—	10.3
Height of spinous process of midcaudal[3]	—	—	9.6

Note: > indicates that the actual value is estimated to be less than 3 mm larger.
[1] Cp refers to the number of preserved cervicals (e.g., Cp_2 is the second preserved cervical).
[2] Based on the first preserved caudal of MACN-N-03.
[3] Based on the single midcaudal preserved in MUCPv-48.

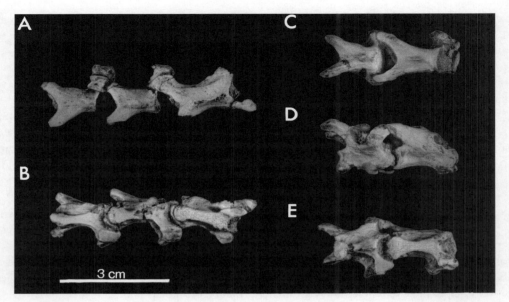

Figure 13.8. Cervical vertebrae of the caudal portion of the neck of *P. deferrariisi* (MACN-N-03). A, B, first three preserved vertebrae in dorsal (A) and ventral (B) view. C, D, E, last two preserved vertebrae in dorsal (C), right lateral (D), and ventral (E) view.

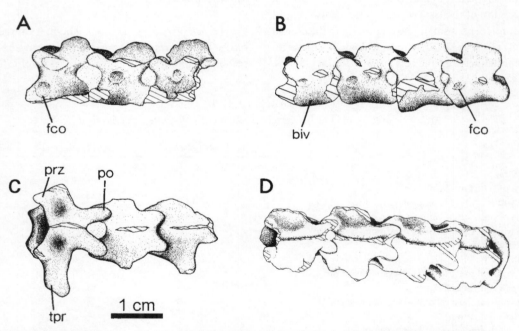

Figure 13.9. Thoracic vertebrae of *P. deferrariisi* (MACN-N-03). A, C, D_2–D_4 in left lateral (A) and dorsal (C) view. B, D, D_5–D_8 in left lateral (B) and dorsal (D) view. Abbreviations: biv, biconvex vertebra (D_5); fco, costal fovea (parapophysis); po, postzygapophysis; prz, prezygapophysis; tpr, transverse process.

tures are typical of cervical vertebrae. These characters, however, are combined with thoracic features such as well-developed transverse processes and ribs articulated—not fused—to the vertebra. The body of this vertebra is broader cranially than caudally, and its articular surfaces show a primitive degree of heterocoely, as do the caudal cervicals.

The bodies of the contiguous two thoracic vertebrae (D_2 and D_3) are even broader cranially. The articular surfaces are transitional from a primitive heterocoelous condition to an opisthocoelous or "pseudopisthocoelous" one. The second thoracic vertebra has a prominent depression in the base of the transverse process (Fig. 13.9C), between the

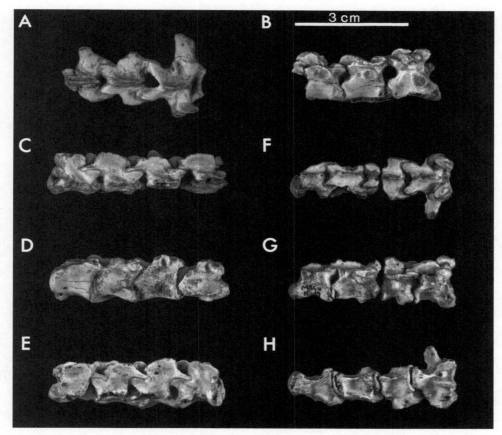

Figure 13.10. Thoracic vertebrae of *P. deferrariisi*. A, B, D_3 and D_4 of MACN-N-03 in dorsal (A) and right lateral (B) view. C, D, E, D_5–D_8 of MACN-N-03 in right dorsolateral (C), right lateroventral (D), and left dorsolateral (E) view. F, G, H, D_2–D_5 of MACN-N-14 in dorsal (F), right lateral (G), and ventral (H) view.

prezygapophyses and the spinous process. This depression is less pronounced in the first thoracic vertebra and absent in the third (Figs. 13.9C, 13.10A).

The parapophyses of these three vertebrae are large and well defined (Figs. 13.9A, 13.10B). In the first thoracic vertebra, they are located in a lower position on the cranial border of the centrum, while in the second thoracic vertebra, they have a slightly more dorsal position. In the third thoracic vertebra, they are located somewhat underneath the transverse process.

Midthoracic Vertebrae (D_4–D_7). In these vertebrae, the centrum is essentially the same width cranially and caudally. On the basis of the large ventral process of the fifth thoracic vertebra of MACN-N-11, these processes must have also been present in D_4–D_6, despite not being preserved. The ventral process of the seventh thoracic vertebra, however, is poorly developed. The articular surfaces are opisthocoelous in D_4, biconvex in D_5, and procoelous in D_6 and D_7. The fifth thoracic vertebra (Figs. 13.9B, 13.10D) is transitional between the procoelous condition of the caudal thoracic ver-

tebrae and the opisthocoelous-heterocoelous condition of the preceding elements. The parapophyses of D_5 nearly reach the level of the transverse processes.

In the midthoracic vertebrae, the separation between both pre- and postzygapophyses reaches a minimum, increasing from D_7 backward.

Caudal Thoracic Vertebrae (D_8–D_{11}). The last four thoracic vertebrae are procoelous, as are the two preceding them (Figs. 13.9B, 13.10D). The bodies of these vertebrae are broad and dorsoventrally compressed and kidney-shaped in cranial aspect (Figs. 13.11–13.13). Ventrally, they lack hypapophyses.

As was mentioned previously, the separation between both pre- and postzygapophyses increases with respect to the midthoracic vertebrae (Table 13.1), although it is not larger here than in the cranial thoracic vertebrae. The spinous processes are high and more robust than those of the preceding vertebrae. The transverse processes are slender and of moderate length with respect to the centrum (Fig. 13.13D–F). Thoracic vertebrae D_{10} and D_{11} are situated be-

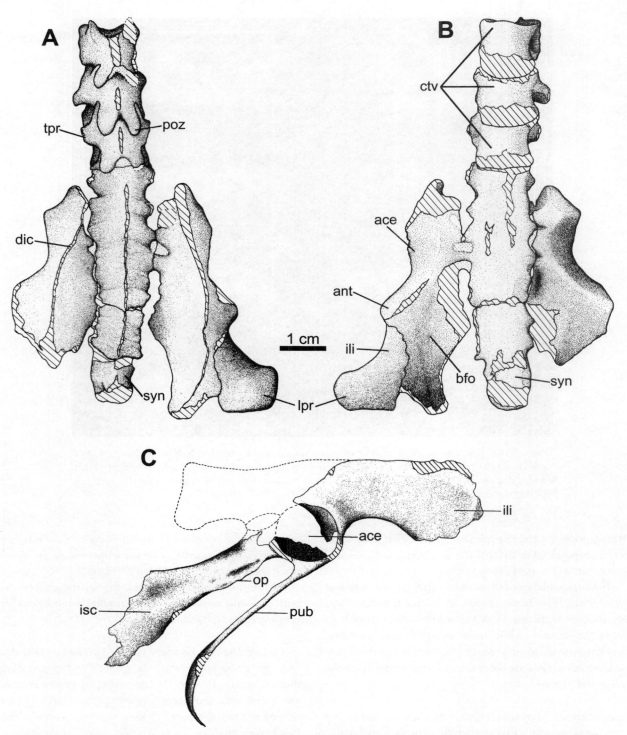

Figure 13.11. Caudal thoracic vertebrae, synsacrum, and ilia of *P. deferrariisi* (MACN-N-03) in dorsal (A) and ventral (B) view. Pelvis of *P. deferrariisi* (MACN-N-11) in right lateral view (C). Abbreviations: ace, acetabulum; ant, antitrochanter; bfo, brevis fossa; ctv, caudal thoracic vertebrae; dic, dorsal iliac crest; ili, ilium; isc, ischium; lpr, lateral projection of the postacetabular wing; op, obturator process; poz, postzygapophysis; pub, pubis; syn, synsacrum; tpr, transverse process.

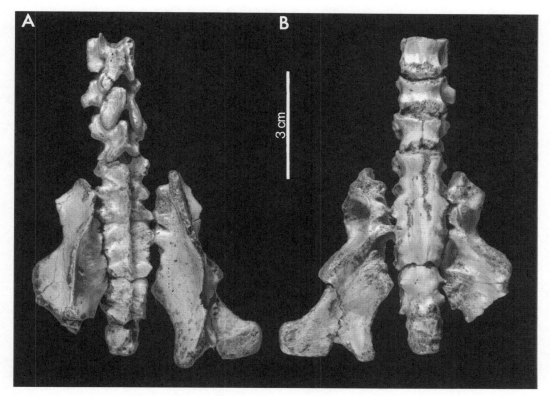

Figure 13.12. Caudal thoracic vertebrae, synsacrum, and ilia of *P. deferrariisi* (MACN-N-03) in dorsal (A) and ventral (B) view.

tween the ilia, although they are not co-ossified to the synsacral vertebrae (Figs. 13.11, 13.12).

Synsacrum. The synsacrum of *Patagopteryx* is formed by nine vertebrae (Figs. 13.11, 13.12). These are completely fused, with no differentiation of either zygapophyses or individual spinous processes. As in the caudal thoracic vertebrae, the synsacral bodies are broad and dorsoventrally compressed (Figs. 13.11B, 13.12B). Caudal to the midpoint, the vertebral centra gradually decrease in size. The cranial articular surface of the synsacrum is concave, while the caudal one is convex. In lateral view, caudal to the acetabulum, the synsacrum curves slightly ventrally.

The dorsal processes of the synsacral vertebrae are fused, forming a longitudinal crest that decreases in height caudalward. Although the cranial part of this crest is not complete in any of the available specimens, the preserved portion of MACN-N-03 suggests that it was as high as the dorsal process of the caudal thoracic vertebrae (Fig. 13.12A).

Ventrally, the vertebral bodies of the cranial half define a nearly flat surface grooved craniocaudally by a shallow broad furrow (Fig. 13.12B) comparable to the ventral sulcus of neornithine birds (Baumel et al., 1979).

At the level of the acetabulum, the transverse processes of S_3 and S_4 are fused to the medial surfaces of the ilia. Un-

fortunately, the remaining transverse processes are missing, and it is not possible to determine whether they were fused to the ilium, as well.

Caudal Vertebrae. Although in MACN-N-11 there is a minimum of five free caudals following the synsacrum, the existence of isolated, free middle caudals in MUCPv-48 and MUCPv-207 suggests a much larger number of free tail vertebrae. Yet their true number remains unknown.

The caudal vertebrae of *Patagopteryx* have well-developed, ventrally directed transverse processes but lack both pre- and postzygapophyses (Fig. 13.13A–C). In the proximal caudals the neural arch is high, defining a triangular vertebral canal. In the middle caudals, the neural arch is lower, and it defines a more arched vertebral canal. The proximal caudals are strongly procoelous, although the degree of their procoelous condition decreases distally. In the middle caudals, as preserved in MUCPv-48 and MUCPv-207, the proximal articular surface is just slightly concave, while the distal surface is convex. No haemapophyses have been found in any of the preserved elements of the caudal series.

In the caudal vertebrae preserved in MACN-N-03 and MACN-N-11, the spinous processes are missing from their base. The middle caudals of MUCPv-48 and MUCPv-207, however, have very high (Table 13.1), laminar spinous

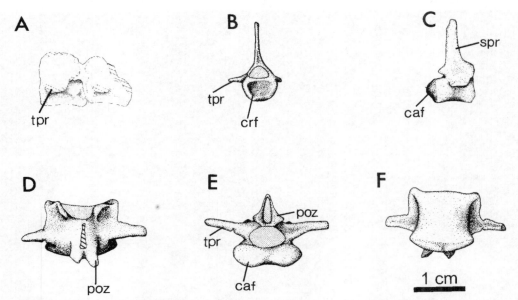

Figure 13.13. Thoracic and caudal vertebrae of *P. deferrariisi*. A, first two caudal vertebrae of MACN-N-03 in left lateral view. B, C, middle caudal vertebra of MUCPv-48 in cranial (B) and right lateral (C) view. D, E, F, caudal thoracic vertebra of MUCPv-48 in dorsal (D), caudal (E), and ventral (F) view. Abbreviations: caf, caudal articular facet; crf, cranial articular facet; poz, postzygapophysis; spr, spinal process; tpr, transverse process.

processes (Fig. 13.13B, C), which taper toward their end. The remarkable development of these processes suggests that the tail of *Patagopteryx* was long. Unfortunately, the available material prevents determination about the presence or absence of a pygostyle.

Thoracic Girdle and Sternum

The thoracic girdle seems to comprise only the scapula and coracoid. These two bones articulate at an angle of approximately 65° (Figs. 13.14B, 13.17M). The glenoid cavity defined by these two bones is deep and narrow. The presence of an ossified furcula is unlikely. At least in MACN-N-11, in which the elements of the thoracic girdle were articulated, there are not even vestiges of a furcula. The two elements interpreted as portions of clavicles by Alvarenga and Bonaparte (1992) are in fact the shoulder ends of the coracoids.

Coracoid. The coracoid is straight, triangular, and proportionally short (~65% of the scapula; Table 13.2; Figs. 13.14, 13.15). The acrocoracoid process is very reduced, and the procoracoid process is absent (Figs. 13.16A–C, 13.17C, D, G, H). The proximal end is completely different from that of other birds. The glenoid facet is located at the shoulder end of the coracoid. It is wide and flat, and it slopes laterally (Figs. 13.16A, 13.17C, D). Its borders are parallel, and its lateral margin is round. The glenoid facet overhangs the lateral border of the coracoid. On the medial border of this facet there is an inflated area that seems to define the ventral limit of the articular surface with the

scapula. This area is barely distinct from the glenoid facet, however.

The medial surface of the shoulder end of the coracoid is flat (Figs. 13.16B, 13.17G), whereas its lateral surface is concave. Distal to the shoulder end, the coracoid narrows, and below the midshaft it widens to form the broad sternal end (Fig. 13.14). In dorsal view, the coracoidal shaft is flat, without the groove present in the fragment reported by Alvarenga and Bonaparte (1992) as the sternal portion of the coracoid. The fragment described by these authors and interpreted as part of the holotype is likely not of *Patagopteryx*, since no bone with this shape has been found in any of the known specimens.

The ventral surface of the coracoid is convex. In the shoulder half, a ventral border separates a medial from a lateral surface, which, combined with the flat dorsal surface, gives a subtriangular cross section to the coracoid. In ventral view, the lateral margin is strongly concave, and there is no supracoracoidal nerve foramen or notch (Figs. 13.14, 13.15B). The sternal end is broad and dorsoventrally compressed.

Scapula. The scapula is laminar and relatively broad, especially in its distal half (Figs. 13.14B, 13.15A). The glenoid facet is large and subcircular (Fig. 13.16D, E). Above this facet, proximally, there is a tubercle that topographically corresponds to the coracoidal tubercle of the scapula of neornithine birds. Ventral to this tubercle, bordering the ventroproximal margin of the glenoid facet, there is a flat and elongate surface situated in an oblique plane with re-

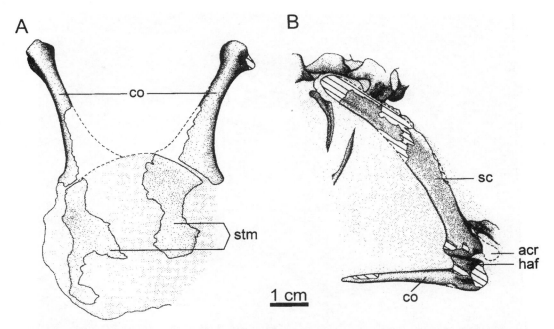

Figure 13.14. Thoracic girdle and sternum of *P. deferrariisi* (MACN-N-11). A, coracoids and sternum in ventral view; B, right scapula and coracoid in lateral view. Stipple pattern indicates matrix, and cross-hatch pattern indicates broken or poorly preserved bone. Abbreviations: acr, acromion; co, coracoid; haf, humeral articular facet; sc, scapula; stm, sternum.

spect to the latter facet (Fig. 13.16D, F). This surface abuts with another surface located medially to the glenoid facet of the coracoid (see previously). The role of the tubercle mentioned earlier in the scapulocoracoid articulation is uncertain. In neornithine birds as well as in *Ichthyornis,* a comparable tubercle (coracoidal tubercle) articulates in a facet of the coracoid; in the coracoid of *Patagopteryx,* however, there is no trace of this facet.

The acromion is large (Figs. 13.16D–F, 13.17A, B, E, F). It is remarkably expanded dorsoventrally, having a small proximal projection. In proximal view, it is separated from the glenoid cavity by a more slender neck. Interestingly, the presence of a well-developed acromion suggests the presence of clavicles, although, as pointed out earlier, the completely articulated MACN-N-11 lacks any sign of these elements.

The scapular blade is sagittally curved, with the ventral border concave (Figs. 13.14B, 13.15A). Most basal birds have straight scapular blades (e.g., Chiappe et al., 1999; Sanz et al., Chapter 9 in this volume; Chiappe and Walker, Chapter 11 in this volume). *Patagopteryx* appears to be the most basal bird for which the scapula is sagittally curved; this derived feature is interpreted as a synapomorphy of *Patagopteryx* plus Ornithurae (Chiappe, 1996a; Chapter 20 in this volume). In the proximal (shoulder) half the width of the shaft increases distally, whereas in the distal half it decreases toward the apex. In the midshaft, the dorsal border forms a slender ridge that projects dorsally (Fig. 13.15A). Hence, the lateral surface of the midshaft is concave.

The scapular apex is not preserved in any of the available specimens; however, the right scapula of MACN-N-11 preserves the impression of it showing that the apex had a round end (Fig. 13.15A).

Sternum. The sternum is thin and slightly convex ventrally (Figs. 13.14A, 13.15B). The cranial border is round. The articular facets for the coracoids are well separated from each other, as is typical for flightless birds. Unfortunately, the median portion is badly preserved, preventing confidence about the presence or absence of a carina (Figs. 13.14A, 13.15B). The slight convexity of the area near the median portion, however, suggests that a carina was probably absent. No other information (e.g., dimensions, lateral processes) is available considering the state of preservation of the single known sternum (MACN-N-11).

Thoracic Limb

Humerus. The humerus is robust and longer than the ulna (Table 13.3; Fig. 13.18). In dorsal view, it has a sigmoid aspect, with the proximal half convex and the distal half concave toward the cranial face. It lacks a pneumotricipital fossa and foramen (Figs. 13.17I, 13.18B). As in other nonornithurine birds, the humeral head is cranially concave and caudally convex, and it meets dorsally with the deltopectoral crest (Fig. 13.18C–E). In cranial view, immediately distal to the head and slightly displaced dorsally, there is a small and shallow circular depression that recalls the con-

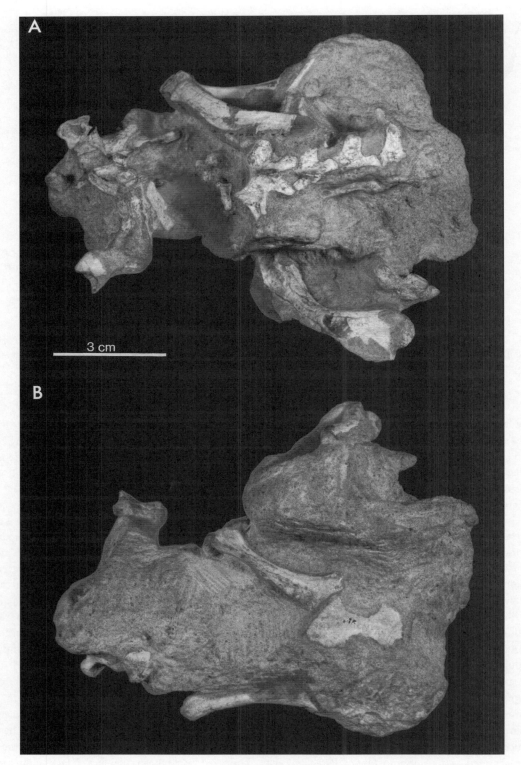

Figure 13.15. Thoracic girdle and sternum of *P. deferrariisi* (MACN-N-11). A, right scapula in lateral view, cranial thoracic vertebrae in dorsal view, left humerus in caudal view; B, coracoids and sternum in ventral view.

TABLE 13.2
Measurements (in mm) of the thoracic
girdle of *Patagopteryx deferrariisi*

	MACN-N-11
Coracoid[1]	
Total length	38.0
Maximum width of shoulder end	8.9
Scapula[2]	
Distance from shoulder margin of humeral articular facies to apex	60.1
Maximum width of blade	7.7

[1]Based on left element.
[2]Based on right element.

dition present in Enantiornithes (Chiappe and Walker, Chapter 11 in this volume). The deltopectoral crest is robust and extends through the proximal third of the humerus (Table 13.3; Figs. 13.17I, J, 13.18). The bicipital area is poorly developed. Unfortunately, the area of the ventral tuberosity is badly preserved in all available specimens, preventing determination of whether it was projected caudally.

The humeral shaft is somewhat craniocaudally compressed, and it has a concave ventral border. Caudally, there are two pronounced ridges (Fig. 13.18B). The most proxi-

mal, the strongest, originates on the dorsal border, distal to the midshaft. This ridge runs proximoventrally until it nears the deltopectoral crest, where it disappears. The second ridge, situated distal to the latter, also has its origin on the dorsal border of the humerus, near the distal end. These ridges are interpreted as intermuscular lines separating different areas of muscular attachment.

The distal end of the humerus has a width equivalent to that of the proximal end (Figs. 13.17I, J, 13.18A, B). Unlike in more primitive birds (e.g., *Archaeopteryx,* Enantiornithes), the main axis of the distal end is oriented in the same plane as that of the proximal end. The dorsal condyle is well developed. It has an elliptical outline with its major axis oriented 30° to the axis of the shaft (Figs. 13.17J, 13.18A). This condyle is far from the dorsal margin of the humerus and is strongly projected distally in such a way that it is visible in caudal view. The distal projection of the dorsal condyle is much stronger than the typical condition in neornithine birds. The ventral condyle is circular, although it is less bulbous than the lateral one. A well-defined brachial depression is absent, although M. brachialis anticus might have originated in a depressed area proximal to the ventral condyle. Both dorsal and ventral epicondyles are poorly developed (Fig. 13.18A, B). The caudal surface of the distal end is flat, lacking tricipital sulci. Proximal to the ventral condyle there is a smooth depression forming a small olecranon fossa.

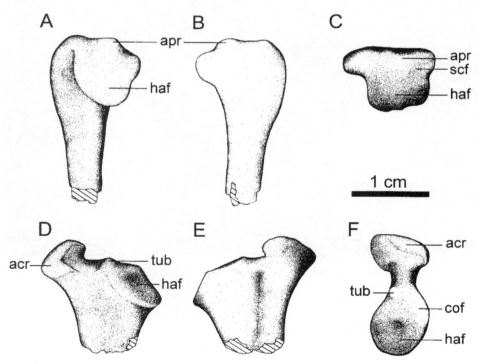

Figure 13.16. Thoracic girdle of *P. deferrariisi* (MACN-N-03). A, B, C, shoulder end of left coracoid in lateral (A), medial (B), and end (C) view; D, E, F, shoulder end of right scapula in lateral (D), medial (E), and end (F) view. Abbreviations: acr, acromion; apr, acrocoracoidal process; cof, coracoidal facet of scapula; haf, humeral articular; scf, scapular facet of coracoid; tub, tubercle facet.

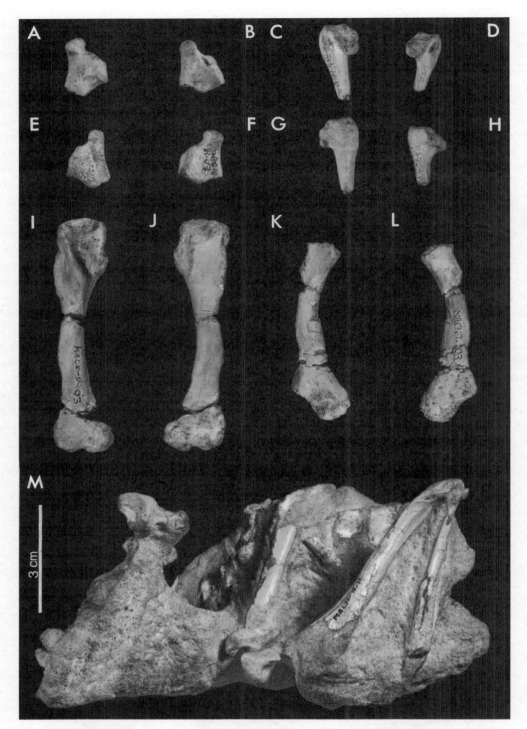

Figure 13.17. Thoracic girdle and humerus of *P. deferrariisi*. A, B, E, F, shoulder end of right (A, E) and left (B, F) scapula of MACN-N-03 in lateral (A, F) and medial (B, E) view; C, D, G, H, shoulder end of left (C, G) and right (D, H) coracoid of MACN-N-03 in lateral (C, D) and medial (G, H) view. I, J, K, L, left (I, J) and right (K, L) humerus of MACN-N-03 in caudal (I, L) and cranial (J, K) view. M, lateral view of the scapula and humerus of MACN-N-11.

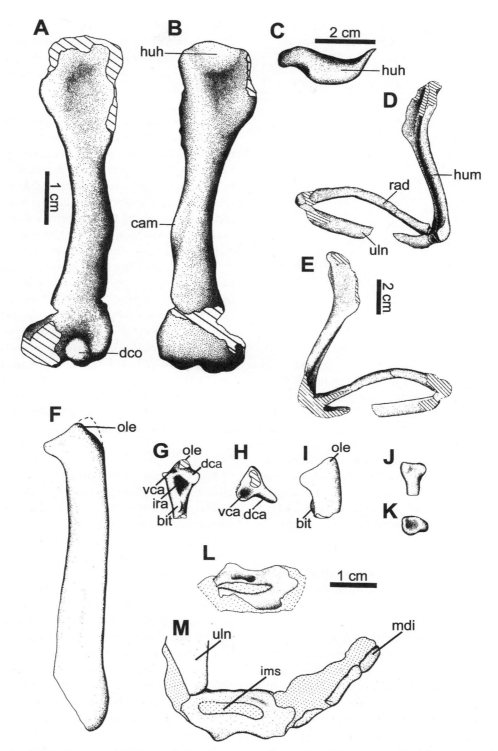

Figure 13.18. Forelimb of *P. deferrariisi*. A, B, left humerus of MACN-N-03 in cranial (A) and caudal (B) view; C, right humerus of MACN-N-11 in proximal view; D, E, humerus, ulna, and radius of MACN-N-11 in ventral (D) and dorsal (E) view. F, left ulna of MACN-N-11 in caudal view. G, H, I, proximal end of left ulna of MACN-N-03 in interosseal (G), proximal (H), and caudal (I) view. J, K, proximal end of left radius of MACN-N-03 in interosseal (J) and proximal (K) view. L, M, left carpometacarpus and phalanges of major digit (II) of MACN-N-11 in ventral (L) and dorsal (M) view. Abbreviations: bit, bicipital tubercle; cam, caudal intermuscular line; dca, dorsal cotyla; dco, dorsal condyle; huh, humeral head; hum, humerus; ims, intermetacarpal space; ira, incisura radialis; mdi, major digit (digit II); ole, olecranon; rad, radius; uln, ulna; vca, ventral cotyla.

TABLE 13.3

Measurements (in mm) of the forelimb of *Patagopteryx deferrariisi*

	MACN-N-03	MACN-N-11
Humerus		
Total length	66.3(l)	59.8(r)
Maximum width of distal end	17.2(l)	>14.8(r)
Length of pectoral crest	19.7(l)	19.5(r)
Ulna		
Total length	—	>51.5(l)
Maximum width of proximal end	9.6(l)	>9.1(l)
Carpometacarpus		
Total length	—	>22.7(r)
Midshaft width	—	8.9(r)

Note: (l) and (r) refer to left and right elements, respectively. > indicates that the actual value is estimated to be less than 3 mm larger.

Ulna. The ulna is craniocaudally compressed, especially in its distal half, in which it forms a slender lamina. In the proximal end (Fig. 13.18F–I), the olecranon is well developed and the cotylae well defined. The dorsal cotyla is located distal to the ventral one, being projected dorsally. The articular surface of this cotyla is nearly flat. The ventral cotyla is round. Both cotylae are separated by a low intercotylar area. In proximal view, these cotylae define a concave cranial border that corresponds to the proximal margin of the radial proximal incision. This well-developed depression is triangular, and its distal vortex ends in a prominent bicipital tubercle. Ventrally, there is no evidence of the impression for M. brachialis anticus. The caudal face is flat, and it has no papillae for the insertion of secondary feathers (Fig. 13.18F). In caudal view, the proximal half of the shaft has a uniform width, while the distal half broadens distally.

Radius. The radius is significantly more gracile than the ulna. The shaft has a subcircular cross section and a ventral curvature that is increased distally. The proximal end is broad with respect to the shaft and in proximal view is subtriangular (Fig. 13.18G, K). The humeral cotyla is subcircular. In MACN-N-03, there is a projection of the proximal end directed toward the ulnar side, a feature comparable to the capital tubercle of neornithine birds and *Ichthyornis*. Distally to the proximal end there is a weak bicipital tubercle. This tubercle, however, is in a more distal position with respect to that in Neornithes.

There is almost no information about the distal end. Only in MACN-N-11 is this end preserved, although badly. The remarkable torsion of the distal end toward the ulnar side (Fig. 13.18D, E) is notable, although it might be exag-

gerated by distortion. In the dorsal portion of the distal end, there is a condylar structure that is tentatively interpreted as the aponeurosis tubercle, the insertion of the tendon that fans out the flight feathers of the wrist region (Baumel and Witmer, 1993).

Carpometacarpus and Phalanges. The carpometacarpus is very short (Table 13.3; Fig. 13.18D, E). The major metacarpal (II) is less robust than the minor one (III) (Fig. 13.18L, M). In its cranial border, proximal to the midshaft, metacarpal II bears a laminar projection directed cranioventrally. Major and minor metacarpals abut proximally and distally, forming broad symphyses and enclosing an elongated intermetacarpal space. Unfortunately, both distal and proximal carpals are not preserved.

Only the major digit (II) preserves anatomical information. This digit is proportionally long (as long as the metacarpals), being formed by three phalanges (Fig. 13.18M). The proximal phalanx of this digit has the greatest depth dorsoventrally. This phalanx is not broad and dorsoventrally compressed as is the proximal phalanx of the major digit of carinate birds (common ancestor of *Ichthyornis* and Neornithes plus all its descendants; Clarke and Chiappe, 2001). The intermediate phalanx is elongate and cylindrical, also being the longest of the three. The distal phalanx is a claw of triangular shape.

Pelvic Girdle

Ilium. The ilia are elongate and broadly separated from each other throughout their length (Figs. 13.11, 13.12, 13.19).

The preacetabular wing, which is slightly longer than the postacetabular (Table 13.4), is vertical and laterally compressed (Figs. 13.11C, 13.19). The vertical portion of the preacetabular wing suggests a poor development of M. iliofemoralis cranialis (= M. iliotrochantericus caudalis), an important muscle in the pelvis of extant birds, the function of which is still unclear (Raikow, 1985). In lateral view, the ventral margin of the preacetabular wing is concave, forming a notch that becomes more pronounced near the acetabulum. Dorsally, the preacetabular wing continues backward as a strong crest that passes above the acetabulum. The relationship between this crest and the other iliac regions suggests its homology with the iliac crest of neornithine birds.

The acetabulum is large (Table 13.4) and round and has thick walls (Fig. 13.11C). MACN-N-11 shows that the acetabulum was not completely perforated but partially obliterated by projections of the ilium, pubis, and ischium. A partial obliteration of the acetabulum has been reported for hesperornithiforms (Marsh, 1880; Martin and Tate, 1976) and *Archaeopteryx* (Martin, 1983, 1991). Caudal to the acetabulum there is a well-developed antitrochanter (Table 13.4). This is below the dorsal margin of the acetabulum, and its main axis points caudoventrally. This condition re-

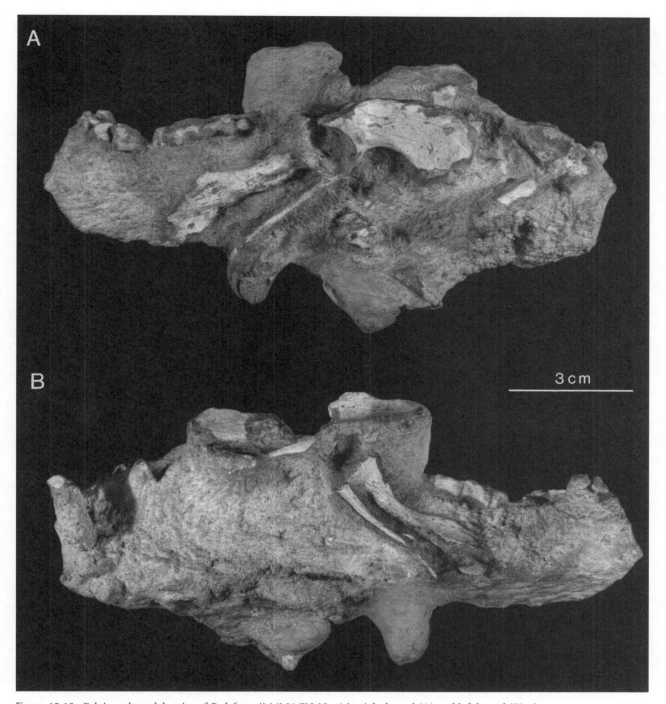

Figure 13.19. Pelvis and caudal series of *P. deferrariisi* (MACN-N-11) in right lateral (A) and left lateral (B) view.

sembles that of alvarezsaurids (Perle et al., 1994; Chiappe, Norell, and Clark, Chapter 4 in this volume), but it differs from the typical neornithine (as well as enantiornithine) condition, in which the antitrochanter is above the acetabular margin and its main axis is caudodorsally oriented.

The postacetabular wing of *Patagopteryx* is transversally broad (Figs. 13.11, 13.12) and lower than the preacetabular wing. Dorsally, it bears a prominent caudal iliac crest that continues cranially in the dorsal border of the preacetabular wing. This crest separates an internal surface, ventromedially slanted, from an external one, sloped ventrolaterally. The caudal iliac crest, high and laterally compressed, has an inflexion point at the level of the caudal third of the acetabulum. In front of this point, continuing in the preacetabular wing, this crest is slightly concave outward, while it is convex behind the inflexion point.

TABLE 13.4

Measurements (in mm) of the pelvic girdle of *Patagopteryx deferrariisi*

	MACN-N-03	MACN-N-11	MUCPv-48
Ilium			
Maximum length of postacetabular wing	29.9(r)	—	>28.7(r)
Maximum length of preacetabular wing	—	35.6(l)	—
Maximum length of acetabulum	15.4(r)	14.3(r)	—
Maximum length of antitrochanter	9.7(r)	—	—
Ischium			
Maximum preserved length	—	52.5(r)	—
Dorsoventral width at the level of obturator process	—	8.6(r)	—
Pubis			
Total length	—	>49.6(r)	—

Note: (l) and (r) refer to left and right elements, respectively. > indicates that the actual value is estimated to be less than 3 mm larger.

The external surface of the postacetabular wing is large and projected ventrolaterally (Figs. 13.11, 13.12). Caudal to the acetabulum, the lateral border of this surface forms a strong notch that ends in the caudolateral corner of the ilium. This area has a round, laterally projected border. The caudal margin of the external surface is formed by a thickening that caudally limits a depressed area developed behind the antitrochanter and lateral to the iliac crest. Ventrally, both the external and internal surfaces of the postacetabular wing delimit a broad fossa that originates behind the acetabulum and widens caudally (Figs. 13.11B, 13.12B)—this fossa strongly resembles the brevis fossa of nonavian theropods.

Ischium. The ischium is an elongate, laterally compressed bone (Figs. 13.11C, 13.19). In its proximal half, the ventral margin forms a thin obturator process that extends for about 1.5 cm. Caudal and parallel to this process, on the lateral face, there is a longitudinal groove extending nearly the same length as the obturator process (Fig. 13.11C). It is likely that this groove provided a larger area for the insertion of the ischiopubic ligament, which in extant birds attaches to the obturator process and closes the obturator foramen. Given the ample notch separating the ischium from the pubis, the obturator foramen of *Patagopteryx* must have been much larger than that of ornithurine birds.

The distal half of the ischium is an ample and flat surface that laterally bears a median longitudinal ridge. Although the distal ends of the ischia are not preserved, it is clear that they did not form a distal symphysis. In contrast to the condition in Enantiornithes and other basal birds, the ischium of *Patagopteryx* lacks a proximodorsal process. Hence, if an ilioischiadic fenestra was present, it could not have been caudally closed by bone.

Pubis. The pubis is proximally fused to the ischium. It is a craniocaudally compressed, beltlike bone of uniform width throughout its length (Figs. 13.11C, 13.19). The craniocaudal width is less than half of the transverse width. The straight proximal half of the pubis is oriented caudoventrally, defining an angle of about 45° with respect to the main axis of the synsacrum (Fig. 13.11C). The distal end of the pubis is curved cranially and slightly medially, in a way that it is oriented cranioventrally. Unlike nonavian theropods and most basal birds, the distal ends of the pubes of *Patagopteryx* do not fuse to each other.

Pelvic Limb

Femur. The femur is robust and strongly arched craniocaudally (Figs. 13.20, 13.21). In cranial view, the medial margin is generally concave, although its central third is flat to slightly convex. In this view, the lateral margin is virtually flat with only a slight concavity in the central portion. The femur is hollow, and its walls are thick (approximately 1.7 mm). On the femoral shaft, two strong intermuscular lines are present. The cranial intermuscular line originates above the internal condyle (Fig. 13.20A), gradually passing toward the external margin. In the central portion, this line becomes more pronounced, and proximally it weakens until it vanishes distal to the trochanteric crest. Most likely, this line represents the boundary between the origin of M. femorotibialis medius and M. femorotibialis externus, as occurs in some neornithine birds (Holmes, 1963; McGowan, 1979). The caudal intermuscular line is a prominent ridge that originates as a projection of the medial margin of the popliteal fossa and ascends toward the proximal end as a straight line (Figs. 13.20B, 13.21B). Proximal to the midshaft, this line slopes medially to disappear about 1.5 cm distal to the femoral head. This line is likely the extensive area of attachment of M. pubo-ischio-femoralis (= M. adductor longus et brevis; McGowan, 1979), a major femoral retractor (Raikow, 1985). It must have also formed the limit of the caudal extension of M. femorotibialis externus (Hudson, 1937; Holmes, 1963). Also on the caudal face—lateral to the intermuscular line and somewhat proximal to the midshaft—there is a small foramen similar to the nutrient foramen of neornithine birds. About 4–5 mm proximal to this foramen there is an elliptical tubercle displaced laterally (Fig. 13.21C). This 5- to 7-mm-long tubercle is well developed in all known specimens, and it probably corresponds to the insertion of M. iliofemoralis of neornithine birds (Hudson, 1937; Holmes, 1963; George and Berger, 1966). The central section of the shaft is subcircular, although the intersection of the two intermuscular lines gives it a more irregular outline.

The femoral head is large, robust, and spherical. It faces medially and is separated from the trochanter by a thick, short neck (Figs. 13.20, 13.21). The trochanteric crest is well developed. The trochanter was not projected proximally; it does not exceed the level of the head. The lateral surface of the proximal end shows areas for muscular attachment (Fig. 13.20F). The most pronounced of these are two broad grooves that together form a crescentic line that is cranially convex. It is likely that these were the areas of insertion of the femoral protractors of the iliotrochantericus group (see Rowe, 1986, for a discussion of the different origin of the three muscles constituting this group).

In the distal end, the cranial surface is flat, lacking a patellar groove (Figs. 13.20A, 13.21A). Caudally, there is an ample, triangular, popliteal fossa (Table 13.5; Figs. 13.20B, 13.21B, C). On the base of this fossa, in MACN-N-11, there are two small depressions of different size. On the lateral border of the popliteal fossa, proximal to the external condyle, there is a strong tubercle; on the medial border of this fossa there is another tubercle of smaller size. It is probable that the first tubercle served as the area of attachment of the ligamentous fibers of ansa M. iliofibularis ("biceps loop" of George and Berger, 1966; Baumel et al., 1979), which acts as a pulley for the action of M. iliofibularis (George and Berger, 1966). The other tubercle probably coincides with the insertion of M. puboischiofemoralis (George and Berger, 1966). The distal border of the popliteal fossa is limited by a low, transverse ridge that connects both condyles.

The distal condyles are strongly projected caudally. The medial condyle is flat on its caudalmost portion but convex distally. The lateral condyle is much larger. It possesses a well-developed tibiofibular crest that delimits a large, lateral area for the articulation with the fibula.

Tibiotarsus. The tibiotarsus is a slender, elongate bone (Figs. 13.22–13.24). The shaft is straight and hollow, with walls approximately 0.8–1.0 mm thick. The shaft is slightly compressed craniocaudally (Table 13.5), especially in its distal portion. The fibular crest extends over the proximal third of the lateral face; it is projected laterally and cranially. This well-developed crest rises 25 mm distal to the proximal surface, reaching its maximum development near its distal end.

The proximal articular surface of the tibiotarsus is craniocaudally expanded owing to the cranial extension of the single cnemial crest. The medial articular area was more caudally projected than the lateral one. The cnemial crest is thick and robust throughout its length, decreasing in thickness distally (Figs. 13.22A, 13.23A, 13.24A). In cranial view, it is laterally twisted. Medially, it forms a slightly convex, smooth surface that extends over all of the proximal end of the tibiotarsus. On this surface, proximal to the distal end of the cnemial crest and displaced to the caudal border of the tibiotarsus, is a small tubercle that must have served as an area for muscular attachment. The lateral surface of the cnemial crest is concave, ending on a ridge in its cranialmost portion.

In the distal end, both astragalus and calcaneum are completely fused to each other and to the tibia (Figs. 13.22–13.24). The distal condyles are separated from one another by a relatively deep intercondylar groove. As in the Late Cretaceous Malagasy bird *Vorona* (Forster et al., Chapter 12 in this volume), this groove opens proximally into a circular fossa. A transverse groove projects medially from this fossa and undercuts the proximal margin of the medial condyle.

As in other nonornithurine birds, the medial condyle is much larger and broader than the lateral (Fig. 13.23A). At the proximal border of the medial condyle, on the medial face, there is a medially directed prominence. This feature is comparable to the medial ligamentary prominence of neornithine birds, on which the ligaments of the tibiotarsal-tarsometatarsal articulation are attached (Baumel et al., 1979). The medial face of the medial condyle has a large, circular depression. The lateral condyle projects less cranially than the medial, and its proximal border rises less abruptly from the shaft. On its lateral face there is a small, circular depression (depressio epicondylaris lateralis; Baumel et al., 1979).

On the medial margin of the distal end—proximal to the medial condyle—there is an elongated, proximodistal depression, the development of which varies among different specimens (Figs. 13.22A, 13.23A, 13.24A, B). The medial border of this depression is delimited by a smooth ridge. In MACN-N-03, this depression forms a groove interpreted by Alvarenga and Bonaparte (1992) as the tendinal groove. On the cranial surface, and centered with respect to the condyles, there is a deep circular fossa. *Patagopteryx* does not have a supratendinal bridge.

Fibula. The proximal end of the fibula is expanded craniocaudally and compressed laterally. Somewhat distal to the proximal end, at the level of the proximal half of the

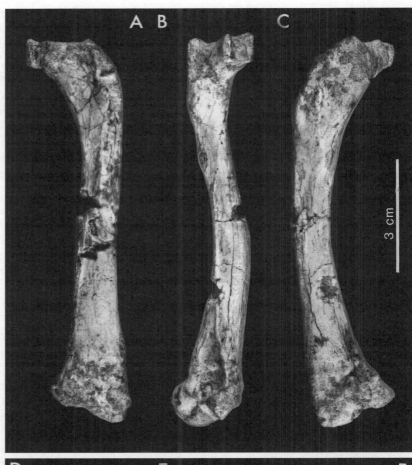

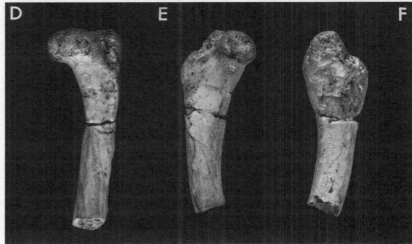

Figure 13.20. Femur of *P. deferrariisi*. A, B, C, left femur of MACN-N-11 in cranial (A), caudal (B), and lateral (C) view; D, E, F, proximal half of right femur of MACN-N-03 in caudal (D), medial (E), and lateral (F) view.

fibular crest of the tibiotarsus, there is a large and robust, laterally directed tubercle (Figs. 13.22D, 13.23, 13.24C, D) for the insertion of M. iliofibularis. The position of this tubercle resembles that of the Late Cretaceous Malagasy *Vorona* (Forster et al., Chapter 12 in this volume) as well as the bizarre alvarezsaurids (Chiappe et al., 1996; Chiappe, Norell, and Clark, Chapter 4 in this volume). Distal to the iliofibularis tubercle, the fibula narrows, forming a slender

bar that progressively becomes craniocaudally compressed and migrates cranially. The distal end of the fibula is almost laminar and is fused to the lateral border of the cranial surface of the tibiotarsus (Figs. 13.23C, D, 13.24C, D). Clearly, the fibula fails to reach the proximal tarsals.

Tarsometatarsus. The tarsometatarsus is much shorter than both the femur (~48% of the femoral length) and the

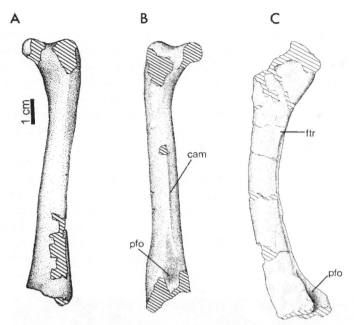

Figure 13.21. Femur of *P. deferrariisi*. A, B, left femur of MUCPv-48 in cranial (A) and caudal (B) view; C, left femur of MACN-N-03 in lateral view. Abbreviations: cam, caudal intermuscular line; ftr, fourth trochanter; pfo, popliteal fossa.

TABLE 13.5
Measurements (in mm) of the hindlimb of *Patagopteryx deferrariisi*

	MACN-N-03	MACN-N-11	MUCPv-48
Femur			
Total length	>98.8(r)	>99.1(l)	>97.6(l)
Height of popliteal fossa	16.0(r)	16.6(l)	—
Mediolateral width at midshaft	10.1(r)	10.2(l)	9.8(l)
Tibiotarsus			
Total length	>137.0(l)	>136.2(l)	>140.0
Length of cnemial crest	21.3(l)	—	—
Craniocaudal length of proximal end	>23.9(l)	—	—
Craniocaudal width at midshaft	—	7.2(l)	7.8(l)
Mediolateral width at midshaft	—	9.2(l)	9.5(l)
Maximum width of distal end	—	14.5(r)	—
Fibula			
Total length	—	—	>111.6(l)
Tarsometatarsus			
Total length	—	50.8(r)	51.0(l)
Maximum width of proximal end	—	>13.5(l)	14.4(l)
Craniocaudal width at midshaft	—	5.1(r)	5.1(l)
Mediolateral width at midshaft	—	11.1(l)	10.6(l)
Total length of metatarsal I	—	14.8(r)	—

Note: (l) and (r) refer to left and right elements, respectively. > indicates that the actual value is estimated to be less than 3 mm larger.

tibiotarsus (~34% of the tibiotarsal length; Table 13.5). As in ornithurine birds, metatarsals II–IV are completely fused with one another and with the distal tarsals (Figs. 13.25, 13.26). In contrast to the condition in these birds, however, longitudinal grooves allow for individualization of each metatarsal, and the medullary cavity of these bones remains individualized (Fig. 13.25G). Metatarsals II–IV are placed on the same transverse plane (Fig. 13.25G), contrasting with the condition present in neornithine birds, in which the proximal end of metatarsal III is plantarly displaced with respect to metatarsals II and IV. Metatarsals II and III are more robust than metatarsal IV, which is slightly slender and more gracile. Metatarsal III is straight. In cranial view, it has the same thickness throughout its length. In contrast, metatarsals II and IV narrow distally. Metatarsal I articulates with the medial border of metatarsal II.

The proximal articular surface is well preserved in MACN-N-10 (Figs. 13.25F, 13.26C). It has a triangular shape with two well-defined, circular cotylae. The medial cotyla is significantly larger than the lateral one. The area between both cotylae is slightly elevated, and a very weak intercotylar prominence is present in the cranial border. In cranial view, the medial cotyla is slightly more elevated than the lateral one. The latter has an elevated lateral ridge.

The hypotarsus is restricted to the most proximal area. It is simple and poorly developed, without any tendinal canals or calcaneal ridges. The hypotarsus is primarily developed behind the medial cotyla, where it is caudally projected. In MACN-N-10, on the proximal surface of this projection, there is a broken triangular surface that indicates the presence of a proximally projected hypotarsal process.

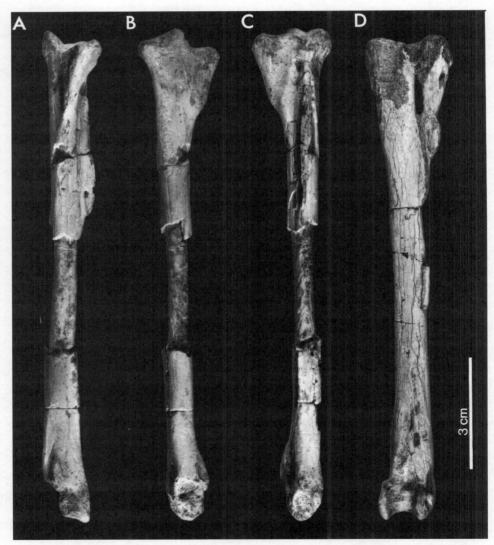

Figure 13.22. Tibiotarsus of *P. deferrariisi*. A, B, C, left tibiotarsus of MACN-N-03 in cranial (A), medial (B), and lateral (C) view; D, left tibiotarsus of MACN-N-11 in cranial view.

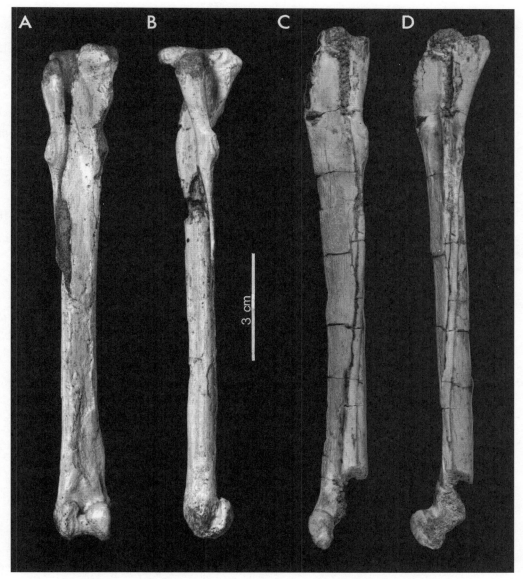

Figure 13.23. Tibiotarsus and fibula of *P. deferrariisi*. A, B, right tibiotarsus and fibula of MACN-N-11 in cranial (A) and lateral (B) view; C, D, left tibiotarsus and fibula of MUCPv-48 in cranial (C) and lateral (D) view.

The tarsometatarsal shaft is straight and strongly compressed dorsoplantarly (Figs. 13.25B, 13.26D). In the mid-portion of MUCPv-48, the thickness is less than 50% of its width (Table 13.5). The distal half of the shaft curves slightly cranially, in such a way that in lateral view the cranial border is slightly concave.

Cranially, somewhat distal to the proximal end, and on the boundary of metatarsals III and IV, there is a small foramen perforating the tarsometatarsus (Figs. 13.25D, 13.26A). This foramen is most likely homologous to the lateral proximal foramen of neornithine birds. At the same level of this foramen, but between metatarsals II and III, there is a small fossa. The position of this fossa is identical to that of the me-

dial proximal foramen of Neornithes, although in *Patagopteryx* this fossa is not perforated by any foramen.

The distal boundary of the proximal medial fossa is formed by swellings of metatarsals II and III. This area may represent the area of attachment of M. tibialis cranialis, the main tarsometatarsal flexor (Hudson, 1937; Raikow, 1985). Distal to these swellings there is a pronounced groove that reaches the midshaft of the tarsometatarsus. This groove is well developed in MUCPv-48. Parallel to it, between metatarsals III and IV, there is another longitudinal groove, although it is much less pronounced. It is in the most proximal portion of this groove that the proximal external foramen is located.

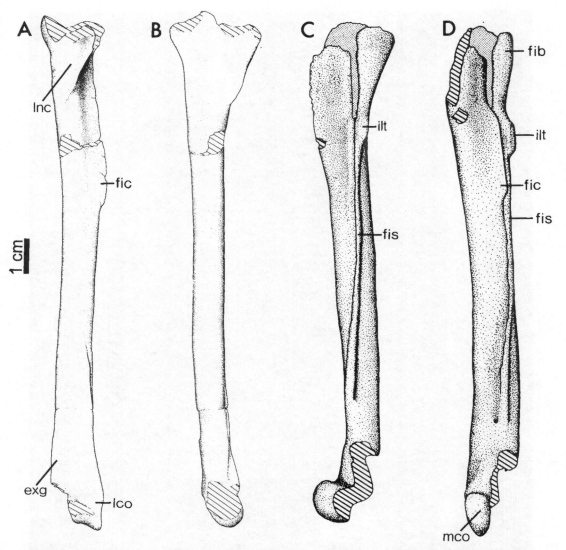

Figure 13.24. Tibiotarsus and fibula of *P. deferrariisi*. A, B, left tibiotarsus of MACN-N-03 in cranial (A) and medial (B) view; C, D, left tibiotarsus of MUCPv-48 in craniolateral (C) and cranial (D) view. Abbreviations: exg, extensor groove; fib, fibula; fic, fibular crest; fis, fibular spine; ilt, tubercle for M. iliofibularis; lco, lateral condyle; lnc, lateral cnemial crest; mco, medial condyle.

In plantar view, the tarsometatarsus is transversely concave, especially in its proximal two-thirds (Figs. 13.25C, E, 13.26B). Metatarsals II and IV are well projected plantarly, defining a broad, triangular depression, the floor of which is formed by metatarsal III. This depression narrows distally down to the level of the articular facet of metatarsal I. This depression must have been the area of origin of several digital muscles (e.g., M. flexor hallucis brevis, M. adductor digiti II, M. abductor digiti IV) as well as of tendons of the flexor digital musculature (George and Berger, 1966). In the proximolateral corner of this depression, the plantar opening of the proximal lateral foramen is visible.

Slightly above the distal end of this depression, at the level of the articulation for metatarsal I, there is a swelling on the plantar face of metatarsal II (this is only slightly vis-ible in MUCPv-48 but well developed in MACN-N-11). A similar structure, although less pronounced, is visible in certain neornithine birds (e.g., *Cathartes aura, Fulica rufifrons*), despite the fact that in these examples the swelling is proximal to the articular facet for metatarsal I. At the level of the swelling of metatarsal II, but on the opposite side of the plantar depression, is the origin of a beltlike groove that is directed laterodistally. This groove borders the lateral surface of the tarsometatarsus, and it reaches the lateral fossa of the lateral trochlea. Most likely, this groove corresponds to the mark left by the tendon of M. abductor digiti IV (George and Berger, 1966). If this interpretation is correct, it suggests a well-developed M. abductor digiti IV in *Patagopteryx*, a muscle that in many groups of neornithine birds is only slightly developed. Metatarsal I is

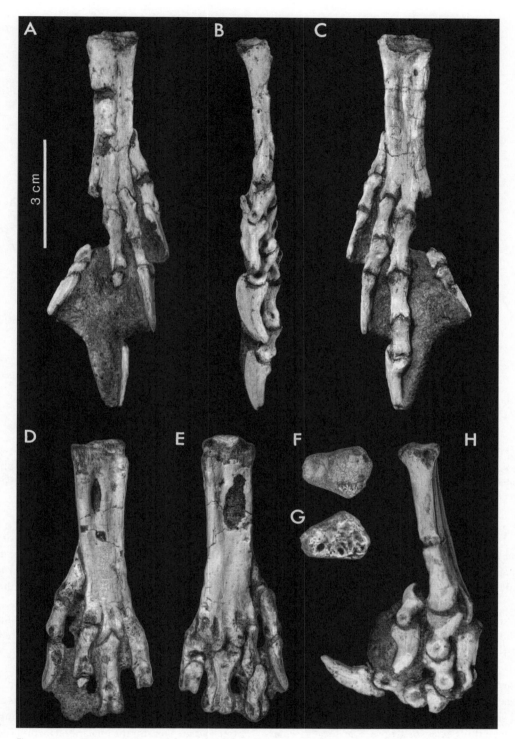

Figure 13.25. Tarsometatarsus and pes of *P. deferrariisi*. A, B, C, right tarsometatarsus and pes of MACN-N-11 in dorsal (A), medial (B), and plantar (C) view; D, E, left tarsometatarsus and pes of MACN-N-11 in dorsal (D) and plantar (E) view; F, G, proximal end of left tarsometatarsus of MACN-N-10 in proximal (F) and distal (cross-section) (G) view; H, tarsometatarsus and pes of MUCPv-48 in medial view.

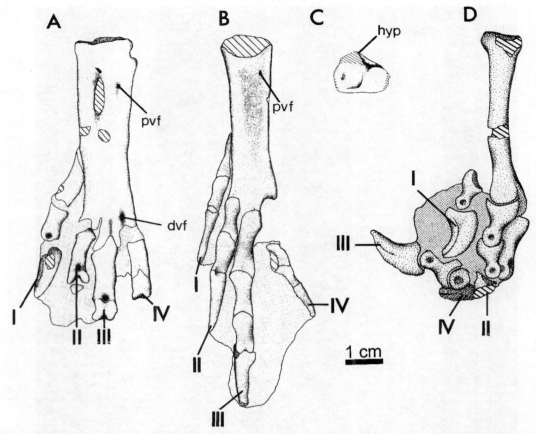

Figure 13.26. Tarsometatarsus and pes of *P. deferrariisi*. A, left tarsometatarsus and pes of MACN-N-03 in dorsal view; B, right tarsometatarsus and pes of MACN-N-11 in plantar view; C, proximal end of left tarsometatarsus of MACN-N-10 in proximal view; D, left tarsometatarsus and pes of MUCPv-48 in medial view. Abbreviations: dvf, distal vascular foramen; hyp, hypotarsus; pvf, proximal vascular foramen; I–IV, pedal digits I to IV.

small (Table 13.5). It articulates by a teardrop-shaped facet to the medial border of metatarsal II, which is slightly displaced cranially. The proximal end of this facet is only slightly distal to the midshaft.

The distal end of the tarsometatarsus is slightly narrower than the proximal end. Trochleae for metatarsals II–IV are nearly in the same transverse plane. The trochlea for metatarsal III is the most projected distally. Trochleae for metatarsals IV and II are slightly less projected, with the latter being shorter than the former. Proximal to the trochlear notch of metatarsals III and IV there is the cranial opening of the distal vascular foramen (Fig. 13.26A). This foramen originates from a short extensor groove. The distal vascular foramen does not open on the plantar surface but rather probably opens on the actual notch between the trochleae; the foramen has not been exposed because of the difficulty of removing the matrix filling the trochlear notch.

The robust trochlea for metatarsal III is the largest of the three, and it projects cranially. This trochlea bears a well-defined central furrow that is very broad plantarly. The lateral and medial trochlear notches, which separate the

trochlea for metatarsal III from the trochleae for metatarsals IV and II, respectively, are very narrow. The trochlea for metatarsal II is somewhat more robust than the trochlea for metatarsal IV. Its medial surface is flat, without a collateral ligamental fossa. This trochlea also lacks a central furrow. The lateral rim of the trochlea for metatarsal I is more distally projected than the medial rim. These borders are separated by a broad, shallow central furrow.

Pedal Digits. The foot of *Patagopteryx* is pamprodactyl (Raikow, 1985), with its four robust, long digits directed cranially (Figs. 13.25, 13.26). The phalangeal formula is 2-3-4-5-*x* (see Padian, 1992, for the phalangeal formula designation), as is typical of theropods (including birds). Digit III, the largest, is as long as the tarsometatarsus. Roughly 30% shorter than this digit is digit II, which is only slightly longer than digit IV. Digit I, although the shortest, is still robust and well developed, being about 50% the length of digit III.

The ungual phalanges are well developed. Except for digit III, the ungual phalanges are the largest phalanges of their respective digits. The ungual phalanges of digits I, III, and IV are only slightly curved. This curvature is more pronounced for the ungual of digit II. Both lateral and medial surfaces bear a longitudinal groove slightly displaced to the ventral margin. The flexor tubercles are weak.

The first phalanx of digit I, which is slightly shorter than the ungual, possesses on its proximal end a tonguelike, ventral projection. Centered on its distal end is a broad, circular collateral ligamental fossa. The proximal phalanx of digit II is very high in its proximal end. The proximal end of the dorsal margin strongly embraces the trochlea for metatarsal II. Toward the distal end, the dorsal border of this proximal phalanx diminishes in height. Slightly proximal to the distal articular facet, the dorsal margin is excavated by a deep depression. A comparable depression is also present in the proximal phalanges of digits III and IV, as well as in some intermediate phalanges, although it may be located more proximally. Phalanx 2 of digit II has a short shaft but a long proximal projection of the ventral border. On its dorsal surface, the fossa described for the proximal phalanx of this digit is not present. The ungual is large and laterally compressed. As in the Late Cretaceous Malagasy *Rahonavis* (Forster et al., 1998), the articular surfaces of the phalanges of digit II are well developed. *Patagopteryx*, however, does not show the raptorial specializations of the latter bird—its digit II is not more robust than the others, and its claw is not strongly curved. Phalanx 1 of digit III is the largest and most robust of all the phalanges. Phalanges 2 and 3 of this digit successively diminish in size. The first four phalanges of digit IV are small. Among these, phalanges 1 and 4 are the largest. The first three phalanges also have a depression on their dorsal surface, although in these phalanges this depression is located in the midshaft portion. The larger ungual phalanx has a pronounced proximal expansion from the ventral margin that embraces the precedent phalanx for more than half of its size.

Phylogenetic Relationships

As pointed out in the introduction, the notion of a close relationship between *Patagopteryx* and extant ratites was stressed in early discussions of this taxon (e.g., Bonaparte, 1986) and was retained in its formal description (Alvarenga and Bonaparte, 1992; see also Alvarenga, 1993). Yet, as shown by Chiappe (1995a), the characters used to support such a claim either are plesiomorphic or have been misinterpreted.

Nessov and Pantheleev (1993) related the Coniacian *Kuszholia mengi* (Nessov, 1992) from the Bissekty Formation at Dzhyrakuduk (Uzbekistan), a taxon erected on the basis of two fragmentary synsacra, to *Patagopteryx*. These authors were followed by Kurochkin (1995b), who mentioned the presence of common features between *Patagopteryx, Kuszholia,* and the Early Cretaceous *Chaoyangia* from northeastern China (Zhou and Hou, Chapter 7 in this

volume), although the alleged common features were not specified. The incomplete and poorly preserved nature of the known material of *Kuszholia* makes comparison to this taxon difficult—indeed, even its avian nature is difficult to support given the available material. Yet the synsacrum of *Kuszholia* differs from that of *Patagopteyx* in the presence of pleurocoels and a concave caudal articular surface.

In another recent paper, Kurochkin (1995a) also supported Alvarenga and Bonaparte's (1992) hypothesis of the palaeognath affinity of *Patagopteryx*, although in this case it is unclear whether *Kuszholia* and *Chaoyangia* were also regarded as closely related to palaeognaths. The characters used by Kurochkin (1995a) to support such a claim are (1) absence of a bicipital crest; (2) open ilioischiadic fenestra; (3) symmetrical lateral and medial ridges of the trochlea of metatarsal III; (4) presence of a deltopectoral crest that begins at the humeral head; (5) thick postacetabular wing of the ilium; (6) hypotarsus lacking canals and (7) proximally raised; (8) large cranial iliosacral crest; and (9) proximally placed medial epicondyle. As pointed out by Kurochkin himself (1995a: Table 1), most of these characters, if not all of them, appear to be primitive—characters 1–6 clearly occur in one or more basal taxa (e.g., *Archaeopteryx, Confuciusornis,* enantiornithines, and nonavian theropods). Furthermore, some of these characters are difficult to divide into discrete character states. For example, Kurochkin's "anterior ilio-sacral crest," which presumably refers to the dorsal iliac crest (see Baumel and Witmer, 1993), is simply the dorsal border of the preacetabular wing of the ilium. Thus, differences between "large" and "small" sizes of this border are rather difficult to distinguish and are highly arbitrary, as well.

None of the aforementioned hypotheses was framed into a cladistic analysis, and they are therefore difficult to evaluate on the basis of parsimony. Nevertheless, various cladistic analyses have supported the sister-group relationship between *Patagopteryx* and Ornithurae (Chiappe, 1992, 1995a,b, 1996b, Chapter 20 in this volume; Chiappe and Calvo, 1994; Fig. 13.27). The hypothesis of ratite affinities has also been challenged by recent histological studies documenting the presence of lines of arrested growth (LAGs) in the compacta of the femora of *Patagopteryx*, which are typically absent in ornithurine birds, including ratites (Chinsamy et al., 1994, 1995; Chinsamy, Chapter 18 in this volume). Consequently, compelling osteological and histological evidence supports the hypothesis that *P. deferrariisi* acquired flightlessness independently from ratites and any other known flightless avian group (Chiappe, 1995a,b, 1996b).

Paleobiology

Patagopteryx was a flightless land bird. Estimates of its body mass based on the diameter of its femur vary significantly

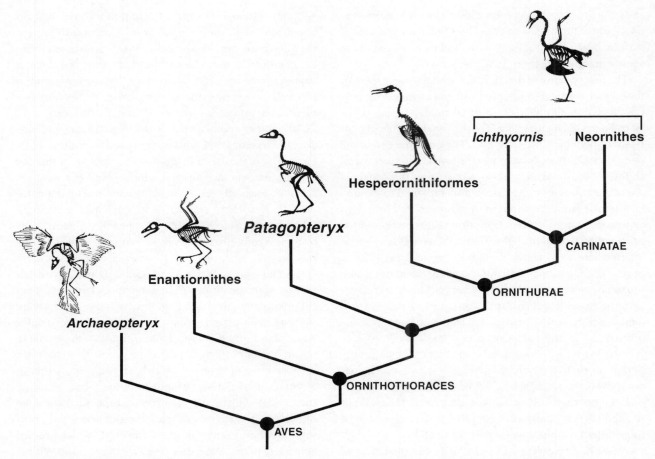

Figure 13.27. Cladogram depicting the relationship of *P. deferrariisi* to other basal birds (modified from Chiappe, 1996b).

depending on the equation used. Yalden's (1984) equation ($W = 13.25 \times D2.353$; D = diameter) provides values of 2,211 and 2,152 g for specimens MACN-N-11 and MUCPv-48, respectively. These values are comparable to the mean mass of the male brown kiwi (*Apteryx australis*; the female is usually heavier; Dunning, 1993). Yet Campbell and Marcus's (1992) standard equation ($W\log = -0.065 + 2.411 \times C\log$; C = circumference) gives values of 3,672 and 3,334 g for these two specimens, respectively. These masses exceed the mean value of female brown kiwis given by Dunning (1993), although they are smaller than the maximum value registered by this author (i.e., 3,850 g) for females of this species.

Flightlessness has been a common theme in the history of birds. At least 12 avian lineages have extant flightless representatives, and a large variety of extinct and recently extirpated birds are known to have been, or have been interpreted as, flightless (Raikow, 1985; Livezey, 1995). The large majority of these are land birds.

Regardless of the group in question and its lifestyle, flightlessness is usually associated with a number of distinct morphological attributes (Diamond, 1981, 1991; James and Olson, 1983; Raikow, 1985; Feduccia, 1996). Most evident among these is the reduced forelimb, which is extreme in some cases, such as hesperornithiforms and certain ratites. In comparison with flighted birds, the forelimb of a flightless bird typically has a substantially smaller carpometacarpus and ulna-radius, and the relative proportions between these midwing elements and the humerus, and of these last two to the entire forelimb, are very different from those of a flighted counterpart (Livezey, 1990). In those birds in which the forelimb is heavily reduced, wing elements bear simple and rudimentary structures. The thoracic girdle also shows radical differences from that of a flighted bird. Flight muscles form a minor portion of the body mass (e.g., 0.13% in the North Island brown kiwi; see McNab, 1996), and their area of origin is highly reduced: the sternal carina and the furcula are absent in a variety of flightless birds (e.g., ratites, *Diatryma*, moa-nalos). The scapula and coracoid articulate to each other in an obtuse angle, and in certain cases (e.g., ratites, *Diatryma*) these bones are fused, forming a scapulo-coracoid. Furthermore, in flightless birds—particularly the terrestrial ones—the hindlimb is robust and proportionally larger than in a flighted counterpart.

In *Patagopteryx,* the ratio between maximum lengths of fore- and hindlimb (1:2.2), and the proportion between wing elements (e.g., humerus larger than ulna-radius) and of the wing to the rest of the skeleton, indicate its incapacity to fly. Moreover, *Patagopteryx* possesses several features present in flightless birds: small carpometacarpus, widely separated sternal articulations of coracoids, and absence of furcula and sternal keel. Yet the forelimb and thoracic girdle of *Patagopteryx* are considerably less reduced than those of ratites and other forms with an extremely simplified flight apparatus. For example, the articulation between the scapula and coracoid forms an acute angle, and these bones are not fused to each other.

The hindlimb and pelvis of *Patagopteryx* are robust and very well developed in proportion to the rest of the skeleton. The hindlimb elements exhibit prominent structures for attachment of muscles (e.g., strong intermuscular lines, robust cnemial crest, prominent tubercle for M. iliofibularis), which point to strong hindlimb musculature. This, combined with the reduction of the forelimb, indicates that the locomotion of *Patagopteryx* was governed by its hindlimbs.

Flightless birds are essentially specialized for two types of locomotion: foot-propelled aquatic diving and terrestrial locomotion. *Patagopteryx* lacks the most obvious specializations of foot-propelled, aquatic birds such as the compression and elongation of the postacetabular portion of the pelvis and synsacrum, the proximal development of the cnemial crests, and the lateral compression of the tarsometatarsus (Raikow, 1985). On the contrary, the general hindlimb anatomy of *Patagopteryx* approaches that of obligate terrestrial birds: the pedal anatomy is generalized, and the foot, although pamprodactyl, is functionally tridactyl. Undoubtedly, *Patagopteryx* was an obligate terrestrial bird. Yet its hindlimb does not show the specializations for high speed present in other obligate terrestrial birds such as certain ratites (e.g., Rheidae, *Dromaius, Struthio*) and most phorusrhacids (Alvarenga and Bonaparte, 1992). For example, in *Patagopteryx,* the tarsometatarsus is significantly shorter than the tibiotarsus. Also, fast runners possess reduced toes (in both size and number) that minimize the surface of contact with the substrate (Raikow, 1985). This is not the case in *Patagopteryx,* in which the toes are large. This suggests that *Patagopteryx* was probably unable to achieve the high running speeds of many flightless birds.

The phylogenetic position of *Patagopteryx* (Fig. 13.27) strongly suggests that this bird evolved flightlessness from flighted ancestors. Indeed, a combined set of functional and phylogenetic analyses indicates that all flightless birds, with the possible exception of the alvarezsaurids, if they are considered avian (Chiappe, Norell, and Clark, Chapter 4 in this volume), are descendants of flighted ancestors. Secondary flightlessness has often been viewed within an adaptational context, as a functional modification shaped by a specific force: natural selection (adaptation is best defined as an apomorphic function—in this case nonfunction—due to natural selection; see Coddington, 1988). The occurrence of flightlessness in many different lineages of birds and, in particular, its recurrence in relatively small monophyletic groups (e.g., more than 25% of rails; see Diamond, 1981; Feduccia, 1996) suggest that in some instances flightlessness may qualify as an adaptation (Andersen, 1995). This is also supported by studies on monophyletic lineages of oceanic islands (e.g., Australasian teals, Hawaiian moa-nalos and geese; see Livezey, 1990; Olson and James, 1991) that have shown a strict correlation between flightlessness and a historical change in environmental conditions (e.g., colonization of a restricted, insular system). Yet caution is advised because even in the best scenarios phylogenetic studies may be insufficient to corroborate a given adaptational proposal (Wenzel and Carpenter, 1994).

Attempts to explain the causes and processes that led to secondary flightlessness have focused more on island lineages (see Olson, 1973; Feduccia, 1980, 1996; Diamond, 1981; James and Olson, 1983; Livezey, 1993; McNab, 1994). It is widely known that most oceanic islands lack aboriginal mammalian predators and that most of them are devoid of major geographic barriers or environmental heterogeneity that complicates the dispersion of their inhabitants (Diamond, 1981). Under these circumstances, secondary flightlessness has been viewed as a by-product of energy conservation, reducing the mass of metabolically expensive thoracic muscles (Olson, 1973; Feduccia, 1980, 1996; Diamond, 1981, 1991; Livezey, 1990, 1992, 1993; McNab, 1994, 1996). McNab (1994, 1996) has shown that in several instances (e.g., rails and kiwis) flightlessness is associated with low basal metabolic rates (average rate of energy consumption under standard conditions), linking the evolution of this functional modification to a reduced rate of energy expenditure. Olson (1973), Feduccia (1980, 1996), Livezey (1990, 1992, 1993, 1995), and others (e.g., Diamond, 1981, 1991) have suggested that heterochronic processes (in particular, paedomorphosis) may have been responsible for the acquisition of flightlessness. In contrast to the case of insular forms, the evolution of flightlessness in continental birds has usually been linked to an increase in size (e.g., Dromornithidae, *Diatryma,* Phorusrhacidae) and sometimes speed (e.g., *Struthio, Rhea, Dromaius*). If factors leading to secondary flightlessness in island lineages are intricate to dissect, those of continental birds appear to be even more complex. Yet, regardless of its causes, it is possible that elimination of the restrictions imposed by the high metabolic demands of flight allowed the increase in size (Raikow, 1985). Again, heterochrony has also been claimed as an important process in the development of flightlessness in continental birds (Feduccia, 1980; James and Olson, 1983). Although neoteny has been highlighted as the specific underlying process in the heterochronic evolution of flightlessness, more than one underlying process (including combinations of paedomorphic and peramorphic processes) is likely to have been at play (Livezey, 1995).

Interestingly, *Patagopteryx* appears to be a singular case in the evolution of secondary flightlessness, not fitting neatly in either of the two models of flightlessness. *Patagopteryx* lived in a continental environment—there are vast deposits of contemporaneous, continental rocks surrounding Boca del Sapo. Yet it neither was highly specialized for running nor achieved large size, despite being sympatric with a variety of predators—dinilysiid snakes, notosuchid crocodiles, small nonavian theropods such as *Velocisaurus*. The morphology of the wing and thoracic girdle of *Patagopteryx* fits well with the predictions expected for paedomorphic skeletal change (see Livezey, 1990, 1993), but such a process theory can be formulated only if close relatives of the lineage under study are known. Unfortunately, this is not the case in *Patagopteryx*. Perhaps, future discoveries of close relatives of this most interesting bird can help us understand the evolution of flightlessness in this ancient lineage.

Acknowledgments

I am grateful to S. Cerruti, S. Copeland, L. Meeker, and A. Rizzetto for preparation of many of the illustrations. S. Orell and S. Copeland edited the final version of this manuscript. Thanks also to J. Bonaparte, L. Salgado, and J. Calvo for letting me study material under their care. Support for this research was provided by grants from the Guggenheim Foundation, the Dinosaur Society, the McKenna Foundation, and the Consejo Nacional de Investigaciones Científicas y Técnicas (Argentina).

Literature Cited

Alvarenga, H. M. F. 1993. A origem das aves seus fósseis; pp. 16–26 *in* M. A. de Andrade (ed.), A Vida das Aves. Editora Littera Maciel, Belo Horizonte, Brazil.

Alvarenga, H. M. F., and J. F. Bonaparte. 1992. A new flightless land bird from the Cretaceous of Patagonia; pp. 51–64 *in* K. E. Campbell (ed.), Papers in Avian Paleontology, Honoring Pierce Brodkorb. Science Series 36. Natural History Museum of Los Angeles County, Los Angeles.

Andersen, N. M. 1995. Cladistic inference and evolutionary scenarios: locomotory structure, function, and performance in water striders. Cladistics 11:279–295.

Baumel, J. J., and L. M. Witmer. 1993. Osteologia; pp. 45–132 *in* J. J. Baumel, A. S. King, J. E. Breazile, H. E. Evans, and J. C. Vanden Berge (eds.), Handbook of Avian Anatomy: Nomina Anatomica Avium, 2nd edition. Nuttall Ornithological Club, Cambridge, Mass.

Baumel, J. J., A. S. King, A. M. Lucas, J. E. Breazile, and H. E. Evans. 1979. Nomina Anatomica Avium. Academic Press, London, 637 pp.

Bock, W. J. 1963. The cranial evidence for ratite affinities; pp. 39–54 *in* C. G. Sibley (ed.), Proceedings of the 13th International Ornithological Congress. American Ornithologists Union, Baton Rouge, La.

Bonaparte, J. F. 1986. History of the terrestrial Cretaceous vertebrates of Gondwana. IV Congreso Argentino de Paleontología y Bioestratigrafía, Actas 2:63–95.

———. 1991. Los vertebrados fósiles de la Formación Río Colorado de Neuquén y cercanias, Cretácico Superior, Argentina. Revista del Museo Argentino de Ciencias Naturales "Bernardino Rivadavia" (Paleontología) 4(3):17–123.

Calvo, J. O. 1991. Huellas de dinosaurios en la Formación Río Limay (Albiano-Cenomaniano), Picun Leufu, Provincia del Neuquén, República Argentina (Ornithischia-Saurischia: Sauropoda-Theropoda). Ameghiniana 28(3–4):241–258.

Campbell, K. E., Jr., and L. Marcus. 1992. The relationship of hind limb bone dimensions to body weight in birds; pp. 395–412 *in* K. E. Campbell (ed.), Papers in Avian Paleontology, Honoring Pierce Brodkorb. Science Series 36. Natural History Museum of Los Angeles County, Los Angeles.

Cazau, L. B., and M. A. Uliana. 1973. El Cretácico superior continental de la cuenca Neuquina. V Congreso Geológico Argentino 3:131–163.

Chiappe, L. M. 1987. Aves mesozoicas. Ciencias Naturales (MACN) 1:1–7.

———. 1989. A flightless bird from the Late Cretaceous of Patagonia (Argentina). Archosaurian Articulations 1(10):73–77.

———. 1991. Cretaceous birds of Latin-America. Cretaceous Research 12:55–63.

———. 1992. Osteología y sistemática de *Patagopteryx deferrariisi* Alvarenga y Bonaparte (Aves) del Cretácico de Patagonia. Filogenia e historia biogeográfica de las aves cretácicas de América del Sur. Ph.D. dissertation. Universidad de Buenos Aires, Argentina, 429 pp.

———. 1995a. The first 85 million years of avian evolution. Nature 378:349–355.

———. 1995b. The phylogenetic position of the Cretaceous birds of Argentina: Enantiornithes and *Patagopteryx deferrariisi*; pp. 55–63 *in* D. S. Peters (ed.), Proceedings of the 3rd Symposium of the Society of Avian Paleontology and Evolution. Courier Forschungsinstitut Senckenberg 181, Frankfurt am Main.

———. 1995c. A diversity of early birds. Natural History 104(6): 52–55.

———. 1996a. Late Cretaceous birds of southern South America: anatomy and systematics of Enantiornithes and *Patagopteryx deferrariisi*; pp. 203–244 *in* G. Arratia (ed.), Contributions of Southern South America to Vertebrate Paleontology, Münchner Geowissenschaftliche Abhandlungen, Reihe A, Geologie und Paläontologie 30. Verlag Dr. Friedrich Pfeil, Munich.

———. 1996b. Early avian evolution in the southern hemisphere: fossil record of birds in the Mesozoic of Gondwana. Memoirs of the Queensland Museum 39:533–556.

Chiappe, L. M., and J. O. Calvo. 1994. *Neuquenornis volans*, a new Upper Cretaceous bird (Enantiornithes: Avisauridae) from Patagonia, Argentina. Journal of Vertebrate Paleontology 14(2):230–246.

Chiappe, L. M., M. A. Norell, and J. M. Clark. 1996. Phylogenetic position of *Mononykus* from the Upper Cretaceous of the Gobi Desert. Memoirs of the Queensland Museum 39:557–582.

Chiappe, L. M., M. A. Norell, and J. Clark. 1998. The skull of a new relative of the stem-group bird *Mononykus*. Nature 392:275–278.

Chiappe, L. M., Ji S., Ji Q., and M. A. Norell. 1999. Anatomy and systematics of the Confuciusornithidae (Aves) from the late Mesozoic of northeastern China. Bulletin of the American Museum Novitates 242:1–89.

Chinsamy, A., L. M. Chiappe, and P. Dodson. 1994. Growth rings in Mesozoic birds. Nature 368:196–197.

———. 1995. Mesozoic avian bone microstructure: physiological implications. Paleobiology 21(4):561–574.

Clarke, J., and L. M. Chiappe. 2001. A new carinate bird from the Late Cretaceous of Patagonia (Argentina). American Museum Novitates 3323:1–23.

Clarke, J., L. Dingus, and L. M. Chiappe. 1998. Review of Upper Cretaceous deposits at Neuquén, Argentina (Bajo de la Carpa Member, Río Colorado Formation). Journal of Vertebrate Paleontology 18(3):34A.

Coddington, J. A. 1988. Cladistic tests of adaptational hypotheses. Cladistics 4:3–22.

Cruz, C. E., P. Condat, E. Kozlowski, and R. Manceda. 1989. Análisis estratigráfico secuencial del Grupo Neuquén (Cretácico superior) en el valle del Río Grande, Provincia de Mendoza. I Congreso Argentino de Hidrocarburos, Actas 2:689–714.

Currie, P. J. 1995. New information on the anatomy and relationships of Dromaeosaurus albertensis (Dinosauria, Theropoda). Journal of Vertebrate Paleontology 15(3):576–591.

Danderfer, J. C., and P. Vera. 1992. Geología. Boletín del Servicio Geológico Neuquino 1:23–45.

Diamond, J. 1981. Flightlessness and fear of flying in island species. Nature 293:507–508.

———. 1991. Twilight of Hawaiian birds. Nature 353:505–506.

Dingus, L., J. Clarke, G. R. Scott, C. C. Swisher III, L. M. Chiappe, and R. A. Coria. 2000. Stratigraphy and magnetostratigraphic/faunal constraints for the age of sauropod embryo-bearing rocks in the Neuquén Group (Late Cretaceous, Neuquén Province, Argentina). American Museum Novitates 3290:1–11.

Dunning, J. B., Jr. 1993. CRC Handbook of Avian Body Masses. CRC Press, Boca Raton, Fla., 371 pp.

Feduccia, A. 1980. The Age of Birds. Harvard University Press, Cambridge, 196 pp.

———. 1996. The Origin and Evolution of Birds. Yale University Press, New Haven, 420 pp.

Forster, C. A., L. M. Chiappe, D. W. Krause, and S. D. Sampson. 1996. The first Cretaceous bird from Madagascar. Nature 382:532–534.

Forster, C. A., S. D. Sampson, L. M. Chiappe, and D. W. Krause. 1998. The theropodan ancestry of birds: new evidence from the Late Cretaceous of Madagascar. Science 279:1915–1919.

Gasparini, Z. 1971. Los Notosuchia del Cretácico de América del Sur como un nuevo infraorden de los Mesosuchia (Crocodilia). Ameghiniana 8(2):83–103.

George, J. C., and A. J. Berger. 1966. Avian Myology. Academic Press, New York, 499 pp.

Heredia, S., and J. O. Calvo. 1997. Sedimentitas eólicas en la Formación Río Colorado (Grupo Neuquén) y su relación con la fauna del Cretácico superior. Ameghiniana 34(1):120.

Holmes, B. E. 1963. Variation in the muscles and nerves of the leg in two genera of Grouse (Tympanuchus and Pedioectes). University of Kansas Publications, Museum of Natural History 12(9):363–474.

Hudson, G. E. 1937. Studies on the muscles of the pelvic appendage in birds. American Midland Naturalist 18(1):1–108.

James, H., and S. L. Olson. 1983. Flightless birds. Natural History 92(9):30–40.

Kurochkin, E. N. 1995a. Morphological differentiation of palaeognathous and neognathous birds. Courier Forschungsinstitut Senckenberg 181:79–88.

———. 1995b. The assemblage of the Cretaceous birds in Asia; pp. 203–208 in Sun A. and Wang Y. (eds.), Sixth Symposium on Mesozoic Terrestrial Ecosystems and Biota, Short Papers. China Ocean Press, Beijing.

Legarreta, L., and C. Gulisano. 1989. Análisis estratigráfico secuencial de la Cuenca Neuquina (Triásico Superior-Terciario Inferior); pp. 221–243 in G. Chebli and L. Spalletti (eds.), Cuencas Sedimentarias Argentinas. Serie Correlación Geológica Volume 6. Universidad Nacional de Tucumán, San Miguel de Tucumán..

Livezey, B. C. 1990. Evolutionary morphology of flightlessness in the Auckland Islands teal. Condor 92:639–673.

———. 1992. Morphological corollaries and ecological implications of flightlessness in the Kakapo (Psittaciformes: Strigops habroptilus). Journal of Morphology 213:105–145.

———. 1993. An ecomorphological review of the dodo (Raphus cucullatus) and solitaire (Pezophaps solitaria), flightless Columbiformes of the Mascarene Islands. Journal of Zoology, London 230:247–292.

———. 1995. Heterochrony and the evolution of avian flightlessness; pp. 169–193 in K. J. McNamara (ed.), Evolutionary Change and Heterochrony. John Wiley and Sons, New York.

Marsh, O. C. 1880. Odontornithes: A Monograph on the Extinct Toothed Birds of North America. Report of Geological Exploration of the 40th Parallel. Government Printing Office, Washington, D.C., 201 pp.

Martin, L. D. 1983. The origin and early radiation of birds; pp. 291–338 in A. H. Brush and G. A. Clark Jr. (eds.), Perspectives in Ornithology. Cambridge University Press, New York.

———. 1991. Mesozoic birds and the origin of birds; pp. 485–539 in H.-P. Schultze and L. Trueb (eds.), Origins of the Higher Groups of Tetrapods: Controversy and Consensus. Comstock Publishing Associates, Ithaca, N.Y.

Martin, L. D., and J. Tate Jr. 1976. The skeleton of Baptornis advenus (Aves: Hesperornithiformes). Smithsonian Contributions to Paleobiology 27:35–66.

McGowan, C. 1979. The hind limb musculature of the brown kiwi, Apteryx australis mantelli. Journal of Morphology 160:33–74.

McNab, B. K. 1994. Energy conservation and the evolution of flightlessness in birds. American Naturalist 144(4):628–642.

———. 1996. Metabolism and temperature regulation of kiwis (Apterygidae). Auk 113(3):687–692.

Musacchio, E. A. 1993. Use of global time scale in correlating nonmarine Cretaceous rocks in southern South America. Cretaceous Research 14:113–126.

Nessov, L. A. 1992. Review of localities and remains of Mesozoic and Paleogene birds of the USSR and the description of new findings. Russian Journal of Ornithology 1(1):7–50.

Nessov, L. A., and B. V. Pantheleev. 1993. On the similarity of the Late Cretaceous ornithofauna of South America and western Asia; pp. 84–94 in R. L. Potapov (ed.), Ecology and Fauna of the Palaearctic Birds. Proceedings of the Zoological Institute 252. Russian Academy of Sciences, Saint Petersburg. [Russian]

Novas, F. E. 1996. Alvarezsauridae, Cretaceous maniraptorans from Patagonia and Mongolia. Memoirs of the Queensland Museum 39:675–702.

Olson, S. L. 1973. Evolution of the rails of the South Atlantic islands (Aves: Rallidae). Smithsonian Contributions to Zoology 152:1–53.

Olson, S. L., and H. F. James. 1991. Descriptions of thirty-two new species of birds from the Hawaiian Islands: Part I. Non-Passeriformes. Ornithological Monographs 45:1–88.

Padian, K. 1992. A proposal to standardize tetrapod phalangeal formula designations. Journal of Vertebrate Paleontology 12(2):260–262.

Perle A., L. M. Chiappe, Barsbold R., J. M. Clark, and M. A. Norell. 1994. Skeletal morphology of *Mononykus olecranus* (Theropoda: Avialae) from the Late Cretaceous of Mongolia. American Museum Novitates 3105:1–29.

Raikow, R. J. 1985. Locomotor system; pp. 57–147 *in* A. S. King and J. McLelland (eds.), Form and Function in Birds. Academic Press, San Diego.

Ramos, V. A. 1981. Descripción geológica de la Hoja 33c, Los Chihuidos Norte. Boletín del Servicio Geológico Nacional, Buenos Aires 182:1–103.

Rowe, T. 1986. Homology and evolution of the deep dorsal thigh musculature in birds and other Reptilia. Journal of Morphology 189:327–346.

Saiff, E. I. 1983. The anatomy of the middle ear region of the Rheas (Aves: Rheiformes, Rheidae). Historia Natural 3(6):45–55.

Sereno, P. C., and F. E. Novas. 1993. The skull and neck of the basal theropod *Herrerasaurus ischigualastensis*. Journal of Vertebrate Paleontology 13(4):451–476.

Simonetta, A. 1960. On the mechanical implications of the avian skull and their bearing on the evolution and classification of birds. Quarterly Review of Biology 35(3):206–220.

Smith Woodward, A. S. 1896. On two Mesozoic crocodilians, *Notosuchus* (genus novum) and *Cynodontosuchus* (gen. nov.) from the red sandstones of the territory of Neuquén (Argentina). Anales del Museo de La Plata 4:1–20.

———. 1901. On some extinct reptiles from Patagonia of the genera *Miolania, Dinilysia* and *Genyodectes*. Proceedings of the Zoological Society of London 1:169–184.

Starck, M. J. 1995. Comparative anatomy of the external and middle ear of palaeognathous birds. Advances in Anatomy, Embryology and Cell Biology 131:1–137.

Uliana, M. A., and D. Dellapé. 1981. Estratigrafía y evolución paleoambiental de la sucesión Maastrichtiano-Eoterciaria del Engolfamiento Neuquino (Patagonia septentrional). Actas VIII Congreso Geológico Argentino 3:673–711.

Verheyen, R. 1960. Considerations sur la colonne vertebrale des oiseaux (non-passeres). Bulletin Institut Royal des Sciences Naturelles de Belgique 36(42):1–24.

Webb, M. 1957. The ontogeny of the cranial bones, cranial peripheral and cranial parasympathetic nerves, together with a study of the visceral muscles of *Struthio*. Acta Zoologica 38:81–203.

Weishampel, D. B., P. Dodson, and H. Osmólska. 1990. The Dinosauria. University of California Press, Berkeley, 733 pp.

Wenzel, J. W., and J. M. Carpenter. 1994. Comparing methods: adaptive traits and tests of adaptation; pp. 79–101 *in* P. Egelton and R. Bane-Wright (eds.), Phylogenetics and Ecology. Academic Press, London.

Yalden, D. W. 1984. What size was *Archaeopteryx?* Zoological Journal of the Linnean Society 82:177–188.

Zusi, R. L. 1985. Muscles of the trunk and tail in the noisy scrub-bird, *Atrichornis clamosus*, and superb lyrebird, *Menura novaehollandiae* (Passeriformes: Atrichornithidae and Menuridae). Records of the Australian Museum 37:229–242.

Zusi, R. L., and G. D. Bentz. 1984. Myology of the purple-throated carib (*Eulampis jugularis*) and the hummingbirds (Aves: Trochilidae). Smithsonian Contributions to Zoology 385: 1–70.

Zusi, R. L., and R. W. Storer. 1969. Osteology and myology of the head and neck of the pied-billed grebes (*Podilymbus*). Miscellaneous Publications Museum of Zoology, University of Michigan 139:1–49.

14

Enaliornis, an Early Cretaceous Hesperornithiform Bird from England, with Comments on Other Hesperornithiformes

PETER M. GALTON AND LARRY D. MARTIN

Lyell (1859:40) noted the discovery of several different bones of a bird, rather larger than those of a common pigeon, in the Cambridge Greensand in 1858 by Lucas Barrett of the Woodwardian Museum of Cambridge University. Owen (1861:327) referred to the distal end of a trifid tarsus found by Barrett of a bird about the size of a woodcock in which the outer toe joint is much higher up than the other two and projects backward above the middle joint. However, Seeley (1876), who collected additional bones for the Wood-wardian Museum (collection now in Sedgwick Museum, Cambridge), noted that these bones were not in the museum in 1859. The new material was named *Palaeocolyntus* [*sic*] *barretti* Seeley, 1864, and *Pelargonis* [*sic*] *sedgwicki* See-ley, 1864, in the title of the paper, but this was all that was published. The name *Pelagornis barretti* Seeley, 1866, was used in a summary paragraph reporting the occurrence of bird remains in the Cambridge Greensand, but because *Pela-gornis* was preoccupied for another bird (Lartet, 1857), this material was later cataloged as *Enaliornis* Seeley, 1869, with two species, *E. barretti* and *E. sedgwicki.* However, Seeley (1869) did not assign specimens to each species or give any diagnoses, so each of these taxa was a nomen nudum.

These did not become valid taxa until 1876, when Seeley finally described and illustrated the material of two species of *Enaliornis* Seeley, 1876: *E. barretti* Seeley, 1876, and *E. sedg-wicki* Seeley, 1876. Included in this description was a well-preserved braincase that was referred to *E. barretti,* SMC B54404, which Seeley (1870a,b) previously described as a pterosaur. Seeley (1876) did not specify any types, an omis-sion partly compensated for by Brodkorb (1963a:221), who made *E. barretti* Seeley, 1876, the type species of *Enaliornis* and designated the lectotypes of *E. barretti* and *E. sedgwicki,* following the description of Seeley (1876), as the distal end of a left tarsometatarsus (collection of T. Jesson, cast as BMNH A1112) and the proximal end of a right tibiotarsus

(Woodwardian Museum), respectively. However, the cast is actually BMNH A112 (Lydekker, 1891:203; Elzanowski and Galton [1991] incorrectly reversed these numbers as A112 and A1112), the original bone is BMNH A477, and the tibio-tarsus is SMC B55314.

The Cambridge Greensand, a lag deposit at the base of the Chalk, was deposited during the Early Cenomanian (lowermost Upper Cretaceous) along with a phosphatized remanié assemblage from the uppermost Albian (upper-most Lower Cretaceous, about 100 Ma; Rawson et al., 1978). The bones of *Enaliornis* (Figs. 14.1–14.8) are undistorted and well preserved (apart from some erosion and a few obscur-ing phosphatic nodules) and include three braincases; sev-eral vertebrae; part of a pelvic girdle; numerous femora, most of which consist of the articular ends with a small part of the shaft; and the end parts of several tibiotarsi and tarso-metatarsi. Seeley (1869) listed several bones from the fore-limb but later noted that there were none (Seeley, 1876).

Lydekker (1891) illustrated the distal end of a referred fe-mur (BMNH A163a) of *E. barretti,* Martin (1984) discussed the hesperornithiform characters of *Enaliornis,* Witmer (1990) partly described the most complete referred brain-case (SMC B54404) in a study of cranial pneumaticity in Mesozoic birds, and Elzanowski and Galton (1991) de-scribed the three referred braincases (SMC B54404, YM 585, YM 586). Apart from these papers, all other discussions of *Enaliornis* appear to be based on the illustrations given by Seeley (1876).

Seeley (1876), who distinguished the species of *Enali-ornis* on the basis of size, referred the larger bones to *E. bar-retti* and the smaller ones to *E. sedgwicki.* However, it is diffi-cult to diagnose these two species because of the following problems:

1. Incongruent lectotypes. Brodkorb (1963a) designated the lectotypes of *E. barretti* and *E. sedgwicki* as figured by

Seeley (1876) as the distal end of a tarsometatarsus and the proximal end of a tibiotarsus, respectively. Thus, they may represent the same species, with *E. sedgwicki* Seeley, 1876, being a subjective junior synonym of *E. barretti* Seeley, 1876.

2. Non-*Enaliornis* bones. Small cervical and thoracic vertebrae (Fig. 14.5A–E) were referred to *E. sedgwicki,* but because these vertebrae are amphicoelous, they were incorrectly referred to *Enaliornis* and indicate the presence of a nonhesperornithiform bird. Other non-*Enaliornis* bones include part of the notarium of a pterodactyloid pterosaur and a caudal vertebra of a turtle (see under vertebrae).

3. Intraspecific variation. The smaller *E. sedgwicki* may represent either juveniles or a sexual dimorph of *E. barretti,* possibilities first suggested by Lambrecht (1933). Intraspecific variation resulting from age differences is present because femora, tibiotarsi, and tarsometatarsi of juvenile individuals are represented (Figs. 14.5M–V, 14.6K–N, 14.7I–K, 14.8A–E). The lectotypes of the two species are not juveniles, but, given the other problems with the material, the possibility of sexual dimorphism is difficult to eliminate.

4. Possibility of three species. Concerning the femora, Seeley (1876:505) noted that "if the bones belonged to three different species, as is probable, [then] they all may have been closely allied." In the caption, femora are identified as *E. barretti, E. sedgwicki,* and another species of bird by Seeley (1876:511). On the basis of the distal ends of the tibiotarsi of adult individuals referred to *Enaliornis,* there are three morphologically distinct forms, which may represent three separate species.

5. Disassociated bones. Femora are the most common bone, and Seeley (1876:505) estimated the minimum number of individuals involved as at least 13. However, as Seeley (1876:497) also noted, "there is no evidence of more than one bird-bone being found at a time; so that every fragment of bone may have belonged to a separate individual bird." Consequently, it is difficult to determine which crania, vertebrae, and hindlimb bones go together.

6. Limb bones as disassociated ends. All the Cambridge Greensand hindlimb bones that manifest hesperornithiform morphology, with the exception of two reasonably complete femora (Figs. 14.5J–L, 14.6C–E), consist only of disassociated ends, so the proportions of the individual leg bones are unknown. This makes it difficult to divide these bones, some of which represent juvenile individuals, between two (or three) species of *Enaliornis,* especially if the average size of males was larger than that of females as occurs in recent diving birds (see measurements in Johnsgard, 1987).

In this chapter, the specific referrals of Seeley (1876) will be followed, but because they were based on size differences, no specific diagnoses are given here. The problems of the number of species represented, which bones (or ends) are referred to which species, and the diagnosis of each species are still being worked on and will be discussed elsewhere (Galton and Martin, in press). The purpose of this chapter is to describe the bones that were referred to *Enaliornis* Seeley, 1876, the earliest foot-propelled diving marine bird and one of the earliest ornithurines described to date, and to discuss its relationships.

The following institutional abbreviations are used in this chapter: BGS, British Geological Survey (Geological Survey Museum), Keyworth, Nottingham, England; BMNH, The Museum of Natural History, London, England; SMC, Sedgwick Museum of Geology, The University, Cambridge, England; SMNH, Saskatchewan Museum of Natural History, Regina, Saskatchewan, Canada; YM, Yorkshire Museum, York, England.

Hesperornithiform Distribution

Hesperornithiformes represent a clade of foot-propelled diving birds that were secondarily flightless and are the best-known Mesozoic ornithurines (Marsh, 1880; Martin, 1983, 1984). If *Enaliornis* is a hesperornithiform, as is proposed here, it constitutes the earliest documented record of this clade (Upper Albian, late Early Cretaceous), although an undescribed hesperornithiform has been reported from the Lower Cretaceous of Antarctica (Kurochkin, 1995). The baptornithid *Pasquiaornis,* from nearshore Cenomanian (early Late Cretaceous) marine deposits of Saskatchewan (Tokaryk et al., 1997), is only slightly more recent than *Enaliornis.* Several genera of hesperornithiforms are known from the marine Upper Cretaceous deposits of western North America that parallel the shoreline of the Late Cretaceous American Western Interior Seaway. *Baptornis* (see Martin and Tate, 1976) and *Parahesperornis* (Martin, 1984) are known from the Smoky Hill (upper) Member of the marine Niobrara Chalk (Coniacian) of Kansas, with *Baptornis* also occurring in the continental Judith River Formation (Campanian) of Saskatchewan (Tokaryk and Harrington, 1992). *Hesperornis* (see Marsh, 1880) is much more widespread (Bryant, 1983: Fig. 2), occurring from the present Gulf of Mexico to above the Arctic Circle in Canada (Russell, 1967) and Alaska (Bryant, 1983) and ranging in age from the Turonian (Greenhorn Formation) to the Campanian (Pierre Shale; Walker, 1967; Martin and Tate, 1976). Outside North America, fragmentary remains of comparable hesperornithiforms have been recorded in Late Cretaceous rocks of Kazakhstan, central Russia, Sweden in the Fenno-Scandian sea straits, and in the Turgai Strait (Nessov, 1992), between eastern Europe and western Asia, as well as in Ukraine and Mongolia (Kurochkin, 1995:53; Feduccia, 1996:159–161). The only record of a purported hesperornithiform from the southern continents is that of *Neogaeornis wetzeli* Lambrecht, 1929, known from an isolated tarsometatarsus from the Campanian/Maastrichtian of Chile (see Olson, 1992). This taxon, however, has recently been referred to the Gaviiformes (Olson, 1992). Hesperornithiform remains have been found in continental (Kurochkin, 1995) and estuarine deposits (Fox, 1974), indicating that these birds also inhabited nonmarine aquatic environments.

Geological Setting

During the nineteenth century, phosphate nodules or "coprolites" from the Cambridge Greensand were used as a source of agricultural phosphate, and there were diggings over an outcrop length of about 65 km from Barton in Bedfordshire northeast through Cambridgeshire as far as Soham, southeast of Ely (Hart, 1973: Fig. 1; Grove, 1976; Norman and Fraser, 1991:31). The exact localities from which *Enaliornis* specimens were collected are unknown. However, Seeley (1876) noted that most of the specimens of *Enaliornis* in the Woodwardian Museum were collected from the neighborhood of Coldham Common or at Granchester (Grantchester), the positions of which are, respectively, about 3 km east and 4 km southwest of the center of Cambridge, Cambridgeshire. Details concerning workings in these areas are given by Penning and Jukes-Brown (1881). Deposition in Cambridgeshire and Bedfordshire was in a narrow seaway along the western margin of the Anglo Brabant Massif, and it linked the East Midlands shelf to the Wessex Basin (Rawson et al., 1978).

The Cambridge Greensand is a thin, condensed bed of glauconitic, arenaceous calcilutite that forms the base of the Chalk (Cenomanian, Upper Cretaceous). It consists of a sandy marl full of green grains of glauconite, and it sits on the uneven and locally bored or holed surface of the Gault (Albian, Lower Cretaceous), in the hollows of which are the phosphatic nodules. A by-product of the phosphatic mining industry was a large number of fossils, mostly invertebrates (Penning and Jukes-Brown, 1881; White, 1932). There has been much discussion concerning the exact age of this fauna (White, 1932; Cookson and Hughes, 1964; Casey, 1965; Hart, 1973; Rawson et al., 1978). The problem is made especially acute because of its proximity to the boundary between the Lower and Upper Cretaceous, but it is now established that the Cambridge Greensand was deposited during the early Cenomanian (*Mantelliceras mantelli* Zone; Rawson et al., 1978). Three specimens of the ammonite *Schloenbachea* cf. *varians subplana* have been identified from the Cambridge Greensand (Cookson and Hughes, 1964), and although the

provenance of these specimens is questioned by Casey (1965), the indigenous microfauna includes seven species of foraminifera (five benthic, two planktonic) that give an unequivocal lower Cenomanian age (upper *Hypoturilites carcitanensis* assemblage subzone; Hart, 1973). However, the majority of the derived phosphatized macrofauna, for which the bed is so famous, represents a remanié assemblage derived from the underlying Upper Gault. The ammonites represent both subzones of the *Stoliczkaia dispar* zone (Upper Albian, Lower Cretaceous; Casey, 1965), and this is probably the age of the *Enaliornis* bones.

Systematic Paleontology

Taxonomic Hierarchy

> Aves Linnaeus, 1758
>> Ornithothoraces Chiappe and Calvo, 1994
>>> Ornithurae Haeckel, 1866
>>>> Hesperornithiformes Fürbringer, 1888

Diagnosis—Fusion of cranial bones caudal to frontoparietal suture; palate with short, broad pterygoids and long, narrow palatines; quadrate pneumaticity reduced or absent; small predentary bone; intramandibular articulation between angular and splenial bones; all thoracic vertebrae heterocoelous; sternum unkeeled; long bones apneumatic; reduction of wing; humerus lacking distinct distal condyles; acetabulum partly closed; postacetabular pelvis greatly elongated; femur relatively short and broad; tibiotarsus with large cnemial crest shaped like an isosceles triangle in cranial view; large and triangular patella with foramen for the ambiens tendon; tarsometatarsus transversely compressed with distinct craniolateral ridge leading to lateral trochlea (IV); lateral trochlea at least as large and distally extending as the middle trochlea (III); fourth toe the longest.

Enaliornis Seeley, 1876

Diagnosis—Antitrochanter on ilium (but not strongly produced laterally); proximally, femur lacks a distinct neck; proximal end of tarsometatarsus with small medial cotyla, distal end arched in distal view with dorsal edge of lateral trochlea (IV) plantar to the prominent dorsal edge of the subequal middle trochlea (III).

E. barretti Seeley, 1876

Lectotype—Distal end of left tarsometatarsus BMNH A477 (Fig. 14.8R–V; this is the original bone for cast BMNH A112 that was incorrectly cited as BMNH A1112 when it was designated as the lectotype by Brodkorb, 1963a:221).

Locality and horizon—Late Albian (Lower Cretaceous) of England. Cambridge Greensand of Upper Gault (*Stoliczkaia dispar* zone, uppermost Albian, uppermost Lower

Cretaceous; Casey, 1965) from the neighborhood of Cold-ham Common or at Granchester, respectively about 3 km east and 4 km southwest of the center of Cambridge, Cambridgeshire, England (localities for all specimens; Seeley, 1876).

Diagnosis—Seeley (1876) referred the more common larger specimens to this species, the detailed study of which will provide a diagnosis (Galton and Martin, in press).

E. sedgwicki Seeley, 1876

Lectotype—Proximal end of right tibiotarsus, SMC B55314, designated by Brodkorb (1963a:221) (Fig. 14.7A–C).

Locality and horizon—Same as given previously for *E. barretti*.

Diagnosis—Seeley (1876) referred the less common smaller specimens to this species, the detailed study of which will provide a diagnosis (Galton and Martin, in press).

Anatomy

The terminology for the osteological structures and the nomenclature for the muscles follow those of Baumel and Witmer (1993) and Vanden Berge and Zweers (1993), respectively, except for cranial structures from Elzanowski and Galton (1991) that are not covered in these references.

Skull

The skull is represented only by three partial braincases: the very well preserved SMC B54404 (Seeley, 1876:498–499, Pl. 26, Figs. 1–4), which is now almost completely free of matrix (Figs. 14.1, 14.2) so that the cranial cavity (Fig. 14.2F) is visible and from which there is a partial endocranial cast (Fig. 14.2G–I), the well-preserved YM 585, and the more fragmentary YM 586 (Elzanowski and Galton, 1991: Fig. 2). Witmer (1990:355–358, Figs. 12, 13) described the features of the braincase of SMC B54404 that were related to the system of pneumatic spaces in the cranium. Elzanowski and Galton (1991) gave a detailed description of these specimens, which is the basis for the following description.

The frontoparietal and the interparietal sutures are un-obliterated (Fig. 14.2A, B), and the parietosupraoccipital suture may be partly so (Fig. 14.2A), but all the other sutures, including those of the laterosphenoid and squamosal, are completely obliterated. Consistent with an adult rather than a juvenile age for the cranial specimens is the full ossification of the dorsum sellae and of the floor of the tectal fossa, which completely closes the trigeminal fossa dorsally (Fig. 14.2F). Thus, the braincases of *Enaliornis* are probably from individuals that completed their ossification, even though the frontoparietal and interparietal sutures probably remained open for most or all of their life, which is, of course, a primitive feature.

The braincase has the airencephalic conformation of Hofer (1952), with the occiput strongly inclined caudo-ventrally so that the plane of the basilar plate is at an angle of 120° to that of the foramen magnum (Fig. 14.2F). Other characters of this type of avian braincase include a high dorsum sellae, a deep medullary fossa, and the occiput inclined caudoventrally, with the cerebellar prominence protruding far behind the level of the occipital condyle (Fig. 14.2F) so that the cerebral hemispheres, optic lobes, and the oblongata are compactly superimposed dorsoventrally (Fig. 14.2G).

In caudal view (Figs. 14.1B, 14.2B), the cerebellar prominence on the supraoccipital is convex (Figs. 14.1B, D, 14.2B, D, F), and the adjacent attachment area for the cranial part of M. biventer cervicis is weakly concave. The attachment area of the medial part of M. splenius capitis, immediately lateral to the groove for the external occipital vein, is strongly concave, whereas that for the lateral part is less distinct. On the base of the zygomatic process are two foramina for branches of the external ophthalmic ramus of the occipital artery, with the medial branch turning dorsally close to the foramen and the lateral branch continuing laterally. Only the lateral margins of the foramen magnum are preserved, the rest being enlarged by the abrasion that removed the occipital condyles (Figs. 14.1B, C, 14.2B, C, F).

The hexagonal fossa on the basilar plate (Figs. 14.1C, 14.2C) is a preservational artifact produced by the collapse of the pneumatic space when the ventral lamina of the base (basioccipital and/or basitemporal) was pressed against the dorsal lamina, the floor of the cranial cavity. The para-sphenoidal rostrum is sectioned just rostral to the pituitary fossa, and a small pneumatic space is visible (Fig. 14.2A). There is no trace of the basipterygoid processes that originally were probably present near the end of the preserved part of the apparently narrow rostrum. The rostral part of the cranial base is flat ventrally, with a pair of asymmetrical openings somewhat enlarged by erosion, the auditory tube foramina of Witmer (1990) that lead to the rostral tympanic recess (Fig. 14.2A, C–E). The rostrum passes laterally into the parasphenoid wings, which are partly damaged and which originally enclosed the rostral tympanic recess (Fig. 14.2C–E).

The sagittal crest, which is almost completely abraded (Figs. 14.1A, B, 14.2A, B), is formed by the parietals between the attachment areas for the superficial part of M. adductor mandibulae externus (Fig. 14.2B, D). On the side wall there is a well-delimited attachment area for M. pseudo-temporalis superficialis (Figs. 14.1D, 14.2D) that extends across the width of the temporal fossa to the base of the zygomatic process (Figs. 14.1B, D, 14.2B, D). Laterally, the parietals are co-ossified to the side wall bones, presumably only to the squamosals. The transverse nuchal crest is prominent on both sides, extending onto the zygomatic processes, but the middle part is abraded (Figs. 14.1B, D, 14.2B, D).

The preserved side wall extends rostrally far beyond the level of the frontoparietal suture, and it is probably formed by both the prootic and laterosphenoid, the latter being represented by the bone rostral to the maxillomandibular foramen (Figs. 14.1D, E, 14.2D, E). The small opening midway between this foramen and the vestibular window is the foramen for the facial nerve (cranial nerve [CN] VII). The orbitocerebral walls converge onto, rather than rostral to, the dorsum sellae, so the hypophyseal (pituitary) fossa is rostral to the cranial cavity (Figs. 14.1A, 14.2A, F). This configuration, which does not occur in neornithines, also occurs in theropod dinosaurs such as *Troodon* (Currie, 1985: Fig. 4) and *Syntarsus* (Raath, 1985: Fig. 1), in which the pituitary fossa opens entirely in front of the cranial cavity. The orbitocerebral wall is pierced by the ophthalmic foramen (Figs. 14.1A, C, E, 14.2A, C, E) and a smaller one, the protractor foramen of Elzanowski and Galton (1991). The latter is on the periphery of a shallow depression, the area of origin for one of the muscles innervated by the protractor nerve, M. protractor pterygoidei et quadrati (Figs. 14.1A, D, 14.2A, D). In *Enaliornis* the dorsum sellae is high, without any shelf protruding over the pituitary fossa (Figs. 14.1A, 14.2A, F). It is perforated by the pair of abducens foramina for CN VI (Fig. 14.2A). Within the pituitary fossa there is one opening for the carotid canal (Fig. 14.2A), so the internal carotid arteries anastomosed without a connecting vessel.

The caudal tympanic wing (= paroccipital process) is poorly developed, and it has no marginal connection to the cranial base through a sphenoccipital bridge. Other primitive characters, which are not artifacts of preservation, include the absence of the following structures: the parabasal fossa, caudal ostia, and canals or grooves for the internal carotid and external ophthalmic (stapedial) arteries. The area of origin of M. columellae (Fig. 14.2B, C) extends medially close to the hypoglossal foramina for CN XII and rostroventral to the vagal foramen for CN X; there is no separate glossopharyngeal foramen for CN XI. The basilar tubercles occupy the extreme lateral position, just rostral to the vagal foramen, so their lateral slopes are continuous with the wall of the tympanic cavity.

The tympanic cavity (Figs. 14.1C–E, 14.2C–E) occupies the caudalmost position in the skull, extending up to the occipital plate. The two quadrate cotylae, squamosal and otic, are separated by the dorsal tympanic recess. However, as Witmer (1990) noted, there is no nonarticular intercondylar depression between the facets (present in hesperornithids), so the dorsal tympanic recess may have extended between the facets, possibly resulting in a small intercapitular incisure to give a "double-headed" quadrate similar to that of many neornithines. The squamosal cotyla occupies most of the ventral surface of the zygomatic process and faces caudoventrally. In the otic cotyla, the caudal portion is entirely sessile, lacking any peduncle or caudal buttress (pila

otica) to block the medial otic capitulum of the quadrate. The rostral portion of the otic cotyla faces laterally, with a relatively slight (less than 45°) ventral tilt and no rostral tilt. The columellar recess is shallow. The vestibular (oval) window is subtriangular and almost oval, with the more acute pole directed ventrally. The cochlear (pseudoround) window is much more elongate. The two fenestrae are separated from the caudal tympanic recess by a broad bony bridge and from each other by a much narrower, but still relatively broad and flat, interfenestral bridge.

The evidence for tympanic pneumatic diverticula was discussed by Witmer (1990:355–358), who concluded that they were similar to those of hesperornithids. In *Enaliornis* and *Hesperornis* the walls of the pneumatic recesses are smooth with little, if any, trabeculation. The dorsal tympanic recess did not communicate with that of the opposite side or with the caudal tympanic recess that pneumatized the paroccipital process, otic capsule, and the basal tubercles. The rostral tympanic recesses were well developed, with contralateral communication (Witmer, 1990:358).

The preserved part of the cranial cavity (Fig. 14.2F, G) includes the fossa for the medulla oblongata, the cerebellar fossa, the tectal fossa, which enclosed the optic lobes, and the auricular (subarcuate) fossa, which enclosed the cerebellar auricle and sinus. The rostral fossa (not preserved, except for the cerebral hemispheres and adjacent structures), which apparently did not extend rostrally beyond the frontoparietal suture and its lateral extensions, was broadly separated from the arcuate eminence by a broad and flat tentorial protuberance and from the tectal fossa by the tentorial ridge, which is formed by the rostral margin of the laterosphenoid.

The fossa for the medulla oblongata is very deep and strongly concave, indicating a prominent ventral bend or sigmoid flexure of the medulla oblongata (Fig. 14.2G). The fossa is widest rostrally, where its width approximately equals its length as well as the width of the cerebellar fossa (Fig. 14.2H, I). The floor of the fossa does not show any trace of a median ridge, which would project into the median fissure of the medulla oblongata (Fig. 14.2H). The trigeminal ganglion fossa (for root ganglion of CN V) opens to the medullary fossa and is entirely separated from the tectal fossa by the ossified floor of the latter (Fig. 14.2F, G). The internal acoustic fossa (Fig. 14.2F) is shallow, rostrocaudally elongate, and poorly delimited ventrally and contains two major pits, well apart from each other, with the rostral pit (Fig. 14.2G, for nn. facialis, cochlearis, and lagenaris) being larger than the caudal one (Fig. 14.2G, for n. ampullaris caudalis). Caudoventral to the internal acoustic fossa is a distinct vagoglossopharyngeal fossa for the root ganglia of CN X and CN XI.

The cerebellar fossa (Fig. 14.2F) extends up to the frontal margin of the parietals. Neither parietals nor supraoccipital

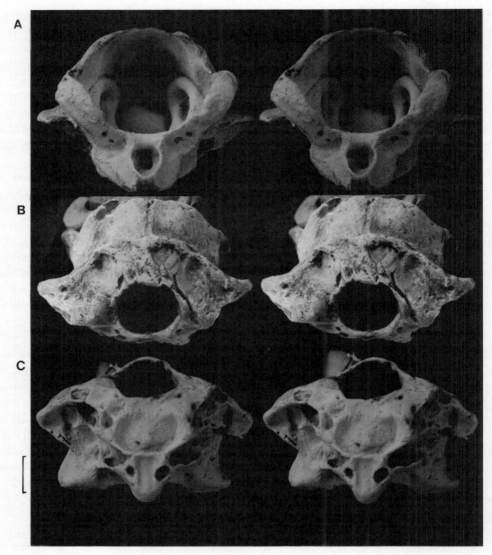

Figure 14.1. *E. barretti*, braincase, SMC B54404: A, rostral; B, caudal; C, ventral; D, right lateral; and E, left rostroventrolateral views (see Fig. 14.2A–E for labeled drawings of comparable views). Scale bars = approximately 5 mm (A–D approximately ×2; E approximately ×3).

(Figs. 14.1A, 14.2A, F) shows any trace of transverse inter-foliar ridges (which would project into the cerebellar fissures of the cerebellum, Fig. 14.2G, I), and there is no trace of a median groove for the occipital sinus on the supraoccipital (Fig. 14.2A, H). The caudalmost portion of the cerebellar fossa was eliminated by erosion.

The tectal fossa (Fig. 14.2F) narrows caudodorsally to a pointed end that comes very close to the cerebellar fossa. The optic lobes were lateral, rather than ventrolateral, relative to the brainstem (Fig. 14.2H). In the horizontally positioned skull, the lobes were only partly covered by the hemispheres in dorsal view (Fig. 14.2I). In contrast to the condition in recent birds, the lobes do not show any appreciable caudalward bending or rotation around their long

axes at the base. Caudally, each lobe closely adhered to the arcuate eminence, as indicated by the flat rostral surface of the eminence, well reflected on the endocast (Fig. 14.2G). In contrast to the condition in *Enaliornis*, the optic lobes of all neornithines are bent or rotated caudally around their basal long axes so they partly overlap the oblongata in lateral view. In *Enaliornis*, the rostral position of the optic lobe is reflected by the bulge of the orbitocerebral wall and the deep caudoventral corner in the orbit (between the orbito-cerebral wall and the cranial base, Fig. 14.2D). A more rostral position of the optic lobe means less space for the cerebral hemisphere, which adhered to the rostrodorsal surface of the lobe as demonstrated by its rectilinear rather than convex outline in side view. The lateral wall of the tectal

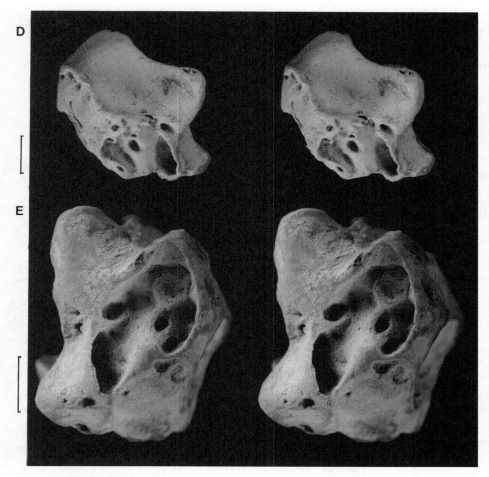

D

E

Figure 14.1. *Continued*

fossa is delimited from the arcuate eminence, which is only slightly embedded in the braincase wall, by the lateral semicircular sulcus (Fig. 14.2F) for the caudal petrosal sinus (Fig. 14.2G–I). This groove encircles the eminence dorsally and rostrally, terminating in the tectal fossa as a slit situated almost exactly internal to the external recess for the obliterated foramen of the middle cerebral vein (Fig. 14.2D, E). Caudally, the groove disappears above the arcuate eminence and under the supraoccipital, beyond which the sinus continues caudally as the external occipital vein (Fig. 14.2B).

The auricular (subarcuate) fossa has a large oval entrance that fills the area enclosed by the bony covering of the anterior semicircular canal, the arcuate eminence (Fig. 14.2F). The ventral part of the bony labyrinth forms the vestibular eminence, with a distinct tubercle at the caudal end perforated by the endolymphatic foramen. There is another small opening caudoventral to this tubercle.

Comparison of the endocranial casts shows that the optic lobes and the medulla oblongata of *Enaliornis* (Fig. 14.2G–I) are comparable in size to those in *Phaethon* (tropic-

bird), although the medulla of *Enaliornis* may be slightly larger. In addition, *Enaliornis* resembles *Phaethon*, Laridae (gulls and terns), Charadriidae (plovers), and Stercorariidae (skuas) by the strong ventral projection of the medulla oblongata. It also approaches *Phaethon* in the shape of the medullary fossa, which, in contrast to the condition in most birds, is not longer than wide. The medulla oblongata is similar in ventral view to that of *Merops* (bee-eater), the brain structure of which is considered to be close to the avian archetype by Portmann and Stingelin (1961). The rostral position of the optic lobes and the caudal position of the orbitocerebral wall, which descends upon rather than being in front of the dorsum sellae, suggest that the cerebral hemispheres of *Enaliornis* were relatively smaller than in neornithines. This is also shown by the lack of any caudal expansion of the hemispheres beyond the frontal. The small hemispheres, lack of caudal rotation of the optic lobes, probably smaller cerebellum, and relatively larger medulla oblongata indicate that the brain of *Enaliornis* was more primitive than that of any living bird.

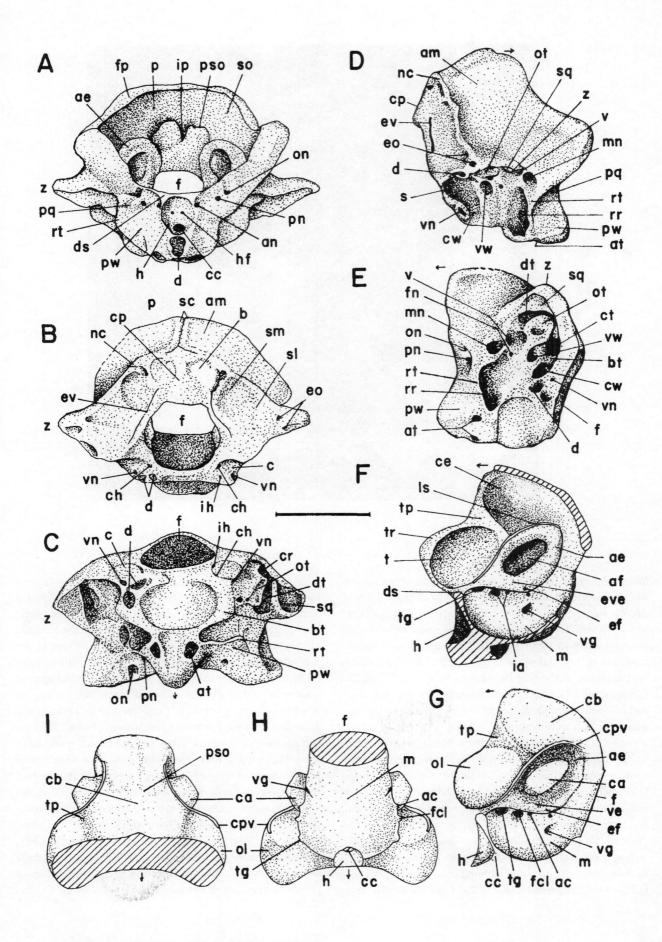

Postcranial Skeleton

Unless stated otherwise, comparisons of the postcranial bones are based on the following references: for *Archaeopteryx*, from the Upper Jurassic of Germany, Ostrom, 1976, and Martin, 1991; for *Baptornis*, Martin and Tate, 1976; for *Hesperornis*, Marsh, 1880; for *Pasquiaornis*, Tokaryk et al., 1997; for *Neogaeornis*, Olson, 1992; for the gaviid "*Polarornis*" from the Upper Cretaceous of Antarctica, Chatterjee, 1997; and for the loon *Gavia*, Wilcox, 1952. A more detailed description of the postcranial anatomy of *Enaliornis* will be published elsewhere (Galton and Martin, in press).

Vertebral Column

A few isolated, incomplete vertebrae are available, and these include a caudal cervical vertebra (YM 507, Fig. 14.3A–E) that is similar to the corresponding parts of cervicals 12 and 13 of *Hesperornis*. The centrum is short, heterocoelous, and subtriangular in ventral view (Fig. 14.3C) with a carotid sulcus that, cranially, is bordered by the bases of the incomplete carotid processes (Fig. 14.3A, C). The cranial central surface is transversely concave (Figs. 14.3C) and mostly convex vertically, except ventrally, where it is slightly concave (Fig. 14.3B). Caudally, the centrum is transversely convex and mostly concave vertically (Fig. 14.3A, E). The spinous process is missing, as are the costal processes.

The strongly heterocoelous thoracic vertebra (SMC B55277, Fig. 14.3F–J) is probably the first free vertebra cranial to the synsacrum, and the preserved parts are very similar to those of this vertebra in *Baptornis* and *Hesperornis*, in which a ventral process is also absent.

Chiappe (1996:225) noted that it is likely that the heterocoelous condition arose only once early in avian history, developing from the cranial cervicals (Marsh, 1880) and extending caudally to a varying degree in different lineages (e.g., hesperornithiforms, neornithines). In *Enaliornis*, cervical vertebra 13 or 14 (Fig. 14.3A–E) and the last free thoracic vertebra (Fig. 14.3F–J) are both strongly heterocoelous, and the preserved parts closely resemble those of *Baptornis* and *Hesperornis*. Because of this, it is reasonable to assume that all the thoracic vertebrae of *Enaliornis* were also strongly heterocoelous, as in *Baptornis*, *Hesperornis*, and most neornithines. Consequently, the small thoracic vertebra SMC B55274 (Fig. 14.5A–E), which does not resemble any of the vertebrae of *Baptornis* or *Hesperornis*, was incorrectly referred to *Enaliornis* by Seeley (1876:501, Pl. 26, Figs. 12, 13) because the centrum is amphicoelous (i.e., concave at both ends), as in *Archaeopteryx* and *Ichthyornis* (Marsh, 1880). A similar but less complete amphicoelous thoracic vertebra (SMC B55280; Seeley, 1876:501) is probably from the same taxon, as is a small and weakly amphicoelous cervical vertebra (YM 584, discussed by Seeley, 1876:500, as being the thoracic vertebra with a sharp ventral ridge to the centrum, in the Reed Collection). The referral of these amphicoelous vertebrae to *Enaliornis* by Seeley (1876) has not been questioned, even though it preceded the detailed description of *Hesperornis* and *Ichthyornis* by Marsh (1880). Indeed, it was noted by Martin and Tate (1976:65) and Martin (1984:147) that *Enaliornis* differs from *Baptornis* in having amphicoelous thoracic vertebrae, and Chiappe (1996) referred to the difference in the caudal extent of heterocoely in hesperornithiforms and neornithines.

Figure 14.2. *E. barretti*, braincase, SMC B54404: A, rostral; B, caudal; C, ventral; D, right lateral; and E, left rostroventrolateral views (Fig. 14.1A–E). F, cranial cavity in left medial view. Partial endocranial cast: G, left lateral; H, ventral; and I, dorsal views. Small arrows point rostrally. (After Elzanowski and Galton, 1991.) Abbreviations: ac, n. ampularis caudalis; ae, arcuate eminence (or impression on cast); af, auricular (subarcuate) fossa; am, attachment area for rostral part of M. adductor mandibulae externus; an, foramen for abducens nerve (VI); at, auditory tube foramen; b, attachment for cranial part of M. biventer cervicis; bt, basilar tubercle; c, attachment for M. columellae; ca, auriculum cerebelli; cb, cerebellum; cc, opening of carotid canal; ce, cerebellar fossa; ch, foramen for caudal branch of hypoglossal nerve (XII); cp, cerebellar prominence; cpv, caudal petrosal sinus; cr, caudal tympanic recess (exposed by damage); ct, caudal tympanic recess; cw, cochlear (pseudoround) window; d, pneumatic space exposed by damage; ds, dorsum sellae; dt, dorsal tympanic recess; ef, endolymphatic foramen; eo, foramina for ophthalmic branch of occipital artery; eve, sulcus for external occipital vein; f, foramen magnum; fcl, nn. facialis, cochlearis, and lagenaris; fn, foramen for facial nerve (VII); fp, frontoparietal suture; h, hypophysial (pituitary) fossa or hypophysis on cast; hf, hypophysiocerebral fontanelle; ia, internal acoustic opening; ih, foramen for intermediate branch of hypoglossal nerve (XII); ip, interparietal suture; ls, lateral semicircular sulcus; m, medulla oblongata (on cast) or fossa; mn, foramen for maxillomandibular nerve; nc, transverse nuchal crest; ol, optic lobe; on, foramen for ophthalmic nerve; ot, otic cotyla for quadrate; p, parietal bone; pn, foramen for a protractor nerve innervating M. protractor pterygoidei et quadrati, M. depressor palpebrae ventralis, and M. tensor periorbitae; pq, attachment for M. protractor quadrati et pterygoidei; ps, attachment for M. pseudotemporalis suprficialis; pso, parietosupraorbital suture (or impression on cast); pw, parasphenoidal wing; rr, rostral tympanic recess; rt, rostral tympanic wing; s, suprarecessal compartment; sc, sagittal crest; sl, sm, attachment areas for lateral and medial parts of M. splenius capitis, respectively; so, supraoccipital bone; sq, squamosal cotyla for quadrate; t, tectal fossa; tg, trigeminal ganglia (on cast) or fossa for it; tp, tentorial protuberance (or impression of on cast); tr, tentorial ridge; v, preotic venous recess, obliterated foramen for middle cerebral vein; ve, vestibular eminence (or impression of it on cast); vg, vagoglossopharyngealis ganglia fossa (or ganglia on cast); vn, foramen for vagus nerve (X); vw, vestibular (round) window; z, zygomatic process. Scale bar = approximately 5 mm (approximately ×2.5).

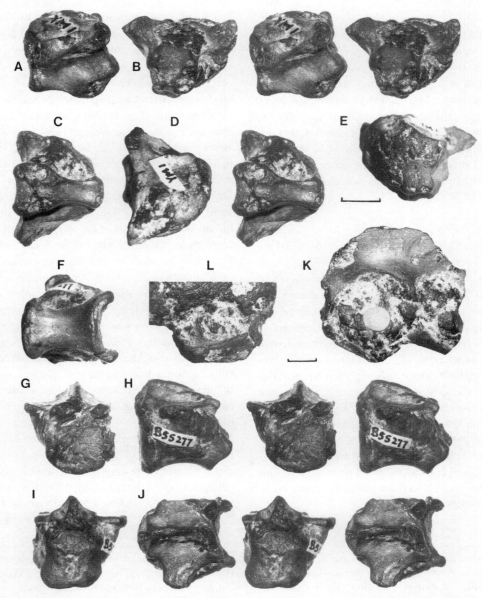

Figure 14.3. *E. barretti*. Cervical vertebra 12 or 13, YM 507: A, right lateral; B, cranial; C, ventral; D, dorsal; and E, caudal views. Last free thoracic vertebra cranial to synsacrum, SMC B55277: F, ventral; G, cranial; H, right lateral; I, caudal; and J, dorsal views. Acetabular part of left pelvic girdle, BGS 87936 (cast as BMNH A5310): K, lateral, and L, medial views. Scale bars = 5 mm (A–J ×2; K–L ×1.5).

The vertebral foramen (Fig. 14.5A) is large, with the ratio of its vertical diameter to that of the cranial articular surface being 0.75. This ratio is comparable to that of hesperornithiforms and ichthyornithiforms (at least 0.7; Marsh, 1880) and well above that of nonavian theropods (much lower than 0.4; Chiappe, 1996). Chiappe (1996:225) noted that, given the present state of knowledge, the "derived condition might be a synapomorphy of the clade composed of Enantiornithes, *Patagopteryx deferrariisi* and Ornithurae (see node below 2, Figs. 54–56), or even of the Ornithothoraces (node 1), or even of Aves." In ornithurines, the cervicals, apart from the atlas, are fully heterocoelous except in

Ichthyornithiformes, in which the heterocoely is restricted to the cranial part of the series (Marsh, 1880; Martin, 1983; Chiappe, 1996). However, the caudal thoracics of *Ichthyornis* have prominent pleurocoels (Marsh, 1880) that are absent in the Greensand specimens, so these vertebrae are probably not from an ichthyornithiform. For the moment, the amphicoelous cervical and thoracic vertebrae (Fig. 14.3A–E) from the Cambridge Greensand can only be identified as Aves incertae sedis.

Part of the preacetabular part of a synsacrum was referred to *Enaliornis* by Seeley (1876). These incomplete vertebrae (SMC B55282, Fig. 14.4H, I) consist of a complete

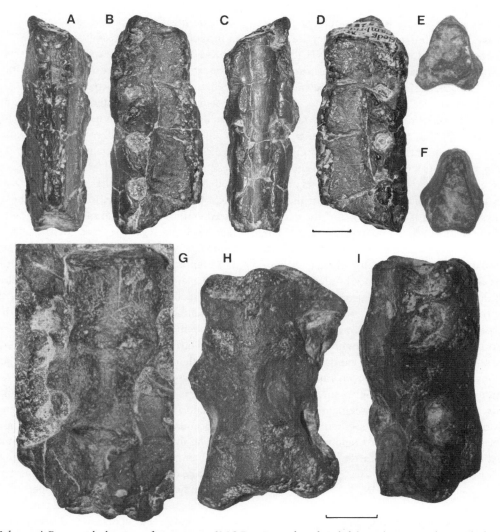

Figure 14.4. *E. barretti.* Postacetabular part of synsacrum, SMC B55283: A, dorsal; B, left lateral; C, ventral; D, right lateral; E, cranial; and F, caudal views. Pterosauria, Pterodactyloidea incertae sedis, cranial part of notarium, SMC B55281: G, ventral view. *E. barretti,* preacetabular part of synsacrum, SMC B55282: H, ventral; and I, left lateral views. Scale bars = 5 mm (A–F ×2; G–I ×2.75).

centrum plus the partial centra of the adjacent vertebrae. The centra are pinched in to form a median longitudinal ridge (Fig. 14.4H). Cranially, part of the dorsal surface of the neural arch is preserved to the right side of the fused but eroded bases of the spinous processes. Large intervertebral foramina are visible in lateral view (Fig. 14.4I). Although the cranial synsacral centra of *Baptornis* and *Hesperornis* are not pinched in, this specimen is tentatively referred to *Enaliornis;* it is not from a pterosaur (Unwin, pers. comm.). The other four fused vertebrae (SMC B55283, Fig. 14.4A–F) are from the postacetabular region of the synsacrum, and, as none has a large sacral rib, they probably represent vertebrae 5 or 6 to 8 or 9. The complete synsacrum probably consisted of 10 vertebrae, as in *Baptornis* (Martin and Tate, 1976). These vertebrae are narrow transversely (Fig. 14.4A, C), indicating that *Enaliornis* had an elongated, compressed

pelvis of the type found in other foot-propelled diving birds. The intervertebral foramina are large and situated between the short sacral ribs (Fig. 14.4B, D). The spinous processes are fused together and rugose dorsally (Fig. 14.4A, B, D). The ventral surface becomes progressively narrower, passing caudally with a central depression that is bordered laterally by a rounded edge (Fig. 14.4C, E, F), as in *Hesperornis* (Marsh, 1880: Pl. 11, Fig. 1)

Seeley (1876) referred the centra of a partial synsacrum (SMC B55281, Fig. 14.4I), consisting of three fused vertebrae exposed in ventral view, to *Enaliornis*. These centra, which are roughly hourglass-shaped and become slightly less massive passing caudally, are very different from those of any in the synsacra of *Baptornis* and *Hesperornis*. Consequently, this specimen was most likely incorrectly referred to *Enaliornis*. These vertebrae probably represent three of a series of fused thoracic vertebrae, the notarium, of a ptero-dactyloid pterosaur (Wellnhofer, 1978, 1991). They are quite similar to the cranial half of a notarium (BMNH R3877), which consists of six fused vertebrae, of "*Ornithodesmus*" *latidens* (Lower Cretaceous, Isle of Wight, England) de-scribed by Hooley (1913).

The centrum of the vertebra SMC B55285 (Fig. 14.5F–I), identified by Seeley (1876) as a caudal, has a strongly pro-coelous centrum (Fig. 14.5F, H) with a markedly concave cranial end (Fig. 14.5G) and a very convex caudal end (Fig. 14.5F–H). The cranial edge of the neural arch is angled craniodorsally at an angle of about 45°, much of the cra-nial zygapophyseal process is preserved, and the caudo-dorsal part of the neural arch is missing (Fig. 14.5F). The transverse process is represented by a rounded edge (Fig. 14.5F) that protrudes very little laterally, whereas in *Bap-tornis* the first four vertebrae, the ones bordered by the caudal end of the elongate postacebular part of the pelvic girdle, have short transverse processes, but the centra are either amphicoelous or amphiplatyan. In *Hesperornis*, all the dorsoventrally flattened caudal vertebrae have large transverse processes. In pterosaurs, the caudals are either all amphicoelous (Wellnhofer, 1978, 1991) or, as in *Pteran-odon* (Bennett, 1987), with a double convex surface at each end (segments 1–4), amphiplatyan (segments 5–8) and fused into a rod (the remaining distal segments). In some chelonians, the caudal vertebra are procoelous, suggesting that this vertebra might be from a turtle (Gaffney, pers. comm.).

Pelvic Girdle

The single fragment of the pelvis (Fig. 14.3K, L) preserves only the area just around the acetabulum and the cranial part of the ilioischiadic fenestra. It does show that the pelvis was opisthopubic and that the pubis and ischium are nar-row, as in *Baptornis* and neornithines. Part of the acetabu-lar margin on the pubis is visible medially (Fig. 14.3L), so the acetabulum is perforate, but laterally the acetabulum is still filled by a phosphatic nodule (Fig. 14.3K) that needs to be removed before the degree of closure of the acetabulum can be determined. The antitrochanter is proportionally smaller than in *Baptornis*. There is also a small pectineal process. On the basis of the form of the postacetabular part of the synsacrum, the postacetabular part of the pelvic girdle of *Enaliornis* was elongated, as it is in the other hesper-ornithiforms.

Pelvic Limb

Femur. The more complete femora (Figs. 14.5J–L, 14.6C–E) are short and broad when compared with the femora of most birds, but they are slender when compared with those of other hesperornithiforms except that of *Pasquiaornis*, which has a femur that is very similar to that

of *Enaliornis*. The proximal end is rounded and lacks detail in those femora from juvenile individuals (Figs. 14.5M–V, 14.6K–N). The femoral trochanter (major) and the head are at the same level and are connected by a continuous articu-lar surface for the antitrochanter (Fig. 14.5Q, R), such that the femur was directed laterally, as in other hesperornithi-form birds. Thus, it seems likely that *Enaliornis* had already lost the power to walk like a normal terrestrial bird. The smaller size of the antitrochanter in *Enaliornis* indicates that the femur could be placed farther under the center of grav-ity than in *Baptornis* so that some sort of a waddling walk might have been possible. In *Baptornis*, the femur was di-rected cranioventrally and markedly laterally (Martin and Tate, 1976: Fig. 19). The fossa for the attachment of the round ligament is present and dorsally situated on the head of the femur (Figs. 14.5P, 14.6F, M). The femur was normally held in a lateral position, but the head is well rounded and slightly inclined proximally (Figs. 14.5M, T, 14.6G, L), and the attachment for the round (= capital) ligament seems to be more medially situated than it is in *Hesperornis* and *Gavia*. The neck of the femoral head is shorter and less con-stricted than it is in the other hesperornithiform birds ex-cept for *Pasquiaornis*.

The cross section of the shaft is subcircular, as in *Bap-tornis* and *Pasquiaornis*, rather than transversely wide and craniocaudally narrow, as in *Hesperornis*. The cranial inter-muscular line, between areas for M. femorotibialis lateralis et medius, runs distally from the femoral trochanter along the craniolateral edge of the femur to the lateral condyle in *Gavia*. In *Enaliornis* and the other hesperornithiform birds, it runs diagonally across the cranial face of the shaft to the medial condyle (Fig. 14.6D), as in neornithines (Baumel and Witmer, 1993: Fig. 4.16A).

The patellar groove is well developed in *Enaliornis* (Fig. 14.6A, B, P, O), and it is not pocketed proximally, as in hes-perornithids, but extends smoothly onto the shaft as a broad shallow groove, as in *Baptornis*. The medial condyle of *Enaliornis* is small compared with the lateral condyle, as in *Baptornis* and *Pasquiaornis*, but in *Baptornis* the medial condyle extends farther distally. The medial condyle is rela-tively much larger in the hesperornithids, in which the me-dial and lateral condyles extend about the same distance dis-tally. The medial surface of the medial condyle is depressed for the origin of M. plantaris, which runs proximally as a de-pression backed by a ridge, much as in *Gavia*. As is charac-teristic of other foot-propelled diving birds, the fibular condyle is enlarged and angled laterally (Figs. 14.5X, 14.6P, Q). The connection between the medial and lateral condyles (Fig. 14.6B, O) is broad and rounded, as in other hesper-ornithiforms, rather than narrow and sharp, as in *Gavia*. In *Gavia*, the caudal surface is excavated for the external artic-ular surface of the tibiotarsus, which is rounded to fit into this area.

Tibiotarsus. The tibiotarsus was probably elongate, as in other foot-propelled diving birds, but not enough of the shaft remains to provide evidence as to the actual length. The cranial cnemial crest in neornithines is an ossification separate from the shaft of the tibia. Several of the proximal tibial ends of *Enaliornis* (Fig. 14.7I–K) are from birds too young for this ossification to be fused to the shaft. In these tibiae, the proximal end is gently rounded with a slightly dimpled surface that in life was covered in cartilage. In the tibiotarsus of adult birds (Fig. 14.7A–H), the large cnemial crest makes essentially a broad isosceles triangle (Fig. 14.7B, E) that is quite different from the more elongate cranial cnemial crests found in grebes and especially in loons. The proximal end of the tibiotarsus of *Enaliornis* is similar to that of *Baptornis,* with a large, lateral cnemial crest that extends proximally into a high, thin cranial ridge (Fig. 14.6A, B, D, E), as in other hesperornithiforms. Distally, the tibiotarsus is continuous with the sharp craniomedial edge of the adjacent part of the shaft (Fig. 14.7A, D), the proximal end of the fibular crest. The smoothly rounded lateral margin of the lateral cnemial crest extends to just beyond the lateral articular surface. The groove between the two cnemial crests is broad and shallow, as in other hesperornithiforms, rather than narrow and deep, as in grebes, "*Polarornis,*" and loons (Cracraft, 1982).

The distal end (Fig. 14.7L–S) has the astragalus and calcaneum completely fused to the tibia in adults, just as in neornithines. The "pretibial" bone (Fig. 14.7S, ascending process of the astragalus of McGowan, 1985) has essentially the same position and shape as in *Baptornis* (Martin et al., 1980). The lateral and medial condyles are small, extending about the same distance both cranially and ventrally, with a very limited extent both craniocaudally and proximodistally (Fig. 14.7L, O–S). This makes the cranial intercondylar fossa very shallow. The condyles are thin and widely separated by the cranial intercondylar fossa (Fig. 14.7L, P, Q, R). There is no supratendinal bridge, but, as in other hesperornithiforms, there is a prominent ligamental prominence above the center of the cranial intercondylar fossa. The intercondylar notch is medial to this prominence, which would have supported a ligament that, like the ligament in certain owls (*Bubo, Otus*), parrots, and ratites (Baumel and Witmer, 1993) and the supratendinal bridge of the other recent birds, restrained the tendon of M. extensor digitorum longus within the extensor canal. *Archaeopteryx* also has small condyles and a shallow cranial intercondylar fossa, so this would seem to be a primitive character state in *Enaliornis*. In other hesperornithiform birds, this fossa is deeper, because of an expansion of the condyles, whereas in enantiornithine birds the whole articular surface, including the shallow intercondylar fossa, is projected cranially (Walker, 1981; Chiappe and Walker, Chapter 11 in this volume).

Tarsometatarsus. Fürbringer (1888:1153) reidentified the proximal portions of two bones identified as fibulae by Seeley (1876: Pl. 26, Figs. 26, 27) as the proximal ends of ulnae, and this was repeated by Lambrecht (1933). However, these bones are actually the proximal portions of tarsometatarsi of adult individuals (Fig. 14.8K–Q), in contrast to the unfused metatarsus of the juvenile individual (Fig. 14.8A–F) illustrated by Seeley (1876). In the latter, there are no distal tarsal bones, and the proximal ends of the metatarsals are unfused (Fig. 14.8A, B, E). However, the metatarsals are fused at the broken distal end (Fig. 14.8F). The metatarsals are fused proximally, and the articular surface is formed by a cap (Fig. 14.8G–Q), much as in *Baptornis* (Martin and Bonner, 1977). There is a slight central ridge (Fig. 14.8G, I, N, O) that probably helped to separate the tendons, much like a hypotarsus in neornithines, but it corresponds to a backward projection of the middle metatarsal, as seen in the juvenile specimen (Fig. 14.8B). Consequently, it is not a genuine hypotarsus (produced by hypotarsal cap) of the sort seen in neornithines. In *Enaliornis* and *Baptornis,* the lateral cotyla is slightly larger than the medial one (Fig. 14.8G, I, J), and the intercotylar eminence is low and gently rounded (Fig. 14.8G, I, J) compared with the condition in *Hesperornis*. Hypotarsal crests are absent plantarly, as are proximal foramina that pass right through the bone (both present in *Neogaeornis*); the proximodorsal face of the shaft is deeply excavated (Fig. 14.8G, N). Proximally, the lateral metatarsal (IV) is the largest, especially dorsoplantarly; the middle one (III) is slightly smaller, and the medial one (II) is quite small and almost as wide as its dorsoplantar length. This results in the medial cotyla being much smaller in *Enaliornis* than it is in the other hesperornithiform birds. The proximal end of the middle metatarsal is wedge-shaped, with the tip oriented dorsally (Fig. 14.8E). This is the normal condition in other hesperornithiform birds and most neornithines, but not in *Archaeopteryx*, in which the proximal ends of the metatarsals are all of about the same size and shape. As in the other hesperornithiform birds, *Enaliornis* as an adult had a long, slender tarsometatarsus that is compressed laterally, with a distinct dorsolateral ridge (Fig. 14.8G, N). This ridge merges with the shaft somewhere along the shaft, as in *Baptornis,* in which this occurs at midlength. The lateral ridge does not continue to the lateral condyle as occurs in hesperornithids.

The tarsometatarsus narrows distally to a long shaft (Fig. 14.8N–Q) and then widens again into the distal end (Fig. 14.8R–V). When viewed distally (Fig. 14.8U), the dorsal edge of trochlea III is dorsal to those of the other trochlea (II, IV); this is the plesiomorphic condition, as is also the case in *Archaeopteryx*. In other hesperornithiforms, the dorsal edge of trochlea IV is dorsal to that of trochlea III (so the leading edges of trochlea II, III, and IV are progressively more dor-

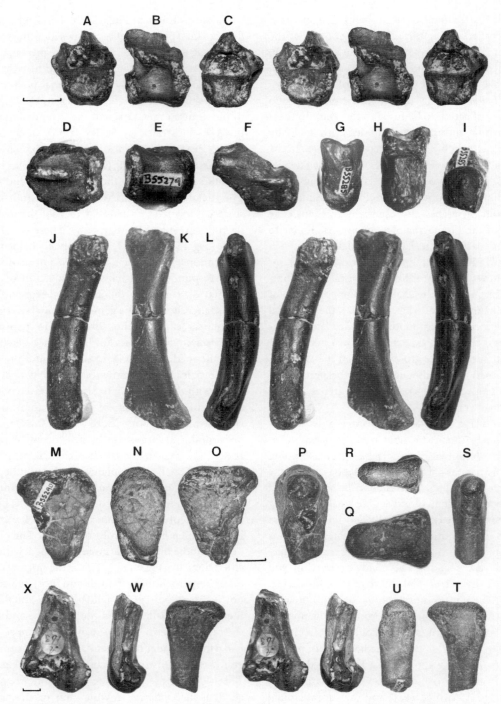

Figure 14.5. Aves, incertae sedis, dorsal vertebra, SMC B55274: A, cranial; B, right lateral; C, caudal; D, dorsal; and E, ventral views. ?Testudines, proximal caudal vertebra, SMC B55285: F, left lateral; G, dorsal; H, ventral; and I, caudal views. *E. sedgwicki,* left femur, BMNH A479: J, lateral; K, cranial; and L, medial views. *E. barretti,* proximal end of left femur of juvenile, SMC B55290: M, cranial; N, lateral; O, caudal; P, medial; and Q, proximal views. *E. sedgwicki,* proximal end of right femur of juvenile, SMC B55287: R, proximal; S, medial; T, caudal; U, lateral; and V, cranial views. *E. barretti,* distal end of right femur, BMNH A163: W, medial, and X, caudal views (also Fig. 14.6A, B; Lydekker, 1891: Fig. 46B). Scale bars = 5 mm (A–I ×2; J–V ×1.5; W–X ×0.9).

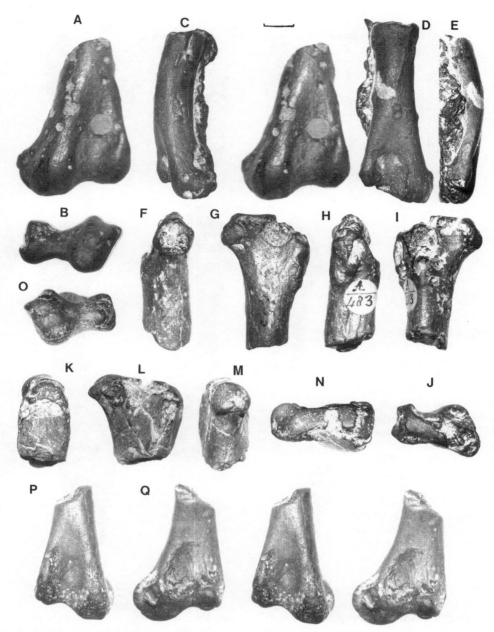

Figure 14.6. *E. barretti.* Distal end of right femur, BMNH A163: A, cranial, and B, distal views (also Fig. 14.5W, X; Lydekker, 1891: Fig. 46B). Left femur lacking proximal end, SMC B55302: C, lateral (and slightly cranial); D, cranial; and E, medial views. Proximal end of left femur, BMNH A483b: F, medial; G, cranial; H, lateral; I, caudal; and J, proximal views. Proximal end of left femur of juvenile, BMNH A483a: K, lateral; L, cranial; M, medial; and N, proximal views. *E. sedgwicki,* distal end of left femur, SMC B55295: O, distal; P, cranial; and Q, caudal views. Scale bar = approximately 5 mm (A–E ×1.4; F–O ×1.5; P, Q ×1.3).

sal in position). This is true for *Baptornis, Parahesperornis,* and *Hesperornis.* It is also the case for *Pasquiaornis* (SMNH P2077.79, distal part of right tarsometatarsus listed but not illustrated by Tokaryk et al., 1997). On the basis of the preliminary description by Tokaryk et al. (1997), no other autapomorphic characters separate this genus from *Enaliornis,* a genus with which they made no comparisons (Seeley, 1876, is not cited).

The medial and lateral trochlear ridges on the medial trochlea are well developed plantarly, but the dorsal face is smoothly rounded, thus permitting the toe to move medially and expand the width of the foot for the power stroke when swimming. The medial ridge is slightly more plantar than the lateral ridge on the medial trochlea, helping to move the inner toe behind the middle and outer toes during the recovery stroke. The medial trochlea is entirely prox-

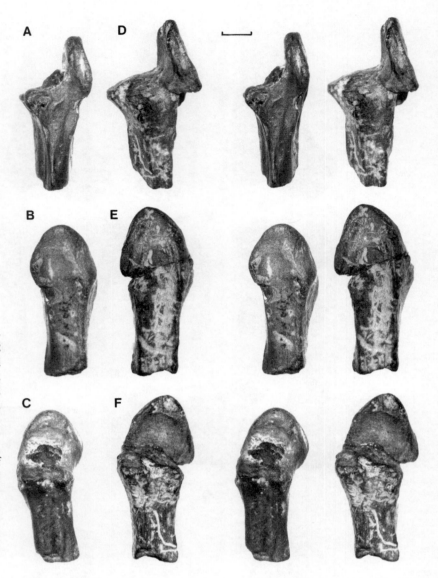

Figure 14.7. *E. sedgwicki* Seeley, 1876, lecto-type, proximal end of right tibia, SMC B55314: A, lateral; B, cranial; and C, caudal views. *E. barretti*, proximal end of right tibia, YM 587: D, lateral; E, cranial; and F, caudal views. *E. barretti*, proximal end of left tibia, BMNH A478: G, lateral, and H, medial views. *E. barretti*, proximal end of right tibia of juvenile, SMC B55312: I, lateral; J, cranial; and K, proximal views. *E. sedgwicki*, distal end of right tibia, SMC B55315: L, distal; M, lateral; N, medial; O, caudal; and P, cranial views. *E. barretti*, distal end of right tibia, SMC B55316: Q, distal, and R, cranial views. *E. sedgwicki*, distal end of right tibia, SMC B55317: S, cranial view. Scale bars = 5 mm (A–F, I–P, S ×1.5; G, H, Q–R ×1.3).

imal to the middle trochlea, facilitating the plantolateral rotation of the inner toe during the recovery stroke. The middle trochlea is slightly larger than the lateral one. This is the normal situation in most birds, but not in the other hesperornithiform birds, and must be considered a primitive condition in *Enaliornis*. When viewed distally (Fig. 14.8U), the middle trochlea is wedge-shaped with the apex in front. Both trochlear ridges are well developed, but the lateral one is longer (more produced dorsally) and extends farther proximally. This would rotate the middle toe plantolaterally in a motion parallel to that of the inner toe on the recovery stroke. The outer trochlea is slightly more distal than the middle trochlea (a derived condition characteristic of hesperornithiforms). The lateral trochlea slants laterally because the medial ridge is much more produced dorsally

than the lateral ridge. This causes the toe to be laterally displaced on the power stroke and expands the surface area of the foot. However, the lateral ridge is more produced plantarly than the medial one and thereby reverses the effect for the recovery stroke.

Phylogenetic Relationships

Although *Enaliornis* is universally accepted as a foot-propelled diver, four very different systematic relationships have been advocated in recent years for *Enaliornis*. It has been considered to be closely related to the foot-propelled diving bird clades (1) Hesperornithiformes, (2) Gaviiformes (loons), or (3) Podicipediformes (grebes) or (4) to be a lineage separate from the other foot-propelled diving birds

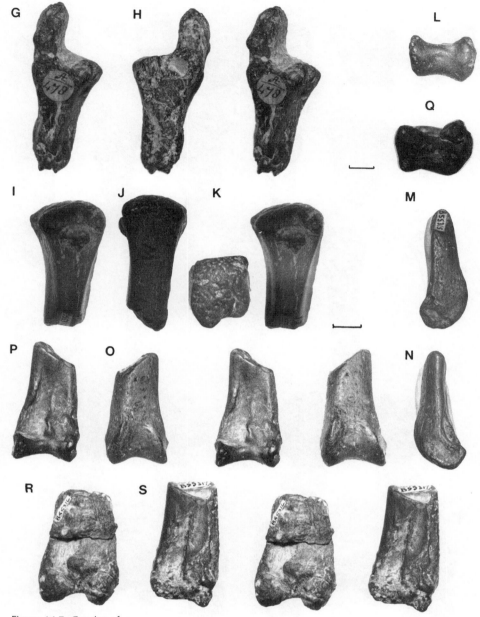

Figure 14.7. *Continued*

(Storer, 1971). However, Cracraft (1982) considered that the hesperornithids probably have close affinities with the gaviids and podicipedids, a position that he has since abandoned (Cracraft, 1986, 1988). Brodkorb (1963b) considered the form of the femur, tibiotarsus, and the distal tarsometatarsus of *Enaliornis* as being similar to that of the loon *Gavia* and hence listed the Enaliornithidae (Fürbringer, 1888) as the basal family of the Gaviiformes. Brodkorb (1971) noted that *Enaliornis* is only slightly less specialized than the Gaviidae and that the Enaliornithidae

appears to be ancestral to the Gaviidae. Although *Lonchodytes* Brodkorb 1963b (Lance Formation, Upper Cretaceous, Wyoming) was removed from the Gaviiformes by Olson and Feduccia (1980), there are two other records of Gaviidae from the Upper Cretaceous: *Neogaeornis* from Chile (Olson, 1992) and "*Polarornis*" from Antarctica (Chatterjee, 1989, 1997). The next earliest record of the Gaviiformes consists of a few isolated bones of *Colymboides anglicus* Lydekker, 1891, from the Late Eocene of England (Harrison, 1976; Harrison and Walker, 1976). Martin and

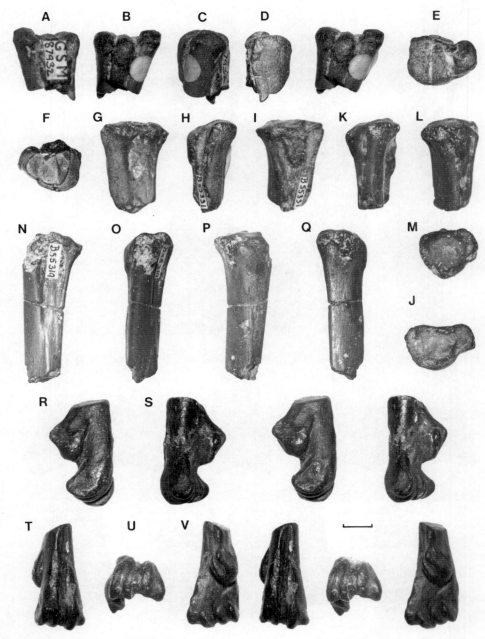

Figure 14.8. *E. barretti,* proximal end of left metatarsus of juvenile, BGS 87932: A, cranial; B, caudal; C, medial; D, lateral; E, proximal; and F, distal views. Proximal end of left tarsometatarsus, SMC B55331: G, cranial; H, medial; I, caudal; and J, proximal views. *E. sedgwicki,* proximal end of left tarsometatarsus, SMC B55318: K, medial; L, caudal; and M, proximal views. *E. sedgwicki,* proximal end of left tarsometatarsus, SMC B55319: N, cranial; O, medial; P, caudal; and Q, lateral views. *E. barretti* Seeley, 1876, lectotype, distal end of left tarsometatarsus, BMNH A477: R, medial; S, lateral; T, cranial; U, distal; and V, caudal views. Scale bar = 5 mm (×1.5).

Tate (1976) referred the Enaliornithidae to the Hesperornithiformes. They considered *Enaliornis* to be more similar to *Baptornis* than it is to any other known bird, and they regarded *Enaliornis* as the primitive basal stock of the Hesperornithiformes.

Elzanowski and Galton (1991) noted three characters of the Cambridge Greensand braincases that are shared with most diving birds and supported the assignment of these braincases to *Enaliornis* (rather than to the other sympatric nonhesperornithiform avian taxon). Reduction of the dorsal tympanic recess (1) reflects the tendency of diving birds to increase the specific density by the overall reduction of pneumaticity (Witmer, 1990), but it certainly may be further enhanced by the expansion of the temporal fossa for the attachment of M. adductor mandibulae externus. The enlargement of the auricular fossa (2) and the lack of soft anatomical features on the roof of the cerebellar fossa (3) are most likely related to the expansion of the dural sinuses, but these characters may not have anything to do with diving. However, despite sharing these three characters, the braincase of *Enaliornis* is not particularly similar to any single group of diving birds, including *Hesperornis*, loons, and grebes.

Elzanowski and Galton (1991) noted that the braincase of *Enaliornis* is most similar to that of *Phaethon,* Diomedeidae, *Fregata,* and Lari, among recent birds. However, most of these similarities are primitive, and this includes the airencephalic conformation of the braincase, with the dorsum sellae high and the medullary fossa deep; the caudal (parietal) part of the skull roof sloping down and the occiput inclined caudoventrally, with the cerebellar prominence protruding far behind the level of occipital condyle; and the corresponding configuration and proportions of the parts of the brain, with the hemispheres, optic lobes, and the medulla compactly superimposed dorsoventrally. Possibly correlated with a strongly airencephalic skull is the separation of the trigeminal fossa from the tectal fossa (for the optic lobes). However, the relationship of *Enaliornis* to the birds with these characters is probably only plesiomorphic, and most of this same set of characters are also present in the braincase of *Hesperornis*. The similarities of the pneumatic recesses of *Enaliornis* and *Hesperornis* described by Witmer (1990) probably also represent the plesiomorphic condition.

Elzanowski (1991) and Elzanowski and Galton (1991) noted that the braincase of *Hesperornis* differs from that of *Enaliornis* in approaching the orthocranial condition as suggested by the perpendicular occiput, the shallow medullary fossa, a probably much lower dorsum sellae, the trigeminal fossa opening into the tectal fossa rather than into the medullary fossa, a conspicuously small tectal fossa of different shape, and much stronger muscular attachments on the skull roof. The orthocranial condition may have evolved in connection with underwater fishing, as it did in cormorants and anhingas, and the dorsal opening of the trigeminal fossae is probably correlated with straightening of the brain. The development of the muscular crests and fossae is well known to be size-related and thus systematically less relevant. However, the differences in the size and shape of the optic lobe are more difficult to reconcile with this hypothesis because there is no obvious reason for a comparable reduction of the optic lobe with progressing diving specialization, and known birds do not offer any parallel. Elzanowski and Galton (1991) noted that the cranial evidence neither contradicts nor supports a closer genealogical relationship between *Enaliornis* and the Hesperornithiformes.

In hesperornithiform birds, the cnemial crest on the tibiotarsus is broad and short as compared with grebes, in which it is moderately long, and with loons, in which it is extremely elongate. The configuration that it has in hesperornithiforms is a derived condition related to the very large patella. The proximal end of the tarsometatarsus of *Enaliornis* has the outer trochlea enlarged, as in other hesperornithiforms, and it extends proximally into a high, thin cranial ridge that is also characteristic of hesperornithiform tarsometatarsi (Martin, 1984). *Enaliornis* also has a wide spectrum of primitive characters, including the absence of a supratendinal bridge on the tibiotarsus and the absence of hypotarsal crests and proximal foramina on the tarsometatarsus. Loons and grebes have these derived features, as do almost all other neornithines, and as such they provide characters that unite neornithines and exclude *Enaliornis* and Hesperornithiformes.

Hesperornithiformes are united by several synapomorphies (see diagnosis). Of these, the fusion of cranial bones caudal to the frontoparietal suture, the heterocoelous articular surfaces of all thoracic vertebrae, the long, non-pneumatic bones, the partially closed acetabulum, the greatly elongated postacetabular pelvis, the relatively short and broad femur, the large tibiotarsal cnemial crest shaped like an isosceles triangle in cranial view, the transversely compressed tarsometatarsus with a distinct dorsolateral ridge leading to the lateral trochlea (IV), and the lateral trochlea that is at least as large and distally extending as the middle trochlea (III) are all present in *Enaliornis*.

It is concluded that *Enaliornis* is a hesperornithiform bird that, with the removal of the incorrectly referred amphicoelous dorsal vertebrae, is quite similar to *Baptornis*. However, it appears to be closest to *Pasquiaornis* from the Cenomanian of Saskatchewan (Tokaryk et al., 1997), but a more detailed description of the specimens of *Pasquiaornis* is needed before more detailed comparisons are possible.

Paleobiology

The source of the phosphatic enrichment of the Cambridge Greensand may have involved an organic-rich runoff from nearby rivers and/or upwellings of phosphate-rich, deep ocean waters into the shallow, warmer water of the epicontinental sea, resulting in high levels of organic production (Norman and Fraser, 1991). The phosphatized remanié fauna includes a large number of marine fossils, mostly invertebrates (listed in Penning and Jukes-Brown, 1881; White, 1932), that include many genera of sponges, echinoderms, annelid worms (a few), mollusks (lamellibranchs, gasteropods, ammonites), and crustaceans. The associated vertebrates (see Lydekker, 1904) include the frequent occurrence of teeth and bones of fish representing several genera of galeoid sharks, chimaeroid holocephalians, and neopterygian teleosts, including pycnodonts, the ichthyodectid *Plethodus*, and the swordfishlike pachycomiformid *Protosphyraena*. There are also many bones of marine reptiles, including Sauropterygia (the pliosaurs *Cimoliosaurus* and *Polyptychodon*,), Ichthyopterygia (the ichthyosaur *Ichthyosaurus*), and Testudines (*Protostega* and *Lytoloma*). The bones of short-tailed flying reptiles or pterodactyloid Pterosauria are well represented (all forms with cranial crests: the lonchodectid *Lonchodectes;* the ornithocheirids *Anhanguera, Coloborhynchus,* and *Ornithocheirus;* and the pteranodontid *Ornithostoma;* Unwin, 1991) and were probably predominantly fish-eating forms, like most extant marine birds. Several records of terrestrial dinosaurs (ornithischians *Acanthopholis, Anoplosaurus, Trachodon*) presumably represent bones from carcasses that washed in from the adjacent land.

Enaliornis was clearly a foot-propelled diver, as is shown by several characters mentioned in the description. The specific density is increased by reducing the dorsal tympanic recess of the skull and in having apneumatic long bones. The postacetabular part of the sacrum is elongate and narrow, indicating an elongate and compressed pelvis. The ventrolaterally directed femur is short and broad, compared with other birds; distally, the patellar groove is prominent, and the fibular condyle is enlarged and angled laterally. The tibiotarsus has a large cnemial crest, and distally the tarsometatarsus shows several specializations for foot propulsion, as detailed previously.

On the basis of comparisons with the braincases of procellariiform and charadriiform birds, Elzanowski and Galton (1991) estimated that the body weight of *Enaliornis* was in the range of 400–600 grams, which is far below that of flightless diving birds. Because of this small size, they suggested that *Enaliornis* was capable of flight. They also cited the extensive pneumatization of the braincase, which does not occur in recent flightless divers, as further evidence for flight. However, the presence of a comparable degree of pneumaticity in the braincase of *Hesperornis* (Witmer, 1990), an obviously flightless form in which the thoracic limb is greatly reduced (Marsh, 1880), shows that this argument is invalid. As mentioned earlier, the proportionally smaller antitrochanter of *Enaliornis* suggests that the hindlimb of this bird could have been placed much more forward than that of *Baptornis*. This inference suggests that a waddling walk might have been possible for *Enaliornis.*

Several of the femora, tibiotarsi, and tarsometatarsi are from juvenile individuals, a rarity in the avian fossil record. Only a few bones of juveniles have been described for Cretaceous aquatic birds, and these are of the hesperornithiforms *Baptornis* from the Niobrara Chalk of Kansas (Martin and Bonner, 1977) and *Hesperornis* from equivalent strata in the Northwest Territories of Canada (Russell, 1967). The state of development of the juvenile metatarsus (Fig. 14.8A–E) is comparable to one of *Baptornis* (Martin

and Bonner, 1977), and both lack many of the grooves and ossified articular surfaces that assist in the functioning of the foot. The juvenile individuals of *Enaliornis* probably did not travel far from the shore, so the nesting sites were most likely nearby, presumably on a still emergent portion of the Anglo-Brabant Massif (Rawson et al. 1978). Although *Enaliornis* was probably less awkward on land than the other hesperornithiforms, owing to the less extreme lateral orientation of the femur, it probably still nested like loons and grebes, as suggested for *Hesperornis* by Lucas (1903), and it was probably never far from water.

Acknowledgments

We thank the following for their assistance with the study and loan of specimens at their respective institutions: H. C. Ivemey-Cook (BGS), C. A. Walker and S. Chapman (BMNH), C. L. Forbes and D. B. Norman (SMC), and B. J. Pyrah and P. C. Ensom (YM). T. T. Tokaryk (SMNH) kindly supplied casts of some of the hindlimb bones of *Pasquiaornis*. The photographs of the braincase (Fig. 14.1) were taken by V. Krantz (Smithsonian Institution), and the remaining photographs were printed by A. Collins (University of Bridgeport). Extensive comments by L. M. Chiappe and L. M. Witmer considerably improved the final manuscript. P.M.G. thanks his wife C. Yoon for drawing Figure 14.2, and E. S. Gaffney (American Museum of Natural History) and D. M. Unwin (University of Bristol, England) for their personal communications.

Literature Cited

Baumel, J. J., and L. M. Witmer. 1993. Osteologia; pp. 45–132 *in* J. J. Baumel, A. S. King, J. E. Breazile, H. E. Evans, and J. C. Vanden Berge (eds.), Handbook of Avian Anatomy: Nomina Anatomica Avium, 2nd edition. Nuttall Ornithological Club, Cambridge, Mass.

Bennett, S. C. 1987. New evidence on the tail of pterosaur *Pteranodon* (Archosauria: Pterosauria). Occasional Papers of the Tyrell Museum of Palaeontology 3:12–17.

Brodkorb, P. 1963a. Catalogue of fossil birds. Part 1 (Archaeopterygiformes through Ardeiformes). Bulletin of the Florida State Museum, Biological Sciences 7:179–293.

———. 1963b. Birds of the upper Cretaceous of Wyoming; pp. 55–70 *in* C. G. Sibley (ed.), Proceedings of the 13th International Ornithological Congress. American Ornithologists Union, Baton Rouge, La.

———. 1971. Origin and evolution of birds; pp. 19–55 *in* D. S. Farner and J. R. King (eds.), Avian Biology, Vol. 1. Academic Press, New York.

Bryant, L. J. 1983. *Hesperornis* in Alaska. PaleoBios 40:1–8.

Casey, R. 1965. Cambridge Greensand. General account; pp. 54–56 *in* E. A. Edmonds and C. H. Dinham (eds.), Geology of the Country around Huntingdon and Biggleswade. Memoirs of the Geological Survey of Great Britain. Her Majesty's Stationery Office, London.

Chatterjee, S. 1989. The oldest Antarctic bird. Journal of Vertebrate Paleontology Abstracts 9(3):16A.

———. 1997. The Rise of Birds. 225 Million Years of Evolution. Johns Hopkins University Press, Baltimore, 312 pp.

Chiappe, L. M. 1996. Late Cretaceous birds of southern South America: anatomy and systematics of Enantiornithes and *Patagopteryx deferrariisi;* pp. 203–244 *in* G. Arratia (ed.), Contributions of Southern South America to Vertebrate Paleontology, Münchner Geowissenschaftliche Abhandlungen, Reihe A, Geologie und Paläontologie 30. Verlag Dr. Friedrich Pfeil, Munich.

Chiappe, L. M., and J. O. Calvo. 1994. *Neuquénornis volans,* a new Upper Cretaceous bird (Entantiornithes: Avisauridae) from Patagonia, Argentina. Journal of Vertebrate Paleontology 14: 230–246.

Cookson, I. C., and N. F. Hughes. 1964. Microplankton from the Cambridge Greensand (Mid-Cretaceous). Palaeontology 7:37–59.

Cracraft, J. 1982. Phylogenetic relationships and monophyly of loons, grebes, and hesperornithiform birds, with comments on the early history of birds. Systematic Zoology 31:35–56.

———. 1986. The origin and early diversification of birds. Paleobiology 12(4):383–399.

———. 1988. The major clades of birds; pp. 339–361 *in* M. J. Benton (ed.), The Phylogeny and Classification of the Tetrapods, Vol. 1, Amphibians, Reptiles, Birds. Clarendon Press, Systematics Association Special Volume 35A, Oxford.

Currie, P. J. 1985. Cranial anatomy of *Stenonychosaurus inequalis* (Saurischia, Theropoda) and its bearing on the origin of birds. Canadian Journal of Earth Sciences 22:1643–1658.

Elzanowski, A. 1991. New observations on the skull of *Hesperornis* with reconstructions of the bony palate and otic region. Postilla 207:1–20.

Elzanowski, A., and P. M. Galton. 1991. Braincase of *Enaliornis,* an early Cretaceous bird from England. Journal of Vertebrate Paleontology 11:90–107.

Feduccia, A. 1996. The Origin and Evolution of Birds. Yale University Press, New Haven, 420 pp.

Fox, R. C. 1974. A Middle Campanian, nonmarine occurrence of the Cretaceous toothed bird *Hesperornis* Marsh. Canadian Journal of Earth Sciences 11:1335–1338.

Fürbringer, M. 1888. Untersuchungen zur Morphologie und Systematik der Vögel, zugleich ein Beitrag zur Anatomie der Stütz- und Bewegungsorgane. Bijdragen tot de Dierkunde, Amsterdam, 1,751 pp.

Galton, P. M., and L. D. Martin. In press. Postcranial anatomy and systematics of *Enaliornis,* a foot-propelled diving bird (Aves: Hesperornithiformes) from the Early Cretaceous of England. Revue de Paléobiologie.

Grove, R. 1976. The Cambridgeshire Coprolite Mining Rush. Oleander Press, Cambridge, 51 pp.

Harrison, C. J. O. 1976. The wing proportions of the Eocene diver *Colymboides anglicus.* Bulletin of the British Ornithological Club 96:64–65.

Harrison, C. J. O., and C. A. Walker. 1976. Birds of the British Upper Eocene. Zoological Journal of the Linnean Society of London 59:323–351.

Hart, M. B. 1973. Foraminiferal evidence for the age of the Cambridge Greensand. Proceedings of the Geological Association of London 84:65–82.

Hofer, H. 1952. Der Gestaltwandel des Schädels der Säugetiere und Vögel mit besonderer Berücksichtigung der Knickungstypen und der Schädelbasis. Verhandlungen der Anatomischen Gesellschaft Jena 50:102–113.

Hooley, R. W. 1913. On the skeleton of *Ornithodesmus latidens,* an ornithosaur from the Wealden Shales of Atherfield (Isle of Wight). Quarterly Journal of the Geological Society of London 69:372–422.

Johnsgard, P. A. 1987. Diving Birds of North America. University of Nebraska Press, Lincoln, 292 pp.

Kurochkin, E. N. 1995. Synopsis of Mesozoic birds and early evolution of Class Aves. Archaeopteryx 13:47–66.

Lambrecht, K. 1929. *Neogaeornis wetzeli* n. g. n. sp., der erste Kreidevogel der südlichen Hemisphäre. Palaeontologische Zeitschrift 11:121–129.

———. 1933. Handbuch der Palaeornithologie. Gebrüder Borntrüger, Berlin, 1,024 pp.

Lartet, E. 1857. Note sur un humérus fossile d'oiseaux, attribué à un très-grande palmipède de la section des longipennes. Comptes-rendus Hebdomadaires des Séances de l'Académie des Sciences Paris 44:736–741.

Lucas, F. A. 1903. Notes on the osteology and relationship of the fossil birds of the genera *Hesperornis Hargeria Baptornis* and *Diatryma.* Proceedings of the United States National Museum 26:545–556.

Lydekker, R. 1891. Catalogue of the Fossil Birds in the British Museum (Natural History). Longmans & Co., London, 368 pp.

———. 1904. Vertebrate paleontology; pp. 51–70 *in* J. E. Marr and A. E. Shipley (eds.), Handbook to the Natural History of Cambridgeshire. Cambridge University Press, Cambridge.

Lyell, C. 1859. Manual of Elementary Geology. Supplement to the 5th edition, 3rd edition, John Murray, London, 40 pp.

Marsh, O. C. 1880. Odontornithes: A Monograph on the Extinct Toothed Birds of North America. Report of Geological Exploration of the 40th Parallel. Government Printing Office, Washington, D.C., 201 pp.

Martin, L. D. 1983. The origin and early radiation of birds; pp. 291–338 *in* A. H. Brush and G. A. Clark Jr. (eds.), Perspectives in Ornithology. Cambridge University Press, Cambridge.

———. 1984. A new hesperornithid and the relationships of the Mesozoic birds. Transactions of the Kansas Academy of Science 87:141–150.

———. 1991. Mesozoic birds and the origin of birds; pp. 485–540 *in* H.-P. Schultze and L. Trueb (eds.), Origins of the Higher Groups of Tetrapods: Controversy and Consensus. Comstock Publishing Associates, Ithaca, N.Y.

Martin, L. D., and O. Bonner. 1977. An immature specimen of *Baptornis advenus.* Auk 94:787–789.

Martin, L. D., and J. Tate Jr. 1976. The skeleton of *Baptornis advenus.* Smithsonian Contributions to Paleobiology 27: 35–66.

Martin, L. D., J. D. Stewart, and K. N. Whetstone. 1980. The origin of birds: structure of the tarsus and teeth. Auk 97:86–93.

McGowan, C. 1985. Homologies in the avian tarsus, McGowan replies. Nature 315:159–160.

Nessov, L. A. 1992. Record of the localities of Mesozoic and Paleogene with avian remains of the USSR, and the description of new findings. Russian Journal of Ornithology 1:7–50. [Russian]

Norman, D. N., and N. Fraser. 1991. Phosphates, fossils and fens. Conservation and the Cambridge Greensand. Earth Science Conservation 29:30–31.

Olson, S. L. 1992. *Neogaeornis wetzeli* Lambrecht, a Cretaceous loon from Chile (Aves: Gaviidae). Journal of Vertebrate Paleontology 12:122–124.

Olson, S. L., and A. Feduccia. 1980. Relationships and evolution of flamingos (Aves: Phoenicopteridae). Smithsonian Contributions to Zoology 316:1–73.

Ostrom, J. H. 1976. *Archaeopteryx* and the origin of birds. Biological Journal of the Linnean Society 8:91–182.

Owen, R. 1861. Palaeontology, 2nd edition. A. & C. Black, Edinburgh, 463 pp.

Penning, W. H., and A. J. Jukes-Brown. 1881. The geology of the neighbourhood of Cambridge. Memoirs of the Geological Survey of Great Britain, London, 184 pp.

Portmann, A., and W. Stingelin. 1961. The central nervous system; pp. 1–36 *in* A. J. Marshall (ed.), Biology and Comparative Physiology of Birds, Vol. 2. Academic Press, New York.

Raath, R. A. 1985. The theropod *Syntarsus* and its bearing on the origin of birds; pp. 219–227 *in* M. K. Hecht, J. H. Ostrom, G. Viohl, and P. Wellnhofer (eds.), The Beginning of Birds. Freunde des Jura-Museums, Eichstätt.

Rawson, P. F., D. Curry, F. C. Dilley, J. M. Hancock, W. J. Kennedy, J. W. Nearl, C. J. Wood, and B. C. Worssam. 1978. A correlation of Cretaceous rocks in the British Isles. Special Report of the Geological Society of London 9:1–70.

Russell, D. A. 1967. Cretaceous vertebrates from the Anderson River, N.W.T. Canadian Journal of Earth Sciences 4:21–38.

Seeley, H. G. 1864. On the fossil birds of the Upper Greensand, *Palaeocolyntus barretti* and *Pelargonis sedgwicki*. Proceedings of the Cambridge Philosophical Society 1:228.

———. 1866. Note on some new genera of fossil birds in the Woodwardian Museum. Annals and Magazine of Natural History (3)18:109–110.

———. 1869. Index to the Fossil Remains of Aves, Ornithosauria and Reptilia, from the Secondary System of Strata Arranged in the Woodwardian Museum of the University of Cambridge. Deighton, Bell & Co., Cambridge, 143 pp.

———. 1870a. The Ornithosauria: An Elementary Study of the Bones of Pterodactyles, Made from Fossil Remains Found in the Cambridge Upper Greensand, and Arranged in the Woodwardian Museum of the University of Cambridge. Deighton, Bell & Co., Cambridge, 130 pp.

———. 1870b. Remarks on Prof. Owen's monograph on *Dimorphodon*. Annals and Magazine of Natural History (4)6:129–152.

———. 1876. On the British fossil Cretaceous birds. Quarterly Journal of the Geological Society of London 32:496–512.

Storer, R. W. 1971. Classification of birds; pp. 149–188 *in* D. S. Farner and J. R. King (eds.), Avian Biology, Vol. 1. Academic Press, New York.

Toraryk, T., and C. R. Harrington. 1992. *Baptornis* sp. (Aves: Hesperornithiformes) from the Judith River Formation (Campanian) of Saskatchewan, Canada. Journal of Paleontology 66:1010–1012.

Toraryk, T., S. Cumbaa, and J. Storer. 1997. Early Late Cretaceous birds from Saskatchewan: the oldest diverse avifauna known from North America. Journal of Vertebrate Paleontology 17:172–176.

Unwin, D. M. 1991. The morphology, systematics and evolutionary history of pterosaurs from the Cretaceous Cambridge Greensand of England. Ph.D. dissertation. Reading University, Reading, 527 pp.

Vanden Berge, J. C., and G. A. Zweers. 1993. Myologia; pp. 189–246 *in* J. J. Baumel, A. S. King, J. E. Breazile, H. E. Evans, and J. C. Vanden Berge (eds.), Handbook of Avian Anatomy: Nomina Anatomica Avium, 2nd edition. Nuttall Ornithological Club, Cambridge, Mass.

Walker, C. A. 1981. New subclass of birds from the Cretaceous of South America. Nature 292:51–53.

Walker, M. V. 1967. Revival of interest in the toothed birds of Kansas. Transactions of the Kansas Academy of Sciences 70:60–66.

Wellnhofer, P. 1978. Pterosauria. Handbuch der Paläoherpetologie. Gustav Fisher Verlag, Stuttgart, 82 pp.

———. 1991. The Illustrated Encyclopedia of Pterosaurs. Salamander Books, London, 192 pp.

White, H. J. O. 1932. The Geology of the Country near Saffron Walden. Explanation of sheet 205. Memoirs of the Geological Survey of Great Britain, London, 125 pp.

Wilcox, H. M. 1952. The pelvic musculature of the Loon, *Gavia immer*. American Midland Naturalist 48:513–573.

Witmer, L. M. 1990. The craniofacial air sac system of Mesozoic birds (Aves). Zoological Journal of the Linnean Society of London 100:327–378.

15

The Mesozoic Radiation of Neornithes

SYLVIA HOPE

The record of Neornithes in the Mesozoic consists almost entirely of fragmentary and dissociated specimens, a legacy of lightweight, fragile bones. Identifications of such limited material have been questioned in the past, often with reason. The difficulty of identification is due not only to the fragmentary nature of the remains but also to the unknown avifauna. While an isolated fragment of an extant bird can be identified readily, even to species, a fragment of a very old bird from an almost unknown avifauna is a formidable challenge.

The problem, then, is to identify in these limited remains the diagnostic features of Neornithes and its major subsidiary groups. Most of these birds were described 40 or more years ago, some well over 100 years ago. Early reviewers often tried to place the fragments based only on general resemblance to an extant group of birds. With the many new Cretaceous birds described in the past 20 years, it is now possible to evaluate these fragments in the context of Aves as a whole.

Since reviews by Elzanowski (1983) and Martin (1983) and the comprehensive treatment of the avian fossil record by Olson (1985), there have been briefer reviews and newer reports of several major groups of Neornithes. Tyrberg's (1986) popular account reflects a middle ground on the reading of the Cretaceous record to that date. Unwin (1993) created an annotated list of the fossil groups of birds, with time of appearance and some critical editing of taxonomic assignments.

Feduccia (1995) discounted all the Cretaceous records of Neornithes except the reports of charadriiformlike birds present in the Maastrichtian. Thus, both Martin (1983) and Feduccia (1996) envisioned these early shorebirds as the group from which the principal radiation of Neornithes originated, and Feduccia dated this radiation to the Early Tertiary. However, foreshadowed by Simpson's assessment of the record (1976:60), both Cracraft (1986, 2000) and Chiappe (1995b) mapped previous reports of Late Cretaceous Neornithes onto a phylogeny of Aves and concluded similarly that Neornithes ("modern birds") had differentiated well before the end of the Cretaceous. Thus, the earliest known fossils of a group do not necessarily indicate its true age (Norell and Novacek, 1992).

In another line of evidence, recent analyses based on molecular data have placed the divergence of several extant lineages of birds in middle or earlier Cretaceous time (Hedges et al., 1996; Cooper and Penney, 1997; Mindell et al., 1997; Kumar and Hedges, 1998). However, it is these earliest reports that have received the least support.

The object of this chapter is to evaluate and update the Mesozoic record of Neornithes on the basis of skeletal specimens.

Phylogenetic Framework for Diagnosis within Neornithes

Comparisons with other kinds of birds make it possible now to evaluate the features of the earliest neornithine in a phylogenetic context. In theory, features derived close to the origin of Neornithes can be distinguished from earlier and later derived features. Those features derived after the origin of the group could then be mapped onto the phylogeny within Neornithes to make nested, phylogenetically based diagnoses for the taxa of living birds. However, the relationships among the traditional orders of birds are poorly understood throughout Neornithes. Thus, if a derived feature occurs in several major groups, it might be due to homology or to convergence. Such a feature is not a basis for a confident referral of a specimen to any of the groups that share it.

Huxley's (1867) analysis of the palate placed ratites basally within Neornithes. Pycraft (1900) proposed further that ratites were a closely related group, Palaeognathae, in which he also included tinamous, based on their paleo-gnathous palate, as he analyzed it. The monophyly of Palaeognathae and their sister-group relationship with Neognathae (all other neornithine birds) are further supported by morphology of the cranium, pelvis, and ankle (Jollie, 1957; Bock, 1963; McGowan, 1985; Cracraft, 1988) and by most molecular data (Sheldon and Bledsoe, 1993; Cracraft, 2000). However, some doubt remains about the inclusion of tinamous. Polarization of characters to reconstruct relationships at this basal level is hindered by insufficient knowledge of the immediate ancestor and stem lineage of Neornithes. The palate, upon which these basal relationships were first reconstructed, is poorly known in other ornithurine birds. Houde and Olson (1981) and Houde (1988) discussed the position of tinamous and the probably near-basal neornithine *Lithornis*.

Many recent studies support the early branching of Galliformes and Anseriformes from remaining Neognathae. Most biochemical data additionally support the monophyly of Galliformes and Anseriformes, as do features of the palate and basicranium (Cracraft, 1988; Cracraft and Mindell, 1989; Sibley and Ahlquist, 1990; Sheldon and Bledsoe, 1993; Livezey, 1997a; van Tuinen et al., 2000). Postcranial osteology, behavior, and ecology do not (Olson and Feduccia, 1980b; Ericson, 1996, 1997). Thus, other reconstructions proposed a close relationship of anseriforms with charadriiforms. Feduccia (1976, 1978) suggested that both these groups were closely related to ibises and flamingos, and Howard (1955) and Ericson (1999) have shown the many similarities of early anseriforms to flamingos. However, the filter feeding mechanisms of flamingos and anseriforms evidently evolved independently (Olson and Feduccia, 1980a). Other discussion centered on the charadriiformlike postcranial skeleton of *Presbyornis* combined with its ducklike skull (Olson and Feduccia, 1980b; Olson and Parris, 1987).

Many studies have grouped charadriiforms instead with other shorebirds and seabirds. These terms refer to similarity of form and habitus, not to monophyletic groups. Classical studies never reached a consensus on the inclusiveness of such an assemblage or the relationships among the included groups of birds (Stresemann, 1959; Raikow, 1985). Recent phylogenetic studies based on molecules, behavior, and morphology, using advanced methods of analysis, have done little better. These analyses are notable for their short internal branches, unresolved polytomies, and lack of concordance (Cracraft and Mindell, 1989; Sibley and Ahlquist, 1990; Harshman, 1994; Hedges and Sibley, 1994; Ericson, 1997; Livezey, 1997a; van Tuinen et al., 2000). Mosaic distributions of morphological features are implied by the results, even if comparisons are restricted to presumably derived features within Neornithes (Hope, 1999). However, several groupings emerge repeatedly:

1. Procellariiforms, loons, and penguins (Simpson, 1976; Olson, 1985; Hedges and Sibley, 1994).
2. Procellariiforms and pelecaniforms. This association is so consistent that Cracraft (1985) used Procellariiformes as an outgroup for his study of Pelecaniformes. In some traditional classifications these groups have appeared adjacent to and immediately after Palaeognathae, to signify their presumed basal position (e.g., Fürbringer, 1888; Brodkorb, 1963b). Sibley and Ahlquist (1990) reversed the order of derivation, placing procellariiforms and pelecaniforms well within a large assemblage that included charadriiforms as the earliest branch; their nomenclature differs, and they did not find either procellariiforms or pelecaniforms in the sense of Wetmore (1960) to be monophyletic. Elzanowski (1995) pointed out the many apparently plesiomorphic resemblances of some procellariiforms and pelecaniforms to non-neornithine seabirds, but he concluded that the similarity is probably convergent.
3. Storks, herons, American vultures, hawks, falcons, and pelecaniforms. Garrod (1874) grouped all these as Ciconiiformes, although he also included owls. Fürbringer (1888) agreed approximately with this, but he excluded owls and added flamingos, ibises, and spoonbills. Gradually this group of ecologically very different birds was pared away until only storks, herons, flamingos, ibises, and spoonbills remained (Sharpe, 1891; Gadow, 1893; Wetmore, 1930b, 1960; Mayr and Amadon, 1951). This "Ciconiiformes" in the sense of Wetmore (1960) is an assemblage of long-legged wading birds with elegant plumes and little else in common. Unfortunately, this twentieth-century usage is deeply embedded in textbooks, popular accounts, and the organization of collections. It persists even in many scientific works. However, there is good evidence for a close relationship of pelicans with storks, and of storks in turn with American vultures (Cottam, 1957; Ligon, 1967; Olson, 1985; Konig, 1988; Sibley and Ahlquist, 1990; Avise et al., 1994; Helbrig and Seibold, 1996). Sibley and Ahlquist (1990) found a close relationship among all the birds of Fürbringer's Ciconiiformes, but procellariiforms and loons were embedded within it paraphyletically. They applied the name Ciconiiformes to a much larger group. The American Ornithologists' Union (1998) redefined Ciconiiformes to include storks, herons, ibises, spoonbills, and American vultures.

This history illustrates the state of basal-level systematics within Neornithes. Phylogenetic relationships are also uncertain among other groups of extant birds, including gruiforms, parrots, columbiforms, and the perching landbirds in the broad sense. In contrast, the systematics of subsidiary groupings is much better understood. Thus, Sibley and Ahlquist (1990) corroborated many of the classical family-level groupings (with the prominent exception of passerine families). It is as though some aspect of the evolution of neornithine birds is distorting analyses of relationships, but only at the most basal levels (Davison, 1998). The solution to diagnosis adopted here was to require a combination of unique features for referral of fragmentary, isolated specimens to the most inclusive groups within Neornithes.

Material and Methods

Material

This review includes specimens previously referred to Neornithes in a well-circulated publication and other material that I have studied sufficiently to comment. However, specimens with minimal diagnostic potential, such as distal phalanges or badly worn fragments, are not included. A list of such specimens is available from the author.

Definition of Taxa

Following Ghiselin (1984, 1997), a taxon is defined by indicating its members. Node-based names are used here, in the sense of de Queiroz and Gauthier (1990). Thus, the name Neornithes applies here to the monophyletic group indicated by extant birds, their most recent common ancestor, and all its descendants. The name Galliformes applies to extant galliforms, their most recent shared ancestor, and all the collateral extinct lineages that shared that ancestor. Most previous workers on Neornithes have applied the same higher taxon names as are used here in the broader, stem-defined sense. That is, Galliformes would be defined as all birds more closely related to the extant members than to its sister group. This procedure does not introduce serious error into studies of middle and later Tertiary birds, which are usually clearly assignable to extant lineages. However, the distinction between node- and stem-defined taxa helps to clarify the diagnosis and thus may show the uncertain phylogenetic status of earlier specimens more distant to the extant groups. In addition, the uncertain basal relationships within Neornithes usually preclude a precise specification of the sister group. The older the specimens being considered, the more problematic these uncertain relationships become to the diagnosis of taxa.

Within Neornithes, inclusive taxon names above the generic level are defined by genera included in Wetmore (1960) unless otherwise indicated. Prominent exceptions to Wetmore's usage are the names Neornithes and Palaeognathae. Genus and species names of North American birds follow the American Ornithologists' Union (1998).

Diagnosis and Referral of Specimens to Supraspecific Taxa

Features of the earliest neornithine were reconstructed through a broadly based review of osteology in Neornithes (Hope, 1989), supplemented with later observations, with emphasis on the well-known Early Tertiary neornithines *Lithornis, Telmabates, Presbyornis,* and *Limnofregata.* Nonneornithine birds studied included *Mononykus, Enantiornis, Avisaurus, Patagopteryx, Hesperornis, Ichthyornis,* and casts of *Lectavis, Ambiortus,* and *Baptornis.* Descriptions of other Cretaceous birds were consulted. For polarizing characters, Enantiornithes was assumed to be the sister group of *Patagopteryx* plus Ornithurae (Chiappe, 1995b). Diagnoses of taxa above the species level, or referral of specimens to an inclusive taxon, required shared derived features of the taxon, except that, for the taxa immediately subsidiary to Neornithes, my operational diagnosis or referral also required that the derived features be unique, as justified in the introductory section of this chapter. However, where unique features were lacking, a specimen was often referred tentatively based on a combination of derived but non-unique features. "Unique" was defined as not occurring in any possibly closely related group within Neornithes, or in the ancestral neornithine, or in another ornithurine. Asterisks in the Systematic Paleontology section flag widely shared derived features, as explained in the introductory remarks preceding Charadriiformes.

Diagnosis of Species and Referral to Species

Diagnosis of species required only differentiating features. Referral of specimens to species required that the material be directly comparable to the holotype or confidently referred material. Comparability was judged by association in situ, identity of form, or good apposition to other confidently referred material. Specimens that were referred previously based on less conclusive evidence were made only tentatively referable to the species but were not removed entirely from such a placement unless there was a compelling reason to do so. Similarly, inferences about the characters of species or monophyly of a group of species were based only on confidently referred material.

Anatomical Terminology

In general, anatomical terminology follows Baumel and Witmer (1993). The terms "brachial fossa" and "brachial depression" are distinguished and "capital tubercle" is defined in the Anatomy section with the discussion of the humerus.

The following institutional abbreviations are used in this chapter: AMNH, American Museum of Natural History, New York, New York, United States; ANSP, Academy of Natural Sciences, Philadelphia, Pennsylvania, United States; BMNH, Natural History Museum, Tring, England; CAS, California Academy of Sciences, San Francisco, California, United States; GPMK, Geologisch-Palaeontologisches Institut und Museum, Universität Kiel, Kiel, Germany; IVPP, Institute of Vertebrate Paleontology and Paleoanthropology, Beijing, China; MLP, Museo de la Plata, La Plata, Argentina; MVZ, Museum of Vertebrate Zoology, Berkeley, California, United States; NJSM, New Jersey State Museum, Trenton, New Jersey, United States; NSMJ, National Science Museum of Japan, Tokyo, Japan; PIN, Palaeontological Institute of the Russian Academy of Sciences, Moscow, Russia; PVL, Instituto Miguel Lillo, Universidad Nacional de Tucumán, San Miguel de Tucumán, Argentina; RTMP, Royal Tyrrell Museum of Palaeontology, Drumheller, Canada; SMNH, Saskatchewan Museum of Natural History, Regina, Canada; TMM, Texas Memorial Museum, Austin, Texas, United States; UCMP, University of California Museum of Paleontology, Berkeley, California, United States; USNM, United States National Museum, Washington, D.C., United States; YPM, Yale Peabody Museum, New Haven, Connecticut, United States; ZISP-PO, Paleornithological Collection, Zoological Institute of the Russian Academy of Sciences, St. Petersburg, Russia.

Geology

The two regions in North America discussed in this section are the source of most of the birds reviewed below. Both regions include controversial Cretaceous-Tertiary (K/T) boundary sections. Authorities for formations in other regions of the world are cited with the appropriate specimens in the reviews below. Throughout this chapter, ages of formations follow Lillegraven and McKenna (1986) unless another author is cited.

Northern Part of the Western Interior, North America

The well-known toothed birds reported by O. C. Marsh and documented in his 1880 monograph were recovered from the floor of the shallow inland Cretaceous sea of North America, probably from a Coniacian level of the Niobrara Chalk. The neornithine bird *Apatornis* is included among these remains. The depositional environment is discussed below with this bird. However, the most abundant Cretaceous record of Neornithes comes from later nearshore sediments, accumulated along the retreating shore of the sea during its long regressive phases. The third regression in mid-Campanian time is associated with a regional fauna defined by its mammalian component as the Judithean North American Land Mammal Age (NALMA). The fourth and final regression in the Late Maastrichtian is associated with the Lancian NALMA (Lillegraven and McKenna, 1986). Neornithine birds are present in Judithean and Lancian collections from the northern part of the Western Interior. Eberth (1990) characterized many Judithean localities in Alberta.

Lance and Hell Creek Formations. Clemens (1964) described the geological setting and the history of exploration of the Lance Formation. The Lance consists of more than

1-km thickness of massive sandstones, interbedded lignites, and lignitic shales deposited along the receding shoreline. At the type area in eastern Wyoming, the Lance Formation lies conformably over the marine Fox Hills Formation and is capped conformably by the Fort Union Formation. Traditionally, the K/T boundary was defined in this region by the latter contact. The gentle dip of the units in this sequence results in an exposure of successive strata going up through time as one travels from east to west. Outcrops of the Lance occur in present-day stream banks cut through the thin overburden or as wind-eroded crags upstanding over the sparsely vegetated plain. These are the indurated remains of ancient streambeds. These paleochannel fills expose an abundant biota, with plant remains showing that the environment was humid and subtropical (Dorf, 1942; Estes, 1964). All the collection sites are well below the contact with the Paleocene, including the nineteenth-century localities of J. B. Hatcher (Clemens, 1964). Since the 1950s, field parties from UCMP and AMNH have explored the paleochannels intensively by quarrying and screen washing. All the bird fossils discussed here were recovered as isolated elements.

The Hell Creek Formation outcrops primarily in eastern Montana. It is probably continuous to the south with the Lance Formation, but there is no surface exposure of a connection. Archibald (1982), Bryant (1989), and Archibald and Lofgren (1990) reviewed the stratigraphy and faunal composition of localities in the Hell Creek Formation. Near the top of the formation, some of the Bug Creek paleochannels cut through iridium-containing coals. The topography in this region is disturbed, and lateral correlation of the several coals close to the K/T boundary has been problematic. It is now clear that the highly fossiliferous channel fills in the Bug Creek region consist of a mix of Paleocene material washed down from higher in the section and Cretaceous material eroded from cutbanks (Lofgren, 1995). This is a time-averaged assemblage of fossils from both sides of the K/T boundary. Thus, the age of these specimens has to be considered undetermined for precise stratigraphic dating (Clemens, 1999).

A few specimens from these channels are included in the reviews below if they are referred to a species present in other localities known to be Cretaceous. However, none of the specimens from the Bug Creek facies is included in the compilation of the Cretaceous specimens at the end of this chapter.

Hornerstown and Navesink Formations— Age of the New Jersey Birds

The greensand marls of New Jersey are part of the offshore record of the Late Cretaceous Atlantic coast of North America. Olson and Parris (1987) and Gallagher and Parris (1996) have reviewed the stratigraphy. The section at the Inversand

Company marl pit in Sewell County is illustrative. Here, the entirely Maastrichtian Navesink Formation is overlain directly by the basal Main Fossiliferous Layer (MFL) of the Hornerstown Formation. This is a narrow, about 0.3-m-thick, well-delimited and densely fossiliferous layer, with Maastrichtian macrofossils (ammonites, mosasaurs) in a matrix that includes typically Paleocene foraminifera. Because of this mix, the age of birds from the MFL has been a subject of protracted debate (e.g., Wetmore, 1930a; Richards and Gallagher, 1974). Baird (1967) traced the historical referents of localities from which fossil birds were recovered during nineteenth-century mining operations in the marls, but for many the origin, whether Navesink or Hornerstown, remains uncertain. Most specimens recovered in recent years are from the MFL.

The 3 m of Hornerstown sediment overlying the MFL is entirely Paleocene. It has not yielded vertebrates and is depauperate for approximately 3 m above the MFL, probably reflecting a massive disturbance in marine food chains in the early Paleocene (Gallagher, 1996; Hallam and Wignal, 1997). Thus, the birds collected from the Hornerstown Formation in the early years, without precise stratigraphic location, are very probably from the MFL rather than higher in the section. However, their age still remains uncertain. One micropaleontological zone spanning the K/T boundary is missing from the section (Gallagher and Parris, 1996). Thus, there are no biostratigraphic dates of the order 100,000 years encompassing the MFL. Gallagher (1992) found iridium elevated within the MFL, but not in high, well-defined peaks. Gallagher (1993, 1996) interpreted this layer as an assemblage that accumulated at some time during the missing 100,000 years, made up of the remains of a protracted faunal die-off following the end-Cretaceous event. Kennedy and Cobban (1996) interpreted the MFL as a latest or terminal Cretaceous lag deposit infilled with Paleocene sands, probably by burrowing invertebrates. Staron et al. (1999) approached the problem by investigating the geochemistry of the fossils. They found that rare earth signatures in all but one of the mosasaurs from the MFL differed from those from the underlying Navesink Formation. They suggested that some specimens in the MFL are reworked from the Maastrichtian, others deposited in earliest Tertiary.

The meaning of the lag interval situation is that material in the MFL may be from any time along the gap of 100,000 years. More battered specimens, such as some ammonites, are probably from early in that interval (and are surely Cretaceous). Others, such as the articulated skeletons and other unworn material, including most of the birds, are more probably not reworked but part of a new phase of deposition in the earliest Paleocene (D. Parris, pers. comm.). It seems best for the present to consider the age of birds from the MFL unresolved, either latest Maastrichtian or earliest

Tertiary. As with the Bug Creek birds, specimens from the Hornerstown Formation are included in accounts here only if the species is reported in earlier sediments.

Anatomy

Many innovations in Neornithes are related to advanced flight. In all birds, the bones are lightweight; the bone walls have a geometrical sandwich structure, with air cells in the bone wall. These are slightly developed even in *Archaeopteryx* but pervasive in Neornithes (Bühler, 1992). Pneumaticity, the channeling of the air sac system into bones (Witmer, 1990), contributes to lightweight bones. Beginning early in avian evolution and continuing in the ornithurine lineage, bones of the postcranial skeleton were extensively co-ossified, reduced, or lost. These changes included fusion of pelvic bones, fusion of sacral vertebrae, and fusion of the synsacrum with the pelvis; loss of tail vertebrae and formation of a rigid pygostyle; formation of a carpometacarpus, tibiotarsus, and tarsometatarsus; and loss or fusion of manual and pedal digits. Extensive fusion of bones in Ornithurae makes the skeleton more rigid, probably to prevent it from dissipating the force of the wing stroke. This becomes more extreme in the skull of Neornithes.

The following abbreviated account places the morphology of Neornithes in phylogenetic perspective to clarify the diagnostic features of Neornithes and its subgroups. The emphasis is on skeletal elements that are most common in the Mesozoic record.

Skull

Bones of the skull are extensively co-ossified only in Neornithes. Most of the cranial and jaw elements are fused together, the maxilla is greatly reduced or entirely eliminated from the exterior surface of the jaw, and teeth are completely lost. Throughout the jaws there are new articulations and bending zones within bones, and the palate is rearranged. The resulting jaws of Neornithes are versatile and often powerful manipulative tools, even though lightweight and completely edentulous. The quadrate is the most intricate part of the jaw articulation, but its variation is poorly documented. Early studies of the morphology of the jaw in extant birds laid the foundations of current understanding (Pycraft, 1900, 1901; Jollie, 1957; Huxley, 1867). Contributions in the past 40 years have emphasized functional aspects, interpretative methods, and new fossil material (Bock, 1963, 1964; Bühler, 1981, 1987; Zusi, 1984; Witmer and Martin, 1987; Bühler et al., 1988; Elzanowski, 1990, 1991, 1999; Witmer and Rose, 1991; Dzerzhinsky, 1995, 1999; Witmer, 1995).

The ratite or so-called paleognathous palate may be in part plesiomorphic in Aves, in part derived. In Ratitae and *Lithornis,* the pterygoids articulate with the palatines well lateral to the parasphenoid rostrum, but this apparently primitive position is associated with other features of the jaw that are part of a design for far rostral jaw kinesis, clearly an unusual condition in Aves (Bock, 1963). The neognathous palate is a major design innovation, in which the pterygoids converge on the parasphenoid rostrum far rostral to the basicranium. Tinamous approach this design, although their pterygoids do not quite reach the midline, and, as in ratites, the pterygoids are separated from the parasphenoid rostrum by the caudally elongated vomer. Jollie (1957) cited and interpreted the numerous nineteenth-century publications of W. K. Parker and those of Pycraft (1900, 1901) on ontogeny of the palate in birds. In Neognathae except Galliformes, the pterygoid divides early in ontogeny, and its rostral part fuses partially or completely with the palatines. The rostral part of the divided pterygoid ("hemipterygoid") is often indistinguishable in the adult. In galliforms and anseriforms, the caudal part ("posteropterygoid") articulates just lateral to the parasphenoid rostrum on rostrally shifted basipterygoid processes of the parasphenoid. In other neognaths, the posteropterygoid articulates directly with the parasphenoid rostrum. During development in galliforms, the hemipterygoid is never connected to the posteropterygoid. It first appears rostral to and separate from the posteropterygoid and then fuses directly with the palatine. Jollie (1957) suggested that this is an ontogenetic abbreviation of the process in other neognaths. In anseriforms the ontogeny is slightly different: a rostral process of the pterygoid appears during development and later contacts the palatine. Elzanowski (1991) concluded that *Hesperornis* had a divided pterygoid, but apparently it does not converge toward the parasphenoid rostrum. The ontogenetic and phylogenetic significance of the different forms of the palate still needs clarification.

Vertebrae

The number of cervical vertebrae increased in Aves from 9 in the earliest birds to up to 12 or 13 in Enantiornithes. Within Ornithurae, the number is both lowest and highest in Neornithes, ranging from 11 in parakeets to 25 in swans (Welty and Baptista, 1988). Heterocoely (saddle-shaped intervertebral articulations) began in the cervical vertebrae before the branching of ornithurine birds (Chiappe, 1995a, 1996b; Sanz et al., 1995). It is well developed in all vertebrae of *Hesperornis* but apparently absent in *Ichthyornis* (Marsh, 1880). Heterocoely is complete in many neornithines but absent from the last few thoracic vertebrae in some neognaths. Lateral depressions in the centra (pleurocoels) are present in some nonornithurine birds; they are large, deep, and very distinct in *Ichthyornis* and *Hesperornis* but shallow, indistinct, or absent in Neornithes. Heterocoely appears to permit a wide range of mobility between vertebrae (Welty and Baptista, 1988). Together with a long neck, it may facilitate manipulative use of the jaws and retraction of the head

over the back; this and the lightweight skull in Neornithes seem to be part of the redistribution of weight in the upright stance, summarized below.

Shoulder Girdle

Hesperornithiformes and Ratitae are excluded from reconstruction of shoulder elements in the earliest neornithine because of their extreme reduction in these flightless birds.

In *Enantiornis* (Fig. 15.1A) the humeral facets of the coracoid and scapula each form half of a rounded, cup-shaped glenoid facet. The acrocoracoid process of the coracoid is very small and blunt. The articulation with the scapula is highly apomorphic and cannot be compared closely with that of ornithurine birds. There is no procoracoid process, but there is a large foramen for N. supracoracoideus located centrally on the shaft (illustrated in Chiappe, 1996a). The shaft is stout and rounded at the neck, long and narrow for its entire length, and deeply excavated over the entire dorsal surface, and there is no lateral process (Chiappe and Calvo, 1994; Sanz et al., 1995).

In *Ichthyornis* (Fig. 15.1B), the acrocoracoid process is small, and there is a narrow, linear clavicular facet. A procoracoid process is present; it is a small, flat flange that extends distally along the shaft of the coracoid, merging gradually with the shaft. The shaft, including the neck, is flat and only moderately elongate, and the lateral process is very large. The humeral facet is flat, ovate, tilted medially, and very large. The humeral facet is attached to the scapular cotyla but extends cranially well beyond onto the enlarged acrocoracoid process and sternally slightly beyond the scapular facet. The scapular margin of the procoracoid process bears a very large, deep, mediolaterally elongate cotyla for articulation with the coracoidal tubercle of the scapula. The foramen for N. supracoracoideus is small and very close to the scapular facet, entering the ventral surface of the coracoid just distomedial to the scapular facet and exiting dorsally at the same level.

Many neornithines have a coracoid similar to that of *Ichthyornis;* the shared features present in *Ichthyornis, Lithornis,* and *Presbyornis* (Figs. 15.2A, 15.4C–F) are presumed plesiomorphic for Neornithes. One difference is that medial tilting of the humeral facet is absent in most neornithines, including *Lithornis* and *Presbyornis,* and therefore tilting is presumed derived where it occurs within Neornithes. The acrocoracoid also differs. *Lithornis* (Fig. 15.2A) and *Eudromia* (Fig. 15.3F) have a smaller acrocoracoid process than *Ichthyornis* or most neognaths. In well-flying neognaths, the acrocoracoid process is larger than in *Ichthyornis,* and it is peaked with a stout bicipital protuberance that supports large sites for M. biceps brachii and the acrocoracohumeral ligament (Fig. 15.4). The acrocoracoid is discussed further below in connection with weak flight and with *Ambiortus*. In Neornithes, the humeral facet of the

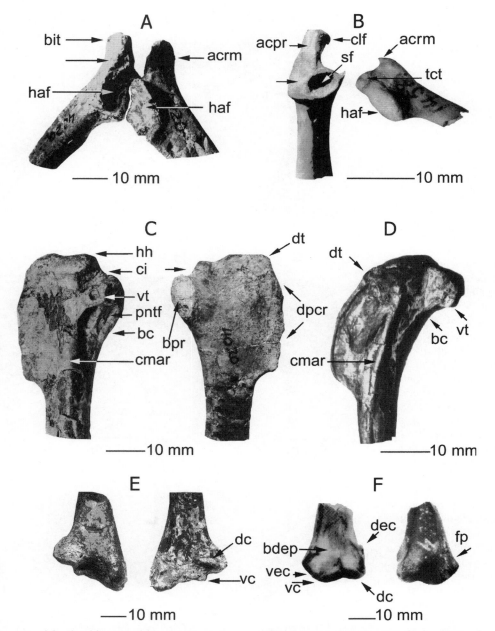

Figure 15.1. Elements of the shoulder articulation in some non-neornithine birds. A, *Enantiornis leali*, shoulder ends of left coracoid and scapula apposed, in lateral view, PVL 4020. B, *Ichthyornis* sp., proximal end of left coracoid in caudal view, YPM 1459, and left scapula in lateral view, YPM 1458. C, *E. leali*, proximal end of left humerus in caudal and cranial views, PVL 4020. D, *I. dispar*, proximal end of right humerus, reversed for comparisons, in cranial and caudal views, TMM 31051-24. E, *E. leali*, distal end of left humerus in caudal and cranial views, PVL 4020. F, *Ichthyornis antecessor*, distal end of left humerus in cranial and caudal views, USNM 22820, reproduced from Olson (1975), permission of the Wilson Ornithological Society. Abbreviations for forelimb structures: acpr, acrocoracoid process; acrm, acromion process; bc, bicipital crest; bdep, brachial depression; bit, bicipital tubercle; bpr, bicipital prominence; ci, capital incisure of humerus; clf, clavicular facet of coracoid; cmar, caudal margin of humerus; dc, dorsal condyle; dec, dorsal epicondyle; dpcr, deltopectoral crest; dt, dorsal tubercle of humerus; fp, flexor process; haf, humeral articular facet; hh, humeral head; pntf, pneumotricipital fossa; sf, scapular facet of coracoid; tct, coracoidal tubercle of scapula; vc, ventral condyle; vec, ventral epicondyle; vt, ventral tubercle. Arrows not labeled indicate unnamed features discussed in the text. Asterisks following a label indicate that the structure is broken off at the tip or badly abraded.

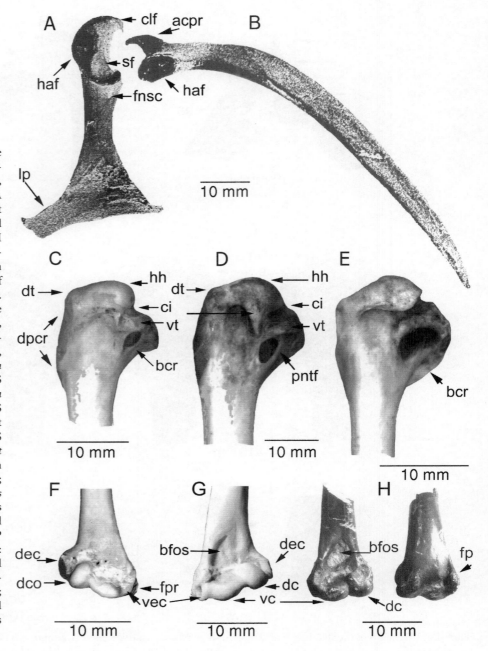

Figure 15.2. Elements of the shoulder articulation in Neornithes: *Lithornis,* Tinamiformes, Galliformes, and Anseriformes. A and B, *Lithornis promiscuus,* left coracoid in caudal view and scapula in lateral view, USNM 336535, holotype, Willwood Formation, Eocene, reproduced from Houde (1988) by permission of the Nuttall Ornithological Club. C–E, proximal end of the humerus in caudal view; F–H, distal end of the humerus: C, *Eudromia elegans* (Tinamiformes), CAS 85344, extant; D, *Acryllium vulturinum* (Galliformes), CAS 57239, extant; E, *Aix sponsa* (Anatidae: Anseriformes), CAS 57843, extant; F, *E. elegans,* left humerus in cranial view, CAS 85344; G, *Megapodius freycinet* (Galliformes), left humerus in cranial view, MVZ 90031, extant; H, *Presbyornis* sp. cf. *P. pervetus* (Anseriformes), right humerus reversed for comparisons, cranial and caudal views, one of UCMP lot 136541, Eocene. Abbreviations: bfos, brachial fossa; dco, dorsal condyle of humerus; fnsc, foramen for N. supracoracoideus; fpr, flexor process; lp, lateral process of the coracoid; others as in Figure 15.1.

coracoid never extends sternally farther than the sternal border of the scapular facet, and the coracoidal humeral facet may be cranial to the entire scapular facet (e.g., Galliformes, Fig. 15.3E). These associated features appear to result from elongation of the coracoid in this region. This separates the two facets and contributes to the elongate glenoid fossa. Diomedeidae (albatrosses) and *Pandion* (osprey) are a near exception to this statement. These features are presumed synapomorphic for Neornithes.

The coracoid is often similar in distantly related, poorly flying neornithines. One or more of the following structures is small or absent: the acrocoracoid process, procoracoid process, foramen for N. supracoracoideus, and the lateral process. The shaft is often stout, narrow, and elongated. Such birds include tinamous, galliforms, grebes, rails, turnicids, and others. Individually, such conditions do not necessarily indicate poor flight, and some occur in well-flying birds (e.g., Piciformes, Passeriformes, with their elongate coracoids). Some of these conditions occur in *Enantiornis,* suggesting that developmental truncation may be responsible for the commonalities. In ratites, the reduction of shoulder elements is much more extreme.

Figure 15.1A shows the coracoid and scapula of *Enantiornis* as they may have apposed in the articulated shoul-

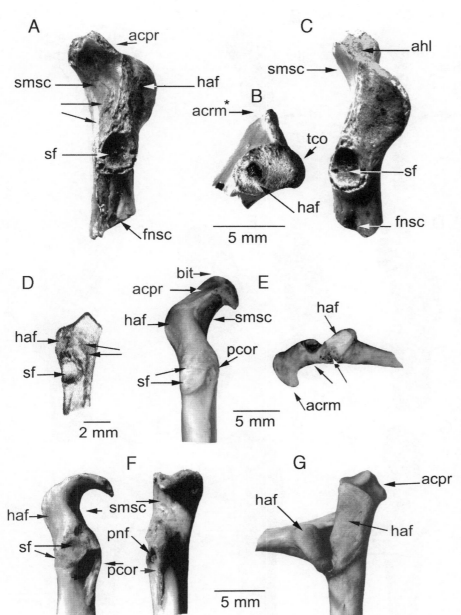

Figure 15.3. The coracoid and scapula in gallinaceous birds and a tinamou. A and C, *Palintropus* sp., right coracoid in dorsomedial and dorsal views, RTMP 88.116.1. B, *Palintropus* sp., right scapula in lateral view, RTMP 86.146.11. A–C from the Judith River Group, Campanian. D, *P. retusus,* left coracoid, YPM 513, holotype, Lance Formation, Maastrichtian, reproduced from Shufeldt (1915). E–G, extant birds: E, *M. freycinet* (Galliformes), left coracoid in dorsal view, MVZ 90031; F, G, *E. elegans* (Tinamiformes), left coracoid in dorsal and medial views and the scapula and coracoid apposed in lateral view, CAS 38670. In *Palintropus,* note absence of the procoracoid process, the size and distolateral position of the scapular facet of the coracoid, and the far lateral position of the coracoidal tubercle of the scapula; the acromion process is broken off. In the coracoid of the extant galliform and the tinamou, note the distolateral extension of the scapular facet of the coracoid and the fracture surface (unlabeled arrow) at the cranial end of the humeral facet of the scapula. This surface indicates the partial synostosis that existed between the scapula and the coracoid just distal to the humeral facet of the coracoid. Abbreviations: ahl, impression for the acrocoracohumeral ligament; pcor, procoracoid process of coracoid; pnf, pneumatic fenestra; smsc, sulcus for M. supracoracoideus; tco, coracoidal tubercle; others as in Figure 15.1.

der. The acromion process of the scapula is short, flattened mediolaterally, slightly recurved laterally, and bluntly rounded at the tip. In *Patagopteryx,* the acromion process is projected cranially, and there is a coracoidal tubercle centrally located on the coracoidal margin of the scapula. In *Ichthyornis* (Fig. 15.1B), the acromion process is styloid and very short. There is a large coracoidal tubercle on the cranial margin of the procoracoid process adjacent to but not impinging on the humeral facet. The humeral facet of the scapula is entirely recessed caudal to the coracoidal tubercle. It is as large as the humeral facet of the coracoid, but in contrast to *Enantiornis,* the scapular humeral facet is flat

and elongate in the axis of the scapula. Its articular face is oriented toward the anatomically ventral border of the scapula, but in ornithurine birds the scapula lies far dorsal and flat along the dorsolateral surface of the thorax. This repositions the scapula so that its narrow ventral border is oriented ventrolaterally with respect to the body axes.

In varied neornithines, the scapula is similar to that of *Ichthyornis.* Again, features shared with *Ichthyornis* are considered plesiomorphic for Neornithes: presence of an ossified coracoidal tubercle, the scapular humeral facet recessed cranially from the coracoidal tubercle, and a cranially projected acromion process. These features are also present in

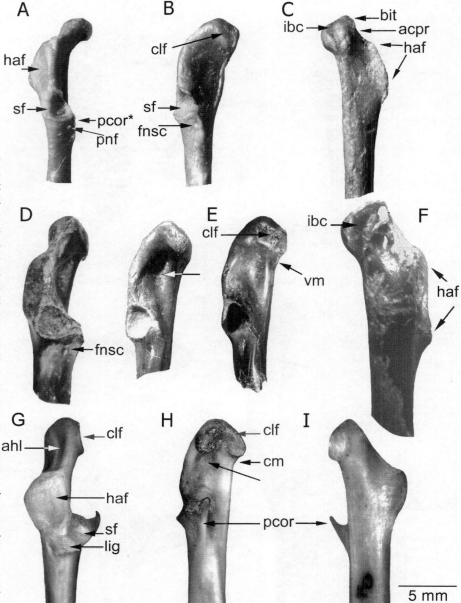

Figure 15.4. The coracoid in Neognathae: *Apatornis* and Anseriformes. A–C, *A. celer,* proximal end of the left coracoid in dorsal, mediodorsal, and ventral views, cast of referred specimen YPM 1734, Niobrara Formation, Coniacian–earliest Campanian. D–F, Presbyornithidae. D and E, undetermined species, left coracoid in dorsal and two medial views, ANSP 15866, from the Bug Creek facies of the Hell Creek Formation, latest Maastrichtian or earliest Paleocene; F, *Presbyornis* sp. cf. *P. pervetus,* right coracoid reversed for comparisons, in ventral view, UCMP 174046, Eocene. G–I, *A. sponsa* (Anatidae), left coracoid in lateral, medial, and ventral views, CAS 57843, extant. In Presbyornithidae and Anatidae, note the columnar ventral border of sulcus M. supracoracoideus, its bifurcation at the clavicular facet, and the well-defined sulcus within sulcus M. supracoracoideus. *Apatornis* lacks some of these features but has diagnostic features of Neornithes and Neognathae. Abbreviations: cm, caudal margin of humerus; ibc, impression for M. coracobrachialis cranialis; lig, impression for Lig. Coracoscapulare ventralis; vm, ventral margin of humerus; others as in Figure 15.1.

Patagopteryx. The absence or far lateral position of the coracoidal tubercle in *Lithornis, Eudromia,* and Galliformes is problematic for such an interpretation and is discussed further at the end of this chapter. In *Lithornis* and Galliformes, the acromion process is dorsoventrally flattened. In *Lithornis,* it is bent ventrally and styloid at the tip; in Galliformes, it is often hooked medially (Figs. 15.2B, 15.3G). This suggests that a dorsoventrally flattened acromion process was present in earliest neornithines, possibly plesiomorphically. In *Lithornis,* the flattened acromion process forms the medial wall of the triosseal canal (Houde, 1988). The status of this feature is discussed further, below with *Ambiortus.*

In Neornithes, the humeral facet of the scapula faces laterally or craniolaterally. This orientation may be considered synapomorphic for Neornithes. In Tinamiformes and Neognathae, the humeral facet of the scapula is smaller than the humeral facet of the coracoid; this is derived within Neornithes and characterizes most members.

Forelimb

Many characters used for phylogenetic analyses to reconstruct the history of birds reflect the redesign of the forelimb (Figs. 15.1, 15.2) in the early lineages of Aves (Cracraft, 1986; Sereno and Rao, 1992; Chiappe, 1995a,b; Sanz et al.,

1995). Beginning in nonavian theropods, the articulation for the forelimb changed from a ventrally situated and caudoventrally oriented ball-and-socket design into a dorsally situated, laterally oriented, saddle-shaped structure that enhanced powered flight in Ornithurae (Jenkins, 1993; Poore et al., 1997; Ostrom et al., 1999). Only a few further innovations in the forelimb were present in the earliest neornithine (Figs. 15.1, 15.2). Changes assignable to the earliest neornithine include enlargement and pneumatization of the head of the humerus and probably also the inflection of the deltopectoral crest. The new design of the forelimb and shoulder girdle allows greater excursion of the wings than in *Ichthyornis*. At the extreme, the wings can be lifted completely over the back during the vertical power stroke (Dial et al., 1991).

In *Enantiornis* the head of the humerus is small, with a globose enlargement adjacent to the very distinct capital incisure (Fig. 15.1C). There is a short but well-developed ventral tubercle, a large pneumotricipital fossa, and a prominent knoblike bicipital prominence. The shaft is short, robust, and moderately curved. The condyles are poorly developed, and there is no distinct brachial fossa or depression. The ventral epicondyle is very large and extends well distal to the ventral condyle (Fig. 15.1E).

In *Ichthyornis,* the head of the humerus and ventral tubercle are small and the capital incisure indistinct, but the deltopectoral crest is very large. The bicipital crest, bicipital prominence, and tricipital fossa are all rudimentary (Fig. 15.1D). The shaft is short, stout, and moderately curved. The condyles are well developed, and the ventral epicondyle extends well ventral to the condyle but not distal to it. The illustrations of the distal end of the humerus of *Ichthyornis* by Marsh (1880) are not fully consistent with the holotype of *I. dispar* (Olson, 1975). In Olson's photograph of the holotype (reproduced here as Fig. 15.1F), the brachial depression is very deep and extensive, excavating the entire distal end of the humerus, including the condyles and the ventral epicondyle. As used here, "brachial depression" refers to the large depression in the entire cranial surface of the distal end of the humerus present in some birds; "brachial fossa" refers to a more limited depression in many neornithines, primarily proximal to the supracondylar prominences, usually shallow, distinctly defined, and obliquely oriented (Fig. 15.2F–H).

In *Lithornis*, tinamous, galliforms, and many other neornithines, the shaft of the humerus is short and moderately curved, and the ventral epicondyle is stout and extends well ventral and often distal to the ventral condyle. These features are here assumed plesiomorphic in Neornithes. A large ventral epicondyle is more marked in *Enantiornis* and neornithine landbirds (e.g., tinamous, galliforms, passerines, coraciiforms) than in *Ichthyornis* and neornithine waterfowl, seabirds, and shorebirds. In *Lithornis,* tinamous, galliforms, and many other neornithines, the head of the humerus is large and globular; there is a capital incisure and a distinct

sulcus for the transverse ligament; the pneumotricipital fossa is pneumatized by a large foramen; and the deltopectoral crest is inflected cranially. These features may be considered synapomorphic for Neornithes. In the articulated shoulder, the ventral lip of the humeral facet of the scapula fits into the capital incisure of the humerus. In *Lithornis,* tinamous (Fig. 15.2C), galliforms (Fig. 15.2D), and columbiforms, a shelf walls the distal end of the capital incisure from the ventral tubercle to the head of the humerus midway between the dorsal tubercle and the capital incisure. Its position and extreme rugosity show this shelf to be an attachment site for one or more ligaments or muscles. A possible rudiment of this shelf appears in many other neornithine birds as a poorly defined crest (some gruiforms, some charadriiforms, e.g., *Scolopax*) or a scar in the capital incisure (ibises, *Graculavus,* Fig. 15.7A). This shelf is probably synapomorphic for Neornithes and lost in most members of the group. In addition, in Neornithes, the bicipital crest and prominence are at least slightly enlarged, there is a well-defined brachial fossa, and the condyles are well rounded. The equivocal status of these and other features of the humerus is discussed below with *Eurolimnornis* and *Ambiortus.*

Hindlimb

As both forelimb and tail were transformed for a role in flight, birds became fully bipedal. The tail skeleton shortened, and hindlimb retractors moved from the tail to the enlarging pelvis. In Ornithurae, the femur lies almost horizontally, with the knee forward under the new center of gravity assumed in the upright, bipedal posture. Thus, the principal axis of motion of the hindlimb moved from hip to knee (Chiappe, 1995b; Gatesy, 1995; Padian and Chiappe, 1998). In Neornithes, changes in the hindlimb are relatively minor, but increases in ossification of ligamentous attachment sites, tendinal channels, tendinal restraining loops, or entire tendons are very common. However, none of these innovations can be ascribed with confidence to the earliest neornithine. Many such changes are so widespread that they do appear to have arisen early in Neornithes, some perhaps more than once, few in a clearly defined phylogenetic pattern. These include increased co-ossification of thoracic vertebrae as a notarium, co-ossification of the synsacrum with the ischium, enlargement of the lateral trochanteric crest of the femur, ossification of the supratendinal bridge of the distal end of the tibia, and ossification of tendinal canals of the hypotarsus. At least two characters occur in a pattern associated with large groups within Neornithes. In neognaths, the ilioischiadic foramen is closed (Cracraft, 1988), and in tinamous and neognaths, the ascending process of the astragalus is reduced or absent and the calcaneum enlarged (McGowan, 1985).

Thus, there was no major reorganization of the hindlimb in Neornithes. This is consistent with the more general finding in Aves that evolution of the hindlimb has been slower than evolution of the forelimb (Cracraft, 1986; Sanz et al., 1988; 1997; Chiappe, 1991a; Sereno and Rao, 1992; Hou et al., 1996).

Elements of the hip, ankle, and toe joints are also represented in collections. Background comparisons for elements of the hip, knee, and ankle joints, including the proximal end of the tarsometatarsus, are included in species accounts below, especially with *Horezmavis, Gansus,* Procellariiformes, and Gaviiformes. Chiappe (1995a) noted that metatarsal V had been lost in the stem lineage of Ornithurae. The distal end of the tarsometatarsus of the earliest ornithurine is reconstructed from features shared by *Ichthyornis* and many neornithines, as follows. Metatarsal I attached far distally on the plantar or medioplantar surface of the metatarsus. Metatarsals II and IV were slightly retracted plantarly compared with metatarsal III. Metatarsal II was shortest, III longest, and IV intermediate. Trochlear incisures were well developed. Because these features also occur in varied neornithines, they are presumed plesiomorphic in Neornithes. In Neornithes, the incisure between metatarsals III and IV is closed by a bridge, forming the distal intermetatarsal foramen, but apparently it was also partly closed in some hesperornithiforms. Thus, no synapomorphic features of Neornithes are positively identified here.

Early Cretaceous Birds with Neornithine Similarities

Ambiortus, Otogornis

Ambiortus dementjevi Kurochkin, 1982, is from the Early Cretaceous (Berriasian-Valanginian) Khurilt Beds of Mongolia (see also Kurochkin, 1985, 1988). The material includes articulated vertebrae and part of the pectoral girdle and forelimb. Based on new preparation, Kurochkin (1999) redescribed *Ambiortus* and compared it closely with *Otogornis genghisi* Hou, 1994, known from less complete forelimb material from probably the earliest Cretaceous Yijinhuoluo Formation, Inner Mongolia, China. He found these birds to be similar to the flying palaeognath *Lithornis,* known from the Paleogene of Europe and North America (Houde, 1988).

The wing feathers of *Ambiortus* had tightly bonded barbs (Kurochkin, 1985), but the two preserved wing feathers of *Otogornis* were not tightly arranged, suggesting that they lacked barbules and that *Otogornis* was a flightless bird (Hou, 1994). Its larger size and more robust bones suggest the same (Kurochkin, 1999). At least the neck vertebrae in *Ambiortus* and *Otogornis* are heterocoelous and bear lateral processes similar to those of palaeognaths. *Ambiortus* had large pleurocoels.

In *Ambiortus,* the apex of the acrocoracoid process is stout and triangularly peaked. It is less clearly shown in

Otogornis but appears to be much smaller. Otherwise, according to Kurochkin (1999), the coracoid and scapula in both birds are similar to those of *Lithornis.* The humeral facet of the coracoid is evidently large and flat, the scapular facet is cotylar and slightly elliptical, and the humeral facet of the scapula is flat, ovate, large, and laterally oriented. However, as a result of crushing and overlap of parts, the scapula-coracoid articulation is not clearly seen in the material or illustrations of either species; thus, the glenoid fossa is hard to evaluate. Both *Ambiortus* and *Otogornis* apparently have a flattened, ventrally directed acromion process, as in *Lithornis.* Both species have a coracoidal tubercle on the coracoidal margin of the scapula, unlike *Lithornis.* The shaft and distal end of the coracoid are similar to these structures in *Ichthyornis* and *Lithornis.*

In both *Ambiortus* and *Otogornis,* the head of the humerus is a small, egg-shaped tubercle, and the ventral tubercle is poorly developed. Although the deltopectoral crest appears to be uninflected in *Ambiortus,* Kurochkin (1999) found that it had been flattened postmortem from a slightly inflected configuration in life. The capital incisure is shallow in *Otogornis* and absent in *Ambiortus.* The tricipital fossa is a simple streak on the bone, and there is no bicipital crest. On the cranial surface, there is a small, pitted tubercle, which appears to correspond with the bicipital prominence, and a small pit that is probably the site for the acrocoracohumeral ligament.

Based on features shared with *Lithornis,* Kurochkin (1999) argued that *Ambiortus* and *Otogornis* are palaeognaths. Several anomalous character distributions in Aves might be explained if this is true. Compared with *Ambiortus* and *Otogornis, Enantiornis,* which is clearly far outside Ornithurae, has a better developed ventral tubercle, tricipital fossa, bicipital prominence, bicipital crest, and capital incisure, features that are well developed in Neognathae but small or absent in Palaeognathae, *Ambiortus,* and *Otogornis.* The small size of these structures might be interpreted as shared, derived reduction uniting *Ambiortus* and *Otogornis* with Palaeognathae. This seems a more parsimonious interpretation than an independent origin of a large number of similar complex features of the head of the humerus in *Enantiornis* and Neognathae (Figs. 15.1A, 15.2C–E). However, most of the humeral structures that are poorly developed in *Ambiortus* and *Otogornis* are also poorly developed not only in Palaeognathae but also in *Ichthyornis* and Galliformes. This suggests an alternative: these features were rudimentary in the earliest ornithurine and remained so in the earliest neornithine. Either explanation implies homoplasy. Thus, whether these forelimb features arose before or after the origin of Neornithes is uncertain, and the phylogenetic positions of *Ambiortus* and *Otogornis* are not resolved.

Gallornis

Gallornis straeleni Lambrecht, 1931, is based on a very worn proximal end of a femur and a humerus fragment from the Neocomian at Auxerre, France. Lambrecht (1931) referred *Gallornis* to Anseriformes. Brodkorb (1963a) considered it close to flamingos and referred it to the extinct taxon Scaniornithidae. Olson (1985) considered it insufficient for identification.

The humerus is badly fragmented and probably beyond evaluation. The femur is very worn, but its general contours are intact. The head of the femur is rounded and the neck short. The antitrochanteric facet is large and widely extended both cranially and caudally from the neck of the femur. The lateral trochanteric crest is elevated above the antitrochanteric facet and slightly recurved over it. Following Chiappe (1995a), a lateral trochanteric crest is a synapomorphy of Ornithurae. An elevated trochanteric crest is known only in the Neornithes. It is widespread in the group, occurring at least in tinamous, galliforms, anseriforms, ibises, flamingos, some gruiforms, and some charadriiforms. In most of these groups it is also recurved. A craniocaudally extended antitrochanteric facet is also restricted to Neornithes as far as is known, but it occurs in many of these groups and several others, so it is not useful for differentiating them.

The distribution of these features suggests that they originated very early in Neornithes rather than outside the group or at its origin, and thus they suggest referral of *Gallornis* to Neornithes. However, the hindlimb is too poorly known in other ornithurine birds to rule out an earlier origin of these features. As for an anseriform assignment, Howard (1964:237) stated, "This very fragmentary material seems quite inadequate . . . as a basis for a positive Mesozoic record of the order [Anseriformes]."

Horezmavis

Horezmavis eocretacea Nessov and Borkin, 1983, is based on the proximal part of the shaft of a tarsometatarsus from the Lower Cretaceous (Albian) Khodzhakul Formation of Karakalpak. The articular facets are missing, and the hypotarsus is damaged. The authors referred *Horezmavis* only to Aves, incertae sedis, although Nessov (1992a) noted a resemblance to gruiforms. Kurochkin (1995b) considered *Horezmavis* to be a gruiform based on the position of the tubercle for attachment of M. tibialis cranialis, presence of an intercotylar prominence, presence of two proximal intermetatarsal foramina, remnants of ossifications suggesting a hypotarsus, and an elongate shaft. However, the intercotylar prominence is present in other ornithurines (Chiappe, 1995a), and hypotarsal ossification without ossified canals is present in *Patagopteryx* (Chiappe, 1996b), in *Hesperornis* (Marsh, 1880; pers. obs.), and slightly in *Ichthyornis* (Marsh, 1880; Martin, 1983). Ossified hypotarsal ridges and canals are known only in Neornithes, but too little of the proximal

end of the tarsometatarsus remains in *Horezmavis* to determine whether these were present.

The proximal intermetatarsal foramina in Neornithes are homologous with the gaps between the metatarsals. The homology is evident in a juvenile domestic chicken (*Gallus gallus*) when development is nearing completion. At this time, the metatarsals are fused distally, and the spaces between them are narrowed proximally but already clearly identifiable as intermetatarsal foramina. Thus, these gaps are the remnants of distal to proximal fusion of metatarsals. This direction of fusion is diagnostic for the larger ornithurine lineage (Martin, 1983; Cracraft, 1988; Chiappe, 1995a), but the developmental process indicates that absence rather than presence of these foramina is the derived state in an ornithurine bird. These considerations undermine the evidence for referral of this limited specimen to Neornithes.

Gansus

Gansus yumenensis Hou and Liu, 1984, is based on an articulated foot skeleton from the Early Cretaceous Xiagou Formation, Gansu Province, China. The authors placed it near the base of a radiation that included *Ichthyornis* and the neornithine shorebirds and seabirds. There are conflicting indications in the description as to the presence or absence of an ossified supratendinal bridge of the distal end of the tibia (Hou and Liu, 1984, compare pp. 1296 and 1300). This structure is known only in Neornithes, and although it is absent in some members of the group (e.g., *Lithornis*, some members of Ratitae, Strigiformes), it is present in most, including all neornithine diving birds. Several reviewers have referred *Gansus* to Neornithes based on a presumption that there was such a bridge or else on the similarity of the foot to that of neornithine swimmers and divers. Kurochkin (1995b) referred this bird to Palaeognathae based in part on the presumed absence of the supratendinal bridge.

The tarsometatarsus is short and strongly flattened mediolaterally, trochlea II is strongly retracted plantarly, and trochlea IV lies almost in the same transverse plane as trochlea III but does not appear elongated. The toes are slender and long; toes II–IV increase in length from medial to lateral. The presence of an intercotylar eminence in *Gansus* is consistent with referral to Ornithurae. Although the metatarsals are closely appressed, the authors noted that they are not fully fused distally. This suggests that the metatarsals fused proximally to distally rather than the reverse; however, Marsh (1880) noted that in *Ichthyornis victor*, lines of fusion remain on the plantar surface between metatarsals II–III and III–IV. The implication is that fusion was less complete in some ornithurine birds than Neornithes, so that incomplete distal fusion in *Gansus* does not necessarily argue against inclusion in Ornithurae. However, this and the small, shallow infracotylar fossa and possible absence of proximal intermetatarsal foramina in *Gansus* (Hou and Liu, 1984:1298) are inconsistent with referral to Neornithes. In summary, *Gansus* is probably an ornithurine bird, but the foot seems to be convergently similar to that of neornithine divers (see also Zhou and Hou, Chapter 7 in this volume.)

Palaeocursornis and Eurolimnornis

For the history and status of the name *Palaeocursornis* and the related history of *Eurolimnornis*, the reader is referred to the meticulous analysis by Bock and Bühler (1996). Briefly, the material was collected from a bauxite lens in the Lower Cretaceous (Wealdian) of Romania. All the material was first described under the name *Limnornis corneti* Kessler and Jurcsak, 1984. Because the name *Limnornis* is preoccupied, the authors ultimately replaced it. The new name *Palaeocursornis corneti* Kessler and Jurcsak (1984) applies only to the holotype of "*Limnornis*" *corneti*, a fragment of a femur that they now considered to have come from a ratite. However, the referred material of "*Limnornis*" *corneti* did not appear to come from a ratite, and it was not found directly associated with the femur. Therefore, they redescribed the referred material under another new name, *Eurolimnornis corneti* Kessler and Jurcsak, 1986. In several publications these authors proposed inclusive taxon names for each species, summarized by Bock and Bühler (1996).

Material of *Palaeocursornis* consists of a badly worn distal end of a femur. There is a deep, distinct, popliteal fossa bounded distally by a crest, a large tibiofibular crest, and a shallow, trochlear fibular condyle. The cranial surface is not illustrated, but a line drawing of the distal surface (Kessler and Jurcsak, 1984) indicates a very narrow, well-defined intercondylar sulcus, which appears to continue proximally on the cranial surface, suggesting the presence of a patellar sulcus. Following Chiappe (1995a), a patellar sulcus and a large tibiofibular crest indicate only an ornithurine bird. Among neornithines, tinamous and lithornithids have a narrow, deep, and distinctly defined intercondylar sulcus and patellar fossa.

The material of *Eurolimnornis* consists of the distal end of a right humerus with a small part of the shaft, a fragment of the shaft of an ulna, and the distal end of a carpometacarpus. Material of both *Palaeocursornis* and *Eurolimnornis* came from a paleokarstic sinkhole, a "bone cemetery" with marks of predation. The material was among "approximately sixty bird-like bone fragments. . . . [T]heir condition is very bad [and] only about six specimens can with certainty be attributed to birds" (Kessler, 1984:119). This account suggests that the bones referred to *E. corneti* may have come from more than one species. Be that as it may, an illustration (Kessler and Jurcsak, 1984) shows a distinct brachial fossa, rounded condyles of the humerus, traces of distally fused

metacarpals, and a distinct tendinal sulcus of the carpometacarpus. Kurochkin (1995a) referred *Eurolimnornis* to Neornithes based on these features. However, fused metacarpals and a tendinal sulcus also occur in *Ichthyornis*. A distinct brachial fossa is known only in Neornithes, but except for the huge brachial depression, the distal end of the humerus of *Ichthyornis* is very similar to that of several early Neornithes, for example, *Presbyornis* and *Torotix*. Deep excavation of the entire brachial depression and the condyles may be autapomorphic in *Ichthyornis*, and a delimited brachial fossa and fully rounded condyles may predate Neornithes. These features in *Eurolimnornis* are suggestive but insufficient evidence for confident referral of so ancient a bird more definitive than to Ornithurae.

Although none of the Early Cretaceous birds reviewed in this section is referred here with confidence to Neornithes, the assignment is not ruled out, and for some, the minimal extent of the material may be the major problem. In general, these Early Cretaceous birds confirm an early origin of neornithine features but tend only to blur any clear-cut characterization of the earliest neornithine. This is a sign that the fossil record is becoming more complete. As the record becomes more dense around the origin of a group, it becomes evident that its new features did not all arise at the same time (McKenna and Bell, 1997).

Mesozoic Neornithes: Systematic Paleontology

Taxonomic Hierarchy

Ornithurae Haeckel, 1866
  Neornithes Gadow, 1893

Limited diagnosis—Specimens are placed with Neornithes based on the following derived features, presumed to have originated close to the origin of the group: premaxilla fused with the maxilla; the maxilla very reduced and almost entirely restricted to the palatal region; the mandibular symphysis fused; the dentary fused with the surangular; teeth absent; the coracoidal humeral facet not extended sternally beyond the scapular facet of the coracoid; the coracoidal humeral facet separated slightly to completely from the scapular facet of the coracoid; the scapular humeral facet oriented laterally or craniolaterally; the humeral head large; a sulcus for the transverse ligament of the humerus distinct; the deltopectoral crest inflected cranially; and the pneumotricipital fossa of the humerus with a pneumatic foramen.

?Galliformes (Temminck, 1820). Landfowl: megapodes, cracids, pheasants

Remarks—Specimens below are placed close to or within Galliformes by the following derived features, but the placement is tentative because these features are not unique: the procoracoid process of the coracoid is reduced; the scapular facet of the coracoid is entirely distal to the humeral facet; the neck of the coracoid is stout and triangular in cross section; the shaft of the coracoid is elongate; the sternal end of the coracoid is narrow; and the lateral process is rudimentary.

The smaller acrocoracoid process of Galliformes further differentiates the group from Anseriformes and most other neognaths. The larger acrocoracoid process and absence of a pneumatic foramen close to the procoracoid process of Galliformes differentiate the group from Tinamiformes.

?Quercymegapodiidae Mourer-Chauviré, 1992

Remarks—*Palintropus* is placed close to or within Quercymegapodiidae by the following derived features: the humeral facet of the coracoid has a large, free lateral flange, and the procoracoid process of the coracoid is further reduced. Quercymegapodiidae, Paraortygidae, and Gallinuloididae are further differentiated from extant galliforms by retention of a large, cotylar scapular facet of the coracoid in the former taxa (Mourer-Chauviré, 1992; Mayr, 2000).

A long, raised, ragged scar within the sulcus M. supracoracoideus and parallel to the scapular facet (Fig. 15.3A) also suggests placement of *Palintropus* within Quercymegapodiidae. Brodkorb (1963a) identified this scar in *P. retusus* (Fig. 15.3D); it is present in the new species and in a newly described Tertiary species of *Ameripodius*, referred to Quercymegapodiidae (Mourer-Chauviré, 2000). However, uniquely in the galliform lineage, *Palintropus* retains a foramen for N. supracoracoideus; this argues against placing *Palintropus* within Quercymegapodiidae.

Palintropus Brodkorb, 1970

Included species—*P. retusus* (Marsh, 1892) and an undescribed new species.

Revised diagnosis—Derived features of *Palintropus* include the complete absence of the procoracoid process of the coracoid; the very large size of the foramen for N. supracoracoideus, and its central position on the coracoidal shaft well distal to the scapular facet.

In *Palintropus* (Fig. 15.3A–D), the coracoidal tubercle of the scapula is placed far laterally, where it caps the cranial end of the humeral facet. The scapula is unknown in *Quercymegapodius* and *Ameripodius,* but the same unusual morphology is predicted for these birds, based on the similar morphology of the coracoidal scapular articulations in all three taxa. Based on a new *Lithornis*-like scapula from the Hornerstown Formation (Parris and Hope, MS), it appears that the coracoidal tubercle may have been in the same far lateral position in *Lithornis.* The taxonomic distribution of this feature needs clarification.

Remarks—Marsh (1889, 1892) named and briefly specified the material of two small species of *Cimolopteryx,* evi-

dently intending a more complete description to follow. Marsh himself made the case for separating the two: "Another species [*C. retusa*], about twice the size of the first, is indicated by various remains, among them the coracoid. This bone lacks the strong inner [procoracoid] process near the pit for the scapula, which is characteristic of the smaller form" (Marsh, 1892:175). A long history of inconclusive and conflicting referrals followed. Sharpe (1899) referred *Cimolopteryx* to Ichthyornithiformes. Hay (1902) listed *Cimolopteryx* at the end of Passeriformes, but this was evidently an error of omission of the heading, because it is clear that his intent was to include it with birds of uncertain position. Shufeldt (1915) considered both species of indeterminate position, certainly not passerine and not closely related to each other. Lambrecht (1933) kept *Cimolopteryx* in Ichthyornithiformes. Wetmore (1956) placed it in Aves, incertae sedis. Finally, Brodkorb (1963a) separated the two species. For the subsequent history of *Cimolopteryx rara,* see the discussion herein with Charadriiformes. Brodkorb (1963a) referred *C. retusa* Marsh, 1892, to *Apatornis retusus,* which he considered to be within Ichthyornithiformes. Later, Brodkorb (1970) referred *A. retusus* to a new genus as *P. retusus,* assigning it to Cimolopterygidae in Charadriiformes, thus reuniting it with *Cimolopteryx.*

The small, scrappy coracoid that is the holotype of *P. retusus* is at first sight very different from that of extant galliforms. Referral was suggested only by new, larger and more complete specimens of another much larger species of *Palintropus* from the Campanian of Alberta, Canada. This material showed that *Palintropus* is similar to Quercymegapodiidae, known from several species in the Early Tertiary of France, Germany, and Brazil and identified as galliforms based on more extensive material (Mourer-Chauviré, 1992; Alvarenga, 1995; Mayr, 2000). The referral of *Palintropus* to Quercymegapodiidae is only tentative because *Palintropus* retains a very large procoracoid foramen. Brodkorb's (1963a) description indicates the presence of this foramen in *P. retusus* by his noting, without identifying it, the proximal end of a deep channel leading to the foramen. The holotype of *P. retusus* is broken away through this channel. This structure is lost in Quercymegapodiidae and extant galliforms. Retention of a foramen for N. supracoracoideus in *Palintropus* is at face value a more plesiomorphic condition than its absence in *Quercymegapodius, Ameripodius,* and extant galliforms. Thus, the phylogenetic interpretation is not straightforward.

The most striking feature of *Palintropus* is the huge size and far lateral position of the coracoidal tubercle, capping the cranial end of the humeral facet of the scapula. Rather than contradicting a relationship with Galliformes, the far lateral position of the coracoidal tubercle in *Palintropus* calls attention to lateralization of the scapular articulation in all galliforms. In extant galliforms, the absence of a coracoidal tubercle of the scapula is perhaps misleading. The cranial margin of the humeral facet of the scapula is truncated as a straight or invaginated margin. Extending from it is a protuberant mass of fibers that attaches to the lateral part of the scapular facet of the coracoid. This may be the unossified remnants of a far lateral coracoidal tubercle. The scapular facet of the coracoid is elongated toward the lateral surface, but its lateralmost part is flat or a very shallow basin (Fig. 15.3E). This basin is probably homologous with the huge, cotylar scapular facet of *Palintropus*. Tinamous have a lateralized articulation very similar to that of extant galliforms (Fig. 15.3F).

P. retusus (Marsh, 1892)

Synonyms—*C. retusa* Marsh, 1892; *A. retusus* Brodkorb, 1970.

Holotype—YPM 513, left coracoid, shoulder end and part of shaft (Fig. 15.3D).

Locality and horizon—Converse (= Niobrara) County, Wyoming; Lance Formation, Late Maastrichtian. Collector J. B. Hatcher, 13 April 1890.

Revised diagnosis—As for the genus in the features preserved.

Remarks—Marsh's (1892) remarks, Shufeldt's (1915:12) comments and photograph, and Brodkorb's (1963a, 1970) moderately detailed description confirm the presence of features given in the diagnosis and description of *Palintropus* above. *P. retusus* was about two-thirds larger than the domestic pigeon, *Columba livia*.

Palintropus, Undetermined Species

Material—RTMP P88.116.1, right coracoid, shoulder end; RTMP 86.36.126, left coracoid, shoulder end; RTMP 86.146.11, right scapula, shoulder end (Fig. 15.3A–C).

Locality and horizon—Dinosaur Provincial Park, Alberta, Canada; RTMP 88.116.1 and RTMP 86.36.126 from the Dinosaur Park Formation, RTMP 86.146.11 from the Foremost Formation. Campanian (Eberth, 1990).

Measurements—See Table 15.1.

Remarks—The coracoids came from a bird in the size range of a mature male and female peacock (*Pavo cristatus*),

TABLE 15.1
Measurements (in mm) of the coracoid in *Cimolopteryx* and other birds

Taxa	Specimen	Humeral Facet Length	Humeral Facet Width	Scapular Facet Length	Scapular Facet Width	Foramen Recess	Neck Width
P. retusus	YPM 513	6.4	4.2	2.6	2.5	—	4.0
Palintropus, undet. sp.	RTMP 86.36.126	14.3	9.0	9.2	8.7	—	8.8
	RTMP 88.116.1	10.5	6.8	5.5	5.8	12.1	6.7
C. rara	YPM 1805	5.5	3.2	2.0	2.3	1.3	2.6
	UCMP 53962	4.6	3.0	2.0	2.2	1.5	2.6
	UCMP 53963	5.4	3.3	2.0	2.3	1.7	2.8
	SMNH 1927.963	4.7	3.2	2.3	2.6	1.6	2.6
C. petra	AMNH 21911	4.4	2.4	1.9	2.2	1.1	1.9
C. maxima	UCMP 53973	8.7	4.7	3.7	3.7	2.6	—
	UCMP 53957	8.9	5.0	3.7	4.2	—	—
	ANSP 15867 (?Paleocene)	—	—	3.9	3.4	1.9	4.1
C. minima	UCMP 53976	3.0	1.8	1.5	1.6	0.8	2.1
Cimolopteryx sp.	RTMP 93.116.1	2.7	1.9	1.6	1.8	—	—
cf. Charadriiformes, undet. sp.	USNM 181923	4.4	3.0	2.9	2.6	1.3	2.9
	ANSP 15868 (?Paleocene)	4.3	2.8	2.5	2.8	1.2	2.6
	AMNH 13011	4.0	2.4	2.2	2.3	1.5	1.9
Presbyornithidae	ANSP 14866 (?Paleocene)	5.2	3.7	3.4	3.4	1.5	4.5
A. celer	YPM 1734 cast	5.0	3.6	2.5	2.3	0.7	2.9
C. major	UCMP 53959	7.9	4.4	3.0	3.2	1.8	—
Neornithes, undet. sp.	RTMP 93.19.1	5.4	3.7	2.9	3.2	2.1	N/A

Sources: YPM and UCMP specimen measurements are from Brodkorb (1963a); SMNH specimen measurements are from Tokaryk and James (1989).

Note: Foramen recess = distance from sternal border of scapular facet of the coracoid to foramen for N. supracoracoideus. Neck width = least mediolateral dimension of the shaft of the coracoid just distal to the scapular facet.

RTMP 88.116.1 being the smaller specimen. The scapula articulates well with the smaller specimen. Other material in the collections of RTMP, not listed here, includes five similar, more fragmentary coracoids in the same two size ranges, apparently referable to the same species, all from localities in the Judith River Group of formations. This collection may represent two species or a single species with extreme sexual size dimorphism, as is common in extant galliforms.

Anseriformes (Wagler, 1830). Screamers and waterfowl

Limited diagnosis—Specimens are placed with Anseriformes by the following features: the capital incisure undercutting the head of the humerus (unique); the humeral shaft strongly S-shaped (derived but not unique).

The coracoid of Anseriformes is further differentiated from that of Charadriiformes principally by the following presumably less derived features in Anseriformes: the acrocoracoid process is more bulbous and erect, the clavicular facet is not extended ventrally, and the foramen for N. supracoracoideus is not recessed caudally from the scapular cotyla of the coracoid.

Remarks—Phylogeny within Anseriformes here assumes the concordant aspects of reconstructions by Woolfenden (1961), Ericson (1997) and Livezey (1997a), with inclusion of *Presbyornis* strongly supported by features of the skull. Taxonomy within Anseriformes follows Livezey (1997b).

Anseres Wagler, 1831. Waterfowl
Anatoidea (Leach, 1820). Presbyornithids, ducks, geese, swans

Limited diagnosis—Specimens below are placed with Anatoidea by the following combination of derived but not unique features: the clavicular facet of the coracoid undercut at its ventral end by a small cavity (Presbyornithidae) or the furcular surface eroded, exposing a shallow sulcus (Anatidae); the ventral margin of the sulcus M. supracoracoideus of the coracoid stout and columnar, increasing in diameter toward the clavicular facet; the tip of the acromion process of the scapula straight and slightly flared dorsally; and the sulcus M. supracoracoideus with a large included central sulcus.

Remarks—The following features in Anseriformes differentiate the coracoid from that of Charadriiformes: a more erect acrocoracoid and absence of ventral extension of the clavicular facet (Fig. 15.4). The long acromion process of the scapula in Anatoidea distinguishes it from the scapula of Charadriiformes. The long acromion process may be plesiomorphic in Neornithes. In the phylogenetic context assumed here, its occurrence in Anatoidea requires homoplasy. In Anatidae, the articular surface at the ventral end of the clavicular facet is concave and appears eroded. This concavity may be homologous with the undercut furcular facet

in Presbyornithidae (Fig. 15.4E). Flamingos and some charadriiforms (*Burhinus, Stercorarius*) also have a small fossa undercutting the ventral end of the clavicular facet of the coracoid, similar to that in Presbyornithidae. In these birds, the ventral margin of sulcus M. supracoracoideus is less stout and more distinctly bifurcated than in Anatoidea. In other charadriiforms, the entire clavicular facet is undercut by a huge fossa. The subclavicular and central sulci of the coracoid are not related morphologically. The central sulcus is large and well defined in presbyornithids (Fig. 15.4E), huge in some anatids (*Chen*), but shallow and poorly defined in others (*Anas,* Fig. 15.4I). *Burhinus* also has a central sulcus within sulcus M. supracoracoideus similar to that in Presbyornithidae.

Presbyornithidae Wetmore, 1926

Limited diagnosis—Specimens here are placed with Presbyornithidae based on the following derived features: the median apical crest of the sternum absent; two apical crests of the sternum present, with a sulcus between, the crests being very stout; the apex carinae of the sternum highly asymmetrical, the right apical crest merging with the external spine, the left apical crest merging into the pila carinae; the distal end of the humeral facet of the scapula strongly protruding dorsally; the humeral facet of the scapula attenuated to a point distally on lower shelf surrounding the facet; a deep, elongate, ragged scar for M. serratus superficialis present along the narrow ventral margin of the scapula just distal to the humeral facet; the humeral shaft very long; and the distal end of the humerus strongly tilted with the condyles recurved proximally.

The following features of Presbyornithidae further differentiate it from other groups of Anseriformes: the humeral facet of the coracoid is larger, flatter, and more ovate; the humeral and scapular facets of the coracoid are less separated; and the scapular facet of the coracoid is slightly larger and more rounded. Another feature of Presbyornithidae further differentiates it from Anatidae: there is no cranial protrusion of the humeral facet of the scapula. In Anatidae, the humeral facet and coracoidal tubercle of the scapula impinge slightly on each other, and the humeral facet extends cranially, creating a bilobed projection at the craniolateral corner of the scapula (Fig. 15.5H). Olson and Parris (1987) gave features differentiating the humerus of *Presbyornis* from that of *Graculavus* (?Charadriiformes).

Remarks—All these differentiating features in presbyornithids would be assumed to be relatively plesiomorphic within Neornithes and require homoplasy for their interpretation as derived features in Anseriformes. The description above of the humeral facet of the scapula in Anatidae indicates the scapular articulation is slightly lateralized. The articulation is similar in gruiforms and is discussed briefly at the end of this chapter.

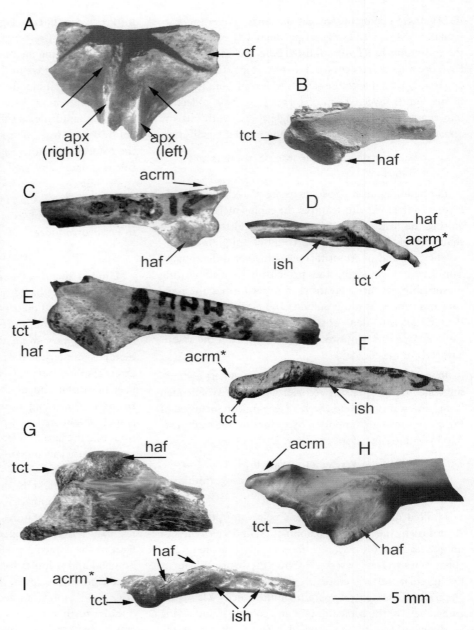

Figure 15.5. The coracoid in Neognathae: *Apatornis* and Anseriformes. A–C, *A. celer,* proximal end of the left coracoid in dorsal, mediodorsal, and ventral views, cast of referred specimen YPM 1734, Niobrara Formation, Coniacian–earliest Campanian. D–F, Presbyornithidae. D and E, undetermined species, left coracoid in dorsal and two medial views, ANSP 15866, from the Bug Creek facies of the Hell Creek Formation, latest Maastrichtian or earliest Paleocene; F, *Presbyornis* sp. cf. *P. pervetus,* right coracoid reversed for comparisons, in ventral view, UCMP 174046, Eocene. G–I, *A. sponsa* (Anatidae), left coracoid in lateral, medial, and ventral views, CAS 57843, extant. In Presbyornithidae and Anatidae, note the columnar ventral border of sulcus M. supracoracoideus, its bifurcation at the clavicular facet, and the well-defined sulcus within sulcus M. supracoracoideus. *Apatornis* lacks some of these features but has diagnostic features of Neornithes and Neognathae. Abbreviations: cm, caudal margin of humerus; ibc, impression for M. coracobrachialis cranialis; lig, impression for Lig. Coracoscapulare ventralis; vm, ventral margin of humerus; others as in Figure 15.1.

Undetermined Species 1

Material—MLP 93-I-3-1; MLP 93-I-3-2, varied postcranial remains of two individuals in concretions, evidently of the same species (Noriega and Tambussi, 1995).

Locality and horizon—Maastrichtian section of the Cape Lamb strata, correlated to the López de Bertodano Formation, Vega Island, Antarctica.

Remarks—Noriega and Tambussi (1995) referred this material only to Presbyornithidae. According to their description and illustrations, referral to Neornithes is indicated by the globular head of the humerus and the slight separation of the humeral and scapular facets of the cora-coid. The coracoid is not seen well enough to assess it further from the illustration. Referral to Anseriformes is suggested by the "impression of M. brachialis anticus . . . deep at its distal margin" (Noriega and Tambussi, 1995:60). This feature is present in both anseriforms and some charadriiforms. Derived features of Presbyornithidae are (1) the very long humerus and (2) the fact that the "internal distal margin [of the humerus] slopes upward . . . consequently the entepicondyle is proximal to the internal [ventral] condyle" (Noriega and Tambussi, 1995:60). In other anseriforms the humerus is short. The humerus is long in many charadriiforms but never strongly curved. The morphology of the ventral epicondyle is unique to Presbyornithidae and appears to be an extreme of the strongly S-shaped shaft of the humerus present in all anseriformes. Olson and Parris (1987) discussed the differentiation of the proximal end of the humerus of Presbyornithidae from that of *Graculavus*.

Undetermined Species 2

Material—AMNH 22602, sternal rostrum including apex carinae, coracoidal facets, and a short segment of pars cardiaca (Fig. 15.5A).

Locality and horizon—Near Lance Creek, Niobrara County, Wyoming; Lance Formation, UCMP locality V5711; Late Maastrichtian. Collected by M. C. McKenna and party, 1981.

Measurements—Coracoidal facets, maximum dorsoventral dimensions: left 3 mm, right 2.4 mm; overlap of facets 3.9 mm.

Remarks—The coracoidal facets overlap slightly in the midline, left over right, as in other birds with a broad sternal end of the coracoid. The pars cardiaca is very deep just distal to the coracoidal facets, and there is no large central pneumatic foramen. The presence of this foramen is highly variable in specimens of *Presbyornis* and among species of anseriforms and other birds having a deep pars cardiaca (flamingos, charadriiforms). The carina was evidently large. Asymmetry of the sternal rostrum occurs also in some specimens of the extant flamingo *Phoeniconaias minor,* but the asymmetry is less than in Presbyornithidae. This fragment is also in other ways closely similar in form and size to 15 specimens of the sternum in UCMP Lot 136538 referred to *Presbyornis pervetus*. In Anatidae, the coracoidal facets of the sternum are short, dorsoventrally deep, and well separated, and there is a median apical crest.

Undetermined Species 3

Material—AMNH 21929, right scapula, shoulder end and part of shaft (Fig. 15.5C, D); YPM 868, left scapula, shoulder end (Fig. 15.5B).

Locality and horizon—Both from Niobrara County, Wyoming, Lance Formation, Late Maastrichtian. YPM 868 from "'P. B. Q' ?Peterson's or Beecher's Quarry?" (exact locality unknown). Collected by J. B. Hatcher and party, April 1890. Hatcher's collections in this region were within the type area of the Lance Formation and well below the K/T boundary (Clemens, 1964). AMNH 21929 from UCMP locality V5711, collected by M. C. McKenna and party, July 1982.

Measurements—AMNH 21929: greatest diameter, humeral facet to base of acromion process, 4.7 mm; greatest depth through humeral facet, 2.1 mm; shaft dorsoventral diameter just distal to humeral articulation, 2.9 mm.

Remarks—YPM 868 is cataloged as *C. rara*. This is probably the "other material" that Marsh (1892) mentioned in his

description of *C. rara* without describing or identifying it or stating his criteria for the association. Shufeldt (1915) gave the specimen number and considered this scapula not diagnosable, but the new scapula, AMNH 21929, is more complete. It is closely similar in all shared details including size to YPM 868. The acromion process is completely broken off in YPM 868, but all but the tip is preserved in AMNH 21929, showing its stout base and tapered shape, as in *Presbyornis*. The two specimens correspond closely in other details to the scapula of *P. pervetus* except for being about one-third smaller.

These scapulae came from a bird about the same size as the charadriiformlike bird *C. rara,* a species that seems to have been abundant in Lance strata. However, it appears that the scapula AMNH 21929 would not appose well to the coracoid of *C. rara*. With the coracoidal tubercle in the scapular facet, the acromion process would project dorsally and well away from the procoracoid process. The scapulae of flamingos and *Presbyornis* are similar, but in flamingos the dorsal margin of the acromion process is not flared at the tip (Ericson, 1999), and although a scar for M. serratus superficialis is present, it is much shallower and shorter. Some rails (Gruiformes) have a distinct, elongate impression for this muscle, but it is much narrower and shallower than in presbyornithids.

Undetermined Species 4

Material—AMNH 22603, left scapula, shoulder end and part of shaft (Fig. 15.5E, F).

Locality and horizon—Near Lance Creek, Niobrara County, Wyoming; Lance Formation, UCMP locality V5711, Late Maastrichtian. Collected by M. C. McKenna and party, 1960.

Measurements—Greatest diameter, humeral facet to base of acromion process, 6.3 mm; greatest depth through humeral facet, 2.5 mm; shaft dorsoventral diameter just distal to humeral articulation, 3.6 mm.

Remarks—The acromion process is broken off. In preserved details, this scapula is very similar to that of the Undetermined Species 3, above, but one-third larger. It apposes well to a coracoid of an undescribed presbyornithid about one-third smaller than *P. pervetus* (ANSP 15866, Fig. 15.4D–F), from the Bug Creek channels of the Hell Creek Formation.

?Charadriiformes (Huxley, 1867). Gulls, auks, plovers, sandpipers, and others

Remarks—Marsh (1870, 1872) and Shufeldt (1915) described 4 genera and 12 species of small neornithine birds from the Late Cretaceous or earliest Paleocene greensands of New Jersey (Hornerstown and Navesink Formations). Marsh (1889, 1892) and Brodkorb (1963a) described 8 more such species in 5 genera from the Late Cretaceous of the

Western Interior of North America. All the material is fragmentary, and most was dissociated when found. Systematic placements have always been problematic. Over the years, these birds were compared to birds as diverse as geese, flamingos, cranes, herons, rails, gulls, snipes, cormorants, pheasants, and turkeys (Marsh, 1880; Hay, 1902; Shufeldt, 1915; Lambrecht, 1933; Howard, 1955; Wetmore, 1956; Brodkorb, 1963a, 1970; Cracraft, 1972, 1973). Each genus was given its own family or subfamily name; the most senior of these is Graculavinae Fürbringer, 1888, for Marsh's (1872) genus *Graculavus* of supposed cormorants.

Olson (1985) focused instead on the similarities among these birds. He included most of them under the heading "Transitional Charadriiformes," to convey their overall resemblance to charadriiforms combined with features also found in gruiforms. Within this assemblage he referred the genera represented in the New Jersey marls to Graculavidae, elevating the rank of that name and including *Graculavus* Marsh, 1872, *Telmatornis* Marsh, 1870, *Palaeotringa* Marsh, 1870, and *Laornis* Marsh, 1870. Olson and Parris (1987) also noted many features of anseriforms in this material and grouped them on the basis of similarity as Form Family Graculavidae. Their intent was not to define a monophyletic group but rather to assemble similar specimens when the evidence for associating them was insufficient. Even with newly referred specimens, material is still not adequate to assess the monophyly of a group Graculavidae.

Among the birds from the New Jersey greensand marls, only *Telmatornis* includes material that is from the Navesink Formation and unquestionably Cretaceous in age (see "Geology"). *Graculavus* and "*Palaeotringa*" *vetus* are also included in this review, because new Cretaceous material of these taxa is known, but the other New Jersey species are from the Hornerstown Formation and are not within the province of this review. Nevertheless, other Cretaceous charadriiformlike birds reviewed here exhibit the same problems of diagnosis. This is true also of material from the Early Paleogene (Nessov, 1992a:473, Fig. 4m–t; Benson, 1999; Boles, 1999b). There are many derived features in the postcranial skeleton of charadriiforms, but none found in material reviewed here is unique to Charadriiformes. All these features are present in other groups of possibly closely related shorebirds and seabirds, often in a further modified form. It might be argued that a unique combination of such derived features is diagnostic for Charadriiformes, but the material reviewed here is too limited to permit such an exception to the guidelines for diagnosis adopted here (see introductory section and "Material and Methods").

For this reason, no diagnosis for Charadriiformes is given. Features suggesting referral are discussed with each species, but all referrals are tentative. Asterisks (*) in the species accounts below flag derived features shared by Charadriiformes with other shorebirds; a double asterisk

(**) indicates derived features present in Charadriiformes and present in a more extreme state in Procellariiformes.

Graculavus Marsh, 1872

Synonym—*Limosavis* Shufeldt, 1915. Inappropriate substitution, Brodkorb (1967).

Included species—*G. velox* Marsh, 1870, type species, designated by Hay (1902); *G. augustus* Hope, 1999.

Revised diagnosis—Derived features of *Graculavus*: *head of the humerus small compared with that of extant charadriiforms; dorsal tubercle very far from the head of the humerus; *dorsal tubercle projected well away from the shaft; a broad, flat surface present between the ventral tubercle and the dorsal tubercle of the humerus; capital incisure of the humerus terminated caudally by a shallow sulcus; *head of the humerus undercut by a transversely oriented sulcus; deep impressions for the fascicles of M. humerotriceps within the pneumotricipital fossa and a raised scar near the base of the ventral tubercle, evidently the site for attachment of the central tendon of the muscle (in part after Olson and Parris, 1987; Hope, 1999).

Remarks—Marsh (1872, 1873c, 1877) described five species of *Graculavus*, considering this a group of cormorants. "*Graculavus*" refers to "*Graculus*," a name formerly used for living cormorants. A sixth species, *G. idahensis*, is an erroneous entry by Shufeldt (1915:68). In his treatise on the toothed birds, Marsh (1880) referred *G. anceps*, *G. lentus*, and *G. agilis* to *Ichthyornis*, presumably because these species came from earlier sediments close to the horizon and locality where *I. dispar* was first discovered in the Western Interior of North America and so would be geographically and stratigraphically distinct. Shufeldt (1915) regarded the material of *G. anceps* and *G. agilis* as indeterminate. It consists of a few bone splinters. He referred *G. lentus* to Galliformes, where it may or may not belong, and to an extant species of grouse, *Pedioecetes* (= *Tympanuchus*) *phasianellus*, where it certainly does not belong. *G. lentus* is listed with Neornithes, incertae sedis. Fürbringer (1888) placed *Graculavus* in his Graculavinae within Phalacrocoracidae. Wetmore (1956) and Brodkorb (1967) continued to list *Graculavus* with Phalacrocoracidae. However, Shufeldt (1915), Lambrecht (1933), and Olson and Parris (1987) agreed that *G. velox* and *G. pumilus* were charadriiforms yet not closely related. Olson and Parris (1987) synonymized *G. pumilus* with *Telmatornis priscus*, and thus *Graculavus* became monotypic until the description of *G. augustus* (Hope, 1999).

Graculavus (Fig. 15.7A) is placed close to or within Charadriiformes based on the following derived features: *humeral impression for the transverse ligament deep; *bicipital prominence and bicipital crest of the humerus large; *pneumotricipital fossa of the humerus not pneumatized; **muscle impressions surrounding the pneumotricipital fossa very rugose; *cranial wall of the pneumo-tricipital fossa very thin, the fossa divided proximodistally by a tumescence that is the obverse side of the sulcus for the transverse ligament; *ventral tubercle of the humerus thin, pointed; **ventral tubercle long; *caudal margin of the humerus dorsal to the ventral tubercle; *caudal margin of the humerus well marked; *humeral attachment for M. latissimus dorsi caudalis situated dorsal to caudal margin (in part after Shufeldt, 1915; Olson and Parris, 1987; Hope, 1999).

Olson and Parris (1987:6) identified three features shared by *Graculavus* and *Telmatornis* that are absent from Burhinidae, a group that appears to be primitive within Charadriiformes: the head of the humerus small, the dorsal tubercle far from the head of the humerus, and the tubercle less projected proximally. In addition, in *Graculavus* and *Telmatornis* the caudal margin deviates abruptly toward the capital tubercle, and the pneumotricipital fossa is oriented directly caudally. These features are also present in the anseriform *Presbyornis*. Such features should be included in any investigation of the phylogenetic positions of *Graculavus* and *Telmatornis*.

A unique combination of so many derived features shared with Charadriiformes might appear to be sufficient for referral to Charadriiformes, where others have placed *Graculavus* (including Hope, 1999). However, each of these features is shared much more broadly with other shorebirds and seabirds. The revised diagnosis here reflects this uncertainty by indicating shared features with asterisks (see above). Features that have figured prominently in previous arguments for the charadriiform assignment of *Graculavus* are the deep sulcus for the transverse ligament, nonpneumatized and double-basined pneumotricipital fossa, long ventral tubercle, lack of pneumatization of the tricipital fossa, and well-marked caudal margin. These features also occur in *Presbyornis* (Anseriformes) and in Oceanitidae (Procellariiformes). The robustness of the caudal margin and its position dorsal to the ventral tubercle are widespread among shorebirds and seabirds.

Other features of the humerus that have appeared to be typically charadriiform need reexamination. The undercut head of the humerus is probably not homologous with the dorsal pneumotricipital fossa present in many charadriiforms (e.g., *Larus*) and in very similar form in some procellariiforms (Oceanitidae). The dorsal pneumotricipital fossa is an excavation for the greatly enlarged dorsal head of M. humerotriceps (Bock, 1962). In contrast, the small fossa that excavates the head of the humerus in *Graculavus* is oriented transverse to the long axis of the humerus, cuts directly under the head of the humerus, and is restricted to that region. This transverse fossa of *Graculavus* resembles a shallower or deeper fossa present in *Presbyornis, Telmatornis,* and some procellariiforms other than Oceanitidae. Thus, this sulcus is not evidence for referral to Charadriiformes.

Resemblance to cormorants is due to the broad, flat surface at the proximal end of the humerus, with the opening of the pneumotricipital fossa in the same plane as this surface, and to the well-defined caudal margin, but there are no other features suggesting this placement.

The two species of *Graculavus* were large birds with thin-walled, highly sculpted bones, no pneumatization of the pneumotricipital fossa, a large bicipital prominence and crest, and very distinct, rugose impressions for flight muscles and shoulder ligaments. These features suggest well-flying, probably far-ranging shorebirds.

G. velox Marsh, 1872

Synonyms—*G. velox* Marsh, 1872; *Limosavis velox* (Marsh), Lambrecht (1933).

Holotype—YPM 855, left humerus, proximal end with a short length of the shaft (illustrated by Shufeldt [1915]; Olson and Parris [1987]).

Locality and horizon—Hornerstown, Monmouth County, New Jersey, Navesink Formation or basal part of the Hornerstown Formation (Baird, 1967), Maastrichtian or earliest Paleocene.

Tentatively referred material—NJSM 11854, right major metacarpal with fragments of the distal symphysis of the carpometacarpus, very worn and abraded; tentatively referred to *G. velox* by Olson and Parris (1987). From the Inversand Company marl pit, near Sewell, Gloucester County, New Jersey, basal layer of the Hornerstown Formation; latest Maastrichtian or earliest Paleocene.

Remarks—*G. velox* was in the size range of moderately large gulls (*Larus*).

G. augustus Hope, 1999

Holotype—AMNH 25223, left humerus, proximal end (Fig. 15.7A).

Locality and horizon—Near Lance Creek, Niobrara County, Wyoming, Lance Formation, UCMP locality V5711, Late Maastrichtian; collected by M. C. McKenna and party, 1985.

Remarks—*G. augustus* was a very large bird, about one-third larger than *G. velox*. It can be differentiated from *G. velox* by the position of the dorsal tubercle much farther from the head of the humerus. This species extends the range of *Graculavus* from the Atlantic to the nearshore of the Western Interior sea of North America.

Telmatornis Marsh, 1870

Included species—Type species only.

Remarks—*Telmatornis* has been compared consistently with rail-like gruiforms or the smaller inshore and upland charadriiforms, especially snipes (*Scolopax*), although *Telmatornis rex* Shufeldt, 1915, is now referred to Anseriformes as *Anatalavis rex* (Olson and Parris, 1987; Olson, 1999). Shufeldt (1915), Cracraft (1972), and Olson and Parris (1987) concurred in the synonymy of *T. affinis* Marsh, 1870, with *T. priscus* Marsh, 1870. Based on a detailed analysis of *T. priscus*, Cracraft (1972) referred *Telmatornis* to Charadriiformes in his new taxon Telmatornithidae. Olson and Parris (1987) also considered it to be charadriiform but placed it in Graculavidae. In addition, they synonymized *G. pumilus* and *P. vetus* Marsh, 1870, with *T. priscus*.

The synonymy of *G. pumilus* with *T. priscus* is based on comparison of adjacent but not quite overlapping fragments of the proximal and distal ends of the humerus; nevertheless, the arguments for synonymy are compelling. However, the synonymy of *P. vetus* with *T. priscus* is questioned here based on a new specimen from the Lance Formation that calls into question the referral to Charadriiformes. This material is discussed herein with Neornithes, incertae sedis.

Telmatornis is placed close to or within Charadriiformes based on the derived features noted above for *Graculavus* and, in addition, the fact that **the shaft of the humerus is almost straight (Olson and Parris, 1987). Muscle attachment sites on the distal end of the humerus agree with those in two groups of charadriiforms (Cracraft, 1972). There is little to add to these analyses, except that referral to Charadriiformes is also suggested by features not present in material of *Graculavus*, including *the recurved flexor process of the humerus, bent toward the olecranon fossa; **a long distal symphysis of the metacarpals; *a deep, well-defined tendinal sulcus of the metacarpus, with a triangular sulcus at its distal end; and **the shortened facet for the minor digit of the manus. *Telmatornis* is distinguished from *Graculavus* by the less undercut head of the humerus; absence of a sulcus at the caudal end of the capital incisure of the humerus; the shorter ventral tubercle of the humerus; position of the dorsal tubercle closer to the head of the humerus; and lesser projection of the dorsal tubercle away from the shaft (Olson and Parris, 1987). All these are less derived states of features present in *Graculavus*. Compared with the condition in many charadriiforms, in *Telmatornis* the shaft of the humerus is shorter and more curved, the brachial fossa shallower, and the ventral epicondyle more widely extended ventrally. All these are less derived states of features found in more extreme form in most charadriiforms, for example, gulls (Laridae), and in procellariiforms. Thus, this best known of the Cretaceous charadriiformlike birds has no feature that is uniquely charadriiform and a large complement of relatively plesiomorphic features.

T. priscus Marsh, 1870

Synonyms—*T. affinis* Marsh, 1870; *G. pumilus* Marsh, 1872.

Holotype—YPM 840, left humerus, distal end (illustrated by Shufeldt [1915]; Olson and Parris [1987]).

Locality and horizon—Cream Ridge Marl Company pits, near Hornerstown, New Jersey, Navesink Formation, Maastrichtian.

Referred material—YPM 845, right humerus, distal end, holotypical material of *T. affinis*, synonymized by Shufeldt (1915). Same locality as the holotype. YPM 850, holotypical material of *G. pumilus*, including the proximal end of the right humerus, distal end of the right carpometacarpus, and other forelimb fragments, apparently associated when collected; from near Hornerstown, Monmouth County, New Jersey (Marsh, 1870); probably from the Navesink Formation (Baird, 1967); Maastrichtian. Synonymized by Olson and Parris (1987). ANSP 15360, left humerus, distal end and shaft; from the Inversand Company marl pit, near Sewell, Gloucester County, New Jersey; MFL of the Hornerstown Formation, latest Maastrichtian or earliest Paleocene (Olson and Parris, 1987).

Identity of YPM 850: Marsh (1870) did not include a specimen number for the holotype of *G. pumilus*, but he gave the locality as indicated above in New Jersey. Shufeldt (1915:19) gave the holotype of *G. pumilus* as YPM 1209 and its collection locality as Kansas, but his discussion and illustration correspond with Marsh's description. YPM 850, which I examined in the collections at Yale, corresponds to the descriptions of both authors and is labeled with the New Jersey locality indicated by Marsh (1870).

Tentatively referred material—NJSM 11900, right ulna, proximal end, from spoil piles near junction of Routes 537 and 539 near Hornerstown, Upper Freehold Township, Monmouth County, New Jersey; presumably from the MFL of the Hornerstown Formation; latest Maastrichtian or earliest Paleocene. NJSM 11853, left tarsometatarsus, distal end and part of shaft, from Sewell, Gloucester County, New Jersey, Inversand Company marl pit, MFL of Hornerstown Formation; both latest Maastrichtian or earliest Paleocene. ANSP 15541, right pedal phalanx 1 of digit II, from Sewell, Gloucester County, New Jersey, Inversand Company marl pit, presumably basal Hornerstown (MFL) or lower strata; Maastrichtian or earliest Paleocene. All referred tentatively by Olson and Parris (1987).

Remarks—*T. priscus* was much smaller than either species of *Graculavus* but similar in size and robustness to the approximately contemporaneous *C. rara*, known from the Lance Formation, Wyoming. The two species share none of the same confidently referred elements, so direct comparisons have been impossible.

Volgavis Nessov and Jarkov, 1989

Included species—Type species only.

V. marina Nessov and Jarkov, 1989

Holotype—ZISP-PO 3638, lower jaw lacking the articular regions (illustrated by Nessov and Jarkov [1989] and by Nessov [1992a:473, Fig. 4m–t]).

Locality and horizon—Belly Jar Valley, Volga Basin, in the region of the Berdeja and Don Rivers, Dubovsky Region, Volgograd District, Russia, Malaja Ivanovka locality, Krasnaja Derevnja Spring, in a greenish quartz glauconite sand 0.3–0.4 m below the Paleocene contact. Late Maastrichtian.

The age is given as Danian by Nessov and Jarkov (1989) and by Tatarinov (1998) in a review of Nessov's (1997) posthumously published review of localities, but Nessov (1992a) gave the age as latest Maastrichtian, and Averianov (1999) has confirmed that specimens from this locality are latest Cretaceous. The environment of deposition was a warm or temperate sea.

Remarks—Placement with Neornithes is justified by fusion of the mandibular symphysis, fusion of the dentary with the surangular, and lack of teeth. Based on its "similarity to primitive charadriiforms," Nessov and Jarkov (1989) and Nessov (1992a) referred *Volgavis* tentatively to Charadriiformes. They also noted the downward orientation of the hooked tip of the mandible, in which it resembles frigate birds (Fregatidae: Pelecaniformes). A slightly downward-hooked mandible is also present in jaegers and skuas (Stercorariidae: Charadriiformes). *Volgavis* was the size of a large gull.

Cimolopteryx Marsh, 1892

Synonyms—*Cimolopteryx* Marsh, 1889 (nomen nudum); *Timolopteryx* Ogilvie-Grant, 1912 (copying error).

Included species—*C. rara* Marsh, 1890, type species (designated by Hay, 1902); *C. maxima* Brodkorb, 1963a; *C. minima* Brodkorb, 1963a; *C. petra*, new species; and another unnamed species discussed below.

Revised diagnosis—Derived features of *Cimolopteryx*: coracoid robust; neck of the coracoid stout and subtriangular in cross section; scapular cotyla of the coracoid slightly elongated transverse to the long axis of the coracoid; lateral process of the coracoid small.

Remarks—Marsh (1889) mentioned *C. rara* in a footnote, but the valid name dates from Marsh (1892), where he specified and illustrated the material and described a further species, *C. retusa*. For the early history of *Cimolopteryx* and subsequent history of *C. retusa,* see *Palintropus* in Galliformes. Brodkorb (1963a) described two new species of *Cimolopteryx* and erected the taxon Cimolopterygidae, in which he also included *Ceramornis major* Brodkorb, 1963a. *Ceramornis* is referred here to Neornithes, incertae sedis.

Cimolopteryx is placed close to or within Charadriiformes based on the following derived but not unique features: **the acrocoracoid process of the coracoid strongly tilted medially at an angle of 60–65° to the humeral facet; *the coracoid massive at the level of the humeral facet, its diameter almost equaling that of the humeral facet; *the coracoidal impression for the acrocoracohumeral ligament short, deep, and ovate; **the humeral facet of the coracoid tilted medially from the axis of the shaft; **the shaft of the coracoid tilted medially in relation to the sternal margin of the coracoid; **the foramen for N. supracoracoideus recessed distally from the scapular facet of the coracoid about one-half the diameter of the facet; and *the foramen for N. supracoracoideus angled in its course through the coracoid, emerging dorsally farther toward the sternal end of the bone.

Other features of *Cimolopteryx* that have suggested the previous referral to Charadriiformes are plesiomorphic (see above). *Cimolopteryx* lacks **ventromedial extension of the clavicular facet of the acrocoracoid, a feature that is present in all extant charadriiforms examined for this review, though minimal in *Burhinus* and *Stercorarius.*

The species of *Cimolopteryx* were very small to moderately large birds close to the size of a small gull.

C. rara Marsh, 1892

Holotype—YPM 1805, left coracoid.

Locality and horizon—Converse (now Niobrara) County, Wyoming, Laramie Deposits, Ceratops Beds (Marsh, 1892; Shufeldt, 1915; = Lance Formation). Collected by J. B. Hatcher, June 1889, exact locality not certain but within the type area of the Lance Formation (Clemens, 1964). Late Maastrichtian.

Referred material—UCMP 53962 (right coracoid, shoulder end) and UCMP 53963 (left coracoid, shoulder end) both from near Lance Creek, Wyoming; UCMP locality V5620, Lance Formation; SMNH P1927.936, left coracoid, shoulder end, from south of Shaunavon, Saskatchewan, Canada, SMNH locality 72F08-0012 (Gryde Locality); tentatively referred by Tokaryk and James (1989), confirmed here from a cast provided by T. Tokaryk. All Late Maastrichtian.

Tentatively referred material—UCMP 53964, left carpometacarpus, distal end, from near Lance Creek, Niobrara County, Wyoming, Lance Formation, UCMP locality V5620. Late Maastrichtian (Brodkorb, 1963a).

A quadrate referred to this species by Brodkorb (1963a) is discussed below with Neornithes, incertae sedis.

Revised diagnosis—As for the genus, and about one-third smaller than *C. maxima*.

Remarks—*C. rara* (Fig. 15.6A) was about the same size as *T. priscus,* known from the Navesink and Hornerstown Formations, New Jersey, and also referred tentatively to Charadriiformes. Although the two species share none of

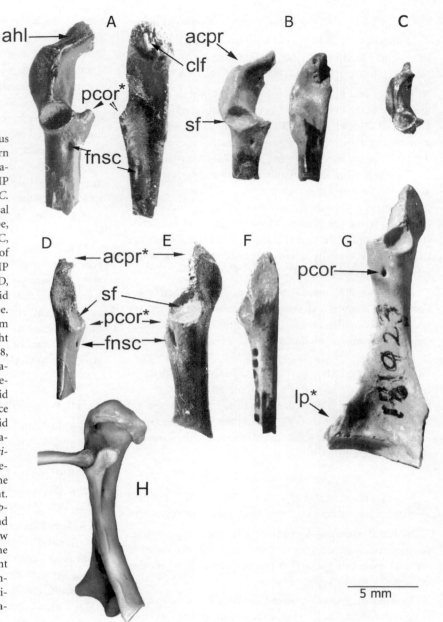

Figure 15.6. The coracoid in Cretaceous charadriiformlike birds and a modern charadriiform. A, *C. rara,* referred left coracoid in dorsal and medial views, UCMP 53962, Lance Formation, Maastrichtian. B, *C. petra,* proximal end of left coracoid in dorsal and medial views, AMNH 21911, holotype, Lance Formation, Maastrichtian. C, *Cimolopteryx,* unnamed sp., proximal end of the left coracoid in dorsal view, RTMP 93.116.1, Judith River Group, Campanian. D, *C. minima,* proximal end of the left coracoid in dorsolateral view, UCMP 53976, holotype. E–G, an unnamed group distinct from *Cimolopteryx.* E, proximal end of the right coracoid in dorsolateral view, ANSP 15868, Bug Creek facies of the Hell Creek Formation, latest Maastrichtian or earliest Paleocene; F, proximal end of the right coracoid in dorsolateral view, AMNH 13011, Lance Formation, Maastrichtian. G, left coracoid in dorsal view, USNM 181923, Lance Formation, Maastrichtian. H, *Stercorarius pomarinus* (Charadriiformes), left coracoid in medial view with the proximal end of the scapula in articulation, CAS 69535, extant. Group E–G is more gracile than *Cimolopteryx* and differs from it in proportions and other details. *C. minima* is similar to the new group but smaller. In most specimens, the acrocoracoid is broken off. In the extant charadriiform *Stercorarius,* note the enlarged and ventromedially extended clavicular facet (compare A with H). Abbreviations as in Figure 15.2.

the same elements in confidently referred material, the carpometacarpus tentatively referred to *C. rara* is similar to that of *T. priscus.* However, the carpometacarpus referred to *C. rara* was not associated in situ with any material comparable to the holotype coracoid. Because another charadriiformlike species of appropriate size occurs in the Lance Formation (Undetermined Species 1, below), the assignment of this carpometacarpus to *C. rara* is very uncertain.

C. maxima Brodkorb, 1963a

Revised diagnosis—About two times the size of *C. rara;* neck of the shaft flatter and broader; the foramen for N. supracoracoideus less recessed from the scapular facet and much less steeply angled in its course than in *C. rara.*

Holotype—UCMP 53973, left coracoid, shoulder end.

Locality and horizon—Near Lance Creek, Niobrara County, Wyoming, Lance Formation, UCMP locality V5711; Late Maastrichtian.

Referred material—UCMP 53957, left coracoid, shoulder end, same locality as the holotype, Late Maastrichtian (Brodkorb, 1963a). ANSP 15867, right coracoid fragment including the distal end of the humeral facet, the scapular facet, and a short length of the shaft, from McCone County,

Montana, Hell Creek Formation, Bug Creek channel facies, probably the Bug Creek West locality; latest Maastrichtian or earliest Paleocene. Matrix collected by R. Sloan, collection date uncertain, donated to Anne C. Dillon and Sidney Hostetter; given by them to ANSP; the specimen recovered from matrix at ANSP.

Remarks—*C. maxima* (Fig. 15.7C, D) was a robust species the size of a small gull. Too little material is preserved in any of the specimens to assess the appropriateness of the referral of this species to *Cimolopteryx*.

C. minima Brodkorb, 1963a

Holotype—UCMP 53976, right coracoid, shoulder end (Fig. 15.6D).

Locality and horizon—Near Lance Creek, Niobrara County, Wyoming, UCMP locality V5003, Lance Formation; Late Maastrichtian.

Measurements—See Table 15.1.

Remarks—*C. minima* is about the same size as the new species *C. petra,* below. For comparisons see that description. Although about one-third smaller, *C. minima* is very similar to specimens listed below as Undetermined Species 1. With more and better material, *C. minima* may prove to belong with these as a new species group.

C. petra, new species

Holotype—AMNH 21911, left coracoid, shoulder end (Fig. 15.6B).

Locality and horizon—Near Lance Creek, Niobrara County, Wyoming, Lance Formation, UCMP locality V5711; Late Maastrichtian. Collected by M. C. McKenna and party, 1983.

Diagnosis—As for *C. rara* but about one-third smaller.

Measurements—See Table 15.1.

Etymology—From Latin *petra*, stone, because of the tiny but robust, rocklike appearance of the specimen.

Description—The holotype includes the glenoid and scapular facets of the coracoid, part of the acrocoracoid process, and perhaps a short fragment of the shaft. The clavicular facet is broken off. Except for its smaller size, the coracoid of *C. petra* is very similar to that of *C. rara* but differs from the coracoid of *C. minima*. The following features differentiate the two. In *C. petra*, the humeral facet is broadest midway between its shoulder and sternal ends, and it tapers gradually toward the two ends, making it more ovate than in *C. minima*. In *C. minima,* the humeral facet remains wide until it tapers abruptly close to the ends. In *C. petra,* the scapular facet of the coracoid is slightly elongated transverse to the long axis of the coracoid. In *C. minima* it is entirely round. In *C. petra,* the shaft is robust, and the neck of the shaft is triangular in cross section; in *C. minima* the shaft is

gracile and flattened dorsoventrally at the neck, making the neck ovate in cross section. In all these features *C. petra* agrees with *C. rara*. Other than size, there do not appear to be any significant differences between *C. rara* and *C. petra*.

Cimolopteryx sp.

Material—RTMP 93.116.1, left coracoid, shoulder end (Fig. 15.6C).

Locality and horizon—Alberta, Canada, Dinosaur Park Formation, RTMP Onefour Car Park site; Middle Campanian (Eberth, 1990).

Measurements—See Table 15.1.

Remarks—This specimen includes the base of the acrocoracoid, the humeral facet, scapular facet, procoracoid process, and a short length of the shaft. Despite its minimal extent, the preservation is excellent, and the specimen includes critical features in the diagnosis of *Cimolopteryx*. However, the material is insufficient as a holotype of a new species. This bird was about one-third smaller than *C. petra* and would have been in the size range of the smallest sandpipers (*Calidris:* Charadriiformes). This material extends the temporal range of *Cimolopteryx* back to the mid-Campanian and the geographic range north to Alberta, Canada.

Undetermined Species 1

Material—USNM 181923, right coracoid; from the Eugene West Ranch, near Lance Creek, Niobrara County, Wyoming, UCMP locality V5622, Lance Formation, Late Maastrichtian; collected 5 September 1977 by F. V. Grady. AMNH 13011, right coracoid, shoulder end; from near Lance Creek, Niobrara County, Wyoming; Lance Formation, UCMP locality V5711, Late Maastrichtian; collected by M. C. McKenna and party, 1985. ANSP 15868, right coracoid, shoulder end and about one-half of the shaft, from McCone County, Montana, Hell Creek Formation. This specimen is from the Bug Creek facies, probably the Bug Creek Anthills locality; whether latest Maastrichtian or earliest Paleocene is uncertain. Matrix collected by R. Sloan about 1985, donated to Anne C. Dillon and Sidney Hostetter and given by them to ANSP; the specimen recovered from matrix at ANSP.

Remarks—Among the new collections of coracoids from the Western Interior, this small series (Fig. 15.6E–G) resembles *C. rara* in size and general form, but these coracoids are more gracile and have other features suggesting that they represent another species complex to which *C. minima* also belongs. The new group differs from *Cimolopteryx* as follows: the coracoid is more gracile; the humeral facet is relatively shorter and less medially tilted; the scapular facet is entirely round; the neck of the coracoid is flatter and more ovate in cross section; the shaft is less medially tilted; and a small foramen is present on the cranial margin of the procoracoid process at its bifurcation for the scapu-

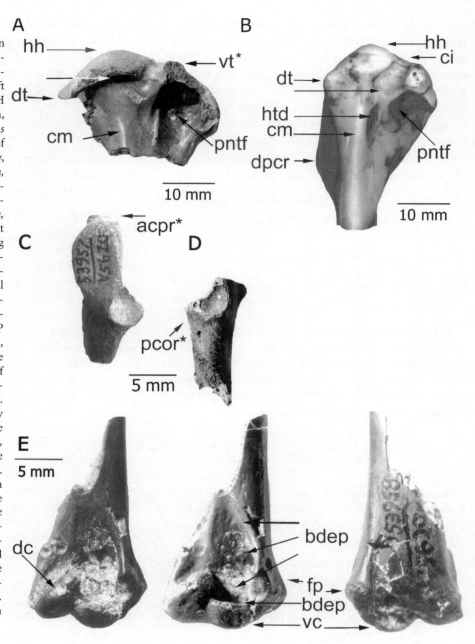

Figure 15.7. Forelimb elements in some Cretaceous birds, with comparison. A, *G. augustus* (?Charadriiformes), proximal end of the left humerus in caudal view, AMNH 25223, holotype, Lance Formation, Maastrichtian. B, *S. pomarinus* (Charadriiformes), proximal end of the left humerus in caudal view, CAS 69535, extant. C, *C. maxima*, proximal end of a referred left coracoid, UCMP 53957, Lance Formation, Maastrichtian. D, *C. maxima*, newly referred fragment of a right coracoid, ANSP 15867, from the Bug Creek facies of the Hell Creek Formation, latest Maastrichtian or earliest Paleocene. E, *T. clemensi*, distal end of the left humerus in two caudal views with differing illumination, and in cranial view, UCMP 53958, holotype, Lance Formation, Maastrichtian. In *Graculavus*, note the large scar at the caudal end of the capital incisure, and in *Stercorarius*, the low ridge in this position. In *Graculavus*, note the transversely oriented sulcus undercutting the humeral head, and in *Stercorarius*, the axially oriented sulcus for the dorsal head of M. humerotriceps. The two are not homologous. In *Graculavus*, the small sulcus at the base of the ventral tubercle may be the site for M. humerotriceps dorsalis. In *Torotix*, note the very narrow ventral border of the brachial depression, its large size, and the very short flexor process. Abbreviations: htd, impression for M. humerotriceps dorsalis; others as in Figure 15.1.

lar facet. The differences are borne out by consistent differences in proportions (Table 15.1). A larger series is needed to diagnose the new group adequately.

Undetermined Species 2

Material—PVL 4730, left tibiotarsus, proximal half (Powell, 1987; Chiappe, 1996a).

Locality and horizon—Salitrál Moreno, Río Negro Province, Argentina, Allen Formation: Maastrichtian.

Remarks—This material is placed tentatively close to or within Charadriiformes based on the following combination of derived features: **cranial and lateral cnemial crests of the tibiotarsus both enlarged; *cnemial crests symmetrical where they join, with no notch in the lateral crest; *cnemial crests projected cranioproximally from the shaft of the tibia at about 135°. The specimen is very similar to the tibia of a present-day long-legged wading charadriiform. Enlarged cnemial crests are known only in the ornithurine lineage. There is a single greatly enlarged cnemial crest in *Patagopteryx* (Chiappe, 1996b). In Hesperornithiformes, both cnemial crests are greatly enlarged, and the two crests are symmetrical where they join, but the lateral crest flares laterally (Marsh, 1880; Martin and Tate, 1976). In *Ichthyornis* the crests are not enlarged. In Neornithes, the crests are usually small, but both crests are conspicuously enlarged in anseriforms, gruiforms, and many shorebirds and seabirds. This distribution suggests that enlargement of cnemial crests has occurred more than once in the ornithurine lineage and is derived within Neornithes. In most neornithines with enlarged crests, the crests are projected proximally at about 135° to the axis of the shaft, and the crests are not symmetrical: the lateral crest is notched near its junction with the cranial crest, and thus it is the smaller of the two. Symmetrical crests occur in presbyornithids, most charadriiforms, procellariiforms, grebes, and loons.

Variations in angle of projection and size of the cnemial crests differentiate these groups. Charadriiforms is the only group that includes birds with symmetrical crests projected cranioproximally at about 135° to the axis of the tibia. In *Presbyornis,* the crests project directly cranially. In auks (Alcidae: Charadriiformes), procellariiforms, loons, and grebes the crests project directly proximally and are further enlarged. In procellariiforms and loons, the crests are twisted so that the intercnemial sulcus is oriented craniolaterally. These conditions in auks, procellariiforms, loons, and grebes appear to be further derived states of the moderately enlarged, symmetrical crests and cranially oriented intercnemial sulcus that occur in many charadriiforms. As a less derived state, the charadriiform condition may have occurred in a common ancestor of one or more of the other groups. Thus, confident referral to Charadriiformes is precluded without more detailed analysis.

Undetermined Species 3

Remarks—J. Case and C. Tambussi (1999) have studied a tarsometatarsus, now in the collections of St. Mary's College, California, recovered from the base of the Cape Lamb strata, Vega Island, Antarctica, below and close to a level geochemically dated at 71 Ma (Early Maastrichtian). They have identified the specimen as charadriiform.

?Procellariiformes Fürbringer, 1888. Albatrosses and petrels

Lonchodytes Brodkorb, 1963a

Included species—Type species only.

Remarks—Brodkorb (1963a) described two species of *Lonchodytes* in his new taxon Lonchodytidae, considering this a group of loons. *L. pterygius* Brodkorb, 1963a, based on the distal end of a carpometacarpus, is not loonlike (Olson and Feduccia, 1980a). This and a distal portion of the shaft of a tarsometatarsus that Brodkorb (1963a) referred with reservations to *L. pterygius* are listed herein with Neornithes, incertae sedis.

Referral of *Lonchodytes* (Fig. 15.8B, F, I) to Procellariiformes is suggested by the following derived but not unique features: (1) the distal intermetatarsal foramen is well proximal to the intertrochlear incisure between metatarsals III and IV; (2) the distal intermetatarsal fora-

men is very large; (3) metatarsal trochlear grooves are very well defined; and (4) the apposed trochlear crests of metatarsals III and IV are longer than the nonapposed crests. Referral to a common lineage of petrels in the broad sense (Oceanitidae, Procellariidae, and Pelecanoididae) is suggested by the following additional derived features: (5) metatarsal trochlea II is very short and retracted far plantarly; (6) trochlea IV is elongated and shifted dorsally to lie almost coplanar with trochlea III (in the same transverse plane); and (7) trochlea IV is closely appressed to trochlea III. *Lonchodytes* is most similar to petrels in Procellariidae (Fig. 15.8F, G), but too little of the shaft is preserved to determine if it was strongly flattened mediolaterally or had other features of that group.

All the foregoing features are highly homoplastic. Feature 4 occurs in charadriiforms (Fig. 15.8A, G), flamingos, presbyornithids, and anatids; features 1–4 occur in auks (Alcidae: Charadriiformes); and features 5–7 are widespread in less extreme form among swimming and wading birds. Features 2 and 5–7 occur in more extreme form in grebes (Podicipediformes; Fig. 15.8H), and all occur in loons, many in a more extreme state. Thus, many of the features listed here are present in birds not closely related to each other but sharing a swimming and diving habit, a subject that has been discussed in detail by Storer (1960). Compared with Procellariidae and with loons (Gaviiformes), *Lonchodytes* and petrels are less modified toward a diving foot form.

L. estesi Brodkorb, 1963a

Holotype—UCMP 53954, right tarsometatarsus, distal end (Fig. 15.8B, F, I).

Locality and horizon—Near Lance Creek, Wyoming, Lance Formation, UCMP locality V5620; Late Maastrichtian.

Remarks—Clearly, *Lonchodytes* is not a loon. The evidence suggests that it derived from the petrel group, but the limited material and homoplasy with other groups of birds preclude a confident referral. Figure 15.8 indicates it was a large bird, perhaps twice the size of the petrel *Puffinus griseus.*

?Diomedeidae Gray, 1840
Undetermined Species

Material—Kurochkin (1995a,b) reported a fragment of a clavicle, which he considered to be from an albatross (Diomedeidae). My brief observation of the specimen was consistent with this determination but did not include detailed comparisons. The specimen is from the Nemegt Formation, Mongolia, probably Early Maastrichtian in age.

Gaviiformes Wetmore and W. D. Miller, 1926.
Loons

Limited diagnosis—Specimens are placed with Gaviiformes based on the following derived features, which in-

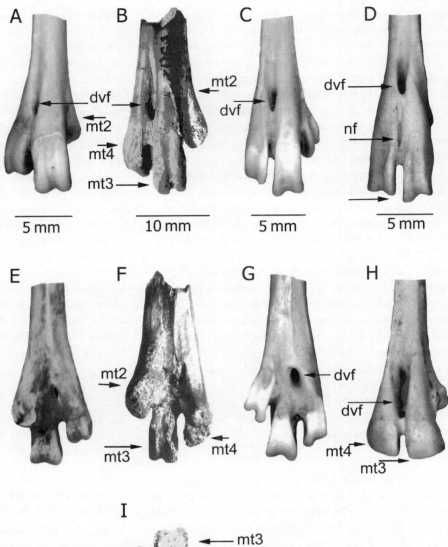

Figure 15.8. The tarsometatarsus in *Lonchodytes,* with comparisons. A and E, *Recurvirostra americana* (Charadriiformes), in dorsal and plantar views, CAS 58390, extant; B, F, and I, *L. estesi* in dorsal, plantar, and distal views, UCMP 53954, holotype, Lance Formation, Maastrichtian; C and G, *Puffinus griseus,* an extant shearwater (Procellariidae), in dorsal and plantar views, CAS 58736; D, *Gavia arctica,* an extant loon (Gaviiformes), in dorsal view, CAS 57859; H, *Aechmophorus occidentalis,* an extant grebe (Podicipediformes), in dorsal view, CAS 71313. *Lonchodytes* compares most closely to the procellariid petrels and shearwaters, but features they share are highly homoplastic in Neornithes. Abbreviations: dvf, distal vascular foramen; mt, metatarsal; nf, nutrient foramen.

clude further derived states of features listed above for Charadriiformes, Procellariiformes, and Procellariidae. As in Charadriiformes and Procellariiformes, both cnemial crests of the tibiotarsus are large and symmetrical. As in Procellariiformes, the cnemial crests project proximally in the axis of the tibial shaft; the intercnemial sulcus is oriented craniolaterally; the distal intermetatarsal foramen is large and well proximal to the intertrochlear incisure; the distal intermetatarsal foramen is very large; metatarsal trochlear grooves are very well defined; and the apposed trochlear crests of metatarsals III and IV are longer than the non-apposed crests. As in Procellariidae, the cnemial crests of the

tibia are further enlarged; the tarsometatarsus is strongly compressed mediolaterally; medial and lateral crests of the tarsometatarsus are well defined; metatarsal trochlea II is strongly shortened and retracted plantarly; and trochlea IV is elongated, shifted dorsally, and closely appressed to trochlea III.

Compared with the condition in Procellariidae, in Gaviiformes the cnemial crests of the tibia are further elongated; the tarsometatarsus is further narrowed; the distal intermetatarsal foramen is much larger and located farther proximally; metatarsal I is lost; metatarsal II is shorter, entirely on the plantar surface, and so strongly compressed mediolaterally that it is flat and bladelike; and metatarsal trochlea IV is slightly longer and situated farther dorsally, with its lateral trochlear crest also slightly longer, although still shorter than the medial crest of trochlea III (Fig. 15.8D).

The following features of the tarsometatarsus of loons differentiate it from that of grebes (Fig. 15.8): a slightly shorter trochlea IV (trochleae III and IV equally long in grebes); the far proximal position of the distal intermetatarsal foramen (far distal in grebes); and the deep incisures between the trochlear crests (incisures absent in grebes). The following features of loons differentiate them from hesperornithiforms: presence of two proximal intermetatarsal foramina and lesser enlargement of metatarsal IV (after Martin and Tate, 1976; Olson, 1992).

Remarks—Other elements not considered here are known in the Cretaceous record, as indicated below.

Neogaeornis Lambrecht, 1929b

Included—Type species only.

N. wetzeli Lambrecht, 1929b

Holotype—GPMK 123, right tarsometatarsus.

Locality and horizon—Southern point of the Tumbes Peninsula, west end of San Vicente Bay, Concepción Province, Chile; Arauco Group; Quiriquina Formation. Campanian-Maastrichtian, following Riccardi (1988). The Quiriquina Formation is a thick intertidal to offshore marine unit of mixed conglomerates, coquina, and sandstones, with an abundant Campanian-Maastrichtian fauna.

Remarks—Lambrecht (1933), Martin and Tate (1976), and Martin (1983) assumed that *Neogaeornis* was close to Hesperornithiformes. After further preparation of the material, Olson (1992) identified *Neogaeornis* as a neornithine bird, based on the presence of two proximal intermetatarsal foramina and indications of ossified hypotarsal ridges, and as a loon based on the shape of the trochleae.

Undetermined Species

Material—Skull and articulated partial skeleton (Chatterjee, 1989, 1997).

Locality and horizon—López de Bertodano Formation, Seymour (Marambio) Island, Antarctica, Campanian-Maastrichtian. The environment of deposition was an offshore marine shelf to nearshore facies (Medina et al., 1989, cited by Chiappe, 1996a).

Remarks—Chatterjee (1997) presented line drawings and a discussion of several skeletal elements but no formal name. This was a large loon very similar to extant *Gavia* (Olson, 1992; pers. obs.). The skull and rostrum are distinctly loonlike. It is of interest that compared with those of extant *Gavia*, the cnemial crests are shorter and more rounded and thus more like those of procellariid petrels and of Tertiary specimens placed in *Colymboides*.

Pelecaniformes Sharpe, 1891. Tropicbirds, pelicans, frigate birds, boobies, anhingas, cormorants

Limited diagnosis—Specimens are placed in Pelecaniformes based on the following feature: iliac facet of the femur entirely convex (unique). In addition, the lack of elevation of the trochanteric crest of the femur differentiates Pelecaniformes from many other birds.

Phalacrocoracidae (Bonaparte, 1853). Cormorants and shags

Limited diagnosis—The specimens are placed in Phalacrocoracidae based on the following derived features: coracoidal tubercle of the scapula absent (also absent in Sulidae, Anhingidae); scapular impression for M. deltoideus minor deep and recurved around the tip of the acromion process; acromion process of the scapula elongate, broad, flat, and recurved at the tip.

Proximal end of the femur very wide, with the medial margin descending directly from the acetabular condyle and the lateral part of the trochanteric crest bowed laterally; shaft of the femur very stout, strongly bowed cranially, strongly narrowed mediolaterally in midshaft, abruptly bent caudally near the distal end, abruptly widened mediolaterally just proximal to the condyles; lateral condyle very large; impression for M. gastrocnemius lateralis a long, very heavy, raised scar situated on the lateral surface of the shaft (unique); finely spaced, diagonally oriented, raised striations present over most of the cranial surface of the femur, interrupted regularly by cross striations, evidently the impressions for fascicles of Mm. femorotibiales; the striations more regularly spaced and more distinct than in the few other birds having such markings (unique).

The scapula resembles only that of Pelecanidae and is differentiated by the following features of pelicans: the acromion process of the scapula projects much farther cranially and dorsally; the scapular impression for M. deltoideus minor is straight and longer, and the coracoidal tubercle is retained. The femur most resembles that of Anhingidae and Plotopteridae. The femur in anhingas is longer and lacks indications of specialization associated with a diving habit (rugose muscle scars; wide proximal end; short, narrow, and strongly bowed shaft, large lateral condyle). The femur in plotopterids is straighter, the trochanter narrower, and the lateral condyle is smaller than in cormorants. For further comparisons see Owre (1967), Hasegawa et al., (1977), and Olson and Hasegawa (1996).

Undetermined Species 1

Material—Scapula in the collection of PIN (Kurochkin, 1995a,b).

Locality and horizon—Nemegt Formation, Mongolia; Late Campanian or Early Maastrichtian.

Remarks—Kurochkin (1995a,b) determined that this specimen came from a very large cormorant. My comparisons with diverse cormorants and other pelecaniforms are consistent with this determination. This bird was larger than any living species of cormorant but near the size of the recently extinct, goose-size Pallas's cormorant, *Phalacrocorax perspicillatus* (= *Compsohalieus perspicillatus,* following Siegal-Causey, 1988). The specimen is not yet described.

Undetermined Species 2

Material—AMNH 25272, right femur (Fig. 15.9A).

Locality and horizon—Near Lance Creek, Niobrara County, Wyoming; Lance Formation, UCMP locality V5711, Late Maastrichtian. Collected 7 July 1985 by M. C. McKenna and party.

Measurements—Shaft mediolateral diameter just distal to the trochanter, 14.1 mm; least midshaft mediolateral diameter, 6.7 mm; depth at the same level, 7.3 mm; iliac facet to base of lateral condyle, 39.5 mm; estimated total length, 47 mm.

Remarks—In the absence of known derived features of the femur in earliest neornithines, the present specimen is referred to Neornithes based on unique features of Phalacrocoracidae. The head of the femur is damaged, and the shaft is broken off just proximal to the condyles, with a fragment remaining that indicates the very large size and lateral orientation of the lateral condyle; part of the popliteal fossa remains. This specimen comes from a bird about one-fifth smaller than the extant cormorant *Phalacrocorax pelagicus*. A small but probably significant difference from extant cormorants is the angular truncation of the trochanteric crest. This feature precludes referral to either of the two major extant lineages of Phalacrocoracidae recognized by Siegal-Causey (1988).

Pelecaniformes?
Torotix Brodkorb, 1963a

Included species—Type species only.

Remarks—Brodkorb (1963a) described *Torotix* as a flamingo in his taxon Torotigidae, which he assigned to Phoenicopteri (= Phoenicopteriformes). Olson and Feduccia (1980a) and Olson (1985) stated that *Torotix* was not a flamingo and suggested that it might have been an early charadriiform. Fragments of two vertebrae together in a vial in the collections of UCMP are accompanied by a note in Brodkorb's hand saying "Same species as humerus?" evidently referring to *Torotix*. This material is referred to *Torotix clemensi* in the UCMP catalog, but Brodkorb did not refer it in any publication. It is discussed here with Neornithes, incertae sedis.

In the absence of unequivocally derived features of the distal end of the humerus in Neornithes, *Torotix* (Fig. 15.7E) is referred to Neornithes based on a feature apparently unique to Pelecaniformes: (1) the ventral rim of the brachial depression of the humerus is very narrow, almost bladelike. Referral to Pelecaniformes is further suggested by the following rare but nonunique derived features that occur in many pelecaniforms: (2) the site for the ventral collateral ligament of the humerus is very small; (3) the flexor process of the humerus is broad, bluntly rounded, and much shorter than the ventral condyle (Sulidae, Pelecanidae); (4) the sites for Mm. flexor carpi ulnaris and pronator profundus are oriented ventrally; and (5) the ventral epicondyle is extended only slightly ventrally. The following further presumably derived feature is present in *Torotix*: (6) the brachial depression of the humerus is very deep and extensive, excavating the humerus almost to the cranial wall and extending distally to excavate the condyles deeply. In extant pelecaniforms, the shallower brachial depression is divided by crests into three or four included fossae. In *Torotix,* crushing obscures the internal contours of the depression.

Feature 2 is absent in some pelecaniforms. Regarding feature 5, the ventral epicondyle is even less extended ventrally in procellariiforms and laridlike charadriiforms. Regarding 6, such deep and extensive excavation of the brachial depression is absent in extant pelecaniforms but is approached in procellariiforms and especially in laridlike charadriiforms. However, in the latter two groups the condyles are not deeply excavated. Possibly, a narrowed ventral epicondyle and a large, deep brachial depression are retained features of a common ancestor.

In flamingos, the ventral rim of the brachial depression is stout; the site for the ventral collateral ligament of the humerus is large; the brachial depression is extensive and subdivided but shallow; and, although the flexor process is broad and blunt, it is only slightly shorter than the ventral condyle.

T. clemensi Brodkorb, 1963a

Holotype—UCMP 53958, right humerus, distal end and part of the shaft (Fig. 15.7E).

Locality and horizon—Near Lance Creek, Niobrara County, Wyoming; Lance Formation, UCMP locality V5620. Late Maastrichtian.

Remarks—At one-half to two-thirds the size of the small cormorant *P. pelagicus, T. clemensi* would have been small for a pelecaniform. The bone has the inflated appearance and rounded contours of the more volant pelecaniforms, rather than the finely sculpted contours of a diving bird.

Psittaciformes? Wagler, 1830. Parrots

Limited diagnosis—Unique features of the mandible: K shape of channels within the bone close to the symphysis;

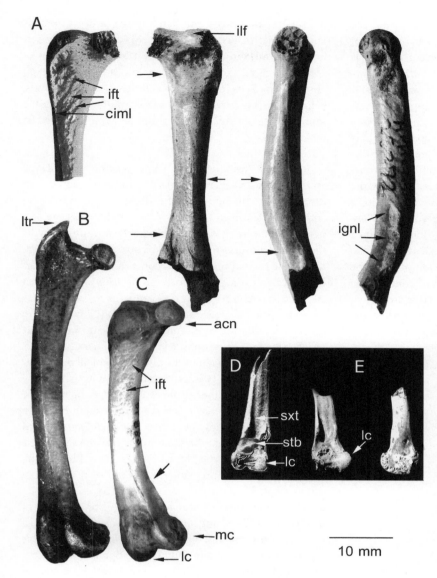

Figure 15.9. Femur and tibiotarsus in some Cretaceous birds, with comparisons. A, undetermined species referred to Phalacrocoracidae, right femur in cranial, caudal, medial, and lateral views, AMNH 29272, Lance Formation, Maastrichtian. B, *E. elegans,* right femur in cranial view, CAS 38670, extant. C, *P. pelagicus,* right femur in cranial view, CAS 68870, extant. In the cormorants, note shape of the shaft (arrows), abrupt broadening at the base of the lateral condyle, elongate impression for M. gastrocnemius lateralis, and prominent impressions for M. femorotibiales. Inset, D and E, distal ends of two left tibiotarsi, probably of the same species, with similarities to Presbyornithidae and Gruformes: D, ANSP 13361, holotype of *P. vetus* Marsh, Hornerstown Formation, Maastrichtian or earliest Paleocene (line editing indicates poorly exposed structures); E, AMNH 25221, Lance Formation, Maastrichtian. Abbreviations: acn, acetabular condyle; ciml, cranial intermuscular line; ift, impressions for Mm. femorotibiales; ignl, impression for M. gastrocnemius lateralis; ilf, iliac facet; lc, lateral condyle; ltr, lateral trochanter; mc, medial condyle; stb, supratendinal bridge; sxt, sulcus for extensor tendon.

dorsal position of entry to the parasagittal canals; very thin bone; tomial crests S-shaped or abruptly truncated at the angle of the mandible; the lower jaw much shorter than the upper jaw (in part after Stidham, 1998c).

Undetermined Species

Material—UCMP 143274, rostrum of the mandible (Stidham, 1998c) (Fig. 15.10A, B).

Locality and horizon—Near Lance Creek, Niobrara County, Wyoming, Lance Formation, UCMP locality V65238; Late Maastrichtian.

Remarks—This specimen lacks the apomorphic features of other beaked archosaurs, thus the fused symphysis is interpreted as a diagnostic feature of Neornithes (Stidham, 1998c). Several of the unique features of the psittaciform lower jaw are due to thinness of the bone and erosion of the

tomial margin. In many neornithine birds, the parasagittal canals enter the lower jaw as large tubes through a stout caudal wall of the symphysis. In parrots, the blade-thin caudal end of the symphysis leaves no space for such entry; instead, the canals enter dorsally through the very thin bone, as in the present specimen. The psittaciform condition is probably a consequence of the extremely thin bone of the entire mandibular rostrum.

The following interpretation comes in part from a conversation with T. Stidham and also from my review of the specimen. The tip of the mandibular rostrum has an unfinished, friable appearance, with spicules of bone protruding, as in parrots. Evidently the entire tomial margin is eroded or incompletely ossified. As a result, the bony part of the lower jaw is much shorter than the upper jaw in all extant parrots. In life, the horny sheath of the bill is continuous with this truncated bone. In the present specimen, the tip of the mandibular rostrum has been broken off, but not quite transverse to the long axis of the jaw. The bone is thinnest at the slightly more distal end of the break, showing that it became thinner toward the tip, as in parrots. It appears that the break severed a protruding tip, which was probably a short tongue-shaped process, such as in many extant parrots (e.g., lories). This tongue-shaped tip is part of a shallow S-shaped profile of the tomial crest in lorylike parrots. The sinuous shape appears to result from only partial excavation of the crest. More extreme excavation of the tomial crest in other parrots such as macaws (*Ara*) results in blunt truncation of the mandible at the angle of the jaw, evidently a more highly derived state. The tongue-shaped process appears to be a remnant of lesser excavation of the rostrum and thus a less highly derived condition than the abrupt truncation in macaws. The extremes of minimal and maximal excavation of the mandible in parrots are shown in Figure 15.10C, D. In addition to the psittaciform diagnostic features noted above, the lower jaw seen in frontal view has a barrel-shaped profile, broader ventrally than at the tomial crests. Such a profile is rare among neornithine birds but is also found in lories.

Thus, the specimen corresponds in several design features to the mandibular rostrum of a lorylike parrot. It probably came from a parrot, but its minimal extent limits confidence in the referral.

Neornithes, incertae sedis
Apatornis Marsh, 1873a

Included species—Type species only.

A. celer (Marsh, 1873a)

Synonym—*Ichthyornis celer* Marsh, 1873a.

Holotype—YPM 1451, partial synsacrum.

Tentatively referred material—YPM 1734, including a coracoid, scapula, partial clavicle, sternum, radius, carpometacarpus, wing phalanges, tibiotarsus, fibula (Marsh, 1880).

Material of the shoulder articulation was examined for this review, including a cast of the referred coracoid (Fig. 15.4A–C). The scapula is evaluated from Marsh's (1880) illustrations and moderately detailed description and from Howard's (1955) comparisons of *Apatornis* with the Early Tertiary presbyornithid *Telmabates*.

Locality and horizon—All from western Kansas, along the Smoky Hill River, Smoky Hill Member of the Niobrara Formation, from a "yellow chalk of the *Pteranodon* Beds" (Marsh, 1880:192); Late Cretaceous.

The Niobrara Formation was formed on the floor of the warm, shallow Cretaceous inland sea of North America. The Smoky Hill Chalk is the upper part of the formation. In Kansas, it spans lower Coniacian through earliest Campanian time (Hattin, 1982), about 89–84 Ma. Unfortunately, yellow beds occur throughout these strata, and *Pteranodon* occurs throughout except at the very lowest levels. However, "Pteranodon Beds" traditionally referred to the lower levels, below the first occurrence of hesperornithiforms. Thus, the minimum age of *Apatornis* is about 84 My, but it may be much older.

Remark—The holotype of *A. celer* is a tiny, crushed synsacrum. It was recovered from the same beds and close to the site where the holotype of *I. dispar* was found. Marsh (1873b:162) described this synsacrum as the holotype of a new species, *I. celer*, distinguishing it from *I. dispar* by its more slender proportions and the "cup of the posterior vertebral face [being] more deeply concave." Soon after, without stating further differences from *Ichthyornis*, Marsh (1873b) referred *I. celer* to a new genus, as *A. celer*. The referred material, cataloged subsequently as YPM 1734, is from the same locality and strata and forms the main body of material of the bird or composite known as *A. celer*.

Questions remain about the association of all this material. The holotypical synsacrum has no counterpart in the referred material, so there is no evidence that the holotype and the later discovered referred material came from the same species. Regarding the referred material, Marsh stated only that it came from the same strata near the site of collection of the holotype (Marsh, 1873b, 1880; Elzanowski, 1995). Thus, there is no evidence that any of this material was associated in situ.

Marsh remained certain about the distinctiveness of *Ichthyornis* from extant birds, but he was also convinced that *Apatornis* was closely related to *Ichthyornis*. Part of his argument for this relationship was stratigraphic co-occurrence, and the rest seems to have been only the presence of biconcave vertebrae. However, the only vertebral remains of *Apatornis* are dorsal vertebrae incorporated into the holotypical synsacrum, and in many neornithine birds the last few dorsal vertebrae are biconcave. Fürbringer (1888) referred *Apatornis* to his taxon Apatornithidae in Ichthyornithes, which he placed near Charadriiformes. The association of *Apatornis* with *Ichthyornis* persisted in most systematic treatments as late as Brodkorb (1967), although Howard (1955:21) had already evaluated the material of *Apatornis* critically and showed its many similarities to flamingos and geese, summarized as follows: "Of the few available bones of *Apatornis*, all except the tibiotarsus have characters which compare favorably with *Telmabates*, although in no instance are the elements identical." She noted that the referred tibiotarsus lacks an ossified tendinal bridge and probably did not belong to the same species as the other material of *Apatornis*. Martin (1983) appears to have assumed that *Apatornis* was closely related to *Ichthyornis*. Olson

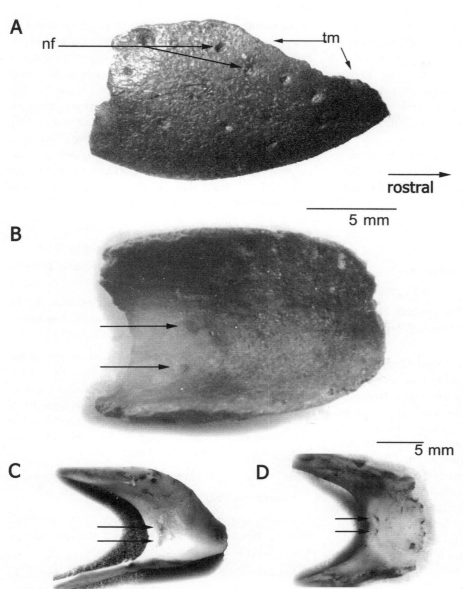

Figure 15.10. Psittaciformes and a parrotlike bird from the Lance Formation. A and B, undetermined species, rostral end of the lower jaw in A, lateral, and B, dorsal views, UCMP 143274, Lance Formation, Maastrichtian. C, D, rostral end of the lower jaw of two extant psittaciforms in dorsal view, showing the lesser and greater extremes of excavation of the tip of the mandible. C, *Domicella garrula*, MVZ 134062; D, *Platycercus adscittus*, CAS 42613. Note the eroded cranial margin in all, and in A and C note the recurved profile of the tomial crest and the tongue-shaped tip of the rostrum. The double arrows point to the dorsal entrances to channels paralleling the symphysis. Abbreviations: nf, nutrient foramen; tm, tomial margin.

(1985) noted that *Apatornis* was similar to the Maastrichtian and Paleogene birds that he characterized as transitional charadriiforms.

Regarding the coracoid and scapula, the humeral and scapular facets are well separated, and the humeral facet of the scapula is small and oriented craniolaterally; if the bones are associated correctly, it is much smaller than the humeral facet of the coracoid. Thus, unquestionably, *Apatornis* belongs with Neornithes. The well-developed acrocoracoid process is consistent with further referral to Neognathae.

A derived feature present in some anseriforms but also in other neornithine birds is the short, angular humeral facet of the coracoid. A presumably derived feature present in *Apatornis* and Anatoidea (Anseriformes) is the long, pointed acromion process of the scapula. Other more cer-

tainly derived features that *Apatornis* shares with birds in Anatoidea are a dorsally protruding distal end of the humeral facet of the scapula with a surrounding shelf tapered distally (Presbyornithidae) and a columnar ventral margin of sulcus M. supracoracoideus of the coracoid, increasing in diameter toward the clavicular facet (Presbyornithidae, Anatidae). The columnar shape of the ventral margin is less pronounced than in Anatoidea. The clavicular facet of the coracoid is neither undercut, as in presbyornithids, flamingos, and the charadriiform *Burhinus*, nor eroded, as in Anatidae. Thus, the coracoid lacks detailed features that would place it securely within Anseriformes or any other group. The scapula corresponds closely with that of Presbyornithidae in all features described and illustrated by Marsh (1880). However, neither Marsh (1880) nor Howard (1955) noted in *Apatornis* the deep, conspicuous scar for M. serratus cranialis present in Presbyornithidae (Fig. 15.5D, F, I). When Ericson's (1999) criteria is used for differentiating *Presbyornis* from the very similar Eocene flamingo *Juncitarsus,* the scapula compares more closely with that of *Presbyornis.* In *Juncitarsus* the acromion process of the scapula is smaller, with a straight lateral margin.

In summary, at least part of the material of *Apatornis* is neornithine and part of an early radiation of Neognathae, possibly close to Anseriformes. *Apatornis* needs a thorough review, including records of its collection.

Ceramornis Brodkorb, 1963a

Included species—Type species only.

C. major Brodkorb, 1963a

Holotype—UCMP 53959, right coracoid, shoulder end.

Locality and horizon—Near Lance Creek, Niobrara County, Wyoming, Lance Formation, UCMP locality V5620; Late Maastrichtian.

Remarks—Brodkorb (1963a) described *Ceramornis* in Cimolopterygidae, which he referred to Charadriiformes. *Ceramornis* was close to the size of a moderately large gull. As in charadriiforms, the humeral facet is large and ovate, and the scapular cotyla is large and rounded. *Ceramornis* lacks any derived features that suggest referral to Charadriiformes. The foramen for N. supracoracoideus is not recessed from the scapular facet; neither the humeral facet nor the shaft is tilted medially. The acrocoracoid process is massive in the region of the humeral facet, and the impression for the acrocoracohumeral ligament is short and deep, but these features, which are present in charadriiforms, are shared with large gruiforms and are too few to suggest a definite referral.

Elopteryx Andrews, 1913

Included species—Type species only.

E. nopcsai Andrews, 1913

Holotype—Left femur, proximal end with a short length of the shaft, identity uncertain but probably BMNH A1234.

Paratype—Left femur, proximal end and about two-thirds of the shaft, probably BMNH A1233.

Locality and horizon—Sinpetru Beds, Szentpeterfalva, Hateg Basin, Transylvania (Romania); Maastrichtian. The environment of deposition was a braided river system on a piedmont, apparently grading into a delta. The age is established as Maastrichtian by superposition on the Campanian and by pollen content (Grigorescu, 1983).

Remarks—Andrews (1913) discussed two femora, implying that he thought they came from the same species of very large cormorant. He did not specify the holotype and gave no specimen numbers, but he included a very detailed description and illustration that correspond closely with the femur of Phalacrocoracidae. In subsequent papers the identification of this and other material referred to *Elopteryx,* and the associated specimen numbers, have been inconsistent or controversial (Lambrecht, 1929a, 1933; Harrison and Walker, 1975; Brodkorb, 1978; Grigorescu and Kessler, 1980; Elzanowski, 1983; Martin, 1983; Olson, 1985; Kessler, 1987). Pending clarification, no comment on the material of *Elopteryx* will be given here, except as indicated in the following section, "Revisions to the Mesozoic Record of Neornithes."

Undetermined Species 1

Material—Specimen in the collections of NSMJ including a scapula, synsacrum, pelvis, fibula, and pedal phalanx, associated in a nodule.

Locality and horizon—Wakkanai, Hokkaido, Japan; from the Campanian section of the Orannai Formation, about 75 Ma.

Remarks—The specimen was referred to Neornithes based on the presence of an obturator foramen. The depositional environment, the thick walled pedal phalanx, and resemblances to several groups of pelecaniforms suggest that it was a marine bird. This material is documented in a masters of science thesis by Y. Kakegawa, University of Tokyo, Japan, completed in 1999 (Y. Kakegawa, pers. comm.).

Undetermined Species 2

Material—UCMP 53969, left quadrate.

Locality and horizon—Near Lance Creek, Niobrara County, Wyoming. Lance Formation, UCMP locality V5620; Late Maastrichtian.

Remarks—Brodkorb (1963a) referred this specimen to *C. rara* in Charadriiformes, but it was not collected in association with other specimens of *C. rara,* and it is not from a charadriiform. Referral to Neornithes is indicated by the small but distinct intercapitular incisure between the squamosal and otic heads of the otic process. Although there are separate squamosal and prootic articulations in some non-neornithine birds (Witmer, 1990; Elzanowski and Galton, 1991), an intercapitular incisure is unknown in Ornithurae except in Neornithes. The incisure is small and shallow, as in a variety of probably near-basal groups of neornithine birds (e.g., galliforms, some anseriforms [*Presbyornis*], some columbiforms, turnicids). The otic process is very tall. The remnant base of the orbital process shows that it was a broad, thin, triangular flange. The mandibular end is very narrow, with a single, transverse condyle, slightly divided by an indistinct sulcus into a very small medial and a slightly larger lateral lobe. There is no caudal condyle. There is a small but distinct pterygoid condyle well separated from the medial condyle. Many of these features occur outside Neornithes and may be plesiomorphic within the group. The intercapitular incisure is probably a more derived state than the undivided head of the otic process in Palaeognathae, but a less derived state than the wide intercapitular incisure of most neornithine birds. In charadriiforms, as in many other neognaths, the squamosal and otic capitula are well separated, the otic process is shortened, the orbital process is elongate and rodlike, the medial and lateral mandibular condyles are large and well separated, and there is a caudal condyle. This specimen is currently under study by Elzanowski and his associates.

Undetermined Species 3

Material—UCMP 53960, two fragments, probably of the same cervical vertebra.

Locality and horizon—Wyoming, Niobrara County, near Lance Creek; Lance Formation, UCMP locality V5620. Late Maastrichtian.

Remarks—The body of the vertebra (or vertebrae) is long and slender; articular facets are similar to those of some neornithine swimming and diving birds. See the remarks above with *Torotix* regarding the specimen.

Undetermined Species 4

Material—RTMP 93.19.1, right coracoid, shoulder end; acrocoracoid process missing.

Locality and horizon—Alberta, Canada, Steveville Railroad Grade; Judith River Formation, Campanian. Collector W. Marshall.

Remarks—As far as preserved, this coracoid is similar to that of *Apatornis* but slightly larger, and the foramen for N. supracoracoideus is situated more distally and medially, so that it is partly emarginate along the medial margin of the procoracoid process. The procoracoid process is deeply incised medially.

Undetermined Species 5

Material—UCMP 53961, left carpometacarpus, distal end, holotypical material of *L. pterygius* Brodkorb, 1963a.

Locality and horizon—Near Lance Creek, Niobrara County, Wyoming; Lance Formation, UCMP locality V5620. Late Maastrichtian.

Remarks—The material of *L. pterygius* was considered undeterminable by Olson and Feduccia (1980a). It is not from a loon, nor was it collected in association with *L. estesi*. The surface of the bone is ceramized, obscuring contours, but the facets for the digits are similar to those of larine charadriiforms, and this specimen may be identifiable if more complete or less worn material is found.

Undetermined Species 6

Material—ANSP 13361, left tibiotarsus, distal end, holotype of *P. vetus* Marsh, 1870 (Fig. 15.9D); AMNH 25221, left tibiotarsus, distal end (Fig. 15.9E).

Locality and horizon—ANSP 13361 from Arneytown, New Jersey, basal layer of the Hornerstown Formation; latest Maastrichtian or earliest Paleocene; AMNH 25221 from near Lance Creek, Niobrara County, Wyoming, Lance Formation, UCMP locality V5711; Late Maastrichtian. Collected by M. C. McKenna and party, 1985.

Remarks—The earliest reviewer of ANSP 13361 (Morton, 1834) compared it most closely with snipes (*Scolopax*: Charadriiformes), as have several subsequent reviewers. Because of many differences from the type species, Olson and Parris (1987) removed this material from *Palaeotringa* and synonymized *P. vetus* with *T. priscus,* which they considered an early charadriiform. Except for being very slightly smaller, the new specimen from Wyoming is almost identical to the first and complements it with a more complete lateral condyle. This calls attention to substantial differences from Charadriiformes. The species name "*vetus*" is available for these tibiotarsi, but the material is insufficient to establish a new genus.

The New Jersey specimen has been fractured and mended, probably with slight displacement of parts in such a way that the intercondylar sulcus appears to be narrower than it actually is. If these specimens are correctly associated, both condyles were small, the intercondylar sulcus was wider than in charadriiforms, and the lateral condyle was a narrow, laterally extended flange. The extensor canal is narrow, very distinctly defined, and placed far medially. Proximal and distal remnants of the channel for M. peroneus brevis suggest that it coursed distally from the cranial surface onto the lateral surface of the tibiotarsus. The supratendinal bridge of the distal end of the tibia is long and exits as an ovate, horizontally elongated opening onto the supracondylar shelf.

Lateral excursion of the lateral condyle is a rare condition found in some gruiforms, anseriforms, and procellariiforms. The complete combination of features found in these specimens occurs in cranes (Gruidae: Gruiformes). The Early Tertiary Messel rails are similar (Idiornithidae: Gruiformes: Mourer-Chauviré, 1992: Figs. 25–28), and except for a shorter tendinal bridge, so is the presbyornithid *Telmabates* (Howard, 1955: Fig. 7). Charadriiforms have a wide, more centrally placed and less well defined tendinal sulcus, a narrower intercondylar sulcus, and larger condyles, with the lateral condyle much larger than the medial condyle and not extended laterally. These features contraindicate referral of the present specimen to Charadriiformes or to *T. priscus*. The occurrence of such similar specimens in both the Lance and the Hornerstown Formations is of interest.

Undetermined Species 7

Material—YPM 1796, left tarsometatarsus, distal end, holotype of *G. lentus* Marsh, 1877 (Shufeldt, 1915: Fig. 127).

Locality and horizon—Near McKinney, Texas; Austin Chalk; Late Cretaceous.

Remarks—The subequal length and short, almost coplanar trochlea, as well as the broad, well-defined intermetatarsal sulcus, suggest a cormorant, as Marsh (1877) evidently thought. Marsh (1880) referred this species to *Ichthyornis*, presumably because it came from earlier strata close to the site where *Ichthyornis* was discovered in Kansas, rather than the Hornerstown locale of *G. velox*. Shufeldt's (1915) photograph indicates no particular morphological reason for the association with *Ichthyornis*. Shufeldt was certain that this species came from the sage grouse of North America, *Pedioecetes* (= *Tympanuchus*) *phasianellus*. In view of its great age, this determination needs reinvestigation.

Neornithes?

Undetermined Species

Material—PO 3434a, left tarsometatarsus, distal end, the trochleae broken off at the bases.

Locality and horizon—Central Kizylkum Desert, Navoi District, Uzbekistan, Dzhyrakuduk locality. Yellowish sandstones of the middle part of the Bissekty Formation, presumably locality CBI-14; Coniacian (Nessov, 1992a: Figs. q–s).

Remarks—Referral to Neornithes is suggested by the long distal intermetatarsal bridge. Uniquely, the third metatarsal is very large and the fourth very narrow.

Revisions to the Mesozoic Record of Neornithes

Palaeognathae

Five birds from Early Cretaceous strata have been referred previously to Palaeognathae: *Ambiortus*, *Otogornis*, *Gansus*, *Palaeocursornis*, *Wyleyia*. For *Ambiortus*, *Otogornis*, and *Palaeocursornis*, the evidence for this placement is insufficient; *Gansus* probably is not neornithine (see the section "Early Cretaceous Birds with Neornithine Similarities" herein). Kurochkin (1995b) thought that *Wyleyia valdensis* Harrison and Walker, 1973, was a palaeognath, but Brodkorb (1978) and Olson (1985) have suggested it might not be avian. The material is probably too damaged to assess and was not reviewed here. From the Late Cretaceous, *Gobipteryx minuta* Elzanowski, 1974, is based on a small, crushed skull, which Elzanowski initially compared to Palaeognathae, and some tentatively referred bird embryos from the same locality (Elzanowski, 1981). Elzanowski (1983) placed *Gobipteryx* separately from Neornithes in its own group, Gobipterygiformes. Martin (1983) referred *Gobipteryx* to Enantiornithes (see Chiappe and Walker, Chapter 11 in this volume). *Patagopteryx deferrariisi* Alvarenga and Bonaparte, 1992, was described as a member of Palaeognathae, but Chiappe (1992a, 1995a, 1996b, Chapter 13 in this volume) has analyzed it as a flightless bird on a stem lineage of Ornithurae. Thus, there are no confirmed records of Palaeognathae in the Cretaceous.

Galliformes

See *G. lentus* below, with Pelecaniformes.

Anseriformes, Flamingos

Lambrecht (1931) described *Gallornis straeleni* as an anseriform, but Brodkorb (1963a) considered it close to flamingos and referred it to the extinct taxon Scaniornithidae. It may be neornithine, but the material is too limited to assess with confidence. Brodkorb (1963a) described *T. clemensi* as a flamingo in his taxon Torotigidae, but herein it is tentatively referred to Pelecaniformes. Lambrecht (1933) described *Parascaniornis stensioi* in his Scaniornithidae, placing this group close to flamingos in Ciconiiformes. *Parascaniornis*, however, is based on the vertebra of a small Asiatic hesperornithiform (Nessov and Prizemlin, 1991; and see Lambrecht's [1933:336] illustration). This eliminates the Mesozoic record for flamingos.

Gruiformes

Nessov and Borkin (1983) compared the Early Cretaceous *Horezmavis* to gruiforms, and Kurochkin (1995b) referred it more conclusively to that group. As far as preserved, its features are consistent with this placement insufficient for confident referral (see the section "Early Cretaceous Birds with Neornithine Similarities" herein).

Charadriiformes

Gansus has sometimes been assigned to Charadriiformes, but the intent of its describers (Hou and Liu, 1984) has been overinterpreted. They placed it provisionally close to the base of a broad radiation of neornithine shorebirds and seabirds plus *Ichthyornis*; assessment in this chapter suggests *Gansus* is probably an ornithurine but probably not a neornithine; see also Zhou and Hou, Chapter 7 in this volume. *P. vetus* has been considered a rail or a charadriiform. Recently, Olson and Parris (1987) synonymized it with *T. priscus* in Charadriiformes. Based on new material, it is placed herein with Neornithes, incertae sedis. *Graculavus*, *Telmatornis*, *Cimolopteryx*, and *Ceramornis* have been considered charadriiform by several authors, and *Cimolopteryx* has been referred to Passeriformes, evidently by printer's error. The first three taxa are here considered only tentatively referable to Charadriiformes because of fundamental problems in the diagnosis of that group, discussed below, and *Ceramornis* is assigned to Neornithes, incertae sedis. *C. retusa* has been referred to *A. retusus* in Ichthyornithiformes and to a new genus as *P. retusus* in Charadriiformes; *Palintropus* is here placed close to or within Galliformes.

Gaviiformes, Podicipediformes

Hesperornithiformes has been linked in early systematic treatments to the neornithine grebes and loons (Lambrecht, 1933; Brodkorb, 1963b). Storer (1960) concluded that all three groups originated separately. Martin and Tate (1976) showed that *Baptornis* was not a grebe, and Elzanowski and Galton (1991) showed that *Enaliornis* was not neornithine. Although Cracraft (1982) resurrected the idea of a close relationship of *Hesperornis* with loons and grebes, he was soon persuaded otherwise (Cracraft, 1986). Brodkorb (1963a) described *L. estesi* as a loon in Gaviiformes, but it is not a loon (Olson and Feduccia, 1980a). It may be closely related to Procellariiformes. The material of *L. pterygius* is probably neornithine but indeterminate within the group (Olson and Feduccia, 1980a).

Pelecaniformes

Since it was first described as a group of cormorants, *Graculavus* has been referred to Charadriiformes by several authors but is here referred only tentatively to that group, and more as a taxonomic convenience than as a well-supported conclusion, as is discussed below. *G. lentus* has been referred to Galliformes but is here considered incertae sedis within Neornithes. *G. agilis* and *G. anceps* have been referred to *Ichthyornis* but are based on indeterminate material. *E. nopscai* was described as a cormorant. Previous reviews have questioned the taxonomic provenance of the specimens referred to this taxon, and they also reflect possible confusion among specimens, which needs clarification.

Strigiformes

Heptasteornis andrewsi Harrison and Walker, 1975, and *Bradycneme draculae* Harrison and Walker, 1975, are based on redescribed tibiotarsi originally referred to *E. nopcsai*. These tibiotarsi do not appear neornithine, and several authors have thought they are not avian (Brodkorb, 1978; Elzanowski, 1983; Martin, 1983; Olson, 1985).

Coraciiformes, Piciformes

Alexornis antecedens Brodkorb, 1976, was described as a coraciiform-piciform ancestor. Elzanowski (1983) separated it from neornithines, and Martin (1983) identified it as an enantiornithine (see Chiappe and Walker, Chapter 11 in this volume).

Passeriformes

The only Mesozoic referral to this group was a printer's error: see *Cimolopteryx*, above.

Neornithes, No Placement

Kessler and Jurcsak (1986) considered the Early Cretaceous *Eurolimnornis* to be a neornithine bird, as did Kurochkin (1995b). The evidence for such referral is inconclusive (see the section "Early Cretaceous Birds with Neornithine Similarities" herein).

Age Uncertain, Previously Considered Mesozoic

The following previously reported specimens are known only from the Bug Creek channels of the Hell Creek Formation or the basal part of the Hornerstown Formation. They are excluded from this review because of uncertainty about the age of specimens from these strata (see "Geology"): *A. rex* (Shufeldt, 1915) (= *T. rex*), previously considered charadriiform, recently referred to Anseriformes (Olson, 1999); *G. velox* Marsh, 1870; *Laornis edvardsianus* Marsh, 1870; *Palaeotringa littoralis* Marsh, 1870; *P. vagans* Marsh, 1872; *Laornis* and *Palaeotringa* previously considered gruiform; all previously considered charadriiform (Marsh, 1870; Shufeldt, 1915; Wetmore, 1956; Cracraft, 1973; Olson and Parris, 1987); *Tytthostonyx glauconiticus* Olson and Parris, 1987; ANSP 15713; and NJSM 11302, all referred by Olson and Parris (1987) tentatively to Procellariiformes; and UCMP 117598 and UCMP 117599, considered respectively to be a bird of graculavid aspect and a bird close to Anseriformes (Elzanowski and Brett-Surman, 1995).

Summary and Discussion

The Record through Time

None of the reports of Early Cretaceous birds previously referred to Neornithes is accepted here, not because the morphology rules out this placement but because the evidence is ambiguous or insufficient. An early neornithine bird might be difficult to place from postcranial material alone if it was still very little differentiated from the common ancestor of all Ornithurae. Because the time of origin of Neornithes is fundamentally important for understanding the evolutionary history of birds, these and other early ornithurine birds need careful attention beyond what has been possible here, using advanced imaging and histological techniques. The subsequent record is summarized in Table 15.2.

Coniacian, Santonian. The earliest unquestionable record of Neornithes is included in the collection of material referred to *Apatornis*. The age is not closely constrained by the published records of collection but is between about

TABLE 15.2
The Mesozoic record of Neornithes

Taxon	Elements	N	Region	Formation	Age	Principal References
?Galliformes						
P. retusus	Coracoid	1	N. America	Lance	Maastrichtian	Brodkorb (1963a)
Palintropus sp.	Coracoid	4	N. America	Dinosaur Park	Campanian	
	Scapula	1	N. America	Foremost	Campanian	
Anseriformes						
Presbyornithidae						
Undet. sp. 1	Postcranial, misc.	2	Antarctica	López de Bertodano	Campanian–Maastrichtian	Noriega and Tambussi (1995)
Undet. sp. 2	Sternum	1	N. America	Lance	Maastrichtian	
Undet. sp. 3	Scapula	2	N. America	Lance	Maastrichtian	Shufeldt (1915); this chapter
Undet. sp. 4	Scapula	1	N. America	Lance	Maastrichtian	
?Charadriiformes						
G. augustus	Humerus	1	N. America	Lance	Maastrichtian	Hope (1999)
T. priscus	Humerus, Carpometacarpus, ulna	3	N. America	Navesink	Maastrichtian	Marsh (1870); Shufeldt (1915); Cracraft (1972); Olson and Parris (1987)
V. marina	Dentaries, surangular	1	Asia	Volgograd District*	Maastrichtian	Nessov and Jarkov (1989); Averianov (1999)
C. rara	Coracoid	3	N. America	Lance	Maastrichtian	Marsh (1892); Brodkorb (1963a)
	Coracoid	1	N. America	Frenchman	Maastrichtian	Tokaryk and James (1989)
C. rara?	Carpometacarpus	1	N. America	Lance	Maastrichtian	Brodkorb (1963a)
C. petra	Coracoid	1	N. America	Lance	Maastrichtian	
C. minima	Coracoid	1	N. America	Lance	Maastrichtian	Brodkorb (1963a)
C. maxima	Coracoid	2	N. America	Lance	Maastrichtian	Brodkorb (1963a)
Cimolopteryx sp.	Coracoid	1	N. America	Dinosaur Park	Campanian	
Undet. sp. 1	Coracoid	2	N. America	Lance	Maastrichtian	
Undet. sp. 2	Tibiotarsus	1	S. America	Allen	Maastrichtian	Powell (1987); Chiappe (1996a)
Undet. sp. 3	Tarsometatarsus	1	Antarctica	Cape Lamb strata	Campanian	Case and Tambussi (1999)
?Procellariiformes						
L. estesi	Tarsometatarsus	1	N. America	Lance	Maastrichtian	Brodkorb (1963a)
Undet. sp.	Furcula	1	Asia	Nemegt	Maastrichtian	Kurochkin (1995a,b)
Gaviiformes						
N. wetzeli	Tarsometatarsus	1	S. America	Quiriquina	Maastrichtian	Lambrecht (1929b); Olson (1992)
Undet. sp.	Skull, postcranial	1	Antarctica	López de Bertodano	Maastrichtian	Chatterjee (1989, 1997)
Pelecaniformes						
Phalacrocoracidae						
Undet. sp. 1	Scapula	1	Asia	Nemegt	Maastrichtian	Kurochkin (1995a,b)
Undet. sp. 2	Femur	1	N. America	Lance	Maastrichtian	
?Pelecaniformes						
T. clemensi	Humerus	1	N. America	Lance	Maastrichtian	Brodkorb (1963a)
?Psittaciformes						
Undet. sp. 1	Dentaries	1	N. America	Lance	Maastrichtian	Stidham (1998c)
Neornithes, incertae sedis						
A. celer	Synsacrum and Postcranial	1	N. America	Niobrara	Coniacian–Early Campanian	Marsh (1873a,b, 1880); Howard (1955)
C. major	Coracoid	1	N. America	Lance	Maastrichtian	Brodkorb (1963a)
E. nopcsai	Femur	1	Europe	Sinpetru Beds	Maastrichtian	Andrews (1913); Lambrecht (1929a, 1933)
Undet. sp. 1	Postcranial, misc.	1	Japan	Orannai	Campanian	Kakegawa (1999) (unpubl. M.Sc. thesis)
Undet. sp. 2	Quadrate	1	N. America	Lance	Maastrichtian	Brodkorb (1963a)

(continued)

TABLE 15.2 (continued)

Taxon	Elements	N	Region	Formation	Age	Principal References
Neornithes, incertae sedis						
Undet. sp. 3	Vertebra	1	N. America	Lance	Maastrichtian	Brodkorb (1963a)
Undet. sp. 4	Coracoid	1	N. America	Judith River	Campanian	
Undet. sp. 5	Carpometacarpus	1	N. America	Lance	Maastrichtian	Brodkorb (1963a); Olson and Feduccia (1980a)
Undet. sp. 6	Tibiotarsus	1	N. America	Lance	Maastrichtian	Morton (1834); Marsh (1870); Olson and Parris (1987)
Undet. sp. 7	Tarsometatarsus	1	N. America	Austin Chalk	Late Cretaceous	Marsh (1877); Shufeldt (1915)

Note: The listing does not include possibly Tertiary specimens of the included species. N = number of specimens.
* See text for notes on the source of the specimen.

84 and 89 Ma, probably toward the earlier end of this range. The holotype is a very minimal, worn, broken, and badly repaired synsacrum; the referred material is probably a composite, and it may be that none of it came from the same species as the holotype. This material urgently needs review, including records of its collection. Nevertheless, the material includes parts of one or more neornithine birds with similarities to geese and flamingos, as was established earlier by Howard (1955). Phylogenetic evaluation of shoulder elements here confirms that at least some of the material of *Apatornis* was neornithine, with derived features suggesting a possible relationship with anseriforms. Most significant, the features that suggest referral to Anseriformes place *Apatornis* with Neognathae. This sets the separation of Palaeognathae and Neognathae to at least 84 Ma.

Campanian, Maastrichtian. *Palintropus* was a polymorphic species or group of species present in North America from mid-Campanian through Late Maastrichtian time. *Palintropus* is closely related to Early Tertiary Quercymegapodiidae. These birds are closely related to Galliformes, possibly as stem-derived lineages of the extant group. Specimens of *Palintropus* from the Campanian in Alberta, Canada, date the galliform lineage to at least 74–75 Ma.

Presbyornithid anseriforms are reported from the Late Campanian through Maastrichtian of Antarctica (Noriega and Tambussi, 1995), North America (this study), and in an unconfirmed report from the probably Late Campanian Barun Goyot Formation, Mongolia (Kurochkin, 1995a). New specimens reviewed here come from at least three species of presbyornithids differing in size, from the Late Maastrichtian Lance Formation, Wyoming. These birds are very similar to *Presbyornis*. Presence of Presbyornithidae in-

dicates that its sister group, the extant Anatidae, as well as earlier branching anseriforms, Anhimidae and Anseranatidae, had already differentiated. This implies that anseriforms were ecologically diverse before the end of the Cretaceous (Stidham, 1998b). Presbyornithids very similar to those in the latest Maastrichtian are present in the Early Paleocene and are well known from later Paleogene sediments (Howard, 1955; Feduccia and McGrew, 1974; Olson, 1994; Benson, 1999).

The record of birds closely related to charadriiforms begins in the Campanian of Alberta at about 75 Ma, with a very fragmentary specimen of a tiny, undescribed species of *Cimolopteryx*. A geochemically dated record slightly older than 71 Ma (Early Maastrichtian) is based on a tarsometatarsus from the Antarctic (Case and Tambussi, 1999). An unconfirmed report of fragmentary remains identified as graculavid comes from the Early Maastrichtian Nemegt Formation of Mongolia (Kurochkin, 1995a). By the end of the Cretaceous, such birds were abundant, with records from western Eurasia (Nessov and Jarkov, 1989), South America (Chiappe, 1996a), and many records from North America (Table 15.2).

It is important to restate that, although some of these birds are very similar to charadriiforms today, none that I have seen is clearly referable to the group. The ambiguity is due only in small part to the fragmentary nature of the material. The larger problem is the difficulty of diagnosing Charadriiformes, as is discussed further subsequently.

Cimolopteryx is the most diverse and abundant group of closely related neornithine species known from the Mesozoic. The five species currently assigned to the group ranged in size from the smallest of sandpipers to a moderately large gull. *Cimolopteryx* extended from mid-Campanian through Late Maastrichtian and from Alberta through Wyoming

(Marsh, 1892; Brodkorb, 1963a; Tokaryk and James, 1989; this study). Newer collections of coracoids from these regions indicate the presence of another similar group of species, not yet named. Both species groups are represented in the Bug Creek channels of the Hell Creek Formation; thus, they extended at least to the terminal Cretaceous. Two coracoids from the Paleocene of Asia, illustrated by Nessov (1992a: Fig. 5r, s), appear to be referable to *Cimolopteryx* and further suggest that this was a widespread, long-lived group. *T. priscus* on the paleo-Atlantic shore of North America was contemporaneous with *C. rara* and probably very similar to it, but no direct comparisons are possible because known material of the two species does not include the same bony elements.

The very large charadriiformlike bird *G. augustus* was present in the Lance Formation. Its congener *G. velox* was present on the paleo-Atlantic, but is known only from the questionably latest Maastrichtian or earliest Danian basal layer of the Hornerstown Formation. This species pair also indicates a close connection between shorebirds from the Atlantic and the Western Interior of North America. The abundant charadriiformlike birds from the basal layer of the Hornerstown Formation are the product of long prior evolution that must have taken place in the Cretaceous (Olson, 1994).

The presence of Procellariiformes is indicated by a furcula similar to the distinctive furcula of an albatross, from the probably Early Maastrichtian Nemegt Formation of Mongolia (Kurochkin, 1995a,b). This is a surprising find, because the bones of these great gliders are very lightweight. For this reason, and if they lived and bred on oceanic islands as they do today, one might expect them to be rare in the record. *L. estesi*, from the Lance Formation, Wyoming, is not a loon as formerly supposed. It is most similar to petrels (Procellariidae: Procellariiformes), but because of homoplastic features it is referred here only tentatively to Procellariiformes. However, loons (Gaviiformes) very similar to extant *Gavia* were already present by Early Maastrichtian time in Antarctica and South America (Lambrecht, 1929b; Chatterjee, 1989, 1997; Olson, 1992). This contrasts with their entirely Northern Hemisphere record throughout the Cenozoic and into the present.

Pelecaniformes is represented by cormorants very similar to extant Phalacrocoracidae, from the Early Maastrichtian of Mongolia and the Late Maastrichtian of North America (Kurochkin, 1995a,b; this study). *Torotix* from the Lance Formation, Wyoming, may also be a pelecaniform, but it was not a cormorant.

A fragment of a lower jaw, probably from a parrot with some similarities to the extant lories (Psittaciformes), is reported from the Lance Formation (Stidham, 1998c). The mandible of this bird had highly unusual features associated with the unique jaw mechanics of present-day parrots. Parrots with a less highly derived jaw existed in the Eocene (Mayr and Daniels, 1998; Dyke and Mayr, 1999). Stidham (1999) suggested that early lineages of psittacines persisted into the Tertiary after the divergence of the extant lineage.

How Representative Is the Record?

Taphonomic Bias. The Late Cretaceous marks an increase in records of Neornithes, heavily dominated by shorebirds and seabirds. Yet among terrestrial lineages, there are records only of gallinaceous birds and a probable parrot. However, phylogenetic analyses point to a very early separation of aquatic from terrestrial lineages of Neornithes (Raikow, 1985; Elzanowski, 1995; Cooper and Penney, 1997; Mindell et al., 1997). Thus, one would expect to find the terrestrial lineages well represented at the same time. The record, however, is probably strongly biased by taphonomic factors favoring preservation of the delicate bones of birds in aquatic environments. Nessov (1992a,b) pointed out that most of the birds known from the Mesozoic come from certain kinds of aquatic environments: chalky sediments of warm-water seas, biologically rich phosphatic waters of quiet, upwelling marine margins, inland lakes, or meandering nearshore streams and estuaries. Small terrestrial birds are likely to be poorly represented because their skeletons are very fragile and many collectors may overlook small elements (Harrison, 1979; Unwin, 1988). Ornithologists are infrequent in paleontological field parties; thus, collectors in the past probably did not have a search image for small birds.

The terrestrial record of vertebrates is generally sparse in the mid-Cretaceous. Thus, one cannot assume that the Late Cretaceous increase in numbers and diversity of specimens from aquatic lineages reflects either the earliest radiation of Neornithes or its taxonomic composition during this time. Nevertheless, the much denser fossil record of mammals suggests that mid- or earlier Cretaceous divergence dates for extant mammals that are estimated by some molecular studies may be too old (Allard et al., 1999).

Biogeography. Unwin (1988) pointed out that the avian fossil record was concentrated from a few well-known Lagerstätten. Thus, long traditions of exploration in Europe and the Americas have distorted the reconstructions of avian distribution through time and space. The great majority of Mesozoic neornithine birds come from only two highly productive regional facies in the latter part of the Late Cretaceous. These are the foreland basin and regressive facies of the inland sea of western North America and the offshore sediments of the paleo-Atlantic of North America. A few specimens come from Chile, Argentina, Antarctica, eastern Europe, Mongolia, Japan (Table 15.2), and possibly one from Central Asia (a specimen reviewed herein as "?Neornithes"). There are still no Mesozoic records of Neornithes from Africa, Madagascar, Australia, the Indian subcontinent, or New Zealand. However, Campanian and Maastrichtian records from South America and Antarctica show that the group was not limited to the northern continents. It seems likely that Neornithes had reached all the great continents well before the end of the Cretaceous.

Land Birds. Early theories held that the various groups of ratites spread to the southern continents separately from the north. With new understanding of continental drift, Cracraft (1974) proposed instead that ratites were isolated in the Southern Hemisphere before the breakup of Gondwana and that they broke into separate groups later when Gondwana split apart. Houde and Olson (1981), however, reported flying paleognathous birds in the Early Paleogene of the Palaearctic, citing this as new evidence for possible polyphyly of ratites by multiple northern sources during the Cenozoic. The present review dates the Neognathae and thus also the lineage of Palaeognathae no later than Early Campanian. This dates the possible isolation of a monophyletic ratite lineage in Gondwana to well before the end of the Mesozoic. More to the point, overland connections may not have been necessary for such isolation if the immediate ancestor of extant ratites could negotiate a narrow water gap by flying. Thus, Paleogene palaeognaths in the north may be relics of an earlier radiation and are not inconsistent with a Cretaceous isolation of a ratite lineage or lineages in the south. Nor is the presence of primitive ostriches in the Paleogene of Eurasia (Houde, 1986; Houde and Haubold, 1987) inconsistent with contemporaneous lithornithids. Another new theory for ratite distributions suggests spread to the Southern Hemisphere across the proto-Antilles (van Tuinen et al., 1998).

However, the absence of ratites from the Mesozoic record of South America is hard to explain in the austral isolation scenario. Antarctica was cool but temperate, and trans-Antarctic connections to Australia and South America persisted into the Late Cretaceous. If they were present, the robust bones of ratites should have been found after the more than 100 years of extensive paleontological exploration in Patagonia. It may be that the diverse Enantiornithes in southern South America (Walker, 1981; Chiappe, 1991a,b, 1992b, 1996a,b, Chapter 13 in this volume; Chiappe and Calvo, 1994) restricted ratites ecologically or geographically.

Cracraft (1974) noted the evidence for a Gondwanan origin of Galliformes. The earliest division in extant galliforms is generally recognized as the split between Australian–New Guinean megapodes and other living galliforms. The avian record in the Early Tertiary includes closely related, probably poorly flying galliforms (Quercymegapodiidae) that occurred both in Europe and South America. Thus, the occurrence of even more primitive gallinaceous birds (*Palintropus*) in North America is of some interest.

Other landbirds have the imprint of an early austral radiation. Most notably, basal disjunctions in Passeriformes are among Australia, New Zealand, and South America (Raikow, 1978, 1987; Feduccia and Olson, 1982; Sibley and Ahlquist, 1990). Passerines are not known until the Eocene, but they appear first in Australia (Boles, 1999a). Their ancestors may lie under Antarctic ice.

Regarding seabirds, procellariiforms are most diverse in the far southern seas, and penguins also have an austral distribution. Thus, it has seemed plausible not only morphologically but also biogeographically that penguins and procellariiforms are closely related. Loons, however, also appear to be closely related to penguins and procellariiforms (Simpson, 1976; Hedges and Sibley, 1994), but loons have been known only from the Northern Hemisphere throughout the Tertiary and into the present. The Cretaceous record shows that they were indeed present in the far south early in their history, and thus a close relationship with penguins and procellariiforms seems plausible geographically.

Systematic Implications

Early Branches in Neornithes. Many features discussed in this review suggest that Anseriformes is more closely related to other neognaths than to Galliformes. In this reconstruction, an ecomorphological transition from Galliformes to Anseriformes within a clade Galloanserae (Livezey, 1997a) would be only partly right. The transition may have been from Galliformes to other neognaths, with an intermediate branch for Anseriformes (Fig. 15.11). Sparse information about the early terrestrial lineages is a major problem for reconstructing these basal relationships within Neornithes. The palaeognaths of today are morphologically and genetically far from the living neognaths and from each other. *Apatornis* sets the time for separation of Palaeognathae from Neognathae to more than 83 Ma. *Palintropus* sets the separation of Galliformes from Anseriformes to at least 75 Ma. This may well have been long enough to distort the genetic evidence for basal relationships, through differing rates of evolution or accumulated noise in the signal, in such a way as to incorrectly unite very old, widely unrelated groups of birds in a tree reconstruction: the so-called long-branch effect (Felsenstein, 1978; Hillis, 1996; Huelsenbeck, 1997).

Reconstruction of the earliest neornithine is complicated not only by poor knowledge of the palate in close outgroups of Neornithes but also by extreme reduction of the flight apparatus in extant palaeognaths. Thus, the reconstruction of shoulder elements done for this review emphasized *Lithornis*, a palaeognath with a shoulder articulation more like that of most neognaths and *Ichthyornis*. Even so, there are many enigmatic character distributions that are hard to interpret phylogenetically. First, the shoulder articulation in many poorly flying neornithines has

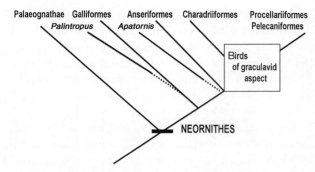

Figure 15.11. Phylogeny of several basal neornithine groups suggested by this study.

similar features that appear plesiomorphic because they occur in Enantiornithes. This may be due to developmental truncation, but it complicates the reconstruction of shoulder elements in the earliest neornithine, because not only Palaeognathae but also other major near-basal groups of Neornithes are poor flyers (Galliformes, Turnicidae).

Second, a plausible interpretation of Kurochkin's (1999) new information on *Ambiortus* (not his conclusion) is that this bird branched from the stem lineage of Neornithes and that in the early neornithine lineage many features associated with flight were rudimentary. In this view, the poor flight of tinamous, ratites, galliforms, and perhaps rails (Rallidae: Gruiformes) may be plesiomorphic or derived by relatively minor losses of flight ability from poorly flying earliest neornithines. The hypothesis that poorly flying early birds lost flight ability several times is not new (Witmer, 1999). This loss is especially likely to have happened in early lineages if they had not yet achieved the most advanced design for flight.

Third, the peculiar, far lateral scapular articulation in the early gallinaceous bird *Palintropus* may suggest an independent line of evolution of flight ability or perhaps a novel function of the forelimb in a flightless bird. Extant galliforms and tinamous also have a lateralized scapular articulation, but it is less extreme and less obvious because the coracoidal tubercle is not ossified.

The similarities among weakly flying birds profoundly affect basal phylogenetic reconstruction within Neornithes by suggesting a paraphyletic relationship of Galliformes with other members of Neognathae. Two kinds of morphological information might illuminate this problem: comparative ontogeny of the palate and shoulder elements in extant basal groups, using advanced techniques that were not available to earlier workers, and a denser fossil record for terrestrial lineages of Neornithes in the Mesozoic.

Anseriformes. The position of *Presbyornis* has long been debated because of its charadriiformlike postcranial skeleton combined with a ducklike skull (Wetmore, 1926; Feduccia and McGrew, 1974; Feduccia, 1976, 1978; Olson and Fe-

duccia, 1980b). Skull morphology strongly supports embedding *Presbyornis* deeply within Anseriformes (Ericson, 1997; Livezey, 1997a), but in the present study this reconstruction results in several discordant, apparently reversed features in Presbyornithidae and other derived features shared by Presbyornithidae with Charadriiformes. Moreover, some of the latter are more widely present in Anseriformes, and some also occur in flamingos. Together with strong evidence supporting the monophyly of Charadriiformes with the seabird groups, the pattern of shared derived features found here suggests that the association between anseriforms, charadriiforms, flamingos, and ibises is basal within some larger assemblage.

Charadriiformes: Mosaicism. Even though none of the specimens reviewed here is referred conclusively to Charadriiformes, some are very charadriiformlike, and it is premature to conclude that the extant group had not yet differentiated. The work of Olson and Parris (1987) called attention to a potentially important body of material that they assembled in Graculavidae. They referred this assemblage to Charadriiformes but recognized that these birds might be placed elsewhere when more complete material was found (Olson, 1999). An obvious obstacle to diagnosing early charadriiformlike birds is the fragmented material, but this is not the whole problem, because lesser fragments from other Cretaceous birds can be diagnosed with relative ease. Even working with the extant members, Strauch (1978), Bjorklund (1994), and Chu (1995) noted the difficulty of diagnosing Charadriiformes.

Two relevant observations have emerged from this and earlier reviews. First, some charadriiforms (e.g., Burhinidae) retain a high proportion of apparently relatively primitive features (Cracraft, 1972; Olson and Parris, 1987). Many of these features are shared at a more inclusive level within Neornithes or were present in the earliest neornithine. These features occur less frequently in other shorebirds or in seabirds—hence, possibly, the often cited resemblance of the postcranial skeleton of Charadriiformes to that of Anseriformes. Thus, a fundamental problem is plesiomorphy of charadriiforms within some larger assemblage of shorebirds and seabirds.

Second, all derived features in the postcranial skeleton of Charadriiformes identified in this review are present in some of the other shorebirds and seabirds. Such features are flagged with asterisks in the systematic reviews above, as explained in the "Material and Methods" section and in introductory remarks to Charadriiformes. The taxonomic distribution of derived features is irregular and overlapping. In addition, some but not all charadriiforms (e.g., gulls: Laridae) share a large suite of derived features exclusively with Procellariiformes, in the same or a more extreme state.

Mosaicism has sometimes been interpreted as evolutionary intermediacy, but seabirds and shorebirds are an entire group of mosaics. Intermediate to what? Livezey (1997a) articulated the view that to interpret its mosaic features, one must locate the animal on a phylogenetic tree. To apply this view to an entire assemblage of mosaics is another matter. If the divergence within a parent lineage occurred over a very short time, the progeny lineages will be similar. Each will retain a different sample of features from the parent lineage, little changed because of the short intervals of time between the separation of lineages.

The mosaicism seen today among shorebirds and seabirds may be the result of rapid divergence. We can guess that it occurred in the latter part of the Late Cretaceous and that the birds of graculavid aspect are a sample of that early avifauna, including, probably, some lineages now extinct and others extant (Fig. 15.11). A better taxonomic arrangement than simply referring these birds tentatively to Charadriiformes would define a more inclusive taxon to which the birds of graculavid aspect might be referred, incertae sedis. The analysis necessary to define such a taxon is beyond the scope of this chapter.

The difficulty of diagnosing Charadriiformes probably indicates that charadriiforms branched early and thus retained a large suite of the common ancestral features. This is consistent with their position as the first branch of the radiation of seabirds and shorebirds found by Sibley and Ahlquist (1990).

Procellariiformes, Pelecaniformes, Gaviiformes. Olson and Parris (1987) described *Tytthostonyx,* from the Hornerstown Formation, as having a combination of features found now in Procellariiformes and Charadriiformes, but with the humerus more strongly curved than in extant members of either taxon. Similarly, *Torotix,* from the Late Maastrichtian Lance Formation, has a unique combination of presumably derived features present in Charadriiformes, Procellariiformes, and Pelecaniformes. These birds also show mosaicism, in this case suggesting they represent an early seabird assemblage analogous to the more charadriiformlike birds of graculavid aspect. Many studies have suggested that loons are closely related to or part of Procellariiformes. This study contributes further evidence by polarizing morphological features in order to diagnose loons. Loons share all the derived features of the tibiotarsus and tarsometatarsus found in procellariid petrels, but in a more extreme form. The coracoid and femur are not reviewed herein but also appear to be consistent with such a placement. The meager fossil support for procellariiforms is buttressed by the clear evidence of loons in the Maastrichtian.

Bradytely. Contemporaneously with possibly extinct lineages of procellariiforms and pelecaniforms, the extant lineage of Pelecaniformes was already represented by cormorants, and loons almost like those of the present had appeared. The difficulty of referring specimens to Charadriiformes contrasts with the easy recognition of Cretaceous loons and cormorants. The inference of slow morphological change, over 65 million years and more, is especially strong because of the relatively complete skeleton of a Maastrichtian loon (Chatterjee, 1997). In addition, the scapula and femur of Late Cretaceous cormorants have the unique and highly unusual morphology of these elements in living cormorants. This indicates that both forelimb and hindlimb were already designed as in the highly modified limbs of living cormorants. These are unequivocal examples of bradytely in birds.

End of the Mesozoic

By Campanian time, the enantiornithine *Avisaurus* shared the terra of North America with gallinaceous birds. By Early Maastrichtian, hesperornithiforms dived beside cormorants in the Turgai Strait and the Niobrara Sea. Enantiornithes and Hesperornithiformes persisted into the Maastrichtian (Kurochkin, 1995a; Stidham, 1998a; Chiappe and Walker, Chapter 11 in this volume; pers. obs.), but they were not alone. Telling numbers come from the abundant collections in the Lance Formation in Wyoming. At sites well below the K/T boundary, neornithine birds were the most diverse element in this nearshore avifauna. A tabulation for hesperornithiforms is still forthcoming, but in collections from the Lance Formation reviewed for this chapter, 5 specimens were enantiornithine and 25 neornithine. This suggests that other birds may have followed the course of many Mesozoic vertebrates, which declined in diversity before the end of the Cretaceous (Hallam and Wignal, 1997; Archibald, 1999; Clemens, 1999). From the present reading of the record, loons, cormorants, presbyornithid anseriforms, and close relatives of galliforms were present from Campanian time on and survived into the Tertiary little changed. This spectacular record of survival is evident even from a very limited record.

Acknowledgments

For the opportunity to describe new Cretaceous specimens, I thank M. C. McKenna, S. L. Olson, and C. Coy. This work is much improved by the efforts of those who brought early neornithine birds from all over the world to a workshop at the 1996 meeting of the Society for Avian Paleontology and Evolution. H. F. James and S. Olson made this workshop at the Smithsonian Institution possible. For access to specimens in their care, I also thank B. Bailey, G. Barrowclough, L. M. Chiappe, W. A. Clemens, T. Daeshler, P. Holroyd, C. Holton, J. H. Hutchinson, N. K. Johnson, E. N. Kurochkin, J. H. Ostrom, D. C. Parris, T. Stidham, and T. Tokaryk. Others helped in ways too numerous to list; I especially thank J. Schonewald, L. Baptista, A. Elzanowski, and D. C. Parris. For editorial help I thank S. Orell.

For expert translations from Russian I am grateful to O. Titelbaum, and for helpful comments on an earlier version of the manuscript I thank K. Padian and D. Bell. I especially thank the editors, L. M. Chiappe and L. M. Witmer, for their persistent and useful comments in the course of this study.

Literature Cited

Allard, M. W., R. L. Honeycutt, and M. J. Novacek. 1999. Advances in higher level mammalian relationships. Cladistics 13:213–219.

Alvarenga, H. M. F. 1995. Um primitivo membro da ordem Galliformes (Aves) do Terciário Médio da Bacia de Taubaté, Estado de São Paulo, Brasil. Anais da Academia Brasileira de Ciencias 67:33–44.

Alvarenga, H. M. F., and J. Bonaparte. 1992. A new flightless landbird from the Cretaceous of Patagonia; pp. 51–64 in K. E. Campbell (ed.), Papers in Avian Paleontology, Honoring Pierce Brodkorb. Science Series 36. Natural History Museum of Los Angeles County, Los Angeles.

American Ornithologists' Union. 1998. Check-list of North American Birds, 7th edition. American Ornithologists' Union, Washington, D.C., 829 pp.

Andrews, C. W. 1913. On some bird remains from the Upper Cretaceous of Transylvania. Geological Magazine, New Series, Decade V, 10:193–196.

Archibald, J. D. 1982. A study of Mammalia and geology across the Cretaceous-Tertiary boundary in Garfield County, Montana. University of California Publications in the Geological Sciences 22:1–286.

———. 1999. Dinosaur extinction: how much and how fast? Journal of Vertebrate Paleontology 12:263–264.

Archibald, J. D., and D. Lofgren. 1990. Mammalian zonation near the Cretaceous-Tertiary boundary; pp. 31–50 in T. M. Bown and K. D. Rose (eds.), Dawn of the Age of Mammals in the Northern Part of the Rocky Mountain Interior, North America. Geological Society of America Special Paper 243. Geological Society of America, Boulder, Colo.

Averianov, A. O. 1999. [Annotated list of taxa described by L. A. Nessov]. Trudy Zoologicheskogo Instituta Russkii Academia Nauk 277:7–37. [Russian]

Avise, J. C., W. S. Nelson, and C. G. Sibley. 1994. DNA sequence support for a close phylogenetic relationship between some storks and New World vultures. Proceedings of the National Academy of Sciences 91:5173–5177.

Baird, D. 1967. Age of fossil birds from the greensands of New Jersey. Auk 84:160–162.

Baumel, J. J., and L. M. Witmer. 1993. Osteologia; pp. 45–132 in J. J. Baumel, A. S. King, J. E. Breazile, H. E. Evans, and J. C. Vanden Berge (eds.), Handbook of Avian Anatomy: Nomina Anatomica Avium, 2nd edition. Nuttall Ornithological Club, Cambridge, Mass.

Benson, R. D. 1999. Presbyornis isoni and other Late Paleocene birds from North Dakota; pp. 253–260 in S. L. Olson (ed.), Avian Paleontology at the Close of the 20th Century: Proceedings of the 4th International Meeting of the Society of Avian Paleontology and Evolution, Washington, D.C., 4–7 June 1996. Smithsonian Contributions to Paleobiology 89. Smithsonian Institution Press, Washington, D.C.

Bjorklund, M. 1994. Phylogenetic relationships among Charadriiformes: reanalysis of previous data. Auk 111:825–832.

Bock, W. J. 1962. The pneumatic fossa of the humerus in the Passeres. Auk 79:425–443.

———. 1963. The cranial evidence for ratite affinities; pp. 39–54 in C. G. Sibley (ed.), Proceedings of the 13th International Ornithological Congress. American Ornithologists' Union, Baton Rouge, La.

———. 1964. Kinetics of the avian skull. Journal of Morphology 114:1–42.

Bock, W. J., and P. Bühler. 1996. Nomenclature of Cretaceous birds from Romania. Cretaceous Research 17:509–514.

Boles, W. 1999a. The world's oldest songbird. Nature 374:21–22.

———. 1999b. Early Eocene shorebirds (Aves: Charadriiformes) from the Tingamarra Local Fauna, Murgon, Queensland, Australia. Records of the Western Australian Museum 57(supplement):229–238.

Bonaparte, C. L. 1853. Classification ornithologique par séries. Comptes Rendus de l'Académie des Sciences, Paris 37:641–647.

Brodkorb, P. 1963a. Birds from the Upper Cretaceous of Wyoming; pp. 50–70 in C. G. Sibley (ed.), Proceedings of the 13th International Ornithological Congress. American Ornithologists' Union, Baton Rouge, La.

———. 1963b. Catalogue of fossil birds, part 1 (Archaeopterygiformes through Ardeiformes). Bulletin of the Florida State Museum, Biological Sciences 7(3):179–293.

———. 1967. Catalogue of fossil birds, part 3 (Ralliformes, Ichthyornithiformes, Charadriiformes). Bulletin of the Florida State Museum, Biological Sciences 11(3):99–220.

———. 1970. The generic position of a Cretaceous bird. Quarterly Journal of the Florida Academy of Sciences 32(3):239–240.

———. 1976. Discovery of a Cretaceous bird, apparently ancestral to the orders Coraciiformes and Piciformes (Aves: Carinatae); pp. 67–93 in S. L. Olson (ed.), Papers in Avian Paleontology Honoring the 90th Birthday of Alexander Wetmore. Smithsonian Contributions to Paleobiology 27. Smithsonian Institution Press, Washington, D.C.

———. 1978. Catalogue of fossil birds, part 5 (Passeriformes). Bulletin of the Florida State Museum, Biological Sciences 23(3):139–228.

Bryant, L. 1989. Non-dinosaurian lower vertebrates across the Cretaceous-Tertiary Boundary in northeastern Montana. University of California Publications in the Geological Sciences 134:1–107.

Bühler, P. 1981. Functional anatomy of the avian jaw apparatus; pp. 439–468 in A. King and J. McLelland (eds.), Form and Function in Birds, Vol. 2. Academic Press, London.

———. 1987. On the mobility of the upper jaw and the segments of the braincase in the Mesozoic birds; pp. 41–48 in C. Mourer-Chauviré (ed.), Table Ronde Internationale du CNRS, Lyon-Villeurbanne, 18–21 Septembre 1985. Documents des Laboratoires de Géologie de la Faculté des Sciences de Lyon 99. Université Claude-Bernard, Villeurbanne Cedex.

———. 1992. Light bones in birds; pp. 385–394 in K. E. Campbell (ed.), Papers in Avian Paleontology, Honoring Pierce Brodkorb. Science Series 36. Natural History Museum of Los Angeles County Science, Los Angeles.

Bühler, P., L. D. Martin, and L. M. Witmer. 1988. Cranial kinesis in the Late Cretaceous birds Hesperornis and Parahesperornis. Auk 105:111–122.

Case, J., and C. Tambussi. 1999. Maastrichtian record of neornithine birds in Antarctica: comments on a Late Cretaceous radiation of modern birds. Journal of Vertebrate Paleontology 19(3, suppl.): 37A.

Chatterjee, S. 1989. The oldest Antarctic bird. Journal of Vertebrate Paleontology 9(Supplement to 3):16A.

———. 1997. The Rise of Birds: 225 Million Years of Evolution. Johns Hopkins University Press, Baltimore, 312 pp.

Chiappe, L. M. 1991a. Cretaceous avian remains from Patagonia shed new light on the early radiation of birds. Alcheringa 15:333–338.

———. 1991b. Cretaceous birds of Latin America. Cretaceous Research 12:55–63.

———. 1992a. Osteología y sistemática de *Patagopteryx deferrariisi* Alvarenga y Bonaparte (Aves) del Cretácico de Patagonia. Filogenia e Historia Biogeográphica de las Aves Cretácicas de America del Sur. Ph. D. dissertation, Universidad de Buenos Aires, 429 pp.

———. 1992b. Enantiornithine (Aves) tarsometatarsi and the avian affinities of the Late Cretaceous Avisauridae. Journal of Vertebrate Paleontology 12:344–350.

———. 1995a. The phylogenetic position of the Cretaceous birds of Argentina: Enantiornithes and *Patagopteryx deferrariisi*; pp. 55–63 in D. S. Peters (ed.), Proceedings of the 3rd Symposium of the Society of Avian Paleontology and Evolution. Courier Forschungsinstitut Senckenberg 181, Frankfurt am Main.

———. 1995b. The first 85 million years of avian evolution. Nature 378:349–355.

———. 1996a. Early avian evolution in the Southern Hemisphere: the fossil record of birds in the Mesozoic of Gondwana. Memoirs of the Queensland Museum 39(3):533–554.

———. 1996b. Late Cretaceous birds of southern South America: anatomy and systematics of Enantiornithes and *Patagopteryx deferrariisi*; pp. 203–244 in G. Arratia (ed.), Contributions of Southern South America to Vertebrate Paleontology. Münchner Geowissenschaftliche Abhandlungen, Reihe A, Geologie und Paläontologie 30. Verlag Dr. Friedrich Pfeil, Munich.

Chiappe, L. M., and J. O. Calvo. 1994. *Neuquenornis volans,* a new Upper Cretaceous bird (Enantiornithes: Avisauridae) from Patagonia, Argentina. Journal of Vertebrate Paleontology 14:230–246.

Chu, P. C. 1995. Phylogenetic reanalysis of Strauch's osteological data set for Charadriiformes. Condor 97:174–196.

Clemens, W. A. 1964. Fossil mammals of the Type Lance Formation, Wyoming, Part I. Introduction and multituberculata. University of California Publications in the Geological Sciences 48:1–105.

———. 1999. Evolution of the Mammalian fauna across the Cretaceous-Tertiary boundary in Northeastern Montana and other areas of the Western Interior. Abstracts with Programs, Geological Society of America Annual Meeting, October 1999, Denver, Colo. Abstract 14530.

Cooper, A., and D. Penny. 1997. Mass survival of birds across the Cretaceous-Tertiary boundary: molecular evidence. Science 175:1109–1113.

Cottam, P. A. 1957. The pelecaniform characters of the skeleton of the shoe-bill stork, *Balaeniceps rex.* Bulletin of the British Museum (Natural History), Zoology 5:51–72.

Cracraft, J. 1972. A new Cretaceous charadriiform family. Auk 89:36–46.

———. 1973. Systematics and evolution of the Gruiformes (Class Aves), 3: phylogeny of the Suborder Grues. Bulletin of the American Museum of Natural History 151:1–128.

———. 1974. Continental drift and vertebrate evolution. Annual Review of Ecology and Systematics 5:215–261.

———. 1982. Phylogenetic relationships and monophyly of loons, grebes, and hesperornithiform birds, with comments on the early history of birds. Systematic Zoology 31:35–56.

———. 1985. Monophyly and phylogenetic relationships of the Pelecaniformes: a numerical cladistic analysis. Auk 102:834–853.

———. 1986. The origin and early diversification of birds. Paleobiology 12:383–399.

———. 1988. The major clades of birds; pp. 339–361 in M. J. Benton (ed.), The Phylogeny and Classification of the Tetrapods, Vol. 1: Amphibians, Reptiles, Birds. Systematics Association Special Volume 35A. Clarendon Press, Oxford.

———. 2000. Avian evolution, Gondwana biogeography and the Cretaceous-Tertiary mass extinction event. Proceedings of the Royal Society of London B 268:459–469.

Cracraft, J., and D. P. Mindell. 1989. The early history of modern birds: a comparison of molecular and morphological evidence. International Congress Series 824:389–403.

Davison, G. W. H. 1998. Scientific controversy over avian taxonomic changes, based on DNA hybridisation. Raffles Bulletin of Zoology 46:253–269.

de Queiroz, K., and J. Gauthier. 1990. Phylogeny as a central principle in taxonomy: phylogenetic definitions of taxon names. Systematic Zoology 39:307–322.

Dial, K. P., G. E. Goslow Jr., and F. A. Jenkins Jr. 1991. The functional anatomy of the shoulder in the European starling (*Sternus vulgaris*). Journal of Morphology 107:327–344.

Dorf, E. 1942. Upper Cretaceous floras of the Rocky Mountain region. 2: flora of the Lance Formation at its type locality, Niobrara County, Wyoming. Carnegie Institute of Washington Publications 508:83–168.

Dyke, G. J., and G. Mayr. 1999. Did parrots exist in the Cretaceous period? Nature 399:317–318.

Dzerzhinsky, F. Ya. 1995. Evidence for common ancestry of Galliformes and Anseriformes; pp. 325–336 in D. S. Peters (ed.), Proceedings of the 3rd Symposium of the Society Avian Paleontology and Evolution, Frankfurt, 1992. Courier Forschungsinstitut Senckenberg 181, Frankfurt am Main.

———. 1999. Implications of the cranial morphology of paleognaths for avian evolution; p. 274 in S. L. Olson (ed.), Avian Paleontology at the Close of the 20th Century: Proceedings of the 4th International Meeting of the Society of Avian Paleontology and Evolution. Smithsonian Contributions to Paleobiology 89. Smithsonian Institution Press, Washington, D.C.

Eberth, D. A. 1990. Stratigraphy and sedimentology of vertebrate microfossil sites in the uppermost Judith River Formation (Campanian), Dinosaur Provincial Park, Alberta, Canada. Palaeogeography, Palaeoclimatology, Palaeoecology 78:1–36.

Elzanowski, A. 1974. Preliminary note on the palaeognathous bird from the Upper Cretaceous of Mongolia. Palaeontologia Polonica 30:103–109.

———. 1981. Embryonic bird skeletons from the late Cretaceous of Mongolia. Palaeontologia Polonica 42:147–179.

————. 1983. Birds in Cretaceous ecosystems. 2nd Symposium on Mesozoic Terrestrial Ecosystems. Acta Palaeontologica Polonica 28:75–92.

————. 1988. Ontogeny and evolution of the ratites; pp. 2037–2046 in H. Ouellet (ed.), Acta XIX Congressus Internationalis Ornithologici. University of Ottawa Press, Ottawa.

————. 1991. New observations on the skull of *Hesperornis,* with reconstructions of the bony palate and otic region. Postilla 207:1–20.

————. 1995. Cretaceous birds and avian phylogeny; pp. 37–53 in D. S. Peters (ed.), Proceedings of the 3rd Symposium of the Society Avian Paleontology and Evolution, Frankfurt, 1992. Courier Forschungsinstitut Senckenberg 181.

————. 1999. A comparison of the jaw skeleton in theropods and birds, with a description of the palate in the Oviraptoridae; pp. 311–323 in S. L. Olson (ed.), Avian Paleontology at the Close of the 20th Century: Proceedings of the 4th International Meeting of the Society of Avian Paleontology and Evolution, Washington, D.C., 4–7 June 1996. Smithsonian Contributions to Paleobiology 89, Smithsonian Institution Press, Washington, D.C.

Elzanowski, A., and M. K. Brett-Surman. 1995. Avian premaxilla and tarsometatarsus from the Uppermost Cretaceous of Montana. Auk 112:762–766.

Elzanowski, A., and P. M. Galton. 1991. Braincase of *Enaliornis,* an Early Cretaceous bird from England. Journal of Vertebrate Paleontology 11:90–107.

Ericson, P. G. P. 1996. The skeletal evidence for a sister group relationship of anseriform and galliform birds—a critical evaluation. Journal of Avian Biology 27:195–202.

————. 1997. Systematic relationships of the Palaeogene family Presbyornithidae (Aves: Anseriformes). Zoological Journal of the Linnean Society 121:429–483.

————. 1999. New material of *Juncitarsus* (Phoenicopteriformes), with a guide for differentiating that genus from the Presbyornithidae (Anseriformes); pp. 245–251 in S. L. Olson (ed.), Avian Paleontology at the Close of the 20th Century: Proceedings of the 4th International Meeting of the Society of Avian Paleontology and Evolution, Washington, D.C., 4–7 June 1996. Smithsonian Contributions to Paleobiology 89, Smithsonian Institution Press, Washington, D.C.

Estes, R. 1964. Fossil vertebrates from the Late Cretaceous Lance Formation, Eastern Wyoming. University of California Publications in the Geological Sciences 49:1–180.

Feduccia, A. 1976. Osteological evidence for shorebird affinities of the flamingos. Auk 93:587–601.

————. 1978. *Presbyornis* and the evolution of ducks and flamingos. American Scientist 66:298–304.

————. 1995. Explosive evolution in Tertiary birds and mammals. Science 267:637–638.

————. 1996. The Origin and Evolution of Birds. Yale University Press, New Haven, 420 pp.

Feduccia, A., and P. McGrew. 1974. A flamingolike wader from the Eocene of Wyoming. University of Wyoming Contributions in Geology 13:49–61.

Feduccia, A., and S. L. Olson, 1982. Morphological similarities between the Menurae and the Rhinocryptidae, relict passerine birds of the Southern Hemisphere. Smithsonian Contributions to Zoology 366:1–22.

Felsenstein, J. 1978. Cases in which parsimony or compatibility methods will be positively misleading. Systematic Zoology 27:401–410.

Fürbringer, M. 1888. Untersuchungen zur Morphologie und Systematik der Vögel, zugleich ein Beitrag zur Anatomie der Stütz- und Bewegungsorgane. Von Holkema, Amsterdam, 1,751 pp.

Gadow, H. 1893. Vögel. II. Systematischer Theil. *In* Bronn's Klassen und Ordnungen des Thier-Reichs, Vol. 6. C. F. Winter, Leipzig, 303 pp.

Gallagher, W. B. 1992. Geochemical investigations of the Cretaceous/Tertiary boundary in the Inversand Pit, Gloucester County, New Jersey. Bulletin of the New Jersey Academy of Sciences 37:19–24.

————. 1993. The Cretaceous/Tertiary mass extinction event in the Northern Atlantic Coastal Plain. Mosasaur 5:75–154.

————. 1996. Marine ecosystem collapse and recovery across the K/T boundary in New Jersey; pp. 46–51 in W. B. Gallagher and D. C. Parris (eds.), Cenozoic and Mesozoic Vertebrate Paleontology of the New Jersey Coastal Plain. Investigation No. 5, New Jersey State Museum, Trenton.

Gallagher, W. B., and D. C. Parris. 1996. Stratigraphy and paleontology at New Jersey's last marl mine, the Inversand Pit; pp. 42–46 in W. G. Gallagher and D. C. Parris (eds.), Cenozoic and Mesozoic Vertebrate Paleontology of the New Jersey Coastal Plain. Investigation No. 5, New Jersey State Museum, Trenton.

Garrod, A. H. 1874. On certain muscles of birds and their value in classification, Part II. Proceedings of the Zoological Society of London 1874:111–124.

Gatesy, S. M. 1995. Functional evolution of the hind limb and tail from basal theropods to birds; pp. 219–234 in J. J. Thomason (ed.), Functional Morphology in Vertebrate Paleontology. Cambridge University Press, Cambridge.

Ghiselin, M. 1984. "Definition," "character," and other equivocal terms. Systematic Zoology 33:104–110.

————. 1997. Metaphysics and the Origin of Species. State University of New York, Albany, 377 pp.

Gray, G. R. 1840. A List of the Genera of Birds. R. and J. Taylor, London.

Grigorescu, D. 1983. A stratigraphic, taphonomic and paleoecologic approach to a "forgotten land": the dinosaur-bearing deposits from the Hateg Basin (Transylvania-Romania). 2nd Symposium on Mesozoic Terrestrial Ecosystems. Acta Palaeontologica Polonica 28:103–121.

Grigorescu, D., and E. Kessler. 1980. A new specimen of *Elopteryx nopscsai* Andrews, from the dinosaurian beds of Hateg Basin. Revue Roumaine de Géologie, Géophysique, et Géographie. 24:171–175.

Haeckel, E. 1866. Generelle Morphologie der Organismen. Georg Reimer, Berlin. De Gruyter, Berlin, 1988. 2 vols.

Hallam, A., and P. B. Wignal. 1997. Mass Extinctions and Their Aftermath. Oxford University Press, New York, 320 pp.

Harrison, C. J. O. 1979. Small non-passerine birds of the Lower Tertiary as exploiters of ecological niches now occupied by passerines. Nature 281:562–563.

Harrison, C. J. O., and C. A. Walker. 1973. *Wyleyia,* a new bird humerus from the Lower Cretaceous of England. Palaeontology 16:721–728.

———. 1975. The Bradycnemidae, a new family of owls from the Upper Cretaceous of Romania. Palaeontology 18:563–570.

Harshman, J. 1994. Reweaving the tapestry: what can we learn from Sibley and Ahlquist (1990)? Auk 111:377–388.

Hasegawa, Y., Y. Okumura, and Y. Okazaki. 1977. A Miocene bird fossil from Mizunami, Central Japan. Bulletin of the Mizunami Fossil Museum 4:169–171. [Japanese]

Hattin, D. E. 1982. Stratigraphy and depositional environment of Smoky Hill Chalk Member, Niobrara Chalk (Upper Cretaceous) of the type area, Western Kansas. Kansas Geological Survey Bulletin 225:1–108.

Hay, O. P. 1902. Bibliography and Catalogue of the Fossil Vertebrata of North America. Bulletin of the United Stated Geological Survey 179. Government Printing Office, Washington, D.C., 868 pp.

Hedges, S. G., and C. G. Sibley. 1994. Molecules vs. morphology in avian evolution: the case of the "pelecaniform" birds. Proceedings of the National Academy of Sciences, USA 91:9861–9865.

Hedges, S. G., P. H. Parker, C. G. Sibley, and S. Kumar. 1996. Continental breakup and ordinal diversification of birds and mammals. Nature 381:226–229.

Helbrig, A. J., and I. Seibold. 1996. Are storks and New World vultures paraphyletic? Molecular Phylogenetics and Evolution. 6:315–319.

Hillis, D. M. 1996. Inferring complex phylogenies. Nature 383:130–131.

Hope, S. 1989. Phylogeny of the avian family Corvidae. Ph.D. dissertation, City University of New York, 280 pp.

———. 1999. A new species of Graculavus from the Cretaceous of Wyoming (Aves: Charadriiformes); pp. 231–243 in S. L. Olson (ed.), Avian Paleontology at the Close of the 20th Century: Proceedings of the 4th International Meeting of the Society of Avian Paleontology and Evolution, Washington, D.C., 4–7 June 1996. Smithsonian Contributions to Paleobiology 89, Smithsonian Institution Press, Washington, D.C.

Hou L. 1994. A late Mesozoic bird from Inner Mongolia. Vertebrata PalAsiatica 32:258–266.

Hou L. and Liu Z. 1984. A new fossil bird from Lower Cretaceous of Gansu and early evolution of birds. Scientia Sinica (Series B) 27(12):1296–1302.

Hou L., L. D. Martin, and Zhou Z. 1996. Early adaptive radiation of birds: evidence from fossils from northeastern China. Science 274:1164–1167.

Houde, P. W. 1986. Ostrich ancestors found in the Northern Hemisphere suggest new hypothesis of ratite origins. Nature 324:563–565.

———. 1988. Paleognathous birds from the early Tertiary of the Northern Hemisphere. Publications of the Nuttall Ornithological Society 22:1–148.

Houde, P. W., and H. Haubold. 1987. Palaeotis weigelti restudied: a small Eocene ostrich (Aves: Struthioniformes). Palaeovertebrata 17:27–42.

Houde, P. W., and S. L. Olson. 1981. Paleognathous carinate birds from the Early Tertiary of North America. Nature 214:1236–1237.

Howard, H. 1955. A new wading bird from the Eocene of Patagonia. American Museum Novitates 1710:1–25.

———. 1964. Fossil Anseriformes; pp. 233–326 in J. Delacour (ed.), Waterfowl of the World, Vol. 4. Country Life, London.

Huelsenbeck, J. P. 1997. Is the Felsenstein zone a fly trap? Systematic Biology 46:69–74.

Huxley, T. H. 1867. On the classification of birds, and on the taxonomic value of the modification of certain of the cranial bones observable in that class. Proceedings of the Zoological Society of London 1867:415–472.

Jenkins, F. A., Jr. 1993. The evolution of the avian shoulder joint. American Journal of Science 293A:253–267.

Jollie, M. 1957. The head skeleton of the chicken and remarks on the anatomy of this region in other birds. Journal of Morphology 100:389–436.

Kennedy, W. J., and W. A. Cobban. 1996. Maastrichtian ammonites from the Hornerstown Formation in New Jersey. Journal of Paleontology 70:798–804.

Kessler, E. 1984. Lower Cretaceous birds from Cornet (Roumania); pp. 119–121 in W. E. Reif and A. E. Westphal (eds.), 3rd Symposium on Mesozoic Terrestrial Ecosystems, Tübingen, 1984. Attempto Verlag, Tübingen.

———. 1987. New contributions to the knowledge about the Lower and Upper Cretaceous birds from Roumania; pp. 129–131 in P. J. Currie and E. H. Koster (eds.), 4th Symposium on Mesozoic Terrestrial Ecosystems, Short Papers. Royal Tyrrell Museum of Paleontology, Occasional Paper 3. Royal Tyrrell Museum of Paleontology, Drumheller, Canada.

Kessler, E., and T. Jurcsak. 1984. Fossil bird remains in the bauxite from Cornet (Romania, Bihor County). Travaux du Muséum National d'Histoire Naturelle "Grigore Antipa" (Bucharest) 25:393–401.

———. 1986. New contributions to the knowledge of the Lower Cretaceous bird remains from Cornet (Romania). Travaux du Museum National d-Histoire Naturelle "Grigore Antipa" (Bucharest) 28:289–295.

Konig, C. 1988. Los catártidos (Aves, Cathartidae) no son rapaces (Falconiformes) sino parientes de las ciguenas (Ciconiidae). Boletín del Instituto de Estudios Almerienses (Extra):127–137.

Kumar, S., and S. B. Hedges. 1998. A molecular timescale for vertebrate evolution. Nature 392:917–920.

Kurochkin, E. N. 1982. Novi otrjad ptizt iz nizhnego mela Mongolii. Dokladi Akademii Nauk SSSR 262:452–455. Nauka, Moscow. [Russian]

———. 1985. A true carinate bird from Lower Cretaceous deposits of Mongolia and other evidence of Early Cretaceous birds in Asia. Cretaceous Research 6:271–278.

———. 1988. Cretaceous birds of Mongolia and their significance for the study of phylogeny of Class Aves; pp. 33–42 in E. N. Kurochkin (ed.), Fossil Reptiles and Birds of Mongolia. Transactions of the Joint Soviet-Mongolian Paleontology Expedition 34. [Russian]

———. 1995a. The assemblage of the Cretaceous birds in Asia; pp. 203–208 in Ailiing S. and Wang Y. (eds.), Sixth Symposium on Mesozoic Terrestrial Ecosystems and Biota, Short Papers. Paleontological Institute Special Issue, China Ocean Press, Beijing.

———. 1995b. Synopsis of Mesozoic birds and early evolution of Class Aves. Archaeopteryx 13:47–66.

———. 1999. The relationships of the early Cretaceous Ambiortus and Otogornis (Aves: Ambiortiformes); pp. 275–284 in S. L. Olson (ed.), Avian Paleontology at the Close of the 20th Century: Proceedings of the 4th International Meeting of the

Society of Avian Paleontology and Evolution Smithsonian Contributions to Paleobiology 89. Smithsonian Institution Press, Washington, D.C.

Lambrecht, K. 1929a. Mesozoische und Tertiare Vogelreste aus Siebenbürgen; pp. 1262–1267 in E. Csiki (ed.), Comptes Rendus 10ème Congrès International de Zoologie, September, 1927. Imprimerie Stephaneum, Budapest.

———. 1929b. Neogaeornis wetzeli, n. g., n. sp., der erste Kreidevogel der Südlichen Hemisphäre. Palaeontologische Zeitschrift 11:121–129.

———. 1931. Gallornis straeleni n. g. n. sp., ein Kreidevogel aus Frankreich. Bulletin du Musée Royal d'Histoire Naturelle de Belgique 7:1–6.

———. 1933. Handbuch der Palaeornithologie. Gebrüder Borntraeger, Berlin, 1,022 pp.

Leach, W. E. 1820. Eleventh room; pp. 65–70 in Synopsis of the Contents of the British Museum, 17th edition. British Museum (Natural History), London.

Ligon, J. D. 1967. Relationships of the cathartid vultures. Occasional Papers of the Museum of Zoology, University of Michigan 651:1–26.

Lillegraven, J. A., and M. C. McKenna. 1986. Fossil mammals from the "Mesaverde" Formation (Late Cretaceous, Judithian) of the Bighorn and Wind River Basins, Wyoming, with definitions of Late Cretaceous North American Land-Mammal "Ages." American Museum Novitates 2840:1–68.

Livezey, B. C. 1997a. A phylogenetic analysis of basal Anseriformes, the fossil Presbyornis, and the interordinal relationships of waterfowl. Zoological Journal of the Linnean Society 121:361–428.

———. 1997b. A phylogenetic classification of waterfowl (Aves: Anseriformes), including selected fossil species. Annals of the Carnegie Museum 66:457–496.

Lofgren, D. 1995. The Bug Creek problem and the Cretaceous-Tertiary transition at McGuire Creek, Montana. University of California Publications in Geological Sciences 140:1–185.

Marsh, O. C. 1870. Notice of some fossil birds from the Cretaceous and Tertiary Formations of the United States. American Journal of Science, Series 2, 49:205–217.

———. 1872. Preliminary description of Hesperornis regalis, with notices of four other new species of Cretaceous Birds. American Journal of Science, Series 3, 3:360–365.

———. 1873a. Notice of a new species of Ichthyornis. American Journal of Science, Series 3, 5 (Jan.):74.

———. 1873b. On a new sub-class of fossil birds (Odontornithes). American Journal of Science, Series 3, 5:161–162.

———. 1873c. Fossil birds from the Cretaceous of North America. American Journal of Science, Series 3, 5 (March): 229–231.

———. 1877. Notice of some new vertebrate fossils. American Journal of Science, Series 3, 14:251–253.

———. 1880. Odontornithes: A Monograph on the Extinct Toothless Birds of North America. Report of the Geological Exploration of the 40th Parallel, Vol. 7. Government Printing Office, Washington, D.C.

———. 1889. Discovery of Cretaceous Mammalia. American Journal of Science, Series 3, 38:81–92, 5 plates.

———. 1892. Notes on Mesozoic vertebrate fossils. American Journal of Science, Series 3, 55:171–175, 3 plates.

Martin, L. D. 1983. The origin and early radiation of birds; pp. 291–338 in A. H. Brush and G. A. Clark (eds.), Perspectives in Ornithology. Cambridge University Press, Cambridge.

Martin, L. D., and J. Tate Jr. 1976. The skeleton of Baptornis advenus (Aves: Hesperornithiformes); pp. 35–66 in S. L. Olson (ed.), Papers in Avian Paleontology Honoring the 90th Birthday of Alexander Wetmore. Smithsonian Contributions to Paleobiology 27. Smithsonian Institution Press, Washington, D.C.

Mayr, E., and D. Amadon, 1951. A classification of recent birds. American Museum Novitates 1496:1–42.

Mayr, G. 2000. A new basal galliform bird from the Middle Eocene of Messel (Hessen, Germany). Senckenbergiana Lethaea 80:45–57.

Mayr, G., and M. C. Daniels. 1998. Eocene parrots from Messel (Hessen, Germany) and the London Clay of Walton-on-the-Naze (Essex, England). Senckenbergiana Lethaea 78:157–177.

McGowan, K. 1985. Tarsal development in birds: evidence for homology with the theropod condition. Journal of the Zoological Society of London 106:53–67.

McKenna, M. C., and S. K. Bell. 1997. Classification of Mammals above the Species Level. Columbia University Press, New York, 631 pp.

Medina, F. A., R. A. Scasso, R. A. Del Valle, E. B. Olivero, E. C. Malagnino, and C. A. Rinaldi. 1989. Cuenca Mesozoica del margen nororiental de la Península Antarctica; pp. 443–465 in G. A. Chebli and L. A. Spalletti (eds.), Cuencas Sedimentanas Argentinas. Serie Correlaciones Geologicas 6. Universidad Nacional de Tucumán, San Miguel de Tucumán, Argentina.

Mindell, D. P., M. D. Sorenson, C. J. Huddleston, H. C. Miranda Jr., A. Knight, S. J. Sawachuk, and T. Yuri. 1997. Phylogenetic relationships among and within select avian orders based on mitochondrial DNA; pp. 214–247 in D. P. Mindell (ed.), Avian Molecular Evolution and Systematics. Academic Press, San Diego.

Morton, S. G. 1834. Synopsis of the Organic Remains of the Cretaceous Group of the United States. Key and Biddle, Philadelphia, 88 pp.

Mourer-Chauviré, C. 1992. The Galliformes (Aves) from the Phosphorites du Quercy (France): systematics and biostratigraphy; pp. 67–96 in K. E. Campbell (ed.), Papers in Avian Paleontology, Honoring Pierce Brodkorb. Science Series 36. Natural History Museum of Los Angeles County, Los Angeles.

———. 2000. A new species of Ameripodius (Aves: Galliformes: Quercymegapodiidae) from the Lower Miocene of France. Palaeontology 43:481–493.

Nessov, L. A. 1992a. Mesozoic and Paleogene birds of the USSR and their paleoenvironments; pp. 465–478 in K. E. Campbell (ed.), Papers in Avian Paleontology, Honoring Pierce Brodkorb. Science Series 36. Natural History Museum of Los Angeles County, Los Angeles.

———. 1992b. Review of localities and remains of Mesozoic and Paleogene birds of the USSR and the description of new findings. Russkii Ornithologicheskii Zhurnal 1:7–50. [Russian]

———. 1997. Cretaceous Non-marine Vertebrates of Northern Eurasia. Edited by L. B. Golovneva and A. O. Averianov. University of St. Petersburg, Institute of Earth Crust, 218 pp.

Nessov, L. A., and L. J. Borkin. 1983. New records of bird bones from the Cretaceous of Mongolia and Soviet Middle Asia. Proceedings of the Zoological Institute, Leningrad 116:108–110. [Russian]

Nessov, L. A., and A. A. Jarkov. 1989. New Cretaceous-Paleogene birds of the USSR and some remarks on the origin and evolution of the Class Aves. Proceedings of the Zoological Institute, Leningrad 197:78–97. [Russian]

Nessov, L. A., and B. V. Prizemlin. 1991. A large advanced flightless marine bird of the order Hesperornithiformes of the Late Senonian of the Turgai Strait: the first finding of the group in the USSR. Proceedings of the Zoological Institute, Leningrad 239:85–107. [Russian]

Norell, M. A., and M. J. Novacek. 1992. Congruence between superpositional and phylogenetic patterns: comparing cladistic patterns with fossil records. Cladistics 8:319–337.

Noriega, J. I., and C. P. Tambussi. 1995. A Late Cretaceous Presbyornithidae (Aves: Anseriformes) from Vega Island, Antarctic Peninsula: paleobiogeographic implications. Ameghiniana 32:57–61.

Ogilvie-Grant, W. R. (ed.). 1912. General Index to A Hand-list of the Genera and Species of Birds. British Museum (Natural History), London.

Olson, S. L. 1975. *Ichthyornis* in the Cretaceous of Alabama. Wilson Bulletin 87:103–105.

———. 1985. The fossil record of birds; pp. 79–238 *in* D. S. Farner, J. R. King, and K. C. Parkes (eds.), Avian Biology, Vol. 8. Academic Press, New York.

———. 1992. *Neogaeornis wetzeli* Lambrecht, a Cretaceous loon from Chile (Aves: Gaviidae). Journal of Vertebrate Paleontology 12:122–124.

———. 1994. A giant *Presbyornis* (Aves: Anseriformes) and other birds from the Paleocene Aquia Formation of Maryland and Virginia. Proceedings of the Biological Society of Washington 107:429–435.

———. 1999. The anseriform relationships of *Anatalavis* (Anseranatidae) with a new species from the Early Eocene London Clay; pp. 231–244 *in* S. L. Olson (ed.), Avian Paleontology at the Close of the 20th Century: Proceedings of the 4th International Meeting of the Society of Avian Paleontology and Evolution, Washington, D.C., 4–7 June 1996. Smithsonian Contributions to Paleobiology 89, Smithsonian Institution Press, Washington, D.C.

Olson, S. L., and A. Feduccia. 1980a. Relationships and evolution of flamingos (Aves: Phoenicopteridae). Smithsonian Contributions to Zoology 316:1–73.

———. 1980b. *Presbyornis* and the origin of the Anseriformes (Aves: Charadriomorphae). Smithsonian Contributions to Zoology 323:1–24.

Olson, S. L., and Y. Hasegawa. 1996. A new genus and two species of gigantic Plotopteridae from Japan (Aves: Pelecaniformes). Journal of Vertebrate Paleontology 16(4):742–751.

Olson, S. L., and D. C. Parris. 1987. The Cretaceous Birds of New Jersey. Smithsonian Contributions to Paleobiology 63:1–22.

Ostrom, J. H., S. O. Poore, and G. E. Goslow Jr. 1999. Humeral rotation and wrist supination: important functional complex for the evolution of powered flight in birds?; pp. 301–310 *in* S. L. Olson (ed.), Avian Paleontology at the Close of the 20th Century: Proceedings of the 4th International Meeting of the

Society of Avian Paleontology and Evolution, Washington, D.C., 4–7 June 1996. Smithsonian Contributions to Paleobiology 89, Smithsonian Institution Press, Washington, D.C.

Owre, O. T. 1967. Adaptations for locomotion and feeding in the anhinga and the double-crested cormorant. Ornithological Monographs 6, American Ornithologists' Union, Allen Press, Lawrence, Kans., 138 pp.

Padian, K., and L. M. Chiappe. 1998. The origin and early evolution of birds. Biological Reviews 73:1–42.

Poore, S. O., A. Sanchez-Haiman, and G. E. Goslow Jr. 1997. Wing upstroke and the evolution of flapping flight. Nature 387:799–800.

Powell, J. E. 1987. Hallazgo de un dinosaurio Hadrosaurido (Ornithischia, Ornithopoda) en la Formación Allen (Cretácico Superior) de Salitrál Moreno, Provincia de Rio Negro, Argentina. X Congreso Geológico Argentino, San Miguel de Tucumán, Actas 3:149–152.

Pycraft, W. P. 1900. On the morphology and phylogeny of the Palaeognathae (Ratitae and Crypturi) and Neognathae (Carinatae). Transactions of the Zoological Society of London 15:149–290.

———. 1901. Some points in the morphology of the palate of the Neognathae. Journal of the Linnean Society of London 28:343–357.

Raikow, R. J. 1978. The appendicular myology and its taxonomic significance in the passerine suborder Menurae. American Zoologist 18(3):377 (abstract).

———. 1985. Problems in avian classification; pp. 187–212 *in* R. F. Johnston (ed.), Current Ornithology, Vol. 2, Plenum, London.

———. 1987. Hindlimb Myology and Evolution of the Old World Suboscine Passerine Birds (Acanthisittidae, Pittidae, Philepittidae, Eurylaimidae). Ornithological Monographs 41, American Ornithologists' Union, Washington, D.C.

Riccardi, A. C. 1988. The Cretaceous system of southern South America. Geological Society of America, Memoires 168:1–161.

Richards, H. G., and W. B. Gallagher. 1974. The problem of the Cretaceous-Tertiary boundary in New Jersey. Notulae Naturae, Academy of Natural Sciences of Philadelphia 449:1–6.

Sanz, J. L., L. F. Bonaparte, and A. Lacasa. 1988. Unusual Early Cretaceous birds from Spain. Nature 331:433–435.

Sanz, J. L., L. M. Chiappe, and A. D. Buscalioni. 1995. The osteology of *Concornis lacustris* (Aves: Enantiornithes) from the Lower Cretaceous of Spain and a reexamination of its phylogenetic relationships. American Museum Novitates 3133:1–23.

Sanz, J. L., L. M. Chiappe, B. P. Perez-Moreno, J. J. Moratalla, F. Hernandez-Carrasquilla, A. D. Buscalioni, F. Ortega, F. J. Poyato-Ariza, D. Rasskin-Gutman, and X. Martínez-Delclòs. 1997. A nestling bird from the Lower Cretaceous of Spain: implications for avian skull and neck evolution. Science 176: 1543–1546.

Sereno, P. C., and Rao C. 1992. Early evolution of avian flight and perching: new evidence from the Lower Cretaceous of China. Science 255:845–848.

Sharpe, R. B. 1891. A review of recent attempts to classify birds. Proceedings of the 2nd International Ornithological Congress, Office of the Congress, Budapest.

———. 1899. A Hand-list of the Genera and Species of Birds, Vol. 1. British Museum (Natural History), London.

Sheldon, F. H., and A. H. Bledsoe. 1993. Avian molecular systematics, 1970s to 1990s. Annual Review of Ecology and Systematics 24:243–278.

Shufeldt, R. W. 1915. Fossil birds in the Marsh Collection of Yale University. Transactions of the Connecticut Academy of Arts and Sciences 19:1–110.

Sibley, C. G., and J. E. Ahlquist. 1990. Phylogeny and Classification of Birds. Yale University Press, New Haven, 976 pp.

Siegel-Causey, D. 1988. Phylogeny of the Phalacrocoracidae. Condor 90:885–905.

Simpson, G. G. 1976. Penguins: Past and Present, Here and There. Yale University Press, New Haven, 150 pp.

Staron, R. M., D. E. Grandstaff, B. S. Grandstaff, and W. B. Gallagher. 1999. Mosasaur taphonomy and geochemistry implications for a K-T bone bed in the New Jersey coastal plain. Journal of Vertebrate Paleontology 19(Supplement to 3):78A.

Stidham, T. A. 1998a. The most diverse Cretaceous avifauna in the world. Abstracts, Proceedings of the 22nd International Ornithological Congress, Durban, South Africa, August, 1998, Ostrich 69:413.

———. 1998b. Phylogenetic and ecological diversification of waterfowl (Anseriformes) in the Late Cretaceous and Paleocene. Journal of Vertebrate Paleontology 18(Supplement to 3):80A.

———. 1998c. The lower jaw from a Cretaceous parrot. Nature 396:29–30.

——— 1999. Reply to G. J. Dyke and G. Mayr. Did parrots exist in the Cretaceous period? Nature 399:318.

Storer, R. W. 1960. Evolution in the diving birds; pp. 694–707 in G. Bergman, K. O. Donner, and L. von Haartman (eds.). Proceedings of the 12th International Ornithological Congress, Helsinki, 1958. Tilgmannin Kirjapaino, Helsinki.

Strauch, J. G., Jr. 1978. The phylogeny of the Charadriiformes: a new estimate using the method of character compatibility analysis. Transactions of the Zoological Society of London 34:263–345.

Stresemann, E. 1959. The status of avian systematics and its unsolved problems. Auk 76:269–280.

Tatarinov, L. P. 1998. Non-marine Cretaceous vertebrates of the former USSR. [Review of Nessov, L. A. 1997. Cretaceous Non-marine Vertebrates of Northern Eurasia. University of St. Petersburg, St. Petersburg, Russia]. Acta Palaeontologia Polonica 43:468–470.

Temminck, C. J. 1820. Manuel d'ornithologie: Un tableau systematique des oiseaux qui se trouvent en Europe, 2nd edition, Vol. 1. G. Dutour, Paris, 46 pp.

Tokaryk, K. T., and P. C. James. 1989. *Cimolopteryx* sp. (Aves, Charadriiformes) from the Frenchman Formation (Maastrichtian), Saskatchewan. Canadian Journal of Earth Sciences 26:2729–2730.

Tyrberg, T. 1986. Cretaceous birds—a short review of the first half of avian history. Verhandlungen der ornithologischen Gesellschaft in Bayern 24:249–275.

Unwin, D. M. 1988. Extinction and survival in birds; pp. 295–318 *in* G. P. Larwood (ed.), Extinction and Survival in the Fossil Record. Systematics Association Special Volume 34. Clarendon Press, Oxford.

———. 1993. Aves; pp. 717–737 *in* M. J. Benton (ed.), The Fossil Record. Chapman and Hall, London.

van Tuinen, M. 2000. The early history of modern birds inferred from DNA sequences of nuclear and mitochondrial ribosomal genes. Molecular Biology and Evolution 17:451–457.

van Tuinen, M., C. G. Sibley, and S. B. Hedges. 1998. Phylogeny and biogeography of ratite birds inferred from DNA sequences of the mitochondrial ribosomal genes. Molecular Biology and Evolution 15:370–376.

Wagler, J. G. 1830. Natürliches System der Amphibien mit Vorangehender Classification der Säugethiere und Vögel. J. G. Cotta, Munich, 354 pp.

———. 1831. Isis von Oken. Vol. 5. Reprinted in Sclater, P. L. (ed.). 1884. Wagler's Six Ornithological Memoirs from the Isis. Taylor and Francis, London, 137 pp.

Walker, C. A. 1981. New subclass of birds from the Cretaceous of South America. Nature 292:51–53.

Welty, J. C., and L. F. Baptista. 1988. The Life of Birds, 4th edition. Saunders, New York, 136 pp.

Wetmore, A. 1926. Fossil birds from the Green River deposits of Eastern Utah. Annals of the Carnegie Museum 16:391–402.

———. 1930a. The age of the supposed Cretaceous birds from New Jersey. Auk 47:186–187.

———. 1930b. A systematic classification for the birds of the world. Proceedings of the United States National Museum 54:577–586.

———. 1956. A check-list of the fossil and prehistoric birds of North America and the West Indies. Smithsonian Miscellaneous Collection 131(5):1–105.

———. 1960. A classification for the birds of the world. Smithsonian Miscellaneous Collection 139(11):1–37.

Wetmore, A., and W. D. Miller. 1926. A revised classification for the fourth edition of the AOU Check-list. Auk 43:337–346.

Witmer, L. M. 1990. The craniofacial air sac system of Mesozoic birds (Aves). Zoological Journal of the Linnean Society 100:327–378.

———. 1995. The extant phylogenetic bracket and the importance of reconstructing soft tissues in fossils; pp. 19–33 *in* J. J. Thomason (ed.), Functional Morphology in Vertebrate Paleontology. Cambridge University Press, Cambridge.

———. 1999. New aspects of avian origins: roundtable report; pp. 327–334 *in* S. L. Olson (ed.), Avian Paleontology at the Close of the 20th Century: Proceedings of the 4th International Meeting of the Society of Avian Paleontology and Evolution, Washington, D.C., 4–7 June 1996. Smithsonian Contributions to Paleobiology 89, Smithsonian Institution Press, Washington, D.C.

Witmer, L. M., and L. D. Martin. 1987. The primitive features of the avian palate, with special reference to Mesozoic birds; pp. 21–40 *in* C. Mourer-Chauviré (ed.), L'évolution des Oiseaux d'après le Témoignage des Fossiles. Table Ronde Internationale du CNRS, Lyon-Villeurbanne, Septembre, 1985. Documents des Laboratoires de Géologie de la Faculté des Sciences de Lyon 99. Université Claude-Bernard, Villeurbanne Cedex.

Witmer, L. M., and K. D. Rose. 1991. Biomechanics of the jaw apparatus of the gigantic Eocene bird *Diatryma*: implications for diet and mode of life. Paleobiology 17:95–120.

Woolfenden, G. E. 1961. Postcranial osteology of the waterfowl. Bulletin of the Florida State Museum 6(1):1–129.

Zusi, R. L. 1984. A functional and evolutionary analysis of rhynchokinesis in birds. Smithsonian Contributions to Zoology 395:1–40.

16

A Review of Avian Mesozoic Fossil Feathers

ALEXANDER W. A. KELLNER

 Feathers are among the most complex integumentary structures found in vertebrates. They are composed primarily of keratin, which consists of insoluble microscopic filaments that are embedded in a proteinaceous matrix. Keratin is regarded as a long-lasting material (Gill, 1990). The amino acid composition of feather keratin, and its shape and behavior, differ remarkably from those of the keratins found in other hard epidermal tissues such as claws, hair, and scales (Brush, 1996). Therefore, feathers are generally regarded as a unique evolutionary novelty of birds (used here as an informal term for all members of Aves [= Avialae sensu Gauthier, 1986]).

The first Mesozoic fossil feather was found in Solnhofen (southern Germany) in the nineteenth century. Since then, many more have been recovered from fossil deposits around the world, but most specimens remain undescribed. In this chapter, I review the occurrences of feathers in Mesozoic strata, which are presented in temporal sequence. I have emphasized the specimens that are poorly known, particularly the isolated feathers. Whenever possible, I have provided a brief description of these fossils in order to justify their assignment to one of the main structural categories of feathers. The classification and terminology of the feathers and their structures mostly follow Lucas and Stettenheim (1972). I was unable to examine certain specimens firsthand, having to rely on photographs, some of which are already published elsewhere.

The following institutional abbreviations are used in this chapter: IEI, Instituto de Estudios Ilerdenses, Lleida, Spain; LH, Collection of Las Hoyas, Unidad de Paleontologia, Universidad Autónoma de Madrid, Madrid, Spain; MB, Museum für Naturkunde, Humboldt-Universität, Berlin, Germany; MCT, Museu de Ciências da Terra, Departamento Nacional da Produção Mineral, Rio de Janeiro, Brazil; PIN, Paleontological Institute of Russia, Academy of Sciences, Moscow, Russia.

Structure and Basic Types of Feathers

Ornithologists have studied feathers in detail, showing the enormous variation of this integumentary structure (e.g., Chandler, 1916; Lucas and Stettenheim, 1972). Basically, feathers are formed by two broad, flat vanes that are separated from each other by a central shaft (Fig. 16.1). The basal portion of the shaft is called the calamus and attaches the feather to the bird's body through a follicle in the skin. The calamus is comparatively short and is hollow. In the young feather, it has an opening at its basal end (the proximal umbilicus) through which blood is supplied during the feather's development. After the short growth period ceases, this hole is closed by a horny covering, and the feather is sealed off.

The main part of the shaft is formed by the rachis, which is long and solid, supporting the vanes. On the ventral side, the rachis has a longitudinal groove. Laterally, the rachis carries several closely spaced, fine branches, called barbs, that form the primary structure of the vanes. The boundary between the rachis and the calamus is marked by an opening in the calamus (the distal umbilicus) and the presence of the basal barbs. On each side of the barbs there is a row of smaller branches, the barbules, some of which bear tiny hooklets. This kind of arrangement forms the cohesive but very flexible surface that is the main part of the feather. In some cases, a hypopenna (afterfeather) is present, with the hyporachis (aftershaft) branching off the main feather at the boundary between the calamus and rachis. With only a few exceptions (e.g., cassowaries), the texture of afterfeathers is plumaceous.

Feather structures can show wide variation in appearance and function. Five main categories of feathers are recognized today: contour feathers (body feathers, remiges, and rectrices), down feathers, semiplumes, filoplumes, and bristles. It is important to note, however, that there are many

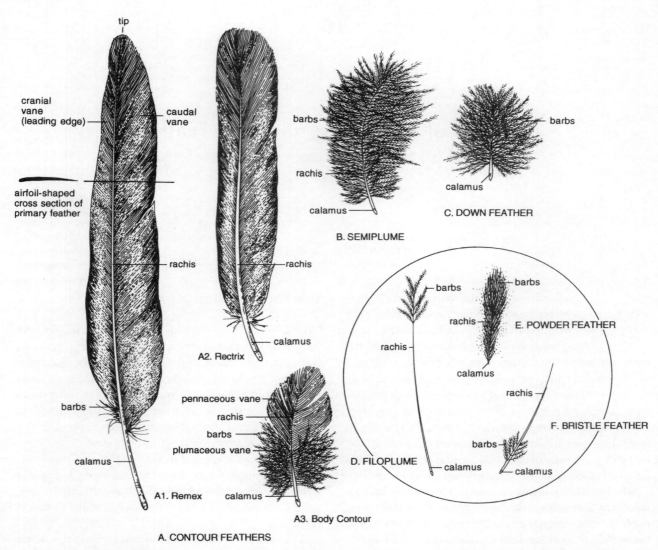

cranial
vane
(leading edge)

caudal
vane

tip

airfoil-shaped
cross section of
primary feather

rachis

barbs

calamus

rachis

calamus

A2. Rectrix

A1. Remex

pennaceous vane

rachis

barbs

plumaceous vane

calamus

A3. Body Contour

A. CONTOUR FEATHERS

barbs

rachis

calamus

B. SEMIPLUME

barbs

calamus

C. DOWN FEATHER

barbs

rachis

rachis

calamus

D. FILOPLUME

barbs

rachis

calamus

E. POWDER FEATHER

rachis

barbs

calamus

F. BRISTLE FEATHER

Figure 16.1. Typical contour feather (from the body), showing the main structures found in feathers, and variations of feather types (from Chatterjee, 1997).

intergrading feathers between most categories, making their classification sometimes very difficult and somewhat arbitrary. Constituting most of a bird's plumage, the contour feathers (Fig. 16.1A) are the ones most commonly known. They include the typical body feathers, the flight feathers (remiges) and tail feathers (rectrices), and the ear coverts. Body feathers are of comparatively medium size, with a distal pennaceous and basal plumaceous texture. Among the several functions advocated for body feathers (Stettenheim, 1976), one of the most important is water repellence. Remiges and rectrices are contour feathers that have been greatly modified for their roles in flight. They are stiff, comparatively large, and mostly pennaceous and lack an afterfeather. Especially in flight feathers, the outer and inner vanes can be larger or smaller, depending on the feather's position in the wing. This asymmetry of the vanes

has also been used to determine the flight capability of some birds (Feduccia and Tordoff, 1979). Ear coverts surround the external opening of the ear and are modified contour feathers thought to improve hearing ability (Stettenheim, 1976). They are characterized by an open pennaceous texture, formed by relatively stiff and widely spaced barbs and very simple barbules. This type of feather is particularly developed in nocturnal birds (e.g., some owls) but is also developed in birds such as hawks.

Down feathers (Fig. 16.1C) are very small and fluffy, forming the underplumage of the bird. The rachis is either shorter than the longest barb or is entirely absent. Barbs are loosely connected, providing the fluffy texture. Further subcategories of down feathers are recognized, such as natal downs, definitive downs, and powder feathers (Stettenheim, 1976). Intermediate between contour and down feathers are the semiplumes (Fig. 16.1B). This type of feather has a large rachis and entirely plumaceous vanes. The size of the rachis is always larger than the longest barb, which is one of the main differences between semiplumes and down feathers (Lucas and Stettenheim, 1972). Functionally, semiplumes fill out the contour of the bird's body and also provide thermal insulation. Very distinctive from other feathers, the filoplumes are hairlike and, once fully developed (Fig. 16.1D), consist of a thin shaft with a cluster of a few short barbs and barbules at the tip. The functions of filoplumes are apparently related to the enhancement of the sense of touch within the plumage (Proctor and Lynch, 1993) and can therefore be regarded as highly specialized structures. Another category of feathers are bristles (Fig. 16.1F). A bristle is identified by a stiff rachis that either completely lacks barbs or has them restricted to the apical region. Similar feathers with more side branches are known as semibristles; these have been regarded as intermediate between contour feathers and bristles (Lucas and Stettenheim, 1972). The functions of bristles and semibristles are apparently of a sensory and protective nature (Stettenheim, 1976).

Mesozoic Feathers

The following section presents in temporal order some of the more important specimens of fossil feathers (see Table 16.1). Given the recent identification of feathers and feather-like structures in nonavian theropods, it is conceivable that some of the feathers discussed here actually pertain to taxa outside Aves. Nevertheless, they all are included here, and it seems likely that most, if not all, derive from birds.

Solnhofen, Southern Germany

The most famous fossil feathers are those attributed to *Archaeopteryx*. These specimens were found in the Solnhofen Limestone of Lower Tithonian age, deposited in a lagoonal environment (Viohl, 1985). The literature discussing several issues on the feathers of *Archaeopteryx,* particularly their preservation, aerodynamics, flight capability, and origin, is extensive (e.g., Rau, 1969; Feduccia and Tordoff, 1979; Hecht et al., 1985; Bock, 1986; Ostrom, 1986; Feduccia, 1993, 1996) and will not be repeated here (but see Elzanowski, Chapter 6 in this volume).

The first feather from this area (Fig. 16.2) was reported by von Meyer (1861a) as essentially a feather that could not be distinguished from a modern bird feather. After careful examination of this feather, von Meyer (1861b) named it *Archaeopteryx lithographica.* Consequently, all the skeletal remains of this early bird are actually referred specimens (see Elzanowski, Chapter 6 in this volume). Von Meyer (1861b) also pointed out the existence of a complete skeleton with the impression of feathers that was found in the same region as the isolated feather. This first *Archaeopteryx*

feather is an almost complete primary flight feather preserved in two limestone slabs that were separated along the bedding plane. The slab with the most complete part of the feather (Fig. 16.2A) is currently housed in the Museum für Naturkunde, Humboldt Universität, Berlin (MB.Av.100), while its counterpart (Fig. 16.2B) is housed in the Bayerische Staatssammlung für Paläontologie und Historische Geologie in Munich. The main part (MB.Av.100) was partially prepared, resulting in the loss of the distal portions of some barbs on the basal section. The counterpart is unprepared and lacks the more basal portion of the feather. The specimen is 58 mm long. The proximal tip of the calamus is missing, while the distal portion is about 2 mm thick. The rachis tapers toward the apex, and the shaft is curved, possibly reflecting the natural curvature of the feather. The vanes are asymmetrical and well preserved and have closely connected barbs. No barbules could be observed.

Several skeletons of *Archaeopteryx* have feathers preserved (Ostrom, 1985). The plumage of some specimens has been described in detail (e.g., Rietschel, 1985). Whereas the feathers associated with the bones are preserved only as impressions, the first *Archaeopteryx* feather is apparently preserved as carbonized traces. This supposition is based on the dark color of this feather, which distinguishes it from the grayish matrix. Similar feathers from other deposits that were chemically examined showed this kind of preservation (Davis and Briggs, 1995).

Other isolated feathers attributed to *Archaeopteryx* from Solnhofen were mentioned in the literature. In a letter published by Shufeldt (1913), C. Schuchert referred to more feathers of this taxon. It is unknown where those specimens are presently housed.

Gurvan Eren, Western Mongolia

Kurochkin (1985a,b) illustrated some feathers that were found in the Gurvan Eren Formation, at a homonymous locality. This unit was deposited in a fluvial/lacustrine environment during the Berriasian (Kurochkin, pers. comm., 1995). Among the material is an incomplete primary feather that lacks the proximal part (Fig. 16.3A). According to the published photo (Kurochkin, 1985a: Fig. 2d), the total preserved length of this feather is approximately 53 mm. The rachis is well preserved and tapers toward the apex. It is straight for most of the preserved part but tends to be slightly curved distally, possibly reflecting the original shape of the feather. Barbs are well developed; barbules are not preserved. Based on the asymmetry of the vanes, this feather is possibly a primary from a volant bird, as Kurochkin (1985a) has already proposed. Another interesting feather from the Gurvan Eren locality is a possible tail feather (Fig. 16.3B). Again, according to the published picture (Kurochkin, 1985a: Fig. 2c), the preserved length of this feather is approximately 28 mm. The proximal part of this specimen

TABLE 16.1
Record of Mesozoic feathers

Deposit	Geological Age	Depositional Environment	Feathers
Southern Alberta, Canada	Campanian	Terrestrial	One remex, one semiplume (?), and two more preserved in amber
Taldysay, Kazakhstan	Santonian–?Campanian	Estuarine	One contour feather
Northern Siberia, Russia	Santonian	Terrestrial	One feather preserved in amber
Kuji, Japan	Santonian	Terrestrial	One incomplete feather preserved in amber
Tjulkeli, Kazakhstan	Upper Turonian-Coniacian	Estuarine	One feather
New Jersey, United States	Turonian	Terrestrial	One semiplume and one fragmentary feather, both preserved in amber
Chaibu-Sumi, Inner Mongolia, China	Early Cretaceous	Lacustrine (?)	At least two remiges associated with one skeleton (*Otogornis*)
Khurit-Ulan-Bulak, Mongolia	Early Cretaceous	Lacustrine	Several isolated feathers, some associated with one skeleton (*Ambiortus*)
Araripe Basin, Brazil	Aptian	Lacustrine	One remex, several semiplumes, body, and down feathers
Koonwarra, Australia	Barremian-Aptian	Lacustrine	Several isolated semiplumes and body feathers
Las Hoyas, Spain	Barremian	Lacustrine/palustrine	Several isolated contours (remiges, body feathers); feathers associated with skeletons (*Concornis, Eoalulavis*)
Transbaikalia, Russia	Hauterivian-Barremian	Lacustrine	Several contour feathers (one remex, body feathers)
Sheen Khuduk, Mongolia	Hauterivian-Barremian	Lacustrine	One contour feather
Jezzine, Lebanon	Hauterivian	Terrestrial	Two contour feathers preserved in amber
El Montsec, Spain	Berriasian-Valanginian	Lacustrine	Contour, semiplumes, remiges; several feathers associated with two skeletons (*Noguerornis*, enantiornithine hatchling)
Gurvan Eren, Mongolia	Berriasian	Fluvial/lacustrine	One remex, one rectrix, one down feather (?), several contour feathers
Liaoning, China	Tithonian-?Berriasian	Lacustrine	Several isolated feathers; many feathers associated with skeletons (*Confuciusornis* and *Protarchaeopteryx*)
Solnhofen, Germany	Lower Tithonian	Lagoonal	One isolated remex; several remiges and rectrices associated with skeletons (*Archaeopteryx*)

Note: Only the confirmed identifications are presented here.

is missing. The rachis is straight, and the vanes are symmetrical. Barbs are well preserved, but no barbules can be recognized. In addition to the feathers mentioned previously, there are several other specimens from Gurvan Eren, most of them contour feathers deriving from unknown parts of the body (Fig. 16.3C). Again, rachis and barbs are preserved, but no evidence of barbules can be found. Kurochkin (1985a: Fig. 2a) also pointed out the presence of a down feather from this locality. If this is a true down feather (not figured here), it has a very reduced rachis and has no preserved barbules on the barbs (Kellner et al., 1994). All these feathers have a dark color and are presumably carbonized.

El Montsec, Lleida, Spain

The area of El Montsec is situated in the northern part of the province of Lleida in Catalonia, Spain. The sedimentary rocks of this area are lithographic limestone of Berriasian-Valanginian age (Brenner et al., 1974) that were deposited in a lacustrine environment. The first isolated feather from this locality was reported by Ferrer-Condal (1955). Subsequently, Pallerola (1979) mistook the remains of a fish for bones of a bird with associated "feathers," which are actually fins (e.g., Nessov, 1984). Pallerola (1979) also mentioned the existence of four isolated feathers but did not provide any information about those specimens.

Additional feathers have been reported by Lacasa-Ruiz (1985, 1986, 1989, 1993), some associated with bird remains. Among the best is one small body feather (27 mm) deposited in the Instituto de Estudios Ilerdenses, Lleida (LP 715 IEI; Fig. 16.4A), which was described in detail by Lacasa-Ruiz (1985). Two more feathers from El Montsec are briefly described here. The first one (Fig. 16.4B) has symmetrical vanes, with the barbs loosely connected. The

Figure 16.2. First isolated feather of *Archaeopteryx*, Late Jurassic, Solnhofen, described by von Meyer (1861a,b): A, main part; B, counterpart.

rachis, preserved only as an impression, is apparently longer than the longest barb. In some areas, barbules are preserved. The calamus cannot be observed. Owing to the described features and its plumaceous aspect, this feather likely represents a semiplume. The second feather (Fig. 16.4C) is incomplete, with calamus and tip missing. The vanes are symmetrical, although some barbs have been moved from their natural position. Unlike in the previous specimens, the rachis of this feather is much longer than any preserved barb, suggesting that it represents a small flight feather. Lacasa-Ruiz (1985, 1989) noted the presence of dotted sequences forming the barbules (and apparently also some barbs) of several feathers from El Montsec. He also pointed out that in neornithines such dotted sequences are the results of tiny deposits of pigment, which might have been the case with these fossil feathers. It is interesting to note that the rachises of the feathers found in the El Montsec area are preserved only as impressions and are observable as white lines between the vanes of the feather. Barbs and barbules, however, can be identified by their dark color and are presumably preserved as carbonized traces.

Figure 16.3. Several feathers found in Gurvan Eren, Early Cretaceous of western Mongolia: A, flight feather; B, tail feather; C, several body feathers.

Amber Deposits, Jezzine, Southern Lebanon

Some of the most interesting specimens of Mesozoic feathers were found in amber deposits in the southern part of Lebanon. Those deposits occur in several horizons near Jezzine and are regarded as being Early Cretaceous (Hauterivian) in age (Schlee, 1973). Only two specimens containing several fragments of feathers are known. A color photograph of one of those specimens was presented by Schlee and Glöckner (1978: taf. 3). The complete specimen is about 7 mm high and 7 mm long, having a depth of 4 mm (Schlee, 1973). One main feather is visible, but incomplete remains

of others are found deeper in the amber. Rachises, barbs, and barbules can be observed in detail. Hooklets are occasionally observed, linking barbules together. The material represents contour feathers, possibly from the trunk of one bird. Other structures were described in detail by Schlee (1973). Unfortunately, no detailed pictures of this extremely well-preserved specimen are available.

Sheen Khuduk, Central Mongolia

One feather from the Sheen Khuduk locality of central Mongolia has been reported by Kurochkin (1985a,b). This specimen was collected from the Hauterivian-Barremian rocks of the Sheen Khuduk Suite (Kurochkin, pers. comm., 1995). Although nearly complete, the specimen is not very well preserved (Fig. 16.5). Its length is about 35 mm, measured along the rachis, which is slightly curved distally. The extension of the calamus cannot be determined. The vanes are essentially symmetrical, with the barbs of one side more loosely connected. On the proximal portion of the feather, only the rachis is preserved; the vanes were lost, possibly during collecting, or are preserved in the counterpart (not figured). No barbules can be identified. Owing to its general characteristics, particularly the pennaceous aspect, this specimen represents a contour feather, possibly from the body.

Bajsa, Chita Oblast, Transbaikalia, Russia

Several feathers have been reported from the Bajsa locality (Bajsa Formation). The rocks of this unit were formed in a lacustrine depositional environment. The age of those layers was once regarded to be Valanginian-Hauterivian (Kurochkin, 1985b) but is now considered as Hauterivian-Barremian (Kurochkin, pers. comm., 1995). Among the

Figure 16.4. Several feathers from the Early Cretaceous of Montsec, Catalonia, Spain. A, body feather. B, semiplume. C, contour feather (flight feather?). Scale bar applies only to A.

Figure 16.5. Body feather from Sheen Khuduk, Early Cretaceous of Central Mongolia.

The locality of Las Hoyas is positioned in the Serrania de Cuenca, in the Spanish province of Cuenca. The lithographic limestone containing the fossil feathers belongs to the "Calizas de La Huārguina" Formation, which was deposited in a lacustrine-palustrine environment during the Barremian (Fregenal-Martínez and Meléndez, 1993). This locality first yielded one isolated feather (Sanz et al., 1988), but more specimens are currently known, including several bird skeletons with associated feathers (Sanz and Buscalioni, 1992; Sanz et al., 1996; Sanz et al., Chapter 9 in this volume). Among the few isolated feathers known from this locality is an incomplete contour feather (Fig. 16.7). This specimen (LH 927003) has not been prepared, and most of its borders are partly covered by matrix. The feather is about 36 mm long, with a straight rachis that tapers toward the apex. The calamus is long, with a length of approximately 9 mm. The vanes are asymmetrical, although one is not very well preserved. The barbs are well developed and tightly connected, but no barbules can be identified. Owing to its characteristics, particularly the asymmetry of the vanes, this specimen represents a flight feather. Associated with the skeleton of *Concornis* (LH 2814), described by Sanz and Buscalioni (1992), there is evidence of feathers, particularly near the right tarsometatarsus and the right wing elements (see Sanz et al., Chapter 9 in this volume). Several flight and body feathers can be observed, some of which overlap. Their preservation, however, is poor. The barbs in several feathers are very clear, but barbules are apparently not preserved.

most interesting feathers reported from this area is a contour feather that shows an indication of the color pattern (Kurochkin, 1985a,b). This specimen is incomplete, lacking both the proximal and distal ends (Fig. 16.6). The preserved length is approximately 37 mm, measured along the rachis, which tapers until it distally fades away. The rachis is only slightly curved, possibly reflecting the original curvature of the feather. The vanes are symmetrical, with most of the barbs closely connected. Distally, the barbs are less marked and tend to fade away, suggesting that they have been worn off. Apparently, no barbules are preserved. The color pattern of this feather is indicated by two irregular light bands that are not entirely symmetrical, for example, darker areas on one vane sometimes corresponding to lighter areas on the other. These bands contrast with the dark color of the remaining parts of the feather. Another light area is found more distally but is confined to one vane. It is formed by irregular dots that are partially connected. Other specimens from the Bajsa locality represent flight feathers (PIN 3064/10598) or body feathers (e.g., PIN 3064/10597), some suggesting the presence of an afterfeather (PIN 3064/10611).

Figure 16.6. Contour feather from Bajsa, Chita Oblast, Transbaikalia, Early Cretaceous of Russia. Observe the color pattern presented by lighter and darker bands and dots.

Figure 16.7. Isolated feather (LH 927003) from Las Hoyas, Early Cretaceous of Spain.

Koonwarra, Southeastern Victoria, Australia

The Koonwarra locality (south Gippsland) has yielded several feathers and is one of two deposits that provides evidence of Mesozoic birds in Australia (Rich, 1991). The sediments of this locality are predominantly claystones that, based on palynological data, were deposited between the Aptian and the Barremian (Dettman, 1986) in a lacustrine environment (Rich, 1991).

The first two fossil feathers from this locality were described in detail by Talent et al. (1966). The best-preserved specimen (Fig. 16.8A) is very curved, possibly as a result of transport prior to fossilization. It is a small feather about 20 mm long. The vanes are symmetrical, and the barbs are loosely connected, exhibiting well-developed barbules. Talent et al. (1966) regarded this specimen as representing a contour feather. However, because of the plumaceous aspect of the vanes, this feather more closely resembles a semiplume. The second specimen figured by Talent et al. (1966: Fig. 1) is not as well preserved. Judging from the published picture, there are two superimposed structures in the same slab that represent either two feathers or one feather with an afterfeather. A third feather from the Koonwarra locality was described by Waldman (1970). This specimen has slightly asymmetrical vanes, a curved rachis, and a well-developed calamus (Fig. 16.8B). The pennaceous aspect of this fossil indicates that it is a contour feather, possibly from the body. Recently, another specimen from this area was found (Fig. 16.8C). It is almost complete, but the proximal end (including the calamus) is missing. The rachis is straight and longer than all barbs. The vanes are symmetrical and plumaceous. Some of the loosely connected barbs exhibit well-developed barbules. This feather is also a semiplume, very similar to the first described by Talent et al. (1966). Other feathers from the Koonwarra locality, mostly semiplumes, are figured by Rich (1991: plate 1).

Santana Formation, Araripe Basin, Northeastern Brazil

The Santana Formation of the Araripe Basin is divided into three members: Crato, Ipubi, and Romualdo (Beurlen, 1971). This unit is known worldwide for its richness and exceptional preservation of fossils (Maisey, 1991). So far, feathers have been found only in the Crato Member. The Crato Member is composed of laminated, light gray, micritic limestones alternating with dark-colored, calcareous, silty shales (Mabesoone and Tinoco, 1973). These sediments were deposited mainly in freshwater conditions (Beurlen, 1971), which is corroborated by the presence of freshwater insect larvae and anurans. Based on palynological data, the age of this unit is regarded as Late Aptian (Pons et al., 1990).

Several different kinds of feathers are known from the Crato Member, all from quarries near the town of Nova Olinda in the state of Ceará. The first one (Fig. 16.9) was reported by Martins Neto and Kellner (1988), and a color picture of this specimen was subsequently published (inverted) in Kellner et al. (1991). It is about 64 mm long with a maximum breath of 8 mm. The rachis is curved, and the vanes are asymmetrical. The barbs are preserved, but no evidence of barbules has been found. Owing to its general configuration, this feather possibly represents a trailing primary or secondary flight feather (Martins Neto and Kellner, 1988; Kellner et al., 1991). So far, this is the only flight feather known from these strata.

A small semiplume (21 mm long) from the Crato Member was reported by Martill and Filgueira (1994). The rachis

Figure 16.8. Several feathers from Koonwarra, southeastern Victoria, Australia. A, isolated feather (semiplume) (counterpart of the one figured by Talent et al., 1966). B, contour feather. C, isolated feather (semiplume).

is curved, and the vanes are formed by loose and plumaceous barbs. Barbules are well preserved and can be observed in several barbs. Kellner et al. (1994) reported a very small (length: 7.5 mm) and fluffy down feather from these sediments (Fig. 16.10). The rachis is approximately 3.4 mm long but shorter than the largest barbs. Barbules are very well preserved and tend to be longer on the proximal part of the barbs, increasing the down density closer to the rachis. More recently, an incomplete contour feather (preserved length: 7 mm) was reported by Martill and Frey (1995). The interesting feature of this specimen is the presence of alternate dark and light bands that indicate the color pattern. Because of its overall configuration, this specimen very likely represents only the apex of the pennaceous portion of a body feather.

A new feather showing color pattern is briefly described here (Fig. 16.11). It is a complete contour feather, possibly from the body. This feather is almost 22 mm long and has a curved rachis. The vanes are symmetrical, and the proximal portion of the calamus is not preserved. The color pattern is represented by five pairs of alternating dark and light bands that are essentially perpendicular to the rachis. Those bands are more visible toward the apex of the pennaceous part of the feather; they fade toward the basal portion, which lacks any color banding. In addition to these, several other feathers were found in the sediments of the Crato Member and are housed in the American Museum of Natural History, New York City, and the Museu Nacional, Rio de Janeiro. They are very small in size (less than 20 mm) and represent either down feathers or semiplumes.

Late Cretaceous of New Jersey, United States

A feather preserved in amber from North America was recently reported by Grimaldi and Case (1995). This specimen comes from the Raritan Formation from central New Jer-

sey. The amber is preserved in lignite and has yielded several insects as well as plant material. The lignite lies over the South Amboy Fire Clay, which, according to palynological data, was formed in the Turonian (Christopher, 1979). Grimaldi et al. (1989) regarded the amber as apparently deriving from an araucarian forest. The site where the amber containing the feather was found represents a lagoonal or deltaic depositional environment where terrestrial plant material was redeposited (Grimaldi and Case, 1995). The preserved part of the feather (Fig. 16.12) is 7.57 mm long. The rachis is very curved, with the distal half bent unnaturally, possibly as a result of being embedded in the plant resin. On each side of the rachis there are more than 200 barbs, none of which bear barbules. The barbs are considerably thinner than the rachis. The proximal region of the feather was broken off, and therefore the total number of barbs and the extension of the calamus are unknown. Because of the plumaceous aspect of the feather and the presence of a well-defined rachis that is larger than all preserved barbs, it has been interpreted as a semiplume (Grimaldi and Case, 1995). Recently, fragments of another feather were found in this same locality. This specimen is also preserved in amber and is currently in a private collection (Grimaldi, pers. comm., 1996).

Taneishi Formation, Kuji, Japan

Grimaldi and Case (1995) mentioned the presence of a feather from the Santonian Uge Member of the Taneishi Formation of Kuji, Japan, which was figured by Sasaki (1996). I had access to a color photograph of this material, which is briefly described here. The material is very fragmentary; it seems to contain one incomplete feather. The area that contains remains of the feather is only 4 mm long (Sasaki, pers. comm., 1996). There are at least seven barbs, with long and widely spaced barbules. The largest preserved

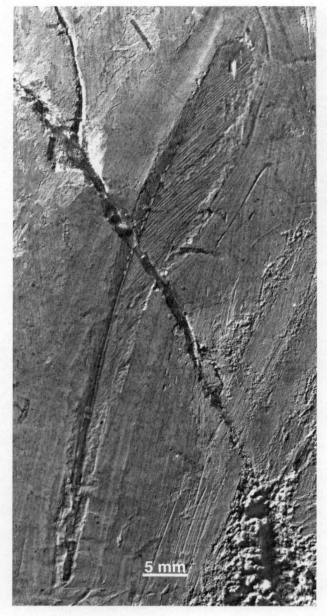

Figure 16.9. Flight feather from the Santana Formation (Crato Member), Early Cretaceous of Brazil.

Figure 16.10. Down feather (MCT 1493-R) from the Santana Formation (Crato Member), Early Cretaceous of Brazil.

portion of a barb is approximately 2.9 mm long (considering the curvature); the length of the smallest does not reach 1 mm. Some of the barbs are superimposed, while others point in different directions inside the amber. All barbs seem to converge to one central axis, suggesting that they all belong to the same feather.

Southern Alberta, Canada

Fragments of feathers have also been found in amber of Late Cretaceous age from southern Alberta, Canada (Davis and Briggs, 1995; Grimaldi and Case, 1995). There are at least

four specimens, which were found in the Foremost Formation (Pike, pers. comm., 1996). None has been previously described or figured. I had access to slides of two specimens, which are briefly described here. The distal portion of one of the specimens shows the preservation of the rachis and barbs but apparently no barbules. The vanes are asymmetric, suggesting that it represents a flight feather. A second specimen is tentatively identified as a semiplume (Fig. 16.13). It consists of several relatively long barbs, some with barbules, pointing in different directions inside the amber. Darker and lighter areas can be observed along the length of some barbs, which are interpreted as evidence of the original color pattern of this feather.

Other Occurrences

There are several other occurrences of Mesozoic feathers in the literature. Rautian (1978) described a supposed feather from the Late Jurassic lacustrine sediments of the Karatau Range, Kazakhstan, naming it *Praeornis sharovi*. Bock (1986) considered that this specimen is not a feather but a leaf of a cycad. Nessov (1992) and Feduccia (1996) examined this material and also concluded that it probably represents plant material. Glazunova et al. (1991), however, maintained the avian origin of this specimen. To my knowledge, no de-

5 mm

Figure 16.11. Complete contour feather (MCT 1509-R) from the Santana Formation (Crato Member), Early Cretaceous of Brazil. Observe the color pattern formed by darker and lighter bands.

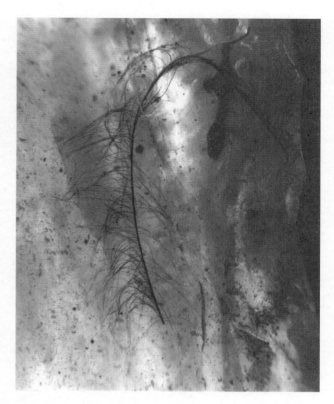

Figure 16.12. Semiplume preserved in amber from the Late Cretaceous of New Jersey.

tailed redescription of this specimen has been done to confirm the interpretation presented by Rautian (1978).

Nessov (1992) mentioned the presence of isolated feathers in two other localities of Kazakhstan, both near the Aral Sea. The first one is a small feather (12 mm), described by P. V. Shilin (apud Nessov, 1992), from the Zhirkindeck Formation (Upper Turonian–Coniacian) found in the Tjulkeli locality (Kzyl-Orda District). The second one consists of a small, asymmetric contour feather from the Bostobe Formation (Santonian-?Campanian) found in the Taldysay locality. This last one was named by V. S. Bazhanov (Nessov, 1992) *Cretaaviculus sarysuensis.* I did not have access to figures or the specimens.

In sediments of the Yixian Formation (?Tithonian), from the Chinese locality of Shangyuan (Beipiao, Liaoning), an incomplete skeleton of a bird was recovered, *Confuciusornis sanctus* (Hou et al., 1995; Zhou and Hou, Chapter 7 in this volume), with several feathers associated with the tibiotarsus. Subsequently, hundreds of specimens have been re-

covered (see Zhou and Hou, Chapter 7 in this volume). Other isolated feathers in this area have been found, as well, most likely from part of the body. Based on their dark color, which contrasts with the matrix, the feathers found in this deposit are possibly preserved as carbonized traces and not as impressions as reported by Hou et al. (1995). See Chiappe et al. (1999) and Zhou and Hou (Chapter 7 in this volume) for description and discussion of these feathers.

Kurochkin (1985a and pers. comm., 1997) reported the existence of a single, tiny feather preserved in the amber deposit from the Yantardakh locality, Hatanga River, in northern Siberia, from Early Cretaceous sediments. Today, these sediments are regarded as Late Santonian (Kurochkin, pers. comm., 1997). Nessov (1992) also reports the existence of small feathers from this locality. However, no detailed information about this material is currently available.

The lacustrine sedimentary rocks of Khurit-Ulan-Bulak (Early Cretaceous) in central Mongolia have also furnished some feathers. Some are preserved as impressions between the wing phalanges of *Ambiortus,* and others have been found isolated (Kurochkin, 1985a: Figs. 1, 3). According to the published illustrations, the feathers known so far from this deposit are not preserved well enough to warrant a detailed description.

Figure 16.13. Feather (semiplume?) preserved in amber from the Late Cretaceous of southern Alberta, Canada. Note the color pattern represented by lighter and darker areas.

Some incomplete but articulated wing bones of an early bird, *Otogornis genghisi* (Hou, 1994), were found in Lower Cretaceous rocks (Yijinholuo Formation) from Chaibu-Sumi, in Inner Mongolia (Zhou and Hou, Chapter 7 in this volume). There are at least two flight feathers associated with the radius and ulna that were figured but not described. Rachises and barbs can be observed, but barbules are apparently not preserved.

Williston (1896) described supposed feathers in one specimen of *Hesperornis* from the Late Cretaceous (Santonian) Niobrara Formation of Kansas. Martin (1983) mentioned the presence of plumaceous feathers in two specimens of *Parahesperornis*. Cracraft (1986) indicates, in his data matrix, the presence of feathers in *Ichthyornis*. To my knowledge, the evidence of feathers in those taxa has not been described and has yet to be confirmed.

Discussion

Mesozoic feathers are uncommon fossils and have rarely been subjected to detailed study. In most cases, the occurrence of this integumentary structure is only briefly men-tioned, sometimes followed with a short description. Nevertheless, the known record provides some interesting information that can contribute to a better understanding of avian evolution. Until very recently, feathers had been reported only in birds. All nonavian theropods in which the skin is fossilized lacked any similar structures. This included one extremely well-preserved specimen discovered in the Santana Formation (Romualdo Member), Brazil; this material contains exceptionally well-fossilized dermis and epidermis, where the evidence of feathers would have been preserved if they were originally present (Kellner, 1996a,b).

New "feathered" fossils, however, could potentially change this picture. The first new taxon, named *Sinosauropteryx prima,* from the Mesozoic strata of China (Ji and Ji, 1996) was originally regarded as an early bird. Contrary to this interpretation, some researchers raised the possibility that *Sinosauropteryx* represents a nonavian theropod, probably related to *Compsognathus* from the Upper Jurassic of Europe (Currie, 1996; Chen et al., 1998). At present, it is also not absolutely clear if the "featherlike" structures of *Sinosauropteryx* are, in fact, feathers, proto-feathers, or something else (Geist et al., 1997; Chen et al., 1998), although a number of recent studies have strongly supported their homology to extant feathers (e.g., Padian et al., 2001; Prum and Williamson, 2001; Xu et al., 2001; see also Witmer, Chapter 1 in this volume). In any case, the structures observed in the dorsal part of *Sinosauropteryx* are filliform and lack the branching pattern observed in avian feathers. Soon after the discovery of *Sinosauropteryx*, other specimens of nonavian theropods were recovered that showed integumentary structures. For example, the therizinosauroid *Beipiaosaurus* and the dromaeosaurid *Sinornithosaurus* (Xu et al., 2001) display filamentous structures similar to those observed in *Sinosauropteryx;* however, the putative basal oviraptorosaur *Caudipteryx* displays true feathers of basically avian form (see Ji et al., 1998; Padian et al., 2001; Witmer, Chapter 1 in this volume; Zhou and Hou, Chapter 7 in this volume; and references therein).

The origin of feathers is still debated (Hecht et al., 1985; Feduccia, 1996). Feathers have traditionally been regarded as derived from the scales present in reptiles. Brush (1996), however, pointed out that the only similarity between these structures is the fact that both are derived from epidermal tissue; in every other feature, such as gene structure, development, and morphogenesis, avian feathers differ from reptilian scales. In extant birds, feathers have a variety of functions, including insulation, flight, and protection (Stettenheim, 1976). The original function of feathers, however, is commonly related to insulation with a later modification for flight (Regal, 1985), but the reverse has also been proposed, that is, feathers first developed in an aerodynamic context and only secondarily achieved an insulation function (e.g., Feduccia, 1985, 1996). The presence in *Caudipteryx* of relatively large feathers restricted to the hand and distal tail may suggest that behavioral display was also important (Ji et al., 1998). So far, there is no clear evidence that can settle this issue, although the insulation hypothesis tends to be favored by some ornithologists (e.g., Bock, 1985). As both the phylogenetic distribution of feathers in Theropoda and the significance of integumentary filaments become better known, hypotheses on the origin of feathers will gain a firmer footing.

No feathers have been found in Triassic strata, and the only confirmed Jurassic feathers are the ones from Solnhofen. In Cretaceous strata, particularly in the lower part, the remains of feathers are more abundant. This is consistent with the known record of early birds, most of which came from strata deposited during this period (Chiappe, 1995).

Undisputed remains of Mesozoic fossil feathers have been unearthed in only a few localities from the following countries: Germany (Solnhofen), Spain (Montsec, Las Hoyas), Mongolia (Gurvan Eren, Sheen Khuduk, Khurit-Ulan-Bulak, Chaibu-Sami), China (Liaoning, Inner Mongolia), Russia (Transbaikalia, Siberia), Kazakhstan (Tjulkeli, Taldysay), Japan (Kuji), Lebanon (Jezzine), Australia (Koonwarra), Brazil (Araripe Basin), the United States (New Jersey), and Canada (Alberta). For some of these deposits, feathers constitute the only avian record currently known, and, given the recent identification of feathers and featherlike structures in nonavian theropods, it cannot be stated with absolute certainty that all the feathers described in this chapter actually do pertain to Aves. So far, no fossil feathers have been found on two continents, Antarctica and Africa.

In *Archaeopteryx,* only flight and tail feathers were identified; no other type, such as down feathers, has been reported. This early bird has been commonly regarded as endothermic, with the feathers used as the primary evidence. Ruben (1991), however, disagreed and presented several arguments favoring the interpretation that this early bird was a flying ectotherm. In fact, some early birds might not have been endothermic, as has been recently suggested based on studies of bone microstructure (Chinsamy et al., 1994; Chinsamy, Chapter 18 in this volume). Therefore, the evolution of feathers and endothermy are not necessarily closely associated (see also Chiappe and Walker, Chapter 11 in this volume).

As mentioned earlier, the record of feathers is comparatively large in the Cretaceous. Flight feathers, including an alula (Sanz et al., 1996), feathers from the tail and body, semiplumes, and down feathers have been reported. The occurrence of down feathers is particularly important, since their function is to provide effective thermal insulation, which was apparently already developed in some basal birds in the Early Cretaceous (Kellner et al., 1994). Of the five main categories of feathers (Lucas and Stettenheim, 1972), only filoplumes and bristles have not been found or recognized as such in the fossil record.

The discovery in Early Cretaceous strata of contour feathers with indications of color patterns presents some interesting information (Figs. 16.6, 16.11, 16.13). Melanin is the most prevalent and durable pigment found in neornithine feathers, producing dark tones such as black, grays, and browns (Gill, 1990). Other pigments, such as carotenoid and porphyrins, produce lighter colors and are less resistant to wear. Similar pigments might have been present in the fossil feathers studied here, since the barbs of the darker bands are better preserved than those in the lighter bands. Although the natural colors in the fossil specimens are unknown, it seems likely that the dark and light bands or dots found in those specimens did represent, in life, respectively darker and brighter colors. So far, only a few fossil feathers with indication of color variation have been found, presenting different patterns. Two of the Araripe feathers show a regular and perpendicular banding (Fig. 16.11). Similar bands, but this time irregular, are found in one Russian specimen (Fig. 16.6). The barbules of one feather from Montsec (Spain) are formed by dotted sequences (Lacasa-Ruiz, 1989: Fig. 3), as observed in some neornithine feathers. A similar condition is found in one feather preserved in amber from Canada (Fig. 16.13), where the barbs are formed by a sequence of dark and lighter "dashes." Color in extant bird feathers plays several roles, particularly in behavior and communication. Its presence in Mesozoic fossil feathers, with different patterns, suggests that birds might have developed similar mechanisms very early in their evolutionary history.

According to Davis and Briggs (1995), fossil feathers can be preserved in one of five main modes: carbonized traces, bacterial autolithification, imprintation, amber, and coprolites (found so far only in one marine Miocene deposit in Maryland, United States). Overall, the best-preserved feathers are found in amber, where all structures can be examined in great detail. Amber preservation, however, is very rare. Most commonly, feathers are found as carbonized traces. This seems to be the case (no chemical tests were performed) for some of the best-preserved Mesozoic material that comes from the Santana Formation (Crato Member, Brazil) and from Koonwarra (Australia). In these, even the barbules on the barbs are commonly preserved.

In terms of the depositional environment, Mesozoic feathers are more commonly found in sedimentary rocks of lacustrine deposits. This is consistent with the hypothesis that the Mesozoic avifauna might have been more confined to terrestrial/inland regions (Kellner, 1994). On the coastal areas, where pterosaur remains far outnumber avian bones, birds were apparently more restricted and represented primarily by flightless diving forms (Hesperornithiformes).

In terms of taxonomy, some authors have based new avian taxa on isolated feathers. Detailed studies of neornithine feathers revealed the presence of ultrastructures (= featherprints) that potentially might be suitable for taxonomical purposes (e.g., Perremans et al., 1992; Laybourne et al., 1994). Except for specimens preserved in amber, such structures are unlikely to be preserved in fossil feathers. Therefore, taxa based solely on isolated feathers should be avoided.

Further comparative studies of fossil feathers with neornithine specimens are warranted in order to provide a better understanding of the development of this important integumentary structure through time. This is particularly true for the fossil specimens found in amber, such as the material from Lebanon, New Jersey, Japan, and Canada, which have the greatest potential for preserving detailed anatomical structures.

Acknowledgments

I thank the following individuals for providing photographs of several feathers: H.-P. Schultze and W. Haare (Museum für Naturkunde, Humboldt-Universität, Berlin) for the picture of the main slab with the first *Archaeopteryx* feather (Fig. 16.2A); L. M. Chiappe (American Museum of Natural History, New York), for slides of the feathers from Montsec (Spain); E. N. Kurochkin (Paleontological Institute, Russian Academy of Sciences, Moscow) for several photographs of feathers from Russia and Mongolia (Figs. 16.3, 16.5, 16.6) and information regarding some Russian specimens; R. Molnar (Queensland Museum, Queensland) for slides of the Australian feathers; E. Frey and V. Griener (Staatliches Museum für Naturkunde Karlsruhe, Karlsruhe) for the photograph of one feather from the Crato Member; J. L. Sanz (Facultad de Ciencias, Universitat Autonoma de Madrid, Madrid) for the photographs of the feathers from Las Hoyas (Fig. 16.7); D. Grimaldi (American Museum of Natural History, New York) for the photograph of the New Jersey feather; K. Sasaki (Kuji Amber Museum, Kuji) for sending a color photograph of one feather preserved in amber from Japan; and T. Pike (Alberta, Canada) for providing some slides of the feathers preserved in amber from Alberta, Canada. D. de Almeida Campos (Museu Ciências da Terra, Departamento Nacional da Produção Mineral, Rio de Janeiro), P. Wellnhofer (Bayerische Staatssammlung für Paläntologie und Historische Geologie, Munich), J. G. Maisey (American Museum of Natural History, New York), and R. G. Martins Neto (São Paulo) provided access to material that is in their care. I am grateful to G. Olshevsky (New York) and P. Currie (Royal Tyrrell Museum of Palaeontology, Drumheller, Canada) for information regarding the newly feathered archosaurs found in China. I thank E. N. Kurochkin, L. M. Chiappe, and L. M. Witmer (Ohio University, Athens) for reviewing drafts of the manuscript. I especially thank L. Meeker and C. Tarka (American Museum of Natural History, New York) for the many photographs used in this chapter, some of which were produced under very difficult conditions.

Literature Cited

Beurlen, K. 1971. As condições ecológicas e faciológicas da formação Santana na chapada do Araripe (Nordeste do Brasil). Anais da Academia Brasileira de Ciências 43(supplement):411–425.

Bock, W. J. 1985. The arboreal theory for the origin of birds; pp. 199–207 *in* M. K. Hecht, J. H. Ostrom, G. Viohl, and P. Wellnhofer (eds.), The Beginnings of Birds. Freunde des Jura-Museums Eichstätt, Eichstätt.

———. 1986. The arboreal origin of avian flight; pp. 57–72 *in* K. Padian (ed.), The Origin of Birds and the Evolution of Flight. California Academy of Sciences, San Francisco.

Brenner, P., W. Goldmacher, and P. Schroeder. 1974. Ostrakoden und Alter der Plattenkalke von Rubies (Sierra de Montsech, Prov. Lérida, N.E.—Spanien). Neues Jahrbuch für Geologie und Paläontologie 8:513–524.

Brush, A. H. 1996. On the origin of feathers. Journal of Evolutionary Biology 9:131–142.

Chandler, A. C. 1916. A study of the structure of feathers, with reference to their taxonomic significance. University of California Publications in Zoology 13(11):343–446.

Chatterjee, S. 1997. The Rise of Birds: 225 Million Years of Evolution. Johns Hopkins University Press, Baltimore, 312 pp.

Chen P.-J., Dong Z., and Zhen S.-N. 1998. An exceptionally well-preserved theropod dinosaur from the Yixian Formation of China. Nature 391:147–152.

Chiappe, L. M. 1995. The first 85 million years of avian evolution. Nature 378:349–355.

Chiappe, L. M., Ji S., Ji Q., and M. A. Norell. 1999. Anatomy and systematics of the Confuciusornithidae (Theropoda: Aves) from the Late Mesozoic of northeastern China. Bulletin of the American Museum of Natural History 242:1–89.

Chinsamy, A., L. M. Chiappe, and P. Dodson. 1994. Growth rings in Mesozoic birds. Nature 368:196–197.

Christopher, R. A. 1979. Normapolles and triporate pollen assemblages from the Raritan and Magothy Formations (Upper Cretaceous) of New Jersey. Palynology 3:73–121.

Cracraft, J. 1986. The origin and early diversification of birds. Paleobiology 12(4):383–399.

Currie, P. J. 1997. "Feathered" dinosaurs; p. 241 *in* P. J. Currie and K. Padian (eds.), Encyclopedia of Dinosaurs. Academic Press, New York.

Davis, P. G., and D. E. G. Briggs. 1995. Fossilization of feathers. Geology 23(9):783–786.

Dettman, M. E. 1986. Early Cretaceous palynoflora of subsurface strata correlative with the Koonwarra fossil bed, Victoria. Memoir of the Association of Australasian Palaeontologists 3:79–110.

Feduccia, A. 1985. On why the dinosaur lacked feathers; pp. 75–79 *in* M. K. Hecht, J. O. Ostrom, G. Viohl, and P. Wellnhofer (eds.), The Beginnings of Birds. Freunde des Jura-Museums, Eichstätt.

———. 1993. Evidence from claw geometry indicating arboreal habits of *Archaeopteryx*. Science 259:790–793.

———. 1996. The Origin and Evolution of Birds. Yale University Press, New Haven, 420 pp.

Feduccia, A., and H. B. Tordoff. 1979. Feathers of *Archaeopteryx*: asymmetric vanes indicate aerodynamic function. Science 203:1021–1022.

Ferrer-Condal, L. 1955. Notice préliminaire concernant la présence d'une plume d'oiseau dans le Jurassic supérieur du Montsech (Province de Lerida, Espagne). Acta XI Congressus Internationalis Ornithologici, Basel: 268–269.

Fregenal-Martínez, M. A., and N. Meléndez. 1993. Sedimentologia y evolution paleogeográfica de la cuberta de Las Hoyas (Cretácico inferior, Serranía de Cuenca). Cuadernos de Geologia Iberica 17:231–256.

Gauthier, J. 1986. Saurischian monophyly and the origin of birds; pp. 1–55 *in* K. Padian (ed.), The Origin of Birds and the Evolution of Flight. California Academy of Sciences, San Francisco.

Geist, N. R., T. D. Jones, and J. A. Ruben. 1997. Implications of soft tissue preservation in the compsognathid dinosaur, *Sinosauropteryx.* Journal of Vertebrate Paleontology 17(supplement 3):48A.

Gill, F. B. 1990. Ornithology. Freeman and Co., New York, 598 pp.

Glazunova, K. P., A. S. Rautian, and V. R. Filin. 1991. *Praeornis sharovi:* bird feather or plant leaf? Materialy 10 Vsesoyuznoi Ornitologicheskoi Konferentzii, Vitebsk: 17–20 sentjabrya 1991. Part 2, 1:149–150. Navuka I tekhnika, Minsk. [Russian]

Grimaldi, D., and G. R. Case. 1995. A feather in amber from the Upper Cretaceous of New Jersey. American Museum Novitates 3126:1–6.

Grimaldi, D., C. W. Beck, and J. J. Boon. 1989. Occurrence, chemical characteristics, and paleontology of the fossil resins from New Jersey. American Museum Novitates 2948:1–28.

Hecht, M. K., J. O. Ostrom, G. Viohl, and P. Wellnhofer. 1985. The Beginnings of Birds. Freunde des Jura-Museums, Eichstätt, 382 pp.

Hou L. 1994. A late Mesozoic bird from Inner Mongolia. Vertebrata PalAsiatica 10:258–266. [Chinese]

Hou L., Zhou Z., Gu Y., and Zhang H. 1995. *Confuciusornis sanctus,* a new Late Jurassic sauriurine bird from China. Chinese Science Bulletin 40(18):1545–1551.

Ji Q. and Ji S. 1996. On the discovery of the earliest bird fossil in China and the origin of birds. Chinese Geology 10(233):30–33. [Chinese]

Ji Q., P. J. Currie, M. A. Norell, and Ji S.-A. 1998. Two feathered dinosaurs from northeastern China. Nature 393:753–761.

Kellner, A. W. A. 1994. Remarks on pterosaur taphonomy and paleoecology. Acta Geologica Leopoldensia 17(39/1):175–189.

———. 1996a. Fossilized theropod soft tissue. Nature 379:32.

———. 1996b. Remarks on Brazilian Dinosaurs. Memoirs of the Queensland Museum 39(3):611–626.

Kellner, A. W. A., J. G. Maisey, and D. A. Campos. 1994. Fossil down feather from the Lower Cretaceous of Brazil. Paleontology 37(3):489–492.

Kellner, A. W. A., R. G. Martins Neto, and J. G. Maisey. 1991. Undetermined feather; pp. 376–377 *in* J. G. Maisey (ed.), Santana Fossils, an Illustrated Atlas. T. F. H. Publications, Neptune City, N.J.

Kurochkin, E. N. 1985a. A true carinate bird from Lower Cretaceous deposits in Mongolia and other evidence of Early Cretaceous birds in Asia. Cretaceous Research 6:271–278.

———. 1985b. Lower Cretaceous birds from Mongolia and their evolutionary significance; pp. 191–199 *in* V. D. Ilyichev and V. M. Gavrilov (eds.), Acta XVIII Congressus Internationalis Ornithologici 1982, Vol. 1. Moscow State University, Moscow.

Lacasa-Ruiz, A. R. 1985. Nota sobre las plumas fósiles del yacimiento eocretácico de "La Pedrera–La Cabrua" en la Sierra del Montsec (Prov. Lleida, España). Llerda 46:227–238.

———. 1986. Nota preliminar sobre el hallazgo de restos óseos de un ave fósil en el yacimiento neocomiense del Montsec. Prov. de Lleida. España. Llerda 47:203–206.

———. 1989. An Early Cretaceous fossil bird from Montsec Mountain (Lleida, Spain). Terra Nova 1(1):45–46.

———. 1993. Hypothetical beginnings of feathers in continental aquatic palaeoenvironments. Terra Nova 5(6):612–615.

Laybourne, R. C., D. W. Deedrick, and F. M. Hueber. 1994. Feather in amber is earliest New World fossil of Picidae. Wilson Bulletin 106 (1):18–25.

Lucas, A. M., and P. R. Stettenheim. 1972. Avian anatomy integument (parts 1 and 2). Agriculture Handbook 362:1–750.

Mabesoone, J. M., and I. M. Tinoco. 1973. Paleoecology of the Aptian Santana Formation (Northeastern Brazil). Palaeogeography, Palaeoclimatology, Palaeoecology 14:97–118.

Maisey, J. G. 1991. Santana Fossils, an Illustrated Atlas. T. F. H. Publications, Neptune City, N.J., 459 pp.

Martill, D. M., and J. B. M. Filgueira. 1994. A new feather from the Lower Cretaceous of Brazil. Palaeontology 37(3):483–487.

Martill, D. M., and E. Frey. 1995. Colour patterning preserved in Lower Cretaceous birds and insects: the Crato Formation of N. E. Brazil. Neues Jahrbuch für Geologie und Paläontologie 1995(2):118–128.

Martin, L. D. 1983. The origin and early radiation of birds; pp. 291–338 *in* A. H. Brush and G. A. Clark Jr. (eds.), Perspectives in Ornithology. Cambridge University Press, Cambridge.

Martins Neto, R. G., and A. W. A. Kellner. 1988. Primeiro registro de pena na Formação Santana (Cretáceo Inferior), bacia do Araripe, nordeste do Brasil. Anais da Academia Brasileira de Ciências 60(1):61–68.

Nessov, L. A. 1984. Upper Cretaceous pterosaurs and birds from Central Asia. Paleontologicheskii Zhurnal 1:47–57. [Russian]

———. 1992. Mesozoic and Paleogene birds of the USSR and their paleoenvironments. Münchner Geowissenschaftliche Abhandlungen (A) 30:465–478.

Ostrom, J. H. 1985. Introduction to *Archaeopteryx;* pp. 9–20 *in* M. K. Hecht, J. O. Ostrom, G. Viohl, and P. Wellnhofer (eds.), The Beginnings of Birds. Freunde des Jura-Museums, Eichstätt.

———. 1986. The cursorial origin of avian flight; pp. 73–81 *in* K. Padian (ed.), The Origin of Birds and the Evolution of Flight. California Academy of Sciences, San Francisco.

Padian, K., Ji Q., and Ji S.-A. 2001. Feathered dinosaurs and the origin of flight; pp. 117–135 *in* D. H. Tanke and K. Carpenter (eds.), Mesozoic Vertebrate Life. Indiana University Press, Bloomington.

Pallerola, J. E. 1979. Una ave y otras especies fósiles nuevas de la biofacies de Santa Maria de Meyá (Lérida). Boletín Geológico y Minero 90(4):333–346.

Perremans, K., A. de Bont, and F. Ollevier. 1992. A study of featherprints by scanning electron microscopy. Belgian Journal of Zoology 122(1):113–121.

Pons, D., P. Y. Berthou, and D. A. Campos. 1990. Quelques observations sur la palynologie de l'Aptian Supérieur et de l'Albien du Bassin d'Araripe (N. E. du Brésil); pp. 241–252 *in* D. A. Campos, M. S. S. Vianna, P. M. Brito, and G. Beurlen (eds.), Atas do 1. Simpósio sobre a Bacia do Araripe e Bacias Interiores do Nordeste, Crato, Brazil. Departamento Nacional da Producão Mineral, Crato.

Proctor, N. S., and P. J. Lynch. 1993. Manual of Ornithology. Yale University Press, New Haven, 340 pp.

Prum, R. O., and S. Williamson. 2001. Theory of the growth and evolution of feather shape. Journal of Experimantal Zoology 291:30–57.

Rau, R. 1969. Über den Flügel von *Archaeopteryx.* Natur und Museum 99(1):1–8.

Rautian, A. S. 1978. A unique bird feather from Jurassic lake deposits in the Karatau. Paleontologicheskii Zhurnal 4:106–114. [Russian]

Regal, P. L. 1985. Common sense and reconstructions of the biology of fossils: *Archaeopteryx* and feathers; pp. 67–74 *in* M. K. Hecht, J. H. Ostrom, G. Viohl, and P. Wellnhofer (eds.), The Beginnings of Birds. Freunde des Jura-Museums, Eichstätt.

Rich, P. V. 1991. The Mesozoic and Tertiary history of birds on the Australian Plate; pp. 721–808 *in* P. V. Rich, J. M. Monaghan, R. F. Baird, and T. H. Rich (eds.), Vertebrate Palaeontology of Australasia. Pioneer, Victoria, Australia.

Rietschel, S. 1985. Feathers and wings of *Archaeopteryx,* and the question of her flight ability; pp. 251–260 *in* M. K. Hecht, J. H. Ostrom, G. Viohl, and P. Wellnhofer (eds.), The Beginnings of Birds. Freunde des Jura-Museums, Eichstätt.

Ruben, J. 1991. Reptilian physiology and the flight capacity of *Archaeopteryx.* Evolution 45(1):1–17.

Sanz, J. L., and A. D. Buscalioni. 1992. A new bird from the Early Cretaceous of Las Hoyas, Spain, and the early radiation of birds. Palaeontology 35(4):829–845.

Sanz, J. L., J. F. Bonaparte, and A. Lacasa. 1988. Unusual Early Cretaceous bird from Spain. Nature 331:433–435.

Sanz, J. L., L. M. Chiappe, B. P. Pérez-Moreno, A. D. Buscalioni, J. J. Moratalla, F. Ortega, and F. J. Poyato-Ariza. 1996. An Early Cretaceous bird from Spain and its implications for the evolution of avian flight. Nature 382:442–445.

Sasaki, K. 1996. The significance of the amber study from the Kuji area. Geology of Iwate 26:29–35. [Japanese]

Schlee, D. 1973. Harzkonservierte fossile Vogelfedern aus der untersten Kreide. Journal für Ornithologie 114(2):207–219.

Schlee, D., and W. Glöckner. 1978. Bernstein. Stuttgarter Beiträge zur Naturkunde (C)8:1–72.

Shufeldt, R. W. 1913. Fossil feathers and some heretofore undescribed fossil birds. Journal of Geology 21:628–652.

Stettenheim, P. 1976. Structural adaptations in feathers; pp. 385–401 *in* H. F. Frith and J. H. Calaby (eds.), Proceedings of the 16th International Ornitholological Congress. Australian Academy of Sciences, Canberra.

Talent, J. A., P. M. Duncan, and P. L. Handby. 1966. Early Cretaceous feathers from Victoria. Emu 66(2):81–86.

Viohl, G. 1985. Geology of the Solnhofen lithographic limestones and the habitat of *Archaeopteryx;* pp. 31–44 *in* M. K. Hecht, J. H. Ostrom, G. Viohl, and P. Wellnhofer (eds.), The Beginnings of Birds. Freunde des Jura-Museums, Eichstätt.

von Meyer, H. 1861a. Vogel-Feder und *Palpipes priscus* von Solnhofen. Neues Jahrbuch für Mineralogie, Geologie und Paläontologie 1861:561.

———. 1861b. *Archaeopteryx lithographica* (Vogel-Feder) und *Pterodactylus* von Solnhofen. Neues Jahrbuch für Mineralogie, Geologie und Paläontologie 1861:678–679.

Waldman, M. 1970. A third specimen of a Lower Cretaceous feather from Victoria, Australia. Condor 72(1):377.

Williston, S. W. 1896. On the dermal covering of *Hesperornis.* Kansas University Quarterly 5(1):53–55.

Xu X., Zhou Z., and R. O. Prum. 2001. Branched integumentary structures in *Sinornithosaurus* and the origin of feathers. Nature 410:200–204.

17

The Track Record of Mesozoic Birds and Pterosaurs
An Ichnological and Paleoecological Perspective

MARTIN G. LOCKLEY AND EMMA C. RAINFORTH

In recent years there has been an exponential increase in the discovery and documentation of tracks attributable to Mesozoic birds (Lockley et al., 1992). There has also been a renewed interest in pterosaur tracks (Lockley et al., 1995, 1997; Mazin et al., 1995, 1997; Unwin, 1997; Wright et al., 1997). The result is that the track record of these two groups of flying vertebrates has increased from virtually nothing to a substantial ichnological record. In addition, owing to a general revival of interest in vertebrate ichnology, it has become clear that the track record often, although not always, represents a component of the fossil fauna that is entirely different from that represented in the skeletal record. Thus, the track record can be particularly instructive in adding information that is not otherwise available from the realm of body fossils. As also demonstrated in recent studies (Lockley and Hunt, 1994, 1995), tracks are often much more common than bones and so facilitate the rapid compilation of a large database. Additionally, tracks are found in situ and so are indicative of the environments actually inhabited by the track makers.

The Mesozoic bird track record provides a good illustration of all these points. We now know of more than 50 Mesozoic bird track assemblages. These all appear to represent "shorebirds" or "waterbirds," in stark comparison to the bone record, in which such groups are quite rare. It has already been noted that "shorebird" track assemblages are sufficiently distinctive and common as to warrant the recognition of a generalized Late Mesozoic to recent "shorebird ichnofacies" (Lockley et al., 1994b) and that the overwhelming majority of ancient bird track sites represent this ichnofacies (Greben and Lockley, 1992). The abundance of tracks in such facies is also remarkable and may measure as high as several hundred per square meter.

It is important to note here that we use the terms "shorebirds" and "waterbirds" in an ecological sense (birds frequenting lakeshore or, in a few cases, seashore environments). While shorebirds in a taxonomic sense (the Charadriiformes) are not known prior to the Campanian (Chiappe, 1995; see Hope, Chapter 15 in this volume), there is ichnological evidence (summarized in this chapter) that shore-frequenting birds (hereafter referred to as shorebirds) existed in the Early Cretaceous. It still remains for paleontologists to address the question of which osteological taxa were responsible for making these tracks or whether, as current evidence suggests, the track makers remain unknown in the skeletal record.

At present, the track record of Mesozoic birds appears to be associated mainly with Cretaceous lacustrine deposits and various coastal plain deposits. This is in contrast to a pterosaur track record associated mainly with Late Jurassic marginal marine and lagoonal settings and Cretaceous terrestrial depositional environments.

The objective of this chapter is to summarize the Mesozoic bird track record in the context of the paleoenvironments frequented or inhabited by the track makers. We extend this paleoecological analysis to include pterosaurs to determine the extent to which the ichnological record of the two groups overlaps in space and time. Such an analysis sheds light on the niches occupied by the two groups. By describing the bird track record, we can also determine something of the overall completeness and bias of the entire avian fossil record (both bones and tracks). It is also possible to evaluate the minimal timing of the diversification of shorebirds or other avian ecological groups that appear convergent with shorebirds (Lockley et al., 1992).

The following institutional abbreviations are used in this chapter: CU-MWC, University of Colorado at Denver and Museum of Western Colorado Joint Ichnological Collection, Denver, Colorado, United States; DGSU, Department of Geology, Seoul University, Seoul, South Korea; KPE, Earth Science Department, Kyungpook National

University, Daegu, South Korea; UCM, University of Colorado Museum, Boulder, Colorado, United States.

Methods

It is important to be able to identify the tracks of birds and not to confuse them with pes tracks of nonavian dinosaurs or manus tracks of pterosaurs (Fig. 17.1). The following criteria are useful, although not foolproof, in determining whether tracks were made by birds or other track makers (Lockley et al., 1992), but, given that almost all fossil bird tracks are those of waterbirds (Greben and Lockley 1992), it is to this group that these criteria apply: (1) general resemblance to the tracks of neornithines; (2) small size; (3) slender digit impressions with indistinct differentiation of the pad impressions; (4) wide divarication angle (approximately 110–120°) between the outer digits; (5) caudally directed hallux (digit I), oriented up to 180° from middle forward digit; (6) slender claws; and (7) claws on the outer digits curving distally away from the middle digit. Lockley et al. (1992) also noted that high track densities are a nonmorphological characteristic of bird track assemblages. In general, pterosaur manus tracks are highly asymmetric, with "fleshy" digits, rendering them distinguishable in most cases from bird tracks. Pterosaur pes tracks have four subparallel toes and are distinctly unbirdlike. Pterosaur trackways also indicate quadrupedal progression.

Although it is not easy to compare the taxonomic composition of ichnofaunas with conventional body fossil taxonomies, an established ichnotaxonomy exists to describe tracks, and we follow its guidelines in assessing diversity of track types. Given the overwhelming predominance of shorebird footprints in the avian track record, ichnotaxonomy is also useful in assessing diversity within this ecological group. We also follow standard ichnological procedures in regarding each discrete track assemblage as a single tracksite or ichnocoenosis. In addition, two or more discrete track-bearing levels in a given formation are regarded as separate sites. An ichnofacies is a consistent recurrence of a particular ichnocoenosis in a given sedimentary facies.

Material

In the mid-nineteenth century, the Reverend Edward Hitchcock described numerous fossil footprints from the Newark Supergroup of the eastern United States. He initially described them as various species within a single new genus (*Ornithichnites*, the "stony bird traces"), forming the basis of modern vertebrate ichnotaxonomy (Hitchcock, 1836). He always considered them to have been made by extinct birds, even though nonavian dinosaurs were discovered in his lifetime (Hitchcock, 1858); later workers

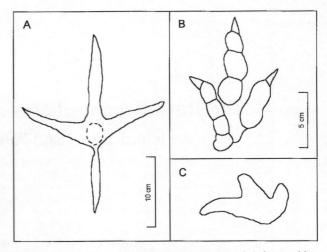

Figure 17.1. A, Neornithine (goliath heron) track (after Lockley et al., 1992). B, *Grallator* track, attributed to a nonavian dinosaur. Late Triassic/Early Jurassic, eastern United States (after Lull, 1953). C, Pterosaur manus track. *Pteraichnus*, Late Jurassic, western United States. CU-MWC 188.8. B and C to same scale.

consider these tracks to have been made by various nonavian dinosaurs.

The majority of known Mesozoic bird tracksite occurrences were reviewed by Lockley et al. (1992), although their stratigraphic distribution was not given precisely—in some cases because of the lack of detailed information available. Since 1992, however, a number of additional sites have been discovered, and ichnotaxonomic descriptions of certain tracks have been published. A brief review of known Mesozoic bird tracksites is presented in stratigraphic order of occurrence (Fig. 17.2).

Jurassic ?Bird Tracks

Currently, there are no reliable reports of pre-Cretaceous bird tracks. Ellenberger (1972) described several new genera and species of small tridactyl tracks from the Late Triassic Stormberg Beds of southern Africa. He noted that they were very birdlike and considered them to have been made by a hypothetical "proavis." The tracks include *Trisauropodiscus aviforma* and *T. superaviforma* (Fig. 17.3A) from the Upper Molteno Beds of the Lower Stormberg Series (Zone A/4), with divarication angles between the outer two digits (d.a.) of 80° and 110°, respectively. *T. galliforma* (d.a. 70°), *T. phasianiforma* (d.a. 70°–100°), *T. levis* (d.a. 90°), and *T. popompoi* (d.a. 90°) all occur in Zone A/6 of the Lower Red Beds of the Lower Stormberg Series, and *Trisaurodactylus superavipes* (d.a. 130°; Fig. 17.3C; Ellenberger, 1974) is found in the lower part of the Upper Stormberg Series (Zone B/3). Although the digit impressions of these tracks are slender, the divarication angles are much smaller than is typical of most bird tracks, and many of these African tracks are large

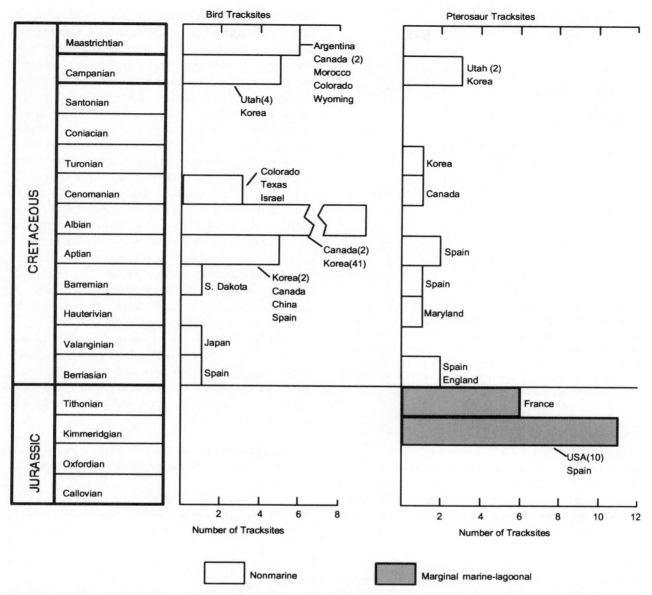

Figure 17.2. Chronostratigraphic distribution of 87 bird and pterosaur tracksites in upper Mesozoic strata. Each discrete track-bearing level in a formation is considered a separate site. Number of tracksites is given in parentheses after geographic locality when there is more than one such locality in a given stage. At least five possible bird track sites, based on *Magnoavipes,* in the Cenomanian of Colorado, are not included. See text for details. Depositional environment of tracksite is indicated as marine or nonmarine (modified after Lockley 1998).

(the type of *T. superaviforma* is 16.5 cm long and 23 cm wide). These observations together lead to the conclusion that these footprints were probably made by relatively small nonavian dinosaurs.

Other small tracks, such as *Trisauropodiscus moabensis* (Fig. 17.3B) from the Lower Jurassic Navajo Sandstone of Utah, are also very birdlike in appearance, especially in size, slenderness of digits, and digit divarication angles (Lockley et al., 1992), but such tracks could be considered similar to *Anomoepus*—a fairly common nonavian dinosaurian ichno-

taxon known from rocks of this age. For example, the Hitchcock collection (Pratt Museum, Amherst College) contains many small *Anomoepus* tracks with relatively wide digit divarication angles. Birdlike tracks have also been reported from the Early Jurassic of Morocco (Fig. 17.3E; Ishigaki, 1985) and are comparable to the ichnogenus *Argoides.*

Enigmatic tridactyl tracks from the Late Jurassic of Spain (Valenzuela et al., 1988), described as "miscellaneous birdlike" tracks (Lockley et al., 1992), are possibly pterosaurian in origin (Lockley et al., 1995) but may not be of either avian

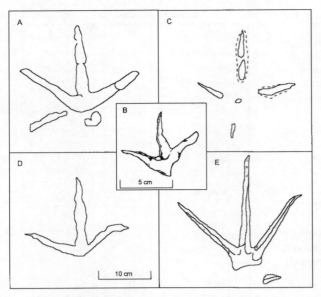

Figure 17.3. Tracks of possible avian affinity (after Lockley et al., 1992). Note the large size of these tracks. A, C, D, and E to same scale. A, *T. superaviforma*, Late Triassic/Early Jurassic, southern Africa. B, *T. moabensis*, Early Jurassic, western United States. CU-MWC 181.3. C, *T. superavipes*, Late Triassic/Early Jurassic, southern Africa. D, Unnamed track from the Early Cenomanian of Israel. E, cf. *Argoides*, Early Jurassic, Morocco.

or pterosaurian origin. Such uncertainty reflects the fact that the tracks have not been carefully described in context (Fig. 17.1C). It is possible that the pterosaurian manus and avian pes could be confused in circumstances in which complete material is not available, as is the case in the Spanish sample.

Lower Cretaceous Bird Tracks

Cretaceous bird tracks are known from Berriasian- through Maastrichtian-age deposits, in nearly all cases apparently associated with nonmarine settings. Moreover, the number of Lower Cretaceous (Berriasian-Albian) reports has increased significantly in recent years (Figure 17.2). All appear to be those of waterbirds and in most cases are essentially indistinguishable from the tracks of neornithines. However, there is sufficient morphological variation to provide some indication of the variety of species represented.

Berriasian Tracks. In a recent paper, Fuentes Vidarte (1996) reports abundant bird tracks from the Wealden of Spain. These are assigned a Berriasian age and have been named *Archaeornithipus meijidei* (Fig. 17.4A, B). They are evidently the oldest confirmed bird tracks yet reported. This track assemblage is interesting for several reasons. First, the tracks are larger than any others previously reported from the Cretaceous, having a maximum length of 16.6 cm and a

mean length of 12 cm. Although the possibility of a nonavian dinosaurian origin was considered by Fuentes Vidarte, it was dismissed on the basis of the high divarication angles between digits II and IV (maximum is 150°; average is 118°) and between digits I and III (maximum is 180°; average is 165°). We agree with this assessment on morphological grounds (cf. Lockley et al., 1992) and note that, except for the shorter hallux in *Archaeornithipus*, these tracks are quite similar to *Jindongornipes* (Fig. 17.4C) from the ?Aptian -?Albian of Korea, which was previously the largest known Lower Cretaceous avian ichnite. In addition, a feature common to *Archaeornithipus*, *Jindongornipes*, *Yacoraitichnus* (Fig. 17.4D), and *Aquatilavipes swiboldae* (Fig. 17.5A) is the cranially concave bend in the medial forward digit (II). However, this feature is not diagnostic for any of these ichnotaxa, because it does not occur consistently in all specimens (compare the holotype and paratype of *Archaeornithipus*, Fig. 17.4A, B); it is probably an artifact of behavior, locomotion, or preservation, although it may be an expression of variation in the flexibility and range of articulation of the pes.

Archaeornithipus is also significant in being the only Mesozoic avian ichnotaxon to exhibit parallel trackways (Fig. 17.6), suggestive of gregarious behavior. This is unusual among bird tracks of any age, previously only demonstrable from a single Oligocene tracksite (Raaf et al., 1965; Yang et al., 1995).

Valanginian Tracks. Tracks from the Tetori Group (Valanginian) of Japan (Lockley et al., 1992) are the oldest avian

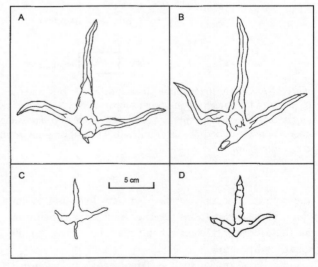

Figure 17.4. A, Holotype of *A. meijidei*, Berriasian, Spain. Left footprint. B, Paratype of *A. meijidei*, Berriasian, Spain. Right footprint. C, Holotype of *J. kimi*, Aptian-Albian, South Korea. KPE 50006. D, *Y. avis*, Maastrichtian, Argentina. All to same scale (A and B after Fuentes Vidarte, 1996; C and D after Lockley et al., 1992).

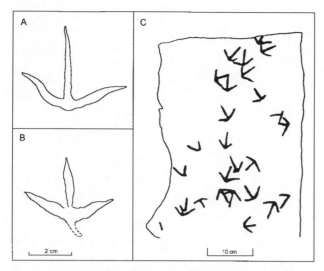

Figure 17.5. *Aquatilavipes* tracks. A, Holotype of *A. swiboldae* Currie, 1981 (after Currie, 1981). Aptian Gething Formation, Canada. The ichnospecies is also known from the Albian Gates Formation of Canada. B, *A. sinensis* Zhen et al., 1987. Drawn from photograph provided by Zhen (after Lockley et al., 1992). C, Unnamed bird tracks comparable to *Aquatilavipes,* from the Valanginian Tetori Group, Japan (after Lockley et al., 1992). A and B to same scale.

ichnites reported from Asia. They are similar to *Aquatilavipes* but have not been formally named (Fig. 17.5C).

Barremian Tracks. Bird tracks of probable Barremian age have recently been discovered in the Lakota Formation of South Dakota and a locality originally reported by Anderson (1939). They occur in association with the tracks of ornithopod and theropod dinosaurs and are considered the oldest bird tracks currently known from North America (Fig. 17.7; Lockley et al., 2001a). They are small (5.0 cm long and wide) with digit divarication angles up to 150–160°. The tracks are tridactyl without hallux impressions and resemble *Aquatilavipes* (Currie, 1981).

Aptian-Albian Tracks. Tracks of undoubted Aptian age were reported from the Gething Formation of western Canada and assigned to the ichnospecies *A. swiboldae* (Currie, 1981). Like the aforementioned tracks from Japan, these footprints are tridactyl and lack any trace of a hallux (Fig. 17.5A). Similar tracks purportedly from the Albian Gates Formation of western Canada assigned to *A. swiboldae* (Lockley et al., 1992: Fig. 8g) have now been reassigned to the Gladstone Formation—a Gething equivalent (Richard Mc-Crea, pers. comm., 2000). Bird tracks, however, are now also known from the Gates Formation of western Alberta (McCrea and Sarjeant, 1999, 2001). Tracks resembling *A. swiboldae* are known from at least three different stratigraphic levels and occur in association with larger bird tracks that

represent a new ichnotaxon (McCrea and Sarjeant, 2001). The tracks are clearly those of waterbirds and often occur in high concentrations. It is reasonable to predict that additional tracksites (levels) are likely to be found in this area.

Tracks of probable Aptian age are found in the Haman Formation of South Korea and in the overlying ?Aptian-?Albian Jindong Formation (these formations have proved hard to date accurately; Lim et al., 1995). Both deposits are remarkable, however, for their abundance of vertebrate footprints, occurring at literally hundreds of stratigraphic levels (Lockley et al., 1992; Lim et al., 1994, 1995). The first Mesozoic bird tracks known from Asia were reported from the Haman Formation by Kim (1969) and named *Koreanaornis hamanensis* (Fig. 17.8C, D). Subsequent studies of the overlying Jindong Formation have revealed that this unit is replete with bird tracks. Conservative estimates put the count of bird track horizons at about 40. In addition to *K. hamanensis,* the Jindong Formation has also yielded a larger bird track named *Jindongornipes kimi* (Fig. 17.4C; Lockley et al., 1995). This large morphotype has since been reported from the Haman Formation (Lim et al., 1995).

Tracks assigned to *Aquatilavipes sinensis* (Fig. 17.5B) have been reported from the Jiaguan Formation of Sichuan Province, China (Zhen et al., 1987, 1994); this formation is of Aptian-Albian age (Hao et al., 1986; Matsukawa and Obata, 1994). Photographs sent to the senior author indicate that this track may have a hallux impression associated with it (Lockley et al., 1992: Fig. 6). If this is the case, then the

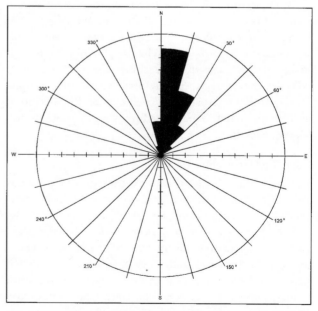

Figure 17.6. Rose diagram showing trends of *Archaeornithipus* tracks (Late Jurassic–Early Cretaceous, Spain). Radius of rose corresponds to 100 tracks. Composite of all localities from data of Fuentes Vidarte (1996).

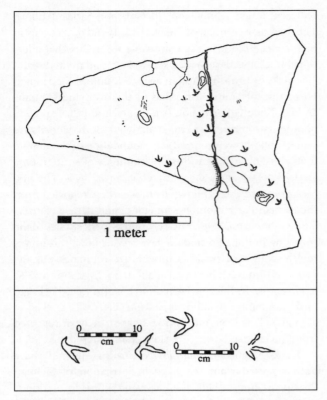

Figure 17.7. Bird tracks from the Lakota Formation of South Dakota (after Lockley et al., 2001a).

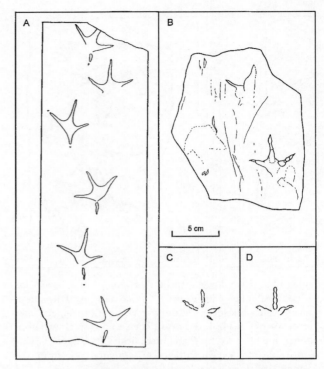

Figure 17.8. Early Cretaceous tracks. Note similarities to *Aquatilavipes* (Fig. 17.5). A, Sketch of type slab of *I. mcconnelli* (after Mehl, 1931; UCM 17614). B, New *I. mcconnelli* track specimen (CU-MWC 203.2) from the Dakota Group, near Golden, Colorado. C, *K. hamanensis* track with hallux impression. DGSU unnumbered specimen. Haman Formation, South Korea (after Lockley et al., 1992). D, *K. hamanensis* track lacking hallux impression. KPE 50001. Haman Formation, South Korea (after Lockley et al., 1992). All to same scale.

track appears morphologically distinct from the type material of *Aquatilavipes.*

A single, small tridactyl track, approximately 2.5 cm long and 3 cm wide with a small impression possibly made by a hallux, was reported from the ?Aptian Enciso Group of Los Cayos, Spain, by Moratella (1993). While it was originally interpreted as an avian ichnite, the present authors consider this trace too poorly preserved to infer a track maker. Tracks interpreted as chelonian and pterosaurian are also found in these strata (Moratella, 1993; Lockley and Meyer, 1999: fig 8.14).

Upper Cretaceous Bird Tracks

Cenomanian Tracks. Bird tracks from the earliest Cenomanian portion of the Dakota Group have special significance as the first Mesozoic bird tracks ever reported (Fig. 17.8A); they were named *Ignotornis mcconnelli* by Mehl (1931). Additional topotype material was reported by Lockley et al. (1989, 1992), and a new specimen is illustrated herein for the first time (Fig. 17.8B); all the material is from a single locality in Colorado. This track type is tetradactyl in most cases, with a prominent hallux. Phalangeal pad impressions are "obscure" in most cases (Lockley et al., 1992), although the new topotype specimen reveals details of two phalangeal pads on what appears to be digit II.

Possible bird tracks may also occur in the Cenomanian of Israel (Fig. 17.3D; Lockley et al., 1992), although these were originally reported as nonavian dinosaur tracks (Avnimelech, 1966; Thulborn, 1990). Lee (1997) erected the new ichnotaxon *Magnoavipes lowei* for large bird tracks from the Woodbine Formation of Texas (Lower to early Middle Cenomanian). These tracks lack hallux impressions, and the step is long; the track maker may have been a cranelike bird. In addition, Lee (1997) claimed this new ichnotaxon to be the largest bird footprint on record in the Mesozoic and noted its similarity to the Cenomanian Israeli tracks. We note the similarity to many tracks in the Dakota Group of Colorado (Fig. 17.9; Lockley et al., 1999). While Lee's claim is intriguing and suggests, on morphological and stratigraphic grounds, that the tracks from Israel and Colorado might be referred to cf. *Magnoavipes*, there are several factors that need to be considered. Are these really bird tracks? Preliminary studies of abundant material from Colorado suggest that digit divarication is between 65° and 100°—considerably less than the 110° reported by Lee (1997) for a single specimen that appears to be compromised by poor

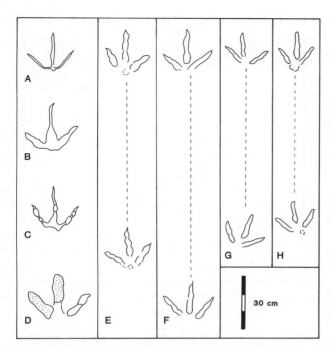

Figure 17.9. *Magnoavipes* tracks from Texas, New Mexico, and Colorado, all drawn to same scale. A, type specimen of *M. lowei* from the Woodbine Formation, Texas (after Lee, 1997), all others from the Dakota Group. B, from Clayton Lake State Park, New Mexico. C, from Gallinas Canyon, Baca County, Colorado (after Lockley et al., 1992). D, from Bandemere site near Morrison, Colorado (after Lockley et al., 1992). E–H, trackway segments (steps) from Dinosaur Ridge (Alameda Parkway) (modified from Lockley et al. 1999).

preservation (M.G.L. pers. obs.). If the divarication angle is consistently less than 110°, the criteria for inferring this as the track of a bird are undermined. In any event *Magnoavipes* is a distinctive ichnotaxon and appears potentially useful in biostratigraphic correlation at the stage and/or substage level. It is also worth noting that the dinosaurian ichnogenus *Caririchnium* co-occurs with *Magnoavipes* in Texas and is also found with *Ignotornis* in Colorado; such evidence suggests that correlations inferred by individual ichnotaxa are often part of a larger picture of correlations that can be deduced from vertebrate track assemblages (Lockley and Hunt, 1995).

Large tracks (approximately 16 cm long, 17 cm wide) somewhat similar to those of *Jindongornipes* have recently been reported from the Dunvegan Formation of western Canada (Plint, pers. comm., 1997). Following the Lower Cenomanian, there appears to be something of a gap in the bird track record until the Campanian. This may be due to the lack of continental sediments of this age in many regions.

Campanian Tracks. In recent years, there have been several reports of bird tracks from the Late Cretaceous Mesaverde Group of northeastern Utah. The first purported

avian tracks were described by Parker and Balsley (1989) from siltstones of the Blackhawk Formation interpreted as representing a nearshore lacustrine environment or possibly very low energy fluvial conditions (Robison, 1991). The original authors interpreted these as tracks of a large bird with a "distinctly asymmetrical foot" (Parker and Balsley, 1989:356). The asymmetry of these footprints (Fig. 17.10F), in particular the elongation of what appears to be the impression of digit IV, was initially taken as an indication of hesperornithiform foot morphology (see editorial footnote in Parker and Balsley, 1989:359). More recent studies of these and similar tracks suggest that these are pterosaur manus tracks resembling the ichnogenus *Pteraichnus,* though considerably larger (Lockley et al., 1995). The tracks are associated with possible feeding traces in the form of oval-shaped "beak" or "prod" marks. The lack of pes traces may be because they are more lightly impressed (Lockley et al., 1995) and thus easy to overlook, especially in this case, in which the locality is underground in a coal mine.

Robison (1991) reported three different types of bird tracks from the lower part of the Blackhawk Formation, at a level about 30 m below the track-bearing layers described by Parker and Balsley (1989). These tracks include a relatively large (approximately 8 cm long by 10 cm wide), slender-toed morphotype from Meetinghouse Canyon (Fig. 17.10A, B). Frog tracks are associated with the track-bearing layer at this locality. Robison reported additional tracks from Straight Canyon and suggested that they represent two different morphotypes. The first (morphotype A) reveals thick-toed footprints about 6.5 cm long by 9 cm wide (Fig. 17.10C), and the second (morphotype B) indicates a slender foot about 5.5 cm long by 9 cm wide (Fig. 17.10D).

A new bird track locality discovered in the Blackhawk Formation near North Horn, Utah (Lockley, 1999b) reveals a set of small tracks (approximately 3 cm long by 3.5 cm wide; Fig. 17.10E) that are clearly different from those reported by Parker and Balsley (1989) and Robison (1991). Thus, at least four bird track types are known from the Blackhawk Formation. This latter locality is important for several reasons. First, this site provides unequivocal evidence of pterosaur tracks (*Pteraichnus*) in the Blackhawk Formation that should be considered in conjunction with the aforementioned pterosaurian reinterpretation (Lockley et al., 1995) of the footprints reported by Parker and Balsley (1989). Second, we believe that this was the first discovery of bird and pterosaur tracks found at the same locality and on the same surface (Fig. 17.11). Third, if we accept that the large tracks are pterosaurian in origin, then the Blackhawk is the only formation producing multiple bird and pterosaur track morphotypes. Present evidence suggests that two pterosaur and four bird track morphotypes are known from this formation.

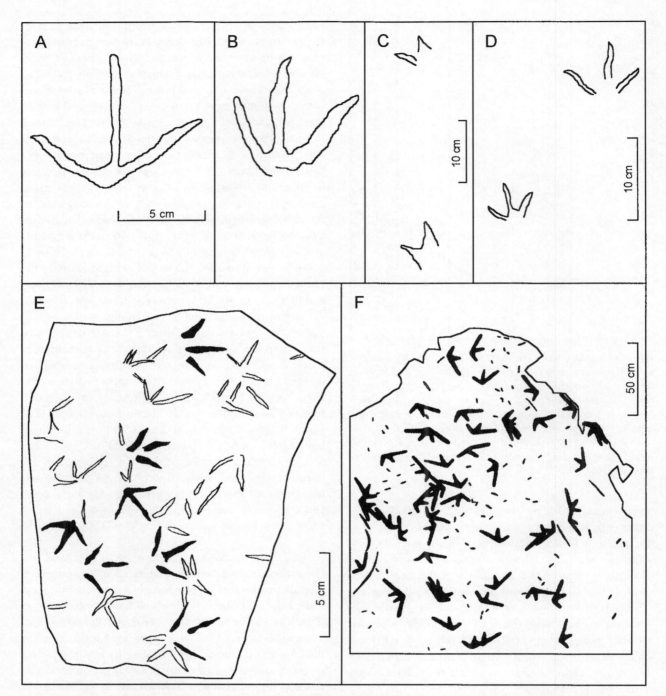

Figure 17.10. Tracks from the Campanian Blackhawk Formation, Utah. A, Meetinghouse Canyon avian morphotype, left footprint. B, Meetinghouse Canyon avian morphotype, right footprint. C, Straight Canyon avian morphotype A. D, Straight Canyon avian morphotype B. E, Tracks of small waterbirds from a newly discovered tracksite near North Horn, Utah. Black shading indicates complete footprint. F, Tracks originally attributed to birds but now believed to be pterosaur manus tracks. A and B to same scale (A–D redrawn from photographs in Robison, 1991; F modified from Parker and Balsley, 1989).

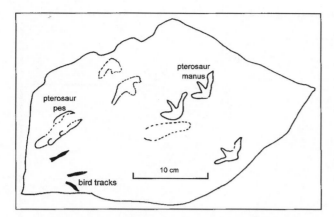

Figure 17.11. First known track-bearing surface with both bird (solid shading) and pterosaur (unshaded) tracks (CU-MWC 217). Blackhawk Formation, Campanian, Utah.

Tracks of possible Campanian age are known from the Late Cretaceous Uhangri Formation of South Korea (Chun, 1990; Lockley et al., 1992; Yang et al., 1995). The tracks occur in strata that are just above the Hwangsan Tuff (85–92 Ma) but older than the Haenam intrusives (63–67 Ma), giving an estimated age range from latest Cenomanian to earliest Campanian. The tracks are unique in the Mesozoic track record in providing evidence of at least two varieties of birds with webbed feet (Yang et al., 1995). The tracks have been assigned to two ichnospecies: a small tridactyl variety, *Uhangrichnus chuni* (Fig. 17.12B); and a larger tetradactyl variety, *Hwangsanipes choughi* (Fig. 17.12A; Yang et al., 1995). These are the first reports of tracks with web impressions from the Mesozoic and also the oldest such tracks known. Recent discoveries have also revealed giant pterosaur tracks at this location (Lockley et al., 1997).

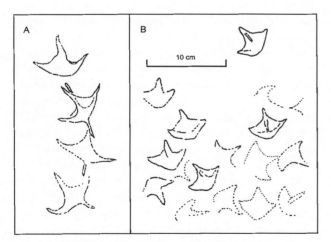

Figure 17.12. Campanian bird tracks from the Uhangri Formation of South Korea. Line drawings of slab KPE 50101 bearing holotypes of both ichnogenera. A, *H. choughi*. B, *U. chuni*. A and B to same scale (modified from Yang et al., 1995).

Maastrichtian Tracks. Maastrichtian bird tracks were first reported from Morocco (Ambroggi and Lapparent, 1954). Little information is available on details of their morphology, and the tracks were poorly illustrated by these authors (Fig. 17.13A). A second Maastrichtian locality was recorded by Alonso and Marquillas (1986), who named *Yacoraitichnus avis* (Fig. 17.4D) from the Yacoraite Formation of Argentina. A third locality, in Río Negro Province, Argentina, was reported by Leonardi (1987, 1994) to have the ichnotaxon *Patagonichornis venetiorum* (Fig. 17.13C) present. A fourth locality was reported by Lockley and Hunt (1995) from the Raton Formation of southern Colorado (Fig. 17.13B). A fifth locality was reported from the St. Mary's River Formation of Alberta, at a site from which hadrosaur tracks with skin impressions were also reported (Currie et al., 1990). The track-bearing layers in the Yacoraite and Raton Formations are associated with the uppermost 10 m of strata below the Cretaceous/Tertiary boundary. In addition, Currie (pers. comm., 1996) has reported the discovery of a sixth Maastrichtian tracksite in the Horseshoe Canyon Formation of western Canada; no further information on this site is currently available.

In contrast to the six aforementioned sites, none of which is known to contain more than a single avian track morphotype, a new fossil footprint site reported from the Lance Formation, on private land in eastern Wyoming, reveals a diverse avian track assemblage. The site is important as the only significant vertebrate tracksite currently known from the Lance Formation. Moreover, the presence of four distinct bird track morphotypes, all from the same bedding surface (Fig. 17.14), is suggestive of a moderate avian diversity, as already seen in Campanian deposits such as the Blackhawk Formation. The four track types are a large morphotype (length 9.5 cm; width 8.3 cm; Fig. 17.14A) with a hallux and partial web; a medium-sized morphotype (length 6.0 cm; width 5.6 cm; Fig. 17.14B) with a hallux; and two smaller-sized footprint types, one (length 4.7 cm; width 3.0 cm; Fig. 17.14D) with a hallux and one (length 2.0 cm; width 3.3 cm; Fig. 17.14C) without. These morphotypes will be described in more detail elsewhere.

The Lance Formation is one of the many Late Cretaceous deposits from which skeletal remains of birds are known (see Hope, Chapter 15 in this volume); for example, Brodkorb (1963) reported eight species. As summarized by Olson (1985:173) these were "described as loons, flamingos, shorebirds and ichthyornithiforms" and included the new genus *Palintropus* (Brodkorb, 1970). According to Olson (1985:173), "the entire known avifauna from the Lance Formation . . . may consist of 'transitional' charadriiforms." However, more recently, the remains of enantiornithine birds have also been found in the Lance Formation. In any event, it is of considerable interest that this diverse avian track assemblage has come to light because it affords us the

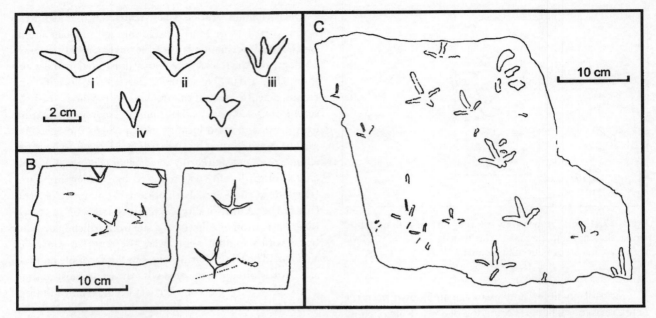

Figure 17.13. Maastrichtian bird tracks. A, Purported bird tracks from Morocco (after Ambroggi and Lapparent, 1954). Tracks (iii) and (iv) could be poorly preserved pterosaur manus tracks. B, Bird tracks on two slabs from the Raton Formation, Colorado, United States, approximately 7 m below the Cretaceous-Tertiary boundary (from Lockley and Hunt, 1995). C, *P. venetiorum*, Upper Maastrichtian of Río Negro Province, Argentina. Slab is housed in the Museo Civico di Storia Naturale, Venice, Italy (modified from free-hand sketch in Leonardi, 1994).

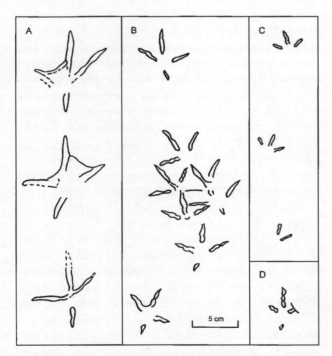

Figure 17.14. Newly discovered bird tracks from the Lance Formation (Maastrichtian) of Wyoming, indicating a high avian diversity. A, Large morphotype with a hallux and partial web. B, Medium-sized morphotype with hallux. C, Small morphotype lacking hallux. D, Small morphotype with hallux. All to same scale.

opportunity to begin to compare the trace fossil record of birds with their body fossil record.

The Pterosaur Track Record

Despite previous debates about the identity of the track maker of footprints assigned to the ichnogenus *Pteraichnus* (Stokes, 1957; Padian and Olsen, 1984), the majority of authors now assert that it is pterosaurian in origin (Lockley et al., 1995, 1996, 1997; Mazin et al., 1995, 1997; Bennett, 1997; Unwin, 1997; Wright et al., 1997). We therefore adopt this position in this discussion. It is now evident that pterosaur tracks are very abundant in the Upper Jurassic and represent volant animals that progressed quadrupedally when on land. Although there is a bias toward the preservation of vertebrate tracks in shoreline paleoenvironments, it is clear that pterosaur tracks are very abundant in marginal marine, carbonate, and clastic-evaporitic lagoonal deposits.

The most important tracksites currently known occur in the Late Jurassic Summerville-Sundance Formations of Arizona, Utah, Colorado, and Wyoming, representing a large, shallow marine embayment (Lockley et al., 1995, 1996, 2001b). Here, in addition to tracks that indicate quadrupedal progression on emergent surfaces, we find evidence that pterosaurs floated in shallow water, touching submerged substrates with their hind feet and probing the substrate with their beaks. Somewhat younger Late Jurassic tracks occur in lagoonal deposits of northern France (Mazin et al., 1995, 1997).

Convincing examples of post-Jurassic pterosaur tracks have been reported from the Lower Cretaceous of England (Wright et al., 1997) and Spain (Lockley et al., 1995), as well as from the Upper Cretaceous of Utah (Lockley et al., 1995) and Korea (Lockley et al., 1997). Recently, pterosaur tracks have also been reported from the Hauterivian-Barremian Arundel Formation of Maryland (Ray Stanford, pers. comm., 1999, and pers. obs. by M.G.L.) and from the Cenomanian Dunvegan Formation of Alberta (McCrea, pers. comm., 1999).

Upper Cretaceous Korean tracks are notable for being the largest known pterodactyloid tracks (up to 33 cm long) and, like the Berriasian tracks from England, were made by individuals larger than any known contemporaneous pterosaur. At several of these sites there are purported beak marks, suggestive of feeding behavior. Unwin (1997:384) has even suggested that high densities of footprints indicate trampling "intended to bring infaunal organisms to the surface where they can be caught and consumed." High densities of pterosaur tracks have been reported from Late Jurassic sites in both France and North America.

Such high pterosaur track density is reminiscent of high avian footprint concentrations at many Mesozoic sites (Lockley et al., 1992). Trace fossil evidence of avian feeding in shoreline settings has been reported by Lockley et al. (1992) for the Mesozoic and by Erickson (1967) and Yang et al. (1995) for the Tertiary. Such feeding behavior represents activity by birds walking on emergent parts of the shoreline or right at the water's edge and is different from the "feeding while floating" evidence preserved in the pterosaur track record (Wright and Lockley, 1999). This is not to say that Mesozoic birds did not feed while floating, as some neornithines do, or that pterosaurs did not feed on emergent substrates, but we must acknowledge that we do not yet have compelling trace fossil evidence for these activities.

Discussion

We readily admit that the identification of bird and pterosaur tracks is sometimes controversial. Recent studies, however, have shown the considerable importance of ichnological evidence in confirming the morphology, functional implications, and distribution of pterosaur tracks (Lockley et al., 1995, 1996, 1997; Mazin et al., 1995, 1997; Bennett, 1997; Unwin, 1997; Wright et al., 1997). See Padian (1998) for dissenting views. There is somewhat less consensus regarding the identification of bird tracks, owing to their similarity to the tracks of bipedal dinosaurs (mainly theropods and ornithopods). Even so, we are unaware of any efforts to refute the claims that most of the bird tracks

reported herein (of Berriasian to Maastrichtian age) are attributable to shorebirds. This seems to be because shorebirds consistently have relatively small, slender tracks with large digit divarications and their tracks look just like the tracks of their Tertiary and recent contemporaries. It is also clear that the Cretaceous shoreline habitats in which these bird tracks are found evidently lack the tracks of diminutive animals with pedal characteristics that indicate theropod or ornithopod affinity, with which waterbird tracks could easily be confused. Moreover, despite the lack of skeletal evidence for Early Cretaceous shorebirds, it is becoming abundantly clear that birds, as a whole, underwent a significant radiation from earliest Cretaceous times onward, as summarized elsewhere in this volume. Not only is the track record chronologically consistent with the timing of this radiation, but it suggests that the track record of shorebirds is currently more complete than the osteological record.

One approach to a further understanding of the problem of differentiating bird and dinosaur tracks is to look at systematic variation in bird track morphology between clades. Preliminary evidence suggests that the difference in digit divarication between songbirds (passerines) and waterbirds is very significant and holds very consistently throughout these clades with few, if any significant, exceptions (Lockley, 1999a,b: Fig. 6.3). It is therefore noteworthy that we find similar distinctions in foot length and digit divarication between theropods (low angles) and ornithopods (high angles) of comparable size. This suggests that systematic patterns of digit divarication can be detected in an extinct clade (nonavian dinosaurs) that are convergent with those known in a living clade (birds). This suggests that digit divarication patterns are not random and may be of some use in identifying track makers within and between groups, especially when looking at large samples and trends through time.

We need not work only with morphological trends. It is known from both the skeletal and ichnological fossil records that pterosaurs reached high diversity levels by the Late Jurassic, the time at which birds first appear. Such concordance of trace and body fossil data is not always apparent and should not be taken for granted, especially in the case of birds and pterosaurs, which are typically not as readily preserved as vertebrates with more robust skeletons. We qualify this statement by noting that birds and pterosaurs can be well preserved in subaqueous lagoonal and lacustrine deposits, where typical terrestrial species are absent, rare, or mixed with abundant remains of aquatic fauna.

As the track record of pterosaurs has improved, it has become evident that the majority of Late Jurassic track occurrences are associated with marginal marine and lagoonal deposits and that Cretaceous tracksites are associated with fluvial and freshwater deposits (Fig. 17.2). Clearly, the pterosaur track record from the Late Jurassic supports the traditional view of pterosaurs as inhabiting marine shoreline habitats. The bird track record appears to show a strong preponderance of occurrences in lacustrine settings, with only a few examples from deltaic environments (Fig. 17.2).

From trackway evidence alone it appears that pterosaurs shifted from marginal marine to nonmarine environments at the Jurassic-Cretaceous boundary. This could be an artifact of preservation and relatively small sample sizes; however, if pterosaurs occupied both these environments throughout their history, it is surprising that their tracks have this distribution. Evidence from the body fossil record indicates that such a shift in habitat may have occurred at the beginning of the Cretaceous (Bakhurina and Unwin, 1995).

Bird tracks appear to predominate in freshwater lacustrine environments, with some occurrences in deltaic and siliciclastic marginal marine sediments. This is perhaps surprising when compared with the body fossil record, which clearly shows that many birds had marine specializations (e.g., the Hesperornithiformes). Perhaps these groups nested on rocky shorelines or small offshore islands, rarely visiting environments with high track-preservation potential. This environmental bias in the track record of birds may reflect a preservational rather than a behavioral bias and/or be a function of discovery as well as relatively small sample size.

It is apparent from the track record that birds and pterosaurs both occupied nonmarine environments during the Cretaceous. However, bird and pterosaur tracks are rarely found together—we know of only two sites where they are found on the same surface. This suggests that birds and pterosaurs may have successfully occupied separate ecological niches in the terrestrial realm.

Such questions can be further investigated by looking at the co-occurrence of bird and pterosaur body fossils in the Cretaceous. To date, however, a number of reported occurrences are controversial owing to the fragmentary nature of remains, which makes the differentiation between birds and pterosaurs difficult in some cases. Moreover, taphonomic processes can cause bird and pterosaur remains from different habitats to be deposited together in some cases (and for body fossils from a single habitat to be deposited in different environments). Such problems do not affect the co-occurrence of bird and pterosaur tracks. Hence, we should continue to compile trackway data in order to flesh out our picture of the spatial and temporal distribution of birds and pterosaurs and to determine what such evidence can tell us about the evolution and ecology of these two groups.

In conclusion, we stress that current trends in Mesozoic vertebrate ichnology are demonstrating that distinctive footprint assemblages are far more ubiquitous than previously supposed. Recent reports of multiple bird tracksites from North America, Europe, and Asia have allowed for the recognition of a "shorebird ichnofacies" (Lockley et al.,

1994b), while it is clear that multiple pterosaur track assemblages in the Upper Jurassic of North America and Europe suggest the existence of distinctive pterosaurian ichnofacies (Lockley et al., 1996; Lockley and Meyer, 1999). It is no longer possible to use the body fossil record of birds or pterosaurs as the only source of data when evaluating the geological history of these groups. Even though the track record may mirror or confirm the body fossil record in certain stratigraphic units, in many cases the footprint record adds substantial new information. Thus, tracks must be taken into consideration for any complete analysis of the vertebrate record.

Acknowledgments

Research on Jurassic tracksites was supported in part by the National Science Foundation. Additional support was provided by the Dinosaur Trackers Research Group, University of Colorado at Denver. Particular thanks go to Julianne Snider for considerable assistance with drafting. We also thank M. Donivan, Salt Lake City, for providing specimens from the Blackhawk Formation. L. Zerbst, Niobrara County, Wyoming, provided access to the Lance Formation tracksite and was responsible for finding many of the bird track specimens that compose that assemblage. Studies in Korea were facilitated by the support of Drs. Yang S.-Y. and Lim S.-K.; Kyungpook National University, Daegu; and Drs. Chun S. S. and Huh M., Chonam National University, Kwangju. We also thank R. McCrea and R. Stanford for providing valuable information on new tracksite discoveries.

Literature Cited

Alonso, R. N., and R. A. Marquillas. 1986. Nueva localidad con huellas de dinosaurios y primer hallazgo de huellas de Aves en la Formación Yacoraite (Maastrichtiano) del Norte Argentino. Actas, IV Congreso Argentino de Paleontologiá y Biostratigrafiá, Mendoza 2:33–41.

Ambroggi, R., and A. F. de Lapparent. 1954. Les empreintes des pas fossiles du Maestrichtien d'Agadir. Service Géologique du Maroc, Notes et Mémoires No. 122, 10:43–57.

Anderson, S. M. 1939. Dinosaur tracks in the Lakota sandstone of the eastern Black Hills, South Dakota. Journal of Paleontology 13:361–364.

Avnimelech, M. A. 1966. Dinosaur tracks in the Judean Hills. Proceedings of the Israel Academy of Sciences and Humanities, Section of Sciences 1:1–19.

Bakhurina, N. N., and D. M. Unwin. 1995. A survey of pterosaurs from the Jurassic and Cretaceous of the former Soviet Union and Mongolia. Historical Biology 10:197–245.

Bennett, C. 1997. Terrestrial locomotion of pterosaurs: a reconstruction based on *Pteraichnus* trackways. Journal of Vertebrate Paleontology 17:104–173.

Brodkorb, P. 1963. Catalogue of fossil birds (Archaeopterygiformes through Ardeiformes), Part 1. Bulletin of the Florida State Museum, Biological Sciences 7:180–293.

———. 1970. The generic position of a Cretaceous bird. Quarterly Journal of the Florida Academy of Science 32:234–240.

Chiappe, L. M. 1995. The first 85 million years of avian evolution. Nature 378:349–355.

Chun S. S. 1990. Sedimentary processes, depositional environments and tectonic settings of the Cretaceous Uhangri Formation. Ph.D. dissertation, Department of Oceanography, Seoul National University, Korea, 328 pp.

Currie, P. J. 1981. Bird footprints from the Gething Formation (Aptian, Lower Cretaceous) of northeastern British Columbia, Canada. Journal of Vertebrate Paleontology 1:257–264.

Currie, P. J., G. Nadon, and M. G. Lockley. 1990. Dinosaur footprints with skin impressions from the Cretaceous of Alberta and Colorado. Canadian Journal of Earth Sciences 28: 102–125.

Ellenberger, P. 1972. Contribution à la classification des Pistes de Vertébrés du Trias: les types du Stormberg d'Afrique du Sud (I). Palaeovertebrata Mémoire Extraordinaire 1972:1–152.

———. 1974. Contribution à la classification des Pistes de Vertébrés du Trias: les types du Stormberg d'Afrique du Sud (II). Palaeovertebrata Mémoire Extraordinaire 1974:1–202.

Erickson, B. R. 1967. Fossil bird tracks from Utah. Museum Observer 5:140–146.

Fuentes Vidarte, C. F. 1996. Primeras huellas de Aves en el Weald de Soria (Espana). Nuevo icnogenero, Archaeornithipus y nueva icnoespecie A. meijidei. Estudios Geologicos 52:63–75.

Greben, R., and M. G. Lockley. 1992. Vertebrate tracks from the Green River Formation, eastern Utah: implications for paleoecology. Geological Society of America, Rocky Mountain Section, Abstracts with Program 24(6):16.

Hao Y., Su D., Yu J., Li P., Li Y., Wang N., Qi H., Guan S., Hu H., Liu X., Yang W., Ye L., Shou Z., and Zhang Q. 1986. The Cretaceous System of China, Vol. 12. Geological Publishing House, Beijing, 301 pp.

Hitchcock, E. 1836. Ornithichnology—Description of the foot marks of birds (Ornithichnites) on new Red Sandstone in Massachusetts. American Journal of Science 29:307–340.

———. 1858. Ichnology of New England. A Report on the Sandstone of the Connecticut Valley, Especially Its Fossil Footmarks. William White, Boston, 220 pp. (Reprinted in 1974 by Arno Press, New York.)

Ishigaki, S. 1985. Dinosaur footprints of the Atlas Mountains I. Nature Study, Osaka 31(1):5–7. [Japanese]

Kim B. K. 1969. A study of several sole marks in the Haman Formation. Journal of the Geological Society of Korea 5(4): 243–258.

Lee Y.-N. 1997. Bird and dinosaur footprints in the Woodbine Formation (Cenomanian), Texas. Cretaceous Research 18:849–864.

Leonardi, G. 1987. Glossary and Manual of Tetrapod Footprint Paleoichnology. Departamento Nacional da Produção Mineral, Brazil, 75 pp.

———. 1994. Annotated Atlas of South America Tetrapod Footprints (Devonian to Holocene). Companhia de Pesquisa de Recursos Minerals, Brazil, 248 pp.

Lim S. K., M. G. Lockley, Yang S. Y., R. F. Fleming, and K. Houck. 1994. Preliminary report on sauropod tracksites from the Cretaceous of Korea. Gaia 10:109–117.

Lim S. K., M. G. Lockley, and Yang S. Y. 1995. Dinosaur trackways from Haman Formation, Cretaceous, South Korea: evidence and implications; pp. 329–336 in Chang K.-H. and Park S.-O. (eds.), Environmental and Tectonic History of East and South Asia, with Emphasis on Cretaceous Correlation (IGCP 350). Proceedings of the 15th International Symposium of Kyungpook National University, Taegu, Korea. Kyungpook National University, Taegu.

Lockley, M. G. 1998. The vertebrate track record. Nature. 396:429–432.

———. 1999a. The Eternal Trail: A Tracker Looks at Evolution. Perseus Books, Reading, **Mass.**, 334 pp.

———. 1999b. Pterosaur and bird tracks from a new locality in the Late Cretaceous of Utah; pp. 355–359 in D. D. Gillette (ed.), Vertebrate Paleontology in Utah. Utah Geological Survey Miscellaneous Publications 99-1. Utah Geological Survey, Salt Lake City.

Lockley, M. G., and A. P. Hunt. 1994. A review of vertebrate ichnofaunas of the Western Interior United States: evidence and implications; pp. 95–108 in M. V. Caputo, J. A. Peterson, and K. J. Franczyk (eds.), Mesozoic Systems of the Rocky Mountain Region, United States. Rocky Mountain Section, Society of Economic Paleontologists and Mineralogists, Denver.

———. 1995. Dinosaur Tracks and Other Fossil Footprints of the Western United States. Columbia University Press, New York, 338 pp.

Lockley, M. G., and C. A. Meyer. 1999. Dinosaur Tracks and Other Fossil Footprints of Europe. Columbia University Press, New York, 323 pp.

Lockley, M. G., M. Matsukawa, and I. Obata. 1989. Dinosaur tracks and radial cracks: unusual footprint features. Bulletin of the National Science Museum, Tokyo, series C, 15(4): 151–160.

Lockley, M. G., Yang S. Y., M. Matsukawa, F. Fleming, and Lim S. K. 1992. The track record of Mesozoic birds: evidence and implications. Philosophical Transactions of the Royal Society of London 336:113–134.

Lockley, M. G., Huh M., Lim S.-K., R. F. Fleming, and K. A. Houck. 1994a. Preliminary report on sauropod tracksites from the Cretaceous of Korea. Gaia: Revista de Geociencias, Museu Nacional de Historia Natural, Lisbon, Portugal 10:109–117.

Lockley, M. G., A. P. Hunt, and C. Meyer. 1994b. Vertebrate tracks and the ichnofacies concept: implications for paleoecology and palichnostratigraphy; pp. 241–268 in S. Donovan (ed.), The Paleobiology of Trace Fossils. Wiley and Sons, New York.

Lockley, M. G., T. J. Logue, J. J. Moratella, A. P. Hunt, R. J. Schultz, and J. W. Robinson. 1995. The fossil trackway Pteraichnus is pterosaurian, not crocodilian: implications for the global distribution of pterosaur tracks. Ichnos 4:7–20.

Lockley, M. G., A. P. Hunt, and S. G. Lucas. 1996. Vertebrate track assemblages from the Jurassic Summerville Formation and correlative deposits; pp. 249–254 in M. Morales (ed.), The Continental Jurassic. Museum of Northern Arizona Bulletin 60. Museum of Northern Arizona, Flagstaff.

Lockley, M. G., Huh M., Lim S.-K., Yang S.-Y., Chun S. S., and D. Unwin. 1997. First report of pterosaur tracks from Asia, Chollanam Province Korea. Journal of the Paleontological Society of Korea, Special Publication no. 2:17–32.

Lockley, M. G., M. Matsukawa, and J. L. Wright. 1999. Is it a bird or is it a . . .? A new scientific name suggests that big bird was around 98 million years ago. Friends of Dinosaur Ridge Annual Report, pp. 18–20 (unpaginated).

Lockley, M. G., P. Janke, and L. Theisen. 2001a. First reports of bird and ornithopod tracks from the Lakota Formation

(Early Cretaceous), Black Hills, South Dakota; pp. 443–452 *in* D. H. Tanke and K. Carpenter (eds.), Mesozoic Vertebrate Life. Indiana University Press, Bloomington.

Lockley, M. G., J. L. Wright, W. Langston Jr., and E. S. West. 2001b. New pterosaur tracks, specimens and tracksites in the Late Jurassic of Oklahoma and Colorado: their paleobiological significance and regional ichnological context. Modern Geology 20:179–203.

Lull, R. S. 1953. Triassic life of the Connecticut Valley. Connecticut State Geological and Natural History Survey Bulletin 81: 1–336.

Matsukawa, M., and I. Obata. 1994. Dinosaurs and sedimentary environments in the Japanese Cretaceous: a contribution to dinosaur facies in Asia based on the molluscan paleontology and stratigraphy. Cretaceous Research 15:101–125.

Mazin, J.-M., P. Hantzpergue, G. Lafaurie, and P. Vignaud. 1995. Des pistes de ptérosaures dans le Tithonien de Crayssac (Quercy, France). Comptes Rendus de l'Académie des Sciences, Paris, t. 321, série IIa:417–424.

Mazin, J.-M., P. Hantzpergue, J.-P. Bassollet, G. Lafaurie, and P. Vignaud. 1997. The Crayssac site (Lower Tithonian, Quercy, Lot, France): discovery of dinosaur tracks in situ and first ichnological results. Comptes Rendus de l'Académie des Sciences, Paris, t. 321, série IIa:417–424.

McCrea, R., and W. A. S. Sarjeant. 1999. A diverse vertebrate ichnofauna from the Lower Cretaceous (Albian) Gates Formation near Grande Cache Alberta. Journal of Vertebrate Paleontology 19:62A.

McCrea, R., and W. A. S. Sarjeant. 2001. New ichnotaxa of bird and mammal footprints from the Lower Createcous (Albian) Gates Formation of Alberta; pp. 453–478 *in* D. H. Tanke and K. Carpenter (eds.), Mesozoic Vertebrate Life. Indiana University Press, Bloomington.

Mehl, M. G. 1931. Additions to the vertebrate record of the Dakota Sandstone. American Journal of Science (5)21:441–452.

Moratella, J. J. 1993. Restos indirectos de dinosaurios del registro Espanol: paleoicnologia de la Cuenca de Cameros (Jurasico superior–Cretacico inferior) y paleoologia del Cretacico superior. Ph.D. dissertation, Universidad Autonoma de Madrid, Facultad de Ciencias, Departamento de Biologia, 727 pp.

Olson, S. L. 1985. The fossil record of birds; pp. 79–256 *in* D. S. Farner, J. R. King, and K. C. Parkes (eds.), Avian Biology, Vol. 3. Academic Press, San Diego.

Padian, K. 1998 Pterosaurs and ?avians from the Morrison Formation (Upper Jurassic, western U.S.). Modern Geology 23:57–68.

Padian, K., and P. E. Olsen. 1984. The fossil trackway *Pteraichnus:* not pterosaurian, but crocodilian. Journal of Paleontology 58(1):178–184.

Parker, L. R., and J. K. Balsley. 1989. Coal mines as localities for studying dinosaur trace fossils; pp. 353–366 *in* D. D. Gillette and M. G. Lockley (eds.), Dinosaur Tracks and Traces. Cambridge University Press, New York.

Raaf, J. F. M. de, C. Beets, and G. Kortenbout van der Sluijs. 1965. Lower Oligocene bird tracks from northern Spain. Nature 207:146–148.

Robison, S. F. 1991. Bird and frog tracks from the Late Cretaceous Blackhawk Formation in east central Utah; pp. 325–334 *in* T. C. Chidsey Jr. (ed.), 1991 Field Symposium. Utah Geological Association, Salt Lake City.

Stokes, W. L. 1957. Pterodactyl tracks from the Morrison Formation. Journal of Paleontology 31(5):952–954.

Thulborn, R. A. 1990. Dinosaur Tracks. Chapman and Hall, New York, 410 pp.

Unwin, D. M. 1997. Pterosaur tracks and the terrestrial ability of pterosaurs. Lethaia 29:373–386.

Valenzuela, M., J. C. Garcia-Ramos, and C. S. de Centi. 1988. Las huellas de dinosaurios del Entorno de Ribadesella. Central Lechera Asturiana, Ribadesella, 35 pp.

Wright, J. L., and M. G. Lockley. 1999. Tracking pterosaurs by land and sea. Journal of Vertebrate Paleontology 19:85–86A.

Wright, J. L., D. M. Unwin, M. G. Lockley, and E. C. Rainforth. 1997. Pterosaur tracks from the Purbeck Limestone Formation of Dorset, England. Proceedings of the Geologists' Association 108:39–48.

Yang S. Y., M. G. Lockley, R. Greben, B. R. Erickson, and Lim S. Y. 1995. Flamingo and ducklike bird tracks from the Late Cretaceous and Early Tertiary: evidence and implications. Ichnos 4:21–34.

Zhen S., Li J., Zhang B., Chen W., and Zhu S. 1987. Bird and dinosaur footprints from the Lower Cretaceous of Emei County, Sichuan; pp. 37–38 *in* First International Symposium, Nonmarine Cretaceous Correlations. Urumqi, Xinjiang, China.

Zhen S., Li J., and Chen W. Z. S. 1994. Dinosaur and bird footprints from the Lower Cretaceous of Emei County, Sichuan, China. Memoirs of the Beijing Natural History Museum 54: 108–120.

Part IV

Functional Morphology and Evolution

18

Bone Microstructure of Early Birds

ANUSUYA CHINSAMY

 Relatively few studies of neornithine bone histology exist, and even rarer are those of fossil birds. Some Tertiary birds have been examined histologically (Amprino and Godina, 1944; Enlow and Brown, 1956; Zavattari and Cellini, 1956; Houde, 1986), but until recently very little was known about the bone microstructure of Mesozoic birds. Prior to 1994, *Hesperornis,* the Cretaceous foot-propelled diver, represented the only Mesozoic bird studied histologically (Houde, 1987). In this study, Houde (1987) compared the bone structure of *Hesperornis* with that of palaeognaths and neognaths (Amprino and Godina, 1944; Amprino, 1947; Zavattari and Cellini, 1956) and deduced that *Hesperornis* was phylogenetically closer to neognaths.

More recently, Chinsamy et al. (1994, 1995) studied the bone microstructure of three basal Cretaceous birds: *Patagopteryx* and two enantiornithine birds. Unlike Houde's (1987) research, these studies utilized the bone microstructure of these early birds to provide insight into questions of growth and biology rather than of phylogeny. The analyses of the microstructure of the bone of these birds have directly contradicted prior notions that early birds were physiologically like their neornithine descendants in having rapid growth rates. On the contrary, the histological findings indicate that these Cretaceous birds grew cyclically and were unable to sustain a rapid, continuous rate of growth (Chinsamy et al., 1994, 1995).

Thus, the use of bone histology for interpreting phylogenetic and paleobiological questions has had a direct impact on our understanding of early birds. More recently, a study of the contemporaneous birds *Hesperornis* and *Ichthyornis* from the Niobrara Formation of western Kansas has provided further insight into histological adaptations of bone with regard to habitat and lifestyles (Chinsamy et al., 1998).

The intention of this chapter is to provide an overview of the histological findings of a variety of previously studied Cretaceous birds (*Patagopteryx* and enantiornithines, Chinsamy et al., 1994, 1995; *Hesperornis* and *Ichthyornis,* Chinsamy et al., 1998) and to include new studies of *Cimolopteryx, Gobipteryx,* and the as yet unpublished Cretaceous loon from Antarctica (Chatterjee, 1989). This chapter thus considers birds basal in the history of the evolution of birds as well as the more advanced ornithurine birds represented by *Hesperornis, Cimolopteryx, Ichthyornis,* and the Antarctic loon.

The methodology used in the thin section preparation of the bird bones is that described by Chinsamy and Raath (1992). In order to make meaningful comparisons between taxa, this study of the bone microstructure of Mesozoic birds was standardized by preferentially examining the bone histology of long bones, mainly femora. This is an important consideration because different bones in a skeleton can show variable histology because of the shape and overall morphology of the bones (Chinsamy, 1995a,b). However, long bones and in particular the midshaft or neutral region of femora are least remodeled and therefore are best for examining growth processes through ontogeny (Chinsamy, 1995a,b). In a recent study, Horner et al. (1999) examined 12 bones of a skeleton of *Hypacrosaurus* and, predictably, found histological variation among the different elements. In addition, they verified that of all the skeletal elements examined, the long bones, that is, the tibia and femora, showed the least Haversian reconstruction and therefore provided the best record of growth during ontogeny. Thus, it appears that multiple skeletal element analyses are appropriate for documenting histological variation in a skeleton, but if the intention is to examine growth processes, the bone microstructure of a growth series of long bones is much more useful (e.g., Chinsamy, 1995a,b; Chinsamy and Dodson, 1995). Since Mesozoic birds are not well represented by different-sized individuals, the current study focused on the long bones of adult birds, although the analyses of the embryonic *Gobipteryx* provide some indication of early ontogenetic growth.

The following institutional abbreviations are used in this chapter: KUVP, Natural History Museum, University of Kansas, Lawrence, Kansas, United States; MACN, Sección Paleontología de Vertebrados, Museo Argentino de Ciencias Naturales, Buenos Aires, Argentina; MGI, Mongolian Geological Institute, Ulaanbataar, Mongolia; PVL, Paleontologí de Vertebrados, Fundación-Instituto Miguel Lillo, Tucumán, Argentina; TTU, Texas Tech University, Lubbock, Texas, United States; ZPAL, Institute of Palaeobiology of the Polish Academy of Sciences, Warsaw, Poland.

Bone Histology of Nonornithurine Birds

Patagopteryx

This hen-sized, flightless, terrestrial bird was recovered from the Bajo de la Carpa Member of the Río Colorado Formation in Neuquen, Argentina (Alvarenga and Bonaparte, 1992; Chiappe and Calvo, 1994; Chiappe, Chapter 13 in this volume). Recent assessments date the Río Colorado Formation as Coniacian-Santonian (Chiappe and Calvo, 1994). In the original description, *Patagopteryx* was described as a ratite (Alvarenga and Bonaparte, 1992), but today it is considered to be a sister group of the Ornithurae (Chiappe and Calvo, 1994; Chiappe, Chapter 13 in this volume).

G. Rougier (formerly of Museo Argentino de Ciencias Naturales, Buenos Aires) made a femoral fragment of the holotype of *Patagopteryx* (MACN-N-03) available for histological examination. Chinsamy et al. (1994, 1995) conducted a detailed analysis of the bone histology of the femur. One of its most distinctive histological features is that the highly vascularized fibrolamellar bone (Ricqlès, 1975) of the compacta is interrupted by the deposition of a single line of arrested growth (LAG), internal to which is a narrow band of lamellated tissue termed the annulus (Fig. 18.1). This LAG signals a pause in the rate of bone formation, while the annulus shows a slower rate of bone formation. These two characteristics directly indicate that *Patagopteryx* was unable to sustain a continuous rapid rate of bone deposition and, hence, growth.

The compacta is intensely vascularized by blood vessels enclosed by osteonal structures (Fig. 18.1). Both primary and secondary osteons are recognized in the compact bone wall. In the perimedullary region several enlarged erosion cavities occur, some of which have been reconstructed as secondary osteons. The centrally located medullary cavity is lined by a narrow band of endosteally formed lamellated tissue (Fig. 18.1) and is free of cancellous bone tissue.

Enantiornithes

Enantiornithines are a diverse volant group of birds that had a worldwide distribution in the Cretaceous. The rela-tionship of Enantiornithes and other avian lineages has been a point of much controversy (Chiappe, 1995a,c; Chiappe and Walker, Chapter 11 in this volume). Some analyses regard them as closely related to *Archaeopteryx* (Martin, 1983), others consider them as within the Ornithurae, and still others consider them as intermediate between *Archaeopteryx* and the ornithurine birds (Chiappe, 1995a–c; Sanz et al., 1995; Forster et al., 1996).

The bone histology of the two enantiornithine femora from the Late Cretaceous of Argentina, MACN-S-01 and PVL-4273, representing two distinct enantiornithine species (Chiappe, 1993), is reported herein. A detailed description of the morphology and histology of these femora is also provided in Chinsamy et al. (1995). From volant birds, both femora are typically lightweight structures having very thin walls and free medullary cavities. The compacta of these bones are noticeably poorly vascularized, if at all, and are essentially composed of a parallel-fibered bone matrix of the lamellar-zonal type (Ricqlès, 1975). The deposition of this tissue is interrupted by the formation of LAGs: in PVL-4273, at least five LAGs can be counted (Fig. 18.2A), whereas in MACN-S-01, four LAGs can be observed. In some areas of the compact bone, enlarged osteocyte lacunae occur with extensive canalicular development (Fig. 18.2B) that probably facilitated the assimilation and distribution of nutrients. A narrow layer of lamellated bone lines the large central medullary cavity.

A third enantiornithine bird examined is a specimen assigned to *Gobipteryx*. Embryonic fragments of the hindlimb skeleton, probably tibia (ZPAL Mgr-I/90) of *Gobipteryx*, were sectioned and prepared by A. Elzanowski (University of Wrocław, Poland). The embryonic material was recovered from the red beds of Barun Goyot Formation at Khermeen Tsav (Elzanowski, 1981).

The histology of the tibia of the *Gobipteryx* embryo is very well preserved. The embryonic nature of the tissue is clearly evident by the fine cancellous nature of its woven bone matrix and the large globular-shaped osteocyte lacunae (Fig. 18.3). The cancellous spaces generally form around vascular canals. A centripetal deposition of bone has not occurred in the spaces around the vascular canals. This would probably have happened at a later stage in development. The uneven peripheral and medullary margin of the bone bears testimony to the remodeling and restructuring (Enlow, 1963) changes that the young bone is undergoing.

Bone Histology of Ornithurine Birds

Hesperornis

Hesperornithiformes are probably the best-known Mesozoic ornithurines. They were toothed, flightless, diving forms that used their laterally compressed feet for propulsion during swimming, as modern loons and grebes do

300 μm

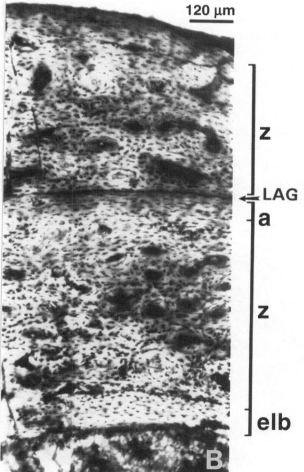

120 μm

z

LAG
a

z

elb

B

Figure 18.1. Transverse sections of a femur of *Patagopteryx deferrariisi* (MACN-N-03). A, the general structure of the bone wall. B, two zones, an annulus, a LAG, and the perimedullary endosteally formed lamellated bone. Note the LAG (arrow) interrupting the deposition of fibrolamellar bone tissue. Abbreviations: a, annulus; elb, endosteally formed lamellated bone; m, medullary cavity; z, zone.

(Martin, 1983). They are generally considered to be a diverse group that had a fairly worldwide distribution in the Cretaceous (Chiappe, 1995a; Feduccia, 1996). Most forms appear to have lived in marine waters, but specimens are known from Late Cretaceous estuarine deposits in Alberta, Canada (Fox, 1974), and freshwater deposits in South Dakota (Feduccia, 1996). Hesperornithiformes are generally regarded as basal ornithurines (Martin, 1991; Chiappe and Calvo, 1994), having secondarily adopted a flightless lifestyle.

Heilmann recognized the ecomorphological similarity of *Hesperornis* to loons as early as 1926, and since then there has been no confusion regarding its way of life. However, the phylogenetic position of *Hesperornis* still remains problematical. Gingerich (1973) proposed that *Hesperornis* was a palaeognath, while Cracraft (1982) placed it with the neognaths and later described it as a sister group of *Ichthyornis* and neornithines (Cracraft, 1986). Houde (1987) attempted to settle this debate by studying the bone histology of *Hesperornis*. In doing so he considered and expanded on the work of Amprino and Godina (1944) and Zavattari and

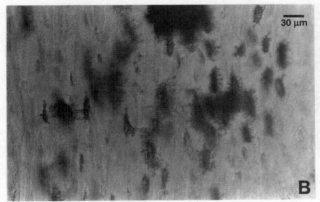

Figure 18.2. Transverse sections of enantiornithine femora (PVL-4273). A, the parallel-fibered, poorly vascularized nature of the compacta and five LAGs. B, the enlarged osteocyte lacunae with extensive canaliculi.

Cellini (1956), who recognized distinct differences in the bone structure of palaeognaths and neognaths.

Houde (1987) agreed with their description of neognaths having vascular canals that were reticular and branched and anastomosed randomly but reevaluated their description of palaeognath bone microstructure. Houde (1987) deduced that palaeognath bone histology could be divided into a tinamou type and a ratite type. He described the tinamou type as having the vascular canals longitudinally arranged and oriented parallel to the longitudinal axis of the bone, whereas the ratite type has a plexiform or laminar arrangement of vascular canals (Enlow and Brown, 1956) that are arranged in closely packed concentric circles in the transverse plane of the bone. Houde (1987) described the bone of *Hesperornis* as neognath type, having a reticular arrangement of the vascular canals, which branch and anastomose randomly, and deduced that *Hesperornis* was therefore more closely related to Neognathae.

In a more recent study, Chinsamy et al. (1998) reported on the bone histology of femora (KUVP 2289 and KUVP 123108) of *Hesperornis* from the Natural History Museum of the University of Kansas. These bones were recovered from the Late Cretaceous Niobrara Formation of western Kansas. Both femora exhibited a fairly thick compact bone wall (Fig. 18.4) that enclosed a rather small central medullary cavity. Thus, both bones exhibited compact, dense characteristics, which are adaptations for an aquatic lifestyle (Buffrénil and Shoevaert, 1989). Erosion cavities and secondary osteons tended to occur more frequently nearer the medullary cavity. Cancellous tissue was observed only in a distal thin section of KUVP 123108.

The primary bone in the *Hesperornis* femora consisted of a fibrolamellar bone tissue in which numerous primary osteons were embedded in the woven framework of the bone. Contrary to Houde's (1987) assessment of *Hesperornis* having a neognath type of tissue, localized areas show his so-called palaeognath tinamou type of histology. Recent studies of femora of *Casuarius* sp. and *Rhea americana* show both Houde's (1987) palaeognath ratite type of bone tissue and the neognath reticular type. Reid (pers. comm., 1994) also mentioned that *Struthio* (Reid, 1984) shows a bone structure intermediate between Houde's (1987) tinamou and neognath types. Thus, Houde's deduction that the palaeognath tissue type is a primitive character state within Aves, while the condition in neognaths is derived, is unsupported by recent studies. Considering the recent work on bone histology of both neornithine (Chinsamy, 1995a; Castanet et al., 1996) and fossil birds (Chinsamy et al., 1998), it appears that the polarization of bird bone histology into palaeognath type and neognath type is not as clear-cut as previously indicated.

Another point worth mentioning with reference to Houde's (1987) study is that he considered bird bone vascularization to be unaffected by ontogenetic factors. Yet juvenile ostriches and secretary birds show the so-called tinamou type of tissue with longitudinally arranged blood vessels, while the adults show a more reticular type of ori-

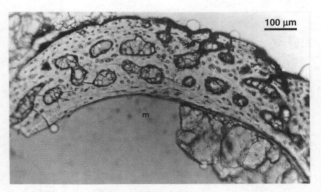

Figure 18.3. Transverse section of a tibia of a *Gobipteryx* embryo (ZPAL Mgr-I/90). Note the cancellous framework of bone with large globular osteocyte lacunae. Abbreviation: m, medullary cavity.

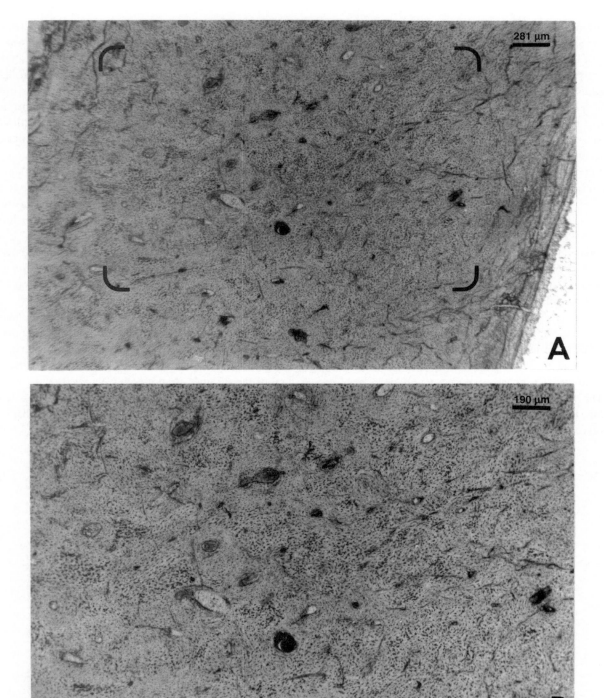

Figure 18.4. Transverse section of a femur of *Hesperornis* (KUVP 2289). A, the general structure of the compact bone wall and the absence of any LAGs. B, higher magnification of bracketed area in A. Indicates the fibrolamellar nature of the tissue. Note also the number of longitudinally oriented primary osteons, that is, the so-called tinamou type of tissue (Houde, 1987).

entation (Chinsamy, 1995a). Furthermore, using image analysis, Chinsamy (1993) determined that the percentage area of bone occupied by blood vessels was higher in juveniles than in adults for both ostriches and secretary birds. In addition, a recent experimental study on mallard ducks (*Anas platyrhynchos;* Castanet et al., 1996) clearly demonstrated that the vascular network in bone is not a characteristic of special taxonomic level or phylogenetic origin. In their study on mallards, Castanet et al. (1996) also found that the structural organization of bone can change in different bones of the same individual as well as in different parts of the same bone at different ontogenetic ages.

Ichthyornis

Ichthyornis is fairly widely distributed in marine Cretaceous deposits of North America (Martin, 1983). It possessed strong wing bones and a well-developed keeled sternum that facilitated powerful flying (Feduccia, 1996). It is generally accepted that it used its long jaw bearing recurved teeth for capturing fish from the water. Originally, the Ichthyornithiformes were regarded as related to Charadriiformes; however, today they are generally considered to be phylogenetically distinct from neornithines (Olson, 1985; Feduccia, 1996).

The humeral fragment described here (KUVP 2294) formed part of a broader study by Chinsamy et al. (1998). It was recovered from the Late Cretaceous Niobrara Forma-

tion of western Kansas. Although the bone is fairly compressed as a result of postmortem damage, histological details are well preserved.

As a specialization for flight, the bone wall is a thin, lightweight structure (Fig. 18.5). It consists of an uninterrupted deposition of fibrolamellar bone tissue. Numerous primary osteons are observed throughout the compacta, although no secondary osteons were recognized. Large erosion cavities occurred in the perimedullary region.

Cimolopteryx

Martin (1991) considered *Cimolopteryx* as a sister taxon of all other neornithines, whereas Olson (1985) and Feduccia (1996) regarded it as "transitional Charadriiformes" in the Graculavidae (see also Hope, Chapter 15 in this volume). A fragment of a distal tibiotarsus (KUVP 126176) of *Cimolopteryx* from the Hell Creek Formation, Montana, was provided for histological analyses.

The bone is richly vascularized. The vessels are mainly longitudinally oriented, although some are radially arranged. An osteonal arrangement of bone can be recognized around the lumen that housed the blood vessels. Several of these canals are noticeably enlarged owing to secondary reconstruction (Fig. 18.6). The vascular canals are located in the woven bone matrix of the fibrolamellar type of bone tissue. There appears to be a thin endosteal lining

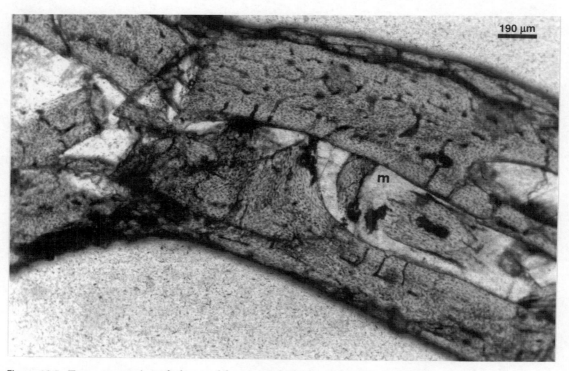

Figure 18.5. Transverse section of a humeral fragment of *Ichthyornis* (KUVP 2294). The bone wall is relatively thin and consists mainly of a fibrolamellar type of tissue. The medullary cavity is lined by a layer of endosteally formed lamellated bone. Abbreviation: m, medullary cavity.

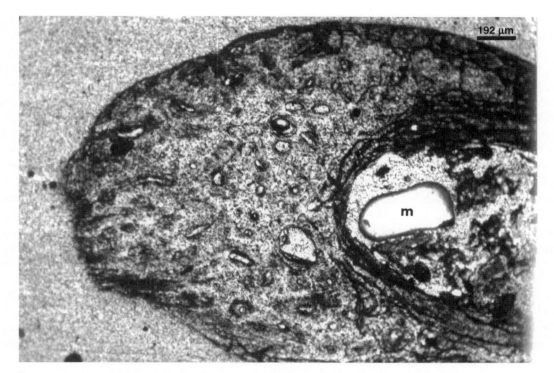

Figure 18.6. Transverse section of a tibiotarsus of *Cimolopteryx* (KUVP 126176). The bone consists of fibro-lamellar tissue without any pauses or interruptions in the rate of bone formation. A large number of vascular canals occur, some enlarged owing to Haversian reconstruction. Abbreviation: m, medullary cavity.

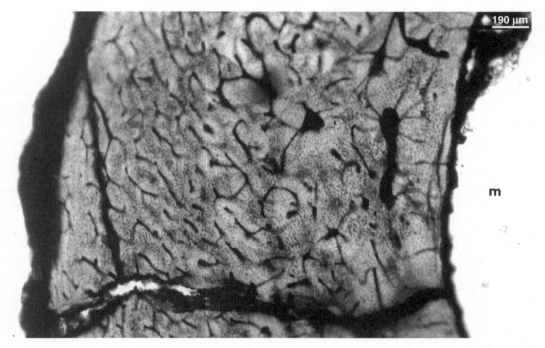

Figure 18.7. Transverse section of a femur of the Cretaceous Antarctic loon (TTU P9265). Note the thick bone wall that consists of fibrolamellar bone tissue. Abbreviation: m, medullary cavity.

of lamellar bone, but this is not clearly defined. No LAGs were recorded in the bone.

Cretaceous Gaviiformes

The fossil loon studied here was recovered from the Upper Cretaceous López de Bertodano Formation of Seymour Island, Antarctica (Chatterjee, 1989). Skeletal morphology suggests that the bird was a foot-propelled diver, since the patella is fused with the inner cnemial crest of the tibia (Heilmann, 1926).

The femoral bone structure of this Antarctic loon (TTU P9265) revealed a thick, compacted bone wall consisting of fibrolamellar bone tissue (Fig. 18.7). The relatively thick, dense nature of the wall suggests that the bird was aquatic and supports the skeletal evidence for an aquatic lifestyle (Chatterjee, 1989).

Numerous primary and secondary osteons are located within the woven bone matrix of the fibrolamellar bone tissue. The blood vessels have a reticular type of orientation. Large erosion cavities are observed in the perimedullary region. A lamellated layer of bone tissue, in which several Volkman's canals are located, lines the free medullary cavity.

Discussion

Anatomical studies of the skeletal remains of fossil birds provide a host of information regarding their diversity, overall morphology, and phylogeny, but knowledge of their biology and physiology is still largely unknown. However, recent studies of the bone microstructure of fossil birds have slowly unraveled some of this mystery and have provided more insight into their biology than ever before. Although bird bone microstructure cannot directly ascertain whether early birds were endotherms or ectotherms (Chinsamy, 1994; Padian et al., 2001), it does allow an assessment of how the bone was formed and, hence, an indication of the rate of bone accretion and overall growth pattern of the bird.

Studies of growth trajectories of early birds are limited by the lack of fossils that allow such deductions. The only apparent growth series of early birds are those of *Archaeopteryx* (Houck et al., 1990; but see Elzanowski, Chapter 6 in this volume) and *Confuciusornis* (Chiappe et al., 1999). In the case of *Archaeopteryx*, it is interesting that all known specimens of this early bird appear to be different in size (Houck et al., 1990; Chiappe, 1995a). This has led to the proposal that *Archaeopteryx*, unlike neornithines, did not attain its adult size rapidly but instead followed a slower growth trajectory similar to that of ectotherms (Chiappe, 1995a; Chinsamy and Dodson, 1995).

Juvenile fossils of early birds are not well known, and while embryonic remains are relatively rare, the unique *Gobipteryx* embryos have provided a glimpse into the prenatal growth of early birds. The embryonic skeletons show well-

ossified shoulder girdles and wing bones (as opposed to poorly developed hindlimbs), which Elzanowski (1981, 1995; see also Chinsamy and Elzanowski, 2001) regarded as indicative of precocial development and advanced flight ability. Although the bone histology of these embryos shows fairly rapid prenatal growth, whether this was sustained through to adulthood is uncertain. Two adult enantiornithines examined show only parallel-fibered (i.e., slowly formed) bone tissue with several cycles of growth rings (Chinsamy et al., 1995). The absence of any fibrolamellar bone tissue could mean that it was never formed, although given the nature of the compacta of the embryonic *Gobipteryx*, it is possible that such tissue was resorbed during the remodeling processes of growth. The bone structure of *Patagopteryx*, the other nonornithurine bird examined, also exhibited a cyclical pattern of bone deposition. However, unlike the enantiornithines, fibrolamellar tissue (indicative of a rapid rate of bone formation) alternated with a LAG and an annulus.

Considering that cyclical bone deposition occurs annually in modern ectotherms (as well as in some endotherms, e.g., Klevezal and Kleinenberg, 1967), it is assumed that the LAGs in these Cretaceous birds were also formed annually and are suggestive of poikilothermy. Hence, unlike neornithine tachymetabolic endothermic birds that are able to sustain a rapid rate of growth to maturity within a single year (Starck, 1993), enantiornithines and *Patagopteryx* required more than a single year to do so (Chinsamy et al., 1994, 1995). Regardless of whether the LAGs are annual or not, the fact remains that the nonornithurine birds studied here grew in an interrupted manner to adult body size. This suggests that they grew at slower rates than neornithines, which grow rapidly to an adult body size, without any pauses in the rate of bone formation. The rest lines formed in the periosteal zone of the cortical bone of mammals and neornithines (Soest and Utrecht, 1971) do not interrupt growth to adult body size but rather are formed once growth has virtually stopped; that is, they are indicative of a determinate growth strategy. The periodic deposition of fibrolamellar bone in *Patagopteryx* suggests that it was physiologically unable to sustain a rapid, continuous rate of bone deposition.

All the Cretaceous ornithurine birds examined (i.e., *Hesperornis*, *Cimolopteryx*, *Ichthyornis*, and the Antarctic loon) showed a rapid rate of bone formation without any periodic interruptions. This suggests a marked physiological advancement of these birds over nonornithurine forms and the probable attainment of homeothermic endothermy, which facilitated rapid, continuous (uninterrupted) growth to maturity.

Physiological Implications for *Archaeopteryx*

The bone histology of birds younger in age than *Archaeopteryx* shows that they grew in a cyclical manner, suggesting that they lacked endothermic homeothermy (Chinsamy et al., 1995). Ruben (1991) proposed an ectothermic physiology for *Archaeopteryx* on the basis of muscle physiology, and the linear size distribution of all known *Archaeopteryx* specimens also hints at a slow (perhaps ectothermlike) postnatal growth rate. Considering the nonornithurine bird bone histology and the recent report by Chinsamy and Elzanowski (2001) of LAGs in *Rahonavis* (potentially a sister taxon to *Archaeopteryx;* see Forster et al., 1998; Chiappe, Chapter 20 in this volume), it seems likely that *Archaeopteryx* would also show a cyclically formed bone cortex. This hypothesis is easily testable, although it is highly unlikely that a specimen of *Archaeopteryx* would be provided for histological analyses. On the basis of the findings of studies on bone histology (Chinsamy et al., 1994, 1995, 1998), muscle physiology (Ruben, 1991), nasal turbinates, and lung structure (Ruben and Hillenius, pers. comm., 1996), it is quite likely that *Archaeopteryx* was not fully endothermic as previously proposed (Bock, 1985; Regal, 1985; Houck et al., 1990).

Conclusion

There appears to be a distinctive difference in the histological character of ornithurine and nonornithurine birds. Nonornithurine birds examined (except *Gobipteryx* embryos, which have typical embryonic bone tissue) show an interrupted pattern of bone deposition, that is, a cycle of rapid growth followed by a cycle of slowed growth (annulus), or a LAG, or both. On the other hand, all ornithurine birds studied, represented by extinct and extant forms, showed a rapid and sustained (uninterrupted) rate of bone deposition irrespective of lifestyle peculiarities. The differing histological patterns among ornithurines and nonornithurines directly suggest marked physiological differences between them. We cannot be absolutely certain about what causes the difference in growth strategy, but it is likely the result of thermal physiology. It appears that ornithurine birds were physiologically capable of a rapid, sustained rate of growth, which, as in neornithines, was quite likely fueled by a high basal metabolic rate characteristic of endothermy. The interrupted growth of nonornithurine birds suggests that these birds were not classic endotherms. Furthermore, there appears to be some degree of difference in the rate of growth between *Patagopteryx* and enantiornithines, since the former is capable of periodically forming fibrolamellar bone. This further suggests physiological differences among early nonornithurine birds. It is quite probable that early birds occupied a range of intermediate physiologies between poikilothermic ectothermy and homeothermic endothermy. Although bone microstructure is not generally considered a reliable taxonomic characteristic, it is noteworthy that for the Mesozoic birds examined the presence or absence of LAGs in their bone suggests differences in growth strategies and physiology that appear to be constrained by phylogeny.

Acknowledgments

I am thankful to S. Chatterjee, L. M. Chiappe, L. D. Martin, M. A. Norell, and G. Rougier for providing specimens for histological analyses. I am grateful to A. Elzanowski for providing the photographs of the *Gobipteryx* embryos. P. Haarhoff and W. J. Hillenius provided comments on a draft of this chapter. I also sincerely appreciate the technical assistance provided by C. Booth and K. von Willig. I thank K. Mvumvu, J. Castanet, M. Schweitzer, L. M. Witmer, and L. M. Chiappe for reviewing the manuscript.

Literature Cited

Alvarenga, H. M. F., and J. F. Bonaparte. 1992. A new flightless land bird from the Cretaceous of Patagonia; pp. 51–64 *in* K. E. Campbell (ed.), Papers in Avian Paleontology, Honoring Pierce Brodkorb. Science Series 36. Natural History Museum of Los Angeles County, Los Angeles.

Amprino, R. 1947. La structure du tissu osseux envisagée comme expression de différences dans la vitesse de l'accroissement. Archives de Biologie 58:315–330.

Amprino, R., and G. Godina. 1944. Osservazioni sui processi di rimaneggiamento strutturale della sostanza compatta della ossa lunghe degli Ucelli corridori. Anatomischer Anzeiger 95:191–214.

Bock, W. J. 1985. The arboreal theory for the origin of birds; pp. 199–207 *in* M. K. Hecht, J. H. Ostrom, G. Viohl, and P. Wellnhofer (eds.), The Beginnings of Birds. Freunde des Jura-Museums, Eichstätt.

Buffrénil, V., and D. Shoevaert. 1989. Données quantitatives et observations histologiques sur la pachyostose du squelette *Dugong dugon* (Müller) (Sirenia, Dugongidae). Canadian Journal of Zoology 67:2107–2119.

Castanet, J., A. Grandin, A. Abourachid, and A. de Ricqlès. 1996. Expression de la dynamique de croissance dans la structure de l'os périostique chez *Anas platyrhynchos*. Comptes Rendus de l'Académie des Sciences de Paris, Série III, 319: 301–308.

Chatterjee, A. 1989. The oldest Antarctic bird. Journal of Vertebrate Paleontology 9(3):16A.

Chiappe, L. M. 1993. Enantiornithine (Aves) tarsometatarsi from the Cretaceous Lecho Formation of Northwestern Argentina. American Museum Novitates 3083:1–27.

———. 1995a. The first 85 million years of avian evolution. Nature 378:349–355.

———. 1995b. A diversity of early birds. Natural History 6: 52–55.

———. 1995c. The phylogenetic position of the Cretaceous birds of Argentina: Enantiornithes and *Patagopteryx deferrariisi*; pp. 55–63 *in* D. S. Peters (ed.), Proceedings of the 3rd Symposium of the Society of Avian Paleontology and Evolution, Courier Forchungsinstitut Senckenberg 181, Frankfurt am Main.

Chiappe, L. M., and J. O. Calvo. 1994. *Neuquenornis volans,* a new late Cretaceous bird (Enantiornithes:Avisauridae) from Patagonia, Argentina. Journal of Vertebrate Paleontology 14:230–246.

Chiappe, L. M., M. A. Norell, and J. M. Clark. 1995. Comment on Wellnhofer's new data on the origin and early evolution of birds. Comptes Rendus de l'Académie des Sciences de Paris, Série IIa, 320:1031–1032.

———. 1996. Phylogenetic position of *Mononykus* from the Upper Cretaceous of the Gobi Desert. Memoirs of the Queensland Museum 39:557–582.

Chiappe, L. M., Ji S.-A., Ji Q., and M. A. Norell. 1999. Anatomy and systematics of the Confuciusornithidae (Aves) from the late Mesozoic of northeastern China. Bulletin of the American Museum of Natural History 242:1–89.

Chinsamy, A. 1993. Image analysis and the physiological implications of the vascularization of femora in Archosaurs. Modern Geology 19:101–108.

———. 1994. Dinosaur bone histology: implications and inferences; pp. 213–227 *in* G. D. Rosenberg and L. Wolberg (eds.), Dino Fest. The Paleontological Society Special Publication no. 7, University of Tennessee Press, Knoxville.

———. 1995a. Histological perspectives on growth in the birds *Struthio camelus* and *Sagittarius serpentarius;* pp. 317–323 *in* D. S. Peters (ed.), Proceedings of the 3rd Symposium of the Society of Avian Paleontology and Evolution. Courier Forschungsinstitut Senckenberg 181, Frankfurt am Main.

———. 1995b. Ontogenetic changes in the bone histology of the Late Jurassic ornithopod *Dryosaurus lettowvorbecki.* Journal of Vertebrate Paleontology 15:96–104.

Chinsamy, A., and P. Dodson. 1995. Inside a dinosaur bone. American Scientist 83:174–180.

Chinsamy, A., and A. Elzanowski. 2001. Evolution of growth pattern in birds. Nature 412:402–403.

Chinsamy, A., and M. A. Raath. 1992. Preparation of bone for histological study. Palaeontologia Africana 29:39–44.

Chinsamy, A., L. M. Chiappe, and P. Dodson. 1994. Growth rings in Mesozoic avian bones: physiological implications for basal birds. Nature 368:196–197.

———. 1995. The bone microstructure of *Patagopteryx* and Enantiornithines. Paleobiology 21:561–574.

Chinsamy, A., L. D. Martin, and P. Dodson. 1998. Bone microstructure of the diving *Hesperornis* and the volant *Ichthyornis* from the Niobrara Chalk of western Kansas. Cretaceous Research 19:225–235.

Cracraft, J. 1982. Phylogenetic relationship and monophyly of loons, grebes, and hesperornithiform birds, with comments on the early history of birds. Systematic Zoology 31:35–56.

———. 1986. The origin and early diversification of birds. Paleobiology 12:383–399.

Elzanowski, A. 1981. Embryonic bird skeletons from the Late Cretaceous of Mongolia. Paleontologica Polonica 42:147–179.

———. 1995. Cretaceous birds and avian phylogeny; pp. 37–53 *in* D. S. Peters (ed.), Proceedings of the 3rd Symposium of the Society of Avian Paleontology and Evolution. Courier Forschungsinstitut Senckenberg 181, Frankfurt am Main.

Enlow, D. H. 1963. Principles of Bone Remodeling. Thomas, Springfield, Illinois, 123 pp.

Enlow, D. H., and S. O. Brown. 1956. A comparative histological study of fossil and recent bone tissue. Texas Journal of Science 8:405–443.

Feduccia, A. 1996. The Origin and Evolution of Birds. Vail-Ballou Press, Binghamton, N.Y., 420 pp.

Forster, C. A., L. M. Chiappe, D. W. Krause, and S. D. Sampson. 1996. The first Cretaceous bird from Madagascar. Nature 382:532–534.

Forster, C. A., S. D. Sampson, L. M. Chiappe, and D. W. Krause. 1998. The theropodan ancestry of birds: new evidence from the Late Cretaceous of Madagascar. Science 279:1915–1919.

Fox, R. C. 1974. A middle Campanian, nonmarine occurrence of the Cretaceous toothed bird *Hesperornis* Marsh. Canadian Journal of Earth Sciences 11:1335–1338.

Gingerich, P. D. 1973. Skull of *Hesperornis* and early evolution of birds. Nature 243:448–462.

Heilmann, G. 1926. The Origin of Birds. Witherby, London, 210 pp.

Horner, J. R., A. de Ricqlès, and K. Padian. 1999. Variation in dinosaur skeletochronology indicators: implications for age assessment and physiology. Paleobiology 25(3):295–304.

Houck, M. A., J. A. Gauthier, and R. E. Strauss. 1990. Allometric scaling in the earliest fossil bird, *Archaeopteryx lithographica*. Science 247:195–198.

Houde, P. W. 1986. Ostrich ancestors found in the Northern Hemisphere suggest new hypothesis of ratite origins. Nature 324:563–565.

———. 1987. Histological evidence for the systematic position of *Hesperornis* (Odontornithes: Hesperornithiformes). Auk 104:125–129.

Klevezal, G. A., and S. E. Kleinenberg. 1967. Age Determination of Mammals from the Layered Structures in Teeth and Bone. Nauka, Moscow. [Russian; English translation by J. Salkind (1969), Israel Program for Scientific Translations, Jerusalem, 128 pp.]

Martin, L. D. 1983. The origin and early radiation of birds; pp. 291–338 *in* A. H. Bush and G. A. Clark Jr. (eds.), Perspectives in Ornithology. Cambridge University Press, New York.

———. 1991. Mesozoic birds and the origin of birds; pp. 485–540 *in* H. P. Schultze and L. Trueb (eds.), Origins of the Higher Groups of Tetrapods. Cornell University Press, Ithaca, N.Y.

Novas, F. E. 1996. Alvarezsauridae, Cretaceous maniraptorans from Patagonia and Mongolia. Memoirs of the Queensland Museum 39:675–702.

Olson, S. L. 1985. The fossil record of birds; pp. 79–252 *in* D. S. Farner, J. R. King, and K. C. Parkes (eds.), Avian Biology, Academic Press, New York.

Padian, K., A. de Ricqlès, and J. R. Horner. 2001. Dinosaurian growth rates and bird origins. Nature 412:405–408.

Perle A., M. A. Norell, L. M. Chiappe, and J. M. Clark. 1993. Flightless bird from the Cretaceous of Mongolia. Nature 62:623–626.

Regal, P. J. 1985. Common sense and reconstructions of the biology of fossils: *Archaeopteryx* and feathers; pp. 67–74 *in* M. K. Hecht, J. H. Ostrom, G. Viohl, and P. Wellnhofer (eds.), The Beginnings of Birds. Freunde des Jura-Museums, Eichstätt.

Reid, R. E. H. 1984. Primary bone and dinosaurian physiology. Geological Magazine 121:589–598.

Ricqlès, A. de. 1975. Recherches paléohistologiques sur les os longs des tétrapodes. VII. Sur la classification, la signification fonctionelle et l'histoire des tissus osseux des tétrapodes. 1ère partie. Annales de Paléontologie (Vertébrés) 61:51–129.

Ruben, J. A. 1991. Reptilian physiology and the flight capacity of *Archaeopteryx*. Evolution 45:1–17.

Sanz, J. L., L. M. Chiappe, and A. Buscalioni. 1995. The osteology of *Concornis lacustris* (Aves: Enantiornithes) from the Lower Cretaceous of Spain and a re-examination of its phylogenetic relationships. American Museum Novitates 3133:1–23.

Soest, R. W. M. van, and W. L. van Utrecht. 1971. The layered structure of bones of birds as a possible indication of age. Bijdragen tot de Dierkunde 41(1):61–66.

Starck, J. M. 1993. Evolution of avian ontogenies; pp. 275–366 *in* D. Power (ed.), Current Ornithology 10. Kluwer Academic Publishers, Dordrecht.

Wellnhofer, P. 1994. New data on the origin and early evolution of birds. Comptes Rendus de l'Académie des Sciences de Paris, Série IIa, 319(11):299–308.

Zavattari, E., and I. Cellini. 1956. La minuta archtetettura delle ossa degli uccelli e il suo valore nella sistematica dei grandi gruppi. Monitore Zoologico Italliano 64:189–200.

Zhou, Z. 1995. Is *Mononykus* a bird? Auk 112:958–963.

19

Locomotor Evolution on the Line to Modern Birds

STEPHEN M. GATESY

 The locomotor system of living birds is dramatically different from that of other extant tetrapods. Nowhere in the design of mammals, lepidosaurs, turtles, or amphibians can the characteristic combination of feathered flight and obligate bipedality be found. Because birds are so divergent, even comparison with their closest living relatives, the crocodilians, reveals few locomotor similarities. This marked distinction of extant birds has made the evolution of their locomotor apparatus one of the most interesting and hotly debated transitions in the history of vertebrates.

My goal in this chapter is to trace aspects of locomotor evolution along the line leading to birds and through the subsequent avian radiation. I integrate morphological information from the fossil record with data from functional studies of living crocodilians and birds. Since this volume concentrates on Mesozoic birds, the locomotor diversity of the approximately 9,700 extant species is not emphasized. Rather, I use a reconstruction of this group's most recent common ancestor to represent neornithines. This ancestor and all its descendants make up the clade Neornithes (Fig. 19.1). I refer to this hypothetical ancestral neornithine as NEO (an abbreviation used as either a noun or an adjective). Although NEO was not identical to any bird alive today, it constitutes my estimate of the primitive neornithine body plan from which all modern forms arose.

The singularity of the neornithine locomotor system is also evident when contrasted with that of its prebird ancestors. A convenient point of comparison on the line leading to birds is the ancestor at the base of the clade Archosauria. Archosauria encompasses the most recent common ancestor of birds and crocodilians, plus all its descendants (Fig. 19.1; Gauthier et al., 1988). I begin this chapter by contrasting the locomotor system in the hypothetical ancestor of Archosauria, hereafter called ARCHO (Gatesy, 1999a), with that in NEO. Such a comparison emphasizes the funda-

mental structural and functional overhaul that occurred between these remote points on the line of descent leading to living birds. This comparison is followed by a brief discussion of the importance of data from fossils and the methodology of reconstructing ancestral states by character optimization. I then reconstruct the morphology of specific intermediate stages between the ARCHO and NEO endpoints and infer transformations in locomotor function. In so doing, I trace the history of the spectacular modifications that took place leading up to the origin and early radiation of birds. Finally, I discuss how character optimization can reduce confusion about ancestors, the origin of birds, and the advent of avian flight.

Methods and Results

ARCHO and NEO Locomotor Structure and Function

Since fossils of lineal ancestors are unlikely to be found and are notoriously difficult to identify, the locomotor system of ancestral forms must be inferred from taxa branching off the main line of interest. For example, the overall design of the ARCHO locomotor apparatus can be reconstructed from the body plan of basal archosaurs and closely related outgroups (Sereno and Arcucci, 1990; Sereno, 1991). Similarly, the general layout and function of the locomotor system can be inferred in NEO at the base of the clade Neornithes. Reconstructing ARCHO and NEO is not without difficulty (see "Reconstructing Ancestral States"), but for this discussion many specific anatomical details are less important than well-supported generalities. The following summary of differences between reconstructed ancestors is restricted primarily to the vertebral and appendicular musculoskeleton. Other regions of the body and other systems contribute to locomotion but are beyond the scope of this chapter.

In the axial skeleton, profound transformations took place along the line between ARCHO and NEO. Two are

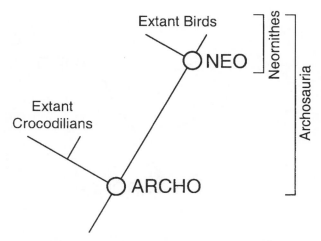

Figure 19.1. A simplified cladogram of Archosauria. The most recent common ancestor of birds and crocodilians, ARCHO, is the hypothetical ancestor of all archosaurs. The most recent common ancestor of all modern birds, NEO, is the hypothetical ancestor of all neornithines. In this chapter, reconstructions of ARCHO and NEO are used to help trace locomotor evolution prior to and following the origin of birds.

particularly significant for locomotion. First, the postcervical vertebral column shortened and stiffened. ARCHO had a relatively long trunk with only two sacral vertebrae (Gauthier, 1984). The tail was a long, heavily muscled structure composed of several dozen vertebral segments. In sharp contrast, NEO had a minute caudal skeleton with very few mobile segments. Fusion of vertebrae into a large synsacrum and a pygostyle gave NEO a much more rigid postcervical body axis than ARCHO. Second, rectrices evolved and formed a variable-sized tail fan (Gatesy and Dial, 1996a).

The NEO forelimb and pectoral girdle can be characterized, relative to that of ARCHO, by four basic changes. First, the pectoral girdle hypertrophied. Modifications of the sternum, coracoids, scapulae, and clavicles (into a furcula) produced an enlarged foundation for the comparatively massive wing musculature. Second, the articular configuration of most joints transformed. The NEO scapula and coracoid came to form a dorsolaterally facing, as opposed to caudolaterally facing, glenoid that permitted greater humeral elevation (Jenkins, 1993). Mobility at the elbow and wrist (Sy, 1936; Vasquez, 1994), unlike that at the shoulder, was reduced. Third, the manus simplified. The NEO hand was composed of three digits, a decrease from the ARCHO count of five. The remaining three metacarpals and several carpals fused into a carpometacarpus. Claws were likely present in NEO but were reduced relative to ARCHO and not found on all digits. Finally, and quite critically, remiges evolved and created a forelimb flight surface.

The NEO hindlimb and pelvis were also highly modified from the ARCHO state. The pelvic girdle enlarged, forming a long contact with many synsacral vertebrae. Elements on each side of the pelvis fused, but the pubic and ischiadic symphyses divided to open the abdomen ventrally. The NEO pubes were oriented nearly caudally, parallel to the ischia, rather than cranioventrally as in ARCHO. The hip joint, augmented by an antitrochanter, restricted the NEO femora to a much more adducted limb posture than in ARCHO (Hutchinson and Gatesy, 2000). Unlike the ARCHO lower leg, the NEO fibula did not reach the ankle and was attached via a fibular crest to the tibia, which fused with the proximal tarsals to form a tibiotarsus. In the foot, digit V disappeared, leaving NEO with only four digits. In NEO the hallux was reduced proximally and was reversed to oppose the other three digits. Metatarsals II–IV fused with distal tarsals to form a tarsometatarsus. Finally, the articulation of the tibiotarsus and tarsometatarsus in NEO formed a single, relatively restricted, mesotarsal hinge joint, in contrast to the complex ankle joint of ARCHO (Sereno and Arcucci, 1990).

These differences between the ARCHO and NEO conditions have profound functional consequences for locomotion (Fig. 19.2A). At the most basic level we can reconstruct ARCHO as a medium-sized, plantigrade, terrestrial quadruped. Facultative bipedalism and aquatic locomotion are difficult to rule out, but the basic design of ARCHO suggests that its forelimbs were used terrestrially. This hypothetical ancestor was clearly more similar, although by no means identical, to living crocodilians than to living birds in its locomotor system. During locomotion the ARCHO hindlimb, like that of living crocodilians and lepidosaurs, was retracted through a large arc by caudofemoral musculature (Gatesy, 1990, 1995). This musculature originated at the base of the tail and inserted on the fourth trochanter of the femur. In sharp contrast, NEO was a small, digitigrade, obligate biped capable of an entirely novel locomotor ability—flight. The feathered forelimbs produced lift and thrust with each wingbeat. Similarly, the novel attachment of feathers to the caudal axis transformed it into a fanning flight surface capable of generating aerodynamic locomotor forces (Thomas, 1993). When walking, NEO employed knee flexion by the hamstring muscles, rather than caudofemoral retraction, to move the limb (Gatesy, 1990, 1999a,b; Carrano, 1998). Thus, the line from ARCHO to NEO underwent dramatic alterations in locomotor structures that entailed a profound functional reorganization.

A Modular Perspective

A construct that has been useful to help clarify the ARCHO-to-NEO transformation is the "locomotor module" (Fig. 19.3; Gatesy and Dial, 1996b). Locomotor modules are anatomical subregions of the body that function as highly integrated units during locomotion. As originally con-

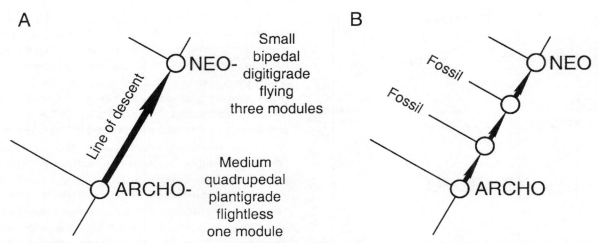

Figure 19.2. A, Dramatic changes in locomotion occurred along the line of descent (arrow) between ARCHO and NEO. Flight evolved somewhere along this line, but data from living taxa are largely uninformative about the nature of this event. B, Only fossils can help resolve the sequence of morphological changes that occurred from ARCHO to NEO. Fossils of taxa branching off this line of descent help reconstruct ancestors (circles) intermediate between ARCHO and NEO, which break the line into a series of smaller transitions.

ceived, modules are identified using data from electromyographic (EMG) analyses to assess each muscle's contribution to a locomotor behavior. We hypothesized that parts of the musculoskeleton that function simultaneously are more tightly integrated than those that do not have temporal overlap in activity. In a modular organism, distant subregions may be more integrated than adjacent ones. Support for the reality of modularity can be found in multivariate morphometric analyses of avian skeletons, which reveal complexes of covarying characters distributed around the body (van den Elzen et al., 1987; Nemeschkal et al., 1992). Such character complexes appear to be united by functional integration rather than anatomical proximity.

Three portions of the musculoskeleton of living birds contribute unequally to different locomotor behaviors; we consider these three to be locomotor modules (Fig. 19.3C; Gatesy and Dial, 1996b). In flight, for example, the pectoral (wing) module functions in close concert with the caudal (tail) module to create aerodynamic forces, but the hindlimb is uninvolved (Gatesy and Dial, 1993). However, when the bird is moving bipedally, the pelvic (hindlimb) module dominates, and the wings and tail are relatively quiescent. Even secondarily flightless birds retain remnants of this modularity, and NEO can be reasonably inferred to have had these same three locomotor modules.

No evidence, however, supports partitioning in ARCHO's locomotor system. ARCHO appears to have retained the primitive tetrapod condition of all four limbs and the axial skeleton functioning as an integrated whole (one module) during terrestrial locomotion (Fig. 19.3A). On the path to NEO, the most basic functional organization of the ARCHO locomotor system—the coordinated activity of the forelimbs, hindlimbs, trunk, and tail—was subdividing into

three parts (Gatesy and Dial, 1996b). Anatomical and functional links primitively present between regions of the body, therefore, must have been broken or reduced in order to form isolated modules that performed more independently. Such decouplings are crucial, but rarely appreciated, events in the history of the NEO locomotor system. Modules allow characters that are treated independently in phylogenetic analysis to be clustered into meaningful groups for subsequent functional investigation. At this level of analysis, the evolution of the flight apparatus can be viewed as the origin of three locomotor modules (Gatesy and Dial, 1996b). Therefore, the evolutionary history of the NEO locomotor system entailed much more than just the origin of a wing. Compared with ARCHO, NEO was thoroughly transformed in its general musculoskeletal layout and neural control.

Despite the insights gained from this modular view, comparison of ARCHO with NEO in isolation is still unsatisfying. The summary of differences focuses on only two points on the line of descent (Fig. 19.2A), thereby leaving important details of the locomotor transformation entirely untouched. All the modifications that distinguish NEO from ARCHO must have occurred, but how and in what sequence? Most important, what were the intermediate morphologies and locomotor capabilities of organisms spanning the gap between earthbound quadrupeds with one locomotor module and flying bipeds with three?

Benefits of Fossils

If we are restricted to information gleaned from living birds and crocodilians, reconstructing the morphology of intermediates between ARCHO and NEO with confidence is almost hopeless. Given that any outline of structural trans-

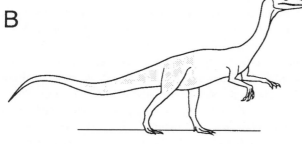

Figure 19.3. Locomotor modules on the line to neornithines. A, In ARCHO, the axial skeleton and all four limbs acted as an integrated unit during locomotion. This condition is retained in modern crocodilians such as *Alligator*, which has one locomotor module (light gray). B, THERO had a single locomotor module consisting of the hindlimb and tail. Such a reduced module was retained by most nonavian theropods, including *Coelophysis* (light gray). C, NEO had three distinct modules: wings, hindlimbs, and tail. The pigeon, *Columba*, retains this unique organization, in which the pectoral and caudal modules create the flight apparatus (dark gray) and the pelvic module (light gray) is used terrestrially (modified from Gatesy and Dial, 1996b).

formation would be unsubstantiated, speculation about the evolution of locomotor behavior would be even more tenuous. Imagine trying to answer even relatively simple questions without the benefit of knowledge from Mesozoic forms. For example, did loss of the pubic symphysis, reduction of the tail, bipedal locomotion, or carpometacarpal fusion come before, during, or after the origin of flight? Since NEO had all these modifications and could fly, testing

whether these novelties are essential for, or even related to, flight is impossible. Without knowing the timing of changes with respect to the rest of the locomotor system, any hypothesis about the historical assembly of the NEO organization is almost entirely conjectural. An infinite number of scenarios could be invoked to account for the differences between ARCHO and NEO (Fig. 19.2A).

Fortunately, the fossil record provides critical information not available in any extant archosaur (e.g., Gauthier et al., 1988). Fossil forms that branch off the line from ARCHO to NEO elucidate the morphology of ancestors along this line. Adding ancestors subdivides the large leap from ARCHO to NEO (Fig. 19.2A, arrow) into many smaller transitions between intermediate ancestors (Fig. 19.2B, arrows). Consider the line between ARCHO and NEO as a string of beads, with each bead representing an intermediate form. Each additional taxon branching off the ARCHO-to-NEO line permits another ancestor (bead) to be reconstructed. The more intermediates we are able to reconstruct, the better the resolution for discerning the sequence of morphological change. Extinct descendants of ARCHO thus help to unravel two important questions. First, from what type of archosaur did birds evolve? Second, what evolutionary paths were taken by birds once they had arisen? Basal birds of the Mesozoic are central to these questions of origin and early evolution. I stress the term "basal" rather than "Mesozoic" because *Archaeopteryx* and other taxa embody a combination of primitive and derived features not found in birds today. Thus, their age is relevant but not the most essential element in their significance for analysis of morphological and functional evolution. The recent discoveries of new basal birds make it strikingly clear how phylogenies, and hypotheses of locomotor history derived from them, must be continually reevaluated.

Reconstructing Ancestral States

How can data from fossil forms, many of which are described in other chapters of this book, be utilized to fill in the missing steps between ARCHO and NEO? Thus far I have used phylogenetic methods very loosely to speak of clades, ancestors, and the line leading to birds. As an evolutionary question, the origin and radiation of the avian locomotor system can be addressed most effectively within a phylogenetic framework. A scenario of locomotor history should be founded on the pattern of evolution revealed by hierarchically arranged monophyletic groups (e.g., Lauder, 1981; Padian, 1987). For this chapter I primarily follow the hypotheses of Gauthier (1986), Sereno (1991), and Chiappe et al. (1996), with additional data incorporated from Novas (1993), Sereno and Arcucci (1993), Sereno et al. (1993), and Sereno (1997). The cladistic analyses of these workers can be combined to span from the origin of Archosauria to the origin of Neornithes (Fig. 19.4). Although such splicing together of cladograms is not ideal, no single analysis currently available covers this entire lineage. By starting with a hypothesis of clade branching, the fate of morphological characters can be traced along the line between ARCHO and NEO. More recent analyses (e.g., Sereno, 1999a) and phylogenetic definitions of nodes and stems (Sereno, 1998; Padian et al., 1999) have yielded nomenclature slightly different from that shown in Figure 19.4, but such details do not affect the overall trends discussed here.

The actual process of reconstructing a hypothetical ancestor, such as ARCHO or NEO, was skirted earlier in the chapter. Ancestral reconstruction uses the distribution of character states in terminal taxa and the cladogram branching pattern to assign states to internal nodes, given certain assumptions about character evolution (Maddison and Maddison, 1992). Thus, the character states of taxa branching off the main line of interest are used to reconstruct hypothetical ancestors at any branch point. For this chapter, data matrices of Gauthier (1986), Novas (1993), and Chiappe et al. (1996) were entered into MacClade 3.03 (Maddison and Maddison, 1992) to reconstruct internal nodes. The vast majority of characters were binary; all transformations were treated as unordered.

For some characters, the reconstruction of ancestral states is relatively straightforward (Fig. 19.5A). Manual digits IV and V, for example, are reduced or absent in all theropods but are present in basal sauropodomorphs, basal ornithischians, and most other archosaurs (Gauthier, 1986). It is most parsimonious to assume that these digits were vestigial in the most recent common ancestor of theropods but not in dinosaurs ancestrally. Similarly, a pygostyle is not found in nonavian theropods, *Archaeopteryx,* or alvarezsaurids yet is known in *Iberomesornis,* enantiornithines, hesperornithiforms, ichthyornithiforms, and most of Neornithes (Chiappe et al., 1996). The simplest explanation for this distribution is to postulate the presence of a pygostyle in the ancestor of Ornithothoraces.

For many characters, however, the history of changes between states is much less clear. Sometimes two or more equally parsimonious solutions can account for the known distribution of states. Such ambiguity results from many factors, such as missing or poorly preserved fossil material, high levels of homoplasy, or a faulty phylogenetic hypothesis. An example of an equivocal character is the presence of a pubic symphysis (Fig. 19.5B). The pubes are known to be in contact in *Archaeopteryx* and other taxa below the Metornithes node; thus, having a symphysis is the primitive condition. Contact is absent in most species of the Alvarezsauridae, *Patagopteryx,* Hesperornithiformes, Ichthyornithiformes, and Neornithes. However, pubic contact is present in Enantiornithes. The state of this character in *Alvarezsaurus* and *Iberomesornis* is not known. Unfortunately, three different scenarios taking a minimum number of

steps can be invoked to account for this distribution. Based on these data, the common ancestor at the base of Aves had pubic contact. Similarly, the hypothetical ancestor of *Patagopteryx* and Ornithurae probably lacked a pubic symphysis. A single loss of the pubic symphysis is most parsimonious, but these data are insufficient to pinpoint this event more precisely.

In this chapter I treat such ambiguities in two ways. First, many uncertainties produced by equivocal characters are avoided by not discussing every known ancestor along the line. I address only six of the ancestors intermediate between ARCHO and NEO for which we have evidence: those of dinosaurs (DINO), theropods (THERO), Paraves (PAR), Aves (AVES), ornithothoracines (THOR), and ornithurines (ORNI) (Fig. 19.6). Ambiguous states at skipped nodes are often unambiguous at higher nodes. As discussed previously, the exact node at which loss of the pubic symphysis is a synapomorphy is not known (Fig. 19.5B). However, despite this ambiguity, the most recent common ancestor of ornithurine birds can be confidently reconstructed as lacking a pubic symphysis. Second, I do not try to explain the functional significance of every change in character state on the line to neornithines. Rather, as with ARCHO and NEO, I make larger-scale statements about the general arrangement of the locomotor system and its transformation between select nodes. Since the relationship between locomotion and many morphological changes is unclear or poorly studied, ambiguities in these characters are not emphasized.

The hypothetical ancestors that follow are reconstructed entirely from unambiguous characters. However, novel character states are synapomorphic only at the level at which they first appear. At each node described, some character states may have arisen as novelties at lower nodes not discussed. For example, a carpometacarpus was likely present in the ancestor of Metornithes (Fig. 19.4). Since this node is skipped (Fig. 19.6), a carpometacarpus is first discussed one node above Metornithes in the ancestor of Ornithothoraces. To avoid excessive citation, I do not credit authors for each character state in my review of hypothetical ancestors. Unless specific citations are given, bird characters are from Chiappe et al. (1996), whereas those of other theropods are from Gauthier (1986) and Novas (1993).

DINO

The hypothetical ancestor of all dinosaurs, DINO (Fig. 19.6), has been reconstructed as a small (body length, ca. 1 meter), primarily bipedal, cursorial carnivore (Gauthier, 1986; Sereno, 1997). It had a mobile, S-shaped neck, a short trunk, three sacral vertebrae, and a long tail (Gauthier, 1986; Sereno et al., 1993; Novas, 1996). Its short forelimbs were used for prey manipulation and possibly during locomotion at slow speeds. DINO's hindlimbs were long (relative to ARCHO's) and highly adducted, with a perforate acetab-

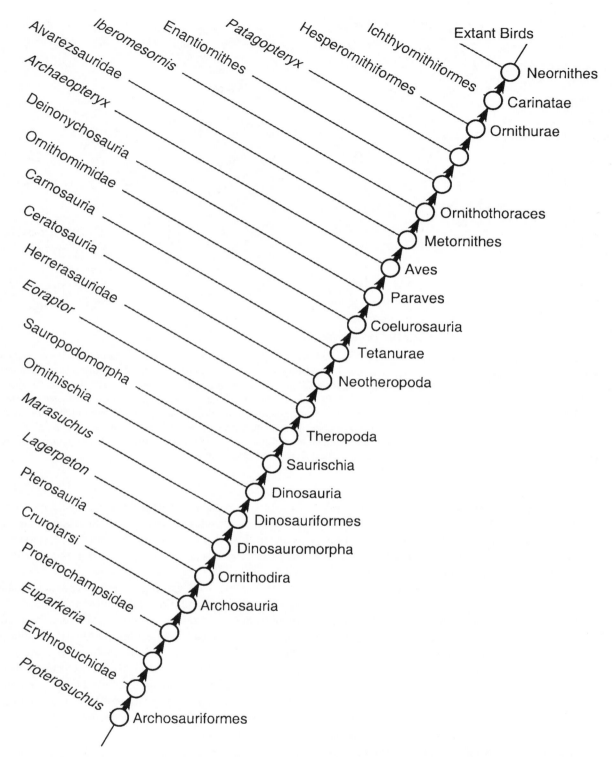

Figure 19.4. A simplified composite cladogram of Archosauriformes emphasizing successive outgroups to neornithines. The results of phylogenetic analyses and terminology by several workers have been combined to reveal the hypothetical ancestors (circles) intermediate between ARCHO (Archosauria) and NEO (Neornithes). By subdividing the transition from ARCHO to NEO into smaller steps, fossils help elucidate the sequence and timing of locomotor evolution on the line to neornithines. Cladogram based on the work of Gauthier (1986), Sereno (1991, 1997), Sereno and Arcucci (1993), Sereno et al. (1993), Chiappe et al. (1996), and Novas (1996).

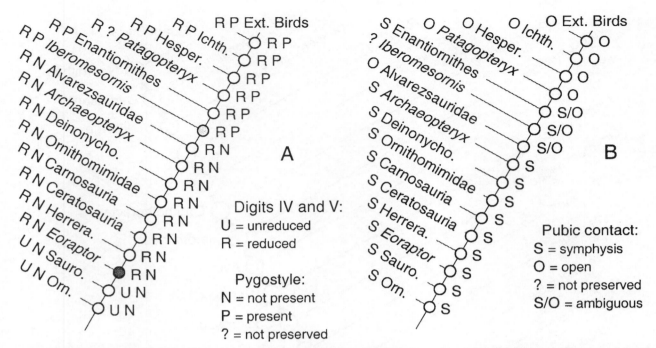

Figure 19.5. Reconstruction of character states in hypothetical ancestors (circles) using character optimization. A, Two traits (size of manual digits IV and V and a pygostyle) are traced on a simplified cladogram of Dinosauria based on the distribution of states on terminal branches. The state of the digits (U or R) and the pygostyle (N or P) is shown for each taxon and for each hypothetical ancestor. Unreduced (U) digits are primitive, but a shift to reduced (R) digits can be unambiguously optimized in the most recent common ancestor of Theropoda (dark-filled circle). Similarly, a pygostyle is not present (N) primitively but was likely present (P) in the ancestor of Ornithothoraces (light-filled circle). B, Not all characters can be unambiguously reconstructed. The presence of a pubic symphysis (S) is clearly primitive, but the distribution of missing data (?) and open pubes (O) make it impossible to unequivocally reconstruct this character in three hypothetical ancestors (S/O). Citations as in Figure 19.4.

ulum and a hingelike mesotarsal ankle joint. The pes was digitigrade and functionally tridactyl.

DINO embodied several of the modifications previously identified as distinguishing NEO from ARCHO. Indeed, many basic features of the hindlimb unique to birds among living tetrapods were already present in DINO and were inherited relatively unchanged. Unlike NEO, however, DINO did not have three locomotor modules and could not fly. The hindlimb retained the primitive caudofemoral retraction system connecting the femur to a long, heavily muscled tail (Gatesy, 1990).

THERO

The hypothetical most recent common ancestor of all theropods, THERO (Fig. 19.6), had a locomotor apparatus similar to that of DINO, but with some important differences. In the axial skeleton the dorsal vertebrae bore accessory (hyposphene-hypantrum) articulations (Gauthier, 1986). The tail was still long but had a "transition point" between distinctively shaped proximal and distal caudal vertebrae. The THERO forelimb was not used for locomotion; the manus was specialized for prey handling (Gauthier, 1986; Sereno and Novas, 1992; Sereno et al., 1993). Digits IV

and V were reduced, whereas the penultimate phalanges and unguals of I–III were enlarged. All long bones had thin walls. In the THERO hindlimb the femur bore a trochanteric shelf, the tibia possessed a fibular crest, and metatarsal V was reduced to a spur (Gauthier, 1986).

THERO was an obligate biped. Vertebral processes made the THERO trunk and distal tail stiffer than in DINO. Although only one locomotor module was present, this module was reduced relative to the primitive ARCHO module in that the forelimbs were freed for nonlocomotor use (Fig. 19.3B; Gatesy and Dial, 1996b). The THERO module consisted of the hindlimb and tail, still linked by the functionally important caudofemoral musculature, and likely still produced significant hip extension during hindlimb movement (Gatesy, 1990).

PAR

The clade defined by the most recent common ancestor of deinonychosaurs (dromaeosaurs and troodontids) and birds, plus all its descendants, has been named Paraves (Sereno, 1997). Deinonychosaur monophyly has been questioned (e.g., Gauthier, 1986; Holtz, 1994), however, leading to a redefinition of Deinonychosauria as a stem-based

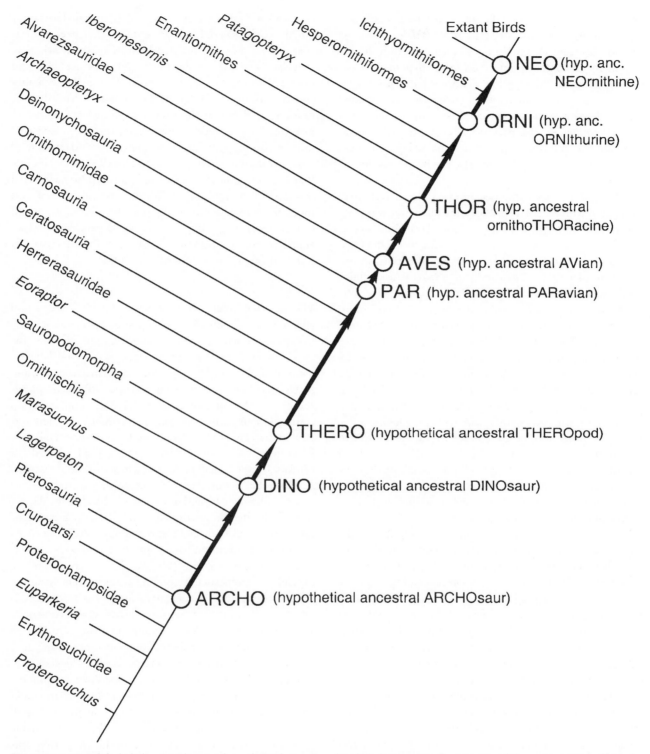

Figure 19.6. Simplified cladogram showing hypothetical ancestors (circles) discussed in the text. Six of the many intermediates for which there is evidence are used to trace the locomotor system on the line of descent between ARCHO and NEO. Citations as in Figure 19.4.

group of theropods more closely related to *Deinonychus* than to birds (Currie and Padian, 1997). The exclusion of troodontids from Deinonychosauria has important implications for character optimization in the hypothetical ancestor of Paraves, PAR (Fig. 19.6; Gauthier, 1986; Holtz, 1994; Sereno, 1997). Although Gauthier (1986) treated troodontids as deinonychosaurs, the condition in dromaeosaurs such as *Deinonychus* was provisionally accepted as ancestral when shared with birds. Since PAR's characters were optimized from Gauthier's data matrix, my reconstruction of PAR retains this bias.

PAR had a trunk composed of 13–14 dorsal vertebrae. Compared with that of THERO, its tail had fewer segments (ca. 35) and a more proximally positioned transition point. Vertebrae at the tail base had boxlike centra, short chevrons, and low neural spines and transverse processes (Ostrom, 1969; Gauthier, 1986; Russell and Dong, 1993). The shoulder girdle was composed of a straplike scapula and a subrectangular coracoid; fused sternal plates and a furcula were likely present in adults (Norell and Makovicky, 1997; Norell et al., 1997). The PAR forelimb was proportionally longer than that of THERO, with a semilunate carpal in the wrist and an elongate manus. In the pelvis, the pubis was oriented caudoventrally, rather than cranioventrally as in THERO, and bore an expanded "foot" distally. The femur had a moundlike greater trochanter, with which the lesser (anterior) trochanter was nearly confluent, but the primitive fourth trochanter present in ARCHO, DINO, and THERO was much reduced. Metatarsal I did not reach the level of the tarsals proximally.

Recent discoveries of nonavian theropod specimens from China preserving integumentary structures (Chen et al., 1998; Ji et al., 1998; Xu et al., 1999a,b; see also Witmer, Chapter 1 in this volume; Zhou and Hou, Chapter 7 in this volume) have important implications for PAR. Although the phylogenetic affinities of *Protarchaeopteryx* and *Caudipteryx* are not fully resolved (Ji et al., 1998; Sereno, 1999a), their likely positions outside the paravian clade make it most parsimonious to infer that PAR was at least partially feathered. At a minimum, PAR can be reconstructed with several remiges and rectrices, as well as a more extensive plumulaceous covering over much of the body.

PAR represents the nonavian theropod ancestor most closely related to birds for which we have good evidence. Its structure and locomotor abilities are therefore of major importance for understanding the origin of birds and flight. The elongate forelimb and hand were presumably predatory features, although these have been interpreted as climbing adaptations (e.g., Chatterjee, 1997, 1999). The wrist, capable of large ulnar deviations but more restricted in flexion and extension than in THERO, may have been modified to permit limb folding (Ostrom, 1969). PAR, like its ancestor THERO, was a biped with a single locomotor module composed of hindlimb and tail. There were significant differences, however, that indicate that this primitive unit was incipiently divided to promote separate pelvic and caudal function. Reduction of both the fourth trochanter and available area of origin at the tail base affirms that the caudofemoral musculature was diminished in size and in functional importance during locomotion. It is likely that muscles running from the pelvis to the tibia and femur were more involved in moving the limb in PAR than in THERO. Partial decoupling of hindlimb and tail can also be inferred from caudal morphology. PAR's narrower, lighter tail was much more mobile proximally than THERO's (Tyson and Gatesy, 1997), suggesting increased independence and an enhanced role in dynamic, and potentially aerodynamic, stabilization during terrestrial locomotion.

AVES

The hypothetical common ancestor of *Archaeopteryx* and all other birds, AVES (Fig. 19.6), is our best perception of the first bird. As with previous character optimizations, the condition in AVES must be reconstructed from the data matrix and cladogram topology. Although *Archaeopteryx* and AVES are often equated, characteristics of *Archaeopteryx* should not be attributed to AVES unless optimization confirms their reconstruction. Surprisingly, very few unambiguous states can be optimized for this node: a tail with fewer than 25–26 vertebrae, short prezygopophyses in distal caudals, a reversed hallux opposing toes II–IV, and feathers. A substantial list of characters are equivocal in AVES because of missing data and homoplasy in basal birds. Yet, despite the limited number of clear novelties, the unambiguous presence of both flight and contour feathers in AVES had major implications for locomotor evolution.

Many features of the AVES locomotor repertoire are difficult to reconstruct. Ambiguity arises from inclusion of the Alvarezsauridae, particularly *Mononykus*, in the cladogram. The position of these clearly flightless birds one node away from *Archaeopteryx*, a presumably flying bird, throws doubt on the locomotor abilities of their common ancestor, AVES (a recent analysis by Sereno [1999b] allies alvarezsaurids with ornithomimosaurs, potentially simplifying this issue; see also Chiappe, Norell, and Clark, Chapter 4 in this volume; Novas and Pol, Chapter 5 in this volume). As noted by Perle et al. (1993), two scenarios are equally parsimonious. First, AVES could have flown, passing this novelty on to *Archaeopteryx* and the ancestor of alvarezsaurids plus all other birds. In this more classic scenario, flight evolved once in birds (flying AVES) and was subsequently lost in the alvarezsaurid line as it would be many times in later bird lineages. However, an equally parsimonious scenario proposes that AVES and alvarezsaurids were primitively flightless. Flight would then have evolved twice, once on the *Archaeopteryx* line and once on the line to Ornithothoraces.

I favor a single origin of flight and a flying AVES because the feathered wings of *Archaeopteryx* are so similar to those of other birds. Such wings were likely present in the ancestral metornithine but were reduced, causing flightlessness, in the alvarezsaurid lineage. If this traditional scenario is provisionally accepted, AVES is the first theropod with three locomotor modules for which we have direct evidence. The wings and tail that bore the feathered flight surfaces also formed pectoral and caudal modules, respectively. The hindlimb acted as its own pelvic module during bipedal locomotion. Although not identical to NEO in many respects, including flight refinements and hindlimb function, a flying AVES would have exemplified the unique modular combination still found in living birds today.

THOR

The clade Ornithothoraces is composed of the most recent common ancestor of *Iberomesornis* and NEO, plus all its descendants. A reconstruction of this hypothetical ancestral ornithothoracine, THOR (Fig. 19.6), reveals the large number of changes that took place in the axial skeleton and appendages on the line from AVES to THOR. THOR had fewer than 13 dorsal vertebrae and 15–25 caudals. Significantly, distal caudal segments were fused into a pygostyle. The pectoral girdle was highly modified relative to AVES, with a strutlike coracoid, scapula with prominent acromion, and large sternum with an ossified keel. The THOR hand skeleton was fused into a carpometacarpus. The diameter of its ulna exceeded that of its radius. The pelvis of THOR had a prominent antitrochanter and fused elements but no ischiadic symphysis. The foot was specialized for perching.

THOR and its descendants (Sanz and Bonaparte, 1992; Chiappe, 1995) possessed many refinements of the flight apparatus relative to the condition in *Archaeopteryx* and, presumably, in AVES. The wing and the tail skeleton of THOR were quite modern in aspect and likely were capable of more modern aerial performance. The sternal keel and strutlike coracoid were associated with large, aerobic pectoralis and supracoracoideus muscles to power the wingbeat cycle. The carpometacarpus formed a solid foundation for the remiges, which also attached to the stout ulna. Most rectrices in THOR were probably supported by soft tissue on either side of the pygostyle. This arrangement would have promoted tail symmetry and allowed the independent movement of the caudal axis without deformation of the tail (Gatesy and Dial, 1996a). An incipient rectricial bulb (Baumel, 1988) was likely present, allowing variable tail fanning and a large range of lift-force production. Fusion of the pelvic girdle of THOR increased the isolation of the pelvic

and caudal modules by providing a stable foundation for hindlimb and caudal musculature. Separation of the ischia moved the origin of several caudal muscles laterally, thereby increasing their ability to rotate the tail fan and deflect it to each side.

ORNI

The most recent common ancestor of Hesperornithiformes and NEO and all its descendants constitute Ornithurae. ORNI (Fig. 19.6), the hypothetical ancestral ornithurine, possessed a large suite of features not present in THOR (note that some of these character states are novelties of ORNI; others first appeared at skipped nodes between THOR and ORNI). The ORNI sacrum was large, encompassing more than eight vertebrae, whereas the trunk was composed of fewer than 11 dorsals and the tail of fewer than 15 caudals. ORNI lacked hyposphene-hypantrum accessory articulations in its dorsals and prezygapophyses in its tail but had ossified uncinate processes. In the shoulder girdle, the sagittally curved scapula formed a sharp angle with the coracoid, which bore a procoracoid process.

Relative to THOR, ORNI had a hindlimb with many modifications. In the pelvis, the laterally compressed pubes, which lacked a terminal foot, no longer formed a symphysis. The pubis, ischium, and caudal ilium were oriented essentially in parallel; the acetabulum was small. The ilium formed a broadened pelvic shelf dorsally (Gatesy and Dial, unpub. obs.). The ORNI femur had a trochanteric crest and a deep patellar groove but no posterior trochanter. ORNI possessed a fully fused tibiotarsus with a cranial cnemial crest, an extensor canal, and a greatly reduced fibula. The completely fused tarsometatarsus lacked metatarsal V and bore a well-developed intracondylar eminence.

ORNI was essentially similar to NEO in all aspects of its locomotor apparatus. The trunk was substantially stiffened by the enlargement of the synsacrum, reduction in mobile dorsal vertebrae, and presence of uncinate processes. In contrast, mobility between caudal vertebrae in front of the pygostyle was no longer hindered by prezygapophyses. Movement of the rectricial tail fan may also have been improved by reorientation of the pubes. With the loss of the pubic symphysis, the laterally spread pubes would have reoriented the pubocaudal musculature to pull the fan ventrolaterally, rather than ventromedially. Broadening of the pelvis by the pelvic shelf also contributed to improved tail control. The femur moved little during walking; the highly reduced caudofemoral retraction system was only employed at high speeds and during takeoff (Gatesy, 1990; Gatesy and Dial, 1993; Carrano, 1998). An essentially modern mechanism of knee flexion by the hamstrings produced most of the limb movement during ground contact.

Discussion

In this chapter I advocate using a phylogeny and its underlying data matrix to trace the evolution of locomotor structures along the line leading to neornithines. Such an approach is not without its weaknesses. The reconstruction of ancestral states, for example, is by no means simple and requires certain assumptions. Missing data and homoplasy will always leave the history of many characters incompletely understood. The discovery of new taxa and better specimens of known taxa may resolve some of these ambiguities, but additional data could yield ambiguities in what had appeared to be unequivocal character-state reconstructions. New information and analyses will also likely alter aspects of the phylogeny underlying my hypotheses of locomotor evolution. Such instability, however inconvenient, is also a strength, since an unalterable view of avian history would be naively unscientific. Finally, interpreting the functional significance of morphological change for locomotor evolution is fraught with difficulty. This concern is not unique to the phylogenetic approach and can only be mitigated through the collection of more data. Continuing study of living archosaurs will help increase our understanding of many aspects of extinct taxa, but we must be willing to accept a lower level of resolution for reconstructions of locomotor behavior. Despite these and other caveats, I still consider the phylogenetic approach superior to alternative methodologies.

One such alternative is to divorce ancestral reconstruction from phylogeny entirely. Tarsitano (1985, 1991) constructed a hypothetical proavis based not on character distributions but solely on aerodynamic principles. By eliminating any historical factors, Tarsitano was free to interpret how flight could have evolved, given general biophysical constraints. This type of analysis appears robust at first. However, a similar investigation of the aerodynamics of preflight based on different initial assumptions arrived at completely different conclusions (Caple et al., 1983). The value of such an approach is difficult to assess, but given the unpredictability of evolution revealed by better-known transformations, a search for ancestors using reverse engineering is dubious. Making functional hypotheses of morphological change independent of phylogenetic hypotheses of morphological change appears to avoid circularity (Lee and Daughty, 1997), but such separation only hampers attempts to integrate all available data (see Witmer, Chapter 1 in this volume).

Confusion about Ancestors: A Call for Optimization

As noted by Witmer (1991, Chapter 1 in this volume), methodologies other than cladistics are often employed by critics striving to disqualify nonavian theropods from a close relationship to birds (Heilmann, 1926; Bock, 1965, 1986; Tar-

sitano and Hecht, 1980; Martin, 1983, 1991; Tarsitano, 1985, 1991; Feduccia, 1996; Hou et al., 1996). Many of the concerns raised against theropod ancestry are founded upon methods of ancestral reconstruction other than character optimization. Even among workers using a cladistic methodology, failure to properly optimize characters can lead to confusion. If character optimization is bypassed, a single data matrix and cladogram can yield very different conclusions about the size, structure, behavior, and age of hypothetical ancestors at each node. The following examples show how disparate interpretations of the same cladogram act as roadblocks to communication and understanding.

Several cladistic analyses place deinonychosaurs (sensu Currie and Padian, 1997) as the closest known relatives of birds (Padian, 1982; Gauthier and Padian, 1985; Gauthier, 1986; Sereno and Rao, 1992; Holtz, 1994; Sereno, 1997, 1999a), lending strong support to the theropod hypothesis outlined by Ostrom (e.g., 1976a). However, the meaning of this branching pattern can be misinterpreted if characters are not optimized. Consider the following criticism: deinonychosaurs are all too large, too terrestrial, and too non-aerodynamic to have given rise to avian flight, and therefore such birdlike theropods are merely convergent upon birds, rather than close to bird origins. This line of reasoning (Tarsitano, 1985; Feduccia, 1996) equates a hypothesis of relationship (deinonychosaurs are the sister group of birds) with a hypothesis of ancestry (deinonychosaurs gave rise to birds). Feduccia (1996), for example, imbued PAR with all the qualities of the oldest well-known deinonychosaur, *Deinonychus* (Ostrom, 1969). PAR likely resembled *Deinonychus* in many respects, but designating features of one terminal taxon as ancestral indicates that a shortcut was taken around optimization. Taxa on terminal branches have novelties (autapomorphies) not present in the lineal ancestors of birds. Including autapomorphies in a reconstructed ancestor is therefore unjustified and a sign that its characters were not optimized. Conclusions based on such an approach do not invalidate the underlying cladogram; they merely confound the study of locomotor evolution.

Another issue raised against a theropod ancestry of birds is stratigraphic disjunction (Padian and Chiappe, 1997, 1998). The reproach that deinonychosaurs are stratigraphically too young to be involved in bird origins (Feduccia, 1996; Hou et al., 1996) is flawed because it equates the age of the oldest known deinonychosaur with the age of PAR. Such a literal reading of the fossil record creates a temporal contradiction within the cladistic hypothesis where none exists (see also Witmer, Chapter 1 in this volume; Clark, Norell, and Makovicky, Chapter 2 in this volume). Proper optimization of PAR places it as at least as old as *Archaeopteryx,* the oldest of PAR's descendants currently well known. Organisms within the interval between PAR and known deinonychosaurs are part of a "ghost lineage" (Norell, 1993) or "missing ancestral lineage" (Sereno, 1997) from which fossils are predicted yet still lacking. The current absence of older definitive deinonychosaurs does not disqualify known forms from a close relationship to birds (Padian and Chiappe, 1997, 1998; Witmer, 1997). New finds and the confirmation of Late Jurassic material provisionally interpreted as dromaeosaurid (Jensen and Padian, 1989; Padian and Chiappe, 1997) may bridge this gap.

Supporters of a theropod ancestry of birds can also interpret a cladogram's topology loosely and forgo character optimization. For example, Chatterjee (1997) envisioned a hypothetical proavian as a small "protodromaeosaur," complete with a rigid tail stiffened by elongate prezygapophyses and chevrons. Such a highly specialized tail morphology is known only in dromaeosaurs; there is no reason that it should be reconstructed in PAR. Chatterjee's justification is made clear when he states, "It is now widely accepted that birds are the descendants of a small, unknown dromaeosaur" (1997:163). Although dromaeosaurs and birds shared a common ancestor, unless there is evidence that some dromaeosaurs are more closely related to birds than they are to other dromaeosaurs, there is no reason to call PAR a dromaeosaur or to give it exclusively dromaeosaurian novelties. Chatterjee, while disagreeing with Feduccia on major issues, falls into the same trap of reconstructing PAR based on the group thought to be most closely related to birds. Character optimization prevents such confusion by forcing the description of each hypothetical ancestor to be an explicit output of the data matrix and cladogram.

Unfortunately, confusing known taxa for lineal ancestors, while methodologically flawed, appears to be quite common. In some cases, terminal taxa are used as surrogates of ancestors to provide a simplified overview of evolutionary change. For example, Dingus and Rowe (1997:114) first present bird evolution as a transition from *Euparkeria* to *Deinonychus* to *Archaeopteryx* to *Columba* (Fig. 19.7). Linking skeletal reconstructions of these taxa with arrows is confusing because it implies an ancestor-descendant relationship the authors do not believe. Even when arrows are omitted, the distinction between illustrating terminal taxa and implying an evolutionary sequence is very slight and depends almost entirely on the sophistication and philosophy of the reader.

The temptation to jump from branch tip to branch tip (rather than from ancestral node to ancestral node) stems from an effort to visualize organisms at internal nodes on a cladogram. ARCHO, NEO, and all ancestors in between remain elusive because they are, as noted, hypothetical. We can conceive of ORNI as a relatively modern flying bird; our intuition leads us to the conclusion that many aspects of *Hesperornis* are diving specializations of its lineage, not features present in ORNI. But what of AVES, PAR, or THERO? As we move backward from NEO to ARCHO, our ability to

discriminate autapomorphy from synapomorphy becomes weakened. Without optimization, judging which character states are primitive features of remote bird ancestors and which are unique attributes of side branches becomes increasingly difficult. Many known taxa are "similar enough" to hypothetical ancestors on the line to birds that differences can be overlooked, but such a relaxation of rigor is dangerously misleading. Efforts must be made to distinguish the true line of descent from the surrogate line of descent (Fig. 19.7).

From PAR to AVES: Evidence and Inference

Given the probability that PAR could not fly and AVES could, the evolutionary changes that took place during this interval are critical to our understanding of the transition from one to three locomotor modules. Fragmentary taxa such as *Avimimus* (Kurzanov, 1987; Vickers-Rich, Chiappe, and Kurzanov, Chapter 3 in this volume) and *Unenlagia* (Novas and Puerta, 1997) may be derived from ancestors in this interval, but for the time being, the intermediates between PAR and AVES in which flight evolved remain almost completely unknown. Unfortunately, many presumptions have been made about what transformations could or could not have occurred between PAR and AVES.

First, no reason exists to assume that PAR itself acquired flight. PAR's size and morphology need not be identical to those of its first flying descendant. PAR could have been too large and not fully equipped for flight, but this does not disqualify PAR from avian ancestry. One of PAR's descendants on the line to AVES could have evolved smaller size and other flight features. The cladistic hypothesis itself makes no statement about the possibility of significant evolutionary change between PAR and AVES. Using such transitions in size, morphology, and function to invalidate the theropod ancestry hypothesis is not warranted. Disqualifying the ancestral vertebrate, ancestral tetrapod, ancestral amniote, and ARCHO from avian ancestry would be equally illogical. None of these ancestors appears well suited to give rise to flying descendants, yet they clearly did.

What of the noncladists' alternative (Heilmann, 1926; Bock, 1965, 1986; Martin, 1983; Tarsitano, 1985, 1991; Feduccia, 1996)? Feduccia's preferred ancestor, which "was surely a small, quadrupedal, arboreal archosaur" (Feduccia, 1996:viii), appeals to many of those not convinced by the theropod hypothesis. Is there any reason to believe that this proavis descended from a line that had always been small, quadrupedal, and arboreal? An examination of the evolutionary history of Archosauria suggests not. As discussed earlier in this chapter, the ancestral archosaur, ARCHO, was of intermediate size and nonarboreal. Therefore, Feduccia's small, arboreal ancestor of birds must have descended from this much larger, terrestrial ancestor. Critics should not reject the possibility of size reduction and

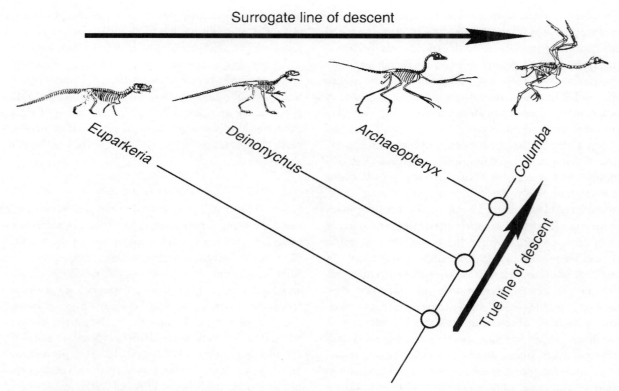

Figure 19.7. Illustrating a cladogram can blur the distinction for some readers between the true line of descent between hypothetical ancestors and a surrogate line of descent between terminal taxa. Autapomorphies of figured taxa were not present in the ancestors on the true line of descent. Mistreating terminal taxa as ancestors creates confusion and misunderstanding in the study of avian evolution (modified from Dingus and Rowe, 1997).

increased climbing activity between PAR and AVES while invoking these very same changes between ARCHO and a hypothetical proavis. If a small descendant of PAR were to enter the trees, it would be described as a small, quadrupedally locomoting, arboreal archosaur. For the cladistic hypothesis adopted here to be correct, PAR need not be *the* proavis, only one of its ancestors. Critics of theropod ancestry may have identified the correct hypothetical proavis but missed its true ancestry.

Some supporters of theropod ancestry have also overlooked the possibility of significant modification between PAR and AVES. Again, a distinction must be made between those ancestors of AVES involved in flight acquisition and more distant ancestors. Returning to the beaded string analogy, we do not know how many beads form the line between PAR and AVES. Just because deinonychosaurs are the closest nonbirds to AVES for which we have relatively complete information, PAR is not automatically the "immediate" ancestor of birds (Padian, 1985:422). PAR and AVES may be adjacent internal nodes on the cladogram, but this relationship does not require PAR to have been the most proximate ancestor (adjacent bead) of AVES or the first flier. The line from PAR to AVES could have included many intermediate forms that evolved features and behaviors unforeseen in PAR.

The behavior of greatest interest is the advent of flight. Can the phylogeny resolve whether flight originated from the ground up or the trees down? Put simply, given current data, does ancestral reconstruction have the power to establish when climbing habits and flight were acquired? Assuming flight evolved only once and was present in AVES, there are three possibilities: (1) PAR and AVES were both exclusively terrestrial; (2) PAR was exclusively terrestrial, and AVES could climb; and (3) both PAR and AVES could climb. Note that only the first hypothesis requires a purely "ground up" origin of bird flight. Difficulty choosing between these alternatives primarily stems from trying to understanding the habits of *Archaeopteryx,* which are critical for reconstructing the terrestrial and climbing abilities of both PAR and AVES. Although data supporting arboreal specializations in *Archaeopteryx* (Yalden, 1985; Bock, 1986; Paul, 1988; Feduccia, 1993) may not be conclusive (Ostrom, 1974, 1985, 1986; Peterson, 1985; Rayner, 1991; Peters and Görgner, 1992; Chiappe, 1997), there is a strong possibility that AVES could both fly and climb. If so, which behavior evolved first?

Arboreal specializations are clearly present in THOR, but when did tree climbing begin? Given that PAR was probably primarily ground-dwelling, some workers advocate a cursorial origin of flight because it appears to be a simpler explanation (Ostrom, 1976b; Padian, 1985; Gauthier and Padian, 1985). However, delaying the advent of climbing until after flight is no more parsimonious than having climbing precede or coincide with flight; each is one step between character states. Taxa stemming from intermediates between PAR and AVES may help discriminate the timing of this transition, but inference of climbing ability is not without difficulty. An alternative that must be considered, therefore, is the evolution of flight *in a theropod* from the trees down (Paul, 1988; Chatterjee, 1997, 1999). Feduccia's concern (quoted in Gibbons, 1996:721) that a cursorial origin of flight is "a biophysical impossibility" may hold true but does not preclude theropods from bird ancestry. The assertion that "a dinosaurian origin of birds is inextricably linked with the cursorial, or ground-up, origin of avian flight" (Feduccia, 1996:viii) simply does not hold (Padian and Chiappe, 1998; see also Witmer, Chapter 1 in this volume).

Conclusions and Future Directions

A theropod ancestry of birds and a cursorial origin of flight are two independent hypotheses. The phylogenetic relationship of birds can be evaluated from morphological characters alone (Chiappe, 1995; Padian and Chiappe, 1997), whereas flight and climbing are not directly preservable. As locomotor behaviors, these must be inferred from the morphology of extinct archosaurs and are therefore less robust. Critics of theropod ancestry should not use evidence for "trees down" flight in their phylogenetic argument against morphological evidence of a close relationship between theropod and birds. At the same time, proponents of theropod ancestry should not assume that tree climbing followed flight just because PAR may have been fully terrestrial. Many of the morphological and even behavioral prerequisites of flight appear to have evolved in terrestrial theropods (Ostrom, 1976a,b; Gauthier and Padian, 1985; Padian, 1987; Novas and Puerta, 1997; Padian and Chiappe, 1997, 1998; Sereno, 1997, 1999a), but perhaps the final steps toward aerial locomotion really were from trees or other elevated structures (Chatterjee, 1997, 1999). Until further evidence can resolve whether theropods first climbed into trees or flew into trees, the possible link between climbing and flight should not be ruled out on phylogenetic grounds alone.

Advances in understanding the evolutionary history of the avian locomotor system will require a synthesis of hypotheses from specialists in different areas. One source will be the discovery of new fossils, which will broaden our rapidly expanding (Chiappe, 1995), yet still poor, knowledge of theropod diversity. Phylogenetic analysis will continually revise data matrices and cladograms to provide the most up-to-date hypotheses. Identification of taxa branching off the line from ARCHO to NEO will increase the number of nodes, thereby increasing the resolution of descriptions of morphological change through time. Understanding the significance of these morphological transformations for locomotor evolution, however, will require more than just character optimization. There is a large gap between proposing a scenario and acquiring data to support it. A cladogram and its data matrix can only go so far.

Ultimately, studies of living archosaurs are required to provide insight into the functional role of characters and character complexes. Many major issues within avian history have yet to be analyzed rigorously by functional morphologists. For example, the predatory forelimb strike of dromaeosaurs has been equated with the flight stroke of birds (Gauthier and Padian, 1985), but the reality of this kinematic pattern has not been verified. Similarly, the functional interaction of the limbs in a climbing theropod has been outlined in great detail (Chatterjee, 1999), but the forces and limb movements proposed are at present entirely hypothetical. Is gliding primitive for birds, or did flapping flight come first? What was the behavioral precursor of flapping? Answering such questions and learning more about the evolution of locomotion on the line from ARCHO to NEO will come from the cooperative interaction of paleontologists, systematists, and functional morphologists.

Acknowledgments

Special thanks go to K. M. Middleton for his tireless work entering data sets, checking optimizations, and charting character evolution. L. M. Chiappe kindly allowed access to unpublished phylogenetic analyses. Discussions with K. M. Middleton, J. E. Gatesy, J. R. Hutchinson, and J. M. Clark were most helpful, as were their comments, and those of both editors, on earlier versions of this chapter.

Literature Cited

Baumel, J. J. 1988. Functional morphology of the tail apparatus of the pigeon (*Columba livia*). Advances in Anatomy, Embryology and Cell Biology 110:1–115.

Bock, W. J. 1965. The role of adaptive mechanisms in the origin of higher levels of organization. Systematic Zoology 14:272–287.

———. 1986. The arboreal origin of avian flight. Memoirs of the California Academy of Sciences 8:57–72.

Caple, G., R. P. Balda, and W. R. Willis. 1983. The physics of leaping animals and the evolution of preflight. American Naturalist 121:455–476.

Carrano, M. T. 1998. Locomotion in non-avian dinosaurs: integrating data from hindlimb kinematics, in vivo strain, and bone morphology. Paleobiology 24(4):450–469.

Chatterjee, S. 1997. The Rise of Birds: 225 Million Years of Evolution. Johns Hopkins University Press, Baltimore, 312 pp.

———. 1999. *Protoavis* and the early evolution of birds. Palaeontographica 254:1–100.

Chen P.-J., Dong Z.-M., and Zhen S.-N. 1998. An exceptionally well-preserved theropod dinosaur from the Yixian Formation of China. Nature 391:147–152.

Chiappe, L. M. 1995. The first 85 million years of avian evolution. Nature 378:349–355.

————. 1997. Climbing *Archaeopteryx*? A response to Yalden. Archaeopteryx 15:109–112.

Chiappe, L. M., M. A. Norell, and J. M. Clark. 1996. Phylogenetic position of *Mononykus* (Aves: Alvarezsauridae) from the Late Cretaceous of the Gobi Desert. Memoirs of the Queensland Museum 39:557–582.

Currie, P. J., and K. Padian. 1997. Encyclopedia of Dinosaurs. Academic Press, San Diego, 869 pp.

Dingus, L., and T. Rowe. 1997. The Mistaken Extinction: Dinosaur Evolution and the Origin of Birds. W. H. Freeman, New York, 384 pp.

Feduccia, A. 1993. Evidence from claw geometry indicating arboreal habits of *Archaeopteryx*. Science 259:790–793.

————. 1996. The Origin and Evolution of Birds. Yale University Press, New Haven, 420 pp.

Gatesy, S. M. 1990. Caudofemoral musculature and the evolution of theropod locomotion. Paleobiology 16(2):170–186.

————. 1995. Functional evolution of the hind limb and tail from basal theropods to birds; pp. 219–234 in J. J. Thomason (ed.), Functional Morphology in Vertebrate Paleontology. Cambridge University Press, New York.

————. 1999a. Guineafowl hind limb function II: electromyographic analysis and motor pattern evolution. Journal of Morphology 240:115–125.

————. 1999b. Guineafowl hind limb function I: cineradiographic analysis and speed effects. Journal of Morphology 240:127–142.

Gatesy, S. M., and K. P. Dial. 1993. Tail muscle activity patterns in walking and flying pigeons (*Columba livia*). Journal of Experimental Biology 176:55–76.

————. 1996a. From frond to fan: *Archaeopteryx* and the evolution of short-tailed birds. Evolution 50:2037–2048.

————. 1996b. Locomotor modules and the evolution of avian flight. Evolution 50:331–340.

Gauthier, J. 1984. A cladistic analysis of the higher systematic categories of the Diapsida. Ph.D. dissertation, University of California, Berkeley, 564 pp.

————. 1986. Saurischian monophyly and the origin of birds. Memoirs of the California Academy of Sciences 8:1–55.

Gauthier, J., and K. Padian. 1985. Phylogenetic, functional, and aerodynamic analyses of the origin of birds and their flight; pp. 185–197 in M. K. Hecht, J. H. Ostrom, G. Viohl, and P. Wellnhofer (eds.), The Beginnings of Birds. Freunde des Jura-Museums, Eichstätt.

Gauthier, J., A. G. Kluge, and T. Rowe. 1988. Amniote phylogeny and the importance of fossils. Cladistics 4:105–209.

Gibbons, A. 1996. New feathered fossil brings dinosaurs and birds closer. Science 274:720–721.

Heilmann, G. 1926. The Origin of Birds. Witherby, London, 210 pp.

Holtz, T. R., Jr. 1994. The phylogenetic position of the Tyrannosauridae: implications for theropod systematics. Journal of Paleontology 68:1100–1117.

Hou L., L. D. Martin, Zhou Z., and A. Feduccia. 1996. Early adaptive radiation of birds: evidence from fossils from Northeastern China. Science 274:1164–1167.

Hutchinson, J. R., and S. M. Gatesy. 2000. Adductors, abductors, and the evolution of archosaur locomotion. Paleobiology 26(4):734–751.

Jenkins, F. A., Jr. 1993. The evolution of the avian shoulder joint. American Journal of Science 293:253–267.

Jensen, J. A., and K. Padian. 1989. Small pterosaurs and dinosaurs from the Uncompahgre Fauna (Brushy Basin Member, Morrison Formation: ?Tithonian), Late Jurassic, western Colorado. Journal of Paleontology 63:364–373.

Ji Q., P. J. Currie, M. A. Norell, and Ji S.-A. 1998. Two feathered dinosaurs from northeastern China. Nature 393:753–761.

Kurzanov, S. M. 1987. Avimimidae and the problem of the origin of birds. Transactions of the Joint Soviet-Mongolian Paleontological Expedition 31:5–92. [Russian]

Lauder, G. V. 1981. Form and function: structural analysis in evolutionary morphology. Paleobiology 7:430–442.

Lee, M. S. Y., and P. Daughty. 1997. The relationship between evolutionary theory and phylogenetic analysis. Biological Reviews 72:471–495.

Maddison, W. P., and D. R. Maddison. 1992. MacClade: Analysis of Phylogeny and Character Evolution. Version 3.0. Sinauer Associates, Sunderland, Mass.

Martin, L. D. 1983. The origin of birds and of avian flight; pp. 106–129 in R. F. Johnston (ed.), Current Ornithology, Vol. 1. Plenum Press, New York.

————. 1991. Mesozoic birds and the origin of birds; pp. 485–540 in H.-P. Schultze and L. Trueb (eds.), Origins of the Higher Groups of Tetrapods: Controversy and Consensus. Cornell University Press, Ithaca, N.Y.

Nemeschkal, H. L., R. van den Elzen, and H. Brieschke. 1992. The morphometric extraction of character complexes accomplishing common biological roles: avian skeletons as a case study. Zeitschrift für Zoologische Systematik und Evolutionsforschung 30:201–219.

Norell, M. 1993. Tree-based approaches to understanding history: comments on ranks, rules, and the quality of the fossil record. American Journal of Science 293:407–417.

Norell, M., and P. J. Makovicky. 1997. Important features of the dromaeosaur skeleton: information from a new specimen. Novitates 3215:1–28.

Norell, M., P. J. Makovicky, and J. M. Clark. 1997. A *Velociraptor* wishbone. Nature 389:447.

Novas, F. E. 1993. New information on the systematics and postcranial skeleton of *Herrerasaurus ischigualastensis* (Theropoda: Herrerasauridae) from the Ischigualasto Formation (Upper Triassic) of Argentina. Journal of Vertebrate Paleontology 13:400–423.

————. 1996. Dinosaur monophyly. Journal of Vertebrate Paleontology 16:723–741.

Novas, F. E., and P. F. Puerta. 1997. New evidence concerning avian origins from the Late Cretaceous of Patagonia. Nature 387:390–392.

Ostrom, J. H. 1969. Osteology of *Deinonychus antirrhopus*, an unusual theropod from the Lower Cretaceous of Montana. Bulletin of the Yale Peabody Museum of Natural History 30:1–165.

————. 1974. *Archaeopteryx* and the origin of flight. Quarterly Review of Biology 49:27–47.

————. 1976a. *Archaeopteryx* and the origin of birds. Biological Journal of the Linnean Society 8:91–182.

————. 1976b. Some hypothetical anatomical stages in the evolution of avian flight. Smithsonian Contributions to Paleobiology 27:1–21.

————. 1985. The meaning of *Archaeopteryx;* pp. 161–176 *in* M. K. Hecht, J. H. Ostrom, G. Viohl, and P. Wellnhofer (eds.), The Beginnings of Birds. Freunde des Jura-Museums, Eichstätt.

————. 1986. The cursorial origin of avian flight. Memoirs of the California Academy of Sciences 8:73–81.

Padian, K. 1982. Macroevolution and the origin of major adaptations: vertebrate flight as a paradigm for the analysis of patterns. Proceedings of the Third North American Paleontological Convention 2:387–392.

————. 1985. The origins and aerodynamics of flight in extinct vertebrates. Paleontology 28:413–433.

————. 1987. A comparative phylogenetic and functional approach to the origin of vertebrate flight; pp. 3–23 *in* M. B. Fenton, P. Racey, and J. M. V. Rayner (eds.), Recent Advances in the Study of Bats. Cambridge University Press, Cambridge.

Padian, K., and L. M. Chiappe. 1997. Bird origins; pp. 71–79 *in* P. J. Currie and K. Padian (eds.), Encyclopedia of Dinosaurs. Academic Press, San Diego.

————. 1998. The origin and early evolution of birds. Biological Reviews 73:1–42.

Padian, K., J. R. Hutchinson, and T. R. Holtz Jr. 1999. Phylogenetic definitions and nomenclature of the major taxonomic categories of the carnivorous Dinosauria (Theropoda). Journal of Vertebrate Paleontology 19(1):69–80.

Paul, G. S. 1988. Predatory Dinosaurs of the World. Simon and Schuster, New York, 464 pp.

Perle A., M. A. Norell, L. M. Chiappe, and J. M. Clark. 1993. Flightless bird from Cretaceous of Mongolia. Nature 362:623–626.

Peters, D. S., and E. Görgner 1992. A comparative study on the claws of *Archaeopteryx;* pp. 29–37 *in* K. E. Campbell (ed.), Papers in Avian Paleontology, Honoring Pierce Brodkorb. Science Series 36. Natural History Museum of Los Angeles County, Los Angeles.

Peterson, A. 1985. The locomotor adaptations of *Archaeopteryx;* pp. 99–103 *in* M. K. Hecht, J. H. Ostrom, G. Viohl, and P. Wellnhofer (eds.), The Beginnings of Birds. Freunde des Jura-Museums, Eichstätt.

Rayner, J. M. V. 1991. Avian flight evolution and the problem of *Archaeopteryx;* pp. 183–212 *in* J. M. V. Rayner and R. J. Wootton (eds.), Biomechanics in Evolution. Cambridge University Press, Cambridge.

Russell, D. A., and Z. Dong. 1993. A nearly complete skeleton of a new troodontid dinosaur from the Early Cretaceous of the Ordos Basin, Inner Mongolia, People's Republic of China. Canadian Journal of Earth Science 30:2163–2173.

Sanz, J. L., and J. F. Bonaparte. 1992. A new order of birds (Class Aves) from the Lower Cretaceous of Spain; pp. 39–49 *in* K. E. Campbell (ed.), Papers in Avian Paleontology, Honoring Pierce Brodkorb. Science Series 36. Natural History Museum of Los Angeles County, Los Angeles.

Sereno, P. C. 1991. Basal archosaurs: phylogenetic relationships and functional implications. Journal of Vertebrate Paleontology (supplement) 11:1–53.

————. 1997. The origin and evolution of dinosaurs. Annual Review of Earth and Planetary Sciences 25:435–489.

————. 1998. A rationale for phylogenetic definitions, with application to the higher-level taxonomy of Dinosauria. Neues Jahrbuch für Geologie und Paläontologie 210(1):41–83.

————. 1999a. The evolution of dinosaurs. Science 284:2137–2147.

————. 1999b. Alvarezsaurids (Dinosauria, Coelurosauria): birds or ornithomimosaurs? Journal of Vertebrate Paleontology 19(3):75A.

Sereno, P. C., and A. B. Arcucci. 1990. The monophyly of crurotarsal archosaurs and the origin of bird and crocodile ankle joints. Neues Jahrbuch für Geologie und Paläontologie 180:21–52.

————. 1993. Dinosaurian precursors from the Middle Triassic of Argentina: *Lagerpeton chanarensis.* Journal of Vertebrate Paleontology 13:385–399.

Sereno, P. C., and F. E. Novas. 1992. The complete skull and skeleton of an early dinosaur. Science 258:1137–1140.

Sereno, P. C., and C. Rao. 1992. Early evolution of avian flight and perching: new evidence from the Lower Cretaceous of China. Science 255:845–848.

Sereno, P. C., C. A. Forster, R. R. Rogers, and A. M. Monetta. 1993. Primitive dinosaur skeleton from Argentina and the early evolution of Dinosauria. Nature 361:64–66.

Sy, M. 1936. Funktionell-anatomische Untersuchungen am Vogelflügel. Journal für Ornithologie 84:199–296.

Tarsitano, S. F. 1985. The morphological and aerodynamic constraints on the origin of avian flight; pp. 319–332 *in* M. K. Hecht, J. H. Ostrom, G. Viohl, and P. Wellnhofer (eds.), The Beginnings of Birds. Freunde des Jura-Museums, Eichstätt.

————. 1991. *Archaeopteryx:* quo vadis?; pp. 541–576 *in* H.-P. Schultze and L. Trueb (eds.), Origins of the Higher Groups of Tetrapods: Controversy and Consensus. Cornell University Press, Ithaca, N.Y.

Tarsitano, S. F., and M. K. Hecht. 1980. A reconsideration of the reptilian relationships of *Archaeopteryx.* Zoological Journal of the Linnean Society of London 69:149–182.

Thomas, A. L. R. 1993. On the aerodynamics of bird tails. Philosophical Transactions of the Royal Society, London, B 340:361–380.

Tyson, A. R., and S. M. Gatesy. 1997. The evolution of tail flexibility in theropods. Journal of Paleontology 17:82A.

van den Elzen, R., H. L. Nemeschkal, and H. Classen. 1987. Morphological variation of skeletal characters in the bird family Carduelidae: I. General size and shape patterns in African canaries shown by principal component analyses. Bonn Zoologische Beiträge 38:221–239.

Vasquez, R. J. 1994. The automating skeletal and muscular mechanisms of the avian wing (Aves). Zoomorphology 114:59–71.

Witmer, L. M. 1991. Perspectives on avian origins; pp. 427–466 *in* H.-P. Schultze and L. Trueb (eds.), Origins of the Higher Groups of Tetrapods: Controversy and Consensus. Cornell University Press, Ithaca, N.Y.

————. 1997. Flying feathers. Science 276:1209–1210.

Xu X., Tang Z.-L., and Wang X.-L. 1999a. A therizinosauroid dinosaur with integumentary structures from China. Nature 399:350–354.

Xu X., Wang X.-L., and Wu X.-C. 1999b. A dromaeosaurid dinosaur with a filamentous integument from the Yixian Formation of China. Nature 401:262–266.

Yalden, D. 1985. Forelimb function in *Archaeopteryx;* pp. 91–97 *in* M. K. Hecht, J. H. Ostrom, G. Viohl, and P. Wellnhofer (eds.), The Beginnings of Birds. Freunde des Jura-Museums, Eichstätt.

20

Basal Bird Phylogeny
Problems and Solutions

LUIS M. CHIAPPE

 Although more than half of the evolution of birds occurred during the Mesozoic era (Chiappe, 1995a), our understanding of this long history focused on the spectacular specimens of the Late Jurassic *Archaeopteryx lithographica* and the more derived Late Cretaceous hesperornithiforms and ichthyornithiforms for more than a century of paleontological research. In the past decade, however, a tremendous burst of new evidence—perhaps unparalleled in the field of vertebrate paleontology—has been uncovered. Indeed, the number of species of early birds described during the 1990s nearly tripled the number of taxa discovered during the previous 130 years, since the discovery of *Archaeopteryx* in the mid-1800s (Chiappe, 1997a; Padian and Chiappe, 1998).

While this new evidence has offered an unprecedented opportunity for better understanding the evolutionary transformations of several remarkable biological attributes of birds (e.g., feathers, uninterrupted growth, active flight), it has also made obsolete the previous phylogenetic analyses of basal birds, which had been based on far fewer taxa (e.g., Cracraft, 1986; Chiappe and Calvo, 1994; Sanz et al., 1995; Chiappe et al., 1996). Clearly, a comprehensive phylogenetic hypothesis of the new diversity of basal avians is needed before the evolutionary history of these attributes can be fully explained. Although this major task is beyond the scope of this chapter and is the subject of ongoing research, this study aims to provide the phylogenetic framework for most of the taxa addressed in this volume and, in doing so, to substantially increase the number of taxa ever to be included in a cladistic analysis of basal birds.

The following institutional abbreviations are used in this chapter: AMNH, American Museum of Natural History, New York, New York, United States; GI, MGI, IGM, Geological Institute, Ulaanbataar, Mongolia; IVPP, Institute of Vertebrate Paleontology and Paleoanthropology, Beijing, China; LH, Las Hoyas Collection, Universidad Autónoma de Madrid, Madrid, Spain; LP, La Pedrera Collection, Institute d'Estudis Illerdens, Lleida, Spain; MUCPv, Museo de Ciencias Naturales, Universidad Nacional del Comahue, Neuquén, Argentina; PIN, Paleontological Institute, Moscow, Russia; PVPH, Museo Municipal "Carmen Funes," Plaza Huincul, Argentina; YPM, Yale Peabody Museum, New Haven, Connecticut, United States; ZPAL-MgR, Institute of Paleobiology, Polish Academy of Sciences, Warsaw, Poland.

Historical Background

Early phylogenetic studies recognized the basalmost status of *Archaeopteryx* (here by definition regarded as the most primitive avian lineage; see Chiappe, 1992, 1996; Witmer, Chapter 1 in this volume) and placed hesperornithiforms and ichthyornithiforms either as closer to Neornithes (e.g., Heilmann, 1926; Howard, 1950) or as basal forms of different extant lineages (e.g., Marsh, 1880). During the 1970s and early 1980s, several important, though incomplete, taxa were recovered. These included the Early Cretaceous *Ambiortus dementjevi* (Kurochkin, 1985a,b) and the Late Cretaceous *Gobipteryx minuta* (Elzanowski, 1977, 1981), both from Mongolia; *Alexornis antecedens* (Brodkorb, 1976), from Mexico; and a variety of mostly isolated bones from the Late Cretaceous of Argentina. These discoveries led to the recognition of a new higher group of basal avians, Enantiornithes (Walker, 1981; Chiappe and Walker, Chapter 11 in this volume).

Martin's (1983) studies of some of these basal birds led to the conclusion that Enantiornithes comprised a diverse group of Cretaceous birds, which included not only the Argentine forms described by Walker (1981) but also *Gobipteryx* and *Alexornis*. Different hypotheses of the relationships of Enantiornithes to other Mesozoic birds were proposed during the 1980s (Chiappe, 1995b). Walker (1981) placed this group in an intermediate position between *Archaeopteryx* and hesperornithiforms. Martin (1983) argued for a close relationship between *Archaeopteryx* and Enantiornithes, classifying them within the alleged clade "Sauriurae." Cracraft (1986), in his pioneering numerical cladistic analysis of basal birds, placed Enantiornithes within Ornithurae (the clade including the common ancestor of Hesperornithiformes and Neornithes plus all its descendants).

Inferences about the evolutionary transformation of different attributes arising from these analyses were restricted by the limited data and constrained by enormous long-branch problems (e.g., Martin, 1983; Cracraft, 1986; Gauthier, 1986). These problems led to some mistaken conclusions—for example, regarding Hesperornithiformes as primitively flightless (Cracraft, 1986). These early phylogenetic hypotheses also shaped the two main current lines of thought intended to explain the interrelationships among the already known and the many newly found species of basal birds.

On the one hand, Martin and followers (e.g., Martin, 1983, 1995; Hou et al., 1995a,b, 1996; Zhou, 1995; Feduccia, 1996) argue for a basal dichotomy of birds: "Sauriurae," including *Archaeopteryx*, Enantiornithes, and the recently described *Confuciusornis sanctus* (Hou et al., 1995a,b, 1996; sometimes included within Enantiornithes), and Ornithurae, including hesperornithiforms, ichthyornithiforms, neornithines, and several poorly known Early Cretaceous Chinese taxa (e.g., *Chaoyangia beishanensis, Liaoningornis longidigitris;* see Hou et al., 1996; Hou, 1997; Zhou and Hou, Chapter 7 in this volume). Within this framework, the vast majority of new finds have been grouped within Enantiornithes—everything that is not an ornithurine is essentially placed here (e.g., Hou et al., 1995a,b, 1996; Martin, 1995; Feduccia, 1996).

On the other hand, Chiappe and his co-workers (e.g., Chiappe, 1991, 1995a,b, 1996, 1997a; Chiappe and Calvo, 1994; Sanz et al., 1995, 1996, 1997; Chiappe et al., 1996; Forster et al., 1996, 1998a; Ji et al., 1999) have supported Walker's (1981; and, to some extent, Cracraft's—see Cracraft, 1986) early suggestion that Enantiornithes are more closely related to neornithine birds than to *Archaeopteryx,* and they have placed a variety of other newly described basal birds (e.g., *C. sanctus, Rahonavis ostromi, Vorona berivotrensis, Patagopteryx deferrariisi*) as successive branches of a pectinate cladogram between *Archaeopteryx* and Neornithes. In this latter view, "Sauriurae" is a paraphyletic group supported only by primitive or inaccurate characters (Chiappe, 1995b; Chiappe and Walker, Chapter 11 in this volume).

This decade of extraordinary findings brought another topic of controversy into the arena of basal birds—namely, the phylogenetic relationships of the Late Cretaceous *Mononykus olecranus* from the Gobi Desert and its alvarezsaurid relatives (Chiappe, Norell, and Clark, Chapter 4 in this volume). In their initial report of *Mononykus,* Perle

et al. (1993a) regarded this bizarre animal as the sister group of all birds except *Archaeopteryx.* Subsequent discoveries (Novas, 1997; Chiappe et al., 1998; Hutchinson and Chiappe, 1998) and the reevaluation of previously described taxa (Novas, 1996) led to the conclusion that *Mononykus* formed part of an early radiation of bizarre birds. Yet the phylogenetic relationships of this peculiar theropod lineage are still unsettled. Methodologically flawed criticisms notwithstanding (see Chiappe et al., 1996, 1997; Novas and Pol, Chapter 5 in this volume for a discussion), serious objections to the avian relationship of the alvarezsaurids stemmed from new cladistic analyses (Sereno, 1999a; Novas and Pol, Chapter 5 in this volume) supporting nonavian coelurosaurian relationships. Other equally exhaustive cladistic analyses, however, are still supporting the initial avian hypothesis provided by Perle et al. (1993a) (e.g., Forster et al., 1998a; Holtz, 2001).

Taxa, Characters, and Coding

The cladistic analysis presented here includes 18 in-group taxa (see Appendix 20.1). With the exception of *Anas platyrhynchos* (mallard duck), all terminals are of Mesozoic age (see Appendix 20.1). Three nonavian theropod taxa (i.e., allosaurids, velociraptorines, and troodontids) were used as out-groups (see Appendix 20.1). All in-groups, with the exception of *Ichthyornis,* constitute distinct species recognized by unique sets of characters. The validity of the different species of *Ichthyornis* still has to be demonstrated, and Clarke (1999) has documented the composite status of the "holotype" of *Ichthyornis victor,* one of the specimens prominently featured in Marsh's (1880) seminal work. Further studies on *Ichthyornis* may demonstrate that some of the specimens used here to score this terminal may be inappropriate.

Scored variables were grouped in 169 characters (see Appendix 20.2); 155 of these are binary, and 14 are multistate. Characters were drawn or modified from previous cladistic analyses (e.g., Cracraft, 1986; Gauthier, 1986; Chiappe and Calvo, 1994; Chiappe, 1995b, 1996; Chiappe et al., 1996), or they derived from subsequent original observations. Approximately 17% of the scored variables (28 characters) are from the skull, 82% (139 characters) are postcranial variables, and 1% (2 characters) are from the plumage. All these characters have at least one derived character state shared by two or more of the in-group taxa; binary characters in which the derived character state is known only in one of the analyzed taxa (autapomorphic characters) were excluded from the analysis a priori. This procedure follows investigations indicating that autapomorphic characters, which are uninformative for the topology of the tree, affect certain statistics of the analysis (e.g., consistency index; see Carpenter, 1988). Aside from obvious preservational artifacts (e.g., presence vs. absence of ossified uncinate processes in *Confuciusornis*), cases of polymorphism (for the character states included in the analysis) were not found among the studied specimens.

The monophyly of the supraspecific taxa used as outgroups (i.e., Allosauridae, Velociraptorinae, Troodontidae) is well documented (Currie and Padian, 1997). When conflicting character states were found within these out-groups (e.g., serrated vs. unserrated tooth crowns in troodontids), the character states were scored on the basis of the condition present in the more basal nonavian theropods (e.g., serrated tooth crowns).

An important aspect of a cladistic analysis is the grouping of the diversity of character states in discrete characters. Given a number of variables within a character, options for character coding range from arranging all of them in a single multistate character (Pimentel and Riggins, 1987; Meier, 1994), to ordering some in separate characters (traditional approach), to treating each variable as a binary (present/absent) character (Pleijel, 1995). Four different problems have been considered at the time of coding observations into a data set: (1) interdependency, (2) hierarchical linkage, (3) missing entries (either unknown or inapplicable data), and (4) information retrieval and testability (see Pleijel, 1995). Unfortunately, none of the proposed approaches is problem-free. Of the two ends of the spectrum for character coding options, the multistate approach minimizes the amount of character interdependency, and the binary (present/absent) practice evades problems of hierarchical linkage. Although both approaches overcome the undesirable addition of inapplicable entries, recent simulation studies have indicated that the strict adoption of any of these extreme approaches may produce spurious results (Strong and Lipscomb, 1999). This simulation suggests using the traditional approach, in which character states refer only to variables of a particular character and not to the absence of the character. This means, for example, that when scoring dental implantation in the jaw, character states are restricted to either being implanted in sockets or being in a groove, excluding a state considering the absence of teeth; absence/presence of teeth is regarded as a different, binary character. This study follows this approach as closely as possible.

Another important, and highly debated, analytical issue is that of treating multistate characters as additive (also known as "ordered") or nonadditive (or "unordered"). While Hauser and Presch (1991) recommended treating all multistate characters as nonadditive (i.e., the length between any two states is invariably one step), Slowinski (1993) suggested using a "mixed" approach in which multistate characters are analyzed as additive when there is confidence, such as in morphoclines (i.e., small-medium-large). In this chapter, both approaches have been followed. Thus, the

analysis was conducted (1) treating multistate characters that clearly represent morphoclines (characters 3, 36, 39, 49, 52, 58, 83, 96, 101, 122, 132, 143, 154, 159) as additive and the remaining multistate characters (i.e., 39, 83, 101, 159) as nonadditive; and (2) treating all multistate characters as nonadditive.

Finally, given that permutation of the out-group order within the same analysis may produce different topologies (Barriel and Tassy, 1998), the data matrix of Appendix 20.2 was analyzed using all possible permutations in the order of the three selected out-groups. Because choice of different out-groups may also result in different topologies of resultant cladograms of a cladistic analysis, the data matrix was also analyzed using two out-groups (all possible combinations of the three initially selected) and then only one out-group (each of the initial three out-groups).

Trees and Statistics

Using the implicit enumeration command of Hennig 86 ("ie" command; see Farris, 1988), the analysis of the data matrix (Appendix 20.3) treating multistate characters as a combination of additive and nonadditive (characters 39, 83, 101, 159) resulted in 28 fundamental (most parsimonious) cladograms. Each fundamental cladogram has a length of 299 steps, a consistency index (CI) of 0.61, and a retention index (RI) of 0.78. The strict consensus tree (Fig. 20.1), the only consensus cladogram that summarizes the information common to all fundamental cladograms (Nixon and Carpenter, 1996), is 304 steps long and has a CI of 0.60 and a RI of 0.77. When the analysis was conducted treating all multistate characters as nonadditive, the result produced the same number of fundamental cladograms with the same number of steps (i.e., 299 steps) and an identical strict consensus tree. The only difference was in the RI, which was slightly lower— and thus, less desirable—in both the fundamental cladograms (RI: 0.77) and their strict consensus tree (RI: 0.76).

Neither treatment of characters 39, 83, 101, and 159 as nonadditive nor permutation of the three out-groups used in this analysis (i.e., Allosauridae, Troodontidae, and Velociraptorinae) produced any difference in the number and topology of the fundamental cladograms or in the resulting strict consensus tree. Values for the statistics of all these trees remained the same.

When the analysis was rooted using two out-groups, the 28 most parsimonious trees (resulting from permutations of Velociraptorinae and Troodontidae or Allosauridae and Troodontidae) were somewhat shorter (i.e., 294 steps, CI: 0.62, RI: 0.76), but the topology of the strict consensus tree (299 steps, CI: 0.61, RI: 0.75) was identical to that obtained by using the three out-groups. Using Allosauridae and Velociraptorinae as out-groups resulted in an equal number of trees but was one step shorter (CI: 0.62, RI: 0.77); the topol-

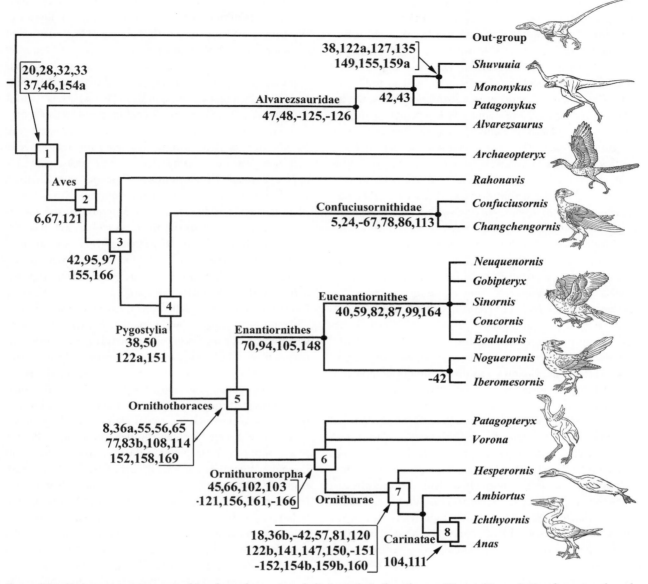

Figure 20.1. Strict consensus tree resulting from the present cladistic analysis (length: 304; CI: 0.6; RI: 0.77). Numbers at each node are unambiguously optimized synapomorphies (see Appendixes 20.2 and 20.3 for character list and matrix).

ogy of the consensus tree remained unchanged (298 steps, CI: 0.61, RI: 0.76). Rooting the analysis with either Velociraptorinae or Allosauridae resulted once again in 28 fundamental cladograms of somewhat shorter length (Velociraptorinae: 287 steps, CI: 0.63, RI: 0.76; Allosauridae: 284 steps, CI: 0.64, RI: 0.76). Once again, the topology of the strict consensus tree (292 steps, CI: 0.62, RI: 0.74 with Velociraptorinae; and 289 steps, CI: 0.63, RI: 0.75 with Allosauridae) remained invariable. Only when the analysis was rooted using Troodontidae as a single out-group was the number of fundamental cladograms increased and the topology of the resulting strict consensus tree modified. This last analysis produced 58 most parsimonious trees shorter than any of the

previous ones (i.e., 278 steps, CI: 0.62, RI: 0.75). The difference between the strict consensus tree (284 steps, CI: 0.61, RI: 0.74) of these fundamental cladograms with respect to those resulting from using more or other out-groups was simply the presence of a trichotomy at the base of Alvarezsauridae, between *Alvarezsaurus calvoi*, *Patagonykus puertai,* and a clade formed by *Shuvuuia deserti* and *M. olecranus.*

Phylogenetic Results

The strict consensus cladogram resulting from the present cladistic analysis (all out-groups included) is shown in Figure 20.1. Following earlier node-based phylogenetic definitions (Chiappe, 1992, 1996), Aves (a term here regarded as interchangeable with "birds") is defined as the common ancestor of *Archaeopteryx* and Neornithes plus all its descendants (in this study, Neornithes is represented by *Anas*). Consequently, *Archaeopteryx* is by definition a member of the most basal lineage of birds (Fig. 20.1). The Late Cretaceous Alvarezsauridae—*Mononykus* and its relatives (Chiappe, Norell, and Clark, Chapter 4 in this volume)—are placed immediately outside Aves, as the sister group to Aves. The Late Cretaceous *Rahonavis* from northwestern Madagascar (Forster et al., 1998a,b) is the next most basal avian lineage after *Archaeopteryx*. *Rahonavis* is the sister taxon to Pygostylia, a name that was previously used as a synonym of Ornithothoraces. Within Pygostylia, the Chinese Early Cretaceous Confuciusornithidae (*C. sanctus* and *Changchengornis hengdaoziensis*; Chiappe et al., 1999; Zhou and Hou, Chapter 7 in this volume) is the sister group of Ornithothoraces. The monophyletic Ornithothoraces, the common ancestor of *Iberomesornis romerali* and Neornithes plus all its descendants (Chiappe, 1996), encompasses two diverse sister taxa: Enantiornithes and Ornithuromorpha. The Early Cretaceous *I. romerali* and *Noguerornis gonzalezi* from Spain (Sanz et al., Chapter 9 in this volume; Chiappe and Lacasa-Ruiz, Chapter 10 in this volume) are included within Enantiornithes—following Sereno's (1998) definition of this clade as all taxa closer to *Sinornis santensis* than to Neornithes. All other enantiornithines are placed in a monophyletic clade, here called Euenantiornithes, and defined phylogenetically as all taxa closer to *Sinornis* than to *Iberomesornis*. The monophyletic Ornithuromorpha include the Late Cretaceous *V. berivotrensis* and *P. deferrariisi* from Madagascar and Argentina, respectively, and Ornithurae in an unresolved trichotomy. Ornithurae includes *Hesperornis regalis* and *A. dementjevi* as successive outgroups of a clade uniting *Ichthyornis* and Neornithes (i.e., *Anas*). Previous hypotheses of relationships of the different in-group taxa, their empirical support, and their comparisons with the hypotheses derived from the present cladistic analysis are discussed subsequently in greater detail.

The Phylogenetic Position and Interrelationships of Alvarezsauridae

M. olecranus (Perle et al., 1993a,b) was the first member of Alvarezsauridae to be recognized. Subsequent discoveries in the Mongolian Late Cretaceous, as well as reexamination of museum collections, documented the existence of close relatives of *Mononykus*, recognizing at least two other Late Cretaceous alvarezsaurids from central Asia—*Parvicursor remotus* (Karhu and Rautian, 1996) and *S. deserti* (Chiappe et al., 1998). Discoveries in the Late Cretaceous of Patagonia recognized the presence of basal alvarezsaurids in South America—*A. calvoi* and *P. puertai* (Novas, 1996, 1997). A fragmentary but diagnostic specimen in Berkeley's Museum of Paleontology proved the existence of this lineage in the Late Cretaceous of North America (Hutchinson and Chiappe, 1998). Although only four alvarezsaurid species were included in the present analysis, the results of this study again support the monophyly of this group (Chiappe, Norell, and Clark, Chapter 4 in this volume). The relationships among these four species are identical to those proposed by earlier cladistic analyses. Namely, the Asian species form a monophyletic group, while the Patagonian species constitute out-groups of this Asian clade (Novas, 1996; Chiappe et al., 1998; Hutchinson and Chiappe, 1998; Chiappe, Norell, and Clark, Chapter 4 in this volume).

Ever since the description of *Mononykus*, the phylogenetic relationships of the alvarezsauridae have been the matter of a heated debate. The initial phylogenetic analysis of Perle et al. (1993a) placed *Mononykus* closer to Neornithes than was *Archaeopteryx*. Such a conclusion triggered a great deal of opposition, although dissent was primarily framed within a noncladistic context (see discussion in Chiappe et al., 1996, 1997; Chiappe, Norell, and Clark, Chapter 4 in this volume). Subsequent cladistic analyses incorporating additional data from the new alvarezsaurid discoveries in central Asia and Patagonia (Chiappe et al., 1996, 1998; Novas, 1996, 1997; Forster et al., 1998a; Holtz, 2001) provided further support for the avian relationship of Alvarezsauridae. Other recent cladistic analyses, however, have supported a nonavian theropod relationship for this group, thus introducing a more serious challenge to the avian hypothesis originally proposed by Perle et al. (1993a). Sereno (1997, 1999a,b, 2001), Clark, Norell, and Makovicky (Chapter 2 in this volume), and Novas and Pol (Chapter 5 in this volume) have argued for a more basal position for the alvarezsaurids among coelurosaurs, and the present analysis supports a sister group relationship between alvarezsaurids and birds (Figs. 20.1 and 20.2). Only Sereno's (1999a) cladistic analysis was critically evaluated at the time this chapter was written (his 1999 data matrix was made available on the Internet). Specifically, Sereno's (1999a,b, 2001) proposal is that alvarezsaurids are the sister group of ornithomimids, both being placed in Ornithomimoidea. Martin and Rinaldi (1994) first entertained this idea, although on the basis of features that are either common to a variety of nonavian theropods and basal birds (e.g., long neck and tail, absence of furcula, short forelimbs, reduced calcaneum) or simply incorrect (e.g., absence of free carpals, unusually short ribs). As with Martin and Rinaldi's (1994) proposal, Sereno's (1999a, 2001) hypothesis remains unconvincing (Suzuki et al., 2002). Even a cursory examination of his data reveals at least one significant caveat: the fact that in his cladistic analysis, Aves is scored as a terminal taxon, which thus prevents the inclusion of alvarezsaurids within Aves.

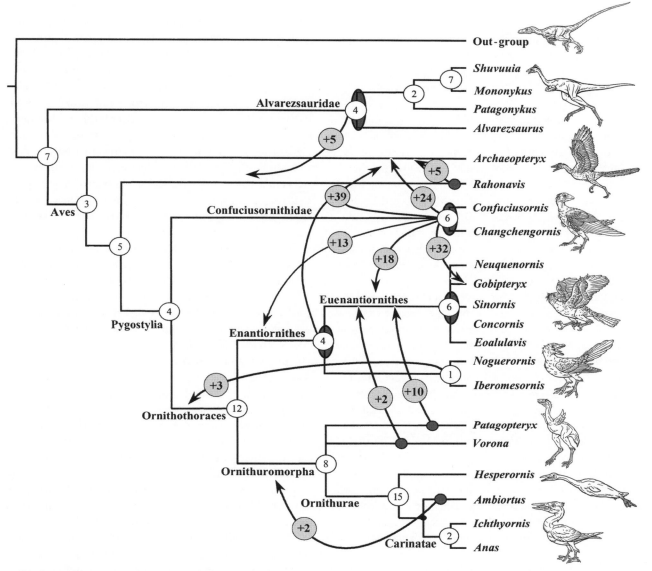

Figure 20.2. Strict consensus tree resulting from the present cladistic analysis (length: 304; CI: 0.6; RI: 0.77). White circles at each node depict the number of unambiguously optimized synapomorphies diagnosing that particular node. Gray circles on arrows indicate the number of additional steps needed to support (more parsimoniously) a different placement for a particular taxon.

Sereno's study (1999a, 2001) fails to appreciate the remarkable similarities between alvarezsaurids and birds (regardless of whether these are homologies). Nevertheless, alvarezsaurids and birds share numerous derived similarities (Chiappe et al., 1996, 1998). Seven unambiguous synapomorphies support the sister-group relationship of Alvarezsaurids and Aves (Figs. 20.1 and 20.2): quadrate articulating with both prootic and squamosal (character 20); teeth with unserrated crowns (character 28); prominent carotid processes in the intermediate cervicals (character 32); weak postaxial epipophyses (character 33); wide vertebral foramen in the midcaudal thoracic vertebrae (character 37); short prezygapophyses of distal caudal vertebrae (character 46);

and fibula with the tubercle of M. iliofibularis laterally directed (character 154a). Unambiguous support for the monophyly of Aves is limited to only three synapomorphies: a caudal margin of the external naris that nearly reaches or overlaps the rostral border of the antorbital cavity (character 6); a prominent acromion in the scapula (character 67); and a pointy and shallow postacetabular wing of the ilium that has less than 50% the dorsoventral depth of the preacetabular wing at the acetabulum (character 121). The relative weak support for the monophyly of Aves is not surprising given the large number of new taxa (Alvarezsauridae among them) documenting a stepwise origin of many of the previously "unique" avian features (Padian and Chiappe, 1998).

Although the present study challenges the avian relationship of Alvarezsauridae proposed by Perle et al. (1993a), the topological shift between these two hypotheses pivots on a handful of steps (Fig. 20.2). Only five additional steps would be required to place alvarezsaurids closer to Neornithes than is *Archaeopteryx*—the original hypothesis of Perle et al. (1993a).

In all probability, the relationships of this spectacular lineage of theropods will continue to be controversial until more complete remains of its basal members are discovered and the interrelationships of nonavian coelurosaur theropods become better understood. The reason for the novel phylogenetic interpretation presented here is not because previously published analyses "did not include any nonavian taxa" as part of the in-group, as stated by Sereno (1999b:75). Several unquestionably nonavian taxa formed part of the in-group in the analyses of Novas (1996), Forster et al. (1998a), Ji et al. (1998), and Holtz (2001). The shift between the original hypothesis of Perle et al. (1993a) and the one presented here is more related to the discovery of new anatomical data for the out-groups and the incorporation into the analysis of confuciusornithids, birds that are now known by hundreds of specimens (Chiappe et al., 1999). As a consequence, these additional data and changes in the discrimination of discrete variables and their scoring led to changes in the optimization of characters and consequently to new topologies for the resulting cladograms. These results as well as others (e.g., Clark, Norell, and Makovicky, Chapter 2 in this volume; Novas and Pol, Chapter 5 in this volume) have made it apparent that the phylogenetic relationships of alvarezsaurids need to be examined within a framework larger than either basal birds (e.g., Chiappe et al., 1996) or nonavian tetanuran theropods (e.g., Sereno, 1999a). Future analyses should look into integrating these data better.

The Phylogenetic Relationships of *R. ostromi*

In 1998, Forster and her associates described a primitive new bird from the Late Cretaceous of northwestern Madagascar—from the same fossil-rich deposits as *V. berivotrensis* (see Forster et al., Chapter 12 in this volume). This bird was first called "*Rahona*" *ostromi* (Forster et al., 1998a), although its generic name was later emended to *Rahonavis* given that "*Rahona*" was found to be in use (Forster et al., 1998b).

The holotype of *Rahonavis* is composed of associated elements (some in articulation) of the trunk, pelvis, hindlimb, and tail, as well as the forelimb and shoulder—all found in an area of 500 cm². *Rahonavis* exhibits pneumatic foramina penetrating the cervicothoracic and thoracic vertebral centra; hyposphene-hypantrum accessory articulations between thoracic vertebrae; a tail composed of more than 13 long caudal vertebrae; an ilium with a preacetabular

wing that is much longer than the postacetabular one; a pubis that is twice the length of the ischium and fused distally into a caudal foot; an ischium with a distinct proximodorsal process; and a reduced fibula that does not reach the proximal tarsals (Forster et al., 1998a). The elongation and distinct quill knobs of its ulna and its laterally oriented glenoid facet suggest feathers and aerodynamic capabilities for *Rahonavis* (Forster et al., 1998a). Clearly the most notable feature of its hindlimb is the highly specialized second digit—a toe much more robust than the other pedal digits, composed of phalanges with expanded distal articular surfaces and ending in an enlarged, sickle-shaped claw. Forster et al. (1998a) highlighted this feature as a striking similarity to the second digit of certain nonavian lineages such as troodontids and dromaeosaurids.

In cladistic analyses including several nonavian coelurosaurs as well a variety of basal avian lineages, the most parsimonious tree placed *Rahonavis* as the sister group of *Archaeopteryx* (Forster et al., 1998a). Yet an alternative hypothesis proposed by Forster et al. (1998a), one supporting the sister-taxon relationship between *Rahonavis* and all avians other than *Archaeopteryx*, was only one step longer. The present study supports the latter hypothesis, one that Forster et al. (1998a:1918) regarded as convincing given that *Rahonavis* "shares a number of characters with more derived birds exclusive of *Archaeopteryx*." Indeed, five unambiguous synapomorphies support the sister-taxon relationship between *Rahonavis* and Pygostylia (new combination): a procoelous synsacrum (character 42); an ulnar shaft that is considerably thicker than the radial shaft (character 95) and with a proximal end bearing a well-defined area for the insertion of M. brachialis anticus (character 97); a fibula that does not reach the proximal tarsals (character 155); and the presence of a distinct tubercle on the dorsal face of metatarsal II (character 166). All these characters are absent in *Archaeopteryx*. Given the present analysis, it would require five additional steps to unite *Archaeopteryx* and *Rahonavis* as sister taxa (Fig. 20.2), re-creating the most parsimonious placement for these taxa of Forster et al. (1998a).

As was emphasized by Forster et al. (1998a), the allocation of *Rahonavis* within birds is strongly supported by what we know of its anatomy. Some critics—essentially those apparently distressed by the support that the discovery of *Rahonavis* provided to the hypothesis of the theropod ancestry of birds—have tried to dismiss the avian relationships of *Rahonavis* by arguing that the holotype was a composite (e.g., Feduccia, 1999). These critics claimed that while the ulna, radius, and scapula were those of a bird, the remaining elements of the holotype were those of a nonavian theropod. As previously mentioned, all bones of the holotype were collected from a very small area. This, along with the identical preservation, texture, and color of all these bones, strongly suggests they belong to a single specimen. Most important is a fact ignored by those arguing the composite status of the holotype: Forster et al. (1998a) ran two separate cladistic analyses, one in which the forelimb and shoulder elements of *Rahonavis* were included and one in which they were excluded from the analysis. In both cases, the resulting cladograms were identical, indicating that even the apparently more primitive morphology of the vertebral, pelvic, and hindlimb elements supports their identification as of those of a bird.

The Phylogenetic Position of Confuciusornithidae

With hundreds of specimens collected during the late 1990s, *C. sanctus* (Zhou and Hou, Chapter 7 in this volume) has become an icon for the unprecedented rate of recent discoveries of primitive birds in China and elsewhere. *Confuciusornis* was first described by Hou et al. (1995a) on the basis of incomplete material from the Chaomidianzi Formation (Lower Yixian Formation of others), an Early Cretaceous lithostratigraphic unit (Swisher et al., 1999) of the northeastern Chinese province of Liaoning. This initial report and subsequent work by Hou and associates (Hou et al., 1996, 1999; Hou, 1997; Zhou and Hou, Chapter 7 in this volume) described the basic features of this bird and provided the first phylogenetic interpretations. Ji et al. (1999) reported on another bird from the Chaomidianzi Formation, *C. hengdaoziensis*, whose morphology closely allies it with *C. sanctus*, and Chiappe et al. (1999) provided a review of the anatomy, systematics, and other biological aspects of these two taxa. Chiappe et al. (1999) discussed the validity of several recently named species of *Confuciusornis*, concluding that *C. sanctus* was the only diagnosable one. *C. sanctus* and *C. hengdaoziensis* are thus the only known members of Confuciusornithidae.

Confuciusornis was first interpreted as an enantiornithine bird and allied with the Late Cretaceous *Gobipteryx* from the Gobi Desert (Hou et al., 1995b). The following year, Hou et al. (1996) removed *Confuciusornis* from Enantiornithes and regarded it as its sister taxon. Neither of these proposals, however, was well substantiated (Chiappe et al., 1999). The inclusion of *Confuciusornis* within Enantiornithes was based on the presence of elongate coracoids, and the sister-taxon relationship between *Confuciusornis* and *Gobipteryx* was supported only by their common lack of teeth (Hou et al., 1995b). The sister-group relationship between *Confuciusornis* and Enantiornithes was argued on the basis of the presence of pygostyles and thoracic pleurocoels (large, lateral excavations) in these taxa (Hou et al., 1996). Elongate coracoids, pygostyles, and pleurocoels in thoracic vertebrae are widespread features among birds. The presence of elongate coracoids and pygostyles in *Confuciusornis* and *Changchengornis* indicates only that the

confuciusornithids are closer to Neornithes than are *Archaeopteryx* or *Rahonavis*. Thoracic pleurocoels are even shared by *Rahonavis*, so their presence would only indicate that confuciusornithids are birds more derived than *Archaeopteryx*. Clearly, all characters used by Hou et al. (1995a, 1996) to ally *Confuciusornis* with Enantiornithes are synapomorphies at a higher level of universality, and therefore they do not support any of the phylogenetic hypotheses entertained by these authors.

Chiappe (1997b) was the first to provide an alternative phylogenetic hypothesis for *Confuciusornis*. Chiappe (1997b) proposed that *Confuciusornis* was the sister taxon of Ornithothoraces, a view corroborated by subsequent studies (e.g., Chiappe et al., 1999; Ji et al., 1999; Sereno, 1999a). The present analysis supports the sister-group relationship of *Confuciusornis* and *Changchengornis* proposed by Ji et al. (1999) in the initial report of the latter—a relationship strongly supported by six unambiguous synapomorphies (Fig. 20.1)—and, once again, Chiappe's (1997b) phylogenetic relationships for *Confuciusornis* (and the Confuciusornithidae). The out-group condition of Confuciusornithidae with respect to Ornithothoraces is strongly supported by the fact that *Confuciusornis* exhibits the primitive condition for the 12 unambiguous synapomorphies diagnosing Ornithothoraces (Fig. 20.1) and that the same can be said about *Changchengornis*, since these characters are not missing in its single known specimen.

The clade formed of confuciusornithids and ornithothoracines is supported here by four unambiguous synapomorphies (Fig. 20.1): absence of hyposphene-hypantrum (character 38); presence of a pygostyle (character 50); a retroverted pubis separated from the main synsacral axis by an angle ranging between 65° and 45° (character 122a); and the presence of a wide and bulbous medial condyle of the tibiotarsus (character 151). A node-based phylogenetic definition and a name—Pygostylia—are provided for this clade. Pygostylia is here defined as the common ancestor of Confuciusornithidae and Neornithes plus all its descendants. Chatterjee (1997, 1998) named a clade that included the most recent common ancestor of the Early Cretaceous *Iberomesornis* and Neornithes plus all its descendants, using the term Pygostylia. The name Ornithothoraces, however, had been previously applied to this same clade (Chiappe, 1996). Consequently, Chatterjee's (1997, 1998) Pygostylia is a synonym of Chiappe's (1996) Ornithothoraces.

The weaknesses of the various phylogenetic hypotheses proposed by Hou and his associates are highlighted by the results of the present cladistic analysis: it would require 13 additional steps to place confuciusornithids as the sister group of Enantiornithes, 18 extra steps to make them share a most recent common ancestor with Euenantiornithes, and many more to cluster them with *Gobipteryx*.

The Phylogenetic Position of Enantiornithes and the Paraphyly of "Sauriurae"

With multiple discoveries in the past decade, Enantiornithes (Chiappe and Walker, Chapter 11 in this volume) has become the most taxonomically diverse clade of Mesozoic birds. Walker (1981) was the first to propose a hypothesis for its phylogenetic position, placing it in an intermediate position between *Archaeopteryx* and Neornithes. Soon after, Martin (1983) proposed his view of a basal dichotomy in avian evolution. On the one hand, he claimed that *Archaeopteryx* and Enantiornithes were sister taxa and classified them under the "Sauriurae." On the other hand, he argued for the sister-group relationship between "Sauriurae" and Ornithurae. Since then, discussions of the phylogenetic relationships of Enantiornithes have been intertwined with those concerning the monophyly of "Sauriurae." Multiple numerical cladistic analyses have consistently dismissed the monophyly of "Sauriurae," supporting the notion that Enantiornithes is phylogenetically closer to Neornithes than to *Archaeopteryx* (e.g., Cracraft, 1986; Chiappe, 1991, 1995a,b, 1996; Sereno and Rao, 1992; Sanz et al., 1995, 1996, 1997; Forster et al., 1996, 1998a; Sereno, 1999a). Comparable objections by authors working outside the framework of cladistics even predated the results of these numerical cladistic analyses (e.g., Steadman, 1983; Olson, 1985).

Martin's (1983) argument of a monophyletic "Sauriurae" was based on five characters, alleged to be exclusive to Enantiornithes and *Archaeopteryx*: (1) scapula with a craniomedial facet abutting against a series of fused cranial thoracic vertebrae; (2) fusion of the cranialmost thoracic vertebrae with a special process on the scapula abutting them; (3) metatarsal bones fused in a straight line; (4) distal tarsal bones either absent or fused as small individual bones; and (5) tarsometatarsus fused proximally but not distally. Martin's (1983) characters 1 and 2 appear to be the same, and most significant, this (apparently single) character is unknown for any of the taxa included within "Sauriurae." Characters 3–5 are not exclusive to Enantiornithes and *Archaeopteryx* but rather are shared by several nonavian maniraptoriform theropods, suggesting that they are primitive. Since the discovery of *Confuciusornis*, the notion of a monophyletic "Sauriurae" became more prominently featured in the literature, although essentially among noncladistic studies (e.g., Hou et al., 1995b, 1996, 1999; Feduccia, 1996; Kurochkin, 1996; Martin and Zhou, 1997). Hou et al. (1995b, 1996) supplemented Martin's characters with three other alleged synapomorphies: (6) broad furcular arms; (7) craniodorsal process of ischium; and (8) crest on the lateral condyle of the femur. Again, character 6 is known in a variety of nonavian theropods (e.g., Chure and Madsen, 1996; Norell et al., 1997; Makovicky and Currie, 1998), and character 7 is known for the maniraptoriform *Unenlagia comahuensis*, apparently a nonavian taxon (Novas and Puerta,

1997; Norell et al., 2001; Clark, Norell, and Makovicky, Chapter 2 in this volume). These two characters are likely plesiomorphic for birds. Character 8 is known for several Enantiornithes (Chiappe and Walker, Chapter 11 in this volume), but it is unknown for *Archaeopteryx,* thus representing a synapomorphy of a subset within Enantiornithes. A review of the evidence in support of a close relationship between *Archaeopteryx* and Enantiornithes suggests that these characters are either plesiomorphic or equivocal.

The present cladistic analysis concurs with many others (Cracraft, 1986; Chiappe, 1991, 1996; Sereno and Rao, 1992; Sanz et al., 1995, 1996, 1997; Forster et al., 1996, 1998a; Sereno, 1999a), indicating the existence of a common ancestor of Enantiornithes and modern avians (Neornithes) not shared by *Archaeopteryx.* Thus, the numerous derived characters shared between Enantiornithes and Ornithuromorpha (including Neornithes) are better explained as synapomorphies of a monophyletic clade uniting these two taxa—Ornithothoraces—than as independently evolved (Figs. 20.1 and 20.2). Among the 12 unambiguous synapomorphies diagnosing Ornithothoraces are the presence of a cup-shaped caudal maxillary sinus (character 8); fewer than 13 thoracic vertebrae (character 36); scapula articulated well below the shoulder end of the coracoid (character 55); scapular caudal end tapered to a sharp point (character 65); alular digit (I) short, not surpassing the length of the major metacarpal (II) (character 108); and the presence of an alula (character 169). To argue for a monophyletic "Sauriurae"—which in its current version includes Confuciusornithidae (Hou et al., 1996)—would require a strict consensus cladogram 39 steps longer than the one resulting from the present analysis (Fig. 20.2).

Not surprisingly, the general inferences drawn from these two very different views of the phylogenetic relationships of Enantiornithes, and of other basal birds, are diametrically opposed. For example, Hou et al. (1996) regarded the growth differences inferred for certain nonornithurine birds (Chinsamy et al., 1994, 1995; Chinsamy, Chapter 18 in this volume) as unique (and thus "synapomorphic") attributes of "sauriurine" birds. Yet, in the framework of a paraphyletic "Sauriurae," these differences are seen not as unique attributes of "Sauriurae" but as primitive conditions inherited from the theropod ancestors of birds (Chiappe, 1995a). Likewise, in the context of a monophyletic "Sauriurae," the sophisticated flight apparatus of Enantiornithes (Chiappe and Walker, Chapter 11 in this volume) requires independent evolution from that of Ornithuromorpha, because many of the flight-correlated features of these taxa are absent in *Archaeopteryx* (e.g., pygostyle, strutlike coracoid, alula). This genealogical corollary has often been ignored by those who, supporting the monophyly of "Sauriurae," have used enantiornithine morphology in constructing interpretations about the evolution of modern avian flight. Nevertheless, if "Sauriurae" were monophyletic, the evidence available in the anatomy of Enantiornithes would not be critical for understanding the evolution of flight in Neornithes.

An interesting difference from most previous cladistic analyses is that the present results support—although not overwhelmingly (Fig. 20.2)—the inclusion of the Spanish Early Cretaceous *Noguerornis* (Lacasa-Ruiz, 1989) and *Iberomesornis* (Sanz and Bonaparte, 1992) within the lineage leading to *Sinornis* and other enantiornithines. These taxa have often been regarded as nonornithothoracines (e.g., Sanz and Bonaparte, 1992; Chiappe, 1995a,b, 1996; Sanz et al., 1995; Forster et al., 1998a), although not universally (e.g., Kurochkin, 1995; Martin, 1995; Feduccia, 1996; Sereno, 1999a), and thus not particularly close to Enantiornithes. The sister-taxon relationship between *Noguerornis* and *Iberomesornis* is supported only by a reversal (absence of a procoelous synsacrum; Fig. 20.1), and their inclusion within Enantiornithes would collapse with only three additional steps (Fig. 20.2). The inclusion of these taxa within Enantiornithes is based only on the presence of a well-developed hypocleideum (character 70); the ulna shorter than the humerus (character 94); the minor metacarpal projecting more distally than the major metacarpal (character 105; this character state is missing in *Iberomesornis*); and a round proximal articular surface of the tibiotarsus (character 148; missing in *Noguerornis*).

Interrelationships within Euenantiornithes

Although the term Enantiornithes was used for almost 20 years to refer to a group of Cretaceous flying birds with a suite of distinct osteological attributes (e.g., Chiappe, 1992, 1996; Chiappe and Calvo, 1994; Sanz et al., 1995, 1996, 1997; Zhou, 1995), it was Sereno (1998) who first applied it to a phylogenetically defined clade: all taxa closer to *Sinornis* than to Neornithes. Under this phylogenetic definition, most attributes used before to diagnose Enantiornithes support a clade that is a subset of Sereno's newly defined Enantiornithes. As previously discussed, *Iberomesornis* and *Noguerornis* may be members of *Sinornis*'s lineage and thus Enantiornithes, but the fact that they lack the majority of the derived characters diagnosing less inclusive enantiornithine clades suggests their placement as basal members of this lineage (Fig. 20.1). All the remaining enantiornithine taxa included in the present analysis cluster in a polytomy. This clade is here named Euenantiornithes and is phylogenetically defined as all taxa closer to *Sinornis* than to *Iberomesornis*. Although only five euenantiornithine taxa are included in the present analysis, several derived similarities between these and other typical enantiornithines (e.g., Chiappe and Calvo, 1994; Zhou, 1995; Chiappe, 1996) strongly suggest the inclusion of most Enantiornithes within Euenantiornithes (Chiappe and Walker, Chapter 11 in this volume).

The interrelationships of Euenantiornithes—a group of more than 15 species—remain one of the main systematic problems of basal avian phylogeny. Chiappe and Walker (Chapter 11 in this volume) attempted to reconstruct the genealogical history of the entire clade, one that in the past was explored only by studies restricted to a small portion of the group's actual diversity (e.g., Chiappe, 1993; Sanz et al., 1995; Varricchio and Chiappe, 1995). Despite analyzing 36 variables across 17 species of enantiornithines, Chiappe and Walker (Chapter 11 in this volume) were unable to find much resolution in the resulting cladograms (see also Sereno, Rao, and Li, Chapter 8 in this volume). At the center of this problem lies the fragmentary nature of most euenantiornithine specimens representing diagnosable species; moreover, specimens of several recently discovered species (e.g., *Boluochia zhengi*, *Eoenantiornis buhleri*, *Otogornis genghisi*) still await detailed study. For example, Chiappe and Walker's analysis (Chapter 11 in this volume) recorded 80% or more of missing data for at least one-third of the in-group species. When this poorly known third was excluded from the analysis, the resulting cladograms supported two sister taxa branching from an otherwise unresolved polytomy. One of these taxa includes *G. minuta* and *B. zhengi*; the other is formed by *Halimornis thompsoni*, *Concornis lacustris*, *Soroavisaurus australis*, *Enantiornis leali*, and *Neuquenornis volans*. To some extent, the latter cluster agrees with previous cladistic analyses of enantiornithine interrelationships (Chiappe, 1993; Sanz et al., 1995). Nonetheless, much more work needs to be done before one can arrive at a phylogenetic hypothesis that portrays the evolutionary history of euenantiornithines with certain confidence.

The Phylogenetic Position of *Vorona*, *Patagopteryx*, and Ornithurae

Two Late Cretaceous birds from Gondwana—the Malagasy *V. berivotrensis* (Forster et al., 1996, Chapter 12 in this volume) and the Argentine *P. deferrariisi* (Alvarenga and Bonaparte, 1992; Chiappe, Chapter 13 in this volume)—form an unresolved trichotomy with Ornithurae (Figs. 20.1 and 20.2). These taxa are included within Ornithuromorpha, a clade here defined as the common ancestor of *Patagopteryx* and Ornithurae plus all its descendants. The monophyly of Ornithuromorpha is supported by eight unambiguous synapomorphies (Fig. 20.1), including the loss of caudal prezygapophyses (character 45); a sagittally curved scapular blade (character 66); metacarpals II and III that are partially to completely fused distally (character 102) and intermetacarpal space that is at least as wide as the maximum width of the shaft of metacarpal III (character 103); and a complete fusion of metatarsals II, III, and IV (character 156).

The phylogenetic placement of *Vorona* indicated by the present analysis is comparable to the one originally proposed by Forster et al. (1996). *Vorona* is incompletely known by two specimens preserving hindlimb elements only (Forster et al., Chapter 12 in this volume), and certain characters—particularly in the tarsometatarsus—are reminiscent of those of some euenantiornithines (e.g., *Avisaurus, Soroavisaurus*). This superficial similarity may have led some authors to place *Vorona* within Enantiornithes (Feduccia, 1996, 1999). At present, however, these similarities are better explained as convergences. Nonetheless, it should be kept in mind that it would take only two additional steps to bring *Vorona* within Enantiornithes, as the sister taxon of Euenantiornithes (Fig. 20.2).

The present analysis also confirms the close relationship between *Patagopteryx* and Ornithurae (Chiappe, 1996, Chapter 13 in this volume), indicating once again that *Patagopteryx* is not closely related to ratites, as initially proposed by Alvarenga and Bonaparte (1992) and later followed by Kurochkin (1995; see Chiappe, Chapter 13 in this volume). *Patagopteryx* is much better known than *Vorona;* it would take a mininum of 10 additional steps to nest it within Enantiornithes. The position of *Patagopteryx* as an immediate out-group of Ornithurae within the Ornithuromorpha appears well established and unlikely to change.

Interrelationships within Ornithurae

Only a handful of workers have ever entertained phylogenetic hypotheses that would compromise the monophyletic status of Ornithurae encompassing *Hesperornis, Ichthyornis,* and all extant birds—Lowe (1935), Holmgren (1955), and Zweers et al. (1997) are among the very few. This is not likely to change because, as shown by the 15 unambiguous synapomorphies that here diagnose this clade (Fig. 20.1), the monophyly of Ornithurae is one of the most robust systematic statements of basal avian phylogeny.

In agreement with a host of previous studies (Martin, 1983; Cracraft, 1986, 1988; Chiappe, 1991, 1995a,b, 1996; Sanz and Bonaparte, 1992; Sereno and Rao, 1992; Sanz et al., 1995), *Hesperornis* is regarded here as a basal ornithurine, outside the common ancestry of *Ichthyornis* and Neornithes—the clade termed Carinatae. In the present cladistic analysis, the Early Cretaceous *A. dementjevi* (Kurochkin, 1985a,b) is nested within Ornithurae, as the sister taxon of Carinatae. Phylogenetic interpretations of the incomplete and only known specimen of *Ambiortus* have varied widely. Kurochkin (1985a,b) regarded it as a carinate bird, allying it with alleged palaeognaths from the Early Tertiary. Kurochkin's interpretation of a close relationship between *Ambiortus* and palaeognaths was reiterated in his 1995 review of Mesozoic birds (Kurochkin, 1995a,b). While highlighting the incompleteness of *Ambiortus* and, thus, his tentative interpretation, Cracraft (1986) placed *Ambiortus* within the Carinatae, in a trichotomy with *Ichthyornis* and Neornithes. Sereno and Rao (1992), however, place it out-

side Ornithurae (as defined here), as its immediate sister taxon.

Although the present analysis depicts *Ambiortus* as an ornithurine, as the sister taxon to Carinatae (Fig. 20.1), there is little support, if any, for this placement. Only 2 of the 15 unambiguous synapomorphies diagnosing the latter clade are present in *Ambiortus* (a procoracoid process on the coracoid [character 57] and a globe-shaped humeral head [character 81])—the remaining are not preserved in the only known specimen. The sister-group relationship between *Ambiortus* and Carinatae is supported by only four ambiguous synapomorphies (i.e., characters 51, 64, 91, 167 of Appendix 20.2)—character states that may be synapomorphic of this node if shown to be present in *Ambiortus*. Furthermore, while removing *Ambiortus* from the Ornithuromorpha would increase the length of the strict consensus cladogram by two steps (Fig. 20.2), no additional steps would be needed to place it as the sister taxon to Ornithurae.

Early Evolution of Avian Locomotion

It has long been known that most birds have two main modes of progression—flying and walking—that are functionally different. While the origin and later refinement of flight in birds has regularly been a topic of intense investigation, the evolutionary development of the mode of terrestrial locomotion in birds remained largely overlooked until the work of Gatesy (1990; Chapter 19 in this volume). This author recognized the profound kinematic differences between the terrestrial progression of extant birds and that of their most immediate nonavian relatives. Chiappe's (1995a) chart of the sequence of transformations (i.e., synapomorphies) of the wing and shoulder versus those of the hindlimb and pelvis demonstrated the early "mosaic" evolution of the two main avian locomotory systems. This latter study suggested that most skeletal structures correlated to enhanced flying capabilities evolved prior to those correlated to the terrestrial locomotion of extant birds. Our understanding of the evolution of the main styles of avian locomotion also benefited from the conceptual recognition of locomotor modules (Gatesy and Dial, 1996)—highly integrated anatomical subregions of the musculoskeletal system that act as different functional units during locomotion. Gatesy and Dial (1996) identified three locomotor modules in extant flying birds: the pectoral module, the caudal module, and the hindlimb module. These authors convincingly argued that the origin of the flight system of birds is equivalent to the evolutionary differentiation of these locomotor modules from the single one (i.e., the combined hindlimb and caudal modules) of nonavian theropods and the functional association of the caudal and pectoral modules.

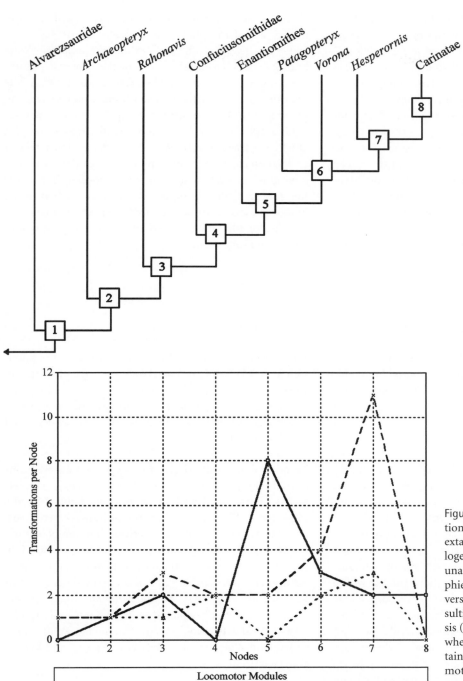

Figure 20.3. Sequence of transformation of the three locomotor modules of extant birds during basal avian phylogeny. The chart shows the number of unambiguously optimized synapomorphies for each node of an abbreviated version of the strict consensus tree resulting from the present cladistic analysis (Fig. 20.1). Because the distinction of whether the pelvis and synsacrum pertain to the hindlimb or the caudal locomotor module is a difficult one, synapomorphies of the pelvis and synsacrum were incorporated into both modules.

Although the work of Gatesy and Dial (1996) laid an important conceptual foundation for understanding the origin of avian flight, the refinements of each avian locomotor module were minimally addressed. The sequence of transformations throughout the evolution of these locomotor modules can be studied by mapping those synapomorphies assumed to belong to a module on a phylogeny of basal birds (see Gatesy, Chapter 19 in this volume). Although there appears to be no reason why every character of a body's subregion should be functionally involved in a particular type of locomotion, it is difficult (if not impossible) to confidently exclude a particular character of any given locomotory module from its potential role in locomotion. Consequently, all skeletal and integumentary unambiguous synapomorphies of the subregions making up each locomotor module were mapped onto an abbreviated version of the strict consensus tree resulting from the present cladistic analysis (Fig. 20.3). Although the mapping of all synapomorphies—regardless of their potential role in locomotion—should be considered a rough estimate, this procedure is possibly the best approximation for understanding the evolution of each of the three locomotor modules.

The sequence of structural transformations of each locomotor module shown in Figure 20.3 is consistent with those derived from phylogenetic analyses using fewer taxa and characters (Chiappe, 1995a). Structures of the caudal locomotor module appear to have evolved almost "progressively." Those of the forelimb and hindlimb modules show distinct peaks of numerous transformations occurring at a single node (e.g., nodes 5 and 7 for the forelimb and hindlimb modules, respectively). It is tempting to conclude that these transformational peaks represent increases in the rate of morphological change. Nevertheless, they could equally be artifacts of taxonomic sampling—long branches between certain nodes of the consensus cladogram that may be subdivided by future discoveries. While the existence of a particular evolutionary process (e.g., an increment in the rate of morphological change) may remain untestable, it is likely that future discoveries may hint at whether these transformational peaks are more related to incomplete taxonomic sampling. Regardless of these questions, the present study indicates once again that the fine-tuning of the forelimb module preceded that of the hindlimb module. Birds enhanced their remarkable flying abilities long before they developed their novel mode of terrestrial progression.

APPENDIX 20.1
Selected Out-Group and In-Group Taxa for the Present Cladistic Analysis

The following list presents selected out-group and in-group taxa (in alphabetical order), specimens, and bibliographic sources used to score the characters listed in Appendix 20.2 into the data matrix of Appendix 20.3.

Out-group

The theropodan ancestry of birds is today well established (Ostrom, 1976; Gauthier, 1986; Holtz, 1994; Favstowski and Weishampel, 1996; Dingus and Rowe, 1997; Padian and Chiappe, 1997). Disagreement, however, centers on the identification of the nonavian theropod clade closer to birds (see Clark, Norell, and Makovicky, Chapter 2 in this volume). Gauthier's (1986) pioneering cladistic analysis of dinosaurs placed Dromaeosauridae and Troodontidae (united within the Deinonychosauria) as the sister group of birds (Avialae sensu Gauthier, 1986). Holtz (1994) has argued that dromaeosaurids are closer to birds than are troodontids, but Currie (1985, 1987), Currie and Zhao (1993a), and Forster et al. (1998a) have suggested the opposite. More recently, Novas and Puerta (1997) have claimed that the nonavian theropod *U. comahuensis* is the sister group of all birds. The fragmentary nature of *Unenlagia* prevents rooting the present cladistic analysis with this taxon, and its precise phylogenetic position awaits further study—Norell and Makovicky (1999), for example, have allocated this species within Dromaeosauridae. Thus, troodontids and velociraptorine dromaeosaurids are used as out-groups. Rooting of the in-group taxa by troodontids and velociraptorine dromaeosaurids has been reinforced with the addition of a more primitive group, Allosauroidea (sensu Padian and Hutchinson, 1997), to the outgroup taxa.

Allosauroidea—Data on this taxon derive from the publications of Madsen (1976) on *Allosaurus fragilis* and that of Currie and Zhao (1993b) on *Sinraptor dongi*. This information has been complemented by personal observations of specimens AMNH-324, AMNH-408, AMNH-666, and AMNH-5753 of *A. fragilis*.

Troodontidae—Data derive from personal observations of the holotypes of *Saurornithoides mongoliensis* (AMNH-6516; Osborn, 1924) and *Saurornithoides junior* (GI SPS 100-1; Barsbold, 1974), as well as on specimens recently collected by expeditions of the American Museum of Natural History–Mongolian Academy of Sciences to the Gobi Desert (see Novacek et al., 1994; Dashzeveg et al., 1995). Additional information was based on the description of Russell and Dong (1993) on *Sinornithoides youngi* as well as on the papers of Barsbold et al. (1987), Currie (1987), Osmólska and Barsbold (1990), Currie and Peng (1993), Currie and Zhao (1993b), Makovicky (1995), and Varricchio (1997).

Velociraptorinae—Data derive from personal observations of several specimens of *Deinonychus antirrhopus* (AMNH-3015, YPM-5204, YPM-5205, YPM-5206, YPM-5210, YPM-5217, YPM-5220, YPM-5226, YPM-5235, YPM-5238; Ostrom, 1969) and *Velociraptor mongoliensis* (AMNH-6518; Osborn, 1924). This information was supplemented by the study of newly collected material of velociraptorines (Norell et al., 1992; Norell and Makovicky, 1997; Brinkman et al., 1998) by expeditions of the American Museum of Natural History–Mongolian Academy of Sciences to the Gobi Desert.

In-group

A. calvoi (**Bonaparte, 1991**)—Information is based on personal observations of the holotype (MUCPv-54), an articulated partial skeleton from the Upper Cretaceous (Coniacian-Santonian) Río Colorado Formation of the province of Neuquén, southern Argentina (see Chiappe, Norell, and Clark, Chapter 4 in this volume). These observations have been complemented by data provided by Novas (1996, 1997).

A. platyrhynchos—Data on this extant species are derived from personal observations of specimens AMNH-10410, AMNH-10726, AMNH-16045, AMNH-16048, and AMNH-16237. Selection of *A. platyrhynchos* as the representative of Neornithes (i.e., modern birds) follows the nearly consensual idea that Anseriformes has a fairly basal position within Neornithes (see Cracraft, 1988; Cracraft and Mindell, 1989; Sibley and Ahlquist, 1990; Sheldon and Bledsoe, 1993; Lee et al., 1997; Mindell et al., 1997; Groth and Barrowclough, 1999).

A. dementjevi (**Kurochkin, 1985a,b**)—Data are based on the description of the holotype (PIN-3790-271, an articulated partial specimen from the Lower Cretaceous [Hauterivian-Barremian] Unduruhinskaja Formation at Hurilt-Ulan-Bulak, southern Mongolia) and a slab with portions of the digits associated with it (PIN-3790-272; Kurochkin, 1985a,b). Additional data have been obtained from the study of a cast of PIN-3790-271 (L. M. Chiappe, pers. coll.).

A. lithographica (**Meyer, 1861**)—Data derive from personal observations of the Berlin, London, Eichstätt, Solnhofen, Haarlem, and Munich (Aktien-Verein) specimens as well as a cast of the Maxberg specimen. For the purpose of the present cladistic analysis, the Munich specimen is considered conspecific to all other specimens of *Archaeopteryx* (see Wellnhofer, 1993, for an argument in support of the specific distinction of this specimen, and see Elzanowski, Chapter 6 in this volume, for a different view on the systematics of these specimens). Additional data were recovered from the extensive literature on these specimens, especially from Witmer (1990), Elzanowski and Wellnhofer (1996), Ostrom (1972, 1976), Wellnhofer (1974, 1992, 1993), and Britt et al. (1999).

C. hengdaoziensis (**Ji et al., 1999**)—The holotype and only known specimen of this species (GMV-2129) comes from the Lower Cretaceous Chaomidianzi (Lower Yixian for others) Formation of the province of Liaoning, China. This nearly complete and articulated specimen was studied in detail by Chiappe et al. (1999).

C. lacustris (**Sanz and Buscalioni, 1992**)—Information is based on the study of the holotype (LH-2814), an articulated partial skeleton from the Lower Cretaceous (Upper Bar-remian) La Huérguina Formation of the province of Cuenca, Spain (see Sanz et al., Chapter 9 in this volume).

C. sanctus (**Hou et al., 1995a**)—Data are derived from personal observations of the holotype (IVPP-V-10918), a skull associated with wing elements from the Lower Cretaceous Chaomidianzi (Lower Yixian for others) Formation of the province of Liaoning, China (see Zhou and Hou, Chapter 7 in this volume), and a great number of referred specimens (see Chiappe et al., 1999). The anatomy of this taxon was summarized by Chiappe et al. (1999).

G. minuta (**Elzanowski, 1974**)—Information is based on the study of the holotype (ZPAL-MgR-I/12), an isolated skull from the Upper Cretaceous (Campanian) Barun Goyot Formation at Khulsan, southern Mongolia (see Chiappe and Walker, Chapter 11 in this volume), as well as several referred specimens. Referred specimens include a second skull (ZPAL-MgR-I/32; Elzanowski, 1977) and a series of embryos (ZPAL-MgR-I/33, 34, 88, 89, 90, 91, 92; Elzanowski, 1981) from the Barun Goyot Formation at Khulsan and Khermeen Tsav, respectively. New specimens referable to this taxon also include a skull (IGM-100/1011), preserving the rostrum and the palate, from the Djadokhta Beds of Ukhaa Tolgod (Chiappe et al., 2001) and a partial skeleton (PIN-4492-1) from the Barun Goyot Formation at Khermeen Tsav. Although PIN-4492-1 was described by Kurochkin (1996) as the holotype of "*Nanantius*" *valifanovi*, this specimen is essentially identical to IGM-100/1011 in the morphology of the rostrum, jaw, and pterygoids, the only overlapping regions of these two specimens. Additional data on this taxon were taken from the publications of Elzanowski (1995) and Kurochkin (1996).

H. regalis (**Marsh, 1872**)—Data derive from personal observations of the holotype (YPM-1200), a partial skeleton from the Upper Cretaceous Niobrara Chalk Formation of Kansas, and a large number of other specimens preserving diverse cranial and postcranial remains. These additional specimens are YPM-1207, YPM-1203, YPM-1476, YPM-1206, USNM-4978, USNM-6622, USNM-13581, USNM-13580, and KUVP-71012. This information was complemented by several specimens labeled as *Hesperornis* sp. (YPM-1479, YPM-1489, YPM-1481, YPM-1480, YPM-1477, YPM-1478, YPM-1491, YPM-1499, YPM-1679, USNM-11640). Data were also supplemented by information available in Marsh (1880), Bühler et al. (1988), Witmer (1990), and Elzanowski (1991).

I. romerali (**Sanz and Bonaparte, 1992**)—Data are based on personal observations of the holotype (LH-022R), a nearly complete, articulated skeleton from the Lower Cretaceous (Upper Barremian) La Huergina Formation of the province of Cuenca, Spain (see Sanz et al., Chapter 9 in this volume).

Ichthyornis—Information is derived from the description of *I. victor* and *I. dispar* by Marsh (1880). Additional information was based on personal observations of the holotype of *I. validus* (YPM-1740), specimens referred to *I. victor* (KUVP-119673, KUVP-25900), and others labeled as *Ichthyornis* sp. (YPM-1208, 1456, 1459, 1731, 1736, 1738, 1743, 1745, 1752, 1753, 1758, 1759, 1765, 1766, 1767, 1768, 1770, 1771, 1772, 1773, 1776, KUVP-2281, USNM-11641). These observations have been supplemented by data from Witmer (1990) and Lamb (1997). A revision of the several species of *Ichthyornis,* known primarily from the Upper Cretaceous of Kansas (Martin, 1983; Feduccia, 1996), has never been conducted (see Clarke, 1999, for some of the problems already detected). For the purpose of the present cladistic analysis, all these specimens are considered to form a monophyletic assemblage.

M. olecranus (**Perle et al., 1993a; see Perle et al., 1993b for a nomenclatural correction**)—Data derive from the study of the holotype (MGI 107/6), an articulated partial skeleton from the Upper Cretaceous (Maastrichtian) Nemegt Formation at Bugin Tsav, southern Mongolia (see Chiappe, Norell, and Clark, Chapter 4 in this volume).

N. gonzalezi (**Lacasa-Ruiz, 1989**)—Information is based on the study of the holotype (LP-1702), a partial skeleton from the Lower Cretaceous (Upper Berriasian–Lower Barremian) La Pedrera de Rúbies Lithographic Limestones Formation of the Montsec Range, Spain (Chiappe and Lacasa-Ruiz, Chapter 10 in this volume).

P. remotus (**Karhu and Rautian, 1996**)—Information is derived from Karhu and Rautian's (1996) description of the holotype (PIN 4487/25), a partial skeleton from the Upper Cretaceous (Campanian) Barun Goyot Formation at Khulsan, southern Mongolia (see Chiappe, Norell, and Clark, Chapter 4 in this volume).

P. puertai (**Novas, 1997**)—Information is based on personal observations of the holotype (PVPH 37), an incomplete dis-articulated skeleton from the Upper Cretaceous (Turonian) Río Neuquén Formation of the province of Neuquén, southern Argentina (see Chiappe, Norell, and Clark, Chapter 4 in this volume). These observations have been supplemented by the papers of Novas (1996, 1997).

P. deferrariisi (**Alvarenga and Bonaparte, 1992**)—Data are based on the study of the holotype (MACN-N-03), a partial skeleton from the Upper Cretaceous (Coniacian-Santonian) Río Colorado Formation of the province of Neuquén, southern Argentina, as well as several other specimens (MACN-N-10, MACN-N-11, MACN-N-14, MUCPv-48, MUCP-207) from the same beds (see Chiappe, Chapter 13 in this volume).

S. deserti (**Chiappe et al., 1998**)—Data are based on the study of the holotype (MGI 100/975), an incomplete skeleton from the Upper Cretaceous (Campanian) Djadokhta Formation at Ukhaa Tolgod, southern Mongolia. Several other specimens (MGI N 100/99, MGI 100/977, MGI 100/1001), some including nearly complete skulls, from the Djadokhta Beds at Ukhaa Tolgod and Tugrugeen Shireh (see Chiappe, Norell, and Clark, Chapter 4 in this volume) have been also studied.

S. santensis (**Sereno and Rao, 1992**)—Data derive from personal observations of the holotype slabs and a cast (positive cast or peel) of the main slab after removal of its bones (see Sereno, Rao, and Li, Chapter 8 in this volume). Sereno, Rao, and Li (Chapter 8 in this volume) convincingly argued for the synonymy of *S. santensis* and *Cathayornis yandica* (Zhou et al., 1992), a species also known by several specimens from the Jiufotang Formation (Zhou and Hou, Chapter 7 in this volume). The type specimen of *C. yandica* (IVPP-V-9769A, B) and positive casts of its two slabs (also after removal of bones) were also studied. Additional observations were conducted on several specimens assigned to *C. yandica* (IVPP-10896, IVPP-10916). Data from Hou (1997) and Martin and Zhou (1997) were also included.

Skull and Mandible

1. Rostral portion of the premaxillae in adults: unfused (0); fused (1).
2. Maxillary process of the premaxilla: restricted to its rostral portion (0); subequal or longer than the facial contribution of the maxilla (1).
3. Frontal process of the premaxilla: short (0); relatively long, approaching the rostral border of the antorbital fenestra (1); very long, extending caudally near the level of lacrimals (2).
4. Premaxillary teeth: present (0); absent (1).
5. Maxilla and dentary: toothed (0); toothless (1).
6. Caudal margin of naris: farther rostral than the rostral border of the antorbital fossa (0); nearly reaching or overlapping the rostral border of the antorbital fossa (1).
7. Dorsal ramus of the maxillary nasal process: present (0); absent (1).
8. Cup-shaped caudal maxillary sinus: absent (0); present (1).
9. Rostral margin of the jugal: away from the caudal margin of the osseous external naris (0), or very close to the caudal margin of the osseous external naris (1).
10. Jugal process of palatine: present (0); absent (1).
11. Ectopterygoid: present (0); absent (1).
12. Squamosal incorporated into the braincase, forming a zygomatic process: absent (0); present (1).
13. Postorbital: present (0); absent (1).
14. Postorbital-jugal contact: present (0); absent (1).
15. Quadratojugal: sutured to the quadrate (0); joined through a ligamentary articulation (1).
16. Quadratojugal-squamosal contact: present (0); absent (1).
17. Lateral, round cotyla on the mandibular process of the quadrate (quadratojugal articulation): absent (0); present (1).
18. Quadrate orbital process (pterygoid ramus): broad (0); sharp and pointed (1).
19. Quadrate pneumaticity: absent (0); present (1).
20. Quadrate: articulating only with the squamosal (0); articulating with both prootic and squamosal (1).
21. Quadrate distal end: with two transversely aligned condyles (0); with a triangular, condylar pattern, usually composed of three distinct condyles (1).
22. Caudal tympanic recess: opens on the rostral margin of the paraoccipital process (0); opens into the columellar recess (1).
23. Basicranial fontanelle on the ventral surface of the basisphenoid (basisphenoid recess): present (0); absent (1).
24. Deeply notched rostral end of the mandibular symphysis: absent (0); present (1).
25. Coronoid bone: present (0); absent (1).
26. Articular pneumaticity: absent (0); present (1).
27. Dentary tooth implantation: teeth in individual sockets (0); teeth in a communal groove (1).
28. Teeth: serrated crowns (0); unserrated crowns (1).

Vertebral Column and Ribs

29. Atlantal hemiarches: unfused (0); fused, forming a single arch (1).
30. One or more pneumatic foramina piercing the centra of midcranial cervicals, caudal to the level of the parapophysis-diapophysis: present (0); absent (1).
31. Cranial cervical vertebrae heterocoelous: absent (0); present (1).
32. Prominent carotid processes in the intermediate cervicals: absent (0); present (1).
33. Postaxial cervical epipophyses: prominent, projecting farther back from the postzygapophysis (0); weak, not projecting farther back from the postzygapophysis, or absent (1).
34. Prominent (50% or more the height of the centrum's cranial articular surface) ventral processes of the cervicothoracic vertebrae: absent (0); present (1).
35. Cervicothoracic vertebrae with parapophyses located at the same level as the prezygapophyses: absent (0); present (1).
36. Thoracic vertebral count: 13–14 (0); 11–12 (1); fewer than 11 (2).
37. Wide vertebral foramen in the midcaudal thoracic vertebrae, vertebral foramen/articular cranial surface ratio (vertical diameter) larger than 0.40: absent (0); present (1).
38. Hyposphene-hypantrum accessory intervertebral articulations in the thoracic vertebrae: present (0); absent (1).
39. Lateral side of the thoracic centra: weakly or not excavated (0); deeply excavated by a groove (1); excavated by a broad fossa (2).
40. Parapophyses: located in the cranial part of the centra of the thoracic vertebrae (0); located in the central part of the centra of the thoracic vertebrae (1).
41. Synsacrum: formed by fewer than eight vertebrae (0); eight or more vertebrae (1).
42. Synsacrum procoelous: absent (0); present (1).
43. Caudal portion of the synsacrum forming a prominent ventral keel: absent (0); present (1).
44. Convex caudal articular surface of the synsacrum: absent (0); present (1).
45. Caudal vertebra prezygapophyses: present (0); absent (1).
46. Distal caudal vertebra prezygapophyses: elongate, exceeding the length of the centrum by more than 25% (0); shorter (1).
47. Procoelous caudals: absent (0); present (1).
48. First caudal with a ventrally sharp centrum: absent (0); present (1).
49. Proximal haemal arches: elongate, at least three times longer than wider (0); shorter (1); absent (2).
50. Pygostyle: absent or rudimentary (fewer than three elements) (0); present (1).
51. Pygostyle: longer than or equal to the combined length of the free caudals (0); shorter (1).
52. Caudal vertebral count: more than 35 (0); fewer than 25–26 (1); fewer than 15 (2).
53. Ossified uncinate processes: absent (0); present (1).

Thoracic Girdle and Sternum

54. Coracoid and scapula: articulate through a wide, sutured articulation (0); articulate through more localized facets (1).
55. Scapula: articulated at the shoulder (proximal) end of the coracoid (0); well below it (1).
56. Humeral articular facets of the coracoid and the scapula: placed in the same plane (0); forming a sharp angle (1).
57. Procoracoid process on coracoid: absent (0); present (1).
58. Coracoid shape: short (0); elongated with trapezoidal profile (1); strutlike (2).
59. Distinctly convex lateral margin of coracoid: absent (0); present (1).
60. Bicipital tubercle (= acrocoracoidal process): present (0); or absent (1).

61. Supracoracoidal nerve foramen of coracoid: centrally located (0); displaced toward (often as an incisure or even without passing through) the medial margin of the coracoid (1).
62. Supracoracoidal nerve foramen opening into an elongate furrow medially and separated from the medial margin of the coracoid by a thick, bony bar: absent (0); present (1).
63. Broad, deep fossa on the dorsal surface of the coracoid: absent (0); present (1).
64. Sternocoracoidal process on the sternal half of the coracoid: absent (0); present (1).
65. Scapular caudal end: blunt and usually expanded (0); tapered to a sharp point (1).
66. Scapular shaft: straight (0); sagittally curved (1).
67. Prominent acromion in the scapula: absent (0); present (1).
68. Dorsal and ventral margins of the furcula: subequal in width (0); ventral margin distinctly wider than the dorsal margin (1).
69. Furcula: boomerang-shaped, with interclavicular angle of approximately 90° (0); U-shaped, with an interclavicular angle of less than 70° (1).
70. Hypocleideum: absent or poorly developed (0); well developed (1).
71. Sternum: subquadrangular to transversely rectangular (0); longitudinally rectangular (1).
72. Distinctly carinate sternum, more prominent than a faint ridge: absent (0); present (1).
73. Sternal carina: near to, or projecting rostrally from, the cranial border of the sternum (0); not reaching the cranial border of the sternum (1).
74. Lateral process of the sternum: absent (0); present (1).
75. Prominent distal expansion in the lateral process of the sternum: absent (0); present (1).
76. Medial process of the sternum: absent (0); present (1).
77. Rostral margin of the sternum broad and parabolic: absent (0); present (1).
78. Wide V-shaped caudal end of the sternum: absent (0); present (1).
79. Costal facets of the sternum: absent (0); present (1).

Thoracic Limb

80. Proximal and distal humeral ends: twisted (0); expanded nearly in the same plane (1).
81. Humeral head: concave cranially and convex caudally (0); globe-shaped, craniocaudally convex (1).
82. Proximal margin of the humeral head concave in its central portion, rising ventrally and dorsally: absent (0); present (1).
83. Ventral tubercle of the humerus: projected ventrally (0); projected proximally (1); projected caudally, separated from the humeral head by a deep capital incision (2).
84. Humerus with distinct transverse ligamental groove: absent (0); present (1).
85. Pneumatic fossa in the caudoventral corner of the proximal end of the humerus: absent or rudimentary (0); well developed (1).
86. Prominent, subquadrangular (i.e., subequal length and width) deltopectoral crest of the humerus: absent (0); present (1).
87. Prominent bicipital crest of the humerus, cranioventrally projecting: absent (0); present (1).
88. Ventral face of the humeral bicipital crest with a small fossa for muscular attachment: absent (0); present (1).
89. Humeral distal condyles: mainly located on distal aspect (0); on cranial aspect (1).
90. Humerus: with two distal condyles (0); a single condyle (1).

91. Well-developed brachial depression on the cranial face of the distal end of the humerus: absent (0); present (1).
92. Well-developed olecranon fossa on the caudal face of the distal end of the humerus: absent (0); present (1).
93. Distal end of the humerus very compressed craniocaudally: absent (0); present (1).
94. Ulna: shorter than humerus (0); nearly equivalent to or longer than humerus (1).
95. Ulnar shaft: considerably thicker than the radial shaft, radial-shaft/ulnar-shaft ratio larger than 0.70 (0); smaller than 0.70 (1).
96. Olecranon process of ulna: relatively small (0); hypertrophied, nearly one-third the length of the ulna (1); one-half the length of the ulna (2).
97. Proximal end of the ulna with a well-defined area for the insertion of M. brachialis anticus: absent (0); present (1).
98. Semilunate ridge on the dorsal condyle of the ulna: absent (0); present (1).
99. Shaft of radius with a long longitudinal groove on its ventrocaudal surface: absent (0); present (1).
100. U-shaped to heart-shaped ulnare (scapholunar): absent (0); present (1).
101. Semilunate carpal and proximal ends of metacarpals: unfused (0); semilunate fused to the alular (I) metacarpal (1); semilunate fused to the major (II) and minor (III) metacarpals (2); fusion of semilunate and all metacarpals (3).
102. Distal end of metacarpals: unfused (0); partially or completely fused (1).
103. Intermetacarpal space: absent or very narrow (0); at least as wide as the maximum width of minor metacarpal (III) shaft (1).
104. Extensor process on alular metacarpal (I): absent or rudimentary (0); well developed (1).
105. Minor metacarpal (III) projecting distally more than the major metacarpal (II): absent (0); present (1).
106. Round-shaped alular metacarpal (I): absent (0); present (1).
107. Alular metacarpal (I) large, massive, depressed, and quadrangular: absent (0); present (1).
108. Alular digit (I): long, exceeding the distal end of the major metacarpal (0); short, not surpassing this metacarpal (1).
109. Alular digit (I) large, robust, and dorsoventrally compressed: absent (0); present (1).
110. Prominent ventral projection of the proximolateral margin of the proximal phalanx of the alular digit (I): absent (0); present (1).
111. Ungual phalanx of major digit (II): present (0); absent (1).
112. Ungual phalanx of major digit (II) much smaller than the unguals of the alular (I) and minor (III) digits: absent (0); present (1).
113. Proximal phalanx of the minor digit (III) much shorter than the remaining nonungual phalanges of this digit: absent (0); present (1).
114. Ungual phalanx of minor digit (III): present (0); absent (1).
115. Proximal phalanx of major digit (II): of normal shape (0); flat and craniocaudally expanded (1).
116. Intermediate phalanx of major digit (II): longer than proximal phalanx (0); shorter than or equivalent to proximal phalanx (1).
117. Alular ungual phalanx with two ventroproximal foramina: absent (0); present (1).

Pelvic Girdle

118. Pelvic elements: unfused (0); fused or partially fused (1).
119. Preacetabular process of ilium twice as long as postacetabular process: absent (0); present (1).

120. Small acetabulum, acetabulum/ilium length ratio equal to or smaller than 0.11: absent (0); present (1).
121. Postacetabular process shallow and pointed, less than 50% of the depth of the preacetabular wing at the acetabulum: absent (0); present (1).
122. Orientation of proximal portion of pubis: cranially to subvertically oriented (0); retroverted, separated from the main synsacral axis by an angle ranging between 65° and 45° (1); more or less parallel to the ilium and ischium (2).
123. Prominent antitrochanter: caudally directed (0); caudodorsally directed (1).
124. Iliac brevis fossa: present (0); absent (1).
125. Pubic pedicel: cranioventrally projected (0); ventrally or caudoventrally projected (1).
126. Supracetabular crest on ilium: well developed (0); absent or rudimentary (1).
127. Supracetabular crest: extending throughout the acetabulum (0); extending only over the cranial half of the acetabulum (1).
128. Ischium with a proximodorsal process approaching, or abutting, the ventral margin of the ilium: absent (0); present (1).
129. Ischiadic terminal processes forming a symphysis: present (0); absent (1).
130. Ischium: two-thirds or less the length of the pubis (0); more than two-thirds the length of the pubis (1).
131. Obturator process of ischium: prominent (0); reduced or absent (1).
132. Pubic apron: one-third or more the length of the pubis (0); shorter (1); absent (absence of symphysis) (2).
133. Pubic shaft laterally compressed throughout its length: absent (0); present (1).
134. Pubic foot: present (0); absent (1).
135. Laterally compressed and kidney-shaped proximal end of pubis: absent (0); present (1).

Pelvic Limb

136. Femur with distinct fossa for the capital ligament: absent (0); present (1).
137. Femoral neck: present (0); absent (1).
138. Femoral anterior trochanter: separated from the greater trochanter (0); fused to it, forming a trochanteric crest (1).
139. Femoral posterior trochanter: absent to moderately developed (0); hypertrophied (1).
140. Conical and strongly distally projected lateral condyle of femur: absent (0); present (1).
141. Femur with prominent patellar groove: absent (0); present (1).
142. Femoral popliteal fossa distally bounded by a complete transverse ridge: absent (0); present (1).
143. Tibiofibular crest in the lateral condyle of femur: absent (0); poorly developed (1); prominent (2).

144. Fossa for the femoral origin of M. tibialis cranialis: absent (0); present (1).
145. Caudal projection of the lateral border of the distal end of the femur: absent (0); present (1).
146. Tibia, calcaneum, and astragalus: unfused or poorly co-ossified (sutures still visible) (0); complete fusion of tibia, calcaneum, and astragalus (1).
147. Cranial cnemial crest on tibiotarsus: absent (0); present (1).
148. Round proximal articular surface of tibiotarsus: absent (0); present (1).
149. Medial border of medial articular facet strongly projects proximally: absent (0); present (1).
150. Extensor canal on tibiotarsus: absent (0); present (1).
151. Wide and bulbous medial condyle of the tibiotarsus: absent (0); present (1).
152. Narrow, deep intercondylar sulcus on tibiotarsus that proximally undercuts the condyles: absent (0); present (1).
153. Proximal end of the fibula: prominently excavated by a medial fossa (0); nearly flat (1).
154. Fibula: tubercle for M. iliofibularis craniolaterally directed (0); laterally directed (1); caudolaterally or caudally directed (2).
155. Fibula: reaching the proximal tarsals (0); greatly reduced distally, without reaching these elements (1).
156. Metatarsals II–IV completely (or nearly completely) fused to each other: absent (0); present (1).
157. Distal tarsals: free (0); completely fused to the metatarsals (1).
158. Metatarsal V: present (0); absent (1).
159. Proximal end of metatarsal III: in the same plane as metatarsals II and IV (0); reduced, not reaching the tarsals (arctometatarsalian condition) (1); plantarly displaced with respect to metatarsals II and IV (2).
160. Well-developed tarsometatarsal intercondylar eminence: absent (0); present (1).
161. Tarsometatarsal vascular distal foramen completely enclosed by metatarsals III and IV: absent (0); present (1).
162. Trochlea of metatarsal II broader than the trochlea of metatarsal III: absent (0); present (1).
163. Completely reversed hallux (arch of ungual phalanx of digit I opposing the arch of the unguals of digits II–IV): absent (0); present (1).
164. Metatarsal IV significantly thinner than metatarsals II and III: absent (0); present (1).
165. Plantar surface of tarsometatarsus excavated: absent (0); present (1).
166. Tubercle on the dorsal face of metatarsal II: absent (0); present (1).
167. Hypotarsus: absent (0); present (1).

Integument

168. Feathers: absent (0); present (1).
169. Alula: absent (0); present (1).

APPENDIX 20.3
Character Matrix Used in the Present Cladistic Analysis

Key: Character state 0 is plesiomorphic; character states 1 or 2 are apomorphic. Entries with "n" are not comparable character states; "?" entries are either not preserved or unknown.

Taxon	1	2	3	4	5	6	7	8	9	10	11	12	13	14	15	16	17	18	19	20	21	22	23	24	25	26	27	28	29	30	31	32	33	34
Allosauroidea	0	0	0	0	0	0	0	0	0	0	0	0	0	0	0	0	0	0	0	0	0	0	0	0	0	0	0	0	0	0	0	0	0	0
Troodontidae	0	0	0	0	0	0	0	0	0	?	0	0	0	0	?	?	?	0	1	0	0	1	1	0	?	?	1	0	?	0	0	0	0	1
Velociraptorinae	0	0	0	0	0	0	0	0	0	0	0	0	0	0	0	0	0	0	0	?	0	0	0	0	0	0	0	0	0	0	0	0	0	0
Archaeopteryx	0	0	1	0	0	1	0	0	0	1	0	0	0	0	1	1	1	0	0	0	?	0	1	?	?	1	?	0	1	0	0	0	?	1
Rahonavis	?	?	?	?	?	?	?	?	?	?	?	?	?	?	?	?	?	?	?	?	?	?	?	?	?	?	?	?	?	?	?	?	?	?
Mononykus	?	?	?	?	?	?	?	?	?	?	?	?	?	?	?	?	?	?	?	?	?	1	?	?	?	?	?	?	1	?	1	0	1	1
Shuvuuia	0	0	0	?	0	0	0	?	0	1	0	?	0	0	1	1	1	0	0	0	1	0	1	0	?	1	?	1	1	0	0	0	1	1
Alvarezsaurus	?	?	?	?	?	?	?	?	?	?	?	?	?	?	?	?	?	?	?	?	?	?	?	?	?	?	?	?	?	0	?	?	1	?
Patagonykus	?	?	?	?	?	?	?	?	?	?	?	?	?	?	?	?	?	?	?	?	?	?	?	?	?	?	?	?	?	0	0	?	?	?
Confuciusornis	1	0	2	1	1	1	0	0	0	?	0	0	0	0	1	1	1	?	?	1	0	?	?	1	?	0	n	n	?	?	0	?	1	1
Changchengornis	1	?	?	1	1	1	0	?	?	?	?	?	?	?	?	?	?	?	?	0	?	?	1	?	0	n	n	?	?	?	?	?	?	?
Noguerornis	?	?	?	?	?	?	?	?	?	?	?	?	?	?	?	?	?	?	?	?	?	?	?	?	?	?	?	?	?	?	?	?	?	?
Iberomesornis	?	?	?	?	?	?	?	?	?	?	?	?	?	?	?	?	?	?	?	?	?	?	?	?	?	?	?	?	?	?	?	?	?	1
Patagopteryx	?	?	?	?	?	?	?	?	?	?	?	?	1	?	?	1	1	1	0	1	1	1	?	1	?	?	0	?	?	1	1	1	1	1
Vorona	?	?	?	?	?	?	?	?	?	?	?	?	?	?	?	?	?	?	?	?	?	?	?	?	?	?	?	?	?	?	?	?	?	?
Concornis	?	?	?	?	?	?	?	?	?	?	?	?	?	?	?	?	?	?	?	?	?	?	?	?	?	?	?	?	?	?	?	?	?	?
Cathayornis	1	0	1	0	0	1	0	?	0	?	?	?	?	?	?	?	?	?	?	0	?	?	?	?	?	?	?	?	0	1	?	?	?	?
Gobipteryx	1	0	2	1	1	1	1	0	1	0	?	?	?	?	?	?	?	?	0	?	?	0	?	?	0	1	0	n	n	?	?	?	?	?
Eoalulavis	?	?	?	?	?	?	?	?	?	?	?	?	?	?	?	?	?	?	?	?	?	?	?	?	?	?	?	?	0	?	?	?	1	1
Neuquenornis	?	?	?	?	?	?	?	?	?	?	?	?	?	?	?	?	?	?	?	?	?	?	?	?	?	?	?	?	?	?	?	?	?	?
Ambiortus	?	?	?	?	?	?	?	?	?	?	?	?	?	?	?	?	?	?	?	?	?	?	?	?	?	?	?	?	?	0	?	1	1	?
Hesperornis	1	1	2	1	0	1	1	1	1	1	1	1	1	1	1	1	1	1	1	0	1	1	0	0	1	0	1	1	?	1	1	1	0	1
Ichthyornis	1	?	2	1	0	?	?	?	?	?	?	?	1	1	1	1	1	1	1	1	?	?	?	?	?	?	1	0	1	1	1	1	1	1
Anas	1	1	2	1	1	1	1	1	1	1	1	1	1	1	1	1	1	1	1	1	0	1	1	0	1	1	n	n	1	0	1	1	1	1

Taxon	35	36	37	38	39	40	41	42	43	44	45	46	47	48	49	50	51	52	53	54	55	56	57	58	59	60	61	62	63	64	65	66	67	68
Allosauroidea	0	0	0	0	0	0	0	0	0	0	0	0	0	0	0	0	0	n	0	0	0	0	0	0	0	0	0	0	0	0	0	0	0	0
Troodontidae	0	0	0	0	0	0	0	?	0	0	0	0	0	0	0	0	0	n	?	1	0	0	0	0	1	0	0	0	0	?	0	?	?	0
Velociraptorinae	0	0	0	0	0	0	0	0	0	0	0	0	0	1	0	n	0	1	0	0	0	0	1	0	0	0	0	0	0	0	0	0	1	0
Archaeopteryx	0	0	?	?	0	0	0	0	?	0	0	1	0	?	1	0	n	1	0	0	0	0	1	0	0	0	0	1	0	0	0	0	1	0
Rahonavis	0	?	1	0	2	0	0	1	0	?	0	1	0	0	1	0	n	?	?	1	?	?	?	?	?	?	?	?	?	?	?	0	1	?
Mononykus	1	?	1	1	0	0	?	1	1	1	?	?	1	1	?	?	n	?	?	0	0	0	0	0	1	0	0	0	0	?	?	?	?	?
Shuvuuia	1	?	1	1	0	0	0	1	1	1	0	1	1	1	0	0	n	?	?	0	0	0	0	0	1	0	0	0	0	?	0	0	0	?
Alvarezsaurus	0	?	?	?	0	0	0	0	0	?	0	1	1	1	0	?	n	?	?	0	0	0	0	?	?	?	?	0	?	?	?	?	0	?
Patagonykus	?	?	1	0	0	0	?	1	1	1	?	?	1	1	?	?	n	?	?	0	0	?	0	0	0	1	0	0	0	?	?	?	?	?
Confuciusornis	?	0	1	1	2	0	0	1	0	0	0	1	0	?	?	1	0	1	1	0	0	0	0	2	0	0	1	0	0	0	0	0	0	0
Changchengornis	?	?	?	?	2	?	?	0	?	0	0	0	1	?	?	?	1	0	1	?	0	0	?	2	0	0	1	0	0	0	0	0	0	0
Noguerornis	?	?	?	?	?	?	?	?	0	?	?	?	?	?	?	?	?	?	?	?	?	?	?	?	?	?	?	?	?	?	?	?	?	?
Iberomesornis	0	1	?	?	0	0	0	0	0	0	0	1	0	0	1	1	0	1	0	?	?	?	?	?	2	0	0	1	?	?	0	1	0	?
Patagopteryx	0	1	1	1	0	0	1	1	0	1	1	n	1	0	?	?	?	?	?	0	1	1	1	0	2	0	0	1	0	0	0	1	1	1
Vorona	?	?	?	?	?	?	?	?	?	?	?	?	?	?	?	?	?	?	?	?	?	?	?	?	?	?	?	?	?	?	?	?	?	?
Concornis	?	?	?	?	1	1	?	?	0	?	?	?	0	0	1	?	?	?	0	1	1	1	0	2	1	0	1	1	1	0	?	0	?	1
Cathayornis	?	?	?	?	1	1	1	?	?	0	?	?	0	?	?	1	?	1	0	1	?	?	0	2	1	0	1	1	1	0	1	0	1	?
Gobipteryx	?	?	?	?	?	?	?	1	?	?	?	0	1	0	?	1	?	1	1	?	0	2	?	0	?	?	1	0	?	?	1	1	1	?
Eoalulavis	0	?	?	?	2	?	?	?	?	?	?	?	?	?	?	0	1	1	1	0	2	1	0	1	1	1	1	0	1	1	1	1	1	1
Neuquenornis	?	?	?	?	1	1	?	?	?	?	?	?	?	?	?	?	1	1	1	0	2	1	0	1	1	1	1	0	?	0	1	1	?	?
Ambiortus	?	?	?	?	?	?	?	?	?	?	?	?	?	?	?	?	1	1	1	1	2	0	0	1	0	?	?	?	1	1	0	?	?	?
Hesperornis	0	2	1	1	2	0	1	0	0	?	1	n	0	0	2	0	0	2	1	1	0	0	1	2	0	n	0	0	0	0	0	0	1	0
Ichthyornis	0	2	1	1	2	0	1	?	0	0	0	0	1	0	?	2	1	1	?	?	1	1	1	2	0	0	1	0	0	1	1	1	0	0
Anas	0	2	1	1	0	0	1	0	0	0	1	n	0	0	2	1	1	2	1	1	1	1	1	2	0	0	1	0	0	1	1	1	1	0

Taxon	69	70	71	72	73	74	75	76	77	78	79	80	81	82	83	84	85	86	87	88	89	90	91	92	93	94	95	96	97	98	99	100	101	102
Allosauroidea	0	0	0	0	n	?	?	?	?	?	0	0	0	0	0	0	0	0	0	0	0	0	0	0	0	0	0	0	0	0	0	0	n	0
Troodontidae	?	?	?	?	?	?	?	?	?	?	?	?	?	?	?	0	?	?	0	?	?	?	?	?	?	?	0	0	0	?	?	?	0	0
Velociraptorinae	0	0	0	0	n	0	n	0	0	0	0	0	0	0	0	0	0	0	0	0	0	0	0	0	0	0	0	0	0	0	0	0	0	0
Archaeopteryx	0	0	?	0	n	?	?	?	?	?	?	?	0	?	0	0	0	0	0	0	0	0	0	0	?	0	0	0	0	0	0	0	0	0
Rahonavis	?	?	?	?	?	?	?	?	?	?	?	?	?	?	?	?	?	?	?	?	?	?	?	?	?	?	1	0	1	0	0	?	?	?
Mononykus	?	?	1	1	0	0	n	0	0	0	0	1	0	0	1	0	0	0	0	0	1	1	0	0	0	0	0	2	0	0	0	?	1	1
Shuvuuia	?	?	1	1	0	0	n	0	0	0	0	1	0	0	1	0	0	0	0	0	1	1	0	0	0	0	0	2	0	0	0	?	1	0
Alvarezsaurus	?	?	?	?	?	?	?	?	?	?	?	?	?	?	?	?	?	?	?	?	?	?	?	?	?	?	?	?	?	?	?	?	?	?
Patagonykus	?	?	?	?	?	?	?	?	?	?	?	?	0	0	1	0	0	?	0	0	1	1	0	0	1	?	0	1	?	0	?	?	1	?
Confuciusornis	0	0	1	0	n	1	0	0	0	1	1	0	0	0	0	1	0	1	0	?	1	0	0	0	0	0	1	0	1	?	0	1	2	0
Changchengornis	0	0	1	0	n	?	?	?	?	?	1	?	0	?	0	?	?	?	1	?	?	?	?	?	?	?	0	1	0	?	?	0	?	0
Noguerornis	?	1	?	?	?	?	?	?	?	?	?	?	?	0	0	?	?	?	0	?	0	?	?	?	?	?	1	0	?	?	?	0	3	?
Iberomesornis	1	1	?	?	?	?	?	?	?	?	?	?	0	0	0	?	0	?	0	0	?	?	0	0	?	?	1	1	0	?	?	?	?	?
Patagopteryx	?	?	?	?	?	?	?	?	1	?	?	1	0	0	?	?	?	?	0	0	1	0	0	0	0	1	0	0	?	0	?	?	?	1
Vorona	?	?	?	?	?	?	?	?	?	?	?	?	?	?	?	?	?	?	?	?	?	?	?	?	?	?	?	?	?	?	?	?	?	?
Concornis	1	1	1	1	1	1	1	1	0	?	0	0	1	2	1	?	0	1	1	1	0	0	?	1	1	1	0	?	?	?	?	?	?	?
Cathayornis	1	1	1	1	1	1	1	1	0	?	0	0	1	2	?	1	0	1	1	1	0	0	1	?	1	1	0	?	1	1	1	1	3	0
Gobipteryx	?	1	?	?	?	?	?	?	?	?	?	?	0	?	?	?	?	?	?	?	?	?	?	0	?	0	?	?	1	1	?	1	1	?
Eoalulavis	1	1	1	0	n	0	n	n	0	0	0	0	1	1	2	1	0	0	1	1	1	0	0	1	1	1	1	0	1	?	1	?	?	0
Neuquenornis	1	0	1	1	0	1	1	?	?	?	?	?	1	?	?	2	2	1	?	?	?	?	?	1	1	1	1	?	?	1	1	?	3	0
Ambiortus	1	0	?	1	1	?	?	?	?	?	?	?	1	0	?	1	0	0	0	0	?	?	?	?	?	?	?	1	?	?	?	1	3	?
Hesperornis	0	0	1	0	n	0	n	0	1	0	1	n	1	0	n	n	0	0	0	0	n	n	0	0	0	n	n	n	n	n	n	n	n	n
Ichthyornis	?	?	1	1	0	?	?	?	1	?	?	1	1	0	2	1	0	0	0	1	1	0	1	1	0	1	1	0	1	1	0	?	3	1
Anas	1	0	1	1	0	1	0	0	1	0	1	1	1	0	2	1	1	0	0	0	1	0	1	1	0	0	1	0	1	1	0	1	3	1

Taxon	103	104	105	106	107	108	109	110	111	112	113	114	115	116	117	118	119	120	121	122	123	124	125	126	127	128	129	130	131	132	133	134	135	136
Allosauroidea	0	0	0	0	0	0	0	0	0	0	0	0	0	0	0	0	0	0	0	0	0	n	0	0	0	0	0	0	1	0	0	0	0	0
Troodontidae	0	0	0	0	0	0	0	0	0	0	0	0	0	0	0	0	0	0	0	0	0	0	1	1	n	0	0	0	0	0	?	?	?	?
Velociraptorinae	0	0	0	0	0	0	0	0	0	0	0	0	0	0	0	0	0	0	1	0	0	1	1	n	0	0	0	0	0	0	0	0	0	0
Archaeopteryx	0	0	0	0	0	0	0	0	0	0	0	0	0	1	0	1	0	n	?	1	1	n	1	?	0	0	0	0	0	0	0	0	0	?
Rahonavis	?	?	?	?	?	?	?	?	?	?	?	?	?	?	?	0	1	0	1	0	0	1	1	1	n	1	1	0	0	0	0	0	?	0
Mononykus	0	0	0	0	1	0	1	1	?	?	?	?	?	?	?	1	1	?	?	?	1	0	?	0	0	1	?	?	?	?	?	?	1	0
Shuvuuia	0	0	0	0	1	0	1	1	?	?	?	?	?	?	?	1	?	0	0	0	1	0	0	0	0	1	0	1	1	2	1	1	1	?
Alvarezsaurus	?	?	?	?	?	?	?	?	?	?	?	?	?	?	0	?	0	0	0	?	?	0	0	0	0	0	?	?	?	?	?	?	?	?
Patagonykus	?	0	?	0	1	0	1	1	?	?	?	?	?	?	?	0	?	?	?	0	0	?	0	0	0	?	?	?	?	1	1	0	0	?
Confuciusornis	0	0	0	0	0	0	0	0	0	1	1	0	0	0	0	1	?	0	?	1	1	1	1	1	n	1	1	0	1	1	0	1	?	1
Changchengornis	0	0	0	0	0	0	0	0	0	1	1	0	0	1	0	?	?	?	?	?	?	?	?	?	?	?	?	?	?	1	0	?	?	?
Noguerornis	0	0	1	1	0	?	0	?	?	?	?	?	?	?	?	?	?	?	?	?	?	?	?	?	?	1	0	?	?	?	?	?	?	?
Iberomesornis	?	?	?	?	?	?	?	?	?	?	?	?	?	?	?	1	?	?	?	?	?	?	?	?	?	1	?	1	1	1	?	?	?	?
Patagopteryx	1	?	?	?	0	?	?	0	?	?	?	0	0	?	?	1	0	0	0	1	0	0	1	1	n	0	1	1	1	2	0	1	0	?
Vorona	?	?	?	?	?	?	?	?	?	?	?	?	?	?	?	?	?	?	?	?	?	?	?	?	?	?	?	?	?	?	?	?	?	1
Concornis	0	?	?	?	0	1	0	?	0	?	?	0	0	1	?	?	?	?	?	?	?	?	?	?	?	1	1	1	0	1	0	?	?	0
Cathayornis	0	0	1	1	0	1	0	0	0	?	0	1	0	1	0	?	0	0	1	1	?	?	1	1	n	1	?	1	1	1	?	?	0	?
Gobipteryx	0	?	1	?	?	?	?	?	?	?	?	0	?	?	?	?	?	?	?	?	?	1	?	?	?	?	?	1	?	1	?	1	?	?
Eoalulavis	1	?	1	?	0	1	0	?	0	?	0	1	0	?	?	?	?	?	?	?	?	?	?	?	?	?	?	?	?	?	?	?	?	?
Neuquenornis	0	0	1	1	0	?	?	?	?	?	?	?	?	?	?	?	?	?	?	?	?	?	?	?	?	?	?	?	?	?	?	?	?	?
Ambiortus	?	0	?	1	0	?	?	?	0	?	?	?	1	1	n	?	?	?	?	?	?	?	?	?	?	?	?	?	?	?	?	?	?	?
Hesperornis	n	n	n	n	n	n	n	n	n	n	n	n	n	n	n	1	0	1	0	2	1	1	1	1	?	0	1	1	1	2	1	1	0	1
Ichthyornis	1	1	0	0	0	1	0	0	1	n	n	?	1	1	1	?	1	?	1	2	1	?	1	1	n	1	1	1	1	2	1	1	?	1
Anas	1	1	0	0	0	1	0	0	1	n	n	1	1	1	1	n	1	0	1	0	2	1	1	1	1	n	1	1	1	2	1	1	0	1

Taxon	137	138	139	140	141	142	143	144	145	146	147	148	149	150	151	152	153	154	155	156	157	158	159	160	161	162	163	164	165	166	167	168	169
Allosauroidea	0	0	0	0	0	0	0	0	0	0	0	0	0	0	0	0	0	0	0	0	0	0	0	0	0	0	0	0	0	0	0	?	?
Troodontidae	0	0	0	0	0	?	0	0	0	0	0	0	0	0	0	0	0	?	0	1	0	0	0	1	0	0	0	0	0	0	0	?	?
Velociraptorinae	0	0	0	0	0	0	0	0	0	0	0	0	0	0	0	0	0	0	0	0	0	0	0	0	0	0	0	0	0	1	0	?	?
Archaeopteryx	1	0	0	0	0	?	0	?	0	0	0	?	0	0	0	0	?	?	0	0	0	0	0	0	0	0	1	0	0	0	0	1	0
Rahonavis	1	1	0	0	0	0	1	0	0	0	0	0	0	0	0	0	0	?	2	1	0	0	?	0	0	0	1	0	0	1	0	1	?
Mononykus	0	1	?	1	0	0	0	0	0	0	0	0	1	0	0	0	1	?	1	0	?	?	1	0	0	0	?	0	0	0	0	?	?
Shuvuuia	0	?	?	?	0	0	?	?	0	0	?	?	1	0	0	0	1	1	1	0	0	0	1	0	0	0	?	0	0	0	0	1	?
Alvarezsaurus	0	?	0	?	?	?	?	?	?	?	0	?	?	?	0	0	0	?	0	0	?	?	0	0	?	0	0	?	0	?	0	?	?
Patagonykus	?	0	0	1	0	0	0	0	0	0	0	0	0	0	0	0	?	0	?	0	0	?	0	0	?	?	?	?	?	?	0	?	?
Confuciusornis	0	1	?	0	?	?	1	?	?	1	?	?	0	0	1	0	?	1	1	0	1	0	0	0	0	0	1	0	1	1	0	1	0
Changchengornis	?	?	?	0	?	?	?	?	?	1	?	?	?	?	?	?	?	1	0	1	0	0	?	?	?	?	1	0	0	?	0	1	?
Noguerornis	?	?	?	?	?	?	?	?	?	?	?	?	?	?	?	?	?	?	?	?	?	?	?	?	?	?	?	?	?	?	?	1	?
Iberomesornis	?	?	?	0	0	?	?	?	?	0	0	1	0	?	?	?	?	?	?	0	0	?	0	0	0	?	1	0	?	?	?	?	?
Patagopteryx	0	?	?	0	0	1	2	?	0	1	0	0	0	0	1	1	1	1	1	1	1	1	0	0	1	0	0	0	1	0	1	?	?
Vorona	0	1	?	0	0	1	1	?	1	0	0	0	0	0	1	1	1	1	1	1	1	0	0	0	1	0	?	0	1	0	0	?	?
Concornis	0	?	?	?	0	?	?	?	1	1	1	0	1	0	?	1	?	1	?	0	1	?	0	0	0	1	1	1	?	?	1	?	?
Cathayornis	0	1	1	0	0	1	1	?	0	1	?	1	0	0	1	1	1	?	0	1	?	0	0	?	?	0	?	1	?	0	1	?	?
Gobipteryx	?	?	?	?	?	?	?	?	?	?	0	1	0	0	1	?	1	?	1	0	1	1	0	0	?	1	?	1	0	1	0	?	?
Eoalulavis	?	1	1	?	?	?	?	?	?	?	?	?	?	?	?	?	?	?	?	?	?	?	?	?	?	?	?	?	?	?	?	1	1
Neuquenornis	?	1	1	?	0	?	?	?	1	?	?	?	?	?	?	?	?	?	?	0	?	?	0	?	0	1	1	1	?	?	?	?	?
Ambiortus	?	?	?	?	?	?	?	?	?	?	?	?	?	?	?	?	?	?	?	?	?	?	?	?	?	?	?	?	?	?	?	1	?
Hesperornis	0	1	0	0	1	1	2	1	0	1	1	0	0	1	0	0	1	2	1	1	1	1	2	1	1	0	0	0	0	0	0	?	?
Ichthyornis	0	1	0	0	1	1	2	?	0	1	1	0	0	1	0	0	?	2	1	1	1	1	2	1	1	0	?	0	0	0	1	?	?
Anas	0	1	0	0	1	1	2	1	0	1	1	0	0	1	0	0	1	2	1	1	1	1	2	1	1	0	1	0	0	0	1	1	1

Acknowledgments

I am grateful to J. Bonaparte, J. Clarke, J. de Vos, K. Fisher, C. Forster, J. Gauthier, Hou L. H., Ji Q., Ji S.-A., E. Kurochkin, A. Lacasa-Ruiz, L. Martin, M. Norell, F. Novas, H. Osmólska, J. Powell, L. Salgado, J. Sanz, D. Varricchio, G. Viohl, P. Wellnhofer, L. Witmer, and Zhou Z. for facilitating access to specimens or for providing discussions that have enriched the results of this investigation. I also thank B. Evans and L. Rhoads for producing the figures and S. Orell for editing the manuscript. This research was funded by the F. Chapman Fund of the American Museum of Natural History, the Natural History Museum of Los Angeles County, and the National Science Foundation (DEB-9873705).

Literature Cited

Alvarenga, H. M. F., and J. F. Bonaparte. 1992. A new flightless land bird from the Cretaceous of Patagonia; pp. 51–64 *in* K. E. Campbell (ed.), Papers in Avian Paleontology, Honoring Pierce Brodkorb. Science Series 36. Natural History Museum of Los Angeles County, Los Angeles.

Barriel, V., and P. Tassy. 1998. Rooting with multiple outgroups: consensus versus parsimony. Cladistics 14:193–200.

Barsbold R. 1974. Saurornithoididae, a new family of small theropod dinosaurs from central Asia and North America. Results of the Polish-Mongolian paleontological expedition, Part V. Palaeontologia Polonica 30:5–22.

Barsbold R., H. Osmólska, and S. M. Kurzanov. 1987. On a new troodontid (Dinosauria, Theropoda) from the Early Cretaceous of Mongolia. Acta Paleontologica Polonica 32(1–2): 121–132.

Bonaparte, J. F. 1991. Los vertebrados fósiles de la Formación Río Colorado de Neuquén y cercanias, Cretácico Superior, Argentina. Revista del Museo Argentino de Ciencias Naturales "Bernardino Rivadavia." Paleontología 4(3):17–123.

Brinkman, D. L., R. L. Cifelli, and N. J. Czaplewski. 1998. First occurrence of *Deinonychus antirrhopus* (Dinosauria: Theropoda) from the Antlers Formation (Lower Cretaceous: Aptian-Albian) of Oklahoma. Bulletin of the Oklahoma Geological Survey 146:1–27.

Britt, B. B., P. J. Makovicky, J. A. Gauthier, and N. Bonde. 1999. Postcranial pneumatization in *Archaeopteryx*. Nature 395:374–376.

Brodkorb, P. 1976. Discovery of a Cretaceous bird, apparently ancestral to the Orders Coraciformes and Piciformes (Aves: Carinatae). Smithsonian Contributions to Paleobiology 27: 67–73.

Bühler, P., L. D. Martin, and L. M. Witmer. 1988. Cranial kinesis in the Late Cretaceous birds *Hesperornis* and *Parahesperornis*. Auk 105:111–122.

Carpenter, J. 1988. Choosing among multiple equally parsimonious cladograms. Cladistics 4:291–296.

Chatterjee, S. 1997. The Rise of Birds: 225 Million Years of Evolution. Johns Hopkins University Press, Baltimore, 312 pp.

———. 1998. The avian status of *Protoavis*. Archaeopteryx 16:99–122.

Chiappe, L. M. 1991. Cretaceous avian remains from Patagonia shed new light on the early radiation of birds. Alcheringa 15(3–4):333–338.

———. 1992. Enantiornithine tarsometatarsi and the avian affinity of the Late Cretaceous Avisauridae. Journal of Vertebrate Paleontology 12(3):344–350.

———. 1993. Enantiornithine (Aves) tarsometatarsi from the Cretaceous Lecho Formation of northwestern Argentina. American Museum Novitates 3083:1–27.

———. 1995a. The first 85 million years of avian evolution. Nature 378:349–355.

———. 1995b. The phylogenetic position of the Cretaceous birds of Argentina: Enantiornithes and *Patagopteryx deferrariisi*; pp. 55–63 in D. S. Peters (ed.), Proceedings of the 3rd Symposium of the Society of Avian Paleontology and Evolution. Courier Forschungsinstitut Senckenberg 181, Frankfurt am Main.

———. 1996. Late Cretaceous birds of southern South America: anatomy and systematics of Enantiornithes and *Patagopteryx deferrariisi*; pp. 203–244 in G. Arratia (ed.), Contributions of Southern South America to Vertebrate Paleontology, Münchner Geowissenschaftliche Abhandlungen, Reihe A, Geologie und Paläontologie 30. Verlag Dr. Friedrich Pfeil, Munich.

———. 1997a. Aves; pp. 23–28 in P. J. Currie and K. Padian (eds.),The Encyclopedia of Dinosaurs. Academic Press, San Diego.

———. 1997b. The Chinese early bird *Confuciusornis* and the paraphyletic status of "Sauriurae." Journal of Vertebrate Paleontology 17(3):37A.

Chiappe, L. M., and J. O. Calvo. 1994. *Neuquenornis volans,* a new Upper Cretaceous bird (Enantiornithes: Avisauridae) from Patagonia, Argentina. Journal of Vertebrate Paleontology 14(2):230–246.

Chiappe, L. M., M. A. Norell, and J. M. Clark. 1996. Phylogenetic position of *Mononykus* from the Upper Cretaceous of the Gobi Desert. Memoirs of the Queensland Museum 39:557–582.

———. 1997. *Mononykus* and birds: methods and evidence. Auk 114(2):300–302.

———. 1998. The skull of a new relative of the stem-group bird *Mononykus*. Nature 392:275–278.

Chiappe, L. M., Ji S., Ji Q., and M. A. Norell. 1999. Anatomy and systematics of the Confuciusornithidae (Aves) from the Late Mesozoic of northeastern China. Bulletin of the American Museum Novitates 242:1–89.

Chiappe, L. M., M. A. Norell, and J. M. Clark. 2001. A new skull of *Gobipteryx minuta* (Aves: Enantiornithes) from the Cretacious of the Gobi Desert. American Museum Novitates 3346:1–15.

Chinsamy, A., L. M. Chiappe, and P. Dodson. 1994. Growth rings in Mesozoic birds. Nature 368:196–197.

———. 1995. Mesozoic avian bone microstructure: physiological implications. Paleobiology 21(4):561–574.

Chure, D. J., and J. H. Madsen. 1996. On the presence of furculae in some non-maniraptoran theropods. Journal of Vertebrate Paleontology 16:573–577.

Clarke, J. 1999. New information on the type material of *Ichthyornis:* of chimeras, characters and current limits of phylogenetic inference. Journal of Vertebrate Paleontology 19(3):38A.

Cracraft, J. 1986. The origin and early diversification of birds. Paleobiology 12(4):383–399.

———. 1988. The major clades of birds; pp. 339–361 in M. J. Benton (ed.), The Phylogeny and Classification of the Tetrapods, Vol. 1: Amphibians, Reptiles, Birds. Systematics Association Special Volume 35A. Clarendon Press, Oxford.

Cracraft, J., and D. P. Mindell. 1989. The early history of modern birds: a comparison of molecular and morphological evidence; pp. 389–403 in B. Fernholm, K. Bremer and H. Jörnvall (eds.), The Hierarchy of Life, Proceedings of the Nobel Symposia. Elsevier Science Publishers, Amsterdam.

Currie, P. J. 1985. Cranial anatomy of *Stenonychosaurus inequalis* (Saurischia, Theropoda) and its bearing on the origin of birds. Canadian Journal of Earth Sciences 22:1643–1658.

———. 1987. Bird-like characteristic of the jaws and teeth of troodontid theropods (Dinosauria, Saurischia). Journal of Vertebrate Paleontology 7(1):72–81.

Currie, P. J., and K. Padian. 1997. The Encyclopedia of Dinosaurs. Academic Press, San Diego, 869 pp.

Currie, P. J., and Peng J.-H. 1993. A juvenile specimen of *Saurornithoides mongoliensis* from the Upper Cretaceous of northern China. Canadian Journal of Earth Sciences 30(10–11): 2224–2230.

Currie, P. J., and Zhao X.-J. 1993a. A new troodontid (Dinosauria, Theropoda) braincase from the Dinosaur Park Formation (Campanian) of Alberta. Canadian Journal of Earth Sciences 30(10–11):2231–2247.

———. 1993b. A new carnosaur (Dinosauria,Theropoda) from the Jurassic of Xinjiang, People's Republic of China. Canadian Journal of Earth Sciences 30:2037–2081.

Dashzeveg D., M. J. Novacek, M. A. Norell, J. M. Clark, L. M. Chiappe, A. Davidson, M. C. McKenna, L. Dingus, C. Swisher, and Perle A. 1995. Extraordinary preservation in a new vertebrate assemblage from the Late Cretaceous of Mongolia. Nature 374:446–449.

Dingus, L., and T. Rowe. 1997. The Mistaken Extinction. W. H. Freeman and Co., New York, 332 pp.

Elzanowski, A. 1974. Preliminary note on the palaeognathous bird from the Upper Cretaceous of Mongolia. Palaeontologia Polonica 30:105–109.

———. 1977. Skulls of *Gobipteryx* (Aves) from the Upper Cretaceous of Mongolia. Palaeontologia Polonica 37:153–165.

———. 1981. Embryonic bird skeletons from the Late Cretaceous of Mongolia. Palaeontologia Polonica 42:147–176.

———. 1991. New observations on the skull of *Hesperornis* with reconstructions of the bony palate and otic region. Postilla 207:1–20.

———. 1995. Cretaceous birds and avian phylogeny. Courier Forschungsinstitut Senckenberg 181:37–53.

Elzanowski, A., and P. Wellnhofer. 1996. Cranial morphology of *Archaeopteryx*: evidence from the seventh skeleton. Journal of Vertebrate Paleontology 16(1):81–94.

Farris, J. 1988. Hennig 86 references. Documentation for version 1.5. Privately published.

Favstowski, D. E., and D. B. Weishampel. 1996. The Evolution and Extinction of the Dinosaurs. Cambridge University Press, Cambridge, 461 pp.

Feduccia, A. 1996. The Origin and Evolution of Birds. Yale University Press, New Haven, 420 pp.

———. 1999. The Origin and Evolution of Birds, 2nd edition. Yale University Press, New Haven, 466 pp.

Forster, C. A., L. M. Chiappe, D. W. Krause, and S. D. Sampson. 1996. The first Cretaceous bird from Madagascar. Nature 382:532–534.

Forster, C. A., S. D. Sampson, L. M. Chiappe, and D. W. Krause. 1998a. The theropodan ancestry of birds: new evidence from the Late Cretaceous of Madagascar. Science 279:1915–1919.

———.1998b. Genus correction. Science 280:185.

Gatesy, S. M. 1990. Caudofemoral musculature and the evolution of theropod locomotion. Paleobiology 16(2):170–186.

Gatesy, S. M., and K. Dial. 1996. Locomotor modules and the evolution of avian flight. Evolution 50(1):331–340.

Gauthier, J. 1986. Saurischian monophyly and the origin of birds; pp. 1–55 in K. Padian (ed.), The Origin of Birds and the Evolution of Flight, California Academy of Sciences, San Francisco.

Groth, J. G., and G. F. Barrowclough. 1999. Basal divergences in birds and the phylogenetic utility of the nuclear RAG-1 gene. Molecular Phylogenetics and Evolution 12(2):115–123.

Hauser, D. L., and W. Presch. 1991. The effect of ordered characters on phylogenetic reconstruction. Cladistics 7:243–265.

Heilmann, G. 1926. The Origin of Birds. Witherby, London, 210 pp.

Holmgren, N. 1955. Studies on the phylogeny of birds. Acta Zoologica 36:243–328.

Holtz, T. R., Jr. 1994. The phylogenetic position of the Tyrannosauridae: implications for theropod systematics. Journal of Paleontology 68(5):1100–1117.

———. 2001. Arctometatarsalia revisited: the problem of homoplasy in reconstructing theropod phylogeny; pp. 99–122 in J. Gauthier and L. F. Gall (eds.), New Perspectives on the Origin and Early Evolution of Birds: Proceedings of the International Symposium in Honor of John H. Ostrom. Peabody Museum of Natural History, New Haven.

Hou L. 1997. Mesozoic Birds of China. Taiwan Provincial Feng Huang Ku Bird Park. Nan Tou, Taiwan, 228 pp.

Hou L., Zhou Z., Gu Y., and Zhang H. 1995a. Confuciusornis sanctus, a new Late Jurassic sauriurine bird from China. Chinese Science Bulletin 40:1545–1551.

Hou L., Zhou Z., L. D. Martin, and A. Feduccia. 1995b. A beaked bird from the Jurassic of China. Nature 377:616–618.

Hou L., L. D. Martin, Zhou Z., and A. Feduccia. 1996. Early adaptive radiation of birds: evidence from fossils from northeastern China. Science 274:1164–1167.

Hou L., L. D. Martin, Zhou Z., A. Feduccia, and Zhang F. 1999. A diapsid skull in a new species of the primitive bird Confuciusornis. Nature 399:679–682.

Howard, H. 1950. Fossil evidence of avian evolution. Ibis 92(1):1–21.

Hutchinson, J. R., and L. M. Chiappe. 1998. The first known alvarezsaurid (Theropoda: Aves) from North America. Journal of Vertebrate Paleontology 18(3):447–450.

Ji Q., P. J. Currie, M. A. Norell, and Ji S.-A. 1998. Two feathered dinosaurs from northeastern China. Nature 393:753–761.

Ji Q., L. M. Chiappe, and Ji S.-A. 1999. A new late Mesozoic confuciusornithid from China. Journal of Vertebrate Paleontology 19(1):1–7.

Karhu, A. A., and A. S. Rautian. 1996. A new family of Maniraptora (Dinosauria: Saurischia) from the Late Cretaceous of Mongolia. Paleontological Journal 30(5):583–592.

Kurochkin, E. N. 1985a. A true carinate bird from Lower Cretaceous deposits in Mongolia and other evidence of early Cretaceous birds in Asia. Cretaceous Research 6:271–278.

———. 1985b. Lower Cretaceous birds from Mongolia and their evolutionary significance. XVIII Congressus Internationalis Ornithologici 1:191–199.

———. 1995. Synopsis of Mesozoic birds and early evolution of Class Aves. Archaeopteryx 13:47–66.

———. 1996. A new enantiornithid of the Mongolian Late Cretaceous, and a general appraisal of the infraclass Enantiornithes (Aves). Special Issue of the Russian Academy of Sciences: 1–50.

Lacasa-Ruiz, A. 1989. Nuevo género de ave fósil del yacimiento neocomiense del Montsec (Provincia de Lérida, España). Estudios Geológicos 45:417–425.

Lamb, J. P., Jr. 1997. Marsh was right: Ichthyornis had a beak! Journal of Vertebrate Paleontology 17(3):59A.

Lee, K., J. Feinstein, and J. Cracraft. 1997. The phylogeny of ratite birds: resolving conflicts between molecular and morphological data sets; pp. 173–211 in D. P. Mindell (ed.), Avian Molecular Evolution and Systematics. Academic Press, San Diego.

Lowe, P. R. 1935. On the relationships of the Struthiones to the dinosaurs and to the rest of the avian class, with special reference to the position of Archaeopteryx. Ibis 5:398–432.

Madsen, J. H., Jr. 1976. Allosaurus fragilis: a revised osteology. Bulletin of the Utah Geological and Mineralogical Survey 109:1–163.

Makovicky, P. J. 1995. Phylogenetic aspects of the vertebral morphology of Coelurosauria (Dinosauria: Theropoda). M.S. dissertation, University of Copenhagen, 311 pp.

Makovicky, P. J., and P. J. Currie. 1998. The presence of a furcula in tyrannosaurid theropods, and its phylogenetic and functional implications. Journal of Vertebrate Paleontology 18(1):143–149.

Marsh, O. C. 1872. Preliminary description of Hesperornis regalis, with notices of four other new species of Cretaceous birds. American Journal of Science and Arts 3:1–7.

———. 1880. Odontornithes: A Monograph on the Extinct Toothed Birds of North America. United States Geological Exploration of the Fortieth Parallel. Government Printing Office, Washington D.C., 201 pp.

Martin, L. D. 1983. The origin and early radiation of birds; pp. 291–338 in A. H. Brush and G. A. Clark Jr. (eds.), Perspectives in Ornithology. Cambridge University Press, New York.

———. 1995. The Enantiornithes: terrestrial birds of the Cretaceous. Courier Forschungsinstitut Senckenberg 181:23–36.

Martin, L. D., and C. Rinaldi. 1994. How to tell a bird from a dinosaur. Maps Digest 17(4):190–196.

Martin, L. D., and Zhou Z. 1997. Archaeopteryx-like skull in Enantiornithine bird. Nature 389:556.

Meier, R. 1994. On the inappropriateness of presence/absence recoding, for non-additive multistate characters in computerized cladistic analyses. Zoologischer Anzeiger 232:201–212.

Mindell, D. P., M. D. Sorenson, C. J. Huddleston, H. C. Miranda Jr., A. Knight, S. J. Sawchuk, and T. Yuri. 1997. Phylogenetic relationships among and within select avian orders based on mitochondrial DNA; pp. 213–247 in D. P. Mindell (ed.), Avian

Molecular Evolution and Systematics. Academic Press, San Diego.

Nixon, K. C., and J. M. Carpenter. 1996. On consensus, collapsibility, and clade concordance. Cladistics 12(4):305–321.

Norell, M. A., and P. J. Makovicky. 1997. Important features of the dromaeosaur skeleton: information from a new specimen. American Museum Novitates 3215:1–28.

———. 1999. Important features of the dromaeosaurid skeleton II: information from newly collected specimens of *Velociraptor mongoliensis*. American Museum Novitates 3282:1–45.

Norell, M. A., J. M. Clark, and Perle A. 1992. New dromaeosaur material from the Late Cretaceous of Mongolia. Journal of Vertebrate Paleontology 12(3):45A.

Norell, M. A., P. J. Makovicky, and J. M. Clark. 1997. A *Velociraptor* wishbone. Nature 389:447.

Norell, M. A., J. M. Clark, and P. J. Makovicky. 2001. Phylogenetic relationships among coelurosaurian theropods; pp. 49–67 *in* J. A. Gauthier and L. F. Gall (eds.), New Perspectives on the Origin and Early Evolution of Birds: Proceedings of the International Symposium in Honor of John H. Ostrom. Peabody Museum of Natural History, New Haven.

Novacek, M., M. A. Norell, M. C. McKenna, and J. M. Clark. 1994. Fossils from the Flaming Cliffs. Scientific American 271(6):36–43.

Novas, F. E. 1996. Alvarezsauridae, Cretaceous maniraptorans from Patagonia and Mongolia. Memoirs of the Queensland Museum 39(3):675–702.

———. 1997. Anatomy of *Patagonykus puertai* (Theropoda, Maniraptora, Alvarezsauridae) from the Late Cretaceous of Patagonia. Journal of Vertebrate Paleontology 17(1):137–166.

Novas, F. E., and P. Puerta. 1997. New evidence concerning avian origins from the Late Cretaceous of Patagonia. Nature 387:390–392.

Olson, S. L. 1985. The fossil record of birds; pp. 79–238 *in* D. S. Farner, J. R. King, and K. C. Parkes (eds.), Avian Biology, Vol. 8. Academic Press, New York.

Osborn, H. F. 1924. Three new Theropoda, *Protoceratops* zone, central Mongolia. American Museum Novitates 144:1–12.

Osmólska, H., and Barsbold R. 1990. Troodontidae; pp. 259–268 *in* D. B. Weishampel, P. Dodson, and H. Osmólska (eds.), The Dinosauria. University of California Press, Berkeley.

Ostrom, J. H. 1969. Osteology of *Deinonychus antirrhopus,* an unusual theropod dinosaur from the Lower Cretaceous of Montana. Bulletin of the Peabody Museum of Natural History 30:1–165.

———. 1972. Description of the *Archaeopteryx* specimen in the Teyler Museum, Haarlem. Proceedings of the Section of Science, Nederlandsche Akademie Wetenschaap (b) 75:289–305.

———. 1976. *Archaeopteryx* and the origin of birds. Biological Journal of the Linnean Society 8:91–182.

Padian, K., and L. M. Chiappe. 1997. Bird origins; pp. 71–79 *in* P. J. Currie and K. Padian (eds.), The Encyclopedia of Dinosaurs. Academic Press, San Diego.

———. 1998. The origin and early evolution of birds. Biological Reviews 73:1–42.

Padian, K., and J. R. Hutchinson. 1997. Allosauroidea; pp. 6–9 *in* P. J. Currie and K. Padian (eds.), The Encyclopedia of Dinosaurs. Academic Press, San Diego.

Perle A., M. A. Norell, L. M. Chiappe, and J. M. Clark. 1993a. Flightless bird from the Cretaceous of Mongolia. Nature 362:623–626.

———. 1993b. Genus correction. Nature 363:188.

Pimentel, R. A., and R. Riggins. 1987. The nature of cladistic data. Cladistics 3:201–209.

Pleijel, F. 1995. On character coding for phylogeny reconstruction. Cladistics 11:309–315.

Russell, D. A., and Dong Z.-M. 1993. A nearly complete skeleton of a new troodontid dinosaur from the Early Cretaceous of the Ordos Basin, Inner Mongolia, People's Republic of China. Canadian Journal of Earth Sciences 30(10–11):2163–2173.

Sanz, J. L., and J. F. Bonaparte. 1992. A new order of birds (Class Aves) from the Lower Cretaceous of Spain; pp. 39–49 *in* K. E. Campbell (ed.), Papers in Avian Paleontology, Honoring Pierce Brodkorb. Science Series 36. Natural History Museum of Los Angeles County, Los Angeles.

Sanz, J. L., and A. D. Buscalioni. 1992. A new bird from the Early Cretaceous of Las Hoyas, Spain, and the early radiation of birds. Paleontology 35(4):829–845.

Sanz, J. L., L. M. Chiappe, and A. Buscalioni. 1995. The osteology of *Concornis lacustris* (Aves: Enantiornithes) from the Lower Cretaceous of Spain and a re-examination of its phylogenetic relationships. American Museum Novitates 3133:1–23.

Sanz, J. L., L. M. Chiappe, B. P. Pérez-Moreno, A. D. Buscalioni, J. Moratalla, F. Ortega, and F. J. Poyato-Ariza. 1996. A new Lower Cretaceous bird from Spain: implications for the evolution of flight. Nature 382:442–445.

Sanz, J. L., L. M. Chiappe, B. P. Pérez-Moreno, J. Moratalla, F. Hernández-Carrasquilla, A. D. Buscalioni, F. Ortega, F. J. Poyato-Ariza, D. Rasskin-Gutman, and X. Martínez-Delclòs. 1997. A nestling bird from the Early Cretaceous of Spain: implications for avian skull and neck evolution. Science 276:1543–1546.

Sereno, P. 1997. The origin and evolution of dinosaurs. Annual Review of Earth and Planetary Sciences 25:435–489.

———. 1998. A rationale for phylogenetic definitions, with application to the higher-level taxonomy of Dinosauria. Neues Jahrbuch für Geologie und Paläontologie 210(1):41–83.

———. 1999a. The evolution of dinosaurs. Science 284:2137–2147.

———. 1999b. Alvarezsaurids (Dinosauria, Coelurosauria): birds or ornithomimosaurs? Journal of Vertebrate Paleontology 19(3):75A.

———. 2001. Alvarezsaurids: birds or ornithomimosaurs?; pp. 69–98 *in* J. Gauthier and L. F. Gall (eds.), New Perspectives on the Origin and Early Evolution of Birds: Proceedings of the International Symposium in Honor of John H. Ostrom. Peabody Museum of Natural History, New Haven.

Sereno, P., and Rao C. 1992. Early evolution of avian flight and perching: new evidence from Lower Cretaceous of China. Science 255:845–848.

Sheldon, F. H., and A. H. Bledsoe. 1993. Avian molecular systematics, 1970s to 1990s. Annual Review of Ecology and Systematics 24:243–278.

Sibley, C. G., and J. A. Ahlquist. 1990. Phylogeny and Classification of Birds: A Study in Molecular Evolution. Yale University Press, New Haven, 976 pp.

Slowinski, J. 1993. "Unordered" versus "ordered" characters. Systematic Biology 42(2):155–165.

Steadman, D. W. 1983. Commentary [on Martin, 1983]; pp. 338–344 *in* A. H. Brush and G. A. Clark Jr. (eds.), Perspectives in Ornithology. Cambridge University Press, New York.

Strong, E. E., and D. Lipscomb. 1999. Character coding and inapplicable data. Cladistics 15:363–371.

Suzuki, S., L. M. Chiappe, G. J. Dyke, M. Watabe, Barsbold R., and Tsogtbaatar K. 2002. A new specimen of *Shuvuuia deserti* from the Mongolian Late Cretaceous with a discussion of the relationships of alvarezsaurids to other theropod dinosaurs. Contributions in Science 494:1–17.

Swisher, C. C., III, Wang Y.-Q., Wang X.-L., Xu X., and Wang Y. 1999. Cretaceous age for the feathered dinosaurs of Liaoning, China. Nature 4000:58–61.

Varricchio, D. J. 1997. Troodontidae; pp. 749–754 *in* P. J. Currie and K. Padian (eds.), The Encyclopedia of Dinosaurs. Academic Press, San Diego.

Varricchio, D. J., and L. M. Chiappe. 1995. A new enantiornithine bird from the Upper Cretaceous Two Medicine Formation of Montana. Journal of Vertebrate Paleontology 15:201–204.

von Meyer, H. 1861. *Archaeopteryx lithographica* (Vogel-Feder) und *Pterodactylus* von Solnhofen. Neues Jahrbuch fuer Mineralogie, Geologie, und Palaeontologie 1861:561.

Walker, C. A. 1981. New subclass of birds from the Cretaceous of South America. Nature 292:51–53.

Wellnhofer, P. 1974. Das fünfte Skelettexemplar von *Archaeopteryx*. Palaeontographica A, 147:169–216.

———. 1992. A new specimen of *Archaeopteryx* from the Solnhofen limestone; pp. 3–23 *in* K. E. Campbell (ed.), Papers in Avian Paleontology, Honoring Pierce Brodkorb. Science Series 36. Natural History Museum of Los Angeles County, Los Angeles.

———. 1993. Das siebte Exemplar von *Archaeopteryx* aus den Solnhofener Schichten. Archaeopteryx 11:1–48.

Witmer, L. M. 1990. The craniofacial air sac system of Mesozoic birds (Aves). Zoological Journal of the Linnean Society 100:327–378.

Zhou Z. 1995. Discovery of a new enantiornithine bird from the Early Cretaceous of Liaoning, China. Vertebrata PalAsiatica 33(2):99–113.

Zhou Z., Jin F., and Zhang J. 1992. Preliminary report on a Mesozoic bird from Liaoning, China. Chinese Science Bulletin 37(16):1365–1368.

Zweers, G. A., J. C.Vanden Berge, and H. Berkhoudt. 1997. Evolutionary patterns of avian trophic diversification. Zoology 100:25–57.

Contributors

Angela D. Buscalioni
Unidad de Paleontología
Departamento de Biología
Facultad de Ciencias
Universidad Autónoma de Madrid
28049 Madrid
Spain

Luis M. Chiappe
Department of Vertebrate Paleontology
Natural History Museum of Los Angeles County
900 Exposition Boulevard
Los Angeles, California 90007
USA

Anusuya Chinsamy
South African Museum
P.O. Box 61
Cape Town 8000
South Africa
and Zoology Department, University of Cape Town
Rondebosch 7700
South Africa

James M. Clark
Department of Biological Sciences
George Washington University
Washington, D.C. 20052
USA

Andrzej Elzanowski
Institute of Zoology
University of Wrocław
Ulica Sienkiewicza 21
50335 Wrocław
Poland

Catherine A. Forster
Department of Anatomical Sciences
Health Sciences Center
State University of New York at Stony Brook
Stony Brook, New York 11794
USA

Peter M. Galton
College of Naturopathic Medicine
University of Bridgeport
Bridgeport, Connecticut 06601
USA

Stephen M. Gatesy
Department of Ecology and Evolutionary Biology
Brown University
Providence, Rhode Island 02912
USA

Sylvia Hope
Department of Ornithology and Mammalogy
California Academy of Sciences
Golden Gate Park
San Francisco, California 94118
USA

Hou Lianhai
Institute of Vertebrate Paleontology and
    Paleoanthropology
Chinese Academy of Sciences
P.O. Box 643
Beijing 100044
People's Republic of China

Alexander W. A. Kellner
    Museu Nacional/Universidade Federal do Rio de Janeiro
    Departamento de Geologia e Paleontologia
    Quinta da Boavista s/n
    São Cristovão, Rio de Janeiro
    Brazil

David W. Krause
    Department of Anatomical Sciences
    Health Sciences Center
    State University of New York at Stony Brook
    Stony Brook, New York 11794
    USA

Sergei Kurzanov
    Paleontological Institute
    Russian Academy of Sciences
    Profsojuznaja Street, 123
    Moscow 117647
    Russia

Antonio Lacasa-Ruiz
    Sección Paleontología
    Institut d'Estudis Ilerdencs
    Plaza La Catedral s/n
    25002 Lleida
    Spain

Li Jianjun
    Beijing Natural History Museum
    126 Tian Ciao Street
    Beijing
    People's Republic of China

Martin G. Lockley
    Department of Geology
    University of Colorado at Denver
    P.O. Box 173364
    Denver, Colorado 80217
    USA

Peter J. Makovicky
    Department of Geology
    Field Museum of Natural History
    1400 South Lake Shore Drive
    Chicago, Illinois 60605
    USA

Larry D. Martin
    Museum of Natural History
    University of Kansas
    Lawrence, Kansas 66045
    USA

Mark A. Norell
    Division of Paleontology
    American Museum of Natural History
    Central Park West at 79th Street
    New York, New York 10024
    USA

Fernando E. Novas
    Museo Argentino de Ciencias Naturales
    Avenida Angel Gallardo 470
    Buenos Aires 1405
    Argentina

Bernardino P. Pérez-Moreno
    Unidad de Paleontología
    Departamento de Biología
    Facultad de Ciencias
    Universidad Autónoma de Madrid
    28049 Madrid
    Spain

Diego Pol
    Division of Paleontology
    American Museum of Natural History
    Central Park West at 79th Street
    New York, New York 10024
    USA

Emma C. Rainforth
    Mesalands Community College
    911 South Tenth Street
    Tucumcari, New Mexico 88401
    USA

Rao Chenggang
    Beijing Natural History Museum
    126 Tian Ciao Street
    Beijing
    People's Republic of China

Scott D. Sampson
    Utah Museum of Natural History
    University of Utah
    1390 East Presidents Circle
    Salt Lake City, Utah 84112
    USA

José L. Sanz
    Unidad de Paleontología
    Departamento de Biología
    Facultad de Ciencias
    Universidad Autónoma de Madrid
    28049 Madrid
    Spain

Paul C. Sereno
    Department of Organismal Biology and Anatomy
    University of Chicago
    1027 East 57th Street
    Chicago, Illinois 60637
    USA

Patricia Vickers-Rich
    Earth Sciences Department
    Monash University
    Clayton (Melbourne) 3168
    Australia

Cyril A. Walker
    Bird Group
    Natural History Museum, Tring
    Hertfordshire HP23 6AP
    United Kingdom

Lawrence M. Witmer
    Department of Biomedical Sciences
    College of Osteopathic Medicine
    Ohio University
    Athens, Ohio 45701
    USA

Zhou Zhonghe
    Institute of Vertebrate Paleontology and
        Paleoanthropology
    Chinese Academy of Sciences
    P.O. Box 643
    Beijing 100044
    People's Republic of China

Index

fossil feathers of, 389–402
Hesperornithiformes in, 319
hind limbs of, 350
locomotor modules of, 226
monophyly of, 33–34
node-based definition of, 6
origin of flight in, 14–17
origins of, 4, 432–445
palaeognathous palate among, 343–344
Patagopteryx deferrariisi in, 282
as polyphyletic, 21–22
post-Mesozoic radiation of, 339
Protarchaeopteryx in, 11
Protoavis in, 8
Rahonavis ostromi in, 454–455
as reptiles, 33–34
Sinornis santensis in, 198–199
Sinosauropteryx in, 10
Spanish fossil birds in, 209, 231
specimens referable to, 330
taxonomy of, 184–185, 211
theropod characters and, 48
in theropod cladograms, 40–41
Vorona berivotrensis in, 270
AVES (common avian ancestor), in locomotor
 evolution, 436, 439, 440–441, 443–445
Avialae
 alvarezsaurids and, 122, 124
 Archaeornithoides and, 39
 in avian taxonomy, 185
 in coelurosaur cladistics, 35, 39, 42
 definition of, 6, 34
 Dromaeosauridae and, 35
 evolution of features of, 46–47
 feathered dinosaurs outside, 38
 temporal extent of, 45
 in theropod cladograms, 40–41, 43
 time problem and, 44
 Unenlagia comahuensis in, 37
Avian evolution, successive waves in, 21–22
Avimimidae, archaeopterygids and, 151
Avimimus, 38, 65–85
 in avian evolution, 21, 443
 as bird, 5
 in coelurosaur cladistics, 42
 historical perspective on, ix, x
 metatarsals of, 275
 phylogenetic position of, 65
 secondary flightlessness of, 66
 taxonomy of, 38
 theropod/bird classification of, 22
 in theropod cladograms, 40–41
Avimimus portentosus, 65–85
 birdlike features of, 5
 character matrix for, 53–56
 descriptions and redescriptions of, 65–66
 femur of, 80–81
 fibula of, 82–83
 humerus of, 76
 ilium of, 78

mandible of, 70
metatarsus of, 84
pedal digits of, 85
pubis of, 79
skeleton of, 68
skull of, 69–70
taxonomy of, 38
in theropod cladograms, 43
tibiotarsus of, 82–83
ulna of, 77
vertebrae of, 71–75
"Avimorph thecodonts," bird origins and, 18
Avisauridae
 in enantiornithine cladogram, 204
 femur of, 255
 geographical distribution of, 243
 historical perspective on, 258–259
 metatarsals of, 257, 277
 thoracic girdle of, 249
 wings of, 253
Avisaurus
 in avian phylogenetic systematics, 458
 in enantiornithine cladograms, 204, 259
 neornithines contemporary with, 381
 taxonomy of, 206, 341
 theropod/bird classification of, 22
Avisaurus archibaldi
 geographical distribution of, 243
 geological setting for, 241
 hind limbs of, 255
 historical perspective on, 258–259
 metatarsals of, 258
 tarsometatarsus of, 256
 taxonomy of, 244
Avisaurus gloriae
 character matrix for, 263
 geographical distribution of, 243
 metatarsals of, 256, 275
 tarsometatarsus of, 176
 taxonomy of, 244
Axial centrum, of *Avimimus*, 66. See also Axis
Axial skeleton. See also Ribs; Vertebrae
 of alvarezsaurids, 97–101
 of archaeopterygids, 138–139
 avian locomotion and, 432–433, 438
 of *Cathayornis yandica*, 189
 characters of theropod, 50–51
 of *Enaliornis*, 325–328
 of euenantiornithines, 247–249
 measurements of *Patagopteryx deferrariisi*,
  289
 of ornithurines, 441
 of *Sinornis santensis*, 188, 190–191
Axis, of *Patagopteryx deferrariisi*, 285, 287. See
 also Axial centrum; Cervical vertebrae

Baca County, fossil footprints from, 411
Bagaraatan, femur of, 109
Baja California, euenantiornithines from, 243
Bajo de la Carpa Member

alvarezsaurids from, 89, 90
dating of, 282
Patagopteryx deferrariisi from, 281–282
Patagopteryx from, 422
Bajsa Formation, fossil feathers from, 394–395
Bambiraptor, bird origins and, 17
Bambiraptor feinbergi, discovery of, 3
Bandemere site, fossil footprints from, 411
Baptornis
 Enaliornis versus, 325, 334, 335
 femur of, 328
 geographical distribution of, 318
 locomotion of, 336
 pelvic girdle of, 328
 tarsometatarsus of, 329, 331
 taxonomy of, 341, 375
 tibiotarsus of, 329
 vertebrae of, 325, 327, 328
Baptornithidae, geographical distribution of, 318
Barbs, of feathers, 389, 390
Barbules, of feathers, 389, 390
Barcelona, 231
Barremian stage
 Chinese fossil birds from, 184
 coelurosaurs from, 47
 euenantiornithines from, 242
 fossil feathers from, 392, 394, 396
 fossil footprints from, 409
 Noguerornis from, 230, 231
 Spanish Mesozoic birds from, 209, 211–215
Barrett, Lucas, *Enaliornis* collected by, 317
Barsbold Rinchen, on theropods as bird
 ancestors, 5
Barton, *Enaliornis* from, 319
Barun Goyot Formation
 alvarezsaurids from, 87, 88–89, 91
 Avimimus from, 65
 euenantiornithines from, 243
 Gobipteryx from, 422
 presbyornithids from, 377
Baryonyx, in theropod cladograms, 40–41
Basal archosaurs, as bird ancestors, 3, 17–18. See
 also Archosauria
Basicranium, in neornithine taxonomy, 340.
 See also Braincase; Palate; Skull
Basisphenoid, of alvarezsaurids, 96–97. See also
 Braincase; Skull
Bastard wing. See Alula
Bathonian stage, coelurosaurs from, 47
Bayan Mandahu Formation, 88
 oviraptorosaur nests from, 36
Bayerische Staatssammlung für Paläontologie
 und historische Geologie, *Archaeopteryx*
 specimen in, 129
 first *Archaeopteryx* fossil feather in, 391
Bayin Obo Formation, coelurosaurs from, 47
Bayn Dzak, alvarezsaurids from, 89
BCF. *See* "Birds Came First" (BCF) theory
Beak. *See* Rostral morphology
Bedfordshire, *Enaliornis* from, 319

forthcoming descriptions of, 180
historical perspective on, 161
locality map for, 162
paleobiology of, 178
Patagopteryx versus, 311
phylogeny of, 175–177
taxonomy of, 164
Chaoyangia beishanensis, 164
in avian phylogenetic systematics, 449
historical perspective on, 161
skeleton of, 176
taxonomy of, 164
thoracic girdle of, 176–177
Characters
in avian phylogenetic systematics, 449–451
in cladistic analysis, 31–33, 463–468
of coelurosaurs, 53–56
of theropods, 48–52
Charadriidae, endocranial cast of, 323
Charadriiformes
Anseriformes versus, 355
Apatornis versus, 370, 371
braincase of, 336
Ceramornis in, 372
Cimolopteryx in, 353, 361, 363
coracoid of, 354, 362
forelimb elements of, 364
fossil footprints of, 413–414
Gansus yumenensis in, 160
Graculavus in, 358–359
Mesozoic, 339, 377–378
in neornithine cladogram, 380
Noguerornis versus, 231
Palaeotringa vetus versus, 373–374
phylogenetic systematics of, 380–381
revisions to fossil record of, 374–375
table of Mesozoic, 376
tarsometatarsus of, 366
taxonomy of, 340, 355, 357–358, 366, 426
Telmatornis in, 360
undetermined neornithine species in, 364–365
Volgavis marina in, 361
Charophytes
in dating *Noguerornis,* 230
in dating Spanish Mesozoic birds, 209
in Las Hoyas Konservat-Lagerstätte, 210
Chatterjee, Sankar, 3
on avian evolution, 443
on *Avimimus,* 66
on bird origins, 16–17
Protoavis and, 7–9
Cheirolepidaceae, *Archaeopteryx* and, 131
Chen, taxonomy of, 355
Chevrons, of *Sinornis santensis,* 191. *See also*
Caudal vertebrae; Tail
Chiappe, Luis M., 65, 209, 230, 240, 268, 281, 448
on avian evolution, 449
on avian taxonomy, 184–185

on *Avimimus,* 66
Chicken, metatarsals of, 351
Chile
Mesozoic neornithine fossil record from, 379
Neogaeornis wetzeli from, 367
China, 162, 185
alvarezsaurids from, 88–89, 110
Chaoyangia from, 311
coelurosaurs from, 42, 46, 47
early therizinosauroid from, 9
euenantiornithines from, 242, 245, 248, 251, 257
feathered dinosaurs from, 4, 6, 9–14, 38, 440
fossil feathers from, 392, 399, 400
fossil footprints from, 409–410
Gansus yumenensis from, 351
Mesozoic birds from, ix, 8, 160–181, 241, 449, 452
neornithine-like birds from, 350
oviraptorosaur nests from, 36
Sinornis santensis from, 184–206
smuggling fossils from, 162–163
therizinosauroids from, 36
Chinese Academy of Sciences, fossil smuggling and, 163
Chinsamy, Anusuya, 421
Chirostenotes
taxonomy of, 36
in theropod cladograms, 40–41
Chirostenotes pergracilis
character matrix for, 53–56
in theropod cladistics, 41, 43
Chita Oblast, fossil feathers from, 394–395
Cicadophytes, in Las Hoyas Konservat-Lagerstätte, 210
Ciconiiformes, taxonomy of, 340, 374
Cimoliosaurus, from Cambridge Greensand, 336
Cimolopterygidae
Ceramornis in, 372
taxonomy of, 353, 361
Cimolopteryx, 361–363
bone microstructure of, 421, 426–428
coracoid of, 354, 362
diversity of, 377–378
fossil record of, 376
growth rate of, 428
taxonomy of, 353, 374, 375
Cimolopteryx maxima, 362–363
coracoid of, 354
forelimb elements of, 364
fossil record of, 376
taxonomy of, 361
Cimolopteryx minima, 363
coracoid of, 354, 362
fossil record of, 376
Cimolopteryx petra n. sp., 363
coracoid of, 362

fossil record of, 376
Cimolopteryx rara
Cimolopteryx petra n. sp. versus, 363
coracoid of, 354, 362
fossil record of, 376
taxonomy of, 353, 357, 361, 372
Telmatornis versus, 360
Cimolopteryx retusa, taxonomy of, 353, 361, 375
Cipoletti, locality map of, 283
Cladistics. *See also* Phylogenetic systematics
of alvarezsaurids, 112–115
avian flight and, 14–15, 16–17
of birds, x
of *Caudipteryx,* 13
of enantiornithines, 202–205, 257–260, 263
group names in, 33–34
locomotor evolution and, 435–441, 442–445
methodology of, 31–33
non-theropod hypotheses of bird origins and, 17–18
stratigraphic information and, 44–46
taxonomy and, 33–34
and theropod fossil record, 44–46
theropod origin of birds via, 19–22, 31–56
Cladograms
of alvarezsaurids, 114
of Archosauria, 433, 434
of Archosauriformes, 437, 439, 444
of Aves, xi, 236, 312
from cladistic analysis, 32
of coelurosaur theropods, 40–41, 43, 123
of Dinosauria, 438
of Enantiornithes, 203, 204, 258, 259, 260
explanatory power of, 33
locomotor modules and, 459
of Neornithes, 380
of Spanish Mesozoic birds, 225, 236
testing of, 15–16
Clark, James M., 31
Classes, in cladistic analysis, 34
Clavicles, furcula and, 5
Claws. *See* Manual unguals; Pedal unguals;
Ungual sheaths
Clayton Lake State Park, fossil footprints from, 411
Climate
for Chinese Mesozoic birds, 177–178
for Solnhofen archaeopterygids, 131
Climbing, by archaeopterygids, 153
Cloverly Formation, coelurosaurs from, 47
Cockroaches, in archaeopterygid diet, 154
Coding, in avian phylogenetic systematics, 449–450. *See also* Characters
Coelacanthiformes, in Las Hoyas Konservat-Lagerstätte, 210
Coelophysis
locomotor modules of, 435
in theropod cladograms, 40–41
Coelurosauria, 39–44
alvarezsaurids and, 113, 115, 122–123

postorbital of, 12
pygostyle of, 191
quadratojugal of, 12
ribs of, 249
Shuvuuia versus, 94
Sinornis santensis versus, 190
Sinosauropteryx versus, 179
smuggling specimens of, 162–163
squamosal of, 284
in strict consensus tree, 451, 453
taxonomy of, 163
in theropod cladograms, 43
thoracic girdle of, 13
tibiotarsus of, 256
Confuciusornis sanctus, 163
anatomy of, 164–167
in avian phylogenetic systematics, 449,
452, 455
character matrix for, 53–56
cladistic analysis of, 461
fossil feathers of, 399
historical perspective on, 161, 184
locality of, 185
phylogenetic systematics of, 277
postorbital of, 8
Sinornis santensis taxonomy and, 186
taxonomy of, 163
in theropod cladograms, 43
Confuciusornithidae, 163
archaeopterygids and, 150
in avian phylogenetic systematics, xi, 203,
236, 452, 454, 455, 456, 459
taxonomy of, 163
Coniacian stage
alvarezsaurids from, 89, 90
Apatornis from, 370
avian coracoids from, 348, 356
coelurosaurs from, 48
euenantiornithines from, 242–243
fossil feathers from, 399
hesperornithiforms from, 318
Kuszholia mengi from, 311
neornithines from, 375–377
in North America, 342
Patagopteryx from, 282, 422
table of neornithines from, 376
undetermined neornithine species from,
374
Coniferales, in *Noguerornis* deposits, 231
Conifers
Archaeopteryx and, 131
in Las Hoyas Konservat-Lagerstätte, 210
Contextual evidence, cladistic analysis and, 31
Continental deposits, hesperornithiforms
from, 319. *See also* Terrestrial deposits
Contour feathers, 389, 390, 391
of archaeopterygids, 149, 150
of *Eoalulavis hoyasi,* 223
Convergence
in dinosaur/bird evolution, 44

in neornithine systematics, 339
Converse County
Cimolopteryx rara from, 361
Palintropus retusus from, 354
Cooper Member, *Protoavis* from, 7
Coprolites, in Las Hoyas Konservat-
Lagerstätte, 210
Coraciiformes, revisions to fossil record of, 375
Coracoid. *See also* Thoracic girdle
of alvarezsaurids, 101, 102
of Antarctic presbyornithid, 356
of archaeopterygids, 140
avian locomotion and, 433
of *Cathayornis,* 169
of *Cathayornis yandica,* 170
of *Caudipteryx,* 13
of *Ceramornis,* 372
of *Chaoyangia,* 173
of *Chaoyangia beishanensis,* 177
of charadriiform birds, 362
of *Cimolopteryx,* 354
of *Concornis lacustris,* 218
of *Confuciusornis sanctus,* 165
of *Enantiornis leali,* 250
of *Eoalulavis hoyasi,* 222–223
of euenantiornithines, 249–250
of *Iberomesornis romerali,* 217
of *Liaoningornis,* 171–173
of neognath birds, 348, 356
of neornithines, 344–348, 376, 377
of *Otogornis,* 171
of *Palintropus,* 354
of *Patagopteryx deferrariisi,* 294, 295, 296,
297, 298
of *Protoavis,* 9
of *Sinornis santensis,* 192
of Spanish Mesozoic birds, 224
of undetermined neornithine species, 357,
363–364, 373
Cormorant
fossil record and, 378
Graculavus as, 358
taxonomy of, 367, 368, 375
Coronoid
of archaeopterygids, 137
of *Shuvuuia,* 97
Cosesaurus, bird origins and, 17
Covert feathers, of archaeopterygids, 147–148
Cranial base, of archaeopterygids, 133
Cranial cavity
in defining birds, 7
of *Enaliornis,* 321–323
Cranial kinesis
of archaeopterygids, 136–137
of *Cathayornis,* 168
in *Shuvuuia,* 94–96
Cranial nerves
of *Enaliornis,* 321
of *Mononykus olecranus,* 96
of *Patagopteryx deferrariisi,* 285–286

in neornithine systematics, 339
Craniofacial kinesis, understanding, 4
Cranium, of euenantiornithines, 246. *See also*
Skull
Crato Member, fossil feathers from, 396–397,
398, 399, 401
Cream Ridge Marl Company pits, *Telmatornis
priscus* from, 360
Creationism, *Archaeopteryx* and, 129
Cretaaviculus sarysuensis, 399
Cretaceous period. *See also* Early Cretaceous
epoch; Late Cretaceous epoch
absence of Madagascan mammals from, 269
Antarctic loon from, 427, 428
archaeopterygids and birds from, 150
Argentine birds from, 240
birdlike theropods from, 19
bird origins and, 45
birds from, 8, 240, 269, ix
Chinese birds from, 160
enantiornithines during, 260–262
enantiornithines from, 422
euenantiornithine phylogeny during,
257–260
fossil feathers from, 401
fossil footprints from, 405, 407, 415–416
fossil pterosaur tracks from, 405
geology of bird-bearing formations of,
342–343
Ichthyornis from, 426
neornithines from, 339
preservational bias against feathers from,
146
taphonomic bias against neornithines from,
378
Crocodylia
Archaeopteryx and, 131
in archosaur cladogram, 433
and evolution of birds, 46
in Las Hoyas Konservat-Lagerstätte, 210
locomotor modules of, 435
Mesozoic Madagascan, 268
in *Patagopteryx* deposits, 281, 314
as reptiles, 33
Crocodylomorpha, as bird ancestors, 3, 4, 17
Crown groups
Aves as, 6, 34, 185
in neornithine systematics, 341
in phylogenetic systematics, 34
Crurotarsi, in archosauriform cladogram,
437, 439
Crus
of alvarezsaurids, 111
of archaeopterygids, 144
Crustaceans
in Mesozoic avian diet, 227
in *Noguerornis* deposits, 231
Cuculidae, scleral ring of, 137
Cuenca province
euenantiornithines from, 242
fossil feathers from, 395–396

Enaliornithidae, taxonomy of, 333–334

Enantiornis
  in avian cladogram, 204
  body size of, 261
  in enantiornithine cladograms, 259
  forelimb of, 349
  neornithine-like features of, 350
  Noguerornis versus, 236
  shoulder girdle of, 344, 346–347
  taxonomy of, 341
  thoracic girdle of, 249, 250
  wings of, 251–253

Enantiornis leali
  in avian phylogenetic systematics, 457
  carpometacarpus of, 253
  character matrix for, 263
  coracoid of, 250
  Eoalulavis hoyasi versus, 222
  geographical distribution of, 243
  humerus of, 252, 345
  shoulder elements of, 345
  taxonomy of, 244
  ulna of, 253

Enantiornis martini
  geographical distribution of, 242
  taxonomy of, 244

Enantiornis walkeri, taxonomy of, 244

Enantiornithes, 163–164, 241–245
  Alexornis in, 375
  anatomy of Chinese, 167–171, 172, 173, 174
  anatomy of Spanish, 215–223
  archaeopterygids and, 150, 151
  in archosauriform cladogram, 437, 439
  in Aves, 21
  Avimimus and, 66
  in avian phylogenetic systematics, xi, 203, 204, 225, 236, 312, 448–449, 455, 456–457, 459
  axial skeleton of, 190–191
  bone microstructure of, 421, 422, 424
  cervical vertebrae of, 287
  Chaoyangia in, 161
  from China, 160–162
  cladograms of, 258, 259, 260
  definition of, 240
  in dinosaur cladogram, 438
  diversity of, 240, 261–262
  Euenantiornithes in, 240
  evolution of alula among, 180
  femur of, 271–272
  flight of, 176–177
  forelimbs of, 192–195
  Gobipteryx in, 374
  growth rate of, 429
  hind limbs of, 195–198
  historical perspectives on, ix, 160–162, 240, 258
  humerus of, 297
  ilium of, 301
  ischium of, 302

metatarsals of, 275, 276, 277
  most primitive known, 180
  neornithines contemporary with, 381
  in *Noguerornis* deposits, 231
  Noguerornis in, 232, 235–237
  Otogornis genghisi in, 161
  paleobiology of, 205–206
  paleobiology of Chinese, 177–179
  paleobiology of Spanish, 224–227
  in *Patagopteryx* deposits, 281
  pectoral girdle of, 191–192
  pelvic girdle of, 195, 196, 197, 436
  perching by, 205–206
  phylogenetic systematics of, 202–205, 223–224, 235–237, 311, 380
  phylogeny of Chinese, 175–177, 198–205
  postorbital of, 8, 12
  pygostyle of, 191, 436
  quadratojugal of, 12
  recognition of, 240
  Shuvuuia versus, 94
  Sinornis santensis in, 184–206
  skull of, 189–190
  from Spain, 209, 230–237
  squamosal of, 284
  taxonomy of, 185, 206, 211–212, 231, 240, 341
  taxonomy of Chinese, 163–164
  in theropod cladograms, 40–41
  thoracic girdle of, 168–169, 249–251
  tibiotarsus of, 273
  vertebrae of, 326
  Vorona and, 277
  wings of, 170

Endocranial cast
  of *Enaliornis*, 322, 323, 324–325
  of *Patagopteryx deferrariisi*, 283

Endothermy. *See also* Thermoregulation
  of archaeopterygids, 151–152, 428–429
  of *Archaeopteryx*, 401
  avian phylogeny and, 5
  bone microstructure and, 428–429
  of euenantiornithines, 262
  evolution of avian, 46
  feathered theropods and, 10

England. *See* Great Britain

Eoalulavis, 212
  in avian cladogram, 225
  body size of, 261
  cervical vertebrae of, 247–248
  character matrix for, 466–468
  diet of, 227
  in Enantiornithes, 240
  flight of, 237
  furcula of, 250
  gastralia of, 191
  historical perspective on, x, 180, 209
  manual digits of, 254
  Noguerornis versus, 231, 232, 236
  ontogeny of, 220
  paleobiology of, 205, 224, 226, 227

pelvic girdle of, 254
  pes of, 261
  sternum of, 251
  in strict consensus tree, 451, 453
  taxonomy of, 206
  thoracic girdle of, 249, 250
  thoracic vertebrae of, 248
  wings of, 192, 251, 254

Eoalulavis hoyasi, 212
  anatomy of, 219–223
  character matrix for, 263
  feathers of, 210, 223
  geographical distribution of, 242
  geological setting for, 241
  historical perspective on, 209
  Noguerornis gonzalezi and, 230
  phylogenetic systematics of, 205
  skeleton of, 214–215, 219–220, 221
  sternum of, 222, 251
  taxonomy of, 206, 212, 244

Eocene epoch
  absence of Madagascan mammals from, 269
  avian coracoids from, 348, 356
  gaviiforms from, 333

Eoenantiornis
  cervical vertebrae of, 247
  furcula of, 250
  historical perspective on, 161, 180
  mandible of, 246
  manual digits of, 254
  pelvic girdle of, 254
  ribs of, 249
  rostral morphology of, 245
  sternum of, 251
  thoracic girdle of, 249, 250
  wings of, 251, 253, 254

Eoenantiornis buhleri
  in avian phylogenetic systematics, 457
  character matrix for, 263
  geographical distribution of, 242
  historical perspective on, 161
  skeleton of, 257
  skull of, 245
  taxonomy of, 244

Eolian deposits. *See* Aeolian deposits

Eoraptor
  in archosauriform cladogram, 437, 439
  in dinosaur cladogram, 438

Eosestheria, in Chinese Mesozoic bird deposits, 184

Ephemeropsis, in Chinese Mesozoic bird deposits, 184

Ephemeroptera, in Las Hoyas Konservat-Lagerstätte, 210

Equisetales, in *Noguerornis* deposits, 231

Erlikosaurus, *Shuvuuia* versus, 93, 96, 97

Erlikosaurus andrewsi
  character matrix for, 53–56
  taxonomy of, 36

in coelurosaur cladistics, 42
feathers in, 112
skull of, 92, 93
in theropod cladograms, 40–41, 43
Mantelliceras mantelli, in dating Cambridge
    Greensand, 319
Manual digital homologies, in bird origins,
    20–21
Manual digits. *See also* Alula
    of archaeopterygids, 142–143, 148
    of *Cathayornis,* 169
    of *Cathayornis yandica,* 189
    of *Concornis lacustris,* 218
    of *Confuciusornis sanctus,* 165–166
    of *Eoalulavis hoyasi,* 223
    of euenantiornithines, 254
    evolution of avian, 438
    of *Noguerornis gonzalezi,* 233
    of *Patagopteryx deferrariisi,* 300
    of *Sinornis santensis,* 188, 193–195
Manual unguals
    in archaeopterygid climbing, 153
    of archaeopterygids, 142–143
    of *Cathayornis,* 170
    of *Eoalulavis hoyasi,* 223
Manus. *See also* Forelimbs; Wings
    of alvarezsaurids, 105–106
    of archaeopterygids, 141–143
    avian locomotion and, 433
    in bird origins, 20–21
    of *Cathayornis,* 169, 170
    of *Caudipteryx,* 12, 180
    of *Concornis,* 193
    of *Concornis lacustris,* 218
    of *Confuciusornis sanctus,* 165–166
    of *Eoalulavis,* 193
    of *Eoalulavis hoyasi,* 222–223
    footprints of pterosaur, 406
    of *Iberomesornis romerali,* 217
    of *Protoavis,* 9
    of *Sinornis santensis,* 193–195
Marambio Island, undetermined neornithine
    species from, 367
Marasuchus, in archosauriform cladogram,
    437, 439
Marine deposits
    euenantiornithines from, 241, 260
    fossil footprints from, 416
    hesperornithiforms from, 423
Marine reptiles, from Cambridge Greensand,
    336
Marovoay, 269
Marsh, Othniel Charles, x–xi, 258
    toothed birds collected by, 342
Martin, Larry D., 317
    on alvarezsaurids, 113, 122
    on avian evolution, 448, 449
    on cladistics, 16
    on cursorial origin of avian flight, 15
    on *Noguerornis,* 235–236

on non-theropod origins of birds, 18, 19, 46
on significance of *Archaeopteryx,* 4–5
Maryland, fossil footprints from, 415
Masarobe Member, 269
Masiakasaurus knopfleri, historical perspective
    on, 268
Massecaps, euenantiornithines from, 243
"Maxberg" *Archaeopteryx,* 129. *See also* Third
    specimen of *Archaeopteryx*
    taxonomy of, 132
Maxilla. *See also* Skull
    of archaeopterygids, 136
    of *Caudipteryx,* 13
    of *Confuciusornis sanctus,* 165
    of *Shuvuuia,* 91
    of *Sinornis santensis,* 190
Maxillary sinus, in defining birds, 7
MAX* program, 42
McCone County, *Cimolopteryx maxima* from,
    362–363
McKinney, Texas, *Graculavus lentus* from, 374
Meckelian groove, of archaeopterygids, 137
Medulla oblongata, of *Enaliornis,* 321, 323. *See
    also* Brain
Meetinghouse Canyon, fossil footprints from,
    411, 412
Megalancosaurus
    bird origins and, 15, 17–18, 46
    temporal extent of, 45, 48
Megalancosaurus preonensis, temporal extent
    of, 48
Megapodius freycinet, shoulder elements of,
    346, 347
Melanin, in feathers, 401
Merops, endocranial cast of, 323
Mesaverde Group, fossil footprints from, 411
Mesokinesis, of *Shuvuuia* skull, 95–96
Mesozoic birds
    in Aves, 6–7
    bone microstructure of, 421–429
    of China, 160–181
    cladogram of, xi
    diversity of Madagascan, 277
    enantiornithines as, 260
    evolution of Chinese, 175–177
    feather fossils of, 389–402
    footprints of, 405–416
    forthcoming descriptions of Chinese,
        180–181
    historical perspectives on, ix, 3, 160–163, 240,
        281
    from Madagascar, 268–277
    metatarsals of, 275
    neornithine, 339–381
    new discoveries of, xi
    oldest confirmed footprints of, 408
    paleobiology of Chinese, 177–179
    table of fossil feathers of, 392
    table of neornithine, 376–377
    taxonomy of Chinese, 163–164

of uncertain age, 375
Mesozoic era
    bird origins and, 45
    birds at end of, 381
Messel rails, *Palaeotringa vetus* versus, 373
Metabolic rate. *See also* Ectothermy;
        Endothermy; Homeothermy
    of archaeopterygids, 151–152
    flightlessness and, 313
Metacarpus
    of archaeopterygids, 142
    in bird origins, 20–21
    of *Cathayornis,* 170
    of *Cathayornis yandica,* 189
    of *Concornis lacustris,* 218
    of *Confuciusornis sanctus,* 165, 166
    of *Eoalulavis hoyasi,* 222–223
    of euenantiornithines, 253
    of *Eurolimnornis,* 352
    of *Noguerornis gonzalezi,* 233
    of *Patagopteryx deferrariisi,* 300
    of *Sinornis santensis,* 193, 194
Metatarsus. *See also* Tarsometatarsus
    of alvarezsaurids, 110–111
    of *Alvarezsaurus calvoi,* 110
    of archaeopterygids, 144–145, 151
    avian locomotion and, 433
    of *Avimimus,* 67
    of *Avimimus portentosus,* 84
    of *Avisaurus archibaldi,* 258
    of *Boluochia,* 171
    of *Cathayornis,* 171
    of *Cathayornis yandica,* 189
    of *Chaoyangia,* 173
    of *Concornis lacustris,* 218–219
    of *Confuciusornis sanctus,* 167
    of *Enaliornis,* 329
    of euenantiornithines, 256–257, 258
    of *Gansus yumenensis,* 351–352
    of hesperornithiforms, 336
    of *Iberomesornis romerali,* 217
    of *Liaoningornis,* 173
    of neornithines, 350, 351
    of Paraves, 440
    of *Patagopteryx deferrariisi,* 304–310
    of *Shuvuuia deserti,* 112
    of *Sinornis santensis,* 188, 198, 199, 200
    of *Vorona,* 273–277
    of *Vorona berivotrensis,* 270
Metornithes
    alvarezsaurids and, 90, 114, 122–123
    in archosauriform cladogram, 437
    in evolution of avian locomotion, 436
Metotic strut, in coelurosaurs, 9
Mexico
    Alexornis antecedens from, 240, 448
    Early Jurassic coelurosaurs from, 45
    euenantiornithines from, 243
MFL. *See* Main Fossiliferous Layer (MFL)
Miadana Member, 269

bird origins and, 7
character matrix for, 466–468
Dromaeosauridae and, 35
growth rate of, 429
historical perspective on, xi, 268–269
ilium of, 195
metatarsals of, 275
Noguerornis gonzalezi versus, 235
pedal digits of, 311
pelvic girdle of, 144, 254–255
in strict consensus tree, 451, 453
tail of, 11
tarsals of, 273
in theropod cladograms, 40–41, 43
Rahonavis ostromi
in avian phylogenetic systematics, 449,
454–455
character matrix for, 53–56
as composite, 454–455
historical perspective on, 268–269
locality map for, 269
in theropod cladistics, 41, 43
Rails
flightlessness of, 313
Telmatornis and, 359
Rainforth, Emma C., 405
Rallidae, phylogenetic systematics of, 380
Rao Chenggang, 184
Rapetosaurus krausei, historical perspective
on, 268
Raritan Formation, fossil feathers from, 397
Ratitae
in Aves, 185
bone microstructure of, 424
Confuciusornis and, 165
embryology of, 261
flightlessness of, 312
Gansus yumenensis and, 351
geographical distribution of, 379
palate of, 343–344
parasphenoidal lamina of, 286
Patagopteryx and, 281, 311
shoulder girdle of, 344
squamosal of, 284
taxonomy of, 339–340
tibiotarsus of, 329
Raton Formation, fossil footprints from, 413,
414
Rb-Sr dating, of Chinese Mesozoic birds, 162
Rectrices (rectrix), 389, 390. *See also* Tail
feathers
Recurvirostra americana, tarsometatarsus of,
366
Reed Collection, *Enaliornis* specimens in, 325
Regurgitated pellets, fossil birds in, 209
Remiges (remex), 389, 390. *See also* Feathers
of *Eoalulavis hoyasi*, 223
Renal fossa, of *Protoavis*, 9
Reproduction, of archaeopterygids, 154
Reptilia

in Chinese Mesozoic bird deposits, 160
feathers and, 400
in Mesozoic bird deposits, 241
as monophyletic group, 33–34
in *Noguerornis* deposits, 231
Respiratory system, evolution of avian, 46
Retroarticular process, of *Shuvuuia*, 97
Retroverted pubis, in defining birds, 6
Rhamphotheca, of archaeopterygids, 145–146
Rhea, flightlessness of, 313
Rhea americana, bone microstructure of, 424
Rhynchocephalians, *Archaeopteryx* and, 131
Ribs
of archaeopterygids, 138–139
of *Chaoyangia*, 173
in cladistic analysis, 463
of *Concornis lacustris*, 218
of *Confuciusornis sanctus*, 165
of *Eoalulavis hoyasi*, 221
of euenantiornithines, 249
of *Shuvuuia*, 97
of *Sinornis santensis*, 191, 196, 197
Rinaldi, C., on alvarezsaurids, 113, 122
"*Rinchenia*" *mongoliensis*, in theropod
cladograms, 43
Río Colorado Formation
alvarezsaurids from, 89
dating of, 282
euenantiornithines from, 243
Patagopteryx deferrariisi from, 281–282
Patagopteryx from, 422
Rio de Janeiro, fossil feathers in, 397
Río Limay Formation, geology of, 282
Río Negro Province
euenantiornithines from, 243
fossil footprints from, 413, 414
undetermined neornithine species from,
364–365
Río Neuquén Formation
alvarezsaurids from, 89, 90
coelurosaurs from, 48
Riverine deposits
euenantiornithines from, 241
Patagopteryx deferrariisi from, 281–282
Roadrunners, limb proportions of, 153
Romania, *Elopteryx nopcsai* from, 372
Romualdo Member, 396
absence of feathered theropods from, 400
Rose diagram, of *Archaeornithipus* tracks, 409
Rostral, of archaeopterygids, 137
Rostral fenestra, of *Shuvuuia*, 92
Rostral morphology, of euenantiornithines,
245
Rostroventral process, of archaeopterygids,
134
Ruben, John A.
on avian lungs, 21
on non-theropod origins of birds, 19
Rubidium-strontium dating, of Chinese
Mesozoic birds, 162

Running speed, of archaeopterygids, 153
Russell, Dale A., on segnosaur cladistics, 39
Russia
fossil feathers from, 392, 394–395, 401
hesperornithiforms from, 318

Sacral vertebrae. *See also* Synsacrum
of archaeopterygids, 138
of *Concornis lacustris*, 218
of *Iberomesornis romerali*, 216
of *Sinornis santensis*, 191, 196, 197
Sacrum. *See also* Synsacrum
avian locomotion and, 433
of *Cathayornis yandica*, 189
of *Iberomesornis romerali*, 216
of *Sinornis santensis*, 188
Sage grouse, *Graculavus lentus* versus, 374
St. Mary's College, undetermined neornithine
species at, 365
St. Mary's River Formation, fossil footprints
from, 413
Salamanders, in Las Hoyas Konservat-
Lagerstätte, 210
Salitrál Moreno, undetermined neornithine
species from, 364–365
Salta province
enantiornithines from, 240
euenantiornithines from, 243
Madagascan birds and, 268
Saltasaurus, enantiornithine fossils found
with, 240
Saltasaurus loricatus, enantiornithine fossils
found with, 240
Sampling bias, fossil record and, 44
Sampson, Scott D., 268
Sandpipers, *Cimolopteryx* versus, 363
Sandstone Basin, alvarezsaurids from, 89
Santa María de Meiá, *Noguerornis* from, 230
Santana Formation
absence of feathered theropods from, 400
fossil feathers from, 396–397, 398, 399, 401
Santonian stage
alvarezsaurids from, 90
fossil feathers from, 392, 397, 399, 400
neornithines from, 375–377
Patagopteryx from, 282, 422
San Vicente Bay, *Neogaeornis wetzeli* from,
367
Sanz, José L., 209
Sao Khua Formation, tyrannosaurids from, 47
SAPE. *See* Society of Avian Paleontology and
Evolution (SAPE)
Saskatchewan
Cimolopteryx rara from, 361
Pasquiaornis from, 318
Saurischia
in archosauriform cladogram, 437
manual digits of, 254
"Sauriurae"
alvarezsaurids and, 113

Shuvosaurus inexpectatus
  as coelurosaur, 46
  temporal extent of, 47
Shuvuuia, 91
  in alvarezsaurid cladogram, 90, 114
  axial skeleton of, 97
  caudal vertebrae of, 100–101
  cervical vertebrae of, 97–98, 247
  character matrix for, 466–468
  coracoid of, 102
  discovery of, 87
  distal tarsals of, 110
  feathers of, 14
  femur of, 108
  forelimb of, 103–104, 105–106
  historical perspective on, x
  ilium of, 106
  ischium of, 107
  metatarsus of, 111
  mononykine skeleton and, 116
  paleobiology of, 115
  parasphenoidal lamina of, 286
  Patagonykus versus, 90
  pelvic girdle of, 106–107
  phylogenetic systematics of, 112–115, 121
  plumage of, 112
  provenance of, 88–89
  pubis of, 107
  quadrate of, 135, 246
  scapula of, 102
  skull of, 91–97
  in strict consensus tree, 451, 453
  synsacrum of, 100
  taxonomy of, 91
  in theropod cladograms, 40–41
  thoracic girdle of, 101–103
  thoracic vertebrae of, 99
  tibiotarsus of, 109–110
Shuvuuia deserti, 91
  in avian phylogenetic systematics, 451, 452
  caudal vertebrae of, 101
  cervical vertebrae of, 98
  character matrix for, 53–56
  cladistic analysis of, 462
  discovery of, 87
  fibula of, 110
  metatarsus of, 112
  paleobiology of, 115
  pelvis and synsacrum of, 107
  provenance of, 88
  skull of, 91–97
  taxonomy of, 91
  theropod characters and, 43, 48
Siamotyrannus, taxonomy of, 37
Siamotyrannus isanensis, temporal extent of, 47–48
Siberia, fossil feathers from, 392, 399, 401
Sichuan Province, fossil footprints from, 409–410

Sieben-Lumpen-Schicht, archaeopterygids from, 132
Sierra del Montsec. *See also* Montsec Range
  geology of, 230
  Noguerornis from, 231
Sierra del Portezuelo, alvarezsaurids from, 89, 90
Sihetun
  Confuciusornis from, 161
  Confuciusornis sanctus from, 163
  euenantiornithines from, 242
  Liaoningornis longidigitris from, 164
Sihetun basalts, dating of, 162
Sihetun-Jianshangou, 185
Sinamia, in Chinese Mesozoic bird deposits, 161
Sino-Canadian Dinosaur Expedition, Mesozoic birds found by, 161
Sinornis, 163, 185–188, 232
  in avian phylogenetic systematics, 203, 204, 452, 457
  bird origins and, 7, 198–205
  body size of, 261
  Boluochia versus, 171
  as *Cathayornis* synonym, 185–188, 232
  caudal vertebrae of, 249
  cervical vertebrae of, 247
  Confuciusornis sanctus versus, 167
  Confuciusornis versus, 165, 166
  in Enantiornithes, 240, 259
  Euenantiornithes and, 257
  feeding habits of, 261
  femur of, 255
  furcula of, 250
  historical perspectives on, x, 160, 161
  locality map for, 162, 185
  long bone ratios in, 190
  mandible of, 246
  manual digits of, 254
  manual phalanges of, 170
  Noguerornis versus, 232, 236
  Otogornis versus, 171
  paleobiology of, 226
  pelvic girdle of, 254
  perching by, 261
  pes of, 261
  phylogenetic systematics of, 175, 176, 223
  pubis of, 170
  ribs of, 249
  rostral morphology of, 245
  skull of, 245
  sternum of, 251
  in strict consensus tree, 451, 453
  synsacrum of, 248
  tarsometatarsus of, 257
  taxonomy of, 163, 167, 185–188, 206
  thoracic girdle of, 249
  thoracic vertebrae of, 248
  wings of, 251, 253, 254
Sinornis santensis, 163, 184–206
  anatomy of, 188–198, 199, 200, 201

as *Cathayornis yandica* synonym, 186–188
  caudal vertebrae of, 187
  character matrix for, 263
  cladistic analysis of, 462
  Euenantiornithes and, 244
  femur of, 255
  geographical distribution of, 242
  geological setting for, 241
  historical perspective on, 160, 184
  locality of, 185
  long bone ratios in, 190
  mandible of, 246
  manus of, 193–195
  paleobiology of, 205–206
  pes of, 199, 200, 201
  phylogeny of, 198–205, 452
  pygostyle of, 187
  skeleton of, 188, 189, 202
  skull of, 245, 246
  sternum of, 251
  taxonomy of, 163, 186–188, 206, 244
  vertebrae of, 248
Sinornithoides
  semilunate carpal of, 20
  taxonomy of, 36
Sinornithoides youngi
  character matrix for, 53–56
  cladistic analysis of, 460
  discovery of, 3
  semilunate carpal of, 20
  taxonomy of, 35–36
  temporal extent of, 47
  in theropod cladograms, 43
Sinornithosaurus
  feathers of, 14
  taxonomy of, 35, 38
  time problem and, 19
Sinornithosaurus millenii
  character matrix for, 53–56
  discovery of, 3
  feathers of, 10, 14
  taxonomy of, 35, 38
  in theropod cladograms, 43
Sinosauropteryx, 179–180
  Chinese Mesozoic bird phylogeny and, 175
  in coelurosaur cladistics, 42
  feathers of, 10–11, 13–14, 18, 179, 400
  paleobiology of, 179
  taxonomy of, 38
  thermoregulation in, 151
Sinosauropteryx prima
  Chinese Mesozoic birds and, 161
  feathers of, 10, 400
  taxonomy of, 38
Sinovenator, time problem and, 19
Sinovenator changii
  character matrix for, 53–56
  discovery of, 3
  taxonomy of, 36
  in theropod cladograms, 43

Sinpetru Beds, *Elopteryx nopcsai* from, 372, 376
Sinraptor, Shuvuuia versus, 93
Sinraptor dongi
    character matrix for, 53–56
    cladistic analysis of, 460
    in theropod cladograms, 43
Sinraptoridae, in theropod cladograms,
    40–41
Sixth archaeopterygid specimen, 129, 131. *See
    also* Solnhofen archaeopterygid;
    Wellnhoferia grandis
    body mass of, 151
    taxonomy of, 132
Skeleton
    of alvarezsaurids, 115, 116
    of archaeopterygids, 150
    of *Archaeopteryx lithographica*, 130
    of *Archaeopteryx siemensii*, 130
    of *Avimimus portentosus*, 68
    of *Boluochia zhengi*, 172, 173
    of *Cathayornis yandica*, 168, 169, 189
    of *Chaoyangia beishanensis*, 176
    of *Concornis lacustris*, 213, 219
    of *Eoalulavis hoyasi*, 214–215, 219–220, 221
    of *Eoenantiornis buhleri*, 257
    of *Iberomesornis romerali*, 211, 212
    of *Liaoningornis longidigitris*, 175
    of *Neuquenornis volans*, 251
    of *Noguerornis gonzalezi*, 232, 233, 234
    of *Otogornis genghisi*, 174
    of *Patagopteryx deferrariisi*, 282
    of *Sinornis santensis*, 188, 189, 202
Skin impressions
    of *Celtedens*, 210
    of *Pelecanimimus*, 210. *See also* Integument
Skuas, endocranial cast of, 323
Skull
    of alvarezsaurids, 91–97
    of archaeopterygids, 132–138
    of *Avimimus portentosus*, 69–70
    of *Boluochia*, 171
    of Catalan hatchling, 246
    of *Cathayornis*, 167–168
    of *Cathayornis yandica*, 189
    of *Chaoyangia*, 173
    characters of theropod, 48–49
    in cladistic analysis, 463
    of *Confuciusornis sanctus*, 165, 166
    of *Enaliornis*, 320–325
    of enantiornithines, 189–190
    of euenantiornithines, 245–247
    length of archaeopterygid, 133
    of *Mononykus olecranus*, 91, 95
    of neornithines, 343–344, 376
    of *Neuquenornis volans*, 247
    of *Patagopteryx deferrariisi*, 283–286
    of *Protoavis*, 8–9
    of *Shuvuuia deserti*, 91–97
    of *Sinornis santensis*, 188, 189–190, 246
Slabs, with *Archaeopteryx* specimens, 131

Smoky Hill Chalk, *Apatornis* from, 370
Smoky Hill Member
    Apatornis from, 370
    hesperornithiforms from, 318
Smoky Hill River, *Apatornis* from, 370
Smuggling, of Chinese fossils, 162–163
Snakes
    Mesozoic Madagascan, 268
    in *Patagopteryx* deposits, 281, 314
Snipes
    taxonomy of, 373
    Telmatornis and, 359
Society of Avian Paleontology and Evolution
    (SAPE)
    Fourth International Meeting of, 163
    meetings of, 3
    Second International Symposium of, 209
Soft tissues, evolution of avian, 46–47
Solenhofer Aktien-Verein quarry,
    Archaeopteryx from, 132
Solnhofen
    archaeopterygids from, 129, 132
    Archaeopteryx fossil feathers from, 391,
    392
    coelurosaurs from, 48
    first fossil feather from, 389, 391, 393
    fossil feathers from, 401
    landscape of, 7, 130–131
Solnhofen archaeopterygid, 129. *See also* Sixth
    archaeopterygid specimen; *Wellnhoferia
    grandis*
    metatarsals of, 275
Solnhofen Lithographic Limestone
    archaeopterygids from, 132
    Archaeopteryx fossil feathers from, 391, 393
    geology of, 130–131
"Solnhofen specimen" of *Archaeopteryx*, 129
Songlingornis linghensis, taxonomy of, 164
Soroavisaurus
    in avian phylogenetic systematics, 458
    in enantiornithine cladograms, 204, 259
    metatarsals of, 275, 276
    perching by, 261
    taxonomy of, 206
    tibia of, 255
    tibiotarsus of, 256
Soroavisaurus australis
    in avian phylogenetic systematics, 457
    character matrix for, 263
    geographical distribution of, 243
    hind limbs of, 255
    historical perspective on, 259
    phylogenetic systematics of, 277
    tarsometatarsus of, 256
    taxonomy of, 244
    tibiotarsus of, 255
South Amboy Fire Clay, 397
South America
    alvarezsaurids from, 115, 452
    euenantiornithines from, 241

    locality map of, 283
    Mesozoic birds from, 268, 379
    Mesozoic neornithines from, 376, 379
    in plate tectonics, 269
South Dakota
    fossil footprints from, 409
    hesperornithiforms from, 423
Southern Hemisphere
    Mesozoic birds from, 281
    ratites in, 379
South Gippsland, fossil feathers from, 396
South Gobi Aimak
    alvarezsaurids from, 90–91
    euenantiornithines from, 243
South Korea, fossil footprints from, 408, 409,
    410, 413
Spain, 210, 231
    coelurosaurs from, 47
    euenantiornithines from, 242, 245, 251
    fossil feathers from, 392–393, 394, 395–396,
    401
    fossil footprints from, 407–408, 409, 410
    Mesozoic birds from, ix, 8, 160, 178, 209–227,
    230–237, 241, 246
Species, diagnosing neornithine, 341
Spermatophytes, in *Noguerornis* deposits, 231
Spheniscidae, scleral ring of, 137
Sphenosuchia
    avian evolution and, 46
    temporal extent of, 45, 48
Splenial
    of archaeopterygids, 137
    of *Cathayornis*, 168
    of *Confuciusornis sanctus*, 165
    of *Sinornis santensis*, 190
Spoonbills, taxonomy of, 340
Squamosal
    of archaeopterygids, 134
    of *Caudipteryx*, 12
    of *Confuciusornis sanctus*, 165
    in defining birds, 8–9
    of euenantiornithines, 245
    of *Patagopteryx deferrariisi*, 284
    of *Protoavis*, 8–9
    of *Shuvuuia*, 94
Stalling speed, of archaeopterygids, 153
State University of New York at Stony Brook,
    Madagascan paleoexpeditions by, 268
Stem-based taxa, 341
Stercorariidae
    endocranial cast of, 323
    Volgavis marina versus, 361
Stercorarius
    Cimolopteryx versus, 361
    coracoid of, 362
    forelimb elements of, 364
    taxonomy of, 355
Stercorarius pomarinus
    coracoid of, 362
    forelimb elements of, 364

Taxonomy *(continued)*

cladistic analysis and, 33–34

of *Sinornis santensis*, 186–188

of theropods, 35–38

Teal, flightless, 313

Tectal fossa, of *Enaliornis*, 321, 322–323

Teeth

of archaeopterygids, 137, 153–154

of *Archaeopteryx bavarica*, 154

of *Avimimus*, 66

of *Boluochia*, 171

of *Cathayornis*, 167

of *Chaoyangia*, 173

characters of theropod, 50

of *Confuciusornis sanctus*, 165

in defining birds, 7

of euenantiornithines, 245

of Jurassic dromaeosaurids, 19

of Portuguese archaeopterygids, 150

of *Protarchaeopteryx*, 11

of *Sinornis santensis*, 190

of *Sinovenator changii*, 36

of *Wellnhoferia*, 137

Teleosteii

in Chinese Mesozoic bird deposits, 160–161, 184

in Las Hoyas Konservat-Lagerstätte, 210

in *Noguerornis* deposits, 231

Telmabates

Apatornis versus, 370

Palaeotringa vetus versus, 373

taxonomy of, 341

Telmatornis, 359–360

historical perspective on, 358

taxonomy of, 359, 374

Telmatornis affinis, taxonomy of, 360

Telmatornis priscus, 360

Cimolopteryx rara versus, 361–362

fossil record of, 376, 378

taxonomy of, 360, 373, 374

Telmatornis rex, taxonomy of, 359–360, 375

Terrestrial deposits, fossil feathers from, 392. *See also* Continental deposits

Terrestrial locomotion, of birds, 432–445, 458–460

Tertiary period

avian distribution during, 379

fossil footprints from, 415

fossil footprints older than, 413

Hornerstown Formation and, 343

loons from, 367

Messel rails from, 373

neornithines from, 339

taxonomy of birds from, 341

Telmabates from, 370

Testability, in avian phylogenetic systematics, 450

Testudines

from Cambridge Greensand, 336

vertebrae referable to, 330

Tetanurae

alvarezsaurids and, 113

in archosauriform cladogram, 437

Avimimus in, 66

manual digital homologies among, 20–21

Tetori Group, fossil footprints from, 408–409

Texas

fossil footprints from, 410, 411

Graculavus lentus from, 374

Protoavis from, 7

Teyler Museum, *Archaeopteryx* specimen in, 129

Thailand, tyrannosaurids from, 47

"Thecodonts," as bird ancestors, 17

Therizinosauridae

alvarezsaurids and, 113, 122

in theropod cladistics, 41

in theropod cladograms, 40–41, 123

Therizinosauroidea, 36

alvarezsaurids and, 122

birdlike features of, 6

in coelurosaur cladistics, 42

feathers of, 13–14, 38

first occurrences of, 47

origins of, 9

pubis of, 144

taxonomy of, 36

temporal extent of, 45

thermoregulation in, 151

in theropod cladograms, 40–41, 43

Therizinosaurus, taxonomy of, 36

Thermocoel, in archaeopterygids, 151

Thermoregulation, of archaeopterygids, 151–152. *See also* Ectothermy; Endothermy; Homeothermy; Poikilothermy

THERO (common theropod ancestor), in locomotor evolution, 435, 436, 438, 439, 440, 443

Theropoda, 34–39. *See also* THERO (common theropod ancestor)

alternatives to bird ancestry within, 17–18

alvarezsaurids in, 90, 115

in archosauriform cladogram, 437

in avian evolution, 442–443

in avian phylogenetic systematics, 454

as bird ancestors, 3–23, 445

birdlike features of small, 16–17

cervical ribs of, 97

character list for, 48–52

Chinese feathered, 9–14

Chinese Mesozoic birds and, 161

cladistic analysis of, 31–56, 460–468

distinguishing footprints of, 406, 415

feathers of, 400, 401

fibula of, 273

forelimb of, 348–349

fossil footprints of, 409

fossil record of, 44–46, 47–48

furcula of, 6, 237

humerus of, 251

ischium of nonavian, 255

in Las Hoyas Konservat-Lagerstätte, 210

locomotor modules of, 226, 435

from Madagascan Mesozoic, 268

metatarsals of, 257, 275

Mononykus and, 121

Noguerornis versus nonavian, 232, 235

Norian phylogeny of, 9

origin of flight among, 14–17

palate of, 135–136

paleobiology of Chinese feathered, 179–180

parasphenoidal lamina of, 286

in *Patagopteryx* deposits, 281, 314

pelvic girdle of, 143

phalangeal formula in, 142

phylogenetic systematics of nonavian, 311

proatlas of, 287

pubis of, 302

pulmonary air sacs of, 252

pygostyle and, 436

scapula of, 140

secondarily flightless, 22

squamosal of, 134

taxonomy of, 35–38

temporal extent of, 45

Third specimen of *Archaeopteryx*, 129, 131. *See also* "Maxberg" *Archaeopteryx*

contour feathers of, 149

flight feathers of, 149

hind limbs of, 145

pubis of, 144

taxonomy of, 132

thoracic girdle of, 140

THOR (common ornithothoracine ancestor), in locomotor evolution, 436, 439, 441, 444

Thoracic girdle. *See also* Coracoid; Furcula; Pectoral girdle; Scapula; Scapulocoracoid; Shoulder girdle; Sternum

of alvarezsaurids, 101–103

of archaeopterygids, 139–140

of *Cathayornis*, 168–169

of *Caudipteryx*, 13

of *Chaoyangia*, 173

of *Chaoyangia beishanensis*, 177

in cladistic analysis, 463–464

of *Concornis lacustris*, 218

of *Confuciusornis sanctus*, 165

of enantiornithines, 205

of *Eoalulavis hoyasi*, 221–222

of euenantiornithines, 249–251

of flightless birds, 312

of *Iberomesornis romerali*, 216–217

of *Liaoningornis*, 171–173

of *Noguerornis gonzalezi*, 232

of *Otogornis*, 171

of *Patagopteryx deferrariisi*, 294–295, 296, 297, 298

Thoracic limb. *See* Forelimbs; Wings
Thoracic ribs
    of archaeopterygids, 139
    of *Sinornis santensis,* 191
Thoracic vertebrae
    of alvarezsaurids, 97, 99
    of archaeopterygids, 138
    of *Avimimus,* 66–67
    of *Avimimus portentosus,* 71–74
    of *Cathayornis,* 168
    of *Concornis lacustris,* 217–218
    of *Enaliornis,* 325–326
    of *Enaliornis barretti,* 326
    of *Eoalulavis hoyasi,* 220–221
    of euenantiornithines, 248
    of *Iberomesornis romerali,* 216
    of *Noguerornis gonzalezi,* 231
    of *Patagopteryx deferrariisi,* 289–293, 294
    of *Sinornis santensis,* 190–191
Thulborn, R. Anthony
    on *Archaeopteryx,* 5
    on *Avimimus,* 66
Tibia
    of alvarezsaurids, 110
    of *Alvarezsaurus calvoi,* 110
    of archaeopterygids, 142, 144, 153
    avian locomotion and, 433
    in bone microstructure studies, 422, 424
    of *Enaliornis barretti,* 332–333
    of *Enaliornis sedgwicki,* 332–333
    of euenantiornithines, 255–256
    of *Iberomesornis romerali,* 217
    of neornithines, 349
    of *Noguerornis gonzalezi,* 235
    of *Patagopteryx deferrariisi,* 303
    of *Vorona,* 273
Tibiotarsus. *See also* Tarsus; Tibia
    of alvarezsaurids, 109–110
    of archaeopterygids, 144, 149
    of *Avimimus,* 67
    of *Avimimus portentosus,* 82–83
    of *Boluochia,* 171
    in bone microstructure studies, 426, 427
    of *Bradycneme draculae,* 375
    of *Cathayornis,* 171
    of *Cathayornis yandica,* 189
    of *Chaoyangia,* 173
    of *Concornis lacustris,* 218
    of *Confuciusornis sanctus,* 167
    of *Enaliornis,* 329
    of euenantiornithines, 255–256
    of *Gansus,* 175
    of *Gansus yumenensis,* 178
    of *Heptasteornis andrewsi,* 375
    of hesperornithiforms, 335
    of *Lectavis bretincola,* 256
    of *Liaoningornis,* 173
    of *Mononykus olecranus,* 111
    of neornithines, 376, 377

    of *Parvicursor remotus,* 109
    of *Patagopteryx deferrariisi,* 303, 304, 305, 306, 307, 308
    of *Sinornis santensis,* 188, 195–198
    of *Soroavisaurus australis,* 255
    of *Vorona,* 270–271, 272–273
    of *Vorona berivotrensis,* 270, 274, 275
Time problem
    secondarily flightlessness and, 22
    in theropod ancestry of birds, 19–20, 44–46
Timolopteryx, taxonomy of, 361
Tinami, in Aves, 185
Tinamiformes
    coracoid of, 348
    limb proportions of, 153
    shoulder elements of, 346, 347
Tinamous
    bone microstructure of, 424
    forelimb of, 349
    hind limbs of, 349
    neornithine-like features of, 352
    in Palaeognathae, 340
    palate of, 344
    Palintropus versus, 354
    phylogenetic systematics of, 380
Titanosauria, enantiornithine fossils found with, 240
Titanosauridae, historical perspective on Madagascan, 268
Tithonian stage
    archaeopterygids from, 132
    Archaeopteryx fossil feathers from, 391, 392, 393
    Archaeopteryx from, 130
    coelurosaurs from, 44, 48
    fossil feathers from, 392, 399
Tjulkeli, Kazakhstan, fossil feathers from, 392, 399, 401
Todus, scleral ring of, 137
Toolebuc Formation, euenantiornithines from, 242
Torotigidae, taxonomy of, 368, 374
Torotix, 368
    forelimb elements of, 364
    fossil record of, 378
    humerus of, 352
    phylogenetic systematics of, 381
    taxonomy of, 373
Torotix clemensi, 368
    forelimb elements of, 364
    fossil record of, 376
    taxonomy of, 374
Trachodon, from Cambridge Greensand, 336
Trackways. *See also* Footprints
    of *Archaeornithipus,* 408
    at Ukhaa Tolgod, 88
Trading fossil vertebrates, from China, 162–163
Transbaikalia, fossil feathers from, 392, 394–395, 401

Transylvania, *Elopteryx nopcsai* from, 372
Tree bisection and regraphing (TBR) algorithm, 42
Trees
    archaeopterygids and, 154
    cladograms as, 32, 449–451
    strict consensus, 451, 453
"Trees down" theory. *See* Arboreal origin of flight
Tremp, 231
Trennende krumme Lage, Archaeopteryx from above, 130
Trialestes romeri, temporal extent of, 48
Triassic period. *See also* Late Triassic epoch
    absence of feathers from, 401
    bird origins in, 17, 45
    Protoavis from, 7–8, 9
Triosseal canal, of enantiornithines, 205
Triosseal foramen
    of archaeopterygids, 140
    of euenantiornithines, 250–251
    of Spanish Mesozoic birds, 224
Trisaurodactylus superavipes, footprints of, 406, 408
Trisauropodiscus aviforma, footprints of, 406–407
Trisauropodiscus galliforma, footprints of, 406
Trisauropodiscus levis, footprints of, 406
Trisauropodiscus moabensis, footprints of, 407, 408
Trisauropodiscus phasianiforma, footprints of, 406
Trisauropodiscus popompoi, footprints of, 406
Trisauropodiscus superaviforma, footprints of, 406–407, 408
Trochilidae, scleral ring of, 137–138
Troodon
    in bird origins, 16
    braincase of, 321
Troodon formosus
    character matrix for, 53–56
    new specimens of, 3
    in theropod cladograms, 43
Troodontidae, 35–36
    in alvarezsaurid phylogenetics, 124
    alvarezsaurids and, 113, 116, 122, 123
    Archaeornithoides and, 39
    in Aves, 21
    in avian phylogenetic systematics, 449, 450, 451
    birdlike features of, 6
    in bird origins, 16, 19, 21
    as birds, 5
    character matrix for, 466–468
    cladistic analysis of, 39, 42, 460–468
    in Deinonychosauria, 438–440
    eggs of, 36
    evolution of features of, 46
    feathers among, 38
    first occurrences of, 47

Troodontidae (continued)
Middle Jurassic, 45
Ornithomimosauria and, 37
Scipionyx samniticus and, 37–38
as secondarily flightless, 22
Shuvuuia versus, 97
taxonomy of, 35–36
teeth of Jurassic, 19
temporal extent of, 45
in theropod cladograms, 40–41, 43, 123
tibiotarsus of, 273
Tropicbird, endocranial cast of, 323
True line of descent, in reconstructing
evolution, 443–445
Tugrugeen Shireh, 88
alvarezsaurids from, 89, 91
Tumbes Peninsula, *Neogaeornis wetzeli* from,
367
Turgai Strait
hesperornithiforms from, 318
neornithines from, 381
Turkey, aepyornithids from, 269
Turonian stage
coelurosaurs from, 47, 48
fossil feathers from, 392, 399
hesperornithiforms from, 318
South Amboy Fire Clay in, 397
Turtles
in Las Hoyas Konservat-Lagerstätte, 210
Mesozoic Madagascan, 268
Two Medicine Formation
euenantiornithines from, 243
troodontid eggs from, 36
Tympanic cavity, of *Enaliornis*, 321
Tympanic recess
of archaeopterygids, 133
in coelurosaurs, 9
in defining birds, 7
Tympanuchus phasianellus, Graculavus lentus
versus, 358, 374
Tyrannosauridae, 37
Avimimus and, 67
as birds, 5
in coelurosaur cladistics, 42
taxonomy of, 37
teeth of Jurassic, 19
temporal extent of, 45, 47–48
in theropod cladistics, 39–41, 43, 123
Tyrannosauroidea
taxonomy of, 37
in theropod cladograms, 40–41
Tyrannosaurus, 5
bird origins and, 16
sacral vertebrae of, 138
Tyrannosaurus rex
character matrix for, 53–56
in theropod cladograms, 43
Tytthostonyx, phylogenetic systematics of,
381
Tytthostonyx glauconiticus, taxonomy of, 375

Udan Sayr, *Avimimus* from, 65
Uge Member, fossil feathers from, 397
Uhangrichnus chuni, footprints of, 413
Uhangri Formation, fossil footprints from,
413
Ukhaa Tolgod
alvarezsaurids from, 88, 89, 91
euenantiornithines from, 243
oviraptorosaur nests from, 36
skull of *Shuvuuia* from, 91–97
troodontid eggs from, 36
troodontids from, 39
Ukraine, hesperornithiforms from, 319
Ulna. *See also* Forelimbs; Wings
of alvarezsaurids, 104
of archaeopterygids, 141, 142, 149
of *Avimimus*, 67
of *Avimimus portentosus*, 77
of *Beipiaosaurus inexpectus*, 14
of *Cathayornis*, 170
of *Cathayornis yandica*, 189
of *Concornis lacustris*, 218
of *Confuciusornis sanctus*, 165, 166
of *Enaliornis*, 329
of *Enantiornis leali*, 253
of *Eoalulavis hoyasi*, 222
of euenantiornithines, 253
of *Eurolimnornis*, 352
of *Iberomesornis romerali*, 217
of *Liaoningornis*, 173
of *Noguerornis gonzalezi*, 232
of *Otogornis*, 171
of *Patagopteryx deferrariisi*, 299, 300
of *Protoavis*, 9
of *Sinornis santensis*, 188, 192
Ultraviolet-induced fluorescence photograph
of *Eoalulavis hoyasi*, 214
of *Iberomesornis romerali*, 212
Uncinate processes
of archaeopterygids, 139
of *Cathayornis*, 168
of *Chaoyangia*, 173
of *Concornis lacustris*, 218
of *Confuciusornis sanctus*, 165
of *Eoalulavis hoyasi*, 221
of euenantiornithines, 249
Uncinatum, of *Archaeopteryx*, 135
Undercovert feathers, of archaeopterygids,
147–148
Undetermined Neornithes, 356–357, 363–365,
367–368, 369–370, 372–374, 376–377
Unenlagia, 37
alvarezsaurids and, 122, 124
archaeopterygids and, 150
bird origins and, 7
cladistic analysis of, 460
in coelurosaur cladistics, 42
femur of, 271
Noguerornis gonzalezi versus, 235

pelvic girdle of, 143, 144, 255
taxonomy of, 37
temporal extent of, 45, 48
in theropod cladograms, 40–41, 43, 123
time problem and, 44
Unenlagia comahuensis
in avian phylogenetic systematics, 456
character matrix for, 53–56
discovery of, 3
ischium of, 12
taxonomy of, 37
temporal extent of, 48
in theropod cladograms, 43
Ungual sheaths
of archaeopterygids, 145–146
of *Sinornis santensis*, 198, 199, 200, 201
Unidad de Paleontologia, Mesozoic birds in,
211–215
United States
alvarezsaurids from, 88–89
euenantiornithines from, 243, 248
fossil feathers from, 392, 397, 401
fossil footprints from, 406, 408, 414
stratigraphy in, 44
Universidad Autónoma de Madrid
Mesozoic birds in, 211–215
Universidad Nacional del Comahue (UNC)
Patagopteryx collected by, 281
Universidad Nacional de Tucumán
enantiornithine fossils at, 240
Université d'Antananarivo, Madagascan
paleoexpeditions by, 268
Upper Cretaceous. *See* Late Cretaceous epoch
Upper Gault
Cambridge Greensand and, 319
Enaliornis species from, 319–320
Upper Jurassic. *See* Late Jurassic epoch
Upper Stormberg Series, fossil footprints from,
406
Upper Triassic. *See* Late Triassic epoch
Urvogel, Archaeopteryx as, 7, 13, 19
Utah
coelurosaurs from, 47
euenantiornithines from, 243, 249
fossil footprints from, 407, 411–413, 414–415
Utahraptor
in Dromaeosauridae, 35
time problem and, 19
Utahraptor ostrommaysorum
character matrix for, 53–56
temporal extent of, 47
in theropod cladograms, 43
Uzbekistan
coelurosaurs from, 47
euenantiornithines from, 242–243
Kuszholia mengi from, 311
undetermined neornithine species from, 374

Vagal canal, in coelurosaurs, 9
Valanginian stage

Compositor:	Princeton Editorial Associates, Inc., Scottsdale, Arizona
Text:	Minion
Display:	Rotis SemiSans
Printer and binder:	Everbest Printing Co. Ltd., Hong Kong, China